THE ORIGIN OF LANDSCAPES

THE
ORIGIN OF LANDSCAPES

A Synthesis of Geomorphology

H. F. Garner

Rutgers University, Newark, New Jersey

With Drawings by Mary Craig Garner

New York
OXFORD UNIVERSITY PRESS
London 1974 Toronto

Permission to use material from the following sources for the figures herewith is gratefully acknowledged:

1.26: George Kukla, from *Boreas*, vol. 1, No. 1 (1972).
2.17: © Crown Copyright Reserved.
2.21 2.22: Published by the Geological Society of America. Copyright © 1964 by Bruce C. Heezen and Marie Tharp. Reprinted by permission.
2.42, 10.26, 10.27, 10.29: Theodore M. Oberlander, *The Zagros Streams*. Syracuse Geographical Series, No. 1 (1965). Reprinted by permission of Syracuse University Press.
2.47, 8.59, 8.60: Robert V. Ruhe, *Quaternary Landscapes in Iowa*. Copyright © 1969 by Iowa State University Press. Reprinted by permission of Iowa State University Press, Ames, Iowa.
2.57: H. B. Willman and John Frye, "Pleistocene Stratigraphy of Illinois." *Illinois State Geological Survey Bulletin* 94 (1970). Reprinted by permission of Illinois State Geological Survey.
3.64: Published by permission of McIlhenny Company, Avery Island.
3.87: William G. Pierce, U. S. Geological Survey, Professional Paper 400-B (1957).
3.90: *Eltanin* cruise 15, NSF Photo.
4.2: From *U. S. Standard Atmosphere*, p. 9, Washington, D. C. (1962).
4.3: From *Aerographer*, p. 20, Washington, D. C. (1958).
4.4: George R. Rumney, *The Geosystem*. William C. Brown Co., Dubuque, Iowa (1970).
4.5: From *Meteorology for Naval Aviators*, pp. 2–9, Washington, D. C. (1958).
4.8: From *Weather for Air Crews*, pp. 6–22, Washington, D. C. (1962).
4.11: George R. Rumney, *Climatology and the World's Climates*. Copyright © 1970 by Macmillan Publishing Co., Inc. Reprinted by permission of Macmillan Publishing Co., Inc.
4.13: From *Climate and Man, Yearbook of Agriculture, 1941*, p. 104, Washington, D. C.
5.9: From Oscar Meinzer, ed., *Hydrology*. McGraw-Hill (1942).
5.14: R. E. Moreau, "Vicissitudes of the African Biomes in the Late Pleistocene." Zoo-

logical Society of London, *Proceedings* (1963). Reprinted by permission of the Zoological Society of London.
6.41: Reproduced from Print laydown, Nigeria 83 (1958) by permission of the Controller of H. M. Stationery Office.
8.4: Richard Foster Flint, *Glacial and Quaternary Geology*. Copyright © 1971 by John Wiley & Sons, Inc. Reprinted by permission of John Wiley & Sons, Inc.
9.33: A. L. Bloom and C. W. Ellis, *Journal of Sedimentary Petrology*, v. 34, p. 601 (September 1964). Reprinted by permission of Society of Paleontologists and Mineralogists.
10.5: A. K. Lobeck, *Geomorphology*. Copyright 1939 by McGraw-Hill Book Co. Reprinted by permission of McGraw-Hill Book Co.

2.10, 6.15: Douglas S. Cherkauer, "Longitudinal Profiles of Ephemeral Streams in Southeastern Arizona." *Geological Society of America Bulletin*, v. 83 (1972).
5.28: B. P. Ruxton and L. R. Berry, "Weathering of Granite and Associated Erosional Features in Hong Kong." *GSA Bulletin*, v. 68 (1957).
5.29: Ernest E. Wahlstrom, "Pre-Fountain and Recent Weathering on Flagstaff Mountain near Boulder, Colorado," *GSA Bulletin*, v. 59 (1948).
7.9: Arthur N. Strahler, *GSA Bulletin*, v. 69, p. 297 (1958).
7.16: K. J. Tinkler, "Pools, Riffles and Meanders." *GSA Bulletin*, v. 81 (1970).
7.24–7.27: S. A. Schumm and H. R. Khan, "Experimental Study of Channel Patterns." *GSA Bulletin*, v. 83 (1972).
7.34: Stephen Born and Dale F. Ritter, "Modern Terrace Development near Pyramid Lake, Nevada and its Geologic Implications." *GSA Bulletin*, v. 81 (1970).
7.35: George T. Moore and D. O. Asquith, "Delta: Term and Concept." *GSA Bulletin* (1972).
8.47: W. C. Bradley, R. K. Fahnestock, and E. T. Rowekamp, "Coarse Sediment Transport by Flood Flows on Knik River, Alaska." *GSA Bulletin*, v. 83 (1972).
8.73: Karl Fritz Bruder and Erk Reimnitz, "River Discharge into an Ice-Covered Ocean

and Related Sediment Dispersal, Beaufort Sea, Coast of Alaska." *GSA Bulletin* (March, 1972).
8.75: Noel Potter, Jr., and J. H. Moss, "Origin of Blue Rocks Block Field and Adjacent Deposits, Berks County, Pennsylvania." *GSA Bulletin*, v. 79 (1968).
8.76: Noel Potter, Jr., "Ice Cored Rock Glacier, Galena Creek, Northern Absaroka Mountains, Wyoming." *GSA Bulletin*, v. 83 (1972).
9.35: William A. White, "Drainage Asymmetry and the Carolina Capes." *GSA Bulletin* (1966).
9.36: John Hoyt and Vernon Henry, "Origin of Capes and Shoals along the Southeastern Coast of the United States." *GSA Bulletin*, v. 82 (1971).
9.44, 9.45: Paul D. Komar, "Nearshore Cell Circulation and the Formation of Giant Cusps." *GSA Bulletin*, v. 82 (1971).
10.32: Rodger T. Faill, "Tectonic Development of the Triassic Newark-Gettysburg Basin in Pennsylvania." *GSA Bulletin*, v. 84 (1973).
10.43: Clyde Wahrhaftig, "Stepped Topography of the Southern Sierra Nevada." *GSA Bulletin*, v. 76 (1965).
Material from the above articles is reprinted by permission of The Geological Society of America.

3.62: G. I. Atwater and M. J. Forman, *American Association of Petroleum Geologists Bulletin*, v. 43 (1959).
7.23: Paul Edwin Potter, "Sand Bodies and Sedimentary Environments: A Review." *AAPG Bulletin*, v. 51, No. 3 (March, 1967).
Material from both articles is reprinted by permission of The American Association of Petroleum Geologists.

3.4: J. R. Heirtzler, X. LePichon, and J. G. Baron, *Journal of Geophysical Research*, v. 71, No. 2 (1966).
3.15: B. Isacks, J. Oliver, and R. L. Sykes, "Seismology and the New Global Tectonics." *JGR*, v. 73, No. 18 (1968).
3.66, 3.68, 3.69, 3.70, 3.71, 3.78, 3.80: John F. Dewey and John M. Bird, "Mountain Belts and the New Global Tectonics." *JGR*, v. 75, No. 14 (1970).
5.50: John C. Frye and S. L. Schoff, "Deep-Seated Solution in the Meade Basin and Vicinity." *Kansas and Oklahoma, Transactions of the American Geophysical Union*, p. 37 (1942).
Copyright by American Geophysical Union for the above material. Reprinted by permission of American Geophysical Union.

1.2, 2.23, 5.6, 7.29, 7.32, 7.40: Luna B. Leopold, M. Gordon Wolman, and John P. Miller, *Fluvial Processes in Geomorphology.* Copyright © 1964 by W. H. Freeman and Company.
5.18: Charles B. Hunt, *Physiography of the United States.* Copyright © 1967 by W. H. Freeman & Co.
9.20, 9.21: John S. Shelton, *Geology Illustrated.* Copyright © 1966 by W. H. Freeman & Co.
All material reprinted by permission of W. H. Freeman & Co.

7.39: James M. Coleman, "Deltaic Evolution." *Encyclopedia of Geomorphology*, ed. Rhodes Fairbridge. Copyright © 1968 by Van Nostrand Reinhold Co.
8.70, 8.74: Arthur H. Lachenbruch; 2.44, 2.48: J. Büdel, *Encyclopedia of Earth Sciences*, ed. Rhodes Fairbridge. Copyright © 1968 by Van Nostrand Reinhold Co.
3.9, 3.14: Bruce C. Heezen; 9.6, 9.17: F. P. Shepard, "Coastal Classification." *Encyclopedia of Oceanology*, ed. Rhodes Fairbridge. Copyright © 1966 by Litton Educational Publishing, Inc.
All material from the above encyclopedias reprinted by permission of Van Nostrand Reinhold Co.

1.25: Cesare Emiliani, "Quaternary Paleotemperatures and the Duration of the High-Temperature Intervals." *Science*, v. 178 (1972). Copyright © 1972 by the American Association for the Advancement of Science.
4.26: Jurgen Haffer, "Speciation in Amazon Forest Birds." *Science*, v. 165 (1969). Copyright © 1969 by the American Association for the Advancement of Science.
8.71: Roger Bélanger, "Pingo Rising out of the Flat Landscape near the Village of Tuktoyoktuk." *Science*, Cover Photo, v. 174 (November 19, 1971). Copyright © 1971 by the American Association for the Advancement of Science.
8.78: J. R. MacKay. "Ice Wedges Developed in Lake Silts Truncating Glacially Deformed, Ice-Rich Pleistocene Sediments more than 40,000 Years Old." *Science*, Cover Photo, v. 176 (June 23, 1972). Copyright © 1972 by the American Association for the Advancement of Science.
Material from the above articles is reprinted by permission of the American Association for the Advancement of Science.

9.49: F. S. MacNeil, *American Journal of Science*, v. 252, pp. 402–427 (1954).
12.7: R. B. Daniels et al., *American Journal of Science*, v. 258 (1960).
Both reprinted by permission of *American Journal of Science*.

To my wife, Mary, and to my children, Ana, Craig, Mark, Bruce, Raquel, Noel, and Michelle, who all helped in so many ways and who always understood what I was trying to do, even when I did not

Preface

It would be presumptuous to undertake the writing of a book in a field so vast and complex as geomorphology without the encouragement and willing assistance of colleagues with expertise in the various subdisciplines. If this work has merit, it is largely to the extent that I have been able to benefit from the very best these scientists had to offer. Indeed, without such help much of what follows would not have been possible. Needless to say, responsibility for any shortcomings lies solely with the author.

I should first like to thank Dr. Lawrence K. Lustig, formerly of the U. S. Geological Survey and now Senior Editor of the Encyclopedia Britannica. He has carefully read the entire manuscript, and I am most grateful for his many invaluable contributions in thought, word, and deed. Professor Stephen C. Porter of the University of Washington also most willingly read and criticized the first five chapters and additionally gave freely of his special technical knowledge to revision of the eighth chapter on glacial and cryergic geomorphology. Professor Robert F. Black, formerly of the University of Wisconsin and now at the University of Connecticut, read several early chapters and offered much thoughtful assistance and encouragement. So too did Professor Lee Wil-

son of Columbia University, who made many helpful suggestions and comments. Professor Rhodes W. Fairbridge of that same institution has also been most considerate. Professor William N. Gilliland of Rutgers University in Newark also kindly read and criticized early drafts and made particularly valuable suggestions on structural geomorphology and geotectonics in the third chapter and in other ways in the fifth and seventh chapters, as well as in later sections. He also helped in long thoughtful discussions and by his boundless enthusiasm for the undertaking. Dr. John Gibbons, recently of Rutgers University in New Brunswick and now of Dames and Moore, Environmental Consultants, was also generous with his time and knowledge of structural geology and petrofabrics in attempts to improve early drafts of several chapters and especially Chapter 3. Special thanks also go to Professor Warren E. Yasso of Columbia University who read and offered much valuable assistance in finalizing the treatment of coastal geomorphology in Chapter 9 and to Drs. Victor Goldsmith of the Virginia Institute of Marine Science and W. W. Wiles of Rutgers University, Newark, N.J., who served in a similar capacity. I am also most grateful to Mrs. Barbara Ryan of Upper Montclair, New Jersey, for her early volunteer copy editing of Chapters 1 and 2. In addition, I am most happy to acknowledge my gem of an editor, Ellis Rosenberg, Leslie Phillips, who processed illustrative material, Joyce Berry, Jean Shapiro, and the many other helpful people on the Oxford staff.

I should also point out that several researchers have been most free with hitherto unpublished information and have supplied data from several papers yet in press. These generous souls include notably R. H. Konig and Al Chin, John Gibbons, L. K. Lustig, K. L. V. Ramana Rao, and W. N. Gilliland.

This book includes some 600 illustrations, a good many of which have not previously been published, and I would like to acknowledge the many who especially assisted in the monumental task of assembling these materials. Those photos not credited in the figure legends are by the author. Special thanks go to my wife, Mary, who did final drafting on the vast majority of my line drawings. Illustrations otherwise acquired or prepared are individually acknowledged within. Joyce Berry of Oxford University Press was most helpful in securing a number of crucial photos from obscure sources. Thanks are also due Professor William N. Gilliland of the Department of Geology, Rutgers University, Newark, New Jersey, who generously gave me complete access to his photo collection. I undertook a special photographic trip throughout the United States and parts of Canada to acquire original photocopy during the summer of 1973. Mr. John Zalkowski of the Geology Department of Rutgers University in Newark was also most helpful in illustration preparation, particularly in processing of several hundred photographs. In that same department, Miss Murial Meddaugh did much of the final typing of the first six chapters. Miss Jean Shapiro of Oxford University Press and Mrs. Phyllis Lennon of Upper Montclair, New Jersey did three, and my son, Mark, completed the job. To the many others who have contributed to the completion of this work, my sincere thanks.

Though this text is designed to be used at the college level, it is my experience that persons taking such courses and others with an interest in the subject of landscape origin have varied backgrounds, and many lack more advanced geology courses. With this in mind, and to assist anyone who occasionally forgets things, an extensive glossary of basic geomorphic terms (italicized in the text) is provided at the end of the text. In general I have attempted to use a dual numerical system throughout this volume (English and metric). Through use of some illustrations from other sources and because of overly cumbersome wording which would locally result from this practice, in certain instances only one or the other system is employed. There is little doubt that the world is gradually converting to the metric system, but that seems insufficient reason to penalize those who have not as yet so adjusted, and there are many.

In writing this book I have made some considerable effort to be comprehensive in terms of subject matter, broadly inclusive in terms of source citations and references, and meticulous in seeking out errors of fact or omission. I have no doubt that any success in these endeavors is a matter of degree and that I will be informed of my fallibility in due course.

H. F. G.

Rutgers University
Newark, New Jersey

March 1974

Contents

INTRODUCTION

1 CONCEPTS OF GEOMORPHIC THEORY   1

DEVELOPMENT OF GEOMORPHIC THOUGHT   1

Beginnings—Catastrophism 1   Uniformitarianism and Gradualism 2
Environmental Passivism 3; *The Geographical Cycle, The Geographical Cycle
Summarized, The Treppen Concept and the Pediplanation Cycle*   Environmental
Dynamism 17; *The Relict Problem, General Systems Theory, Geomorphic Systems,
Dynamic Equilibrium*   Relict Destruction 27   Geomorphology in a
Theoretical Vacuum 32

GEOCHRONOLOGY AND QUATERNARY PALEOCLIMATOLOGY   33

Geologic Time Scales 34   The Quaternary Paleoclimatic Record 35

UNIFORMITARIANISM REVISITED   38

TWO BASIC PRINCIPLES   40

Principle 1. Uniformitarianism 41   Principle 2. Environmental Dynamism 42

REFERENCES   43

2 GEOMORPHIC OBJECTIVES AND METHODS   46

THE GENESIS OF LANDFORMS AND LANDSCAPES   46

The Problem 46   Description and Measurement 47; *External Form, Internal
Aspects, Problems and Analysis*   Landform Chronologies and Quaternary
Stratigraphy 83; *Chronologic Principles and Methods, Relative Geomorphic
Sequences*   Quaternary Stratigraphic Techniques 92; *Geomorphic-Stratigraphic
Settings*   Terminology and Nomenclature 95

PALEOGEOMORPHOLOGY AND PALEOCLIMATOLOGY   97

PHYSICAL ENVIRONMENT AND MAN   99

REFERENCES   101

3   TECTONIC TOPOGRAPHIC ELEMENTS   104

CLASSIFICATIONS, CONSTRAINTS, AND CONCEPTS   104

Endogenic Topographic Effects 104   Development of Plate Tectonic
Concepts 106; *Early Wondering and Wandering, Later Drifting and Dreaming,
Where Do We Go from Here?, Rocks and Topography*   Ocean Floors 114;
*The Mid-Ocean Ridge—An Orogenic Zone, Ridge and Rift Topography—An
Epeirogenic Realm, Oceanic Trenches—Orogenic Zones*   Continents 125;
Epeirogenic Continental Settings, Orogenic Continental Settings   Mesogenic
Topographic Effects 189; *Gravity Tectonics and Topography, Volcanic Geomorphic
Mechanisms, Subaqueous Volcanic Topography, Subaerial Volcanic Topography*

HOW IT ADDS UP   203

REFERENCES   207

4   SURFICIAL GEOMORPHIC SYSTEMS   211

EXOGENIC ENVIRONMENTAL PHENOMENA   211

General Atmospheric Parameters 215; *Atmospheric Circulation*   Circulatory
Interaction—Atmosphere-Hydrosphere 218; *Atmospheric Moisture Transfer*
Climates, Geomorphic Environments, and Interfaces 225   Geomorphic System
Components 231   Orography, Slope, and Geomorphic Adjustments 233
Recent Geomorphic System Displacements 237   Use of the Geomorphic
Space-Time Concept 246

REFERENCES   248

5   HUMID GEOMORPHIC SYSTEMS AND LANDFORMS   251

THE VEGETATED SETTING   251

HUMID GEOMORPHIC PARAMETERS   256

Precipitation and Temperature Variants 256   Vegetal Cover 263

WEATHERING MODES AND EFFECTS   269

Soils and Weathering Profiles 271; *Soil Development*   Primary Structures in
Soils 281

FLUVIAL EROSION AND MASS WASTING   285

Drainage System Development 286; *Inclined Granular Surfaces, Drainage and
Slope Development, Drainage Density, Litho-Structural Drainage Effects,
Trans-Environmental Rivers*

SUMMARY   306

REFERENCES   307

6 ARID GEOMORPHIC SYSTEMS AND LANDFORMS   310

THE DESERT SETTING   311

ARID GEOMORPHIC PARAMETERS   312
Precipitation-Temperature Effects 312

WEATHERING AND MASS WASTING   316

DESERT FLUVIAL ACTIVITY   327
X_c-E_c-D_c—The Arid Fluvial Event 327; *Accounts of Desert Water and Mud,
Erosion: Transportation? Deposition*   Incisional Realms 335   Deposition—
When, Where, and How 340

PEDIMENTS AND PLANATION   343
The Role of Water 344   Base Levels and Scarp Retreat 344

WIND, SAND, AND SCULPTURE   349
Moving Rock 350   Armors and Crusts 350   Dunes 353   Deflation 356

ARID MORPHOGENESIS   362
Climax Arid Landforms 362; *Desert Plains and Uplands: Theory and Process,
Desert Plains and Uplands: Actuality*   Climax Arid Landforms—
An Assessment 374

SUMMARY   376

REFERENCES   377

7 ALTERNATING ARID-HUMID GEOMORPHIC SYSTEMS   381

THE ARID-HUMID ENVIRONMENTAL CONTINUUM   381
Humid-to-Arid Semiaridity 385; *Setting of the Stage—Humid Geomorphic Relicts,
Humid-to-Arid Disequilibrium, Streamfloods and Stripped Surfaces*
Arid-to-Humid Semiaridity 398; *Setting the Stage—Arid Geomorphic Relicts,
Moving Water and Sediment*   Arid-to-Humid Drainage Adjustments 406;
*Channel Redevelopment in Drainage Nets, Alluviated Channel Evolution and
Maintenance, Arid-Humid Deranged Drainages, Arid-Humid Geomorphic Analysis*

REFERENCES   444

8 GLACIAL AND CRYERGIC GEOMORPHIC SYSTEMS
AND LANDFORMS   448

GLACIAL ENVIRONMENTS   449
Existing Glacial Regimes 449   Past Glacial Fluctuations 451

GLACIAL GEOMORPHIC EFFECTS   453
Glacier Movement 453   Patterns of Erosion and Deposition 459; *Glacier
Erosion, Glacier Transport, Mass Wasting, and Deposition*   Alpine Glacial
Landforms 471; *Cirque Development, Cirque-Shaped Landscapes, Valley*

Glaciation, Valley Glacier Deposition Continental Glaciers and Landforms 497;
Erosional Terrain Morphology, The Depositional Halo, Glacial Fringe Benefits

CRYERGIC GEOMORPHIC EFFECTS 517
*Permafrost, Cryergic Process Associations, Cryergic Weathering, Cryergic Mass
Wasting, Eolian and Fluvial Activity*

SUMMARY 530

REFERENCES 531

9 POLYGENETIC LANDSCAPES/COASTS 536
COASTAL GEOMORPHIC SYSTEMS—ENVIRONMENTS IN CONTACT 537
Coastal Environments 537; *Water Conditions, Organisms, Subaerial Land
Conditions, Sedimentary Effects, Tectonic and Volcanic Effects* Coastal
Development and Classifications 541 Coastal Waves and Currents 546;
Waves of Translation, Near-Shore Currents Marine Geomorphic Evolution 556;
*Terrigenous Coasts of Primary Morphologic Disequilibrium, Terrigenous Coasts of
Primary Morphologic Equilibrium, Organic Coasts*

SUMMARY 584

REFERENCES 585

10 POLYGENETIC LANDSCAPES/MOUNTAINS 589
OROGENIC ELEVATION AND RELIEF 591
Character and Maintenance of Mountainous Uplift 592 Early Orogenic
Denudation 594 Environments of Orogenic Uplift 596; *Humid Orogenesis,
Arid Orogenesis, Arid-Humid Orogenesis, Climatic Orogenesis at Moderate
Elevations—A Synopsis* Uplift Morphogenesis: A Theoretical Synthesis 609;
*Low-Level Planation and Aggradation, Intermediate-Level Fluvial Incision, High
Montane Environments* Orogenic Landscape Case Studies 617
Covermasses and Transverse Drainages 617; *Montane Geomorphic Asymmetry*
Terminal Aspects of Orogenic Topography 633

SUMMARY 635

REFERENCES 636

11 ANCIENT LANDFORMS AND LANDSCAPES 640
UNCONFORMITY AND DEPOSIT ANALYSIS 640
Buried Paleozoic Landscapes of the Ozark Region 646; *Basal St. Peter
Unconformity, Post-Boone Unconformity, Mississippian-Pennsylvanian Unconformity,
Bloyd-Winslow Unconformity* Early Pennsylvanian Geomorphology of the
Mid-Continent 651; *Quartz and Quartzite Provenances, Space-Time Drainage
Relationships, Environmental Relationships* Alan B. Shaw and the
Cyclothems 659; *The Tideless Sea and the Strand, Cyclothemic Landscapes and
Environments*

PALEOCLIMATOLOGY, TECTONISM, AND TOPOGRAPHY   667

Trans-Latitudinal Continent Shifts   668; *Migration and Breakup of Gondwanaland*
Ocean Basin Desiccation 672

SUMMARY   674

REFERENCES   675

12 ENVIRONMENTAL GEOMORPHOLOGY   677

MODERN ENVIRONMENTAL PROBLEMS   677

History of Environmental Complexity 677   Environments, Relicts, and
Disequilibria 680; *Accelerated Erosion Problems, Water Problems, Permafrost
Problems*

SUMMARY   693

REFERENCES   693

GLOSSARY   695

INDEX   711

Introduction

Of all the myriad aspects of the solid Earth, perhaps the first to catch and hold our eyes is the sculptured surface. The land has form. The form confronts man as scenery and building site, as route and obstacle, as smooth, fertile slope or rough, barren wasteland. From the encounter man emerges variously pleased, puzzled, enriched, impoverished, or even intrigued. From any or all of these experiences our curiosity may lead us to study the origin of the landscape.

Geomorphology is the study of landforms. At the very least, the study goes beyond mere description and geographic location to an abiding concern about the "why" of landforms and landscapes. Our approach must also be historically scientific (in the geologic tradition) to ascertain the "when" of the things recorded. Moreover, if we observe processes, their settings, and their effects very carefully, we may find out the "how" and "where" of landforms. Most importantly, the study of all these landscape aspects can usually create in the student a unique and satisfying awareness of the role of environment. Potential applications of this knowledge extend far beyond the direct concerns of geomorphology and geology.

The rightful concerns of geomorphology are those process associations that give form to the outer surface of the lithosphere plus the shapes that result. These shapes are principally in the nature of sculpted surfaces and deposits. But continents are, in a large degree, landforms. Their generation and modification necessarily directs our geomorphic interests, at least passingly, beneath the seas and elsewhere to crustal changes whose interplay is seemingly causal.

Geomorphology is not an elementary geology course. In a sense it is applied geology. A broad background in basic sciences and particularly in physical and historic geology and some structural geology and petrology are assumed. A passing acquaintance with organisms, climate, soils, and fossils would be helpful. The reasons are simple. Some landforms, for example those dependent on solutioning of bedrock, mainly reflect particular lithologic variations (Fig. I-1). Others, for instance those dependent on certain bedrock stratal inclinations, owe their form largely to special structural effects (Fig. 3.73). Still others take their shape "in the grasp" of some particular agency and therein mostly express the special influences of process (Fig. 8.55). Some unique landforms are brought about by environment through a special sequence, a particular dominance, or an unusual duration. Commonly there are environmental

Fig. I-1. Natural Bridge, Virginia, a product of solutioning of limestone followed by cavern collapse. (Virginia State Travel Service.)

combinations. The complexities of these relationships are developed at length here. Over each landform is soil or regolith; only rarely, is there nothing. Additionally, there are the ever-present problems of relative landform ages and the consequent need for telling of time with fossils, radioactivity, or physical relationships.

Geomorphology is very special. It ranks as applied geology in that some mastery of each of the geologic subdisciplines is necessary for one to work on the more complex geomorphic problems. Alternatively, geology is in a very real sense applied geomorphology. This relation emerges because geomorphologists combine the historic perspective so dear to geologists with an accentuated awareness of contemporary process. Geologists have the difficult task of unraveling the history of the Earth, its ancient environments, processes, and their effects. The geologist is sorely in need of analogs developed in a "uniformitarian" context by which to gauge the earth's past. And one potential by-product of the geomorphologist's efforts to comprehend landforms is commonly some understanding of environment-process-product relations which can serve as geologic analogs. Geomorphology is rewarding in its own right. Most geomorphologists are content to learn more about the manner in which the earth is shaped. The generation of geologic analogs is merely an added bonus.

The intent here is to communicate about landforms and landscapes. Where pictures, drawings, or diagrams assist, they have been used. The same is true of purely descriptive prose. Where increased understanding seems to come from numbers, quantification is employed. But we will strive for a happy blend of rationalism and empiricism by eliminating both "liars that figure" and "figures that lie." Since we have an intrinsic interest in cause and effect, there is some selectivity in our choice of numbers. And in each case where a question is answered by a quantitative answer, an effort is made to verbalize the significance of the quantity. I hope that the various media of expression have been blended

in a complementary fashion. Private jargons, whether terminologic or numerical, are left to the special interest groups for whom they were devised.

The presentation of geomorphic subject matter here is not quite traditional. We have changed the "Physical Geology" textbook subject sequence which gives each geologic process separate treatment in a chapter or section. Landforms are rarely created by a single process working by itself. Processes, like men, are best known by the company they keep. Our discussion will therefore emphasize process associations. Furthermore, there is reason to believe that processes significantly vary in effect in different environments. With this possibility in mind, we will treat a particular process in its several environments; effects of running water, for example, will be discussed in humid, arid, and glacial contexts and again with respect to alternating environmental conditions.

Growing awareness of the common impermanence of Earth environments has also influenced our rather "non-traditional" presentation of material. Growth of that awareness is charted in Chapter 1 where we also plot the development of geomorphic theory. Chapter 2 is devoted to geomorphic techniques of taking data and recording and processing it during landform analysis. Chapters 3 and 4 deal with the general mechanisms that appear to govern internal and external earth environments, respectively. There we discover that environments appear to induce landform-making systems whose effects where intensively developed are unique. The tectonic and geophysical discoveries discussed in Chapter 3 and accomplished in recent decades are found to strike deep at the theoretical roots of much early geomorphic thinking. G. H. Dury has noted the resulting theoretical vacuum in geomorphology in which recent researchers have been operating.

Our approach to geomorphologic subjects has been influenced by the aforementioned theoretical vacuum and by the character of existing general geomorphic texts. An effort is made here to provide a central unifying theoretical theme reflective of our more recent findings: in geotectonics, in environment-process relations, and in environment-space-time effects. Existing texts set few viable over-all patterns. Entire books are devoted to such subjects as *coasts, tectonic landforms, fluvial processes,* and *volcanic landforms*. These efforts lack little in detail. But they hang separately rather than together. The most widely used and thoroughly integrated general geomorphic text differs little from its initial version composed two decades ago on the dominant theoretical theme devised by W. M. Davis near the turn of the last century.

Our general theme herein reflects genesis of landforms by associations of processes within environmental systems and through the substitution of one system for another in time and space. The driving forces are seen to be both internal and external with respect to the earth's surface. Following our consideration of landforms relating to interal mechanisms in Chapter 3 and an over-all look at surficial phenomena in the next chapter, we embark on a detailed consideration of the specific geomorphic effects of particular surficial conditions, especially as typified by certain types of groundcover and amounts or states of moisture. Chapter 5 treats the humid, vegetated environment and its peculiar geomorphic effects insofar as we understand them. Chapter 6 gives a similar exposure to arid, non-vegetated settings, and in Chapter 7 we attempt an analysis of the consequences when arid and humid systems have mutually replaced one another in time and space. The concern in the latter is with environmental relicts and what happens to them in new settings.

Chapter 8 surveys glaciation and glacial landforms as additional aspects of another kind of surficial environment, plus cryergic conditions and their effects. Chapter 9 treats the great group of polygenetic landscapes exemplified by coasts where environments are in continual contact, and Chapter 10 considers mountains as a second major group of polygenetic landforms where internal and external Earth processes vie most directly with one another. The two closing chapters (11 and 12) consider the application, respectively, of geomorphologic techniques and principles to ancient landforms and deposits (paleogeomorphology) and their essential antithesis —the current environment and its problems. Thus, we intend to come full circle, from the processes and conditions of the present to the interpretations of the geologic record, and back again to a better understanding of our present surroundings.

An interesting consequence of the present synthesis is that it is at once new and not new. The volume of geomorphic literature utilized here, and not previously incorporated in a general geomorphology text, is vast indeed. There, much is new, in both fact and theory. But the older ideas and observations are examined and compared and contrasted as we proceed and are not found entirely wanting. Far from it. Indeed, many of the original observations and conclusions made by W. M. Davis around the end of the 1800's and long before by many others about such things as the nature of humid erosion, the character of glaciation, or the effects of prolonged aridity have an essential validity, often even in detail. Our placement of these and related environments and their associated processes in a space-time continuum hardly alters their basic character. But it should be emphasized that most landscapes are products of the environmental sequence, and this is the essential theoretical thread that binds otherwise detached concepts together into what we hope is a meaningful whole.

THE ORIGIN OF LANDSCAPES

1 Concepts of geomorphic theory

DEVELOPMENT OF GEOMORPHIC THOUGHT

BEGINNINGS—CATASTROPHISM

Geomorphic theory is rooted in the origins of geology, but not all the tendrils are traceable through the obscurity of the past; in any case, space limitations forbid the effort. Interest in the general subject of landforms widened about the beginning of the 1700's when the exchange of ideas necessary to science began. I do not intend to ignore the early and rare brilliance of such giants as Leonardo da Vinci (1452–1519), who apparently understood much of what we are still trying to find out. But da Vinci stood almost alone. We should therefore briefly examine the conceptual climate of that time before noting individual contributions.

First, several factors influenced the thinking of the eighteenth-century naturalists who sketched the beginnings of geomorphology. It was a time of strict religious orthodoxy; in Western civilization this usually related to some form of Christianity. The most widely quoted book was the Bible; it was accepted literally by many, and, apart from The Creation, the biblical event of greatest geomorphic import was Noah's Flood. In keeping with the emotional climate of the times, major natural happenings were frequently attrib-uted to willful acts of various deities. In a population generally unskilled in detailed observations of minor changes it is under-standable that catastrophes were most im-pressive as agents of change to the early earth scientists.

Second, the vast majority of early earth scientists developed their concepts in Eng-land, Europe, and somewhat later, in east-ern North America. These regions have abundant rainfall, vegetation, and running water. Not surprisingly, the early schemes of landscape genesis reflect this. Even now it is commonplace to read that running water is the most important agent of land sculpture. Perhaps it is. But you might have difficulty convincing an Arabian or a Greenlander of this.

Third, keep in mind that early theoreti-cal formulations were made relative to a vastly foreshortened earth-time scale. Until at least 1750 it was heresy in the Christian world to assume that the earth was more than some 6,000 years old. Relative geo-logic dating was in its infancy. The entire 4,500-million-year (plus) history of the earth had to be telescoped. Individual geo-logic events were thus effectively acceler-ated in the minds of the theorists and were, to that degree, rendered somewhat catastrophic. Radiometric dating of geologic events did not become important until the 1950's.

Fourth, one should be aware that notions of climate change hardly entered into early geomorphic thinking. This probably relates in part to the biblical tradition of a special creation where everything in paradise was "just so." Apart from glaciation, which was considered accidental, subaerial environments were regarded as geographically fixed and therefore as passive elements in landscape genesis. Landform-making was attributed to subaerial processes after appropriate homage had been made to factors involved in bedrock formation.

With some major aspects of the intellectual atmosphere of the 1700's in mind, let us now consider the more outstanding ideas of early geologists and geomorphologists. We cite both, although initially there were neither, at least by those names. Rather, there were a surprisingly large number of "naturalists" who did not hesitate to do simultaneous research in such areas as medicine, botany, hydrology, mineralogy, zoology, and law in a list of activities that were often of overpowering diversity. The mind boggles. Considering the state of the geological arts and scientific techniques of the time, their accomplishments were formidable.

UNIFORMITARIANISM AND GRADUALISM

Ideas of geomorphic significance go back at least to ancient Greek, Arabic, and Asiatic cultures. But modern geomorphic approaches date little farther back than the work of James Hutton (1726–79). This is not because earlier workers lacked valid notions of natural processes. On the contrary. However, in 1785 Hutton developed and expounded what has come to be known as the "Law of Uniformitarianism." Its essence is that *the present is the key to the past*. Formulated in direct opposition to the then-popular doctrine of catastrophic

earth change, Hutton's thesis related changes in the earth to natural, observable, commonplace processes. And all subsequent geologic and geomorphic theories of wide acceptance and application have been framed with due respect to Hutton's "law."

The Law of Uniformitarianism focused attention upon agents of change that might have caused various geologic features. Those who continued to emphasize the earth surface (*surficial*) effects of process became the first geomorphologists. Hutton himself related granite to an igneous origin in opposition to the Wernerian* notion that it had been formed chemically in water. Playfair (1748–1819), who voiced Hutton's views with a fluency never matched by their originator, described in a most logical fashion the manner by which running water was responsible for the formation of river valleys. He further noted (1802) that river tributaries enter trunk streams at the same level, presumably because the drainage system develops as a unit—a concept now incorporated into the "Law of Accordant Junctions." And countless others expressed their views about the effects of ocean waves and currents, volcanism, glacier ice, weathering changes, and mass wasting. Not surprisingly, there were numerous points of disagreement. There are even now.

The most striking aspect of Hutton's "law" is that papers discussing its validity and applicability continue to appear annually almost two hundred years after its proposal. One gathers from this that, at least, the law is not an easy rule to apply. This difficulty of application should be examined in depth since conclusions based on the law have become the very substance of geomorphology and geology. One of the difficulties relates to the Huttonian implication of gradualism, which was later advocated with great vigor by Charles Lyell (1872).

In effect, geologic gradualism implies that earth changes generally are slow, that they occur as small increments, and that

---

* Abraham Gottlob Werner (1749–1817), German mineralogist.

they involve relatively weak forces. Only rarely do changes occur rapidly, on a large scale, with great force. Since time is an abundant geological commodity, the earth has become what it is through the prolonged action of weak forces. Following this thesis of gradualism, it has been relatively easy for geomorphologists to document many of the slow changes that occur, and to demonstrate that these changes can, in time, be very effective.

The reality of slow natural changes in no way negates other observations that force us to review gradualism as an interpretive approach in geomorphology. The record of rare but severe earthquakes, volcanic eruptions, floods, hurricanes, and the like shows that events with recurrence frequencies measurable in hundreds or even thousands of years can produce tremendous changes. For example, decades of measured sediment transport in a stream may pass prior to a flood that modifies an entire channel configuration in a matter of hours (Schumm and Lichty, 1963; Leopold, Wolman and Miller, 1964, pp. 81–85, 208). Elsewhere, a variety of shoreline marine processes, operating over scores of years, can produce a configuration of hooks, spits, offshore islands, bars, beaches, and inlets that vanish before the force of a major storm in a few hours (Wanless, 1965, pp. 76–79). The natural catastrophes are a part of the present, and thus one key to the past, but uniformitarianism as a concept requiring gradualism is indeed difficult to apply.

A second difficulty confronts the uniformitarian analyst. He must be able to relate a geologic process to its effects. The effects in this case comprise the landforms and deposits being created. A not uncommon response to this statement is, "What is the problem?" The question should be in the plural. For one thing, the process literally MUST BE CAUGHT IN THE ACT. It is easy enough to go see a river flowing in Pennsylvania or France (though this is not equivalent to knowing WHAT the flow accomplishes or *how*). But consider the difficulties confronting one with a desire to see it rain in Egypt, or even Arizona. Chances are

a person who tried this in the more moist of these deserts would return to report "After several weeks there finally was a shower near Flagstaff but I was in Winslow at the time." Similar difficulties attend the efforts of those who desire to study the effects of running water in Algeria or Arabia or who wait by a steep slope or cliff to observe the results of a rock fall. Thus, there is the problem of evaluating the significance of rare events and THE RELATED PROBLEM OF EVEN RECORDING THE NATURE AND EFFECTS OF SUCH EVENTS.

There are at least three other difficulties related to Hutton's "law" that merit attention here. One such difficulty is encountered when the effort to record the behavior of a given geomorphic agent alters that behavior. The classic example of this is, of course, the so-far futile attempts to measure stream *traction loads* when insertion of the sampling device distorts bottom conditions. The quantification of fluvial erosion rates has been notably hampered by this problem. A similar difficulty attends those who would record the nature and effects of processes that function very slowly, though more or less continuously. Weathering and certain types of mass wasting are particularly good examples of this, as are the more gradually expressed effects of erosion on consolidated bedrock. In some cases these difficulties can be overcome by the establishment of long-term Vigil sites where periodic measurements are made (see, for example, Washburn, 1967). Finally, there is the problem created by the fact that ours is a historic science. Each scene down to the present is formed from ruins (*relicts*) of the previous scene. These relicts often become confused with things now forming. The many complications of this relict problem will be detailed in the section on environmental dynamism.

ENVIRONMENTAL PASSIVISM

Hutton's law and its logical extension in Lyell's gradualism found expression in much of the geomorphic thought of the nineteenth and early twentieth centuries. Its culmination appeared in the geomorphic rationale of W. M. Davis (1899), who classified glaciation as little more than a climatic accident. The same view continued to be shared by other members of the Davisian school of geomorphology for almost half a century (Cotton, 1942). Consequently, geomorphic syntheses were commonly stated in terms of static subaerial environmental systems. More precisely, apart from the "accidents," climate was not merely not discussed. It was ignored. Reasons for landform variations were sought elsewhere—mainly in differences of geologic structure, geomorphic process, and stage of development.

Davis described the erosion of lands as a function of decreasing potential energy expended mainly as the kinetic energy of water that runs on an uplifted segment of the earth's crust. Although other geomorphic ideas existed at the turn of the century, none were as pervasive and encompassing as those of Davis. He borrowed extensively and was an excellent synthesizer of the materials he chose. He developed several subordinate hypotheses that have had almost as much effect on later thinking as his general theory. This last is widely known as The Normal Erosion Cycle, The Geographical Cycle, or The Theory of Peneplanation.

The Geographical Cycle

The geographical cycle (Davis, 1899; 1902) had its basis in Huttonian concepts, but Davis also used the ideas and observations of early surveyors of the North American West. Notably among these were three men. J. W. Powell (1834–1902) is best known for his explorations of the Grand Canyon of the Colorado River and for his support of the idea of an erosional base level. G. K. Gilbert (1843–1918) is especially acclaimed for his hydrologic researches which culminated in a statement about the load-carrying abilities of streams; since termed the "Law of Stream Capacity" (Garner, 1959). He is also well known for his classic geologic studies of the Henry Mountains. C. E. Dutton (1841–1912) explored much of the Colorado Plateau and developed the basic concepts of widespread denudation that have subsequently been applied elsewhere.

The Davisian landform synthesis can be summarized as follows. An uplifted land remains structurally stable while it passes through a series of time-significant erosional stages during its progressive lowering and leveling. These stages, by analogy to man's growth were designated YOUTH, MATURITY, and OLD AGE, a time framework that was readily grasped and has been immensely popular down to the present (Fig. 1.1). The presumed long-term landform culmination was the low-relief, low-elevation erosion surface Davis called a *peneplain*. The assumed landscape and landform changes depended on a series of secondarily postulated interrelationships. These were largely included in Davis' original synthesis but most were elaborated upon in many subsequent papers up to the time of his death. The following aspects of the geographical cycle appear to be most basic.

1. Constancy of Environment. Though it is rarely expressly stated, Davis was clearly speaking of changes occurring in a humid, temperate environment characterized by abundant rain, vegetation, chemical modes of weathering, and river drainage systems. He called such an environment "normal" and described the stream's load as

> . . . *small at the beginning and rapidly increases in quantity and coarseness during youth . . . it continues to increase in quantity, but probably not in coarseness, during early maturity. . . .*

The stream sedimentary loads postulated are thus contradictory to the chemical weathering mode in a humid land (Thornbury, 1969) which normally produces fine-grained soils with particle dimensions that progressively diminish with time (cf. Chapter 5).

Later Davis (1905) developed a separate scheme to accommodate the landscape changes that he believed occurred accidentally, elsewhere, under desert conditions. In no instance did he indicate any belief that subaerial environments were NORMALLY interchangeable.

2. Fluvial Erosion. The main erosional mechanism discussed by Davis was running water in streams, supplemented by the downslope creep of soil under the pull of gravity. This is certainly compatible with the humid climate he took to be the world norm. Drainage was analyzed by Davis in terms of kinetic energy to the extent that the efficiency of the fluvial system was usu-

A. **Youth**: v-Shape valleys, few or no floodplains, extensive interfleuves, many falls and rapids plus some lakes and swamps; incising watercourses

B. **Maturity**; Well-drained terrain, all in slopes except floodplains; trunk and some tributary streams meander; maximum relief

C. **Old Age**; Broad, open valleys with widely meandering streams, indistinct divides, erosion remnants of resistant lithologies, surface near erosional base level

Fig. 1.1. Three-stage diagram of the classical Normal Erosion Cycle of W. M. Davis (1899).

ally expressed in terms of slope/velocity/ load relations. Energy, as potential energy, was introduced initially by land uplift and ultimately consumed by degradation. The effects of other variables including discharge volume were also considered, but Davis generally expressed the erosional power of the water directly as a function of slope. Since watercourses typically exhibit steep headwaters reaches and more nearly flat downstream areas (Fig. 1.2), a stream's sediment-carrying capacity was thought to decrease downslope or downstream as its velocity decreased. From this, one can conclude that Davis either knew no riverboat pilots or was not guided by their remarks on this subject. Any of them could have told him that river currents pushed down more strongly against their side-wheelers near the river's mouth than upstream. Some pilots were aware of this from travel-time records; others knew some of the relations between river cross section, *wetted perimeter*, slope, and velocity as early in 1882. Samuel Clemens records both types of information (Mark Twain, 1944, pp. 110–12, 336–68). Curiously, G. K. Gilbert (1880) had already worked out the same relationship more technically by the time Davis wrote. More recently, Leopold, Wolman, Maddock, and others have confirmed the technical implications of Gilbert's Law of Stream Capacity (cf. Chapters 5 and 7).

In summary, we now know that the volume of a normal humid river increases downstream at a rate and in a manner that more than compensates for decreasing slope. The manner is examined in Chapter 7, where it is noted that wetted perimeters of channels actually decrease in sum as channel cross sections increase (cf. Atkinson [in] Twain, 1944, p. 368). As a result, velocity actually increases downstream in many cases,* and erosive power is augmented. Thus, a typical humid river cannot induce more than local, sporadic aggradation of its own channel (Carlston, 1968).

3. The Graded Stream. This concept appears to have developed directly from the

---

* See **Alluviated Channel Evolution and Maintainance, Chapter 7.**

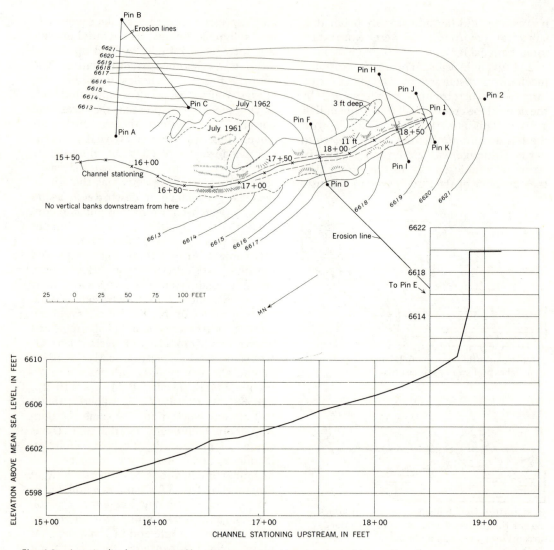

Fig. 1.2. Longitudinal stream profile of an arroyo in New Mexico with correlative topographic map show-ing pin locations for an erosion study. (From Leopold and others, 1964.)

rationale employed by Davis in the previous concept. As noted by Dury (1966), *"By its very persistence, debate suggests that the concept of grade offers persistent difficul-ties."* In brief summation, according to Davis (1902) a *graded stream* is a mature or old river, one with a balance between ero-sion and deposition (e.g., between capacity to do work and quantity of work to be done), the condition being first established in downstream reaches (cf. Fig. 1.1). In a re-cent redefinition that is widely quoted, it is assumed that the water of the graded stream is moving over the lowest slope and with just the velocity required for the sedi-ment load to continue to move (Mackin, 1948). Stream action via the graded state was inferred by Davis and his followers to result ultimately in a longitudinal slope con-figuration tied to base level and termed the *profile of equilibrium* (Fig. 1.3).

The most exhaustive treatment of the graded stream concept of recent date (Dury, 1966) includes several very significant observations. Davis apparently adopted the equilibrium idea from Gilbert (1880). In this regard Kesseli (1941) observed that as long as a stream's load is in motion the bed undergoes *corrasion*. Thus, since underloaded and fully loaded streams alike erode, and overloaded streams deposit, *a neutral condition of balance (grade) is impossible*. In fact, *overloading*, cannot occur except on an instantaneous or a local basis, for this would imply that work is being accomplished in excess of the energy available to do it.

A protracted search for streams in the graded state has been fruitless, though short reaches of various rivers have been cited as being in a state of quasi-equilibrium. And most recent discussions proceed in this vein. Leopold and Maddock (1953) conclude that the average system of river channels tends to develop so as to produce an approximate equilibrium between the channel, on the one hand, and the water and sediment in it, on the other. Wolman (1955) points out that a smooth longitudinal profile of descent need not be present (i.e., there may be *knickpoints*) along a stream with reaches in quasi-equilibrium. Incidentally, a REACH is rather like a league as a unit of distance measure. In Mexico it is about the distance a man can ride a horse in an hour. It varies some. Dury (1966) sums up his discussion to the effect that ". . . because the concept of grade necessarily involves the converse idea of an ungraded state, it is to be recognized as unserviceable both in the study of actual terrains and in the theoretical analysis of landforms generally."

4. The Base Level of Erosion. This concept is based largely on ideas presented by Powell (1875), though both Leonardo da Vinci and James Hutton recognized the principle. Conceptually, *base level* has both *regional* and *local* aspects. In the local sense, to be described, it has scarcely been questioned. In the sense used by Davis, regional base level is essentially mean sea level, with the implication that downward erosion by running water stopped there and

Fig. 1.3. Idealized longitudinal regional profile of a stream in relation to the theoretical regional base level and sea level.

that the lower end of a stream's profile of equilibrium was fixed at that level. *Base leveling* is the term frequently applied to the erosional reduction of a land to a surface that is essentially a plane, representing the landward extension of regional base level (Fig. 1.3). The ultimate, low-relief erosion surface resulting from base leveling is the *peneplain* of the Davisian scheme.

Peneplains were presumed to slope with respect to regional base level and to be controlled erosionally by that level through continuous slopes formed by graded streams. Local (temporary) base level has been variously employed (1) for trunk channels of major rivers at grade relative to all superjacent tributaries and slopes, (2) for erosionally resistant obstructions (knickpoints) within streams, and (3) for local depressions in deserts. Presumably, local base levels induce breaks in longitudinal stream profiles and, if expressed in a series of adjacent drainages, could induce regional breaks in slope.

Discussions of the regional base level concept have proceeded along several lines. Three or four of these appear to have greater geomorphic significance than the others. One theory, advanced by A. Penck (1894), is that glacial ice cap (sheet) growth and wastage frequently produce changes in sea level and, hence, regional base level. This has been extensively verified by several converging lines of evidence. Such glacio-eustatic sea level changes of a first magnitude number at least eight and possibly ten during the Pleistocene Epoch (the past ±2.5 million years).* Moreover, such base level shifts do not seem to be con-

---

* Deep-sea core records of eight fluctuations in the past 700,000 years may or may not coincide with and reflect the continental glacial sequence.

fined to that epoch. Under these conditions, it is difficult to explain why a regional, low-relief erosion surface developed over a period of millions of years should exhibit a single general slope attitude. In effect, the presumed gradational target (base level) would be shifting vertically, up and down, as much as three or four hundred feet every few tens of thousand of years during the over-all erosional episode. If the control were effective, one would expect the resulting erosion surfaces to have several hundred feet of relief. Many planar landforms in central Australia, and on the Colorado and Bolivian plateaus lack such relief.

The second development of note on regional base level concerns numerous observations that show erosion does not stop there. Tidal currents, marine *turbidity (density) currents* and major river channels all erode well below mean sea level. The Mississippi River channel is locally more than 100 ft (30 m) below sea level and the Amazon channel several hundred feet (100+ m). These and several related agencies demonstrate that gravitationally directed water movement and erosion are not limited to the subaerial realm, nor are they limited by base level (Garner, 1968). Widespread and frequent structural movements are acknowledged in some areas and also enter the theoretical picture of base level control there.

Finally, if our understanding of the proposed mechanism of base level control of regional erosion surfaces is correct, such control would only apply to lands where environments sponsor *exhoric drainage* (i.e. runoff flowing normally to the oceans) and therein only to the extent that graded streams exist in the Davisian sense (Garner, 1965). In this context, L. C. King states (personal communication to C. A. Cotton, 1960) that he accepts regional base level as a factor controlling pediplanation (see next section and also Wheeler, 1964).

5. Downwasting. This is a slope reduction process corollary to the progression of Davis's normal erosion cycle. It emphasizes the pull of gravity on weathered material as a mechanism to supply streams with sedi-

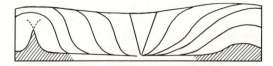

Fig. 1.4. Valley widening and slope variation during erosional cycles. *Right*: downwearing, youth to old age, according to W. M. Davis. *Left*: backwearing, with parallel scarp retreat, according to W. Penck. (After W. M. Davis, 1930.)

ment. It involves the vertical lowering of land slopes, principally by creep as opposed to the back-wearing of slopes advocated later by King (1953) and others (Fig. 1.4). The process was presumed to slow down as slopes became lower. Davis's discussion of hillslope development implies that initial erosional slopes steepen as valleys deepen and that steepness continues or even increases until maturity and a graded state is achieved, after which slopes are reduced.

To this writer, what would happen to slopes in the late stages of the Geographical Cycle is a moot point. If, as postulated by Davis, drainage lines achieve a maximum depth at one stage when slopes are also maximal, there can be little question but that an eventual reduction in relief and elevation would cause a reduction of slopes. These issues will be considered at greater length in the section on pediplanation and in Chapter 5.

6. Lateral Stream Erosion. According to Davis, a secondary erosional process involving stream meander swinging occurs and is assumed to lead to base leveling of a crustal segment during the "normal" cycle. Davis infers that this process proceeds in conjunction with graded stream development and downwasting. Effective vertical incision was attributed by him to all ungraded streams that had yet to develop an equilibrium profile. Meandering was assumed to mark the attainment of a graded state when some (unspecified) increment of energy became available for lateral cutting. The flat bottom of his "mature" river valley was considered to be a product of this process (cf. Fig. 1.1).

The hydraulic relationships that seem to

Fig. 1.5. A completely recurved meander developed in unconsolidated alluvium in a small tributary of the Yellowstone River in Yellowstone Park. (W. N. Gilliland.)

govern the cause, size, and shape of meanders will be considered in a later section. We will here restrict ourselves to some observations on meander occurrence and the apparent effectiveness of lateral meander swing as an instrument of erosional planation. The most effective lateral erosion by meanders that migrate would appear to be in unconsolidated sediment, usually alluvium (Fig. 1.5). This judgment is based on the incidence of *scars*, principally *ox-bow lakes, sloughs,* and arcuate *point-bar deposits.* Occasionally, where meanders impinge on consolidated bedrock, there is evidence of effective lateral erosion, the precise nature of which (solutional, corrasional, etc.) is often uncertain. In other instances where bedrock is involved there appears to be little or no erosion.

Studies of entrenched meanders particularly illuminate the modes of bedrock erosion. In a set of entrenched meanders along the Buffalo River in Arkansas there are *slip-off slopes,* occasional *point bars,* and distinct undercutting of cliffs along the outside of meander bends. All these features suggest some lateral erosion during downcutting (Fig. 1.6). In contrast, in most of the awesomely entrenched meanders of the Colorado and San Juan rivers, the average distance is often the same from the mid-line of the stream to the inside and outside valley wall at a given elevation above the river (Fig. 1.7). Thus, there appears to have been little effective undercutting of many outer meander banks during thousands of years of vertical incision.

In the above examples it may be signifi-

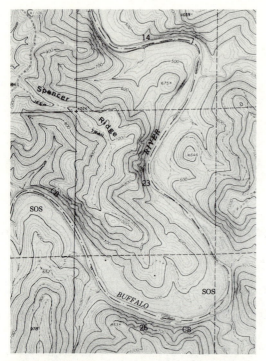

Fig. 1.6 Portion of a 7.5' quadrangle map (Big Flat, Arkansas) showing a reach of the Buffalo River which displays entrenched meanders with slip-off slopes (SOS) and opposing steep cut banks (CB) on the outer sides of bends. Countryrock is middle and lower Paleozoic limestone and sandstone.

cant that the eastern area includes much limestone, friable sandstone, and shale country rock. Rivers in the western area have been working on generally more resistant sedimentary lithologies and at least some igneous and metamorphic types. The relation between meandering and grade postulated by Davis is yet to be empirically demonstrated. Streams that possess straight or nearly straight channels and that are incised directly into consolidated bedrock show little sign of meander development and lateral erosion (Fig. 1.8). Wide-swinging, large-scale meanders are evident in the broad river valleys that exhibit evident alluvial fills. But, as Tanner (1968) correctly notes, the Davisian inference of a particular age ("mature," "old") is incorrect, and

streams of all sizes meander at a variety of elevations, slopes, and discharges. Davis, himself, refuted his original contentions here. There is a growing body of evidence that meandering probably will not occur in the absence of granular bed materials.* In resistant rocks, lateral erosion by meanders appears to be so inhibited as to be inappropriate for planation of large land areas (Fig. 1.8).

7. Rejuvenation. In the original Davisian scheme of landscape evolution, a relative downward displacement of erosional base level (sea level drop or land uplift) was presumed to induce renewed incision by streams. This renewed fluvial erosion was termed *rejuvenation.* The inception of rejuvenation was thought to mark an interruption in the geographical cycle; the increased vertical erosion was attributed to increased potential energy expressed in higher elevations and steeper slopes. More recently, Thornbury (1969) designated rejuvenation as DYNAMIC, EUSTATIC, or STATIC on the basis of assumed cause.

It is generally assumed by members of the Davisian school that rejuvenation is expressed by changes in established stream profiles of equilibrium and changed behaviors of graded streams. Regrading of slopes presumably follows establishment of a new base level of erosion. The new base level is taken to be situated above or below any earlier datum (the amount of effective land uplift and/or sea level depression). Eventually, multiple erosion surfaces are assumed to be possible, one for each erosional datum produced by rejuvenation followed by a period of quiescence and a protracted period of erosion (Fig. 1.9; see also Fig. 10.25).

It seems clear that the identification of a naturally rejuvenated stream depends upon our ability to recognize a graded stream, a profile of equilibrium or some other sign of fluvial NON-EROSION in retrospect (i.e. after rejuvenation). Rejuvenation would be documented for a stream that has cut into a particular erosion level, incised its valley flats,

* Meanders developed in glacier ice were probably initiated on firn, a granular material; see Chapter 8.

Fig. 1.7. Wide-swinging entrenched meanders expressed in the "goosenecks" of the San Juan River, southern Utah. Note the manner in which differential weathering and erosion has etched out beds of varying resistance. (W. N. Gilliland.)

or entrenched its meanders, only if the erosion level, valley flats, or meanders could be proved to result from a graded condition. To date, such proofs are not readily forthcoming. Induction weaves us a tangled web. Rejuvenation is thus a difficult phenomenon to document, particularly since the behavior of a normally incising stream and the behavior of a rejuvenated one otherwise should be the same.

Of interest here is the observation that glacially induced sea level changes have eustatically rejuvenated the world's rivers at least four times by sea level drops during the Pleistocene Epoch and canceled that rejuvenation by sea level rises an equal number of times. The resulting record is complex.* Perhaps rejuvenation should be saved to describe what we all hope will happen to us when we are old. However, it is widely applied to anything that initiates or accentuates stream incision where presumably there was previously little or none.

8. The Peneplain: A Terminal Erosion Surface. This is the theoretical erosional resultant postulated by Davis for the geographical cycle, from a prolonged erosional confluence of a series of graded streams

* Certainly, recent studies (Shepard, 1972, among others) show that changes in stream regimen can account for fluvial effects otherwise usually attributed to crustal movements in controlled studies where movements were definitely absent.

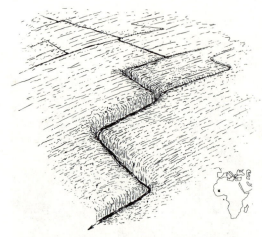

Fig. 1.8. Diagrammatic sketch of right-angle jointing which controls drainage developed on quartzite in the upper basin (Fouta Djallon) of the Senegal River (Bafing), Guinea (map inset). Though incised many tens of feet, the channels continue in right-angle bends, since they only carry solutional loads. (See also Figs. 5.40 and 5.41.)

supported by slope downwearing. The *peneplain* is Davis's "old age" landscape. It has been called an imaginary landform. Perhaps it is. At one point the term was applied to almost every low-relief land area. In Davis's declining years, he and a good many others were reluctant to cite actual examples. Even so, at this writing, one can encounter enthusiasts who are eager to point to remnants of peneplains, usually in the form of more or less accordant upland summits (Fig. 1.10). One thing seems certain. Peneplain is a term with strong genetic overtones. If peneplains do exist, they were formed in an essentially humid environment under the conditions stipulated by Davis. Presumably, surfaces formed under other conditions should be called something else.

The Geographical Cycle Summarized

A theory is no stronger than the major postulates upon which it is based and the Geographical Cycle of W. M. Davis is no exception. The documented reality of linear

fluvial erosion in humid lands is widely acknowledged. The normality and permanency of the requisite environment are much less certain. Soil formation and mass wasting in humid lands are also well documented. However, the actual detailing of slope reduction mechanisms through observation was not part of the Davisian synthesis. He seldom related theory, actual process, and resulting hillslope form, even though he was one of the first geomorphologists to attempt an over-all landscape analysis and to postulate the staged evolution of landforms. The long-term tendency of a drainage system to evolve toward increased efficiency is indicated in a number of studies but the extent to which graded streams or idealized equilibrium profiles play a role in this evolution is uncertain. Many of Davis's ideas and analyses deserve further consideration in the context of more recent findings and apart from the theoretical framework synopsis presented here. Consequently, the relationship between landscape change and time (under a humid environment) will be reconsidered again in greater detail relative to the constraints of Davisian and other theories in Chapters 5 and 7.

The Treppen Concept and the Pediplanation Cycle

The first major alternative to the Davisian Geographical Cycle was formulated a quarter of a century later by Walther Penck (1924). Penck's landscape synthesis incorporated observational techniques and geomorphic quantification, neither of which characterized earlier approaches. Included were the ideas (1) that under a given set of conditions hillsides would assume a particular (stable) angle of slope and thereafter recede at that angle (*parallel scarp retreat*); (2) that the recession of hillslopes produced subjacent concave surfaces of planation (Penck used the term peneplain for these surfaces, most of which today would be described as terraces, pediments, or piedmonts); (3) that subsequent uplift could create subjacent scarps in a series that

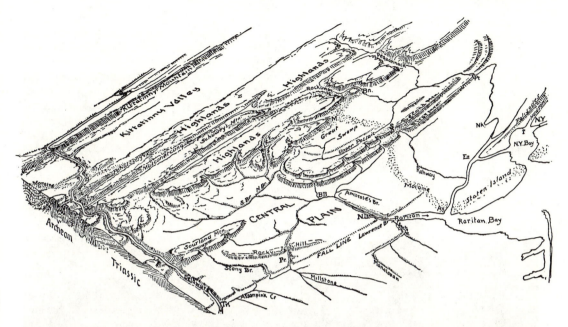

Fig. 1.9. Bird's-eye view of the topography of northern New Jersey, drawn by W. M. Davis, portraying his Schooley Peneplain reflected presumably by Kittatinny and associated mountains with accordant summits, plus lower erosion levels, which together produce a stepped landscape. (From W. M. Davis, 1899.)

would also recede parallel to one another to produce a *piedmont benchland* (Fig. 1.11); (4) that the character of the resulting erosional scarps and flats could be used to interpret the form of crustal movements; and (5) that the general mechanism functioned without regard to subaerial climatic variations in more than degree. The long-term landform culmination of the treppen mechanism was said to be the *inselberg* (island mountain) *landscape*, a low-relief surface dotted with scattered, steep-sided erosional remnants (Fig. 1.12).

Penck's theory has come to be known as the *Treppen Concept* from the stair-stepped landscape which presumably was one possible result. The idea was first published in German in 1924 in a rather difficult style (translated 1953 by Hella Czech and Katharine C. Boswell) so initially it was not widely read or perhaps even understood. Penck's hypothesis was quite naturally attacked by Davis (1932), but survived to serve as the basis for the Pediplanation Theory of King (1953). As one might expect, W. M. Davis's ideas were well enpect, W. M. Davis's ideas were well en-

trenched indeed by the time King's "Canons of Landscape Evolution" appeared. This is true even though varying degrees of disaffection with Davisian schemes had been expressed from time to time.

The King theory has as its basis the work of several early researchers as well as Penck (1924), notably the desert erosional observations of McGee (1897), the Arid Geographical Cycle idea of Davis (1905), and the pediment studies of Bryan (1923, 1935).

In his original presentation, King (1953) goes to great lengths to establish what he terms semiaridity as the normal climate of the world. In part this appears to have been done to counter Davis's claims that humidity was the norm, though at several points in his discussion there is the impression that King believed that the type of semiaridity he favored best accounted for the postulated processes. However, in no instance does he designate the meteorologic or other parameters of semiaridity. Other terms (arid and subhumid) appear more or less interchangeably, and the reader gains the distinct impression that King did not

Fig. 1.10. Accordant summits (*gipfelflur*) developed west of the Front Range crest near Canyon City, Colorado. Some apparent middle-ground declevity relates to the Royal Gorge.

believe it greatly mattered. King even says as much, but leaves one to wonder why he bothers to praise semiaridity elsewhere.

The final point is reinforced by the fact that, at the outset (1953), and even more pointedly later (1957), King argues that the basic elements of hillslopes are present in all environments in varying degrees, concluding that differences in environmental effect are not geomorphically significant. This was also a part of Penck's thesis (1924). In this regard, Frye (1958) was quick to point out that the ubiquitous occurrence of certain hillslope elements could just as well signify the general interchangeability of environments. In any event, it is clear that environment comprises an essen-

tially passive ingredient in King's theory of landscape genesis as it does in Penck's.

According to King, scarp retreat resulting in the generation of pediments (*pedimentation*) is the standard mode of cyclic landscape evolution. He maintains that the process

> comes into operation after the rivers and streams of an area have incised themselves following substantial tectonic uplift,

and further that the process,

> operates to a general continental base level . . . but local base levels are also valid within the concept.

Hillslopes (scarps) are assumed to retreat more or less parallel to one another, reducing uplands and leaving ever-broadening *pediments* at a lower elevation. *Sheet erosion* (water flowing in planar form) is cited as a major formative mechanism for pediment sculpture and scarp retreat. Modes of hillslope retreat are stated to vary with rock type and other factors, an assertion we will examine in some depth later. He concludes that the endpoint of the cycle is a surface composed of coalesced pediments (a pediplain) via the over-all process of *pediplanation*. This landform appears to differ in no significant manner from the inselberg landscape (Fig. 1.12), Penck's terminal landscape morphology (1924).

A glance at the King landscape theory is sufficient to disclose certain similarities and contrasts to the Davisian Geographical Cycle. King's slopes wear back in the Penckian fashion, not down as in the Davisian scheme (cf. Fig. 1.4). King employs base level in terms of a theoretical downward limit that erosion must approach, but with the implication of base level control of erosion surfaces. Clearly, this could be implemented only through the agency of graded streams extending to the sea. Davis emphasized geomorphic *structure, process,* and *stage* in that order, whereas King emphasized *process, stage,* and *structure* in that order. In describing the main process, both emphasize fluvial erosion by water flowing to regional base level. Both relate the effectiveness of that erosion to slope variations in the presence of intermittent uplift. Finally, Davis regarded all non-humid climates as accidental, thereby making it possible to disregard environment in most general geomorphic discussions. Similarly,

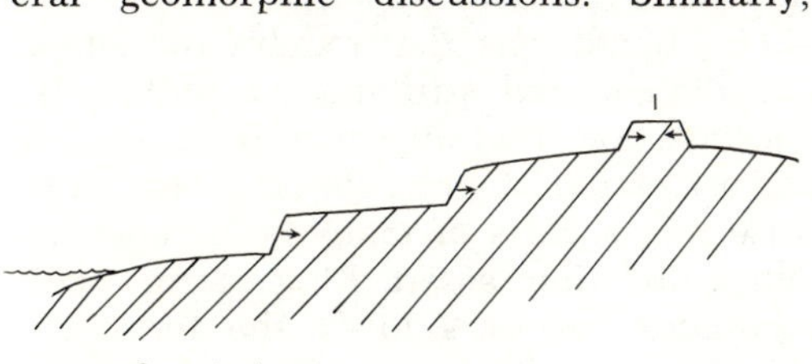

→ Scarp retreat
I. Inselberg

Fig. 1.11. Diagram of piedmont benchland development by parallel scarp retreat according to Penck (1924).

King states, "The difference between landforms of humid-temperate, semiarid, and arid environments are differences only of degree," and he thus effectively ignored climate. Several aspects of the King theory merit additional comment.

(1) King speaks of an early phase of stream incision and states (1953, p. 749): "After the streams are graded the dominant agencies are the processes acting directly upon the hillslopes. . . ." The weakness of the graded stream concept and the problems of base level control of erosion surfaces have been noted in the discussion of Davis's work. The related difficulties of the early theory are clearly shared by the later. The emphasis on hillslope processes (cf. Penck, 1924) evidently reflects a concern with inclined surfaces as focal areas for unconsumed potential energy. Research on hillslope evolution continues.

(2) Tectonic uplift is stated by King to be followed by stream incision, though no reason is given for the presence of sufficient water for effective downcutting. Water is merely assumed to be present.

Schumm (1963) notes average denudation rates in humid drainage basins of 0.1 to 0.3 ft (3–10 cm)/1000 years (maximum average 3 ft or 91 cm/1000 years) as opposed to modern orogenic uplift rates of 25 ft (8 m)/1000 years. He concludes that hillslope form relates to slope and channel erosion rates as opposed to uplift rates (cf. Garner, 1959). Initial aridity on a new landmass would presumably alter or eliminate incisional drainage effects and related denudation as described in Chapters 5, 6, and 10. Since crustal movements can occur without regard to land moisture, fluvially generated relief can hardly encompass a region's structural history. Or, stated somewhat differently, land moisture that can cause stream incision can occur in the absence of crustal movements.

(3) According to King, the pedimentation process can occur in arid, semiarid and humid environments. However, its principal erosive mechanism (rill and sheet erosion) cannot occur where there is an effectively continuous vegetal cover. King's

Fig. 1.12. Ayres Rock, an inselberg of massive, Precambrian rock rises some 1100 ft above the surrounding Australian pediplain some 200 mi (320 km) west-southwest of Alice Springs. (For a closeup view, see Fig. 6.20.) (South Australian Government Tourist Bureau.)

descriptions and illustrations of areas undergoing pedimentation include extensive areas of exposed regolith and bedrock with sparse or nil plant cover.

(4) King states: "A pediment is the ideal landform for the rapid dispersal of surface water, encouraging sheet flow and with a proper hydraulic profile. Pediments are, indeed, moulded under sheet flow." It should also be noted that dispersal MEANS spreading. Sheet flow involves the maximum wetted perimeter of all recorded forms of surface water runoff. Frictional energy loss and resistance to flow and load transport are therefore maximum in sheet flow. However, the bedrock surface under the pedi-

ment is not planar. And the relative flatness of the pediment surface is due, in part, to an alluvial veneer so that both erosion and deposition apparently are involved in the "moulding" process.

(5) Landforms that exhibit the shape and dimensional attributes of pediments, and hillslopes that appear to have receded and maintained appreciable steepness, comprise real aspects of many landscapes, as Penck and King assert. King's term "pedimentation" appears to fit the formative processes involved, and these processes seem to have been operative in many parts of the world. Details of the pedimentation mechanisms—what conditions they oper-

ate under, to what controls and erosional data they respond—are currently being studied intensively. One implication is that earlier work failed to answer questions in these areas. We will consider several aspects of these problems in depth in Chapter 6.

ENVIRONMENTAL DYNAMISM

The idea of climate as a factor in landform genesis emerged subtly more than a half a century ago, though it was not until about 1950 that climate became a significant factor in geomorphic theory. The expression environmental dynamism implies a concern with possible changes in climate, apart from glaciation, and with the geomorphic implications of these changes. As noted earlier, several basic geologic concepts were developed at a point in history when climate was regarded as geographically fixed and temporally stable. Steno's Law of Superposition (1669) is perhaps least dependent on this view. Playfair's Law of Accordant Stream Junctions (1802) is more dependent; and the Law of Uniformitarianism (1785) already under consideration here, is especially dependent upon the rationale of static subaerial environments.

Though the Law of Uniformitarianism was proposed in 1785, the initial notions of continental glaciation did not appear until after the work of Venetz (1829), Bernhardi (1832), Charpentier (1836), and the summary publication of Agassiz in 1840. Opposition to the Glacial Theory continued to be voiced until after 1900. Furthermore, the idea that growth and wastage of continental ice sheets markedly affected subaerial environments in other latitudes was developed much later Jamieson, 1873; Russell, 1889 —see Morrison, 1968). With that idea came recognition that increased precipitation— related by some to a cold climate associated with an expansion of ice sheets and glaciers and by others to increased evaporation associated with a warm climate and a recession of ice sheets and glaciers—could explain prehistoric lake formation in presently arid regions. Thus, the seeds of what can be termed the "relict problem" or climatic leftovers were planted with the continental glaciation concept. Nearly a century lapsed before the geomorphic implications of the climatic "wild oats" were noted.

Davis (1899), Penck (1924), and King (1953) avoided the climatic issue (1) by arguing for environmental constancy (if they mentioned it at all) and by pleading the normalcy of particular climates (humid in Davis' case; semiarid in King's), thus inferring the accidental nature of variations from their norms (glaciation, aridity) and (2) by repeatedly insisting (in the case of King and Penck) that the geomorphic effects of various environments differed only in degree, not in kind.

The doctrine of environmental passivism has already been discussed. It would be seriously challenged by documented proofs of repeated climate change and will be noted in this regard shortly. A related concept of *climatic ineffectualness* was originally adopted by the latter two theorists; the basic ideas of this concept were still held by Penck at the time of his death and, to date, King has not modified his position. Typical of the arguments presented are those of the latter writer (1953, p. 724):

> One can understand that a characteristic cycle and set of landforms should be generated in frigid zones where water is frozen into ice and the whole mechanism of abrasion alters, . . . but that mere differences in amount and incidence of rainfall, evaporation, and like factors should have such far-reaching effects as to result in two entirely distinct cycles requires further investigation. The primary agent moulding the landscape in both humid and arid cases is water flow, and this should produce comparable results in both types of region.

Rather curiously, it was W. M. Davis, himself, who dismayed King by refusing to accept the ". . . *identity of landscape evolution under humid and under arid influences* . . . ." Near the close of his career (1930, p. 147), Davis wrote:

> . . . *although arid erosion resembles humid*

erosion in many respects the two kinds of erosion nevertheless differ so much in process and in product that they cannot be clearly understood if they are briefly brought together as examples of normal erosion, as has been done by several of the German writers . . . . The unlikenesses of the two erosions deserve explicit treatment.

It should be well understood that this statement by Davis does not mean that he accepted any condition other than humidity as climatically normal and constant.

King has commented that Davis's wrong last view kept him from the final unifying geomorphic concept. But being wrong in science and philosophy is like dying on the field of battle. It is something that always happens to someone else. One may well ask, "How wrong was Davis at the end?" After all, he had spent more than a quarter of a century looking for differences between aridity and humidity? More importantly, if climates change, and if different climates cause different kinds of landforms and deposits, there should be relics.

The Relict Problem

The relict problem exists if W. M. Davis was correct in 1930; that is, if different climates make different kinds of landscape elements, and if climates change. The relict problem is geomorphically important because features created in one subaerial environment could thereafter be encountered in an opposing environment. Such features would be *relicts* of a prior condition. For proponents of Hutton's Law of Uniformitarianism, this means that the behavior of a present agent or process may be changed by a relict. An agent or a process functions in one manner and with one result in the presence of relics and in a distinct manner and with a different result in the absence of said relics. *It follows that the "present is the key to the past" only to the extent that we can recognize environmental relics and accurately identify their effects and origins.* It seems clear that such relics could consti-

tute a major complication in the use of the uniformitarian concept.

Davis's assertions on the uniqueness of arid and humid climatic effects have no significance with respect to relics unless such climates replace one another from time to time—in space. Otherwise, each environment merely stays put. We must therefore ask precisely what evidence there is that climates do not "stay put"? As mentioned previously, indications that large lakes once existed in now arid regions are one example. Many lakes in southwestern North America are so removed from glaciation of any kind as to absolutely require previously higher "effective" precipitation. Similar evidence is seen in Cotton's description (1968) of a humid tropical soil (laterite) preserved by lava flow burial in the central Sahara Desert. Elsewhere he notes organized valleys and divides of river systems in several deserts including the Atacama in Peru, similar drainage forms exist in Death Valley (Fig. 1.13). These organized drainage forms contrast sharply with the usual discontinuous and poorly defined flow lines with obscure divides encountered in many deserts.

On the other side of the environmental coin there are the sand seas (*ergs*). Extensive, unconsolidated, free-blowing sand bodies are known only on the shores of certain oceans and the midportions of the Sahara and Arabian deserts (Fig. 1.14). The latter are unique. Extensive areas of free-blowing sand in major dune configurations are an unmistakable product of aridity —particularly when the comprising quartz grains are rounded, polished, sorted, and frosted in an obvious nonmarine setting with evident fluvatile and bedrock sources. Two such ergs are well known beyond desert boundaries. The Sand Hills of Nebraska (Fig. 2.46) is an erg now stabilized under vegetal cover in a region that presently qualifies as either subhumid or semiarid in most meteorologic or geographic-climate classifications (Lugn, 1960; Smith, 1965). A similar erg remnant exists in northern Nigeria along the southern edge of the

Fig. 1.13. A relict dendritic drainage net formed on the floor of Death Valley, California and probably related largely to the last pluvial episode in the region. Some headward extension of tributaries is possible under present conditions but downstream reaches are being alluviated by ephemeral runoff in braided form (cf. Chapter 6). (William A. Garnett.)

Sahara (Grove, 1958). It too is largely immobilized by vegetal cover.

Also indicative of arid climates are widespread accumulations of mechanically weathered alluvium (some of it *arkosic*) found at low elevations in the tropics where frost action has not been a factor in past weathering and where most bedrock types presently "dissolve" under accelerated chemical decomposition. Tricart (1959), Tricart and Cailleux (1965), Bigarella et al. (1965) and Damuth and Fairbridge (1970) have described examples in various parts of Brazil; Garner (1966) notes similar deposits in southern Venezuela under a rainforest (Fig. 1.15). In Venezuela there are also extensively deranged drainages which include centripetal flow patterns like those developed in closed desert basins that previously lacked external drainage outlets (Fig. 1.16).*

Each of the geomorphic features cited in the above paragraphs is situated within an environment where it apparently could not have formed. Each can be genetically related to processes operating in opposing en-

* Since this is a shield region, recent structural movements are nil.

Fig. 1.14. A portion of the Grand Erg Occidental in southwestern Algeria taken at about sunrise. Such sand seas in an active form are limited to Africa and Arabia at present although relicts occur elsewhere.

vironments. Each can therefore be presumed to constitute a relict from a prior condition. And each relict necessarily reflects a change of climate in its respective area. A voluminous literature contains dozens of additional examples of climatic relicts, and later chapters deal with many of these. The point has been made—W. M. Davis (1930) seems to have been correct when he maintained that arid and humid climates do different kinds of things. In a separate context he seems to have been incorrect. Climates do change. Otherwise the features cited as relicts would even now be in their parent environments and quite in keeping with their surroundings. Later we

will consider whether the changes were "accidental." Among their other attributes, accidents are infrequent.

If the impression that only climates make relicts has been created in the foregoing discussion, we must correct it. There are many kinds of relicts. We have described some of the special relicts created by humid and arid climates; relicts that appear to have been overlooked because their genesis was not included in any existing theoretical framework. Even Hutton and Lyell recognized elevated marine deposits decomposing in the atmosphere, and they and many other workers have described metamorphic and igneous rocks decomposing and flaking in

Fig. 1.15.   Rainforest on soil on alluvium reflects opposing weathering equilibria developed in sequence. These gravels near Socopo Wells, Venezuela demonstrate former mechanical fragmentation of rock under aridity on the northeast lowlands of South America. The dense forest cover typifies plant response to present humidity and (inset) is environmentally compatible with the surficial fine-grained soil and contemporary chemical rock decomposition. Similar chemical weathering extends to much greater depths on uplands in this region (cf. Fig. 3.43).

the sun and rain, far from the conditions of high pressure and high temperature that created them. Of course, glacial effects were acknowledged as relicts by, for example, Davis and Penck when they found them in non-glacial settings. Landforms and landscapes clearly owe their character to the interplay of many such environments, whether subaerial or not.

Some relicts are more obvious than others and have a way of looking more or less "out of place," like a French horn in a string quartet. This discrepancy is probably one of the best clues available to the geomorphologist whose ability to apply this clue depends in large measure on his conceptual framework and his understanding of the way environments can alter the landscape. Of

course, the greater the contrast between the environments that have replaced one another, the more obvious is the substitution. For more than 200 years, rocks melted or deformed in the earth's depths have presented no major identification problem when observed at the earth's surface; such relicts, however, do not begin to approach the geomorphic subtleties created by subaerial climatic alternations. In our consideration of relicts, we have thus far not attempted to explain how climatic conditions like aridity and humidity alter the landscape, producing effects that, under other conditions would qualify as relicts. We will now.

The mechanism of relict formation is perhaps best seen in the basic concept of the geologic rock cycle. This, of course, is the open-ended igneous-sedimentary-metamorphic sequence (Fig. 1.17). When one of these lithic elements is displaced (by whatever means) into one of the opposing environments, it is altered because it is only stable in its parent environment. The displaced elements are relicts of the former condition, and for a time, their origin may be recognizable. But exposure to opposite environmental conditions of sufficient intensity or duration may destroy the relict. Thus, a metamorphic rock at the surface may become sediment, or in an igneous environment, the rock may melt. Its destruction coincides with and results in the creation of a "thing" unique to the new environment. That special "thing," and perhaps several of the stages that led to it, would be recognizable in any opposing environment as a relict of a particular condition. Let us now be more specific.

There is a growing body of evidence that suggests that special climatic effects occur

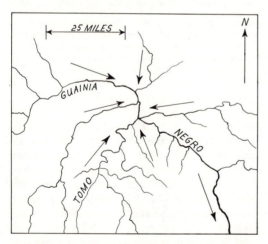

Fig. 1.16. Centripetal drainage pattern in headwaters area of the Rio Negro centered at Lat. 3° N, Long. 68° W. The northwest trending portion of the Rio Negro coincides with the Venezuelan-Columbian border. Note that the main channel flowing away from the pattern center has three barbed tributaries, whereas streams flowing toward the pattern center have normal tributary junctions (see Fig. 7.55 for location).

under conditions that parallel the lithic changes of the rock cycle. Soil scientists (pedologists) have long been aware of one major aspect of the problem; that is, relicts, by their response to a new environment, indicate how special climatic effects come about. As an example, consider the changes occurring in deposits left by the last continental glaciers, which are presently encountered in the relatively warm, moist, and vegetated conditions of central Europe or northeastern North America. The glacial deposits are relicts of what many might not consider a subaerial environment. One type of deposit, glacial *till*, is the product of the mechanical plucking, grinding, rasping action of the ice. It generally is a poorly sorted, commonly non-stratified accumulation of rock and soil debris ranging in size from clay to boulders (Fig. 1.18).

As the glacier ice melted, the rock and mineral fragments in the till were exposed to moisture (unfrozen) and to above-freezing temperatures. Under these conditions, chemical and biological activity may occur as a part of the *weathering* process. We will discuss some of the details of the pedologic changes in Chapter 5. Now we will only note that a major process is the conversion of till materials to *soil;* this starts at the surface and gradually penetrates downward into the relict deposit. In time, the entire till deposit can thus be converted into soil and the glacial relict thereby entirely destroyed if the process is not interrupted.* There are instances on record where this destruction has apparently happened.

Similar changes occur in materials created in other fashions by agencies such as volcanism. Debris erupted in a aura of intense heat is rapidly chilled in the atmosphere, and the resulting minerals are highly unstable at surface temperatures and pressures. Where moisture exists, lava flows and ash falls are rapidly altered from the surface downward and where a series of eruptions has occurred each may be separated from the other by a soil (Fig. 1.19).

From the foregoing discussion it is pos-

---

* Clearly, a thick relict deposit might well escape complete destruction.

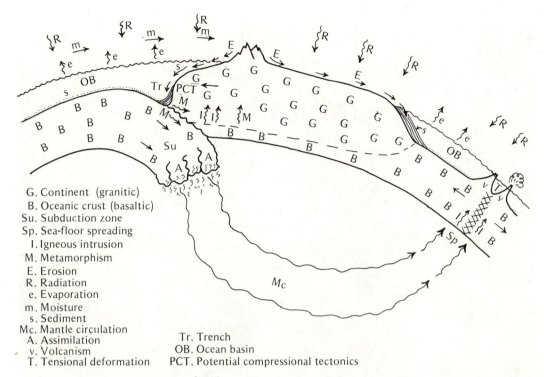

G. Continent (granitic)
B. Oceanic crust (basaltic)
Su. Subduction zone
Sp. Sea-floor spreading
I. Igneous intrusion
M. Metamorphism
E. Erosion
R. Radiation
e. Evaporation
m. Moisture
s. Sediment
Mc. Mantle circulation
A. Assimilation
v. Volcanism
T. Tensional deformation

Tr. Trench
OB. Ocean basin
PCT. Potential compressional tectonics

Fig. 1.17. Idealization of the geologic rock cycle, in terms of the plate-tectonics mechanism (cf. Chapter 3). Certain aspects of the relationship appear to be essentially terminal rather than cyclic. Thus, granitic continental material, once brought to the surface, does not appear to undergo later subduction and assimilation.

sible to make a generalization: *Environments apparently tend to destroy or alter relicts left within them by prior environments which were different.* It is a rare environment that creates nothing new during the destruction or alteration of a relict. Setting aside the detailed mechanisms that are involved for the moment, a major geomorphic consequence can be noted. When we examine the landforms and deposits in a region, they sometimes include environmental relicts other than those of the geologic substrate. If no environmental relicts are found, they may be presumed to have been destroyed or altered beyond recognition. This is because there was only one "first" environment on earth. All subsequent environments have created relicts. This being the case, it is probable that the mechanisms for relict destruction in most contemporary environments that appear to

lack relicts are either very effective or that there have been no environment changes for a very long time. In terms of subaerial environmental effects that naturally tend to dominate at the earth's surface, such a landscape is *monogenetic* (made by one climate). *The landforms and deposits of such a landscape comprise one group of special climatic effects.*

A subaerially formed landscape that includes both newly formed features and relicts reflects more than one environmental condition. It is *polygenetic* (produced by multiple environments). *The partially modified landforms and deposits constitute a second group of special climatic effects.* Such landscape elements exhibit the form modifications of more than one climate; their character apparently depends upon (1) character of environments and processes, (2) duration of alteration processes,

Fig. 1.18. Terminal moraine glacial till of Wisconsin age exposed in a gravel pit along the Pequest River, western New Jersey. Pick (lower right) provides scale.

(3) environment sequences and (4) geologic settings.

We used a simplified example of soil formation to demonstrate how one special climatic effect could be achieved. Such a process and effect is merely a single part of a complex geomorphic system that functions to produce landscapes. In order to grasp the complexities of climatic alternation presented in later chapters, we must consider this type of larger natural system.

General Systems Theory

The environments we have been discussing, together with their contained activities and effects, constitute natural, general systems. The theory of such systems should help us cope with their manifold variables, the

problems that arise when they endure very long or when they alternate, and the implications of these relations to uniformitarianism. According to Howard (1965) general systems theory is more a collection of complementary concepts than a theory. Smalley and Vitafinzi (1969) object to its use in geomorphology as an unnecessary complication, but continued attempts by others to apply systems theory to geomorphic problems suggests that it meets certain needs. We will introduce related ideas at this point for precisely that reason.

Natural systems involve *environments, agencies, processes,* and *products.* We are concerned with their instantaneous state and their interactions and changes through time. Workers in the field of thermodynamics (heat reactions and actions) recognize closed systems and open systems. A

Fig. 1.19. Soil zones developed on pyroclastic materials at Craters of the Moon, southern Idaho. Dark zone, at top, under vegetation is little-altered cinders from a recent eruption. Light interval just beneath is paleosoil (paleosol) developed in older ash and dark zone beneath is unaltered parent material. Light material, lower, is partly altered scree from outcrop face.

closed system may be defined in terms of an initial fixed quantity of energy which is gradually used up during the life of the system. Such a system is thus characterized by a progressive decrease in available energy (increased entropy). "Artificial" chemical reactions occur in these circumstances. And a supposedly classical geomorphic model of this kind is the Geographical Cycle devised by Davis (1899). It can be argued that Davis's was not a closed system. But since one factor (land elevation above sea level) was initially added and ultimately consumed, it can be treated as such.

An *open system* involves the flow of mat-ter and energy through interrelated parts. If inflow and outflow of energy are balanced, an equilibrium or steady-state condition is a theoretically possible development. Landscapes typically form under open-system conditions. This concept was simultaneously applied in a classical river study (Strahler, 1950) and a landscape analysis (Cholley, 1950).

Geomorphic Systems

We are most interested in the natural systems that form landscapes. They may be broken down into subsystems such as the river drainage basins that were Strahler's primary concern (1950), or they may be defined in terms of some other parameter. Cotton (1958) attributes the idea of an erosional landscape system to Cholley (1950) and its modification to include depositional phenomena to Tricart and Cailleux (1955). These workers call it the *morphogenetic* (form-making) *system*. It is defined by Cotton (1958) as a group of agencies or processes interacting under a particular climate to produce a particular landscape. Since geomorphology currently implies a concern for landform genesis, the term morphogenetic is already conceptually implied in the word geomorphic. Hence, we will use the briefer expression *geomorphic system* here. *Morphogenesis* is a useful expression for landform-making by a geomorphic system. For a good general discussion of systems theory, see Chorley (1962).

As already noted, there are many kinds of environments, some within the earth and some without. By substituting the word environment for climate in the definition of a geomorphic system, we can extend the possible scope of such systems. This seems to be a necessary substitution if we are to be able to discuss landscapes created by conditions that are non-subaerial.

Among the various attributes of subaerial geomorphic systems we cite the following as perhaps most significant.

(1) *The limits are environmental.* The events of the system continue *as long as* the parent environment endures and *where* those conditions are imposed on the earth's crust.

(2) *The parts interact.* In doing so they necessarily change one another (*action-reaction-feedback*); the eventual results cannot be readily anticipated by studying a single part, be it agency, process, or product.

(3) *They are historically predisposed.* This is the direct outgrowth of relict inheritance. After the "first" environmental system, other systems followed and have thereby been altered. And, as we shall see, "some acts were pretty hard to follow."

(4) *They are controlled.* This is a direct prerequisite for the creation of special landscape effects. Some element of a given system supersedes other elements enough to impart a distinct developmental bias. Without such control the effects of subaerial conditions would be random and therefore diagnostically meaningless. We would find no relicts.

(5) *They evolve.* This is well documented by the progressive changes that occur in relicts. Like a chemical reaction that goes toward completion where the reactants are either consumed or neutralized, the system changes through time, with respect to the gradual internal modification of its parts and relative to variations in external energy forms and quantities.

(6) *They convert matter and energy.* External energy enters and flows through a geomorphic system, and in the process the energy is converted from one type to another; this can be diagrammed: Potential ⇌ Kinetic ⇌ Heat ⇌ Friction. Where this energy affects matter, there are changes in shape (landforms), consistency (matter states), and behavior (agency responses).

(7) *The parts tend to respond on the threshold principle.* The special effects of a geomorphic system are achieved when certain energy levels are exceeded or when certain physical relationships are created. Thus, water tables are too depressed to support forests or they are not; runoff normally escapes the environment or it does not; natural lubricants facilitate mass wasting or, because they are insufficiently developed, they do not; groundcover is an effective preventative of eolian deflation or it is not; and so forth.

Dynamic Equilibrium

Dynamic equilibrium, or steady state, is an integral part of general systems theory. Axiomatically, the counter notion of *disequilibrium* is implied, and conceptually (relative to the entire system), Equilibrium ⇌ Disequilibrium.* Thus, an agitated body of water can become tranquil as wave energy dissipates, but it also can be reagitated only to become tranquil again. Because of geomorphic relicts, however, the inception of a geomorphic system in their presence ensures an initial state of disequilibrium with the implication that dynamic equilibrium might or might not follow. As previously noted, many of the special effects of subaerial geomorphic systems relate to the destruction of relict landforms and deposits. Presumably, this is the initial disequilibrium phase of the system.

Several writers have described modifications of relict arid and humid landscape elements following changes in climate (e.g. Garner, 1958, 1959; Bigarella *et al.*, 1965; Cotton, 1968). Landforms and deposits plus associated stream loads, drainage configurations, soils (indeed each element of the geomorphic system) appear to pass through recognizable stages of development. Such features would be *time dependent*. As we have seen, Davis (1899) employed the idea of a staged landscape development in his geographical cycles, but by this he meant a development initiated by uplift and terminated by rejuvenation rather than by climate change. The only relicts that generally concerned Davis occurred as a part of the geologic substrate lithology or structure.

In one of the above studies (Garner, 1959), the arid and humid regions cited have been in existence for prolonged periods of time. Hack (1960) cites similar circumstances. In all these examples various landscape elements (e.g. soils, hillslopes) are subject to changes that are so gradual as to defy observation and measurement. These features have become, to a degree, *time independent*, and to that extent they can be said to be in dynamic equilibrium or to exist as steady-state landforms. Davis (1899, 1902) had postulated progressively slower landscape development in cases where his humid and arid geographical cycles were greatly prolonged. But Hack (1960) argues

that after a brief period of adjustment to new environmental conditions a state of dynamic equilibrium is achieved and that most landscapes reflect this. The issues are thus clearly drawn. Have most landscapes adjusted to their co-existing environments or have only a few?

Like many questions, this one has no simple answer, and we are forced to pose two corollary questions. How long does it take to eliminate relict landforms or deposits from a landscape? and How frequently, recently, intensively, and generally have climates changed? The second of these questions will be considered shortly with respect to the Quaternary climatic record. The first forces us to examine in some detail the manner in which geomorphic relicts are destroyed.

RELICT DESTRUCTION

In our initial simplified example of relict modification (glacial till converted to soil) it is apparent that the till would not have changed in its new environment if it had been stable there. In natural systems, therefore, change in any element is *prima facie* evidence of instability (disequilibrium). Because of its importance in this context, we will rephrase a statement made earlier in this section. *The inception of a geomorphic system in the presence of relicts insures an initial state of disequilibrium.* Such an event is analogous to combining some chemicals that are known to react, and in our till-soil example this is essentially what occurred. The following experiment is given as illustration.

Roughly crush and partly pulverize about 50 gm of a 150-gm piece of fossiliferous, oolitic Bedford Limestone from Indiana, place the large fragments and crushed pieces in a beaker, and add 200 ml of con-

---

* At the outset we should note that one can discuss equilibrium of a single element of a landscape (e.g. a particular stream or hillslope) as well as of an entire ensemble.

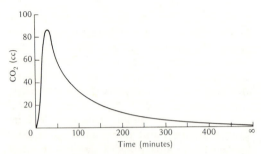

Fig. 1.20. The rate of CO_2 gas evolution from limestone in a solution of HCl as described herewith in relation to natural system disequilibrium and dynamic equilibrium (steady-state) conditions. When the rate of change per unit time can no longer be measured, a steady-state relation may be said to exist.

centrated hydrochloric acid. The calcium carbonate begins to dissolve and carbon dioxide gas is evolved. The acid can be added slowly or quickly to simulate environment changes initiated in a like manner. The quantity of acid added is insufficient to dissolve all the limestone. Consequently, the acid will be neutralized in the following chemical reaction

$$CaCO_3 + 2\,HCl \rightarrow CO_2\uparrow + H_2O + CaCl_2$$

There are a variety of changes that occur in this model micro-environment. But we are particularly concerned with rates of change because they reflect the free energy available at any given instant and hence the degree of disequilibrium in the system. Here, a graph of the rate of gas (CO_2) evolution with respect to time is instructive (Fig. 1.20). One could also plot mean particle diameter changes against time or exothermic heat emission against time. The curve produced is essentially the same as that illustrated previously (Garner, 1968), where one environment is replaced by another in the presence of relicts. In our model, limestone in air becomes limestone in acid; the pulverized and crushed fragments simulate an artificial climate relict and the unbroken undissolved remainder

unweathered bedrock. Any hydration or oxidation of the calcium in air presumably stops when the acid reaction starts. The rates of change induced are functions of several variables, of which the most important seem to be the rate of inception, surface area of solid fragments, concentration of the reagent, and temperature.*

The smaller, pulverized limestone fragments are analogous to a relict climatic landform or deposit. The rate of CO_2 evolution increases more or less abruptly (depending on how fast the acid is added), and the corresponding part of the curve thus rises steeply toward a maximum. The peak rate of gas evolution is of relatively short duration and apparently corresponds to solution of the most finely divided limestone fragments and the corners and edges of the larger pieces. Thereafter, reaction rate decreases gradually and steadily as the larger fragments (relicts) go into solution, and less surface area is exposed to the acid, and as the acid is neutralized. Depending upon their size, the smaller clasts (relicts) may or may not dissolve completely. The gas evolution curve descends and then flattens to approach zero asymptotically (Fig. 1.20). In a natural open system, such as soil formation, the solutions would normally be water plus solutes at several pH values, and in amounts limited only by the norms for the environment. The reaction of such a system continues as long as there are reactants, and this could conceivably be as long as the environment endures.

There are certain obvious differences between the model just described (which is a closed system) and a natural geomorphic system (which is open). Yet certain valid conclusions can be drawn from the analogy. Climatic relict destruction goes through a highly visible (or at least an accelerated), potentially measurable phase, starting with the inception of the environment. In the model, this ends with the complete solution of the small limestone fragments (the climatic relict model) or with the quasi-

---

* In natural geomorphic system substitutions the observed disequilibrium is a function of the environmental differences as expressed by *what was achieved geomorphically by the former* (relicts) and by *what has developed relative to the new potential.*

neutral chemical reaction and flattening of the curve (Fig. 1.20), whichever comes first (cf. Yaalon, 1960). It involves accelerating and decelerating phases, which, when plotted, demonstrate environmental disequilibrium as dependent upon the relicts that comprise the system's historical predisposition. Most of the shift in equilibrium is effected by the presence of finely divided material, the climatic relict model; to a lesser extent it is effected by the large fragments which comprise the symbolic geologic (bedrock) relict model. Initial changes in the relicts are *time dependent* and geomorphically significant to the extent that similar changes can be observed and measured in the field.

The flattened portion of the curve (Fig. 1.20) is not without change through time. But even in our model system, it becomes progressively more and more difficult to observe or record changes in the smallest fragments, if any remain, or to measure the ever smaller and more rare gas bubbles arising in the system. Ultimately, even though it is clear that a reaction (change) is still occurring, it is beyond our ability to detect and measure these changes without excessive error, notwithstanding man's evident ingenuity along these lines. This is analogous to what can probably be assumed in nature. The final rate of change is several orders of magnitude below that occurring in the disequilibrium phase of the environmental sequence. Even though the reaction proceeds, a state of *dynamic equilibrium* may be said to exist since the changes cannot be measured. The reaction is said to have reached a steady-state and is thus *time independent;* the environment should remain unchanged until otherwise disrupted.

According to Howard (1965) a natural system tends to establish dynamic equilibrium as a least-work relationship—a condition where energy expenditure per unit time is minimal and is equally distributed per unit area. The writer concurs (1967) and would add here that the existence of the equilibrium condition in the present context requires the essential absence of geomorphic relicts. Hack (1960) recognizes

dynamic equilibrium where ". . . topography is in equilibrium and erosional energy remains the same"; on p. 89 he equates "equilibrium topography" with erosionally *graded* topography. The mixed blessings that accompany use of "erosional grade" as a gauge of landscape evaluation have already been discussed. Dynamic equilibrium as a concept involving a balance between energy inflow and outflow of a system has also been noted. Both concepts involve severe measurement problems and fall in the realm of pure theory.

Pragmatically, the potential value of landscape and landform *disequilibrium* is as an indicator of climatic relicts as opposed to *dynamic equilibrium* landscapes which lack such relicts. In this regard, Dury (1966) notes: "In some valleys, part at least of what now constitutes coarse bed-material was fed into the valley-bottom in conditions unlike those of today . . . ." Measurements of the extent of some of these deposits and the degree of subsequent removal constitute *bona fide* stages of landscape development (Garner, 1959). Similarly, in moist and vegetated environments exposures of limestone tend to be rounded by solution (Fig. 1.21). Regolith and soil downslope from outcrops that contain limestone clasts freed by mechanical weathering are therefore partly relict. And the dimensions, quantities, and placements of such relict clasts would indicate the degree of adjustment of the regolith to current morphogenesis. A detailed example is useful here.

On the east side of West Mountain in Fayetteville, Arkansas at an elevation of 1320 ft (400 m) the Pitkin Limestone of Mississippian age crops out in a low bluff 2 to 5 ft (0.6–1.5 m) high. It overlies the 200+ ft (70–80 m)-thick Fayetteville Shale. The Pitkin outcrops are rounded (cf. Fig. 1.22), presumably by continuing solution in an environment that features mild winters, 30–40 inches (78–104 cm) annual precipitation and a deciduous hardwood forest vegetal cover. There is no talus, and the single massive bed displays only very widely spaced joints [6–8 ft (2–2.5 m)]. Recent foundation excavations downslope from the

Fig. 1.21. An expanse of etched, fluted, and rounded limestone typical of exposures developed by solutioning under humid conditions as are the associated red "terra rossa" soil remnants. The occurrence of these features and related sinks in Morocco under semi-desert conditions establishes their relict nature. (W. N. Gilliland.)

Pitkin outcrop encountered Fayetteville Shale at depths of 10–12 ft (3–4 m) overlain by *colluvium*. This colluvium has a 7° to 10° surface inclination, contains many angular, ferrugeneous sandstone and siltstone clasts, but is apparently devoid of limestone clasts for an average distance of 130 ft (33 m) downslope from the Pitkin outcrop (Fig. 1.22). From that point on downslope for an average distance of 160 ft (50 m) the colluvium includes pitted and rounded clasts of limestone from 6 inches (15 cm) to four ft (125 cm), median diameter, that are identifiable as originating from the Pitkin Formation upslope. No limestone was observed in the colluvium more than about 300 ft (100 m) from the Pitkin outcrop. There is abundant evidence that the colluvial mass is presently undergoing downslope movement in a variety of forms including *creep* and local solifluction; the latter is seen especially during rainy seasons and wetter years.

It is possible to draw several conclusions from this example. (1) The outcrop does not now seem to be breaking up mechanically by weathering or even yielding blocks by joint separation. (2) The colluvial interval immediately downslope from the limestone outcrop that is devoid of limestone clasts is demonstrably moving and, hence, documents an interval of time during which no fragmental limestone has been produced —the time involved being equal to the average colluvial downslope movement rate divided into that distance [130 ft (33 m)]. (3) The limestone clasts in the colluvium can reasonably be attributed to a former episode when bedrock was fragmented mechanically rather than dissolved so that in terms of current conditions the limestone clasts are relict. (4) Since the mechanical-weathering episode ended, the uppermost clasts have moved downslope an average of 130 ft (40 m). (5) Since this episode began the lowermost clasts have moved an average of 300 ft (92 m). (6) The mechanical weathering interval can reasonably be related to a time of cold and/or aridity. The latter condition has been suggested for the area during the early Altithermal (3,000–5,000 years B. P. ≠ Jewell interval of Ruhe, 1969; 3,000–7,000 B.P.) and during proximal glacio-environmental effects of uncertain character (correlative to the Wisconsinan Glacial Stage which last peaked out in Iowa about 14,000 years B.P. according to Ruhe, 1969); these tapered off farther north about 8,500 years ago (see Expanded Quaternary time scale, Fig. 1.24).

In the absence of absolute dates for the Arkansas events, one cannot determine which earlier climatic episode is responsible for the relict limestone clasts. Recently dated desert faunal remains in the same area show a C_{14} age of 4,290 years B.P. which is strongly suggestive of an Altithermal age.* In any event, if the clasts of limestone were formed during the Altithermal, the clasts formed last have averaged 0.5 inches/year (1.3 cm/year) movement downslope and the clasts formed first averaged 0.75 inches/year (1.8 cm/year). If the clast origin was of Wisconsin Glacial Age, the beginning and ending times are less certain, but mean mass wasting movement rates would then be on the order of about one third as fast. The setting appears to me to favor an arid Altithermal origin.** It seems evident that the movement rates need not have been constant nor need the movement conditions have been the same as those at present. Also, microclimatic effects should be considered, for slopes with other directional exposures might disclose contrasting effects.

The artificial model used to illustrate

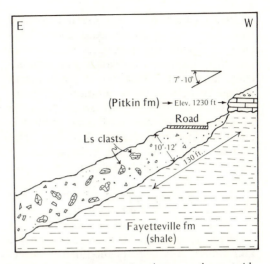

Fig. 1.22. East-west section drawn on the east side of West Mountain, Fayetteville, Arkansas (cf. Fig. 1.22). Slope angle illustrated is exaggerated above value cited.

relict alteration indicates that landscape development in a given environment tends to occur in two phases, relative disequilibrium and relative equilibrium. Little is known about the time magnitudes involved for similar changes in nature. This comes, instead, from a consideration of actual relicts including those already discussed. The quartz sand dune concentrations (ergs) are extremely durable features, both in composition and as landforms. Since sand-size material does not generally blow out of deserts, such masses of granular quartz are almost indestructible. They are virtually insoluble except in the humid tropics, they comprise an effective "sponge" where there is no runoff erosion since direct precipitation soaks straight in, and they are subject to notable fluvial erosion only where flow on a relatively impervious surface encounters the sand masses. This last is relatively rare since the ergs are concentrated on uplands (Fig. 1.14). Currently, evidence, from the French Saharan Research Center, suggests that the Algerian ergs are now being replenished with sand from *oueds* (dry water-

---

* *See* Davis's study of a cave deposit (1969).
** For corroborative data, see glacio-environmental discussions in Chapter 8.

courses) which have not carried surface water in recorded history. And it would appear that arid-humid climate fluctuations otherwise indicated for this region periodically cause soil A horizons to yield quartz grains. In light of these relations the principal erg sand depletion would appear to be by granular attrition until size reduction permits clast removal by deflation. The over-all erg budget is not known, but the general sand seas may well be several million years old.

The continental glacial deposits of the Pleistocene are probably less ancient than the desert ergs. The glacial deposits are widespread in the northern hemisphere, and many (only slightly modified by alternative environments) range in age from a few thousand years to perhaps two million years or more. The relict limestone talus cited was at least several thousand years old. Thus, it appears that relict landforms may persist for a geomorphically significant (measurable) period of time. Moreover, relicts tend to fall into two rather distinct categories according to their time response to subaerial geomorphic systems.

One relict group, represented by general bedrock lithology, structure, and land elevation, is introduced to the earth's surface by tectonic and other related environmental systems, many of which function with apparently random periods. The durations of these periods commonly measure tens of millions of years. As will be seen in the next section, subaerial geomorphic systems do not commonly endure so long; hence, most subaerial geomorphic systems, individually, can hardly eliminate the more enduring geologic relicts, though many geomorphic systems, functioning in sequence, may eventually do so. Wilson (in Harris and Twidale, 1968) in a related context observes,

A landscape acting as an open system need not achieve a steady state, as suggested by Hack (1960). Such an unchanging topography can only occur where all external and internal conditions are constant. This is rarely the case. Even assuming a constant climate and no tectonic activity there will be an inevitable reduction of relief and,

hence, a loss of potential energy within the system. The result is a gradual evolution, perhaps similar to that postulated by Davis. The landscape evolution follows the basic laws of allometric growth (in this case the growth is negative).

Wilson, in substance, describes the landscape role of geologic setting relative to landform genesis under subaerial environments. The protracted response of the geologic setting is a low-key background to climatic relict destruction and creation. Later we will consider the extent to which climatic relicts reflect the geologic substrate and the apparent reasons for this. In any case, relicts appear to undergo accelerated modification with the inception of each new subaerial environment. Climatic relicts often have an endurance that is at least an order of magnitude shorter than major elements of the geologic setting. Even so, some climatic relicts do survive—some through several consecutive environmental episodes. At issue here is the extent of this survival, the reasons for it, and the question of the "accidental" nature of climate change as maintained by W. M. Davis and many others of his school. These will be discussed in the section on geochronology and Quaternary paleoclimatology which follows shortly.

GEOMORPHOLOGY IN A THEORETICAL VACUUM

In the wake of recent discoveries in geotectonics (Chapter 3) the diastrophic basis for the theoretical frameworks of Davis, Penck, and King is thrown into serious question as noted by Chorley (1963). Geomorphology currently progresses in what Dury (1972) with some reason designates a theoretical vacuum, characterized by a variety of national preoccupations: ". . . American with processes, French and German with climatic geomorphology, British with denudation-chronology, . . . Polish with the dominance of Pleistocene effects, Russian with applied geomorphology, and so on." The American momentum appears to stem mainly from work in the early 1950's, notably that of Strahler (1952), whose *dynamic basis of geomorphology* proposes to deal with geomorphic processes "as gravitational or molecular shear stresses acting upon elastic, plastic, or fluid earth materials to produce the characteristic varieties of strain, or failure that constitute weathering, erosion, transportation and deposition."

These quasi-national geomorphic preoccupations noted by Dury fortunately are beginning to show signs of idea interchange in isolated instances where processes are studied in a climatic framework for landform analysis, denudation chronologies, and practical applications of the findings (cf. Bigarella *et al.*, 1965; Leopold and

Maddock, 1953; Senstius, 1958). In this text I attempt a thorough amalgamation of these legitimate areas of concern with the hope of delineating a broad, updated theoretical overview—an overview with some balanced consideration of process, environment, chronology, quantification, and application. That the various contemporary schools of geomorphic interest are as yet incompletely aware of developments in all other areas is clear from Scheidegger's (1970) treatment of "Theoretical Geomorphology," within whose 435 process-oriented pages, four paragraphs are directed at the subject of climate and therein only through the perspective afforded by W. M. Davis and his Geomorphological Cycle.

GEOCHRONOLOGY AND QUATERNARY PALEOCLIMATOLOGY

In the two decades since L. C. King's Theory of Pediplanation, records of probable climate changes of many kinds and the techniques for dating them have grown. Extensive age determinations using a variety of radioactive isotopes began in the early 1950's. The early ratio techniques with uranium/lead and other elements having long half-lives were little suited to geomorphologic dating of very recent events. To draw an analogy, a yardstick isn't much use in gauging the eye spacing of a gnat. Later, isotopes of carbon (C_{14}) were used to date deposits containing organic material where the upper limit is about 50,000 years. Isotopes of oxygen (O_{18}) were used to date, and to measure paleotemperatures (cf. Dansgaard and Tauber, 1969), in marine sediments which are ±175,000 years old or more, as noted by Emiliani (1970). Protactinium/thorium ratios (Th_{230}/Pa_{231}) have been used to date some marine carbonate deposits up to several hundred thousand years old (Emiliani, 1969). Also, potassium/argon (K_{40}/A_{40}) ratios have been extensively used to date older rock, and recently attempts have been made to

use these isotopes to date volcanic deposits associated with the remains of ancient hominids (fossil pre-men) which are as recent as 2 to 3 million years. The latter results are somewhat uncertain because the gaseous decay product (argon) is difficult to measure accurately when it is present in extremely small amounts.

Absolute dating techniques have been of inestimable value in dating a number of relatively recent geomorphic events. But they have their limitations. They are currently being used to date igneous rocks for the most part. They are expensive. And they are subject to contamination in some instances as noted by several authors including Ruhe (1969). The spans for dating have notable gaps, particularly for early Pleistocene and late Tertiary continental deposits. Also, materials from which a date could be determined are not always present where needed (i.e. wood upon which C_{14} determinations are commonly based is frequently destroyed by accelerated vegetal decay in the tropics).

Fortunately for geomorphologists, other dating techniques of a relative character have become available to supplement the absolute radiometric record.* A variety of fossil forms including planktonic foraminifera, terrestrial snails, vertebrates, coccoliths, and especially spores and pollen grains have proved very useful. The recorded sequences of the various fossil types have an evident relative dating value. Many of the same fossils are also temperature sensitive and quite a number are excellent environmental indicators. A word of caution should be voiced in at least one regard here. Oceanographers and marine biologists too often imply that dated sequences of marine water temperature changes can be directly equated with specific terrestrial climatic conditions. There is as yet no unanimous agreement that cold oceans directly correlate with a cold earth or atmosphere and glaciation or that warm oceans correlate with a warm subaerial climate assemblage and no continental glaciers. The fit may not be quite as neat as that. There are suggestions of hemispheric inequalities, and, from available evidence, these issues remain in doubt.

Several other dating and environmental detection techniques are proving helpful to the geomorphologist. Borings through the Greenland and Antarctic ice sheets appear to record 100,000 years or more of climate and temperature changes (Dansgaard, Johnsen, Møller, and Langway, 1969; Epstein, Sharp, and Gow, 1970). In addition, a randomly fluctuating sequence of earth magnetic reversals has been detected which provide a paleomagnetic time scale. (Vine, 1966; Heirtzler, LePichon, and Baron, 1966); see Chapter 3. Finally, but by no means of lesser importance, there is the application of stratigraphic techniques to geomorphic problems. Therein, landscape histories can be worked out in terms of the position-in-sequence of features within depositional landforms that were generated while associated erosional landforms were being sculpted (Bigarella, Mousinho and Da Silva, 1965; Ruhe, 1969). Details of these methods are discussed in Chapter 2.

GEOLOGIC TIME SCALES

Two geologic time scales are included (Figs. 1.23; 1.24). The first is a general time scale spanning the interval from the earth's origin down to the present. The second is an expanded scale of the Quaternary Period.

A general time scale (Fig. 1.23) is necessary for any geomorphologic consideration of the earth's first-order relief features (Engeln, 1942). These are the continents and ocean basins, with origins that relate to events initiated several hundreds of millions of years before present (B.P.). Such a time scale is also needed as a chronologic reference for the study of second-order relief features of the crust which modify the continental rises and ocean basins and comprise mountain systems and other structurally or lithologically delineated geomorphic provinces. The origins of these provinces go back a few tens or hundreds of millions of years. Topographic roughness developed in various ways (e.g. erosion; crustal movement) on first- and second-order features can be termed third-order relief; and secondary roughness on third-order features (gullies, terraces, benches) are fourth-order relief features. The usage here departs slightly from that of von Engeln (1942) who cites only three orders of relief features.

The expanded Quaternary scale of time (Fig. 1.24) was devised to account for several geomorphologic problems. Foremost among these is the consideration that the majority of the third- and fourth-order relief features are no older than Quaternary. We are not quite prepared to join those who claim that there are no features older than the Quaternary, though the geomorphologist clearly has a special interest in this time interval. None of the above should be taken to signify that there have been no recent crustal movements reflected in landforms. Far from it—the cases that do exist involve circumstances that should be given special consideration. Recent structural landforms will be considered in Chapter 3.

---

* Volcanic ash studies (tephrochronology) and paleomagnetic dating have been helpful.

GEOLOGIC TIME SCALE

ERA	Periods and Epochs		Orogenic Episodes	Duration and Date B.P.(Millions of Years)
CENOZOIC ERA	Quaternary Period	Pleistocene Epoch	Cascadian-Alpine	1-3 — 3
	Tertiary Period	Pilocene Epoch		8 — 11
		Miocene Epoch		14 — 25
		Oligocene Epoch		15 — 40
		Eocene Epoch		20 — 60
		Paleocene Epoch	Laramide	10 — 70 ±2
MESOZOIC ERA	Cretaceous Period			65 — 135 ±5
	Jurassic Period		Nevadian	45
			Palisade	— 180 ±5
	Triassic Period			45 — 225 ±5
PALEOZOIC ERA	Permian Period		Appalachian-Ouachita / Marathon	45 — 270 ±10
	Pennsylvanian Period			80
	Mississippian Period		Acadian	— 350 ±10
	Devonian Period			50 — 400 ±10
	Silurian Period		Caledonian-Taconic	40 — 440 ±10
	Ordovician Period			60 — 500 ±15
	Cambrian Period			100 — 600 ±20
PROTEROZOIC ERA	Keweenawan "Period"		Killarney ±1100 m.y.	— 1650
	Huronian "Period"		Algomen 1800-2000 m.y.	
ARCHAEOZOIC ERA	Timiskaming "Period"		Laurentian 2400-2700 m.y.	3000
	Keewatin "Period" Onverwacht		>3200 m.y.	
AZOIC ERA	Godthaab "Period"			3980 m.y. 5000 m.y.

Fig. 1.23. A scale of geologic time. Precambrian "periods" reflect arbitrary assignments to rock units of shields in North America, South Africa, and Greenland. For an expanded version of the Quaternary Period, see Fig. 1.24.

THE QUATERNARY PALEOCLIMATIC RECORD

Various details of Quaternary paleoclimatology will emerge in several subsequent chapters. All of the details would fall far beyond the scope of this text (cf. Charlesworth, 1957). All I intend to do here is to survey the evidence indicating (1) the frequency of climate change, (2) the geographic extent of those changes, and (3) something of the intensities of the environments involved.

The frequency of climate change can be adduced in part from the record of glacial changes themselves, as symptomatic of correlative climate shifts elsewhere on every continent (Butzer, 1957; Garner, 1958; Büdel, 1959; Garner, 1959; 1967). It is traditional to think of glacial activity as characteristic of the Pleistocene Epoch, but there is growing evidence that glaciation occurred intermittently throughout much of the later Tertiary Period (Bandy, Butler, and Wright, 1969; Margolis and Kennett, 1970) and at many earlier times when land was present at high latitudes. In any case, frequency has to do with incidence per unit time. The Quaternary Period is currently assigned a span of about 2.0 to 2.5 million years on the average (see, for example, Longwell, Flint, and Sanders, 1969).

During the Quaternary Period there appear to have been at least four major glacial episodes that left records in mid-continental North America (Fig. 1.24) and possibly a fifth indicated elsewhere in alpine settings.* The onset of each episode and its termination can each be taken to reflect a first-order change in earth climate patterns in nonglaciated regions. There have thus been eight and possibly ten or more first-order climate changes in the past 2.5 million years or one such change every 250,000

---

* The several alpine glacial advances are not here individually equated with major continental "stages" in all instances.

Quaternary Time Scale						
Period and Epochs	Radiometric Ages (10^3 yrs) B.P.	Europe Glacial & Inter-	N. American Glacial & Inter-	Paleomagnetic Polarity	N. Amer. Mammalian	European Mammalian
Quaternary — Pleistocene — Holocene	4–7	Flandrian — Altithermal	Holocene — Recent	Brunhes Normal	Rancholabrean	Upper
	10					
	30	Weichsel — Late	Wisconsin — Valders			
			Twocreek			
	50	Middle	Woodford			
			Farmdale			
	70?	Early	Altona			
	100?	Eem	Sangamon	Jarmillo event		
	200?	Saale				
		Holstein	Illinoian			
	400?	Elster				
	600?	Cromer		Reversed Epoch	Irvingtonian	Middle
		Menap	Yarmouth			
	800?					
	1,000?	Waal	Kansan	Matuyama		
		Eburon		Gilsa event	Blancan	Villafranchian (Lower)
	2,000?		Aftonian	Olduvai events		
		Tegelen				
		Brüggen	Nebraskan			
	2,500?			Gauss Normal		
Tertiary — Pliocene	3,000?	K/Ar Dating				
	11,500?					

years on the average.* Averages can be misleading, however, and there is some evidence that the glacial stages were shorter than the interglacial stages, though Emiliani (1972) maintains that really warm episodes were brief. In any event, the major glacial episodes were apparently not times of constant climate. This is demonstrated by pronounced fluctuations in the ice sheet margins which nevertheless fell short of actual disappearance of the parent ice mass.

The best known of the glacial stages in North American mid-continental areas is the Wisconsin, largely because it is most recent, its deposits are best preserved and can be radiometrically dated to a large extent by C_{14}. The Wisconsin appears to be atypical in two aspects, however. It may well be the most brief of the known Pleistocene glacial stages ($\pm75,000$ to $\pm8,500$ years B.P.), and, as measured in terms of its maximum southerly extent in North America, it was the least intense. Even so, there are at least two glacial substages indicated in Iowa (Ruhe, 1969) during the late Wisconsin (29,000 to 14,000 B.P.).

In Illinois, where the Wisconsin record is more nearly complete, there is a sequence of seven separate till deposits (Kempton and Gross, 1970); C_{14} dating suggests that the substages of Iowa cited above may be equivalent to two of the more recent of these. Thus, there appear to be at least seven glacial substages and six recessional substages presumably reflecting some twelve second-order climate shifts during the Wisconsin Glacial Stage (Fig. 1.25). Considering that the Wisconsin was comparatively brief and weak, it does not seem unreasonable to assume that earlier glacial stages were at least equally complex cli-

matically. If we make this assumption and extrapolate back through the Pleistocene Epoch, there are possibly 48 to 60 *glacial stage*, second-order climate changes that affected the earth during the Pleistocene. The apparently longer duration of earlier glacial episodes raises the possibility that there were even more. Emiliani (1972) suggests that there were, when a different basis for calculations is used.

Climate fluctuations during glacial stages may account for less than 50 per cent of those occurring during the Pleistocene Epoch. As noted previously, the interglacial stages may have been longer than the glacial stages. Moreover, the interglacial record was also one of climatic fluctuations (of second-order magnitude), albeit kinds that did not involve low latitude glaciation (cf. Bigarella *et al.*, 1965). Details of both glacial and interglacial climate fluctuations also come from studies of so-called "pluvial" lakes (Morrison, 1965; 1968). Lake level changes suggest a minimum of 12 second-order Wisconsin Age climate changes in agreement with the till-soil records of the Illinois region. These and similar climate-change sequences suggest that some 36 to 48 second-order climate changes can also be assigned to Pleistocene INTERGLACIAL episodes (Fig. 1.25). During postglacial time there seem to have been at least three more nonglacial climate shifts that probably rank as second order (Ruhe, 1968).

In total, some 87 to 111 second-order climate changes probably occurred during the Pleistocene Epoch. This amounts to one climate shift every 28,700 years at a minimum, one every 22,500 years at a maximum or one every 25,250 years on the average. It is interesting to compare some recently calculated climate change fre-

Fig. 1.24.   Expanded Quaternary geologic time scale that incorporates data on radiometric dating, glacial and inter-glacial subdivisions of Europe and North America, mammalian zonations of the same regions, and paleomagnetic polarity reversals. The upper and lower Pleistocene limits are entirely space-time dependent since, on the basis of glacial extent, that epoch still endures in the Antarctic and began there long before 3 million years B.P. Drawn from a number of sources including several cited for the same purpose by Flint (1971). Post-Wisconsin sea-level rise (Flandrian) is equal to about the last 18,000 years (cf. Fig. 9.20). *Subdivisions of the Wisconsin are type localities of the U. S. Geological Survey (Frye *et al.*, 1968).

* Marine records suggest 40,000- to 50,000-year fluctuations.

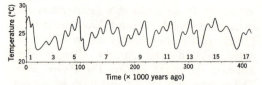

Fig. 1.25. Paleotemperature curves of the Caribbean based on oxygen isotopic analysis for the past 400,000 years. Note comparative brevity of thermal extremes. Numbers above horizontal axis are stages. (After Emiliani, 1972.)

quencies (Veeh and Chappell, 1970), attributed to astronomic causes (*Milankovitch Theory*), which relies in great part on the ±25,800 year period of equinox precession. This precession of the earth's axis would presumably result in variations in solar insolation ranging from 7% to 14% (Emiliani, 1969). It should also be noted that smaller-order climate fluctuations are presumably more numerous than those already cited but are also probably less intense (Fig. 1.26).

In summary, it would appear that changes in climate have been frequent. Even if the foregoing calculations are too high by 50%, which seems most unlikely, changes have been too frequent to be termed accidental. Some of the changes have been very intense and some have been very re-

cent. In geographic scope the areas affected are global. At the extremes are instances of barren deserts being replaced by rain forests and vice versa (Fig. 1.15) cited by Garner (1958; 1959; 1966), Bigarella *et al.* (1965), Cotton (1968) and Damuth and Fairbridge (1970). Elsewhere, grasslands and forests or grasslands and deserts have apparently alternated. Such instability may be contrasted with the rather isolated and local refuges (*climatic nuclei*) where, for reasons outlined in Chapter 4, climatic conditions seem to have remained remarkably stable for very long periods (Garner, 1967; Haffer, 1969). There seems to have been essentially all degrees of gradation between these extremes of stability and intensity, both temporally and geographically.

It would appear that special landform effects of subaerial geomorphic systems are created in an array of environmental changes. Indeed, some of these special effects seem to occur in spite of those changes, whereas others may owe their existence to the fact of those changes. Before considering the precise manner in which these effects are achieved, we must ask, "What has happened to Uniformitarianism in light of such environmental instability?"

UNIFORMITARIANISM REVISITED

We have come full circle, but we can hardly do without a valid rule or continue to countenance one that is invalid. And Hutton's "law" must be reconsidered on this basis alone. The law stands in doubt because of environmental relics and what they might mean. However, "the present will serve as a key to the past" if we can identify environmental relics and distinguish them and their effects from those of present day processes and their products.

After introducing the "law" we considered the notion of geomorphic systems and the actual record of climate changes. From

these considerations, the following points seem valid. (1) If the present is a key to the past, the past is at least equally a key to the present. (2) In varying degrees the vast majority of the world's landforms are relics from recent climate changes. (3) Because of relict effects, a degree of process or agency disequilibrium exists which varies from one geomorphic region to another— in a few regions disequilibrium is near zero, which is to say that landscape changes there take the form of allometric growth, and something approaching dynamic equilibrium exists. (4) Because of relics, there is indeed little Huttonian product uniformity apart from that induced by evolving geomorphic systems—the present is like the past as a man is like an *amoeba;* there is a relationship, one whose understanding requires a knowledge of many developmental stages and a unifying theory. (5) Because of relicts, most landforms and deposits are an ENVIRONMENTAL BLEND (cf. *compound landscapes* of Thornbury, 1969). They can be used as keys to geologic situations where similar environments-in-transition created portions of the geologic record. (6) Because of the occasional long-term climatic situation, we are here and there confronted with landscapes that are essentially in complete adjustment with a particular environment. Botanists use the term *climax* for plant assemblages in this condition, and Senstius (1958) extended the concept to soils.* *Climax landscapes* are those created under a single environment (cf. simple landscapes of Thornbury, 1969). *Climax landforms* and particularly *climax deposits* should also have counterparts in the greater geologic record. (7) The relatively short-term geologic and geomorphic system effects are superimposed upon the long-term matter evolution of the earth. This last is only partly cyclic in the few billions of years of earth history—"nature" thus simulates but never duplicates, and a historical analysis of the earth can only proceed in terms of environmental dynamism in a time continuum.

A significant sidelight of the foregoing summation is its historical implication about modes of geologic analysis. Historically, Hutton formulated the uniformitarian concept on an island near the Gulf Stream that carries a record of only two major subaerial environments for many thousands of years—humidity and glaciation. This statement is made in full recognition of a body of British literature that advocates marine planation of some of the same area. British glaciation was apparently more mild than its continental counterpart, and in the south

of England (where accentuated freezing and thawing were periodic but apparently without glacial ice) virtually nothing but moist, vegetated conditions have prevailed for at least scores of thousands of years. Thus, the landscape in the south of England approximates an example of climax humidity. Small wonder that Lyell and many other early British proponents of Hutton's "law" advocated gradualism if the south of England was their model. What could be more gradual than changes in a dynamic-equilibrium landscape?

It is important to realize that places like the southern part of England which have seemingly escaped profound and recent climate changes seem to be local exceptions. More or less profound disequilibrium apparently characterizes various aspects of the majority of present-day environments, and man's manipulations are frequently cited as the primary cause of such expressions of disequilibrium as accelerated erosion (Schumm, 1963). Manipulations there have been. But man had often merely triggered a condition which had already approached a threshold through natural dynamic change. Many efforts to reconcile current erosion rates with isostatic rebound rates or orogenic tectonic rates largely depend on the premise of widespread dynamic equilibrium and gradualism. The results are, at least, uncertain. And the extrapolation of such rates back through time as averages without regard to environmental setting or stage seems a questionable procedure. The extent to which environmental disequilibrium characterizes the earth cannot be easily expressed. It is pervasive. Yet it is probably normal. And the dynamic equilibrium state is therefore comparatively rare. Paradoxically, therefore, "the key to the past" which was formulated to counteract "catastrophism" is a highly dynamic earth with changes which are intermittently catastrophic, frequently accelerated by disequilibrium, and possibly not too effective where they are truly gradual.

* For a detailed discussion of the climax concept, see Chapters 5 and 6.

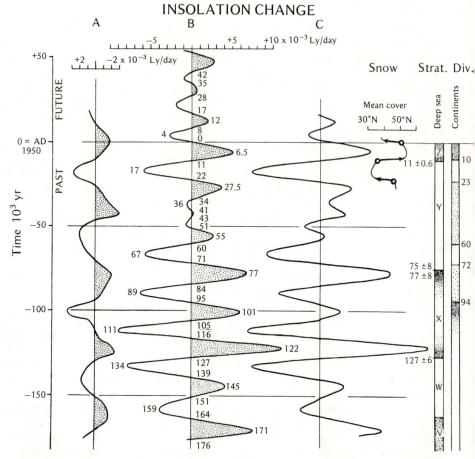

Fig. 1.26. Insolation curves related to correlative geologic events over the past 150,000 years, compiled by Kukla (1972). The relatively warm periods appeared in times of positive winter insolation gradient and vice versa. Thus, shifts between negative and positive trend (B) date the start and end of correlative gross climate modes (±1,000 year accuracy). Highest insolation gradients correspond with peak cold or warm conditions.

(A) Year-to-year departures in annual totals of insolation received at top of atmosphere in a 5° latitudinal belt centered on 65° N. It is inversely related to warming or cooling of the equatorial belt of both hemispheres; (B) seasonal variations in area-weighted averages of insolation received at top of atmosphere beyond the latitudinal belt between 25° and 75° N, in the Northern Hemisphere winter-half years. It basically controls extension of seasonal snow covers and pack ice in that hemisphere, and thereby, Earth's surface albedo; (C) integration of (B) with the converse of (A) (1:1) which approximates the combined effects of two independent effects on northern mid-latitudes in winter. The 6.5-, 6.7-, and 127-thousand year warming peaks are intensified; the mid-Wisconsin section is modulated by the 42,000 B.P. warming.

TWO BASIC PRINCIPLES

I have presented ideas in this chapter that have often been used as fundamental approaches to the analysis and understanding of landforms and landscapes. I hope that these ideas will influence every student to examine each landscape problem from a genetic point of view. The "facts" always count. But those who can observe and record "facts" appear to outnumber those who can interpret them. Facts are always

GEOLOGICAL EVIDENCE

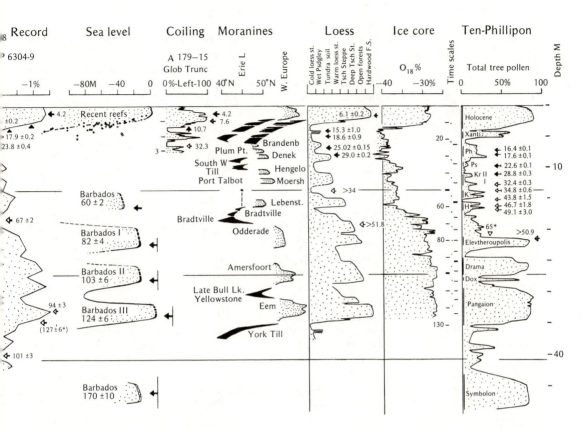

Right diagrams:

1. Northern Hemisphere mean snow line; 2. marine and continental stratigraphic subdivisions, warm zones dotted; 3. sea surface temperature and salinity (O_{18}) in Caribbean core P 6304-9; note Broecker and Van Donk's age difference (in parentheses); 4. sea level fluctuations (mostly east and south U. S.)—unsafe before 15,000 B.P.; 5. sea surface temperature based on foraminifera coiling (*Globorotalia truncatulinoides*); 6. radiometrically dated moraines and other climate-dependent deposits from North America and Europe— warm units to right; 7. climate dependent soils in loess of Czechoslovakia in terms of plant cover— increasing warmth and moisture right; 8. Camp Century (Greenland) ice core (O_{18}) variation (time scales highly tentative); 9. tree-pollen variations (radiocarbon dates).

New time scales based on assumption of regular climatic periodicities are unsafe; older scales calibrated against C_{14} are better according to Kukla (1972).

filtered through a sensory system that has been programmed with a theoretical bias. Things tend "not to exist" if we don't believe in them or if we don't know what they look like. The converse of this is also true. We will begin with two notions; more may emerge later.

PRINCIPLE I.   UNIFORMITARIANISM

An examination of Hutton's "law" has been a theme throughout this chapter. Our conclusion is that we cannot just say, "the present is the key to the past." We can say *the basic physical laws appear to apply to all of geologic time as well as the present.* But an entire new family of environments has evolved over the past four or five billion years, and all the common geologic agencies have been impressively modified during their creation and evolution. Gravity is the old standby Newton took it for as are most of the basic astronomic relations of the earth, though the constancy of solar radiation (effects) and the permanence of Earth-Moon relations and tidal forces has been questioned.

In the beginning there was no atmosphere as we know it; a methane atmosphere has been postulated, but such an atmosphere could hardly have affected the landscape as has our present-day atmosphere of nitrogen, oxygen, and carbon dioxide. Oceans as bodies of water and continents appeared much later; we cannot be sure when (cf. Chapter 3). Eolian influences came later as did the weathering that was influenced by land plant and land animal life. Ice presumably appeared along with water. And since the over-all relief of the earth is in large measure a product of continent-ocean-basin differentiation, rigorous high altitude environments must be a comparatively recent development, possibly not appearing until mid or even late Precambrian.

As the earth's matter evolved, the behaviors of geologic agencies were necessarily affected as were related processes. It is only in part (as uniformitarianists are quick to observe) a matter of varying process intensity. Environments do vary in intensity—in part cyclically. Thus, the events that are happening now also happened in the past. That is the "key." But events happened in the past and were not repeated. That is evolution, a problem that Hutton's "law" has not solved.

PRINCIPLE 2. ENVIRONMENTAL DYNAMISM

This concept focuses attention on the nature of major landscape control. Davis' (1899) genetic emphasis was on *geologic structure, process,* and *stage,* in that order (see also Davis, 1954). King (1953) reversed this order. The doctrine of terrestrial environmental dynamism, however, *deemphasizes preferred sequence. Where environments are interchangeable, controls must also be interchangeable.* We now know that Earth dynamism has several forms. It is not wholly surficial; it is not entirely internal, nor do its events appear to be precisely coordinated in any particular order.

To grasp the global extent of the earth's external form, we must admit that each of the Earth's various environments occasionally dominates the landscape via its respective geomorphic system. Since these environments replace one another in both time and space, we are apparently dealing with a time-environmental continuum. Therein, it is probably rare for the effects of two alternating environments to be equally expressed, and rarely is there no alternation. Thus, *landscapes tend to reflect an environmental dominance and its counterpart.* Furthermore, within a geomorphic system the role of relicts is necessarily determined by the order in which the systems occur. Thus, *landscapes usually express the environmental sequence (a matter of historic predisposition).* Finally, the time-dependent aspects of landforms emerge and ultimately may disappear as the environment persists. Thus, *landscapes commonly reflect the passage of time by relict response to geomorphic system evolution; they may cease to reflect the passage of time on those relatively rare instances where a climax state is attained.* Therein, the past and present would be one.

REFERENCES

Bandy, O. L., Butler, E. A. and Wright, R. C. (1969) "Alaskan Upper Miocene Marine Glacial Deposits and the Turborotalia pachyderma Datum Plane," *Science,* v. 166, pp. 607–609.

Beckinsale, R. P. and Chorley, R. J. (1968) "History of Geomorphology" [in] *Encyclopedia of Geomorphology,* R. W. Fairbridge, ed., Reinhold Book Corp., New York, pp. 410–416.

Bigarella, J. J., Mousinho, M. R. and Da Silva, J. S. (1965) *Processes and Environments of the Brazilian Quaternary,* University of Paraña Press, Curitiba, Brazil, 71 pp.

Bryan, Kirk (1923) "Erosion and Sedimentation in the Papago Country, Arizona," *U. S. Geol. Surv. Bull.,* v. 730, pp. 19–90.

Bryan, Kirk (1935) "The Formation of Pediments," *16th Int. Geol. Cong. Rept.,* pt. 2, pp. 765–775.

Büdel, J. (1959) "The Periglacial Morphologic Effects of the Pleistocene Climate over the Entire World" (trans. by H. E. Wright and D. Alt) *Int. Geol. Rev.,* v. 1, pp. 1–16.

Butzer, K. W. (1957) "The Recent Climatic Fluctuation in lower Latitudes and General Circulation of the Pleistocene," *Geogr. Ann. Stockh.,* v. 39, pp. 91–111.

Carlston, G. W. (1968) "Slope-Discharge Relations for Eight Rivers in the United States," *U. S. Geol. Survey Prof. Pap.* 600-D, pp. D45–D47.

Charlesworth, J. K. (1957) *The Quaternary Era,* Edward Arnold, London, 2 vols.

Cholley, A. (1950) "Morphologie Structural et Morphologie Climatique," *Geogr. Ann.,* v. LIX, pp. 321–335.

Chorley, R. J. (1962) "Geomorphology and General Systems Theory," *U. S. Geol. Surv. Prof. Pap.* 500-B, pp. 1–10.

Chorley, R. J. (1963) "Diastrophic Background to Twentieth-Century Geomorphological Thought," *Geol. Soc. Amer. Bull.,* v. 74, pp. 953–970.

Chorley, R. J., Dunn, A. J. and Beckinsale, R. P. (1964) *The History of the Study of Landforms,* John Wiley and Sons, New York, 678 pp.

Cotton, Sir Charles (1942) *Climatic Accidents,* Whitcombe and Tombs Ltd., Wellington, esp. Chapters 1 and 2.

Cotton, Sir Charles (1958) "Alternating Pleistocene Morphogenetic Systems," *Geol. Mag.,* v. XCV, pp. 125–136.

Cotton, Sir Charles (1960) "The Origin and History of Central Andean Relief: Divergent Views," *Geogr. Jour.,* v. CXXVI, pp. 476–478.

Cotton, Sir Charles (1968) "Relict Landforms" [in] *Encyclopedia of Geomorphology,* R. W. Fairbridge, ed., Reinhold Book Corp., New York, pp. 936–940.

Damuth, J. E. and Fairbridge, R. W. (1970) "Equatorial Atlantic Deep-Sea Arkosic Sands and Ice-Age Aridity in Tropical South America," *Geol. Soc. Amer. Bull.,* v. 81, pp. 189–206.

Dansgaard, W., Johnsen, S. J., Moller, B. and Langway, C. C. Jr. (1969) "One Thousand Centuries of Climatic Record from Camp Century on the Greenland Ice Sheet," *Science,* v. 166, pp. 377–381.

Dansgaard, W. and Tauber, H. (1969) "Glacier Oxygen 18-Content and Pleistocene Ocean Temperatures, *Science,* v. 166, pp. 499–502.

Davis, L. C. (1969) "The Biostratigraphy of Peccary Cave, Newton County, Arkansas," *Ark. Acad. Aci. Proc.,* v. 23, pp. 192–196.

Davis, W. M. (1899) "The Geographical Cycle," *Geogr. Jour.,* v. 14, pp. 481–504.

Davis, W. M. (1902) "Base-level, Grade and Peneplain," *Jour. Geol.,* v. 10, pp. 77–111.

Davis, W. M. (1905) "The Geographical Cycle in an Arid Climate," *Jour. Geol.,* v. 13, pp. 381–407.

Davis, W. M. (1930) "Rock Floors in Arid and Humid Climates," *Jour. Geol.,* v. 38, pp. 1–27.

Davis, W. M. (1932) "Pediment Benchlands and Primärrumpfe," *Geol. Soc. Amer. Bull.,* v. 43, pp. 399–440.

Davis, W. M. (1954) *Geographical Essays,* Dover, New York, repub. 1909 (D. W. Johnson, ed.), 777 pp.

Dury, G. H. (1966) "The Concept of Grade" [in] *Essays in Geomorphology,* American Elsevier, New York, esp. pp. 211–233).

Emiliani, C. (1969) "Interglacial High Sea Levels and the Control of Greenland Ice by the Precession of the Equinoxes," *Science,* v. 166, pp. 1503–1504.

Emiliani, C. (1970) "Pleistocene Paleotemperatures," *Science,* v. 168, pp. 822–825.

Emiliani, C. (1972) "Quaternary Paleotemperatures and the Duration of the High-Temperature Intervals," *Science,* v. 178, pp. 398–401.

Engeln, O. D. von (1942) *Geomorphology,* Macmillan, New York, esp. pp. 17–37.

Epstein, Samuel, Sharp, R. P. and Gow, J. J. (1970) "Antarctic Ice Sheet: Stable Isotope Analysis of Byrd Station Cores and Inter-hemispheric Climatic Implications," *Science,* v. 168, pp. 1570–1572.

Frye, J. C. (1958) "Climate and Lester King's 'Uniformitarian Nature of Hill Slopes'," *Jour. Geol.,* v. 67, pp. 111–113.

Garner, H. F. (1958) "Climatic Significance of Anastomosing Channel Patterns Typified by Rio Caroni, Venezuela," *Geol. Soc. Amer. Bull.,* v. 70, pp. 1568–1569.

Garner, H. F. (1959) "Stratigraphic-Sedimentary Significance of Contemporary Climate and Relief in Four Regions of the Andes Mountains," *Geol. Soc. Amer. Bull.,* v. 70, pp. 1327–1368.

Garner, H. F. (1965) "Base-Level Control of Erosion Surfaces," *Ark. Acad. Sci. Proc.* v. 19, pp. 98–104.

Garner, H. F. (1967) "Geomorphic Analogs and Climax Morphogenesis," *Ark. Acad. Sci. Proc.,* v. 21, pp. 64–76.

Garner, H. F. (1968) "The Form of the Lands," *Yale Scientific Mag.,* v. XLIII, pp. 16–20, 33.

Gilbert, G. K. (1880) *Report on the Geology of the Henry Mountain Region,* 2d ed., U. S. Geogr. and Geol. Survey Rocky Mtn. Region (Powell), 170 pp.

Grove, A. T. (1958) "The Ancient Erg of Hausaland, and Similar Formations on the South Side of the Sahara," *Geogr. Jour.,* v. CXXIV, pp. 526–533.

Hack, J. T. (1960) "Interpretation of Erosional Topography in Humid Temperate Regions," *Amer. Jour. Sci. Bradley Volume,* v. 258-A, pp. 80–97.

Haffer, Jürgen (1969) "Speciation in Amazonian Forest Birds," *Science,* v. 165, pp. 131–137.

Harris, S. A. and Twidale, C. R. (1968) "Geomorphic Cycles" [in] *The Encyclopedia of Geomorphology,* R. W. Fairbridge, ed., Reinhold Book Corp., New York, pp. 237–240.

Heirtzler, J. R. and Le Pichon, Xavier (1965) "Crustal Structure of the Mid-Ocean Ridges, 3, Magnetic Anomalies over the Mid-Atlantic Ridge," *Jour. Geophys. Res.,* v. 70, pp. 4013–4034.

Heirtzler, J. R., Le Pichon, X. and Baron, J. G. (1966) "Magnetic Anomalies over the Reykjanes Ridge," *Deep-Sea Res.,* v. 13, pp. 427–433.

Howard, A. D. (1965) "Geomorphological Systems—Equilibrium and Dynamics," *Amer. Jour. Sci.,* v. 263, pp. 302–312.

Kempton, J. P. and Gross, D. L. (1970) *Stratigraphy of the Pleistocene Deposits in Northeastern Illinois,* 34th Ann. Tri-State Field Conf. Guidebook, pp. 65–72.

Kesseli, J. E. (1941) "The Concept of the Graded River, *Jour. Geol.,* v. 49, pp. 561–588.

King, L. C. (1953) "Canons of Landscape Evolution," *Geol. Soc. Amer. Bull.,* v. 64, pp. 721–752.

King, L. C. (1957) "The Uniformitarian Nature of Hillslopes," *Edinburgh Geol. Soc. Trans.,* v. 17, pp. 82–102.

Kukla, G. J. (1972) *Insolation and Glacials,* Boreas Osleo, v. 1, pp. 63–96.

Leopold, L. B. and Maddock, Thomas (1953) "The Hydraulic Geometry of Stream Channels and Some Physiographic Implications," *U. S. Geol. Surv. Prof. Pap.* 252, 57 pp.

Leopold, L. B., Wolman, M. G. and Miller, J. P. (1964) *Fluvial Processes in Geomorphology,* W. H. Freeman, San Francisco, 522 pp.

Longwell, C. R., Flint, R. F. and Sanders, J. E. (1969) *Physical Geology,* John Wiley and Sons, New York, 685 pp.

Lyell, Sir Charles (1872) *Principles of Geology,* 11th ed. (esp. Chapters 2, 3, 4), D. Appleton, New York.

Lugn, A. L. (1960) "Origin and Sources of Loess in the Great Plains of North America," *Int. Geol. Cong. XXI, Rept.,* pp. 223–235.

McGee, W. J. (1897) "Sheetflood Erosion," *Geol. Soc. Amer. Bull.,* v. 8, pp. 87–112.

Mackin, J. H. (1948) "Concept of the Graded River," *Geol. Soc. Amer. Bull.,* v. 59, pp. 463–512.

Margolis, S. V. and Kennett, J. P. (1970) "Antarctic Glaciation during the Tertiary Recorded in Sub-Antarctic Deep-Sea Cores," *Science,* v. 170, pp. 1085–1087.

Morrison, R. B. (1965) "Quaternary Geology of the Great Basin" [in] The Quaternary of the United States, H. E. Wright and D. G.

Frey, eds., Princeton Univ. Press, Princeton, pp. 265–285.

Morrison, R. B. (1968) "Pluvial Lakes" [in] *Encyclopedia of Geomorphology,* R. W. Fairbridge, ed., Reinhold Book Corp., New York, pp. 873–883.

Penck, Albrecht (1894) *Morphologie der Erdoberfläche,* Bd. 1, pt. 2, Bibliothek Geogr. Handbücher.

Penck, Walther (1924) *Morphological Analysis of Land Forms,* (Translated in 1953 from the 1924 version by H. Czech and K. C. Boswell), Macmillan and Co. Ltd., London, 430 pp.

Playfair, John (1802) *Illustrations of the Huttonian Theory of the Earth,* William Creech, Edinburgh, 528 pp. (Republished in 1858 by Univ. of Illinois Press).

Powell, J. W. (1875) *Exploration of the Colorado River of the West,* Smithsonian Institution, Washington, esp. p. 203.

Ruhe, R. V. (1969) *Quaternary Landscapes in Iowa,* Iowa State Univ. Press, Ames, 256 pp.

Russell, I. C. (1889) "Quaternary History of Mono Valley, California," *U. S. Geol. Surv. 8th Ann. Rept.,* pp. 261–394.

Schumm, S. A. and Lichty, R. W. (1963) "Channel Widening and Floodplain Construction, Cimarron River, Kansas, *U. S. Geol. Surv. Prof. Pap.* 352-D, pp. 67–88.

Senstius, M. W. (1958) "Climax Forms of Rock Weathering," *Amer. Scientist,* v. 46, pp. 355–367.

Shepard, R. G. (1972) "Incised River Meanders: Evolution in Simulated Bedrock," *Science,* v. 178, pp. 409–411.

Smalley, I. J. and Vita-finzi, Claudio (1969) "The Concept of 'System' in the Earth Sciences, Particularly Geomorphology, *Geol. Soc. Amer. Bull.,* v. 80, pp. 1591–1594.

Smith, H. T. U. (1965) "Dune Morphology and Chronology in Central and Western Nebraska," *Jour. Geol.,* v. 73, pp. 557–578.

Strahler, A. N. (1950) "Equilibrium Theory of Erosional Slopes, Approached by Frequency Distribution Analysis," *Amer. Jour. Sci.,* v. 248, pp. 673–696, 800–814.

Strahler, A. N. (1952) "Dynamic Basis of Geomorphology," *Geol. Soc. Amer. Bull.,* v. 63, pp. 423–458.

Tanner, W. F. (1968) "Meandering and Braided Rivers" [in] *Encyclopedia of Geomorphology,* R. W. Fairbridge, ed., Reinhold Book Corp., New York, esp. pp. 957–963.

Thornbury, W. D. (1969) *Principles of Geomorphology,* 2d ed., John Wiley and Sons, Inc., New York, esp. pp. 139–141.

Tricart, Jean and Cailleux, André (1955) *Introduction à la Géomorphologie Climatique,* Cours de Géomorphologie, Centre de Documentation Universitaire, Paris, 228 pp.

Tricart, Jean and Cailleux, André (1965) *Introduction to Climatic Geomorphology* (Eng. trans., 1972), Longman Green Ltd., London, 295 pp.

Twain, Mark (pseudo. & S. L. Clemens) (1944) *Life on the Mississippi,* Heritage Press, New York, 418 pp.

Veeh, H. H. and Chappell, John (1970) "Astronomical Theory of Climate Change; Support from New Guinea," *Science,* v. 167, pp. 862–865.

Vine, F. J. (1966) "Spreading of the Ocean Floor: New Evidence," *Science,* v. 146, pp. 1003–1010.

Wanless, H. R. (1965) *Aerial Stereo Photographs,* T. N. Hubbard Scientific Co., Northbrook, Ill., esp, pp. 76–79.

Washburn, A. L., (1967) "Instrumental Observations of Mass-Wasting in the Mesters Vig District, Northeast Greenland," *Meddelelser Om Gronland,* Bd. 166, 297 pp.

Wheeler, H. E. (1964) "Base level, Lithosphere Surface and Time-Stratigraphy," *Geol. Soc. Amer. Bull.,* v. 75, pp. 599–610.

Wolman, M. G. (1955) "The Natural Channel of Brandywine Creek, Pennsylvania," *U. S. Geol. Surv. Prof. Pap.* 271, 56 pp.

Yaalon, D. H. (1960) "Some Implications of Fundamental Concepts of Pedology in Soil Classification," *7th Int. Cong. of Soil Sci. Trans.,* v. 4, pp. 119–123.

2 Geomorphic objectives and methods

THE GENESIS OF LANDFORMS AND LANDSCAPES

THE PROBLEM

It is reported that Gertrude Stein, on her deathbed, repeatedly murmured, "What is the answer?" And then, just before she died, she asked, "What is the question?" The answers to wrong questions are at best irrelevant and at worst misleading. Science repeatedly pays the price of wrong questions and occasionally reaps the rewards for right ones. A major point of this chapter is the recognition of worthwhile problems in geomorphology.

By definition, geomorphology is the determination of the origin of landforms and the landscapes they comprise. The effort is historically scientific and tends to fall naturally into three phases. Miss Stein's queries apply to phase one, *definition of the problem*. Phase two involves *description*. Phase three involves *data analysis and conclusions*.

A problem in geomorphology usually becomes apparent when a given landscape element is found to be anomalous within a particular theoretical and genetic framework. In addition, there is usually some attribute that makes us want to know more about the element. For example, (1) a landform is unusual because its features are not comparable to descriptions of features supposed to be of a kind, (2) a feature has not been previously described (it belongs to no

known group), (3) there are unique study advantages—relations that somehow reduce the number of genetic variables, or temporary dimensions that are added by either natural or artificial breaches of the regolith; and (4) new theoretical developments suggest that a restudy of a feature would be fruitful. Of course, there are no hard and fast rules that can take the place of unbridled curiosity and experience or experienced guidance.

The following geomorphic studies were undertaken for a variety of reasons. Horton (1945) studied drainage basin and drainage net parameters (including stream order) in an attempt to quantify Playfair's Law of Accordant Junctions. Weathering and mass-wasting phenomena were studied by Washburn (1967; 1968) because they occurred in an area of Greenland with the apparent attributes of a typical Arctic climate. Garner (1959) studied geomorphic process variations under a variety of climates on a single mountain system (the Andes) because relief, elevation, distance to base level, and general tectonic age could all be treated as relative constants in the analysis. Oberlander (1965) examined the relations among fluvial incision, transverse drainage, and geologic structure in the Zagros Mountains of Iran so he could compare the Zagros drainage with Appalachian drainage and generalize about stream dissection of folded mountain systems. Ruhe (1969) detailed modes of study of glacial geomorphic sequences to establish the relationship between age of deposits and development of associated landforms. Black (1969) undertook the study of some exceptional stagnant-ice features, particularly *moulin kames*, which were in danger of being destroyed because of the demand for construction aggregates. The possibilities are clearly manifold.

For most landscape studies, a geomorphologist delineates the problem area and devises a descriptive approach based on a series of initial postulates. These postulates should not be confused with conclusions. Rather they are ideas related to known theory and to the information generally available from previous studies. It is rare indeed to find an extensive region that has not been studied. The preliminary postulates usually indicate the relative importance of the geomorphic role of bedrock, both lithologically and tectonically (passive or active), the significance (if any) of sea level changes or volcanism, and the role of subaerial environments. As description progresses, the geomorphologist tests his initial generalizations against his new data, and modifies his postulates as he determines the sequence and nature of geomorphic events. Analysis actually continues throughout the study. Ultimately, the researcher must analyze the geomorphic system as a whole. Three questions can be asked. (a) What is the current system? (b) What landscape changes is it causing? (c) To what extent, as indicated by relicts, does this system differ from prior systems and the dominant system (if these are not the same)? With these questions in mind we can now consider techniques.

DESCRIPTION AND MEASUREMENT

Though measurement is a form of description, we wish to emphasize it in this work, while giving full rein to other modes of description. Thus, it is meaningful to say a car has a low profile and good lines. It is also meaningful to say it has a 455 cu. inch motor and a four-barrel carburetor. Descriptively, something would be lost if either phrase were left out.

In the early days of geomorphology, researchers based their conclusions almost entirely on the external aspect of landforms and landscapes in descriptions that were largely verbal and qualitative. A notable exception to this was in the area of glacial geomorphology in which deposits had already received considerable attention by 1900. In great part, the emphasis on external form evaluation was the result of a lack of instrumentation and related techniques; these have been widely available at moderate cost only in the years since about 1950. Before that time the internal character of landforms was conjectural except where it was revealed by the occasional natural exposure or by artificial excavation. Intentional digging was understandably uncommon, and drilling is still expensive. Nevertheless, it is important that description and measurement of landforms be extended, wherever possible, to include both external form and internal aspect.

External Form

The outward appearance of landforms can be recorded in a variety of ways. Each way of recording has its uses and shortcomings. Included are: (1) the use of such surveying devices as the plane table and alidade, Brunton compass, hand level, Jacob's staff, or slope pantometer* to record slope dimensions and attitudes, (2) on-the-spot written or taped descriptions, (3) still, time-lapse, or movie photography, (4) sketches and schematics, (5) aerial photographs, (6) areal topographic maps, (7) orthophoto maps (Fig. 2.1). For the best results, these devices should be employed conjunctively. Also, where possible, statistical treatment

---

* The pantometer (Pitty, 1968) is one of the simplest and most promising devices for obtaining multiple slope values that can then be analyzed statistically for morphometric purposes.

of landform and associated process attributes can add information in the study of populations, norms and deviations from norms, frequency occurrences, and distributions with respect to such frameworks of reference as time, space, or environmental limits, to name only a few geomorphic reference systems that can be quantified.

Topographic Maps

It is difficult to geomorphically analyze a landscape using topographic maps alone (see Hartshorn, 1967). Apart from the lack of basic geologic data, in most cases the landform detail is simply inadequate. For example, the most widely used maps in the United States are the 7½-, 15-, and 30-minute* topographic series issued by the U. S. Geological Survey and the Army Map Service, and the most commonly employed contour intervals are 5, 10, and 20 ft.** The contour interval chosen usually delineates second- and third-order relief features of a region adequately, but much of the third- and fourth-order roughness of the relief features (often geomorphically most diagnostic) hardly appears and can only be seen with an order-of-magnitude decrease in contour interval.

To illustrate the above difficulty, the St. Paul Quadrangle, Arkansas, is a 15-minute sheet with a 20-ft contour interval (Fig. 2.2). The major valleys, mountains, valley flats, and drainage lines are clearly visible. According to a widely used physical geology laboratory manual, the larger stream (White River) on this quadrangle is said to meander, which is noted as one expression of the topographic "maturity" of the region. However, every major valley tributary to the White River has an alluvial deposit at its mouth whose topographic details tend to disappear with contour intervals much in excess of 2 ft (Fig. 2.3). These deposits have the external form and internal character (Fig. 2.4) that are generally accepted

for alluvial fans. The over-all relief of these fans rarely exceeds 25 ft so they hardly show on the St. Paul quadrangle map. Yet the White River channel is seen to take its course along the bases of the alluvial fan slopes. This is the flow route that *consequent* runoff would normally take during initial movement down valley if the fans were already present. Therefore the bends of the White River appear to reflect a relict alluvial topography, not meanders. This explains the absence of *meander scars* on air photos of the area (scars often fail to show up on contour maps even when present) and the general lack of arcuate symmetry of the river bends. It should be pointed out that the actual nature of this flow pattern was initially indicated by a statistical analysis of trunk channel-tributary juncture locations relative to valley flat margins that demonstrated a distinctly nonrandom distribution.

Considerable geomorphic utility can be derived from proper use of topographic maps, and in this regard we should note the trend-surface analysis of the Basin and Range Province by Lustig (1969a) in which he utilizes 7½-minute (1:250,000 scale) maps to establish eleven topographic parameters. These serve as the basis on which he was able to quantify regional topographic variations. In a second example (1969b), Lustig applied similar cartographic techniques to the study of individual landforms with excellent results.

Hillslopes

Detailed geomorphic analysis based on adequate measurements of the exterior of landforms has hardly begun. Leopold *et al.*, (1964) rightly lament the "dearth of descriptive measurement of both form and process on slopes." Data on outward form are relatively easy to acquire if one is willing to make the effort. But the manner of presentation usually offers difficulties be-

---

* The latter are rather obsolete; 1° (1:250,000) sheets are the only coverage in much of the world.
** Larger intervals are used in mountainous terrain.

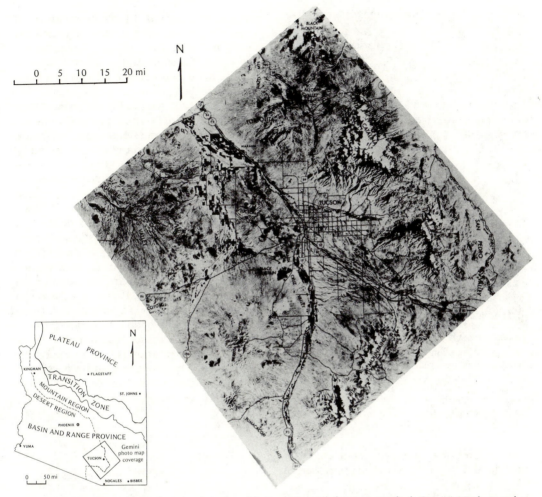

Fig. 2.1. Orthophoto map, Tucson, Arizona vicinity, prepared from a rectified Gemini V photo taken August 21, 1965 by astronauts Gordon Cooper and Charles Conrad, Jr. Features depicted include topography, snow, culture, and drainage. Data compiled by the U. S. Geological Survey in cooperation with NASA. (U. S. Geological Survey.)

cause of various scalar problems, effects of vertical exaggeration, and our uncertainty as to the range of quantitative slope values allowable for various hillslope elements. Following the lead of Wood (1942), King (1953) recognizes four distinct hillslope elements. "From top to bottom . . . these are: (1) the *waxing slope* (convex), (2) the *free face* (rock outcrop), (3) the *talus* or *debris* slope, (4) the *waning slope* [including pediments] (concave)," (Figs. 2.5 and 2.6).

Ruhe (1969) includes a *summit* and *shoulder* essentially in place of King's waxing slope, a *backslope* in place of a free face, a *footslope* that seems roughly equivalent to a talus or debris slope, and a *toeslope* related to a waning slope. It should be noted that King's terms encompass the effects of a bedrock outcrop whereas Ruhe is mainly concerned with unconsolidated regolith materials which generally will not stand steeply enough to create an outcrop. Each writer comments that all the hillslope com-

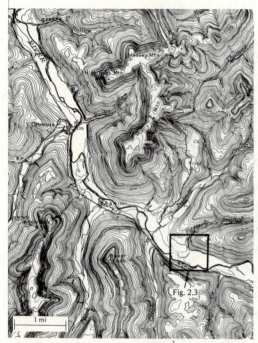

Fig. 2.2. A portion of the St. Paul Quadrangle, Arkansas showing a reach of the White River whose bends have been said to be meanders. Actually the curves are at the bases of alluvial fan slopes situated at the mouths of side valleys as shown by 2-ft contour lines (Fig. 2.3).

highest in a region. In origin it can be almost anything—erosional, depositional, structural, volcanic, etc.—with corresponding values for slope and roughness, none of which have been really adequately documented to date.

The free face seen in Fig. 2.5 is composed of consolidated bedrock and may have a range of slope expressions up to and including vertical or overhanging. As noted by King (1953) the debris slope has an attitude that is a reflection of *repose angle* (Fig. 2.7), in part a function of fragment size and shape. He states, "A talus of boulders 3–4 feet in diameter may have a declivity of as much as 35°." Angular blocks may accumulate even more steeply, in part for the same reason that you can build a vertical wall of bricks and in part, presumably, because of interlocking. It is well known that silt-size angular rock fragments (*loess*) can stand vertically in open faces, in part for the latter presumed reason and in part because of some cementation of grain contacts. Therefore it may not always be possible to evaluate all the factors that affect slope repose angle values (Fig. 2.8).

ponents do not necessarily occur on every hillside and that they may be suppressed in varying degrees.

King is notably vague on the assignment of actual slope values to the waxing slope. This may cause problems for those geomorphologists who wish to distinguish quantitatively between the product of some leveling mechanism (e.g. planar erosion) and a feature created by a relief-generating device (e.g. stream incision or high-angle faulting). In the field the former (flats) are generally distinguishable from the latter (scarps), at least subjectively. Purely statistical cartographic studies of the same features present the analyst with many "gray" areas of uncertainty, particularly since a change in contour spacing actually coincides with a change in slope only by chance. Of course, the waxing slope is, by definition, an upper topographic level, possibly the

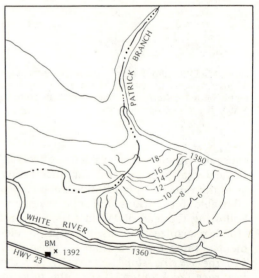

Fig. 2.3. Micro-topography depicted by contour along a reach of the White River, Arkansas. The alluvial fan shown by 2-ft contours is not evident with a greater contour spacing (cf. Fig. 2.2).

Fig. 2.4. Typical alluvium of a small fan situated along the White River southeast of Fayetteville, Arkansas. The clasts display little rounding, and material ranges from a silty sand matrix to boulders.

Repose of sand on the lee side of dunes is close to 34° in air. Rahn (1969) recorded kame terrace sand and gravel talus repose angles averaging 33.08° which was independent of material type (though the clasts were presumably somewhat rounded). The late A. C. Trowbridge (personal communication, 1970) reported seeing sand grains freezing to a dune face south of Lake Michigan to an angle of 37°. Sand grains are said to accumulate to steeper angles on foreset beds under water where particle buoyancy is a factor. Observations of underwater sediment accumulation are in short supply, however. As any professional sand-castle builder knows, water-saturated, fine-grained beach sand can be dribbled into pinnacles with vertical or even overhanging sides as long as the water can drain away freely and quickly. This happens naturally along sandy scarps, though the reason for the increased cohesion is uncertain. It might initially be due to adhered water, but it continues after drying if the material is not disturbed mechanically. It might involve minute amounts of clay or salt acting as a matrix at grain contacts. In a later section we will consider the varying role of lubricants in the repose of nonhomogeneous material.

The waning slope of King (1953) is presumably a surface of low relief and slight

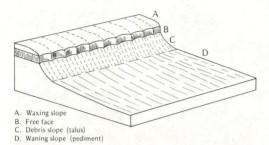

A. Waxing slope
B. Free face
C. Debris slope (talus)
D. Waning slope (pediment)

Fig. 2.5. Diagram of the major elements of a hillslope. (After King, 1953.)

inclination. King notes that it can include *pediments* and states: "Though it rises to a maximum angle of 13 degrees the pediment is essentially a slope of less than 5 degrees over by far the greater part of its (profile) length" (Fig. 2.9). Similar slope values are cited for pediments by other authorities, and there is general agreement on the relative flatness of the lower pediment slopes and the steepness of the upper ones.* It

Fig. 2.7. Block talus of Silurian quartzite developed at the base of an escarpment in the Delaware Watergap between New Jersey and Pennsylvania. The slope (inclined toward the viewer) is in excess of 40° (cf. Fig. 3.44).

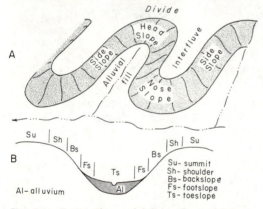

Fig. 2.6. Terminology of hillslopes according to Ruhe and Walker (1968), particularly appropriate for terrain developed in unconsolidated materials. Geomorphic components of hillslope. (A) Slopes in an open system: *Headslope* is at the head of the valley, and slope lengths converge downward. *Sideslope* bounds the valley along the sides, and slope lengths generally are parallel. *Noseslope* is at the valleyward end of interfluve and slope lengths diverge downward; (B) on slope profile, *summit* is upland surface and descent downslope successively crosses *shoulder, backslope, footslope,* and *toeslope.* (Cf. Fig. 2.5.)

Fig. 2.8. An essentially vertical road-cut exposure in loess deposits on the east side of the Mississippi River Valley, Tennessee. At least one soil profile is included in the visible section.

* As noted in Chapter 6, some pediments do not steepen.

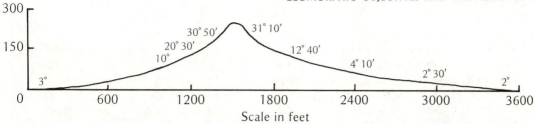

Fig. 2.9.   Cross-section profile of an erosion remnant flanked by pediments in Natal, South Africa, surveyed by T. J. D. Fair. Note slope relationships compared to stream longitudinal profiles (Fig. 2.10) and discussion of pediment development in Chapter 6. (After King, 1953.)

should be pointed out, however, that potentially retreating escarpments that comprise hillslope elements initially may be generated in a variety of ways—as stream banks or valley walls through fluvial incision (Fig. 7.27), as fault escarpments (Fig. 9.3), or as sea cliffs due to marine erosion (Fig. 9.1). Before actual scarp retreat, the subjacent surface (waning slope) could reflect a variety of agencies, and hence display a variety of attitudes and amounts of roughness.

Profiles and Sections

As our discussion progresses some effort will be made to distinguish between the descriptive problems of landforms or distinct portions thereof that (because of their size) can be perceived at a single glance and summarily recorded (perhaps by a sketch or on film) and those larger features that

usually defy such efforts. In the latter category one must assemble a body of data in order to convey an adequate impression of the feature to others. Included here are the actual regional slopes of stream channels as discussed by Fairbridge (1968). These can be shown as *longitudinal profiles* based on numerous slope measurements and field observations. The data can be displayed in section on a suitable grid with certain sacrifices of scale and with some vertical exaggeration (Fig. 2.10). The down-valley slope relations of river terraces offer similar problems and so do a good many cross-valley parameters that can readily be illustrated in *transverse sections* (Fig. 2.11). Sometimes good visual comparisons are achieved by displaying serial sections, possibly superimposed (either longitudinal- or cross-profile), with individualized symbol systems (Fig. 2.12).

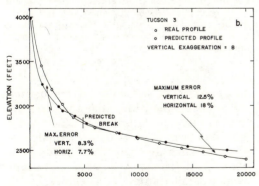

Fig. 2.10. Longitudinal profiles of ephemeral streams (actual and predicted) in southeastern Arizona. Note changes in concavity indicative of two kinds of hydraulic relations along each channel. (From Cherkauer, 1972.)

Remnant Planar Landforms of Low Inclination

Of particular importance, and of some illustrative difficulty, in the aforementioned large-landform category is the detection and portrayal of planar landforms of low inclination, particularly where they only exist regionally as remnants at one or more levels in rather thoroughly dissected country. Possible origins for such features include several types of planar erosion, marine deposition and regional subaerial aggradation in a variety of modes, including glacial, desert, and volcanic. Sometimes the geomorphologist will be uncertain of the existence of such surfaces as he approaches a problem of landscape analysis. More often than not, he will be confronted with the statement that such-and-such a surface "exists." The burden of proof or disproof will usually fall on him. The question then, in terms of external form is, "Is there an accordance of levels at a particular elevation or within a particular range of elevation?" He can ask, "Why?" later.

One of the commonly cited indications (or proofs) of the existence of low inclination land surfaces, and one that may also be very misleading, is visual horizon aspect (Fig. 1.10). There are several pitfalls here, including the fact that the control is local and generally tends to be two dimensional. Also, distance tends to eliminate our visual perception of roughness. A typhoon we experienced in the South China Sea in 1945 is unforgettable as a case in point. The storm produced 65-ft waves (towering well above the ship's main deck, 20 to 30 ft), but the water horizon viewed some 9 miles from the superstructure appeared not only quite smooth but motionless.

To verify visual impressions, one possible approach is to construct serial cross sections from areal topographic maps. These can be superimposed, either with transparent overlays or with an identifying color or symbol system. The well-known Appalachian summit accordance ("Schooley Peneplain," cf. Johnson, 1931) has great "apparent" regularity (Fig. 2.13). It is actually far from perfect. An accordance of levels will show up visually by this method unless the existing surface(s) incline at a high angle to the trend of the sections. The latter possibility can be checked by making one or more transverse control profiles which are geographically tied into the initial series. Alternatively, a topographic model can be constructed by cutting out profiles mounted on cardboard and inserting them into a three-dimensional mockup with proper horizontal and vertical scalar control (cf. Fig. 1.9). Planar landforms thus indicated (if there are any) may be an accordance of summit levels (*gipfelflur* of Penck, 1919), or the

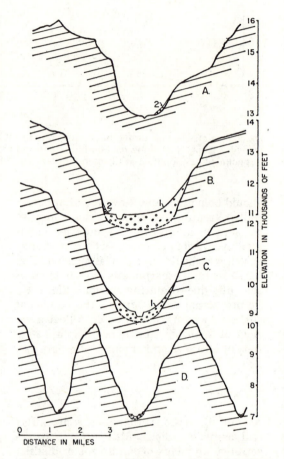

Fig. 2.11. Cross profiles of the Urubamba River Valley, an Amazon tributary in Peru, at four distinct valley-floor elevations showing shape modifications by glaciation and alluvial fill. (A)–(C), U-shaped, glaciated profiles modified by (1) coarse valley fill comprised of alluvial cones and fans which predates most recent high-level glaciation (A) and by (2) minor talus cones developed since recent glaciation; (D), V-shaped valley in an area of protracted humidity at intermediate elevations.

surfaces may be subsummit at one or more elevations. Establishing that a planar landform (or its remnants) exists is not equivalent to determining its nature. Application of terms such as *aggradation plain, peneplain, pediplain,* or *structural plain* must await a detailed study of the feature.

As an alternative approach to the foregoing problem, or as a cross check, elevations secured by a random scan of a region's

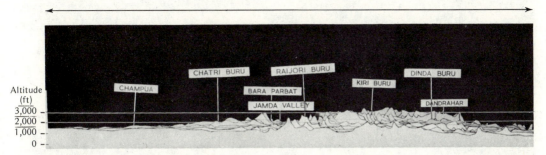

Fig. 2.12. Serial topographic profiles erected in a topographic model of the Keonjhar Region, India; view is toward the south. Three planar levels visible have actual counterparts at altitudes of 3,000–2,800 ft, 2,200–2,000 ft, and 1,700–1,400 ft, and a few peaks rise above 3,000 ft. (K. L. V. Ramana Rao.)

Fig. 2.13. Accordant Appalachian Mountain summit levels in New England. Isolated upland summit (arrow) is W. M. Davis classic example of an erosion remnant, Mt. Monadnock, New Hampshire. (George Theokritoff.)

Fig. 2.14.   View of a rectilinear escarpment along the San Andreas fault where it crosses the Carrizo Plain, California. Similarly straight scarps typify fault zones. Though the foreground stream is "offset" away from the viewer to left center, it is more probably an instance of runoff seeking out zones of easy erosion than actual structural displacement. (William A. Garnett.)

Fig. 2.15. Cathkin Peak, part of the great Drakensberg Escarpment of Natal, South Africa, where gully heads extend into the Jurassic basaltic lavas. The 4,000-ft-thick layer of lavas caps the 10,000-ft-high Basutoland Plateau. (Aircraft Operating Co. Ltd.)

detailed topographic maps can be processed by computer to determine if there is a random vertical scatter or a statistically significant grouping at particular levels (cf. Strahler, 1968; Tanner, 1968). If there is adequate topographic detail, the range of elevations (relief) of the surfaces so located can also be recorded. These figures provide a roughness index which might be useful in differentiating planar surfaces of various origins in conjunction with other data. Where many maps provide the data for an evaluation of a planar landform, the topographic data from each map can be recorded on tape or punch cards and fed into a computer programmed for a comparative regional morphometric analysis.

Scarps

Escarpments (*scarps*) are the topographic opposite of planar landforms of low inclina-

tion, though both types can be planar. The attributes of scarps as hillslope elements have already been noted. They may constitute formidable physiographic barriers, even when they hardly cause a crowding of contours on a topographic map. But, in any case, scarps also exhibit the additional form parameter of areal distribution or pattern. This areal pattern may have genetic significance in some cases. For example, structural escarpments related to faulting tend to be rectilinear in plan form (Fig. 2.14). It is sometimes argued that an escarpment pattern hardly changes during retreat in areas of uniform lithology. Even in the common setting where lithologies vary along the escarpment trend, much of an original areal pattern may be maintained (King, 1956). In the case of the South African Drakensberg Escarpment cited by King (1956), the lithologic variations (granite, quartzite, dolomite, shale) hardly affect the over-all height of the scarp (Fig. 2.15). The same

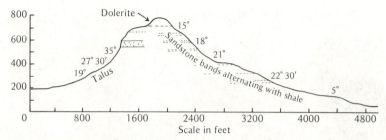

Fig. 2.16. Effects of strata massiveness on eroded slopes involving shale and other lithologies. [After Fair (in King, 1953).]

conditions are seen in the Eureka Springs Escarpment along the northern edge of the Boston Mountains in Arkansas. There, chert, sandstone, limestone, and shale all occur as lithologies in the scarp face and crest in various places without notably af-fecting its height or steepness. It is interesting to note that, in both these instances, bedrock lithology is a constant relative to whatever downwearing and backwearing processes are involved. Nevertheless, King (1953) notes bedrock lithic variations that

Fig. 2.17. A series of nearly straight cliffs produced by headland erosion south-southeast of Newhaven, Sussex, England. Valleys between uplands (Seven Sisters) have been left hanging. For a close-up view of chalk being eroded, see Fig. 9.31. (H. M. Geological Survey.)

favor escarpment development as a hill-slope element (Fig. 2.16).

Cliffs and steeply inclined strands generated by marine erosion have highly variable areal patterns but are rarely rectilinear over long distances. King (1956) maintains that the Drakensburg Escarpment was produced by marine erosion. Its considerable parallelism with the African coast seems to be one of the better arguments in favor of this hypothesis in terms of the theory of parallel retreat of slopes. On irregular coasts, short, nearly straight cliffs often develop on headlands normal to the prevailing winds (Fig. 2.17). Embayment margins are more commonly gentle and curvilinear in areal pattern. As noted by Quinn (1965), irregular scarps formed by the incision of drainage nets tend to straighten out with backwearing, since erosion attacks all scarp surfaces and tends to eliminate narrower secondary and tertiary divides first (Fig. 2.18). This apparently means that scarp straightness (smoothness) is an expression of erosional equilibrium. If this is so, it would presumably apply to scarps of any origin, and the original genesis of scarps that have undergone straightening through extensive backwearing would become increasingly problematical.

Drainage Patterns

Another group of comparatively large landforms that require cartographic or related descriptive techniques are drainage patterns. Their basically two-dimensional form can as readily be taken from air photos as from drainage maps. Scalar distortions are negligible (Howard, 1967). Seasonally occupied flow routes may not appear on photos, and some such flow lines are commonly omitted from topographic maps, particularly where vegetation is very dense. The more common, geometrically regular drainage patterns (Fig. 2.19) are generally presumed to have bedrock structural connotations. Other more randomly variable flow patterns and certain interwoven flow types may reflect particular hydraulic or

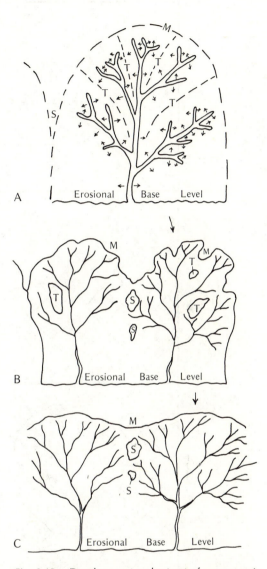

Fig. 2.18. Development and retreat of escarpments by backwasting of valley walls of a drainage net (A) which initially affects tertiary divides within the net (T), with downward erosion limited by erosional base level. In the second stage (B), tertiary divides are almost destroyed and secondary divides (S) are being reduced, with drainage basins beginning to merge. Divide remnants are commonly buttes or mesas. In the final phase, major divides (M) are being eroded and straightened, at least to a sinuous form. In extreme cases, major divides may be reduced to remnants as in eastern South America where the major drainage basins exchange runoff. Based on concepts of Fenneman (1931, 1938) and Quinn (1965).

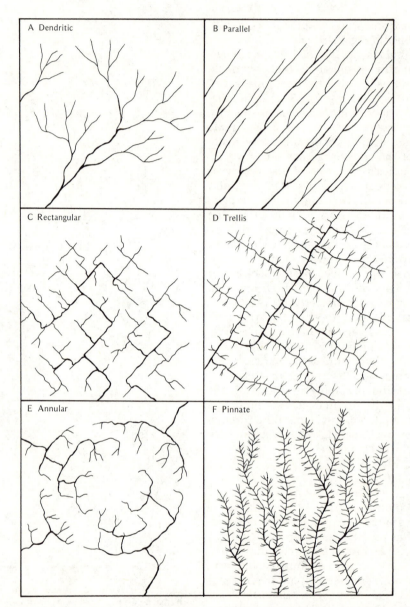

Fig. 2.19. Drainage patterns reflect various terrain compositions or structures. Possible relations include (A) dendritic—granular surface; (B) parallel—elongate sand dune surface; (C) rectangular—intersecting bedrock fractures; (D) trellis—erosion of folded strata; (E) annular—erosion of a domal rock structure; (F) pinnate—very fine-grained granular surface, possibly loess. (See also Fig. 2.20.)

environmental effects. The latter include *braided, meandering, anastomosing,* and *deranged* examples plus the *distributary configurations of deltas,* all usually with accompanying deposits (Fig. 2.20). Preferred geomorphic distributions of the various pattern types will be discussed later.

First- and Second-Order Landforms

The most cumbersome features of the earth's surface are, of course, the largest and the hardest to describe. First- and second-order relief features include the continents and oceanic depressions as first-

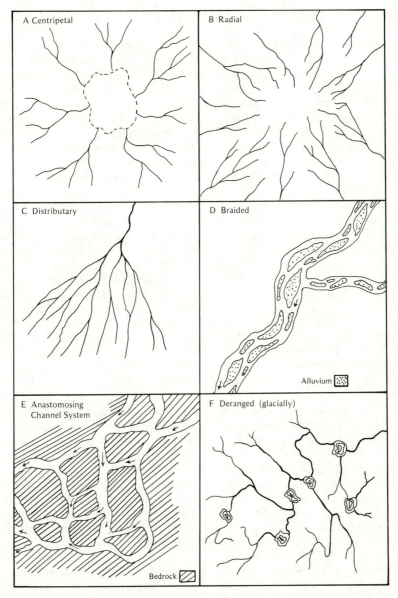

Fig. 2.20. Drainage patterns reflect various terrain compositions or structures. Possible relations include (A) centripetal—flow into closed depression (volcanic, eolian, or structural) commonly occupied by a playa, pan, or lake; (B) radial—drainage from local elevated area, commonly a volcano; (C) distributary—self diversion of channels in deltas, alluvial fans, or riverine plains; (D) braided—alluviation with interwoven flow pattern within a channel; (E) anastomosing channel system—network of channels developed by consequent runoff on irregular (non-drained) terrain; (F) deranged (glacially)—flow patterns common to glaciated bedrock surfaces with numerous lakes which later become swamps.

order features, and the major mountain chains, plus erosional, depositional, and structural elevations and depressions of comparable magnitude as second-order features. It is rare for one geomorphologist's descriptive efforts to adequately encompass any of these. The work that comes closest is perhaps that of Bruce C. Heezen and Marie

Tharp (1961; 1964; 1968) and the various team efforts of major oceanographic institutes. Such description and analysis flirts with the disciplines of geodesy, satellite cartography, submarine bathymetry, global tectonics, infrared and radar scanning techniques, and regional cartography. The results are variously portrayed on relief maps with a diversity of projections, globes, block diagrams, and topographic maps with overlays (Figs. 2.21; 2.22). In recent years great strides have been made (see Fairbridge, 1966), and the geomorphic details of the ocean floors and submerged continental margins, have joined with satisfying regularity the array of subaerial geomorphic features already available.

Third- and Fourth-Order Landforms

We are certainly guilty of neglecting some of the special descriptive problems of various large landforms. The matter will be considered elsewhere and particularly in Chapter 3 where geologic structure is given some emphasis and Chapter 10 where mountains are discussed. Now, however, we will focus on the particular aspects of relatively small landforms and portions thereof. Size is not necessarily a measure of geomorphic importance, for many small landforms reflect the genesis of the over-all associated landscape. Among the scores of categories here, we will cite only a few to illustrate the range of variation. Among the more important features are the channels of drainage systems, both those that currently carry water and those that no longer do so. In the former, one can consider process geomorphology in the hydraulic sense; in the latter one is concerned with progressive modifications of relicts. The general longitudinal topographic aspects of channels have already been noted. We must now consider their hydraulically significant cross sections which many researchers (for example Leopold et al., 1964) believe reflect the intimate interactions of contained water and

sediment. Channel cross sections are usually recorded in series, and actual measurements may require a combination of surface surveying and sounding techniques, depending on flow quantities. Relatively consistent channel shape variations exist with respect to relatively straight reaches and bends (Fig. 7.17). In addition to cross-sectional data, detailed channel flow patterns and sediment distribution maps and diagrams are finding increased use (Leopold et al., 1966; Tinkler, 1970) (Fig. 2.23).

A variety of landforms (mostly third- and fourth-order relief features), apparently generated by running water, are of the same general scalar magnitude as channel cross sections and flow pattern features. These landforms include *point bars, lacustrine deltas, alluvial islands* (in braided streams), *slipoff slopes, sloughs, knickpoints, plunge pools, potholes, channel banks, channel deeps (thalweg)*, and *alluvial cones* and *fans*. Clearly, there are other features of fluvial origin. Of special note as elements of surface roughness on plains and plain remnants are *rills, gullies, arroyos, wadis, oueds*, and *box canyons*. Non-fluvial roughness is also seen in lava flow and ash fall irregularities, marine deposit initial dips, and several types of primary structural flexures. To this we should add eolian roughness features such as *dunes, yardangs, zastruga*, and *sapped cliffs*. We should not omit random slopes on depositional topography generated by sheet floods and mudflows, *playa* desiccation features, and individual clast roughness effects.* There are also mass wasting landscape elements including *landside topography, landslide scars, rotational slump blocks, soil creep ridges, talus* and related *scree piles*, and *solifluction markings*. Were we to add the myriad landform details of karst, glacial, permafrost, and structural landscapes, it would only serve to emphasize our point; each landscape includes many relatively minor landforms to be sought out and described by the geomorphologist. It is not simply that they are a part of the landscape.

---

* Outstanding among these are some enormous glacial erratics (cf. Chapter 8).

Fig. 2.21.   Bathymetric sketch of the Indian Ocean. (B. C. Heezen and Marie Tharp, 1964.)

Fig. 2.22.   The Eastern Indian Ocean portion of the Physiographic Diagram of the Indian Ocean. (B. C. Heezen and Marie Tharp, 1964.) (Cf. Fig. 2.21.)

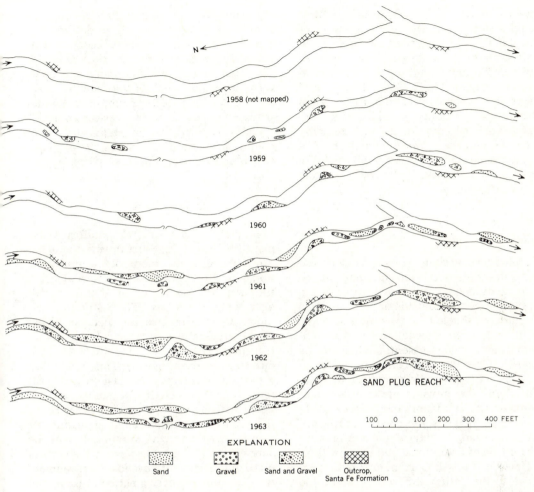

Fig. 2.23.   Sand and gravel bar positions along Arroyo de los Frijoles, New Mexico (1958–63). Bar locations were consistent from year to year, though detritus was being moved. (From Leopold and others, 1966.)

It is also that most of them were formed during the creation of larger associated landforms, and they therefore may furnish clues to the genesis of these large landforms. Formation of the above features and others will be considered in later chapters dealing with structural, fluvatile, volcanic, glacial, and eolian activity.

As previously noted, the form of the "smaller" landforms can often be described adequately by a single photo (e.g. Fig. 1.21) or sketch (Fig. 2.23) or both when one supplements the other. There should also be a written description of the feature's outward

aspect including such traits as apparent roughness (smoothness) particularly for items that are not likely to show on illustrations. Comments on the nature of associated plant cover, any surficial soil or alluvial texture, and apparent regolith moisture would also be appropriate. Bedrock configurations and lithologies should be included when they are visible. The latter should include rock types, bedding characters and attitudes, joint and similar structures, and their directions and spacings, particularly if they contribute to the outward aspect of the landform. This is clearly a "gray" descriptive area, for these lithic features would also relate to the internal aspect of any landform. In addition to other techniques, many smaller landforms can be illustrated to advantage by a micro-topography form map (Fig. 2.3). Of at least equal importance is some notation of the regional cartographic, megatopographic, and general geologic situation of the feature being described and the degree to which the feature might be subject to microclimatic influences including those due to preferred geographic exposures.

Internal Aspects

We have devoted some pages to a discussion of the description and measurement of the external aspects of landforms. In some exceptional cases the outside of a landform is all that is available to the geomorphologist, but it is often the internal aspect of a landform that provides essential clues to its genesis. It is evident that many of the external aspects of a landform reflect internal conditions, the previous descriptive comments on soil, alluvium, and bedrock being cases in point. Thus, there is no sharp transition between a study of the features on the outside of a landform and those within.

However, it is important to distinguish between the rather different geomorphic problems presented by landforms composed largely of consolidated bedrock and those composed largely of unconsolidated material. Again, it must be admitted that the categories are often difficult to define since many landforms combine these characters. However, the distinctions are useful. Some landforms acquire their shape largely through the removal of rock debris; others result from accumulation, often of the same debris. Thus the latter (*depositional landforms*) develop more or less contemporaneously with the former (*erosional landforms*).

Depositional landforms composed of indurated bedrock or lava are perhaps more rare than those of unconsolidated material. Erosional landforms composed of lithified material may be more common than those

of unconsolidated detritus. Or it merely may be that we feel more certain in designating a bedrock landform erosional because lithification is presumed to take time and it is often not clear if an unconsolidated landform is depositional or has simply been formed by the erosion of nonindurated detritus. In any case, despite certain compositional variations and overlaps, the forthcoming discussion is based on depositional versus erosional landform categories.

Depositional Landforms

Examples of depositional landforms are plentiful. Excepting subsequent modifications, the external morphology of depositional landforms reflects the internal structure and the character of the formative agency to some degree (Fig. 8.44). Furthermore, the internal structure necessarily expresses the failure of a geologic agent of transport. The reasons for failure are various. Far too often, however, one reads that a particular category of clast in a deposit represents the activity of some geologic agent with a certain competency. Perhaps the clast did represent such an agent (or agents) WHEN IT WAS MOVING. But in the deposit, we are simply viewing the resting place of objects that have lost their momentum, presumably over a period of time, and an increment of distance for a variety of reasons. The distinction is an important one. In a depositional landform the principal genetic issue is, "What conditions prevailed during creation of the landform?", not merely, "What transported the materials?"

For example, *sinter* accumulations around thermal springs (Fig. 2.24) commonly comprise lithified depositional landforms. They may owe their existence to precipitation from a solution whose hypersaturation is due to cooling. But organisms may induce the actual deposition. Again, *alluvial fan gravels* are often unconsolidated and may reach their destination by *stream flow* or *stream floods* or, ultimately, from the *sheetfloods* or *mudflows* these agents became (Fig. 6.14). *Travertine*

Fig. 2.24. Sinter deposits topped by rising steam in the Mammoth Hot Springs terraces, Yellowstone National Park.

mounds in abandoned watercourses and *organic reef mounds* are lithified depositional forms which reflect, in varying degrees, the depositional intervention of organisms, coupled with the factors that affect the carbon dioxide content of water and carbonate solubility. Cavern *dripstone* (Fig. 5.47) is a third example of depositional landforms that are usually lithified. Carbonate solubility is usually a major factor here as well, but there are exceptions (Fig. 2.25). The suite of depositional landforms that owes its existence to a slackening of air or water currents or the immobilization or melting of ice masses are also important. These include *till plains* (Fig. 8.47), *moraines* (Fig. 8.19), *sand dunes* (Fig. 2.26), *point bars* (Fig. 7.16), and *deltas* (Fig. 7.30). The materials that constitute these features are frequently unconsolidated. The hardened tabular sediment layers of emergent sea floors (*structural plains*) (Fig. 9.37) and desiccation indurated clays of *playa lake beds* (Fig. 2.27) are potentially similar in origin. Mineral encrustations on the latter features more probably reflect evaporation than particle settling. Often other mineral coatings (*duricrust, caliche*) may hardly modify the shape of the various surfaces they encrust. Here the issue may be raised as to whether more material has been added or removed.

When one considers the ways in which depositional landforms can originate, a re-

Fig. 2.25. Salt stalactites developed in the Arapien Salt in an open pit salt mine near Redmond, Utah. (W. N. Gilliland.)

Fig. 2.26. Vertical aerial photograph of "star" dunes in the Namib Desert, South West Africa. (J. J. Bigarella.)

cently overheard remark by a geomorphologist is not surprising. He said, "You have to be something of a sedimentologist in order to do glacial geomorphology these days." Clearly, the word "glacial" could easily be deleted from his statement and it would still be essentially correct (*see* Jackson, 1970). The study of till fabrics has proved to be essential to the understanding of glacial depositional modes, *provenance* and even erosional mechanisms. Moreover, the usual glacial-interglacial geomorphic association includes ice-, water-, and wind-laid materials (often more or less stratified) plus soil (Fig. 5.19). Thus, as noted by Ruhe (1969) the science of soils (pedology) is important in geomorphic analysis along with the principles of stratigraphy (cf. Frye and Willman, 1970; Schafer and Hartshorn, 1965).

Non-glacial deposits, like their glacial counterparts, require a variety of sedimentologic analyses. Channel, valley, and basin fills; alluvial terraces, cones, fans, and

bajadas; pediment veneers and lacustrine sequences are typically involved. Again, many of these deposits are found to comprise geomorphic sequences in which fluvial and/or eolian deposition and erosion alternate with soil formation (cf. Figs. 2.54 and 2.55). For this reason, studies are directed at grain size, grain shape, clast orientation, sorting composition and distribution, often with the data being processed statistically. Other potentially significant factors include weathering depths and modes, heavy and accessory mineral assemblages, and the results of insoluble residue analyses and clay studies (Tricart, 1958; Lugn, 1935; 1960; Bigarella and Ab'Saber, 1964; Lustig, 1966). This body of research and that of many others including, notably, Bigarella *et al.* (1965), Ruhe (1969), and Willman and Frye (1970), suffice to establish the validity of the stratigraphic-sedimentologic-paleopedologic approach to the study of depositional landforms.

In addition to sedimentology and mor-

Fig. 2.27. Sun-baked clay on a dry lake bed (playa or large pan) developed in the western Algerian Sahara. In the western United States, similar ephemeral lake basins are typically situated in intermontane depressions.

phology, physical relationships within deposits are usually time significant. Materials that can be used for dating (wood, animal fossils, and pollen) can usually be recovered from deposits, and as both Bigarella and Ruhe emphasize, the dates so acquired may have pertinence for both the depositional events and for associated erosional landformation events. The proximity of erosional event to depositional site varies greatly in both time and space and depends largely upon agent and environment. Thus, coarse eolian deposits in quantity are rarely moved far from sources (Fig. 2.26). But fine clastic eolian deposits traceable to a particular provenance are usually limited to downwind silt (loess) concentrations (Fig. 2.28). Gravity deposits that have not been re-eroded are usually relatively close

to their points of origin and can often be traced to specific parts of the landforms where they originated (Figs. 2.29 and 2.30). Detritus eroded by waves and currents along open coasts can only rarely be traced to precise place of origin.

Data on the internal composition and structure of depositional landforms are acquired in a variety of ways. Outcrops naturally are preferred because of their dimensional aspects and mega-structures plus lithic samples. Many a geomorphologist has benefited from exposures produced during a highway construction program. However, natural exposures produced by geologic agents are commonplace sources of data, particularly those caused by fluvial incision (Fig. 2.29), faulting, and the like.

In the absence of surface exposures,

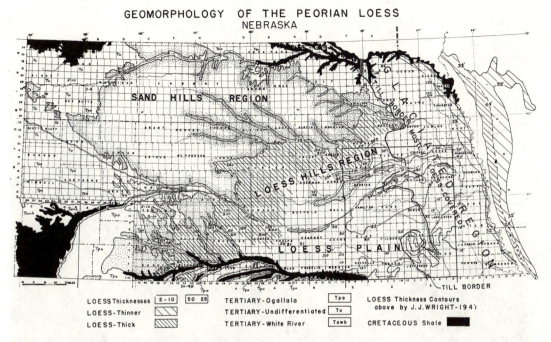

GEOMORPHOLOGY OF THE PEORIAN LOESS
NEBRASKA

Fig. 2.28. Note the association of the thickest portions of loess deposits with margins of a fossil Nebraska erg (Sand Hills Region) and with watercourses such as the Missouri River, which tend to remain vegetated even in dry times. (From Lugn, 1935.)

other approaches are used. Auger drilling has been developed to a fine art by glacial geomorphologists, and small portable drilling rigs are widely used for collecting samples and determining depositional sequences. Since about 1960, two devices are increasingly and effectively employed to establish surficial geologic structure in solid rock, internal regolith configurations, and depths to bedrock. These are *earth resistivity instruments* of several types and single- or multiple-channel *portable seismographs*. Lightweight models are available that are simple to operate and relatively inexpensive. They are very useful; their application includes such operations as differentiating between rock-cut and alluvial terraces (Fig. 2.31) and establishing alluvial fill thicknesses (Fig. 2.32), weathering depths, and pediment-bajada boundary relations (Fig. 6.22). The instrumental data can be plotted directly in contour or cross-sectional form, or it can be processed digitally by a computer to yield bedrock form maps. Apart from geomorphic considerations, such in-

formation clearly has potential economic significance in such areas as groundwater geology (Garner, 1966), engineering geology, and quarry operations.

Erosional Landforms

Erosion can, of course, modify any surficial earth material, whether it is lithified or not. But the special descriptive problems centering on erosional landforms concern mainly those features that are, in large part, consolidated bedrock. Their outward form may reflect a sculptured bedrock character alone as in the case of a cirque-shaped peak (*horn*). More often there is some shape modification by weathered *regolith* with a resultant softening of line.

Depending on the present environment, the regolith is usually under more or less continual form modification by mass wasting. Thus, an escarpment or hillside without a debris slope of talus or colluvium is unusual. It might indicate the age of the

Fig. 2.29. Landslide scar on the north slope of the Venezuelan Cordillera de la Costa due to hypersaturation of the regolith from heavy rains in 1951. Scar is more than 325 ft (100 m) wide and about three times as long.

slope-making event (e.g. very recent?) or perhaps something special about the geomorphic mechanism that caused the stripping.

The external description of erosional landforms essentially parallels that of other types. But the treatment of internal aspects of these landforms can vary considerably, depending upon whether the role of bedrock has been that of an essentially passive lithic substrate or that of a dynamic geomorphic agency such as volcanism or diastrophism. There are exceptions, but some generalizations can be made about the loci of minor or nil crustal, igneous, or tectonic activity versus areas of intense, continuing actions. Signs of notable, continuing crustal adjustments are present (1) in areas of recent deglaciation, (2) in orogenic belts—particularly those less than 70 million years old,

(3) along mid-oceanic ridges, their probable extensions and associated transverse fractures, and (4) in association with oceanic trenches. Areas of notable tectonic and volcanic quiescence include (1) continental segments consisting in large part of Precambrian igneous and metamorphic rocks, some with a relatively thin veneer of Paleozoic and/or younger sediments (*cratons*) and some without (*shields*), and (2) oceanic *abyssal plains* well away from ridges, troughs, and transform fractures. Chapter 3 deals with the tectonic aspects of landforms in some detail. The structural stability of the continental United States is representative of that of recently glaciated lands (Fig. 2.33).

Passive Geologic Substrate—Descriptive Techniques. In the absence of evidence to

Fig. 2.30. Landslide debris on the narrow, arid Venezuelan coastal plain derived from scars like that depicted in Fig. 2.29 in the nearby mountains following heavy rains in 1951. Road (bridge) was originally buried, but excavation and subsequent runoff have removed finer debris. The Caribbean Sea is in the background; the scene is about 0.5 mi (800 m) from the mountain base.

the contrary, geomorphic studies in areas of cratons, shields, and worn down mountain systems may be assumed to involve a structurally passive rock substrate. Where landforms are developed in the absence of notable crustal movement or igneous activity, the major potential geomorphic influence of bedrock is lithologic in the broadest sense of that term, through the mechanisms of differential erosion and/or weathering. It is a rare countryrock that is texturally and structurally homogeneous. Therefore, a major issue in such settings is the extent (if any) to which landforms express substrate inhomogeneities and homogeneities. The problem of why a given character is or is not topographically expressed is related.

From discussions set forth in Chapter 1, it seems clear that if particular subaerial environments can induce particular kinds of landforms, the degree of bedrock expression in landforms can vary. Thus, valleys

etched by linear erosion (particularly in streams) frequently reflect bedrock structure (joint sets or systems or other fractures) where the flow pattern was developed in contact with bedrock (Fig. 2.19). Higgins (1961) has suggested that even some buried fractures may influence drainage. In related contexts, *dendritic drainage patterns* are usually taken to reflect development on some type of non-structured, probably granular covermass. Eolian erosion may etch out rather grotesque physiographies where relief is already present (Fig. 2.34) but is relatively ineffectual in the presence of flat lands with *lag gravel* residues on their surfaces (Fig. 6.4). King (1953) has noted variable expressions of bedrock lithology on hillslope form (Fig. 2.16); he previously noted escarpments with a morphology that appears to be independent of lithology (1966). Many researchers have observed common beveling

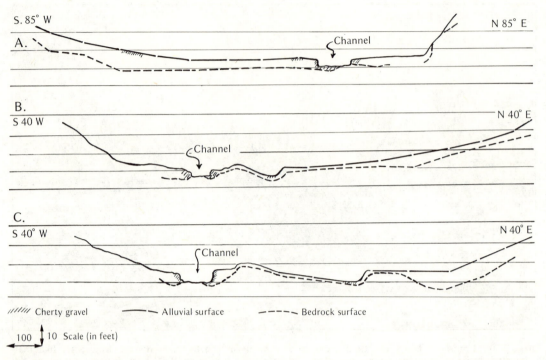

Fig. 2.31. Alluvial cross sections based on resistivity meter traverses along Mountain Fork Creek, northwest Arkansas. Some features are cored by bedrock; others are not. (For location and fuller discussion, see Chapter 7 and Figs. 2.32, 7.10, and 7.21.)

of bedrock structures and various textural lithologies by pediments. Davis, in 1893, described bedrock masses (*monadnocks*) rising above erosion surfaces, and there has been some tendency to regard these and similar forms (*inselbergs*) as being especially resistant to erosion (cf. Figs. 1.12 and 2.13). In many if not most instances it is difficult to establish any notable lithic or structural difference between the rise and surrounding terrain. Many such erosion remnants can be shown to occupy former drainage divides where their preservation would be favored by mere geographic position (Quinn, 1965).

Landscapes influenced by bedrock lithologies often lack the obviousness of erosionally weak rock masses (e.g. shale formations) which have been exploited by such agencies as glacial ice, running water, or the frequent gully or ravine development along fault or shear zones in bedrock. De-

tailed studies of the bedrock geology may be required before a comprehensive geomorphic analysis of a landscape can be made. The evaluation must often extend beyond mere rock-topography associations such as the usual one between limestone and *karst* landscape. For example, the Boston Mountains of northern Arkansas are often cited as classic residual mountains produced by erosion (Figs. 2.35 and 2.36), but they are separated from the Arkansas Valley to the south by a steep monoclinal flexure locally expressed in bedrock by faults. The same mountains generally overlie a broad structural rise of Mississippian Age which helped localize late Paleozoic carbonate deposition (Garner, 1967). And probably equally important, the relief is developed in a Paleozoic shelf lithofacies dominated by sandstone and limestone, whereas the adjacent valley to the south roughly coincides with a southward transition into a shale facies de-

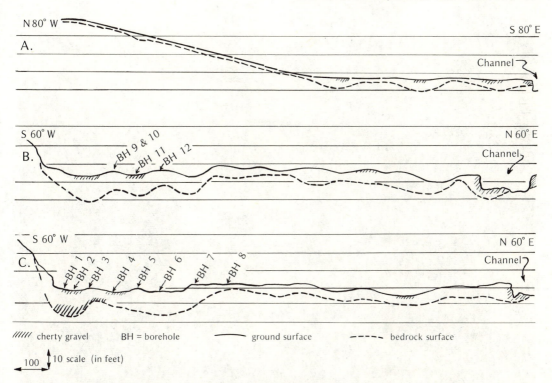

Fig. 2.32. Alluvial cross sections in northwest Arkansas based on resistivity meter traverses, confirmed by auger drilling to bedrock. Cherty deposits are those made by runoff moving down-valley under humid conditions. Chert-free fill was introduced from valley sides. (For location and fuller discussion, see Chapter 7 and Figs. 2.31, 7.10, and 7.21.)

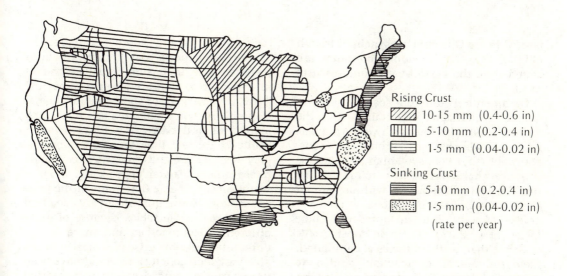

Fig. 2.33. Map depicting degrees of crustal unrest in the United States in terms of uplift or subsidence per year. Based on an illustration of the U. S. Coast and Geodetic Survey.

Fig. 2.34. Delicately sculptured topography formed from fine-grained material on Sunrise Point, Boat Mountain, Bryce Canyon National Monument, Utah. Though some eolian effects are indicated, much of the morphology has probably been achieved by differential weathering and sheetwash. (Grant, National Park Service.)

posited in the Ouachita geosynclinal trough (Fig. 2.37). The Boston Mountains are bounded on the north by erosional escarpments.

On an entirely different scale of observation, many individual mountains within the Boston Mountains in northwest Arkansas reflect subtle bedrock lithic variations. Many of the Pennsylvanian formations there are characterized by patch reefs. The reef limestone lithofacies (with or without some associated sandstone) typically grades laterally into an inter-reef shale facies (Fig. 2.38). There is considerable evidence that, as these facies accumulated, their differential compaction tended to cause reef facies to redevelop on or near older, compactionally emphasized (higher) reef masses, thus causing inter-reef shale

facies to overlie one another. Compaction along these patterns should have produced reef areas that are topographically high and inter-reef areas that are lower whether the latter are emergent or submergent. Thus, it is probably not mere chance that many mountains in the area (e.g. Bloyd, Kessler, White) have reef limestone cores and display reef flank sand and inter-reef shale facies on their outer spurs (Fig. 5.5). It seems most probable that many of the present Boston Mountain valleys are localized on inter-reef shale facies because of early differential compaction and more recent differential erosion. It seems most improbable that such conclusions could have been drawn from a casual description of external landform morphology and passing reference to outcrop lithologies. Indeed, nothing

Fig. 2.35. The Boston Mountain Plateau (looking north) near Heber Springs, Arkansas. This erosion surface has an average elevation near 2,000 ft and is incised by V-shaped valleys, some of which are more than 1,000 ft deep.

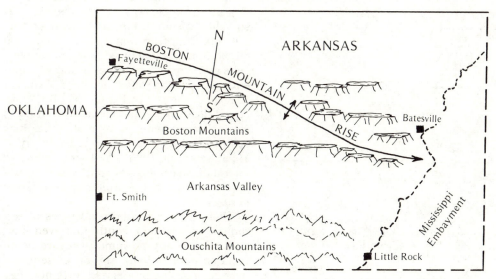

Fig. 2.36. Location of the southern Ozark region including the Boston Mountains and the Mississippian Age flexure discussed here (Boston Mountain Rise). Cross section (Fig. 2.37) is indicated by line N-S.

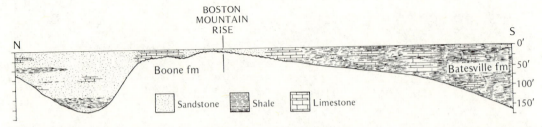

Fig. 2.37. North-south cross section through the Batesville Formation on the Boston Mountain Rise (cf. Figs. 2.36 and 2.38), depicting sandstone, shale, and limestone distribution. The "rise" is a structurally positive area developed at the end of Boone time (mid-Mississippian).

less than a detailed examination of the regional stratigraphy and structure would yield an accurate geomorphic picture of the Boston Mountains. Yet available evidence indicates little tectonism beyond possible uplift in the past 200 million years. So, in its present context the geologic substrate would be considered structurally passive.

The extent to which differential weathering and erosion characterizes various geologic agents and the conditions under which these agents operate will be considered in some detail in the discussions of the several geomorphic systems in Chapters 5, 6, 7, and 8. For the present, and on the strength of the discussion here, it will suffice to note the need for detailed bedrock studies of ero-

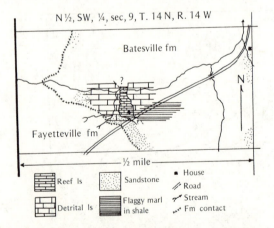

Fig. 2.38. Reef and near-reef deposits of the Mississippian Age Batesville Formation. Shown is a patch reef representative of many on the Boston Mountain Rise (Figs. 2.36 and 2.37). All lithologies indicated are laterally equivalent within the same 20–25 ft (6–7 m) interval of strata.

sional landforms, even in areas where recent volcanic or structural activity is not in evidence and particularly where lateral lithologic continuity is uncertain.

Active Geologic Substrate—Descriptive Techniques. In areas of probably recent structural activity (whether orogenic or epeirogenic, and including recent volcanism), the bedrock roles and the effects of this activity on geomorphic processes must be considered in landform analysis.* Subaerial agencies presumably begin to modify many structural or volcanic landforms even as they develop. Yet there are landforms in both categories (Figs. 2.39; 3.91) that display almost no modification since their formation. Either they are almost resistant to existing geologic agents and, hence, presumably in equilibrium with them, or they are very recent. In many cases, the lateral dissipation of unbalanced stress in orogenic belts and some forms of up-thrusting apparently occur over relatively brief spans of geologic time (cf. Chapter 3). But associated isostatic adjustments resulting in vertical movements in orogenic areas may span scores of millions of years (Garner, 1959; Schumm, 1963). And some indicated lateral displacements of continental masses appear to endure for similar or even larger periods.

There are areas where all types of tectonism appear to be continuing even now, with what many presume to regard as effects on both bedrock and on present and past geomorphic processes. Certainly one's geomorphic treatment of a region of recent

* The effects of eustatic sea level shifts may simulate some epeirogenic effects on geomorphic processes.

Fig. 2.39. View of the anticlinal mountain and synclinal valley that comprise the "Grande Chartreuse," north of Grenoble, French Alps illustrating a structural control of topography. (Swissair-Photo.)

structural activity will be influenced by what one infers the effects of crustal movements on processes such as erosion to be. The same or similar considerations apply to regions that may be modified by sea level changes elsewhere—regardless of origin. We would do well to recognize and advertize the fact that essentially every area of geologic specialization can be directed toward the study of geomorphic aspects of the earth and, in fact, have produced such specialists as volcanic or orogenic geomorphologists. Certainly the former could hardly function without a thorough knowledge of extrusive igneous processes and effects, and the latter would need a working knowledge of tectonic principles, deformational modes and styles, and basic geophysical and isostatic relationships of mountain systems. The sa-

lient aspects of each of these types of tectonic landforms will be discussed in subsequent sections, with the usual assumption that in most instances some subaerial modification has occurred (see Chapters 3 and 10).

Regardless of the probable dynamism of the geologic substrate in a region, the geomorphologist studying an erosional landform related to bedrock characters should record the rock types and their vertical and lateral distributions. He should note their apparent responses to current weathering and erosion processes with some documentation whenever possible. Major partings present should be recorded including joint sets and systems, fissures, faults and shatter zones, stratification characteristics, and any apparent individual or collective reflections of these on topography or drainage. Often, the geomorphologist can do his field work with the aid of an existing geologic map plus some topographic base map. He will commonly find that the geologic data are irrelevant to his purposes or exist in insufficient detail because of the way they are recorded or illustrated.

Problems And Analysis

Suggesting an analytic approach to geomorphic problems is a bit like prescribing a remedy for a patient before determining the nature of the illness. Certainly each problem makes its own special demands on the geomorphologist. But there are some general rules which can be followed in most cases, and these rules govern the order of presentation of material in this chapter and throughout much of the remainder of the text.

The position taken here is that geomorphic analysis is a continuing process. It does not merely occur after data acquisition in the way this section follows the one on description. Or at least, it should not. One tends to see what one excepts to see. It is a matter of programming. Indeed, it is a rare geomorphologist who has not shared the well-known experience of the paleontologist who was unable to find fossils in unfossiliferous, non-marine deposits until word was circulated that some marine fossils had been found. After that it was easy. Thus the initial delineation of the problems tends to determine what theory will be followed and what observations will be made in the course of the geomorphic endeavor. There may be no escaping this, which is why areas studied from one point of view usually yield new data when approached differently. One can only hope for a perenially open mind and a periodic, in-depth re-examination of theoretical bases.

Some detailed over-all analyses and related conclusions will be presented for actual geomorphic problems in several of the following chapters. What follows here are merely examples of preliminary general data on three selected geomorphic settings together with suggested problems and tentative analytical approaches to be used during data acquisition. Each area is viewed as a unit in which the geomorphic findings on one aspect may well influence the approach on a related aspect.

Example 1.   General Information

In southern Ontario and Manitoba, Canada, there is an area of contorted, Precambrian schist, gneiss, and granite, with a mixed coniferous and decideous forest cover. It is drained by rivers and studded with lakes; it has subdued rolling hills with relief averaging less than 300 ft and less than 500 ft elevation. There are patches of poorly sorted gravel containing faceted cobbles and wood fragments radiocarbon-dated as 8,000 years B.P. ±250. The region is now undergoing uplift estimated at a rate of 10 to 15 mm/ year (Figs. 2.33; 2.40). There is no apparent recent volcanism or recent marine submergence.

Problem Analysis. This is an area known to have been glaciated during the Wisconsin Glacial Stage. Specific erosional and depositional effects, either glacial or fluvial, should be noted. Bedrock appears to be consistent with a former orogenic belt. The relation of tectonic forces to present low relief and elevation merits examination. Possible eustatic sea-level influences should be considered with respect to relatively great distances to regional base level. Origin of uplift is uncertain, but post-glacial isostatic rebound is documented in the nearby Great Lakes area. The effects of uplift on the present geomorphic system including drainage should be determined. Low elevation and relief and widespread bedrock exposure suggest that subglacial erosion is dominant over other geomorphic influences in the region, but the crustal thickness here should be studied. Cobble faceting in deposits could be examined to test glacial versus eolian origins. Relicts suggest that the present geomorphic system has not reached a dynamic equilibrium state. Current and prior geomorphic mechanisms should be compared.

Fig. 2.40. Banded Precambrian gneiss exposure near the shore of Lake Superior in southern Ontario, Canada. The surfaces have been rounded by glacial scour and a patch of drift (foreground) has a vegetal cover. See geologic pick for scale.

Chaotic surface water runoff configurations (both nonlinear and nondendritic) suggest glacial drainage derangement, but bedrock lithologic influences should be considered. Pre-glacial, glacial, and post-glacial land-form-deposit relations (including origin of dated wood fragments) should be examined in the context of the over-all setting.

Example 2.   General Information

In southcentral Idaho, there is a low-relief area with many alluvial remnants veneered over gently sloping surfaces. The whole area

is dotted by a series of basaltic, volcanic deposits, averaging 2,150 years old, distributed along an apparent fissure in an intermontane basin setting with barren-to-sparse, grass-sage floral groundcover and no significant associated perennial watercourses (see Bullard, 1971). It is the Craters of the Moon National Monument (Fig. 2.41).

Problem Analysis. This area is broadly orogenic in character and location; evidence of continuing earth movement has not been presented, though such evidence may exist. Low-relief surfaces appear to

Fig. 2.41.   Craters of the Moon National Monument, southcentral Idaho. *Below*: distant view from north-west; dark area is recent eruptive material, lighter foreground is sage growth on older flows. *Above*: two craters and associated pyroclastic debris and lava flows. In the 2,000 years or so since the most recent activity, remarkably few desert plants have become established, and surface material is little altered (cf. Fig. 1.19).

truncate a variety of bedrock lithologies, but associated alluvium is of uncertain depth, and areal extent and should be studied as should genesis of the low-relief land surfaces. A variety of pyroclastic and lava emissions comprise cones and a suite of other eruptive landforms that are not as yet adequately documented. The precise rela-

tion of nonvolcanic slopes to volcanic processes should be investigated. It is stated that extrusive igneous rocks are hardly altered by weathering and erosion. This statement deserves close scrutiny in view of the radiometric ages of the materials. The dominant local geomorphic environment is clearly volcanic, but regional geomorphic history probably resides in the detailed internal and external morphology of plains. Current subaerial morphogenesis would appear to be of a rather uncertain nature in the absence of abundant runoff, but it still deserves detailed study. The relationship of fissure and oceanic type eruptive rocks to plate tectonics and continent-oceanic ridge positions should also be considered (cf. Lipman *et al.*, 1971).

Example 3. General Information

In the Zagros Mountains of Iran (Fig. 10.27) a fold and complex mountain range is seismically active. The range has a negative gravity anomaly suggesting a deep subsidence of a sediment-filled trough; faults and folds affect the most recent sediments, some anticlinal hills and mountains are barely modified by erosion, and there is evidence of a continuing uplift. The Zagros Mountains present a variety of structures that are intricately dissected by a complex river system, though rainfall is notably less and vegetation more sparse in the lowlands than the uplands (see Oberlander, 1965).

Problem Analysis. This area appears to be that of an orogenic landform complex requiring multiple working hypotheses. Much of the region seems to be dominated by tectonic landforms. Karst and allied features show lithologic influences locally and a maze of drainage complexities (including some transverse to structural trends) indicates a long and intricate erosional history with several examples of clear-cut contemporaneity between erosion and deformation (Fig. 2.42). Piedmont areas include some lower relief landforms that truncate structures in younger sedimentary formations (Fig. 2.43). Geomorphic analysis requires examination of structural history, lithostratigraphy, and associated erosional and depositional modifications. Geomorphic systems at low elevations function in low rainfall deserts, but many elevated areas are nearly barren vegetally. A high-elevation frost zone contributes runoff by meltwater,

and mechanical weathering of bedrock characterizes much of region. Possible variations in subaerial environments during Pleistocene merit study (for details, see Chapter 10).

The three examples of geomorphic problems and tentative analytical approaches certainly do not exhaust the possible variety. But they include areas where morphogenesis initially appears to be dominated by subaerial erosion, volcanism, and orogenic tectonism. Current erosional and depositional mechanisms differ in each region, both in character and apparent rate. Each summary therefore merely outlines the major aspects of a distinct geomorphic setting to provide a basis for a detailed landform description and study. As data accumulate, original suppositions about specific physical relationships in the setting should be reappraised and possibly modified. Ultimately, the researcher will have performed the main requirements of scientific procedure by (1) posing a problem based on preliminary information, (2) gathering additional data, analyzing it, and revising initial premises, as necessary, (3) summarizing the results and drawing conclusions.

LANDFORM CHRONOLOGIES AND QUATERNARY STRATIGRAPHY

With the recognition that different environments produce distinct landforms, geomorphologists are now aware that synchronous geomorphic events can differ in kind from place to place. When climates were thought to be essentially fixed in time and space, the timing of geomorphic events from place to place was a tenuous matter indeed. Not until the advent of absolute dating using radioactive decay rates (particularly C_{14} in near-recent deposits) was appreciable progress made (e.g. Black and Rubin, 1967–68). And as previously noted, the scope of this method is notably limited. Before radiometric dating, regional geomorphic correla-

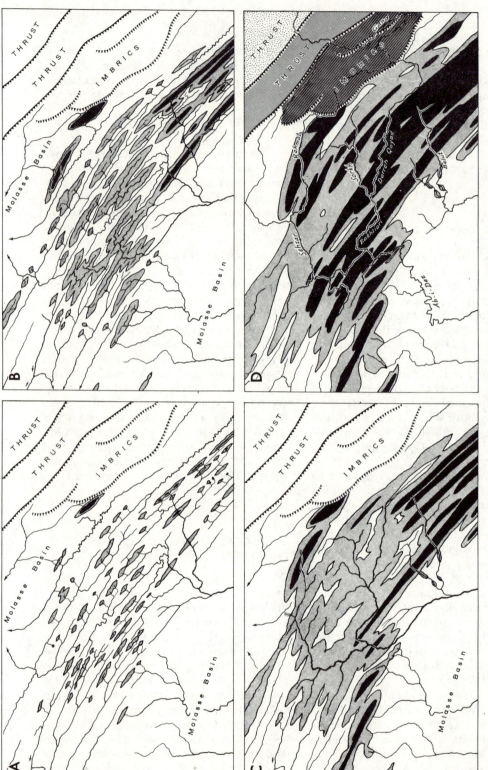

Fig. 2.42. Transverse drainage relations of the Bakhtiari Mountains in the central Zagros Mountain Region. (A) Hypothetical consequent drainage of initial stage, showing development of small axial basins by autogenic denudation of anticlinal ridges; (B) expansion and coalescence of axial basins producing compound subsequent valleys; (C) complete inversion of relief producing transverse subsequent lowlands and facilitating drainage extension across the strike; (D) exposure of Cenomanian fold cores, transected on a massive scale as a result of the superposition of drainage from the transverse subsequent valleys shown in (C). The Ab-i-Balut appears to be an antecedent stream. The absence of other drainage anomalies in the southeast of the region is attributable to the poor development of the erodable flysch in this area. South of the present Bakhtiari River the Cenomanian fold cores were exposed shortly after denudation breached the initial Asmari limestone fold envelopes, precluding any extension of subsequent drainage across the strike. At least some fold development has occurred during drainage development. (From Oberlander, 1965.) (See Chapter 10 for a further discussion.)

Fig. 2.43. Hogback ridges developed in the interbedded salt, anhydrite, and marl of the Miocene, Lower Fars Formation, Iran, near the Masjed-i-Sulaiman oil field. The ridges are products of differential erosion and weathering in an arid environment, but some would readily dissolve under humid conditions. (Aerofilms Ltd.)

tion was largely based on relations between certain structural movements or base-level shifts and, presumably, resultant erosional and depositional events.

In general, and this applies to most existing texts, planar landforms were related to structural quiescence, steeper slopes between planar levels to stream incision caused by land uplift or base level depression, and valley aggradation to land subsidence or sea level rise, all within the theoretical framework of the graded stream and related stream profiles of equilibrium controlled by base level (see Chapter 1). Though some types of structural activity were clearly localized, some workers, at least, sought a form of world-wide structural synchroneity that would permit geomorphic correlations between widely separated regions or even continents. Efforts along these lines are being continued by what has come to be known as the school of structural geomorphology (Beckinsale and Chorley, 1968; Büdel, 1968).

Comparisons by region of geomorphic events on an alternative basis were largely initiated by glacial geomorphologists. They worked largely with the sequence of deposits made by Pleistocene glaciers and the soils formed on these deposits between glacial intervals and associated eolian (loess) and outwash materials. Basically, they followed the principles of traditional stratigraphy, while taking into consideration the landform attributes and temporal aspects of some deposits. Thus, regional comparison of glacio-geomorphic sequences has become commonplace; at the same time, there has been some effort to compare regions intercontinentally.

In many respects, glacial stratigraphic work is comparable to a study of deposits made by transgressing and regressing seas. In the former, sub-glacial conditions alternate with subaerial environments as ice fronts advance and waste away in broad geographic belts. In the latter, submarine conditions alternate with subaerial environments in similarly broad geographic belts. The addition of subaerial environments to the list of conditions capable of inducing unique geomorphic features gives new dimensions to regional geomorphic analysis. Büdel (1961) was AMONG the first to suggest that suites of landforms relate to regional climate zones that roughly parallel the equator and that these zones have been displaced relative to equatoral zones and the poles (Fig. 2.44). There is some question whether such generalized zones meaningfully coincide with well-defined geomorphic systems (cf. Garner, 1968). But at least the potential appears to exist for regional comparisons of landforms and deposits in both glaciated and nonglaciated regions—especially when data from radiometric dating are available.

Chronologic Principles and Methods

The establishment of precise chronologic relations among parts of depositional landforms and associated, usually upslope, erosional landscape elements comprises much of the local geomorphic problem in many instances. Several different approaches are usually possible. Intrinsically, the dating problem is relative, and a perfectly meaningful geomorphic sequence can often be worked out for a region without direct recourse to absolute geologic dates. The relative chronologies permit geomorphic analysis. Eventually, absolute dates for some events are necessary, both to verify the relative chronology of the local area being studied and to compare regional landscapes on bases apart from mere similarity of events in sequence, number, and kind. Technically, only radiometric dating is considered absolute, but fossil dating often supports a chronology obtained by other means.

As noted by Ruhe (1969), even where bedding is horizontal, one should not expect to encounter a layer-cake type sequence in depositional landforms. Following erosional episodes (stream incision), younger alluvial deposits may be inset topographically below older alluvium (cf. Fig. 7.32). Further, the same author notes that an erosional valley hillslope is younger than the youngest (uppermost) deposit within the hill. It is also younger than the hill summit. If the valley has an alluvial fill, the hillslope is the same age as the fill if the erosion of the hillslope provided the fill sediment. However, we should note that where the hillslope is a part of a retreating escarpment, the subjacently forming pediment is youngest next to the superjacent hillslope and progressively older as distance from the scarp increases (Fig. 2.45). The scarp, as a steep landform, predates the entire subjacent pediment. But the actual escarpment face may well be younger than most of the subjacent erosion surface. The debris slope clearly forms as the free face recedes. Ruhe (1969) further affirms the *principles of ascendancy and descendancy* to the effect that a "hillslope is the same age as the alluvial valley fill to which it descends, but is younger than the higher surface to which it ascends."

In the foregoing it seems clear that the Davisian concept of downwearing is being

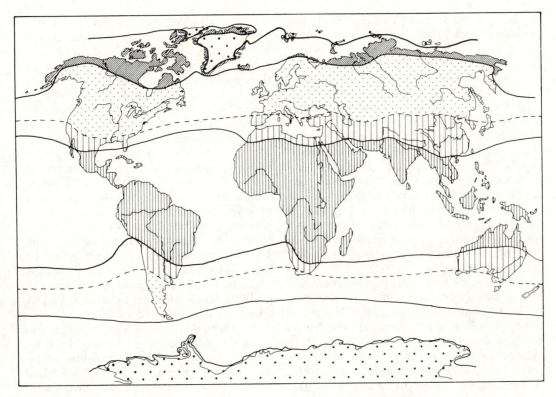

Zone of glaciated areas

Zone of pronounced valley formation

Extra-tropical zone of valley formation

Subtropical zone of pediment and valley formation

Tropical zone of planation surface formation

Fig. 2.44. Latitudinally disposed climate zones related to particular types of landform genesis according to Büdel (1968).

rejected. Thus, if hill summits were reduced by downwearing, they would be significantly younger than valley hillslopes previ-

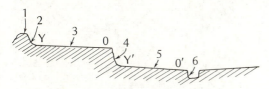

Chronologically, in cross section:

1 was formed first, 6 last
Odd numbers are "flats", even ones scarps
Under parallel scarp retreat, Y and Y' are younger than
O and O', respectively
Under parallel scarp retreat, Y surface and deposits could
equal Y' surface and deposits in age

Fig. 2.45. Chronology of a land surface undergoing planation by parallel retreat of scarps.

ously incised by streams in the area. It would also appear that lateral alluvial emplacement in valleys is incorporated by the principle of descendancy. In instances where valley alluvium is emplaced by water moving downvalley (e.g. by stream aggradation due to downslope volume loss) the valley fill formed could be considerably younger than adjacent valley hillslopes (cf. Garner, 1966).

Trowbridge (1921) cites the following chronologic principles for cross-country erosion surfaces. (1) Any erosion surface is younger than the youngest material it cuts. (2) It is younger than any structure it bevels. (3) An erosion surface is younger than any material of which there are distinguishable fragments or fossils in alluvial deposits on the surface. (4) It is contemporaneous with basal alluvial deposits that lie on it; (5) it is the same age as or older than other terrestrial deposits lying on it. (6) An erosion surface is older than valleys cut below it; (7) it is younger than materials forming erosion remnants above it; and (8) it is older than deposits in valleys below it. (9) An erosion surface is younger than any adjacent erosion surface that stands at a higher level; and (10) it is older than any lower adjacent erosion surface. A comparison of Trowbridge's principles with the age relations illustrated here (Fig. 2.45) for areas of pedimentation and scarp retreat discloses certain contradictions when downwearing is postulated as opposed to backwearing of hillslopes.

We will have numerous occasions to examine and apply the various chronologic principles of erosion surfaces later. It is necessary, however, to note certain restrictions in their use. (1) An erosion surface developed by downwearing would presumably be the same age everywhere; however if such erosion were initiated following stream incision, headward extension of drainage nets would make divide portions relatively younger than other parts. (2) As indicated previously (Fig. 2.45), an erosion surface developed by parallel scarp retreat would necessarily be youngest next to the scarp and oldest along the surfaces distal margins; here, portions of low-level surfaces could, in fact be older than portions of superjacent surfaces. (3) In accordance with the tendency of escarpments to straighten, as noted by Quinn (1965), re-entrants in escarpments would have to recede very slowly or hardly at all for protrusions to "catch up" (cf. Fig. 2.18); the age of a surface along and subjacent to a "straightened" escarpment could vary greatly, being greatest in areas of primary re-entrants.

Conditions under which streams of water vertically incise channels differ materially from the conditions under which water contributes to the shaping of planar landforms. We will consider the precise nature of the differences in Chapters 5, 6, and 7. In Chap-

ter 1, it was pointed out that there was no necessary relationship between crustal movements and atmospheric conditions favoring particular amounts of precipitation, runoff, and fluvial erosion. This would appear to be a basic weakness in any attempt to date all erosional landforms or landscapes through chronologies of structural movements or to relate particular erosional modes (e.g. incision versus planation) to specific directions, amounts, or rates of crustal movement (see Garner, 1959). The seemingly necessary expedient of some theorists to rely on the graded-stream - equilibrium - profile - base - level - control Davisian association (e.g. Johnson, 1931) increases the basic weakness of the logic and theory here. Yet there are many cases of planar landforms (sometimes at multiple levels) incised by stream valleys which require explanation.

An alternative to the theory of structural control of fluvial erosion relates regional incisional episodes to time of *exhoric drainage*. Only then can sediment be flushed from drainage lines as erosion progresses. Clearly, channel bedrock must be more or less continuously exposed to linear erosion mechanisms if valley deepening is to progress rapidly. The associated planar landforms might be of one or more distinct origins, but the landform assemblage would nevertheless be susceptible to relative chronologic analysis with or without relying on the genetic discussions in later chapters. Essentially, this is simply a relative

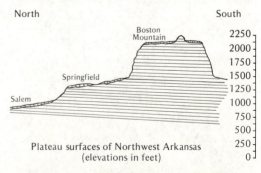

Fig. 2.46. North-south cross section through the southern Ozark region, a terrain possibly generated through the parallel retreat of scarps. (See Fig. 2.45.)

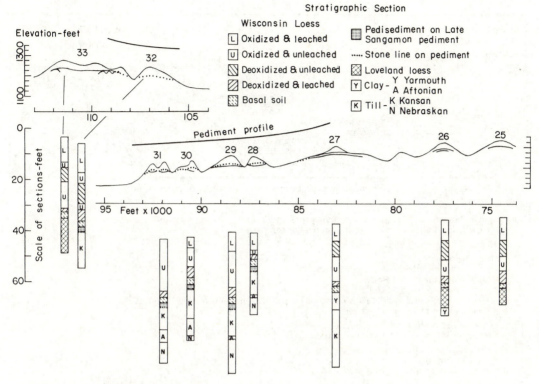

Fig. 2.47. Relationships between land surfaces and soils, eolian deposits and glacial materials, north-central Iowa, according to Ruhe (1969).

scarp-flat chronology that relates each major low-inclination land surface of an area to a leveling event and each escarpment to a relief-producing event (Figs. 2.45; 2.46). Such stair-stepped landscapes may be analyzed chronologically by the rules set forth at the beginning of this section, with the oldest erosional features being generally highest and the youngest lowest. Actual ages (relative and absolute) of such a landform assemblage could only be arrived at by a genetic analysis of individual features coupled with a study of the associated deposits (including contained fossils).

Relative Geomorphic Sequences

No one region provides the range of emphasis necessary to the study and illustration of geomorphic sequences. Consequently, we offer here five contrasting examples from widely separated parts of the world. They include a glacio-depositional setting in the upper midwestern United States, an upland valley in central Europe, a valley in the Boston Mountains of southcentral United States, formerly glaciated valleys on the eastern Andes Mountain flank, and desert valleys and surfaces of the western Andes Mountain flank.

In his summary study of the Quaternary of Iowa, Ruhe (1969) deals most effectively with the geomorphic sequence problems of an area that has been alternately exposed to glacial erosion-deposition effects and subaerial conditions (see also Washburn, 1956). Ruhe describes an association of soil zones, eolian deposits, glacial deposits, and erosion surfaces (Fig. 2.47). In addition, he notes the occurrence of *geomorphic dis-*

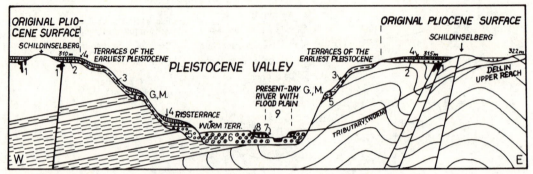

Fig. 2.48. Relative geomorphic sequence of events from an upland valley Central Europe, according to Büdel (1968).

I, Original Pliocene surface, truncating varying rock conditions faults and (present-day) water divides
    1 = Karst fissures filled up with red and brown Pliocene loams
II, Terraces of the earliest Pleistocene
III, Pleistocene valley
    2 = Cryoturbation mantle on planation surfaces
    3 = Solifluction mantle on slopes
    4 = Loess cover overlying 2 and 3
    5 = Gravel of Early and Middle Pleistocene terraces G = Günz, M = Mindel, R = Riss
    6 = Gravel of the older Würm terrace
IV, Late glacial and prehistoric Holocene
    7 = Younger Würm terrace (late-glacial period)
V, Post-Palaeolithic (cultivation)
    8 = Valley loams on 7
    9 = Present-day river with flood plain

cordances between erosion surfaces and associated soil-mineral suites. In this study and in similar studies (e.g. Black, 1962; Black and Rubin, 1967–68), many C_{14} dates and some relative dates anchor the sequences rather firmly into the absolute time scale.

In the Boston Mountains of northern Arkansas there is a series of erosion levels and erosional escarpments carved into a sequence of nearly horizontal Paleozoic sedimentary rocks with some associated gravels and soils plus eolian (loess) deposits. Here the major landforms are clearly erosional in origin whereas, as noted in Chapter 1, some minor features are depositional. Bretz (1962) interprets the resulting sequence in the Davisian manner, whereas Quinn (1958) generally follows the pediplanation concept of King (1953). Few absolute dates are available, apart from those obtained in association with vertebrate remains in a cave developed in the Springfield

Plateau level (Fig. 2.46) and elsewhere in gravel veneers on the erosion levels. This region exemplifies the problem of differentiating erosion surfaces and terraces, possibly created by scarp retreat, from minor benches, created by differential erosion of inter-stratified shale, and more erosionally resistant rock types in nearly flat-lying strata (cf. Fig. 2.46 and Fig. 2.16).

Büdel (1968) illustrates a terraced valley and upland erosion surface in central Europe reminiscent of that illustrated in the previous example (Fig. 2.48). The area was not glaciated, but the series of alluvial terraces at various elevations, and an associated complex of soils, eolian deposits, and permafrost mass-wasting features, are correlated with European glacial or interglacial stages. Such a correlation involves certain assumptions, which may or may not be sound, about the effects of such stages on stream regimen elsewhere. Some archaeologic dating of more recent deposits helps anchor the relative sequence, but few radiometric dates are available.

In the eastern Andes Mountains of Peru, high level plateaus (some erosional, some

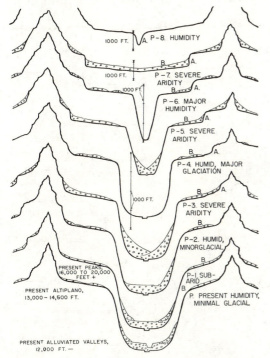

Fig. 2.49. Cross sections of the geomorphic sequence and relief development, Cordillera Oriental, Peru, late Tertiary and Pleistocene. Sequence: oldest to youngest; the apparent events are (P-8) stream incision of an elevated surface of low inclination (montane plateau) having local topographic highs of considerable magnitude, development of scarp A; (P-7) arid pedimentation and aggradation of high-elevation surfaces and retreat of scarp A forming surface B; (P-6) major episode of stream incision under humid conditions, valley depths attained exceed 2,000 ft (610 m); (P-5) prolonged arid episode with valley widening and aggradation; (P-4) major "humid" glaciation, erosion of arid residual alluvium and downslope extension of U-shape valley form; (P-3) severe and widespread episode of arid aggradation, high-level scarp retreat, many talus cones and confluent alluvial fans developed in valleys; (P-2) minor watercourse incision correlative to high-level glaciation; (P-1) minor arid episode, local talus cone formation; (P) contemporary stream incision of relict valley alluvium at present, glaciers confined to high peaks.

depositional surfaces) surmounted by higher peaks have been deeply incised by fluvial and glacial agencies and intermittently aggraded (Fig. 10.6 shows the location). The resulting geomorphic sequence (Fig. 2.49) can be assumed to have a synchroneity of glacial events with those elsewhere in the Southern Hemisphere (subject to certain orographic and geographic variations; compare Chapter 4). Archaeologic dates are available, but absolute radiometric dates are not. Opinions differ as to whether alluvial valley fills are glacial age *cryergic* deposits (Dresch, 1957; 1958; Cotton, 1960) or combined glacial-arid aggradational deposits (Garner, 1959).*

On the western side of the Andes, in southern Peru, a second geomorphic sequence has been recorded (Fig. 2.50) in which glaciation is not particularly significant, though meltwaters have been involved. Here a sequence of volcanic and erosion-deposition events appears to have developed under the dominantly arid conditions that characterize the mountain flank —with periodic subordinate humid episodes. No absolute dating is available, though archaeologic finds indicate the nature and timing of the more recent geomorphic events (Garner, 1959).

These examples of relative geomorphic sequences indicate the possible range of variation in landform development. Some temporal equivalence of events among the regions discussed can be assumed, though, in view of the diverse settings, such correlations could hardly relate to structural controls. And because of geographic-climatic diversity at any one time, there is little reason to expect that landforms of the same *age* in different regions will be the same *kind* of landforms (cf. Fig. 2.51). At the same time it is evident that all the areas discussed exhibit planar landforms and escarpments. To the extent that conditions capable of inducing such features are globally present during a given interval, widely separated regions *could* experience a similar morphogenesis.

* Data presented in Chapter 8 support the latter view.

QUATERNARY STRATIGRAPHIC TECHNIQUES

The geologic subdiscipline, *stratigraphy,* includes a number of areas of special concern here. One such area, paleogeomorphology, will be briefly considered in a later section of this chapter and at greater length in Chapter 9. A second area concerns relatively recent stratified deposits that are also landforms. As pointed out in Chapter 1, a thorough analysis of a given geomorphic setting must include the internal as well as external aspects of landforms. And here the internal aspects of depositional landforms created during the Quaternary Period are in question, since the developmental landform chronologies discussed in the previous section can frequently be refined only through stratigraphic techniques.

The great bulk of the Quaternary stratigraphic literature relates to glacial depositional sequences of the mid-continental Pleistocene of North America and western Europe. But recent stratified deposits are not limited to such situations, nor are glacial deposits necessarily most meaningful in terms of landform genesis. The creation of each erosional landform, of necessity, gives rise to a deposit. The deposit may also be a landform. And it may be closely associated with the correlative erosional feature or so removed or altered as to prohibit our recognition of a common genetic relationship. Reworking of land-derived sediment by lacustrine or marine agents may completely obscure subaerial origins. Or, most of the erosional history of a drainage basin may be incorporated in adjacent deltaic deposits accumulated near or below base level in a lake or sea. The erosional landforms tell one part of the story. The derived deposits tell another part.

Geomorphic-Stratigraphic Settings

It will be to our advantage to examine the three principal geomorphic settings that govern the relative spatial associations of erosional landform and equivalent deposit. The stratigraphic techniques relevant to

each type differ significantly. For the moment, our frameworks of reference will be the subaerial environments and the geomorphic systems they sponsor. Individually, these conditions act on a segment of the crust (*geomorphic area*). And, in this con-

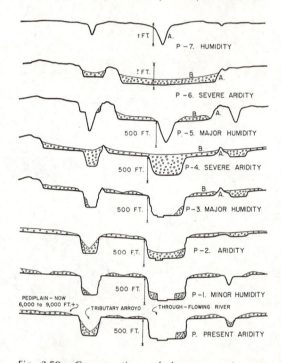

Fig. 2.50. Cross sections of the geomorphic sequence and relief development, Cordillera Occidental, Peru, late Tertiary and Pleistocene. Sequence: oldest to youngest; the apparent events are (P-7) humid stream incision of a surface of volcanic deposition; scarp A results; (P-6) arid pedimentation and scarp retreat with aggradational formation of surface B; (P-5) prolonged humid incision of streams to a maximum depth of 400 ft (120 m) at an elevation of 7,500 ft (2,300 m); (P-4) severe arid aggradation of channels and interfleuves with coincident scarp retreat; (P-3) fluvial removal of alluvial deposits leaving only remnants in valleys; (P-2) minor aggradation of channels to a depth of 80 ft (25 m), essentially complete destruction of scarp A and development of a pediplain; (P-1) incision of channel alluvium along main streams by moisture from humid uplands with terrace development; (P) present arid aggradation of arroyos not receiving flow from humid uplands, these are perched (hanging) above streams carrying snow melt-waters and eroding at high elevations; they alternately erode and deposit at intermediate levels as seasons change and aggrade at low elevations through volume loss downslope.

text, it appears that each type of system tends to retain certain weathering and erosion products within the geographic limits of the geomorphic area (*environmental residuals*) and to yield certain others (*envi-*

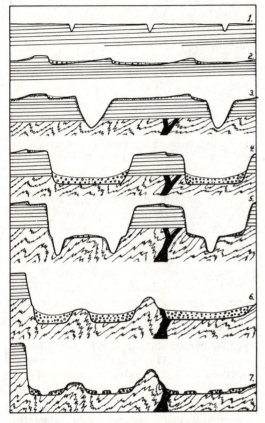

Fig. 2.51. Cross-section profiles of the sequence of geomorphic development, northeastern South American lowlands. The events depicted are (1) humid incision of relatively undeformed Roraima Series (Precambrian) to a depth of about 200 ft (65 m); (2) arid erosion develops cuestas and scarps, with remnants on high plateaus (cf. Fig. 3.36); (3) humid incision of elevated plateau surface; (4) arid aggradation and planation; (5) humid incision into lower surfaces ranging from about 500 ft (185 m), north, to more than 1,000 ft (300 m), south; (6) drainage system destruction (derangement) under prolonged aridity (possibly repeated several times); cumulative local aggradation thicknesses from this phase and older regolith relicts totals 0–400 ft (0–135 m) on lowland and 2,000–3,000 ft (600–900 m) in Andes Mountain valleys to the northwest; (7) recent perennial drainage system development extending to present. Incisional episodes tended to alternate with arid-releveling in each instance.

ronmental extracts) (Fig. 2.52). Environmental residuals may, by definition, have been moved around within the geomorphic area in which they originate by various agencies, as opposed to *weathering residuals* that have developed directly from subjacent bedrock and that are little moved (*in situ*). Environmental extracts have not been just eroded and moved. They have been removed from the parent geomorphic area and environment (see Johnson, 1901; Gignoux, 1955; Garner, 1959). The geographic framework of environment should be kept in mind as we consider examples of differing stratigraphic associations.

(1) Vegetated regions. The soil tends to be protected from erosion by plant cover and consequently waxes in thickness, to a degree, with the passage of time, even while it is being slowly removed by solution and linear fluvial erosion. This was noted long ago by Davis (1902) and many others. Soils thus comprise a common environmental residual in vegetated lands. Usually, some soil is in place and some has moved downslope, but only the former *in situ* material tends to be distinctively zoned in a time-intensity-thickness age relationship that can be of use in stratigraphy (Fig. 1.19). The weathered debris that enters drainage lines in this setting tends to move as far as the water carrying it moves, often for long distances, possibly to lacustrine sites, and commonly out of the geomorphic area, to become an environmental extract. Thus, fluvially eroded landforms in vegetated regions may be a great distance from derivative sedimentary deposits. These deposits, though commonly stratified, are often not landforms. And any tracing of deposit lithofacies back to provenance does not usually permit the identification of the individual erosional landforms that provided the material. Sedimentary sequences that reflect the progressive denudation of an orogenic source in reverse order have been noted (Pettijohn, 1957). And the basis for correlation is usually tied, in part, to attributes acquired during the movement and accumulation of the sedimentary material in water, e.g., fossils.

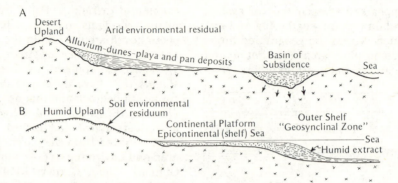

Fig. 2.52. Generalized, theoretical diagrams of sediment deposition under humid and arid environments, based, in part, on a construct of Gignoux (1955). Deposits on continents (A) relate to either arid conditions where agencies able to transport coarse detritus long distances are absent or relate to basins of subsidence. Basin deposits may or may not be made in standing water. Ocean basin deposits and shelf accumulations reflect continental humidity (B) but include eolian desert increments; main residuals are soil.

(2) In deserts, rock fragments small enough to blow away tend to be removed to entrapment sites beyond the geomorphic area limits (usually standing water or vegetation). *Loess* is thus one of the more widely acknowledged desert extracts (cf. Cressy, 1932; Lugn, 1960). Sand, on the other hand, is generally too coarse to be blown completely out of a desert except by exceptional storms. Dune sand is thus a common desert residual (Fig. 1.14). So is alluvium which also includes clasts coarser than sand (cf. Figs. 2.52; 2.53; 6.11). For though sand and gravel may be shifted about by desert runoff, such water and sediment does not consistently or even frequently escape the desert geomorphic area (cf. Gignoux, 1955). Therefore, in deserts, alluvium, talus, and allied coarse residuals are usually closely associated with the hillslopes from which the material was eroded. Many such accumulations comprise distinctive landforms. Moreover, such deposits are often more or less stratified, and the age and genesis of the related landforms, as noted by Bigarella and his associates (1965), may be intimately reflected in the internal character of the deposits (cf. Figs. 2.54; 2.55 and, 2.56).

(3) Regions subjected to glaciation, and examined following recession of ice sheets, show rather well-defined areas of dominant erosion and dominant deposition. By defini-

tion, both would fall within the same geomorphic area. Erosion clearly occurs in and near areas of ice accumulation, where the closest thing to an environmental residual may be a stripped, scoured, polished, and grooved surface with scattered patches of *drift* (Fig. 8.11). Relatively marginal to the glaciated area is the locus of direct ice deposits which, though clearly moved, remain within the glacial regimen, as long as it lasts, to accumulate as environmental residuals (Fig. 8.37). Glacio-fluvial outwash deposits carried by meltwater and iceberg-rafted detritus rank as glacial extracts. The complex of residual drift deposits is often not geographically very close to the erosional landforms that yielded the debris. Yet such precise landform origins have been noted (Shilts, 1970), even though most of the deposits are non-stratified.

The residual elements of each of the foregoing geomorphic systems are the potential *relict* depositional landforms. Stratigraphic techniques are of little use in the study of a single layer of glacial drift lying on bedrock, a deposit of desert alluvium on bedrock, or a soil on bedrock in a vegetated region. But in areas where multiple glacial episodes have alternated with other conditions, the resulting intricately stratified sequences may yield detailed geomorphic histories (Ruhe, 1968; 1969b, Flint, 1971). Such geomorphic histories apparently exist

Fig. 2.53.   Desert alluvium under sparse plant cover, southern end of Owens Valley, California. Note restriction of snowfall to mountain areas, a relation echoing the occurrence of plant cover illustrated elsewhere (Fig. 5.15). (L. K. Lustig.)

where other subaerial environments have alternated, as illustrated (Figs. 2.54 and 2.55).

TERMINOLOGY AND NOMENCLATURE

Extended treatments of this subject are available in any standard textbook of stra-

tigraphy, and most such works refer back to issues of the American Commission of Stratigraphic Nomenclature (A.C.S.N.), particularly the 1961 issue. An excellent review is offered by Wright and Frey (1965) and Willman and Frye (1970). Both the latter emphasize the need for uniformity and the value of multiple classifications that are entirely independent. Of the vari-

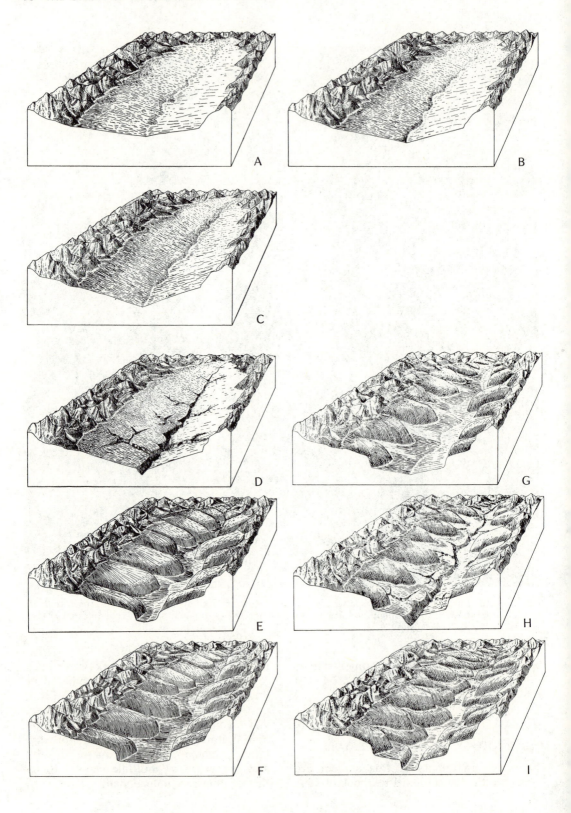

ous classifications cited by Willman and Frye (1970), only the four they actually use appear to have had extensive Quaternary applications to date.

These are listed below, with examples; geologic time units (where appropriate) are in parentheses.

Time-stratigraphic
 Quaternary System (Period)
  Pleistocene Series (Epoch)
   Wisconsinan Stage (Age)
    Altonian Substage
Rock-stratigraphic
 Wedron Formation
  Tiskilwa Till Member
Soil-stratigraphic
 Sangamon paleosol
Morphostratigraphic
 Humboldt end-moraine

Briefly, a time-stratigraphic unit is a lithic layer with a particular time connotation. Rock-stratigraphic units are lithic but with a possibly variable and unspecified time connotation from place to place. Soil-stratigraphic units are contiguous soil horizons and may be developed from a variety of parent materials with a variable time significance from place to place. Morphostratigraphic units are particular lithic units identifiable from outward surface form. In glaciated areas, these are commonly moraines, but, in other regions, they could be terraces, fans, etc. A classification of the A.C.S.N. not used by Willman and Frye is the *Geologic-climatic* unit. It is defined as "an inferred widespread climatic episode defined from a subdivision of Quaternary rocks." The boundaries of such units in different latitudes may differ in age, and the unit might include a variety of facies of rock, soils, or other materials. A geomorphic system within a given geomorphic area where a *climax* expression has been essentially attained would create such a unit (cf. Chapter 1) as would a definite environmental "overprint" on a relict landform-deposit association.*

Typical of depositional landform sequences that can be described by the foregoing stratigraphic terminology are those illustrated by Ruhe (1968) and adapted here (Fig. 2.47) and by Willman and Frye (1970) adapted here (Fig. 2.57). In these examples the deposits are relatively far from their source, but the opposite extreme, with a similar stratigraphic potential, is illustrated by Bigarella *et al.* (1965) (see Figs. 2.54; 2.55).

PALEOGEOMORPHOLOGY AND PALEOCLIMATOLOGY

Fig. 2.54. Basic scheme of slope evolution (Bigarella et al., 1965). Stages depicted are (A) extensive intermontane surface formed by pediplanation under arid climate; (B) and (C) regrading of the planated surface caused by a slight lowering of the local base level of erosion due to small climatic fluctuations toward humid conditions within the arid epoch; (D) generalized dissection of the planated surface due to onset of humidity; (E) valley-widening, alluviation, and colluviation accelerated by short arid episodes within the humid epoch; (F) escarpment retreat and formation of a pediment surface under arid conditions; (G) regrading of pediment slopes during slight humid fluctuations within the arid epoch; (H) generalized incision under a new humid epoch; (I) widening and alluviation of valleys caused by episodic climatic fluctuations trending toward increasing aridity.

The findings of geomorphologists working on contemporary settings can be logically extended to include analysis of ancient landforms (*paleogeomorphology*) and—to the extent they have been involved— ancient climates (*paleoclimatology*). Much of the lithosphere is composed of sedimentary rocks or their somewhat altered equivalents, and, because of their surficial accumulation, each of these rock bodies comprises a series of lithospheric surfaces and depositional increments. The surfaces may be presently expressed by bedding plane partings or unconformities. And a number, if not the majority of these surfaces have undergone appreciable subaerial

---

* For a further discussion of climax associations see the section on vegetal cover in Chapter 5.

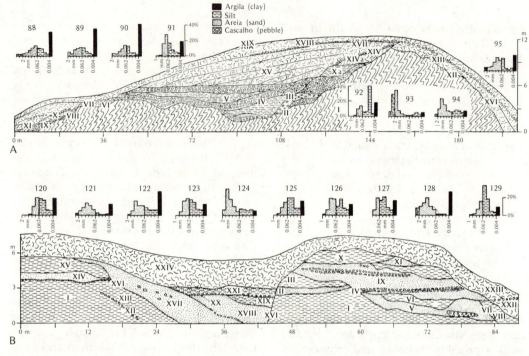

Fig. 2.55. Cross section of two road-cuts, Brazil, that display deposits referable to the erosion surfaces in Fig. 2.56. (A) Section is 5.2 km from Brusque in the state of Santa Catarina, on the road to Gaspar; (B) section is 28 km from Barra Mansa, on the road to Tres Rios in the state of Rio de Janeiro (after Bigarella et al., 1965).

The A section corresponds to the Tp2 terrace and is contemporaneous with pediment P2 formation in the Itajai-Mirim River Valley (Fig. 2.56). Key: I, Weathered phyllite; II, unconformity (channel); III, layers of sub-angular to sub-rounded quartz pebbles (5 cm max diam), cut and fill structure (samples 92, 93, 94); IV, unconformity; V, sub-angular to sub-rounded 5-cm pebbles with inter-stratified sandy, clayey, silty layers (sample 91); VI, unconformity; VII, gravel similar to V; VIII, unconformity; IX, reworked gravel from higher deposits; X, mottled colluvium (sample 88) containing pebbles, topped by paleopavement (stone line); Xa, stratified sandy clay; XI, red-brown colluvium with many quartz pebbles and enclosed paleopavement (sample 89); XII, unconformity; XIII, reddish colluvium; XIV, mottled red colluvium; XIVa, detrital paleopavement; XV, red-brown colluvium (sample 95) containing several stone lines; XVI, reddish colluvium; XVII, discontinuous detrital paleopavement of quartz pebbles and granules following present topography; XVIII, sandy, clayey, silty colluvium (sample 90) with quartz granules; XIX, humic zone.

The B section corresponds to the Tp2 level in Paraiba Valley (I to XV), to Tp1 (SVII to XXI) and to colluviation (XXII to XXIV). Key: I, Banded, red, silty, clayey, sandy material (sample 123); II, light purple, silty sand (sample 126); III, red, banded silty, clayey sand; IV, unconformity; V, yellow, feldspathic sand with quartz granules and pebbles to 5 cm (sample 122) and gravel lenses of angular and sub-angular 5-cm quartz and feldspar pebbles; VI, purplish red, sandy, silty, clayey material (sample 125); VII, unconformity; VIII, unstratified sandy, silty, clayey material with small ferruginous granules (sample 120); IX, yellow, arkosic sands with lines of quartz pebbles (3–10 cm), sub-angular to rounded; X, reddish colluvium; XI, sandy loam; XII–XV, equal parts of above; XVI, major erosive unconformity above previous (older) sequence; XVII, arkosic sand with variously sized pebbles, angular to sub-rounded; XVIII, brown-yellow arkosic sand with small quartz pebbles on a quartz pebble layer; XIX, white, sandy, silty loam (sample 127); XX, gravel layer 20 cm thick of quartz (5–15 cm diam); XXI, light yellow, pebbly colluvium; XXII, brown, pebbly colluvium (sample 121); XXIII, local detrital paleopavement; XXIV, brown colluvium. The major unconformity (XVI) is believed to represent a humid interval, whereas the deposits above and below reflect aridity.

exposure and are, thus, fossil landforms or landscapes. On occasion, previously buried land surfaces have been uncovered (*ex-* *humed landscapes*) (cf. Fig. 11.1). But uncovered or not, fossil landforms would appear to fall within the logical expertise of geomorphologists. Most unconformities are, by definition, erosional landscapes (see Chapter 11).

It seems somewhat less certain where the legitimate interests of geomorphologists lie with respect to ancient deposits. Certainly, where these were formerly landforms there is little question. Traditionally, all such deposits are also studied by sedimentologists and stratigraphers. And, as previously noted, these specialists also work in geomorphology. Indeed, "hardrock" sedimentologists are among the most avid students of present sedimentary settings in their search for analogs to compare with the effects of ancient environments. But more than one "example" in a recent sedimentary situation was only recognized after its discovery in a fossil relationship. The limitations of the uniformitarian approach being what they are, it is not surprising that both present and fossil sedimentary settings, considered separately, exhibit features without recognized counterparts in the other.

It is in the area of environmental systems analysis that geomorphology has perhaps the greatest potential values. Applications to contemporary environmental problems will be discussed shortly. Of immediate concern, however, are the sedimentologic contributions of paleogeomorphic events to ancient deposits. At issue here are the characters of sediment particles or particle associations that can be traced to specific geomorphic events related to geologic setting, weathering, erosion, transportation, and deposition. Sedimentologists (e.g. Pettijohn, 1957) have long been aware that extremes in almost any of the foregoing, or *diagenetic effects* can obscure details of *provenance* conditions. It is therefore not surprising that the most intimately associated ancient landforms and deposits can be thus analyzed.

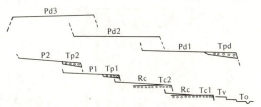

Fig. 2.56. Cross profiles depicting spatial relationships among surfaces of planation and those of aggradation in central and eastern Brazil. (After Bigarella *et al.*, 1965.) Oldest to youngest: pediplains (Pd) 3–1; terrace (Tpd) at a high level, inset with pediments (P) 2 and 1 with corresponding pediment terraces (perisediments) Tp2 and Tp1; colluvial slopes (Rc), a low, gravel-veneered terrace pair, Tc2 and Tc1; flood-plain terrace, Tv; and the present flood-plain, To.

According to uniformitarian theory (Chapter 1), studies of contemporary process-product associations should provide geologic analogs. If environmental dynamism, the relict problem, and the evolution of matter are taken into account, uniformitarianism should even now comprise a valid approach to the interpretation of ancient landforms and deposits. It seems reasonable to accept the present as a valid sample of environmental diversity for geologic episodes with numerous, discrete, relatively emergent continents (one polar) that vary in topography and altitude. Possible atypical aspects of the present include the grouping of lands in relation to Northern and Southern Hemispheres or the positions of lands with respect to particular zones of latitude that exhibit climatic overtones. In Chapter 11 we will examine a number of paleogeomorphic models in some depth, relative to potential modern analogs.

PHYSICAL ENVIRONMENT AND MAN

One weevil in a slice of bread may well go unnoticed. So may two. But there comes a time to even the least perceptive when foreign substances are not only noticed, they are paid attention to. Men in large numbers have recently crossed a threshold of this

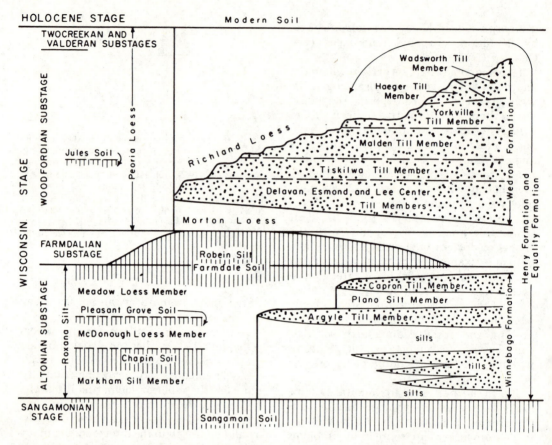

Fig. 2.57. Cross section showing the relations of formations and members of Wisconsin age in northern and western Illinois. (After Willman and Frye, 1970.)

kind relative to their physical environment. There are, of course, many environments, and some of these are also incidentally, life-support systems. From the point of view of man's survival, it may well appear incidental that the same environments contribute to the genesis of landforms. So it is that the contributions made in geology and geomorphology to knowledge of environments should be turned to advantage wherever possible in mankind's environmental dilemma.

Actually, of course, man has a tolerence for only a certain number of environments or *ecologic niches*. And the principal concerns here are problems and manipulations. Existing problems have generally originated in manipulations, whether intentional, man-made or not. Contrary to popular opin-

ion, man did not invent environmental problems. He is responsible for artificial environment problems. He also must be concerned about potential problems resulting from manipulation. In effect, man needs to know what to do and what not to do to this earth. His efforts to date have largely been a matter of trial and error.

Man's initial environmental action usually comprises a manipulation. So does his usual reaction. But in most instances our ability to anticipate the environmental consequences of certain actions has been indeed marginal. It has to do with the state in which we all live. It's called ignorance. We plowed steep hillslopes parallel to their inclination, and achieved catastrophic erosion in one fell stroke. We detected life, documented distributions of kinds both pleasing and displeasing, and reacted—with, for example, DDT. Knowing what is going to happen "in the long run" has not been our long suit. A long-term perspective of environments is one of two major areas where the geomorphologist's contribution may be unique—and invaluable.

Geomorphology may contribute to the solution of contemporary environmental problems via its geologic kinship. Geology (historically speaking) is essentially a study of ancient environments and their effects. Here, then, is the needed historic perspective of environments. There are scores of examples of paleoenvironments that endured for thousands or millions of years. Some ended. Some changed into something else. Associated organisms re-sponded. Some adapted. Some died. Not many were in a position to consciously manipulate their surroundings. In this last, man may simultaneously possess a talent that can lead either to his destruction or his salvation.

The geomorphologist studies the effects of environmental systems, and the manipulations man periodically contemplates parallel, to a degree, natural changes that induce disequilibrium. A manipulation must be considered in terms that mirror the effects of a relict on an open, natural system. Presumably, we are considering changes in systems, some in equilibrium, some not. The final chapter of this book summarizes an approach to certain environmental problems.

REFERENCES

American Commission on Stratigraphic Nomenclature (1961) "Code of Stratigraphic Nomenclature," *Amer. Assoc. Petroleum Geologists Bull.*, v. 45, pp. 645–665.

Bigarella, J. J. and Ab Sáber, A. N. (1964) "Palägeographische und Paläoklimatische Aspekte des Känozoikums in Südbrasilien," *Zeits. für Geomorph.*, Bd. 8, heft. 3, pp. 286–312.

Bigarella, J. J., Mousinho, M. R. and Da Silva, J. X. (1965) *Quaternary Geomorphologic Problems of Brazil* [reports on], Paranaense Geogr. Bols. 16–17, Univ. Paraná Press, Curitiba, Brazil, 200 p.

Black, R. F. (1962) "Pleistocene Chronology of Wisconsin," *Geol. Soc. Amer. Spec. Pap.* 68, esp. p. 137.

Black, R. F. and Rubin, Meyer (1967–68) "Radiocarbon Dates of Wisconsin," *Wis. Acad. Sci., Arts, and Letters, Trans.*, v. 56, pp. 99–115.

Black, R. F. (1969) "Glacial Geology of Northern Kettle Moraine State Forest, Wisconsin," *Wis. Acad. Sci., Arts, and Letters, Trans.*, v. 57, pp. 99–119.

Bretz, J. H. (1962) "Dynamic Equilibrium and the Ozark Land Forms," *Amer. Jour. Sci.*, v. 260, pp. 427–438.

Büdel, J. (1961) "Morphologenese des Festlandes in Abhängigkeit von den Klimazonen," *Naturwissenschaften*, v. 48.

Büdel, J. (1968) "Geomorphology—Principles" [in] *The Encyclopedia of Geomorphology*, R. W. Fairbridge, ed., Reinhold Book Co., New York, pp. 416–422.

Bullard, F. M. (1971) "Volcanic History of the Great Rift, Craters of the Moon National Monument, South-Central Idaho (Abstract), *Geol. Soc. Amer. Bull.*, v. 3, p. 234.

Cressy, G. B. (1932) "Distribution and Source of Chinese Loess," *Geol. Soc. Amer. Bull.*, v. 43, pp. 130–132.

Davis, W. M. (1889) "The Rivers and Valleys of Pennsylvania," *Nat. Geogr. Mag.*, v. 1, pp. 183–253.

Davis, W. M. (1909) *Geographical Essays*, New York (Dover reprint, 1954), 777 pp.

Dresch, Jean (1957) "Les types de relief morphoclimatiques et leurs limites dans les Andes Centrales, *Ass. Geogr. franc. Bull.*, v. 263–4, pp. 2–19.

Fair, T. J. D. (1947) "Slope Form and Development in the Interior of Natal, South Africa," *Geol. Soc. S. Africa Trans.*, v. 51, pp. 105–118.

Fair, T. J. D. (1948) "Hillslopes and Pediments of the South African Karroo," *South African Geogr. Jour.*, v. 30, pp. 71–79.

Fairbridge, R. W. (1955) "Some Bathymetric and Geotectonic Features of the Eastern Part of the Indian Ocean," *Deep-Sea Res.*, v. 2, pp. 161–171.

Fairbridge, R. W. (1968) *The Encyclopedia of Geomorphology*, Reinhold Book Co., New York.

Fairbridge, R. W. (1968) "Profile of Equilibrium" [in] *The Encyclopedia of Geomorphology*, R. W. Fairbridge, ed., Reinhold Book Co., New York, pp. 891–893.

Frye, J. C., Willman, H. B., Rubin, Meyer and Black, R. F. (1968) "Definition of Wisconsinan Stage," *U. S. Geol. Surv. Bull.*, 1274-E, pp. E1–E22.

Frye, J. C. and Willman, H. B. (1970) *Rock Stratigraphy in the Illinois Pleistocene*, Thirty-fourth Ann. Tri-State Field Conf., Guidebook, pp. 60–64.

Garner, H. F. (1959) "Stratigraphic-Sedimentary Significance of Climate and Relief in Four Regions of the Andes Mountains," *Geol. Soc. Amer. Bull.*, v. 70, pp. 1327–1368.

Garner, H. F. (1966) "Groundwater Aquifer Patterns and Valley Alluviation along Mountain Fork Creek, Crawford County, Arkansas," *Ark. Acad. Sci. Proc.*, v. 20, pp. 95–103.

Garner, H. F. (1967) "Moorefield-Batesville Stratigraphy and Sedimentation in Arkansas," *Geol. Soc. Amer. Bull.*, v. 78, pp. 1233–1246.

Gignoux, Maurice (1955) *Stratigraphic Geology*, W. H. Freeman, San Francisco, trans. by G. G. Woodford, 682 pp.

Hartshorn, J. H. (1967) "Geologic Descriptions of USGS Topographic Maps," *Jour. Geol. Ed.*, v. XV, pp. 188–196.

Heezen, B. C. and Tharp, Marie (1961) *Physiographic Diagram of the South Atlantic Ocean, the Caribbean Sea, the Scotia Sea, and the Eastern Margin of the South Pacific Ocean*, Geol. Soc. Amer. Spec. Pub.

Heezen, B. C. and Tharp, Marie (1964) *Physiographic Diagram of the Indian Ocean, the Red Sea, the South China Sea, the Sulu Sea and the Celebes Sea*, Geol. Soc. Amer. Spec. Pub.

Heezen, B. C. and Johnson, C. L. (1968) "Bathymetry" [in] *The Encyclopedia of Geomorphology*, R. W. Fairbridge, ed., Reinhold Book Co., New York, pp. 104–110.

Higgins, C. G. (1951) "Effect of Buried Topography on Drainage in Sonoma County, California" (Abstract), *Geol. Soc. Amer. Bull.*, v. 72, p. 73.

Horton, R. E. (1945) "Erosional Development of Streams and Their Drainage Basins," *Geol. Soc. Amer. Bull.*, v. 56, pp. 275–370.

Howard, A. D. (1967) "Drainage Analysis in Geologic Interpretation: A Summation," *Amer. Assoc. Petroleum Geologists, Bull.*, v. 51, pp. 2246–2259.

Jackson, K. C. (1970) *Textbook of Lithology*, McGraw-Hill Book Co., New York, esp. Chapter III.

Johnson, D. W. (1931) "A Theory of Appalachian Geomorphic Evolution," *Jour. Geol.*, v. 39, pp. 497–508.

Johnson, W. D. (1901) "The High Plains and Their Utilization," *U. S. Geol. Survey Ann. Rept.*, v. 4, pp. 601–741.

King, C. A. M. (1966) *Techniques in Geomorphology*, Edward Arnold, London, 342 pp.

King, L. C. (1953) "Cannons of Landscape Evolution," *Geol. Soc. Amer. Bull.*, v. 64, pp. 721–752.

King, L. C. (1956) "Drakensberg Scarp of South Africa: A Clarification," *Geol. Soc. Amer. Bull.*, v. 67, pp. 121–122.

Leopold, L. B., Emmett, W. W. and Myrick, R. M. (1966) "Channel and Hillslope Processes in a Semiarid Area, New Mexico," *U. S. Geol. Survey, Prof. Pap.* 352-G, pp. 193–253.

Lipman, P. W., Prostka, H. J. and Christiansen, R. L. (1971) "Evolving Subduction Zones in the Western United States, as Interpreted from Igneous Rocks," *Science*, v. 174, pp. 821–825.

Lustig, L. K. (1965) "Clastic Sedimentation in Deep Springs Valley, California," *U. S. Geol. Survey Prof. Pap.* 352-F, pp. 131–192.

Lustig, L. K. (1969a) "Trend-Surface Analysis of the Basin and Range Province and Some Geomorphic Implications," *U. S. Geol. Surv. Prof. Pap.* 500-D, pp. 1–70.

Lustig, L. K. (1969b) "Quantitative Analysis of Desert Topography" [in] *Arid Lands in Perspective*, W. G. McGinnies and B. J. Goldman, eds., Univ. of Arizona Press, Tucson, pp. 47–58.

Lugn, A. L. (1935) "The Pleistocene Geology of Nebraska," *Nebr. Geol. Survey, Bull.*, v. 10, 223 pp.

Oberlander, Theodore (1965) *The Zagros Streams*, Syracuse Geographical Series, No. 1, Syracuse University Press, Syracuse, 168 pp.

Penck, A. (1919) "Die Gipfelflur der Alpen," *Sitzber. Preuss. Akad. Wiss. (Berlin)*, v. 17, 256 pp.

Pettijohn, F. J. (1957) *Sedimentary Rocks*, Harper and Bros., New York, esp. pp. 525–588.

Pitty, A. F. (1968) "A Simple Device for the Field Measurement of Hillslopes," *Jour. Geol.*, v. 76, pp. 717–720.

Quinn, J. H. (1965) "Monadnocks, Divides and Ozark Physiography," *Ark. Acad. Sci. Proc.*, v. 19, pp. 93–97.

Rahn, P. H. (1969) "The Relationship between Natural Forested Slopes and Angles of Repose for Sand and Gravel," *Geol. Soc. Amer. Bull.*, v. 80, pp. 2123–2128.

Ruhe, R. V. (1968) *Iowan Drift Problem, Northeastern Iowa*, Iowa Geol. Survey, Rept. of Investigations, v. 7, 40 pp.

Ruhe, R. V. (1969) *Quaternary Landscapes in Iowa*, Iowa State Univ. Press, Ames, 255 pp.

Ruhe, R. V. (1969b) "Principles for Dating Pedogenic Events in the Quaternary," *Soil Science*, v. 107, pp. 398–402.

Schaffer, J. P. and Hartshorn, J. H. (1965) "The Quaternary of New England" [in] *7th INQUA Congr. Rev. Volume*, Pt. I, H. E. Wright and D. G. Frey, eds., pp. 113–128.

Schumm, S. A. (1963) "The Disparity between Present Rates of Denudation and Orogeny," *U. S. Geol. Survey, Prof. Pap.* 454-H, pp. H1–H13.

Shilts, W. W. (1970) *Pleistocene History and Glacio-Tectonic Features in the Lac Megantic Region, Quebec*, New Eng. Intercol. Geol. Conf. Guidebook, pp. E1–E14.

Strahler, A. N. (1968) "Quantitative Geomorphology" [in] *The Encyclopedia of Geomorphology*, R. W. Fairbridge, ed., Reinhold Book Co., New York, pp. 898–912.

Tanner, W. F. (1968) "Geomorphology—Statistical Analysis" [in] *The Encyclopedia of Geomorphology*, R. W. Fairbridge, ed., Reinhold Book Co., esp. pp. 422–424.

Tinkler, K. J. (1970) "Pools, Riffles and Meanders," *Geol. Soc. Amer. Bull.*, v. 81, pp. 522–547.

Tricart, Jean (1958) "Etudes sur quelques Cailloutis fluviatiles actuels des Pyrénées Orientales et du Massif Central," *Zeits. für Geomorph.*, Bd. 2, heft 4, pp. 278–304.

Tricart, J. and Cailleux, A. (1955) *Introduction a la Géomorphologie Climatique*, Cours de Géomorphologie, Paris, C. D. U., 228 pp.

Trowbridge, A. C. (1921) "The Erosional History of the Driftless Area," *Univ. Iowa Studies Nat. History*, v. 9, pp. 1–127.

Washburn, A. L. (1956) "Classification of Patterned Ground and Review of Suggested Origins," *Geol. Soc. Amer. Bull.*, v. 67, pp. 823–866.

Washburn, A. L. (1969) "Weathering, Frost Action, and Patterned Ground in the Mesters Vig District, Northeast Greenland," *Meddelelser Om Gronland*, Bd. 176, 304 pp.

Willman, H. B. and Frye, J. C. (1970) "Pleistocene Stratigraphy of Illinois," *Ill. State Geol. Survey, Bull.*, v. 94, 204 pp.

Wood, A. (1942) "The Development of Hillside Slopes," *Geol. Assoc. Proc.*, v. 53, pp. 128–138.

Wright, H. E., Jr., and Frey, D. G., eds. (1965) *The Quaternary of the United States*, Internat. Assoc. Quater. Res. 7th Congr., Princeton Univ. Press, Princeton, N.J., 922 pp.

3 Tectonic topographic elements

CLASSIFICATIONS, CONSTRAINTS, AND CONCEPTS

In the earth's crust there are scores of bedrock attributes ranging from textural minutiae to gross partings. Collectively, this *lithology* and *structure* comprise the bedrock fabric, and, to the extent that it is expressed in topography, it is the legitimate concern of geomorphologists. The earth would indeed have little topography on which to express bedrock fabric were it not for internal relief-making mechanisms. Terrestrial inhabitants would presumably face an under-water world, flat above and flat below. Relief and bedrock fabric are imparted to the lithosphere in several distinct manners and places. In terms of manner, of immediate interest is fabric and relief generated by lithospheric movement (*diastrophism*) and melting (*plutonism*), both here grouped under *tectonism*. In our discussion of tectonic elements in topography, the reader will be quick to observe that it is one thing to distinguish between plutonic and diastrophic lithic features and quite another to distinguish between relief created by deformative events as opposed to that caused by igneous or related mechanisms (e.g. mantle phase changes). Often the precise cause of tectonic relief changes remains unknown.

An orderly arrangement of ideas should also aid our discussion of where bedrock fabric and relief originates. For this we will parallel a classification advocated by Amadeus Grabau in 1904 and later modified by Arthur Holmes in 1920. As used here, *endogenic* topographic characters are those created by forces originating within the earth, and they may be contrasted with *exogenic* features produced by agencies acting on the crust from without or *mesogenic* forms generated by mechanisms that function transitionally between these extremes. Most endogenic events are tectonic in nature. Some mesogenic activities but almost no exogenic ones are thus qualified. Mesogenic and exogenic topographic attributes will be considered in detail in several later sections and chapters. They are used here only to clarify other material. This use should not only justify occasional digressions but should compensate for the artificial character of chapter outlines.

ENDOGENIC TOPOGRAPHIC EFFECTS

Like those effects that function to the same end elsewhere, the processes and agencies interacting within the earth to produce topography also comprise geomorphic systems. Most endogenic relief develops so slowly by tectonic mechanisms that it is modified (as it forms) by other geomorphic systems. W. M. Davis's notion of abrupt initial uplift of an entire landmass followed by subaerial sculpturing is hardly mirrored by the bulk of the geologic record. Weathering, erosion, deposition, and allied phenomena occur ALONG WITH most major earth movements. Where this is so, the extent and nature of endogenic topographic attributes are governed by mesogenic and/or exogenic effects to some degree. For this reason, examples of tectonic topographic effects in this chapter are restricted to those fabric and relief aspects that are least modified by non-tectonic agencies. Examples of modified tectonic landforms are used throughout the remainder of this text.

Unmodified endogenic topography is understandably rather rare. On earth, change is a synonym of normality. In some instances, however, usually because of very recent or continuing events or because of the ineffectual nature of associated geologic processes, essentially unaltered endogenic topography is encountered (Figs. 2.39; 3.82; 3.91). Such tectonic terrain should, of course, be closely scrutinized by geomorphologists because it embodies primary litho-structural shapes on which mesogenic and exogenic terrain modifications could be imposed, and generally have been imposed, elsewhere on earth.

We will discuss details of tectonic bedrock fabric as necessary throughout this chapter. Probably no crustal regions are entirely lacking in such fabric, however, so the presence or absence of tectonic bedrock characters can hardly serve as a basis for evaluating a region's geomorphic tectonism. Major fabric variations are tectonically indicative, however, and geomorphic tectonism should be categorized on the basis of these varients and crustal relief, the second main tectonic element of landscapes. *Tectonic relief* is that part of crustal roughness that would theoretically be destroyed by gradational leveling under the force of gravity if opposing internal earth processes ceased.

As one would expect of any intermittently active geomorphic system, tectonism affects topography via relief through both continuing and relict effects. Both sea floors and continents exhibit relief that departs markedly from the norms for these areas that is associated with current earthquake activity, gravity and magnetic deviations, and (frequently) volcanism. In short, there is abundant geophysical and related evidence of relief being generated or maintained by contemporary diastrophism and plutonism, i.e. CONTINUING TECTONISM. Most of the earth's mountain systems that are less than 100 million years old exhibit some of these traits. We will hereafter refer to such tectonic crustal irregularity as *dynamic relief* (cf. Fig. 2.33).

In contrast to dynamic relief, there is notable crustal roughness in excess of regional norms (not specifically associated with other overt tectonic symptoms) which seems to relate to rock distributions of long standing. Thus, the continents rise above the average elevation of the sea floors because of lithic-density differences that probably began to develop early in the earth's history. The causitive differentiation may be continuing, but most of the present relief effects can be regarded as tectonically RELICT. Similarly, nearly *aseismic* portions of continents that rise well above their mean elevations have been found to overlie notably thickened continental crust. The Colorado Plateau is an excellent example. And like the continents as a whole, the extreme highs on them that owe their existence to prior crustal thickening can be considered to exhibit RELICT TECTONIC RELIEF. With due consideration for the probable source of the forces that perpetuate this type of crustal roughness, we feel this expression of tectonism should be termed *isostatic relief*.

The foregoing distinctions between DYNAMIC RELIEF caused by continuing tectonism versus ISOSTATIC RELIEF caused by relict crustal thickening are geomorphically useful even though both are forms of tectonic relief. In areas of dynamic relief, overall above-average elevations are at least being maintained tectonically and may be increasing so that maximum local erosional or structural relief may therefore also increase. However, in areas of relict isostatic relief, elevations are gradually decreasing as isostatic adjustments accommodate the thickened crust to erosional or other mass reductions. Such events now appear to be functioning in such older mountain systems as the Appalachians, Urals, and Ouachitas. Mean elevations of surrounding crustal areas are approached by the higher eroded surface as excess crustal thickness is eliminated—normally with a very low level of seismicity. Generally, in areas of relict isostatic relief the potential duration of a high point (as a high) is proportional to its elevation (see Chapter 10).

The role of tectonism in areas of dynamic relief is clearly that of an active geomorphic agent. The degree of activity is not only expressed by contemporary elevation, structural relief, and slope imparted by diastrophism, but also by apparent stability (e.g. seismicity, isostatic imbalance, vulcanism), proximity and vulnerability to relief-generating tectonism, and deformative style and stage. Because the earth is essentially spheroidal in shape, crustal movements that have dominantly vertical (radial) components do more to change relief than do movements that have dominantly horizontal (tangential) components. Essentially, all mountainous relief appears to be generated by agencies with high radial movement components. Where the mountain-making agencies are tectonic (as opposed to erosional), it seems appropriate to call the formative events an *orogeny*. The effects are both direct and local, and the tectonic-geomorphic agent usually imparts both relief and bedrock fabric. Tangential lithospheric movements may also geographically redistribute crustal units on a regional scale. Where regional crustal elevation changes comprise only a small and indirect consequence of such shifts, the term *epeirogeny* seems appropriate for the events involved.* Bedrock fabric may be

created or somewhat altered in such areas, but tectonism can be regarded as a passive geomorphic agent. In accordance with the foregoing discussion and the theoretical discourse to follow, it is apparent that dynamic relief characterizes developing orogenic zones, whereas isostatic relief dominates epeirogenic regions.

Probably no major aspect of geology has escaped the principal theoretical developments of the 1950's and 1960's. Geomorphology, as a discipline incorporating most if not all geologic subdisciplines, can hardly be an exception. Internal earth processes and effects have been perhaps most subject to theoretical revision during the period in question (cf. Hart, ed., 1969). Since it is these very phenomena (expressed in topography by particular rock types and structures) that constitute the initial geomorphic theme of this chapter, we must review recent theoretical developments and discoveries in geophysics and structural geology as a prelude to further discourse.

DEVELOPMENT OF PLATE TECTONIC CONCEPTS

The creation of rocks within the crust and their relative movements are probably no more independent activities than an animal's digestion and locomotion. Indeed, surprising parallels appear to exist between earth process-product associations and the biologic activities of an organism. Rocks are the figurative bones of the earth. The rest passes well for flesh and when the flesh moves the bones follow. The earth seems to lack the regular tectonic pulse once attributed to it by Grabau (1904), but there is, to say the least, a certain activity. The more or less solid portions appear to be undergoing periods of stress by, at least, tidal, rotational, compactional, and thermal mechanisms. Nor are the more plastic or fluid phases immune to these forces. As early as

---

* Classically, (cf. G. K. Gilbert, 1880) the term epeirogeny has been used to apply to regional vertical movements of the crust, and only recently has it been indicated that these shifts relate to much-larger-scale lateral movements of the lithosphere as noted in the next section.

1839, Hopkins proposed subcrustal convection. We have long been aware that lithic differentiation, earth tremors, volcanic eruptions, and crustal movements of several kinds were somehow related to internal processes and even to one another. But despite numerous attempts to weld these diverse data into a unified whole (e.g. Holmes, 1931), only discoveries and theoretical developments made during and since the 1950's and 1960's appear likely to serve. A brief historic resumé seems warranted.

Early Wondering and Wandering

Before about 1950, most students of topography were, by necessity, students of subaerial landforms, the principal reason being that data on ocean floors were lacking. Also, many geomorphologists regarded the earth's continent-ocean basin relations to be of long standing. The most notable exception to this school that accepted essentially static global geography was, of course, the earth scientists who continued to cling to the 1915 "continental drift" proposal of Alfred Wegener.* To the "antidrift" group, the submarine realm offered little inducement to study. After all, if the ocean floors were permanent, it was widely agreed that several hundred million if not several billion years of sediment accumulation mantled them. Topographic diversity doesn't characterize settling basins. Occasionally the literature included terse accounts of *submarine canyons, guyots, trenches, abyssal plains*, and little else. Actual data were few indeed and descriptions general.

Until the mid-1900's discussions of the origin of continental rises and ocean basins among antidriftists were, for the most part, related to theoretical mechanisms for lithic differentiation. In this context, the issues were mainly petrologic and geochemical. As noted by Bowen (1928), a foremost concern was to explain how the continents became generally silica-aluminum rich (granitic) in composition with an average

specific gravity of ±2.7 whereas the oceanic rocks seemed to be iron-magnesium rich (basaltic) with an average specific gravity of ±2.9. The cosmologic concepts that were in vogue generally required some stage of magmatic differentiation in a primordial earth that was either partly or wholly hot and molten (cf. Hess, 1962) (Fig. 3.1). Purely geomorphic interests in the discussions largely turned on the geophysical-structural suggestion (*Isostasy*) that the continents stand high because they are less dense than the oceanic material upon which they presumably rest. Isostatic Theory, so conceived, relates to the 1855 observations of Pratt and Airy in India. It is open to generalized speculation on processes in the mantle and crustal buoyancy problems, but it does not permit much detailed geomorphic analysis.

Early advocates of continental drift as well as the scientists who believed in a parent landmass that had broken up and partly foundered beneath the sea (Gondwanaland) were certainly interested in the ocean floors if not actually preoccupied by them. However, the deep seas were generally beyond the technologic grasp of scientists in the early 1900's. Be that as it may, ocean floor configurations were regarded by driftists as relatively recent (± post-Mesozoic) and hence rather thinly covered with sediment. Wegener's hypothesis induced many to search for evidence of continental drift on the lands. To members of the Wegenerian school, they generally succeeded. To most others, they failed. As we shall see, the scientists who tried (e.g. Alex du Toit, 1937) were able to anticipate many of the probable endogenic consequences of continental fragmentation and displacement. But drift was initially advanced as a "one-time" phenomenon which began rather late in geologic time. A plausible cause for such an event was not readily forthcoming unless one is willing to so regard such proposals as that of George Darwin, to start things off by tearing the moon out of what is now the Pacific Ocean Basin.

---

* Meyerhoff (1968) credits the actual drift concept to Baron von Humboldt, *ca.* 1800.

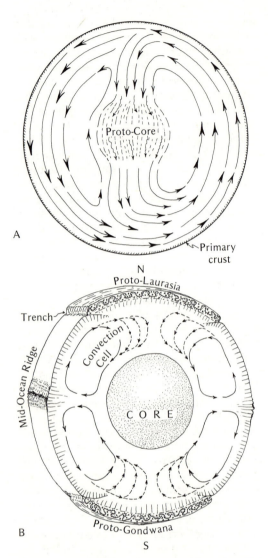

Fig. 3.1. Theoretical cross sections of the primordial Earth showing (A) the possible mode of development of a primary siliceous crust through internal convectional overturn of Earth materials (cf. Hess, 1962) to form a proto-core, circulation shown is modified by a main-stream circulation projected into an equatorial plain; (B) development of polar protocontinents with a multiple-convection cell system (cf. Deitz and Holden, 1970), accretion is continuous until recession of cell margins (dashed arrows) inhibits heat escape and initiates secondary cell development and continental rifting and drift.

The only drift mechanisms initially offered required the *sialic* continents to move through a *simatic* substrate, which is rather like requiring a steel block to sail through a sea of concrete. Needless to say, the formidable resistance encountered by the drift idea in some quarters matched that presumed for the migrating landmasses.

Later Drifting and Dreaming

Potential ingredients for a continental drift mechanism began to appear about 1950 as earth scientists collected geophysical, sedimentologic, and bathymetric data from the ocean floors. One major finding was that few of the sedimentary deposits appeared to be older than Cretaceous. Also, on the land, studies of *remnant magnetism* by F. M. S. Blackette (1961) and later S. K. Runcorn (1962) and others led to a series of reports on changed orientations of earth magnetic fields with time. On the basis of the latter data, it was initially concluded that the magnetic poles had wandered—since, presumably, the earth's crust could not do so. Shortly thereafter, the oceanographic studies verified many physiographic details of the earth's main submarine mountain system (the mid-ocean ridge) and noted the above-average heat flow and seismicity of this zone (Fig. 3.2); bottom core samples established the increasing age of sea floor sediments with increased distance from the mid-ocean ridge. The stage was set for a theoretical revolution of fantastic proportions.

Early in the 1960's a number of researchers including Vine and Mathews (1963) and Heirtzler and Le Pichon (1965) demonstrated through additional remnant magnetic studies (and a chronology based on magnetic field reversals plus paleontologic and radiometric ages) that intrusive, dike-like, basaltic masses were associated with the mid-ocean ridge system. Dikes and associated extrusive volcanic accumulations are clearly visible in the exposed segment of the ridge in Iceland (Cf. Holmes, 1965; Figs. 3.3, 3.5). Then, almost simultaneously, Deitz (1961) and Hess (1962) resurrected an idea probably first advanced by Arthur Holmes (1931) when they suggested that the sea floors were spreading away from the mid-ocean ridge (Fig. 3.3).

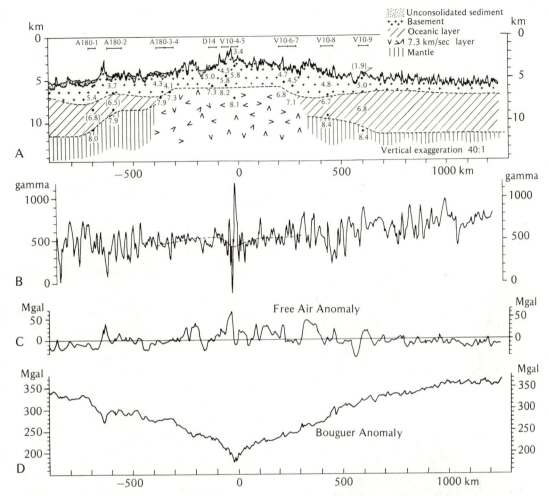

Fig. 3.2. Composite crustal section across the north Mid-Atlantic Ridge (A) and a magnetic profile (B). (From Talwani et al., 1965.) Boundaries of the axial zone anomaly are defined by the low-velocity mantle. (After Heirtzler and Le Pichon, 1965.) Above (C) and (D) are related gravity anomalies; the fact that the free-air gravity anomaly over the entire ridge is close to zero (±50 mgal) and that the Bouger anomaly attains a minimum over the ridge indicates that the ridge is compensated. (After Talwani et al., 1965.)

In agreement with the spreading concept, the dike-like intrusions are apparently youngest next to the ridge and progressively older and more deeply buried away from the ridge (Fig. 3.4). In 1965, J. T. Wilson noted an associated set of unusual sea-floor fractures (*transform faults*). Wilson suggested that the transform faults influenced the pattern of crustal motion; presumably, they would function in conjunction with additions of new material to the lithosphere to permit sea floor segments on opposite sides of the ridge to separate.

It is now widely accepted that intrusive igneous material is being added to the lithosphere along the mid-ocean ridge, and associated tensional stress is causing the adjacent sea floors to spread apart (Fig. 3.5). These ideas of increasing crustal extent, coupled with the suggestion of Kennedy (1959) that ecologite-basalt phase changes in the upper mantle were responsible for mantle-crust rock transitions and isostatic adjustment, led to a brief flury of speculation about an expanding earth. Such ideas culminated in the text by Arthur

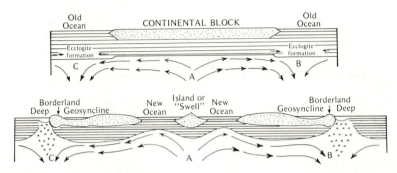

Fig. 3.3.   Probably one of the earliest depictions of convectional circulation of the mantle related to sea-floor spreading and continental drift; (A) ascending currents; (B) and (C) descending currents. (After Holmes, 1931.)

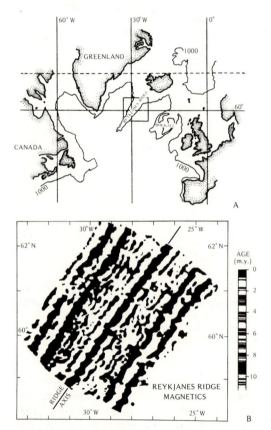

Fig. 3.4.   Magnetic reversal map of the Reykjanes portion of the mid-ocean ridge southwest of Iceland; depth contours in fathoms (A). Details of a segment (B) show magnetic anomalies. Normal (present) polarization is shown in black and reversed polarization in white. Anomalies in this instance, symmetrically disposed on the ridge axis, are presumed to correlate with dike-like intrusions emplaced under sporadically reversed magnetic field conditions. (After Heirtzler, Le Pichon, and Baron, 1966.)

Holmes (1965), but the hypothetical need for extensive earth volume increases had already been largely countered by Hess's 1962 evaluation of the oceanic trenches. Continuing oceanographic studies of the trenches had disclosed several relations. Trenches are the locus for deep-focus earthquakes, and a plot of such quake foci (Fig. 3.6) indicates the presence of an inclined shear zone in relatively rigid material at depths where geothermal considerations would normally require a crust too ductile to yield by fracture. The present consensus is close to the suggestion of Coats (1962) and Hess (1962) that relatively brittle, cool crustal segments are being down-thrust beneath other crustal segments along the deep-focus earthquake shear zones in the ocean trenches. Thus the volume of material presumably being added to the ocean crust in the ridges can supposedly be accommodated by subduction and re-assimilation into the mantle in the trench zones.

Not all geologists have embraced the plate tectonics scheme with equal enthusiasm. Indeed, some question various premises of the theory, and a few (e.g. Meyerhoff and Meyerhoff, 1972) apparently reject the scheme, essentially in its entirety. Many of the points raised by the Meyerhoffs have merit and should lead to causal investigations of such things as (1) the lack of symmetry of magnetic anomalies on either side of some mid-ocean ridge segments, (2) the absence of anomalies along some ridge segments, and (3) the retention of ancient latitudinal climatic patterns by some lands that resemble present-day patterns and imply more east-west movements than north-south. These and other similar points questioned by the Meyerhoffs are of interest to students of global tectonics, but they do not seem to the writer to be basically damaging to the plate tectonics concept. More compelling to me are the geo-historic relations of the earth that require shifts in the latitudinal positions of crustal segments by some mechanism—*sea-floor spreading* being the most acceptable of those so far proposed. Among the geo-historic requirements for crustal shifts, the most outstanding are (1) the growth of trees to provide late Paleozoic coal deposits in the Antarctic and on Spitsbergen (both are now far too polar in position to receive either the required heat or light to support tree growth) and (2) the development of continental glaciers during the Carboniferous on such lands as Africa and Australia (both far too equatorial at present to either grow or maintain a low-elevation ice sheet). In short, the crumpled fenders on the vehicle are more impressive than arguments that it couldn't be wrecked.

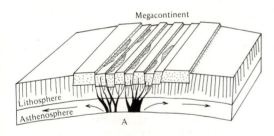

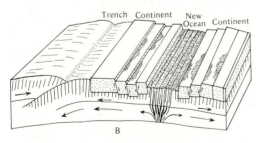

Fig. 3.5. Sequential block diagrams illustrating rifting of a megacontinent over a spreading zone and the development of a new oceanic depression. (A) Initial rifting with extrusion of oceanic (basaltic) lavas analogous to the Triassic Wachung basalt flows, New Jersey (cf. Figs. 3.31 and 10.25), and resultant block faulting and (B) spreading and development of a new pair of continents.

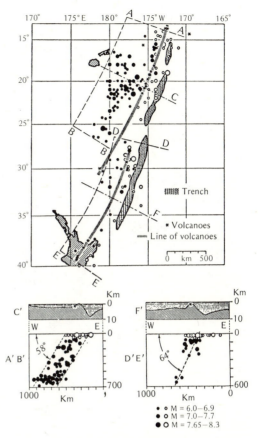

Fig. 3.6. Map and cross profiles of the Tonga-Kermadec trenches and earthquake foci. The zone of deep-focus shocks is believed to coincide with a downward inclined shear between two colliding crustal plates expressed at the surface by the trench —a probable subduction zone. (After Benioff, 1949.)

Where Do We Go from Here?

Continuing geophysical research on oceanic fracture systems and displacement mechanisms by many workers, including Morgan, Isacks, Oliver, and Sykes (see Le Pichon, 1968), tends to substantiate the *plate tectonics* and *sea-floor spreading* schemes which comprise the over-all theoretical result of researches largely accomplished since 1950. Collectively these concepts have been termed "the new global tectonics" (e.g. Oliver *et al.*, 1969), and the apparent mechanism is surely world wide in scope (Fig. 3.7). It is an understatement to say that these developments have revitalized the theory of continental drift by providing

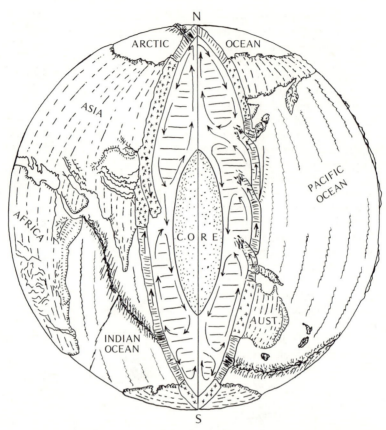

Fig. 3.7. Incised section of the world between approximately 85° and 140° East Latitude, depicting several Pacific trenches and Asian continental plates and the mid-ocean ridge, Indian and South Pacific oceans. Implied convectional circulations within the mantle are purely diagrammatic (arrows) but consistent with trench and ridge distributions.

a plausible mechanism. But, as noted by J. T. Wilson (1968), the theoretical consequences of the theory go far beyond this.

Hammond (1971) rightly comments that the driving mechanism behind what we now term plate tectonics remains uncertain. Convective transfer of heat and material within the mantle has been freely mentioned since tectonic theory was first advanced. But as noted by Knopoff (1970) mantle properties may limit vertical convectional activity to a zone only 600 km thick. Since the horizontal dimension of related convection cells would necessarily equal the vertical dimensions according to Knopoff, lateral plate displacement would be limited to 600 km by this mechanism.

Yet available data suggest that continents have apparently been moved thousands of kilometers, so that an additional driving force or one of greater scope is needed. Radioactive decay and compaction are often-cited heat sources for the plate tectonics process, and certainly gravity and rotational forces are involved as noted by B. F. Howell (1970). But a full understanding of the details must await a more complete knowledge of the earth's mantle. Be that as it may, the geomorphologic implications of sea-floor spreading would be hard to miss. It is clearly unrealistic to limit the morphologic scope of the discipline to emergent portions of the continents. Those continents and the ocean basins comprise the world's largest and most complex, basically endogenic landforms. Not all of the plate tectonics processes and effects are expressed topographically. But indeed few crustal segments can have entirely escaped the morphologic consequences.

Structural geologists now recognize from as few as 6 to as many as 20 lithosphere plates on the earth's surface, each some 50 to 100 km thick and each presumably sliding on the relatively warm and less rigid upper mantle (cf. Hammond, 1971). Continental and plate boundaries do not necessarily coincide, so that some plates include continents and others do not (Fig. 3.7). Clearly, moving plates are therefore not geomorphic units as such. And evidently related phenomena including plutonism, lithic differentiation, fracturing, and mountain building have not yet been fully reconsidered in light of the plate tectonic concept. Even so, it now appears that plate tectonic phenomena are very ancient—possibly dating back to nearly the time of initial compaction and heating. Relatively dense materials have thus presumably long been emplaced in mid-ocean ridge zones. But, as noted by Hess (1962), the lighter continental material that is a probable differentiate of this process is apparently resistant to downthrusting and long-term reassimilation in ocean trench zones (Fig. 3.5). In part this continent resistance to subduction seems to be due to density and in part because of the hyperfusibility (±600° C melting zone) of granitic materials. The conclusion that continents are not appreciably subducted appears to be substantiated by the fact that there seems to be no record of massive sialic intrusions in the mid-ocean ridges where subducted material should return to the surface.

Though the continents seem to be borne along by subjacent oceanic plates, according to most authors, they appear to be neither lithically nor structurally a permanent part of any particular lithosphere plate any more than coal hauled in a dump truck is a part of the truck. In effect, the continents are much more acted upon by sea floor spreading than vice versa. This seems to be true in spite of instances of structural and plutonic interactions between oceanic and continental areas where, for example, a continent has been translated laterally to the margin of a trench or athwart an ocean ridge. Occasionally, a documented interaction between continents similarly appears to be more a result of plate tectonics than a cause. In general, both continents and sea floors exhibit a number of tectonic subprovinces and deformational styles that are expressed in topography. But the sea floors are tectonically much more active than the continents.* Accordingly, we will undertake a discussion of oceanic geomorphology first, so that subsequent references to plate tectonic effects on continent topography can be made with minimum digression.

Rocks and Topography

As a prelude to further consideration of the tectonic elements of terrain it should be noted that such topography in epeirogenic crustal areas is usually limited to an aspect of lithology or some expression of already-formed rock structure exposed at the surface through weathering and *differential erosion* or expressed in a subdued manner through an incomplete sedimentary *covermass*. It is largely if not exclusively in epeirogenic areas that low-relief depositional or erosional terrain, or both, extends for great distances without interruption by structurally created relief.

Our interpretation of bedrock influence on topography necessarily involves two general assumptions: (1) sedimentary burial ultimately tends to have a concealing and leveling effect upon terrain; (2) relatively weak rocks may give way under certain forms of erosion and weathering to form topographic recesses or depressions—in contrast to this, comparatively tough rocks subjected to the same treatment tend to form protrusions or topographic elevations.

---

* This holds simply because there are more plate junctures on sea floors and, historically, because spreading has continued for long intervals without orogenesis.

Classically, topographic depressions coinciding with structural "lows" (i.e. synclines, grabens) or erosionally nonresistant formations are said to be under "structural" (bedrock) control. So, presumably, are topographic rises occurring on structural "highs" (i.e., anticlines, horsts) or erosionally resistant rock types. Terrain found to be lacking in these landform-bedrock relationships is commonly said to be independent of structural control. Since this breakdown of structural categories is man made, it tends to break down. Thus, structural highs with mobile cores often have weak rocks at their centers (Fig. 3.65). There is a great tendency on the part of geologists to make off-hand judgments about relations between structure and topography when what is usually needed is study.

OCEAN FLOORS

Sea floors are interesting. Since most are under water, some marine geophysicists have inadvertently become geomorphologists because much of what they were recording was topography. In accordance with plate tectonics theory, the principal oceanic structural units are relatively rigid lithosphere segments (*plates*) that are being laterally displaced with respect to one another. The fact that the plate movements are dominantly tangential with respect to the earth's surface is of greatest significance geomorphically. It means that plate movements do not directly produce appreciable, short-term crustal relief changes over most of the plate expanse. Thus, as previously defined, tectonism in medial plate areas usually takes the form of epeirogeny. Effectively radial tectonic movements are common to plate margins and result directly in increased crustal relief, of mountainous proportions in many cases. Plate margins therefore tend to be orogenic zones. Nevertheless, for a variety of geologic reasons, neither the borders nor the

medial portions of oceanic plates are topographically uniform. Crustal movements, submarine volcanism, sediment accumulation, and erosion all play a role in the resulting submarinescapes. Clearly only some of these effects are endogenic.

In his summary discussion of plate tectonics, Hammond (1971) lists three types of plate boundaries: (1) spreading zones where new crust is apparently being added by intrusion as along mid-ocean ridges and where the plates are moving away from one another; (2) trenches where one plate is thrust under another; and (3) faults where two plates are sliding past each other. A fourth important structural relationship involves plate corners which join in groups of three in several places on the sea floors.* These triple junctions (as in the central Indian Ocean) apparently influence plate geometry extensively during migration. The topographic aspects of each of the major plate relations must now be considered in some detail.

The Mid-Ocean Ridge—An Orogenic Zone

The mid-ocean ridge comprises one of the two most tectonically active areas of the earth's crust. The deep ocean trenches, also on the sea floor, constitute the other. The ridge is probably best treated in the media used by two of its most avid cartographers, Bruce C. Heezen and Marie Tharp (1965). Their numerous descriptions and illustrations (see, for example, Figs. 2.26; 2.27; 3.8) give us a picture of a form of mountainous submarine topography where geophysical and petrologic associations tend to dictate the geomorphic interpretations (see also K. S. Deffeyes, 1970; Sleep and Biehler, 1970).

As previously noted, the ridge zone exhibits high heat flow (2–6 μcal/cm^2 sec)—the amount varies from place to place along the crest. There is both extrusive and intrusive igneous activity with associated shallow-focus seismic activity and little or

---

* A fifth relation involves long-term hot spots (mantle plumes?) characterized by persistent volcanism.

no sediment cover. Rocks dredged from the ridge zone are predominantly alkali basalt or olivine tholeiite or their derivatives. Bathymetric maps (Fig. 3.8) based on acoustic-sounding profiles show us a mountain range approaching 40,000 mi (56,000 km) in length rising 0.6–2.1 mi (1–3 km) above the adjacent ocean basin floors. Over this distance it varies from rather narrow and high to broad and subdued. Heezen and Fox (1966) give its average width as being in excess of 1,000 mi (1,500 km), but there is great variation. We should actually speak of RIDGES here instead of A RIDGE simply because multiple, parallel ridges comprise the main zone and because there is some doubt about the linear geographic continuity of all ridge segments (e.g. Alpha Rise). The ridge crest is commonly a paired range of mountains with maximum elevations averaging some 1,000 fathoms (2,000 m) below sea level. Heezen describes an axial valley situated between these ranges whose floor averages about 1,000 fathoms (2,000 m) deeper than the crestal mountains (cf. Fig. 3.9). Deffeyes (1970) maintains that the valley is present along only about 50% of the ridge. The crestal mountains also average 600–800 fathoms (1,200–1,600 m) higher than peaks in laterally adjacent ridge-flank areas. Menard (1969) notes the association of the highest ridge centers with slow spreading rates and vice versa. This could be misleading since rates are long-term averages, whereas the topography is current. The ridge-flank area ranges from hilly to mountainous and is separately considered under the designation "ridge and rift" topography in the next section.

As further noted by Menard, the precise relationship between mid-ocean ridge topography and associated geophysical and plutonic phenomena remains somewhat uncertain. Wilson (1968), as mentioned, advanced the idea of opposing ridge sides separating along transform fractures as an accompaniment to igneous intrusion. In this view, the axial valley would constitute an extensional feature, and its presumed subaerial counterpart on iceland locally has the aspect of a broad graben-type struc-

ture (Fig. 3.31). It constitutes an essentially unmodified structural landform. No thoroughly convincing alternative explanation of the valley has been offered. Deffeyes (1970) suggests that this feature is a "steady-state" entity. If so, this is only true for a certain sea-floor spreading condition [a rate of 2 inches/year (5 cm/year) according to Menard]. Heezen observes that there are magnetic anomalies along some portions of the ridge zone (i.e., 60° N to 42° S in the Atlantic Ocean Basin) and indicates that the Atlantic example signifies an intrusive body 7–10 mi (10–15 km) wide and 3–7 mi (5–10 km) deep under the ridge valley. He notes elsewhere that "instead of having one singular fracture along which intrusives have migrated, there are many subsidiary fractures paralleling the main rift along which intrusion has taken place." More recent work by Vine and others (1963) could be taken to indicate that these peripheral intrusions may represent laterally displaced portions of mid-line intrusions, or alternatively, that each type of plutonic activity represents a distinct phase of ridge evolution during spreading (cf. Fig. 3.10).

In spite of steady refinement of geophysical data (i.e. Knopoff, 1970), it is not clear how the central rift valley is modified to form flanking mountain ranges during the sea-floor spreading process. To be more precise, the exact locus of intrusion appears to vary along the ridge, with the result that uncertainties exist about the relation between the intrusive events, mountainous relief, and the spreading process. Menard (1969) maintains that slow spreading accounts for the higher ridge segments by allowing volcanic buildup to higher elevations in the ridge crest area. This does not, however, account for that portion of the relief that is due to high-angle faulting as noted by Deffeyes (1970). Neither does Hess's suggestion (1962) that convectional dilation causes ridge elevation.

Actual emplacement of the magma at the ridge can hardly be expected to push the sea floors apart. Frictional drag by a convecting mantle acting on the bases of lithosphere

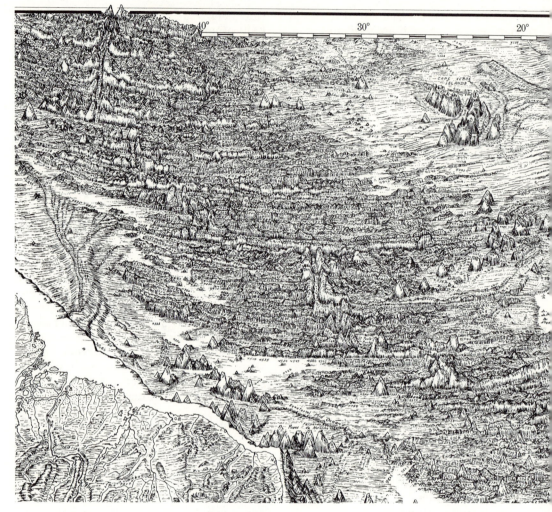

Fig. 3.8. A view of a segment of the mid-ocean ridge in the vicinity of the Romanche Fracture System in the Atlantic Ocean Basin. (From Heezen and Tharp, 1971.)

plates has been suggested as a driving force. But, as noted earlier, the potential lateral scope of such a drag appears to be little more than 600 mi (840 km) (maximum convection cell diameter). This is insufficient to account for the over one thousand-kilometer crustal displacements on record. The answer may in part lie in the intrusive mechanism. It may well be essentially systaltic in nature as noted by Knopoff (1969). If it were, ridge segments would undergo alternate episodes of dilation and contraction, possibly as associated convection cells

"cycle," "precess," or "invert" in the presence of such impulses as the coriolis force (cf. Howell, 1970).

The liquifaction of a linear mass of subcrustal material along a ridge segment presumably requires a thermal buildup. The solid-liquid phase change with accompanying dilation would presumably cause arching and surficial extensional rifting of the ridge segment being affected (Fig. 3.10 a). Intrusion of the resulting fractures could then occur, nominally near the ridge midline, to add to the lithosphere's area and

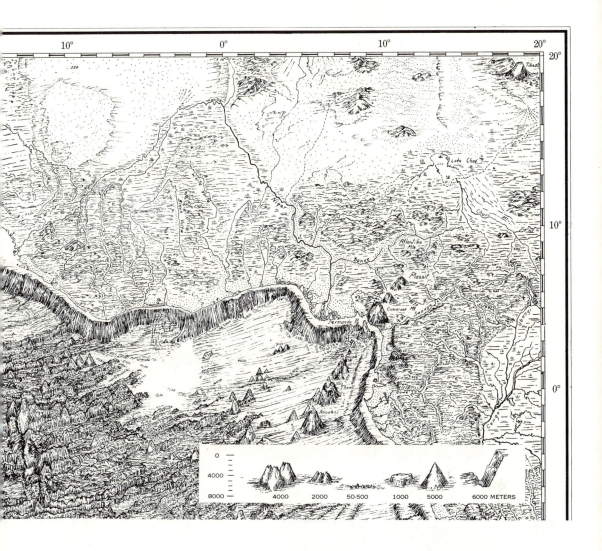

mass in the arched zone. Sea-floor spreading during dilation might notably slow or even stop, though ridge relief would be maximal. Release of molten material and heat from the magma chamber should then cause linear subsidence which would be accentuated by any partial resolidification and contraction of the parent magma prior to the next heat buildup. Axial trenches would presumably relate to this phase (Fig. 3.10 b). Gravitational accommodation of the increased crustal circumference to an isostatically balanced level and shortened crustal radius should then cause extension normal to the ridge axis (cf. Fig. 3.10 c). This type of activity might well account for an appreciable part of the force necessary to move the plates over a low viscosity, upper mantle layer and lend the lateral scope to the spreading mechanism that mantle convection alone may fail to provide. The force needed to cause plate movement has been estimated by Leon Knopoff as 10^{26} ergs/year. Any compressional stress not released by actual sea-floor spreading could cause failure along inclined shear

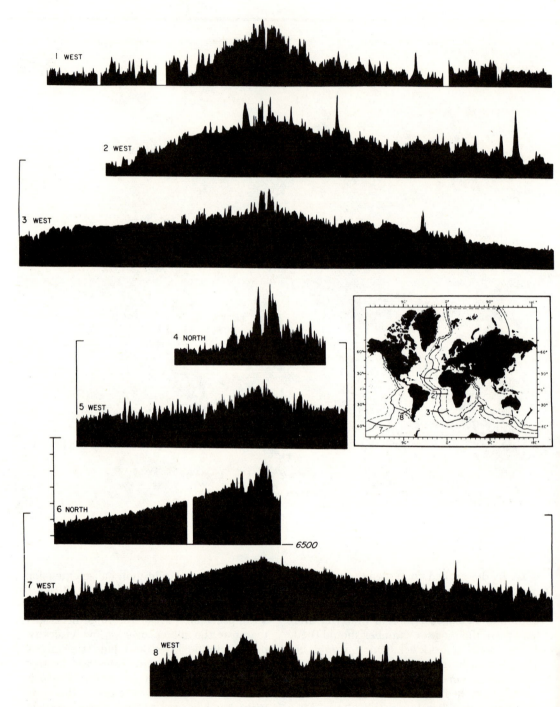

Fig. 3.9. Acoustic cross profiles of the mid-ocean ridge, North Atlantic, South Atlantic, Indian, and South Pacific oceans. Note the variable expression of the axial valley and the more gentle relief of the southern Pacific example. Vertical exaggeration, 100:1. (After Heezen, 1962.)

zones in the ridge and thereby contribute to ridge-crest mountains, either as horsts or upthrust blocks. Deffeyes (1970, p. 213) notes the problems of explaining the ridge as a purely subsidence feature (Fig. 3.11).

Topography of the mid-oceanic ridge presumably reflects the spreading mechanism in some degree—probably in various stages from place to place along its length. Measured spreading rates are cited from 2–12 cm/year, but 0–12 cm/year is probably closer to the actual relationship, since some

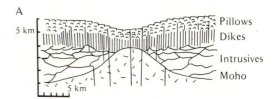

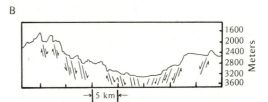

Fig. 3.11. Cross sections of the mid-ocean ridge depicting (A) reconstruction of the axial valley and following outcrops on the island of Cyprus as a guide to lithologies, vertical exaggeration ×2 and (B) profile of the Gorda axial valley obtained by a deep-towed echo sounder by Atwater and Mudie. (After Deffeyes, 1970.)

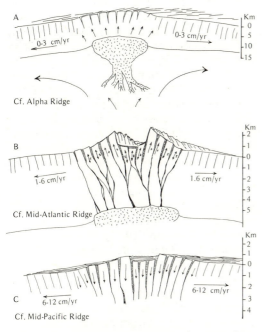

Fig. 3.10. Three-phase schematic of igneous intrusion and sea-floor spreading along a mid-ocean ridge segment. (A) Convectionally induced heat buildup and liquifaction of a portion of the upper mantle under the ridge to cause dilation and extensional rifting; (B) extended magmatic liquifaction and arching accompanied by intrusion near the ridge midline and axial subsidence to form a trench (graben) system; ridge relief would be maximal at this time; (C) resolidification and contraction of the magma chamber with general subsidence of the ridge and crustal extension normal to the ridge axis under the force of gravity; ridge relief would be relatively low. Spreading rates in this model would be low along high-relief ridges and high along low-profile ridges in general agreement with existing relations.

ridge segments appear to be relatively quiescent at present. Heat flow, seismicity, volcanism, and, of course, topography similarly vary along the ridge. Thus, the Mendeleyev Ridge and the Alpha Ridge in the Arctic Basin exhibit essentially all elements of ridge topography recorded elsewhere except for the axial trench. These ridges are inactive seismically, and, as noted by Hunkins *et al.* (1970), they probably represent extinct mid-ocean ridge segments with current spreading rates of zero. Hess (1962) makes the same suggestion about the Trans-Pacific Ridge. It should be kept in mind that spreading rates are calculated on the basis of crustal distance displaced over a period of millions of years. The resulting figures express long-term averages and do not necessarily reflect either current spreading rates or continuous spreading.

Any attempt to relate mid-ocean ridge topography to structural dynamics must be reconciled with the sea-floor spreading requirement that topography of the outer ridge flank is laterally displaced inner-ridge topography. Thus, the apparent topographic increment of displacement is either a linear depression followed by a ridge, a ridge followed by a linear depression, or both together in some order (all essentially paralleling the main ridge). Many reaches of the paired mountain ranges that flank the axial valley have linear depressions to a lesser extent than have the axial valley; they parallel the main ridge zone. The full significance of these features is as yet unknown. Yet it is certain that there is a very long mountain range under the oceans, one whose lack of sediment cover attests to its relative "youth." These mountains also represent primary structural topography reflecting nearly continual growth from within and minimal alteration from without. One can't help but wonder what W. M. Davis would have thought of this kind of "youthful" topography.

Ridge and Rift Topography— An Epeirogenic Realm

Seismic studies of the sea floors demonstrate that rather extensive tracts are seismically tranquil. According to the plate tectonics model, these tranquil areas constitute medial portions of lithospheric plates being displaced. Since the effective component of tectonic movement is tangential, little relief is generated. Hence, tectonism is a passive factor in topographic genesis in this epeirogenic realm. Data from deep sea drilling operations and studies of remnant magnetism (cf. Fig. 3.4) demonstrate that, apart from sediment cover, the sea-floor plates consist of oceanic-type igneous material emplaced as more or less vertical, tabular, dike-like bodies.* These tend to parallel the trend of the mid-ocean ridge segment that presumably generated the plate. Relative to the mid-ocean ridge, the proximal portion of the adjacent plate also comprises the ridge flank. Bathymetric maps of the sea floor prepared by Heezen, Tharp, and others disclose a reticulate topographic pattern in the ridge flank area (Fig. 3.8). One major element of the pattern consists of the

previously mentioned alternating narrow ridges and valleys paralleling the main ridge. Though there is great uncertainty about the genesis of these flank topographic features, they do appear to reflect the theoretical configuration of the intrusives that presumably generate the plate mass as well as the main fracture trends of the axial valley.

The second major element in the ridge-flank reticulate topographic pattern is a series of long narrow depressions disposed at right angles to the ridge crest. These depressions are termed "fractures" on the maps of Heezen and Tharp (1967). They occur at intervals ranging from less than 100 km to more than 500 km. Some are so profound (±2 km in depth) as to merit individual names such as Nansen, Rodriques, Owen, etc. Topographically, many of the fractures are expressed as more or less open rifts. It may be possible to explain this open character by some form of linear subsidence, perhaps as related to the extension of a comparatively rigid plate normal to primary movement directions because of lateral subduction. Alternatively, these fractures could represent rifting related to zonal rotation of the mantle (Fig. 3.12) as suggested by Gilliland (1973). Possibly in jest, the rifts have been called "continental keel marks." But there is little present indication of any Wegenerian type sliding. Since the mid-ocean ridge has clearly been offset along some of these same rifts, there is an evident genetic relation between them and the transform faults postulated by Wilson (1968). We are far less certain as to why these fractures are expressed as open depressions away from the mid-ocean ridge, insofar as many of them are no longer active seismically and therefore presumably are not presently zones of plate dislocation. Many rifts must occur within tectonic plates (if we accept estimates as to the number of such plates), since there are otherwise too many rifts to serve as plate margins only.

Adjacent to the mid-ocean ridge the ridge

---

* One would anticipate the local development of sill-like extensions.

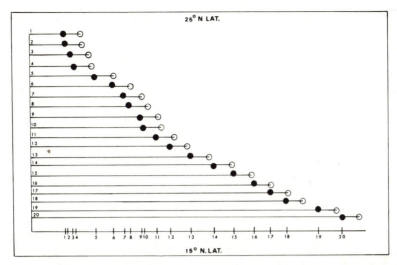

Fig. 3.12. Experimental model of zonal rotation within a fluid-filled sphere approximating mantle circulation as a driving mechanism behind sea-floor spreading. Dot positions represent time-lapse displacements of settling pellets (cf. Gilliland, 1973).

and rift topography of the ridge flank constitutes an essentially unmodified endogenic submarinescape and exhibits essentially no sediment cover. Many researchers have noted an increasing amount of sediment cover on the plates away from the mid-ocean ridge. Acoustic sounding and seismic studies show sediment ponding in depressions (Fig. 3.13). Generally, basal sediment layers appear to be little younger than the basement on which they rest, though as Meyerhoff and Meyerhoff (1972)

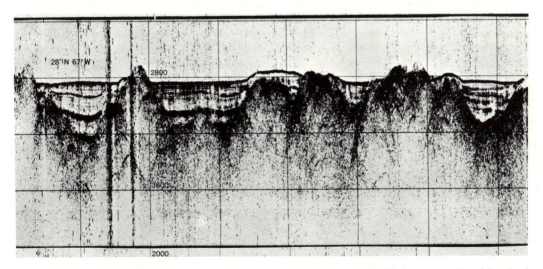

Fig. 3.13. The rugged relief created in the mid-ocean rift valley is gradually being smothered by slowly accumulating deep-sea sediments. Seismic reflection profile, North Atlantic. (From Heezen and Hollister, 1971.)

note, sills may have occasionally been mis-identified as basement. Menard (1969) has pointed out that the plates slowly undergo progressive subsidence as they move away from the mid-ocean ridge. The reasons for this subsidence are not evident, though it is initiated before the acquisition of appreciable sediment cover and its regional extent and magnitude place it within the epeirogenic realm.

The topographic result of ocean plate subsidence, in conjunction with progressively increasing sediment cover, is steadily decreased relief in the ridge and rift area and gradually increased depth below sea level. The distal limits of the ridge and rift zone (away from the ridge) are, in most instances, marked by the margins of one or another of the several abyssal plains. Alternatively, the ridge flank may merge imperceptibly with an oceanic trench zone through subsidence and burial. The abyssal plains are largely depositional and comprise epeirogenic areas of low relief which exhibit little or no tectonic topographic influence in any positive sense.

Heezen and Fox (1966) divide the general mid-ocean ridge flank area into several topographic zones (Fig. 3.14). Counterparts of each of these zones apparently are not present all along the mid-ocean ridge. In part this is because the ridge ranges from narrow and steep (as in the North Atlantic) to broad and rather low (as in the southeast Pacific).

Oceanic Trenches—Orogenic Zones

Approaching the mid-ocean ridge in linear magnitude is a series of elongate, often arcuate depressions in the ocean floor called trenches. Though the term *trench* has been applied to a variety of ocean floor features (Fairbridge, 1966), including the axial valley of the mid-ocean ridge and some of the more pronounced transverse rifts (e.g. Vema Trench, Heezen and Tharp, 1967), it seems best to restrict its usage. Here we use trench to mean seismically active seafloor depressions with commonly asymmetrical cross-profiles where subduction of lithospheric plates appears to be occurring. In short, trenches are terrestrial disposal units. Man needs all of these he can find. Included would be the Ob, Diamantina, Java, Mariana, Aleutian, and Peru-Chile trenches and several others (Fig. 3.15).

Oceanic trenches include the topographically lowest points on earth [−36,198 ft (−11,033 m) in the Mariana Trench] which, together with Mt. Everest [+29,028 ft (+8,665 m)], delimit the earth's maximum relief. Trenches near to or bordering extensive sediment sources (continents and island arcs) contain varying but appreciable amounts of sediment fill and hence must be regarded as more or less modified endogenic topography. Others which contain little or no sediment appear to be essentially unaltered endogenic submarine landforms. As noted by Fisher and Hess (1963),

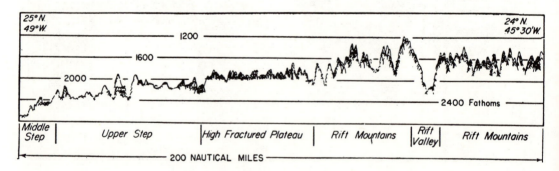

Fig. 3.14. Physiographic subdivisions of the mid-ocean ridge in the Atlantic Ocean based on a profile from a Precision Depth Recorder. (After Heezen et al., 1959.)

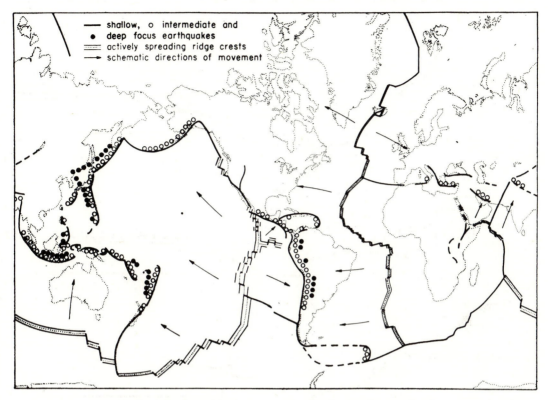

Fig. 3.15.   The seismically defined crustal plates of the Earth (cf. Gutenberg and Richter, 1954) that depict those plates bounded by active ridge crests, faults, trench systems, and zones of compression. Trenches are indicated by those areas of intermediate and deep focus earthquakes. (After Isacks, Oliver, and Sykes, 1968.)

the trench outer slopes are most gentle (4°–8°), being lowest where they merge with the sea floor plains and steepest where they merge with insular or continental slopes (Fig. 3.16). Trench slopes steepen inward up to 45° (e.g. Tonga Trench) and generally produce a V-shape cross-profile, with the V tilted to produce asymmetry. In the Japan Trench, the steep side is terraced, possibly by shearing or slumping, and some sediment is retained on the terraces.

Since some trenches lie along continental margins, it is clear that they can influence the over-all topography of continental rises and slopes as do a number of trench-associated geologic phenomena.

For many years, oceanographers have been aware of several, consistently asso-

ciated geologic phenomena in ocean trench zones. These phenomena include volcanism, notable gravity anomalies, and active seismicity. Recent geophysical studies also suggest a below-average heat flow. Trenches have long been cited as zones of probable crustal subsidence (i.e. Kuenen, 1936, after the suggestion of Vening Meinesz), but a causal relationship involving essentially all the known features has only recently been suggested following the plate tectonics model. The factors to be taken into account, in addition to crustal subsidence, include (1) evidence of extensional fracturing along trench margins; (2) volcanism on the steepest (landward) side of the trench, often with andesitic extrusives as opposed to the usual oceanic basaltic materials, and

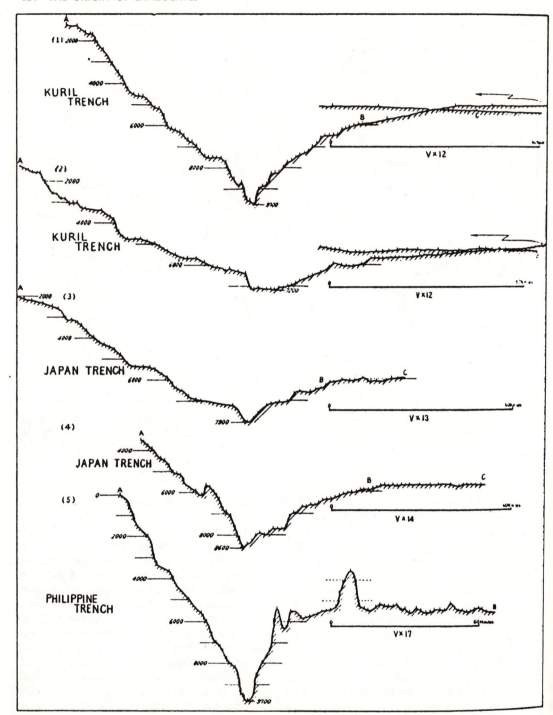

Fig. 3.16. Sounding profiles across several ocean trenches, western Pacific Ocean. (From Dietz, 1970.)

usually producing an arcuately disposed group of islands (*island arc*); (3) strong negative gravity anomalies also centered on the steeper side of the trench; and (4) a locus of deep and shallow earthquake foci along a plane inclined near the surface on the landward side of the trench to nearly 700 km depth farther landward at an angle of about 45°. With this much information, we should have some notion about what is happening in trenches.

We have cited the nearly simultaneous accounts of Robert S. Dietz (1961) and H. H. Hess (1962) in which they attempt to integrate the mid-ocean ridge spreading phenomena and activities in the trenches, thus paralleling Holmes's 1931 model. Long before the work of Deitz and Hess, Benioff (1949) interpreted the plane of earthquake foci in the trenches as an inclined shear surface of enormous dimensions (Figs. 3.6 and 3.17). The spreading concept provides a mechanism that would sustain trench subsidence forces and thus produce a steady-state tectonic relief feature. In this context, extensional fracturing along trench margins is explicable as flexing of a relatively rigid crustal segment as is reduced trench heat flux where a relatively cool, thick lithosphere plate is being thrust downward into and displacing heated upper mantle material. Fracturing and displacements within such a rigid down-thrust mass also provide a reasonable explanation of deep focus earthquake phenomena, and the observed volcanic activity is a logical adjunct to the large-scale extensional fracturing along the landward side of trench margins. The ande-

sitic character of trench zone extrusives is compatible where sea-floor sediments which are often siliceous would be subducted and magmatically assimilated into a parent basaltic melt. The subducted masses of low-density crustal material also help explain the negative gravity anomalies of the trench zones (Fig. 3.2).

In effect, therefore, the topographic asymmetry of the oceanic trenches and their V-shape is a logical consequence of directionally oriented plate tectonics forces and processes. In mid-oceanic isolation, the most obvious subaerial expressions of these trench activities are volcanic island arcs. But the ocean trench contribution to landscape takes on an added dimension where the trenches are adjacent to continents and more heavily sedimented. Both continental and ocean basin morphologies are involved. Materials subducted and assimilated in such trenches may well later put an endogenic stamp on adjacent continental landscapes. It may safely be assumed that a beer can so processed would at least have its trademark obscured. The role of oceanic trenches in such matters as continental accretion, plutonism, and orogeny will be examined in depth—to say the very least—in the following section (see also Chapter 10).

CONTINENTS

It has been theorized that the continents comprise a more or less passive cargo on a

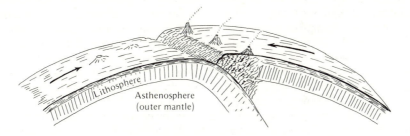

Fig. 3.17.   Block diagram of a segment of the earth transverse to an oceanic trench showing the relation between the colliding plates, the subduction zone, and island-arc volcanism.

system of lithosphere plates which functions much like a conveyor belt. At the same time, cosmological and petrologic considerations require these high-standing masses of low-density, silica-rich material to be some sort of lithic differentiate from a primordial mass of terrestrial matter. The convective circulation of the earth's mantle postulated by Holmes (1931) seems to provide one plausible mode of differentiation operative during at least the latter two-thirds of geologic time. But to the extent that the plate tectonics concept is applicable, the process has not been entirely igneous, for there have apparently been concordant changes in the relative areal magnitudes, elevations, and locations of the continents with direct structural consequences and/or causes, plus changes in the volume of the oceans with a resultant hydrostatic versus lithostatic control over global relief.

As indicated in the previous section, the continents do not appear to occupy a fixed relationship to the several tectonic plates. The present distribution, as summarized by Isacks *et al.* (1968), may not reflect the full range of possible relationships, but a considerable array does exist (cf. Fig. 3.7). North America, for example, appears to rest largely on an oceanic plate that originates along the Atlantic portion of the mid-oceanic ridge (Fig. 3.7). The western part of the same continent may somehow have overridden an extension of the same mid-ocean ridge in the eastern Pacific Ocean Basin and possibly an intervening trench or two before that. South America has apparently been displaced westward relative to the mid-Atlantic Ridge probably while maintaining a position along the margins of the Peru-Chile Trench. This trench may have assimilated an island arc or two since the most recent spreading episode began. Apparent eastward moves of Eurasia and Africa have carried these areas into another ridge area of rifting, and most workers now believe that India and Arabia have undergone northward displacements and consequent collisions with Asia. There is also growing evidence that each of these kinds

of movements and possibly other types as well have been repeated several times in the geologic past. Therefore, the endogenic geomorphic character of the continents must relate in large measure to such plate tectonic events.

An excellent case appears to exist in favor of the continents being accretionary masses with more or less old, "reputedly" stable interiors fringed by younger lithic belts of greater recent tectonic mobility (Fig. 3.18). The concepts involved have been well summarized by Clark and Stearn (1968). See also Dietz (1970) and Fig. 3.1. Hypotheses relating structural stability to rocks of great age go back at least to the work of Charles Schuchert (1910; 1923) but was perhaps most lucidly expressed by Marshall Kay (1951). Kay called his old, stable areas in continents *cratons,* and he classified and made extensive reference to *geosynclines* as structurally mobile belts of sediment accumulation following Dana and Hall's earlier notions of linear sedimentary troughs. It seems clear, however, that the structural stability of a portion of a continent can hardly be a sole function of age insofar as it is subject to the vagaries of plate tectonic events. These events include the potential to disrupt essentially any type of crustal segment, though admittedly the deformative style may alter with lithic nature, time, and position.

Re-examination of the continents in light of geologic history and plate tectonics suggests that structural-geomorphic provinces are established in several distinct ways. (1) Early accretionary processes of the proto-continents probably involved lithic differentiation via igneous activity while these crustal masses were small areally and relatively thin and low (Fig. 3.1). Presumably, the material initially eroded from the proto-continents was ultimately returned to them by plate tectonics mechanisms just as it seems to be at present. The proto-continents may have experienced periods of contact with one another or there may only have been one or two such areas with little interaction (cf. Deitz, 1970). (2) Plate tectonic activity periodically resulted in island arc

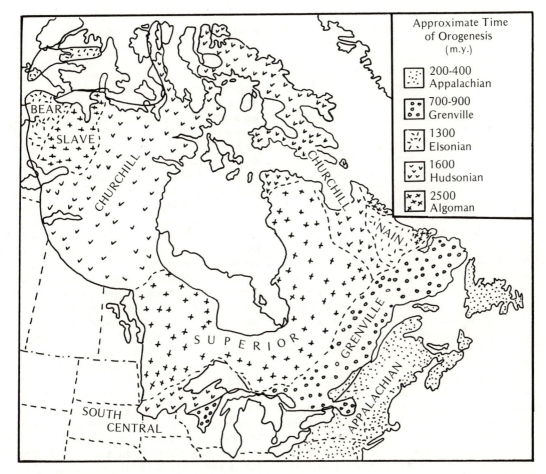

Fig. 3.18.   Distribution of rocks by general age groups, northeastern North America, illustrating the crudely concentric pattern further reinforced to the southwest.

and continental interactions (collisions) along with marginal orogenic activity and resulting accretion (cf. Dewey and Bird, 1970; Bird and Dewey, 1970). (3) From time to time, sea-floor spreading either caused displacement of a continental mass to a position over a mid-ocean ridge, or convection cell re-orientation caused the development of a "ridge" under a continent, thereby causing partial or even complete fragmentation of the continent. (4) Plate tectonic mechanisms apparently caused the displacement of continental masses over adjacent marginal sediment wedges with consequent accretion through plutonism and metamorphism. (5) Displacements

similar to those of (4) caused continents to approach or even override ocean trenches with geomorphic consequences that are far from fully understood (cf. Wilson, 1970). (6) As background music to the orogenic "scenario," sea-floor spreading has apparently produced lateral displacement of continental segments with structural activity thereupon being limited to epeirogenic phenomena such as periodic broad warping or tilting, regional subsidence, or uplift. The precise relationship of such events to the development of localized, vectored stress fields within continental interiors remains uncertain.

The geologic record suggests that each of

the several types of plate tectonic structural activity (and possibly others) occurred several times. Present configurations are merely the most recent and therefore unique largely in that respect. Probable early geographic events are well summarized by Deitz and Holden (1970) and by King (1958).

Portions of continents that are elevated above the sea floor are here termed *continental rises.** The geomorphology of these rises necessarily tends to reflect the tectonic "aging" processes to which these low-density crustal blocks have been subject (cf. Fig. 3.18). The oldest igneous rocks of North America which occupy its approximate heartland near the Great Lakes and another smaller area are about 2,400 to 2,700 million years old. These old areas are fringed by zones with rocks that range from 1,650 to 2,000 million years old which are in turn bordered in some places by rocks averaging about 1,150 million years old; in the more peripheral zones which are the youngest mountain systems, the rocks are some 50–500 million years old. The vaguely concentric pattern that emerges (Fig. 3.18) is far from perfect. It is less perfect on other continents, but it is very informative. We take it as more significant that there are no really young, metamorphic, crystalline bedrock materials in the continental interior than that there are occasionally very old rocks near continent margins. Old rocks near continent margins are quite compatible with the plate tectonics notion that a continent can break up "anywhere." The absence of young rock in the interior must mean that the radiometric clocks there have somehow escaped being reset by the thermal mechanisms usually associated with orogeny and intrusion.

Continental interiors apparently tend to become progressively more immune to the compressive orogenic effects of collisions with other continents or island arcs. In part, there is probably because of progressive marginal crustal thickening and the cush-

ioning effect of the less indurated marginal lithic masses that first become involved in collisions. At the rate of a few centimeter/year movement, the momentum of colliding crustal blocks can hardly be said to be a factor in continental interiors. In addition, the geologic record suggests that marginal rock masses tend to be raised to higher and higher lithic integrity by repeated orogenesis. The marginal crustal rocks are thereby brought closer to the original thickness, homogeneity, and strength of the cratons before they were thinned by erosion. With variations, each continent exhibits some endogenic topography that reflects accretionary growth and related aspects of crustal displacement.

In the previous section dealing with ocean floors it seemed appropriate to consider features in the order in which they apparently form, consistent with plate tectonic theory—mid-ocean ridge to plate to trench. With minor exceptions, a consideration of models of continental structuro-geomorphic provinces (first in terms of tectonically passive areas and then in terms of tectonically active regions) also tends to parallel the formative continental sequence, and this is the approach used hereafter. And in first considering old rocks and then young ones, age is clearly not going before beauty.

Epeirogenic Continental Settings

Vast expanses of each continent presently appear to be situated well away from contemporary oceanic orogenic zones or their probable extensions. These areas are all seismically tranquil and volcanically inactive. Mainly, they seem to be tectonically influenced by lateral continental displacement as a consequence of sea-floor spreading and by such other minor, protracted crustal shifts as accompany tangential plate movements. As previously defined, the regions in question are undergoing more or

---

* A departure from previous usage is recognized here, but the area to which the term was previously applied at the base of the continental slope is not consistently present as an elevated area and is not continental in composition.

less continuous epeirogeny. From a geomorphic standpoint, they are tectonically passive. Such structural and lithologic expressions that may be expressed in landscape are largely matters of differential erosion and weathering or subtly introduced, gentle regional slopes. But just as an older person may exhibit surprising vitality, so old rock areas may become involved in younger activities.

Because the diastrophic activity of continents is seemingly tied to the essentially independent tectonic effects of sea-floor plate activity, each continental tectonic category (active versus passive) includes a variety of bedrock settings. In North America, the epeirogenic realm includes part of an exposed area of old, highly deformed, and altered rocks (*shield*) in eastern Canada; an extension, mainly to the south and west, of similar and somewhat younger rocks with a veneer of mainly flatlying, Paleozoic, and younger sedimentary strata, less than 2,500 ft thick on the average, which, together with the shield, may be termed a *craton* (Fig. 3.19); several areas in the southern and western United States where the surficial sediments are rather deformed but where no appreciable crustal thickening seems to occur (cf. Eardley, 1951).

This apparent lack of crustal thickening is important. There is abundant evidence that compressively tangential forms of orogeny result in crustal thickening and that resulting tectonic elevations are isostatically compensated by a relatively low density root zone which protrudes downward into the mantle (Figs. 10.2–10.4). As pointed out by Woollard (1969)

> ... *there is a regular relation between surface elevation and the depth of the crust-mantle interface defined seismically. It appears that the seismic crust is supported hydrostatically by the mantle in accordance with the Airy concept of isostasy.*

As a consequence, mountainous elevations situated over a low-density crustal root undergo isostatic adjustments (uplift) when the mass above is reduced erosionally or otherwise. Radially directed tectonic forces causing relief changes therefore correlate with the presence of such a root and corresponding elevations. We believe that regions or zones with these tectonic attributes should be regarded as tectonically active geomorphic regions.

By way of emphasizing the fact that older crustal areas are not necessarily diastrophically passive, we note that parts of the central African Shield have been severely rifted

North American Lithosphere and Craton

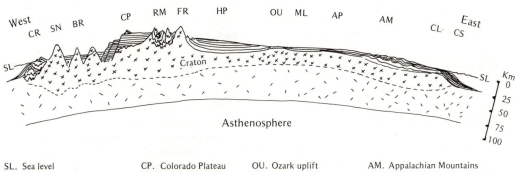

SL. Sea level	CP. Colorado Plateau	OU. Ozark uplift	AM. Appalachian Mountains
CR. Coast Ranges	RM. Rocky Mountains	ML. Mississippi lowland	CL. Coastal lowland
SN. Sierra Nevada	FR. Front Ranges	AP. Appalachian Plateau	CS. Continental shelf
BR. Basin and Range Province	HP. High plains		

Fig. 3.19. Generalized structural cross section through North America in an east-west direction (about the 40th parallel) to show the relative thickness of the continental crust in relation to elevation and relief and particularly the relation between the sedimentary veneer (layered) and the crystalline craton. Vertical scale above sea level is approximately ×5.

in a mid-ocean ridge tectonic style in the comparatively recent geologic past. Also, because of post-glacial isostatic rebound, parts of the Canadian Shield are apparently under compressive stress and undergoing uplift (cf. Woollard, 1966). Mountainous elevations are not a probable consequence in the latter area, however, and the geomorphic effects are undeniable (Fig. 8.57). One must also admit to some uncertainty as to whether the origin of the latter movements is endogenic or exogenic. This classification paradox is intensified by the fact that ice is a mineral that crystallizes in the hexagonal system, an aggregate of minerals is a rock by definition, and glacial ice sheets by logical extension could be regarded as part of the lithosphere undergoing diastrophism. We should think about this.

Tectonic Landscape Elements in Shields

As previously noted, differential weathering and erosion are mainly responsible for any expression of tectonic rock fabric in epeirogenic areas. The exposed portions of cratons (shields) are no exception to this observation. But the several weathering and erosion mechanisms differ in their capacities to exploit bedrock lithic variations. Therefore, landscape can vary with exogenic agencies and processes as well as local nuances of rock type and structure. Here, we will consider the litho-structural landform POTENTIAL of each major type of geologic substrate in each type of epeirogenic setting, beginning with shields. The degrees to which these tectonic elements become expressed in landforms under several surficial geomorphic systems is explored in Chapters 4 through 11.

As a broad generalization, it should be noted that there is a much narrower range of rock TOUGHNESS (resistance to weathering and erosion) in shields than in areas with sedimentary rocks at the surface. Depending upon endogenic conditions, the toughest rock in shields is usually chert or metaquartzite; in sedimentary areas, it is usually chert or orthoquartzite. There is little to choose from in terms of toughness at this end of the lithic spectrum. But the "weakest" rocks in sedimentary provinces would include hardly indurated mudstone, marl, and limestone plus evaporites, whereas their counterparts in shields would include slate, phyllite, mica schist, and possibly marble, the latter often more or less silicified. Most igneous and metamorphic rocks would rank as erosionally resistant in a generally sedimentary rock province.

Though the rock types found in shields are not unique to such tectonic settings, the areal extents of individual rock types (e.g. gneiss, granite, etc.) generally far exceed those of the same rocks elsewhere. As a consequence, in shields, topography is often developed on a single rock type (usually crystalline) over thousands of square miles.

Fabrics of Massive Rocks. The two principal bedrock fabrics in shields are associated with intrusive igneous rocks and metamorphic rocks. There is, to be sure, no sharp line of demarcation between these rock types in the field, since local melting is common within metamorphic rock bodies and incomplete countryrock assimilation and metasomatic replacement is common along borders of larger intrusive masses (*plutons*). In general, the main obligation of the geomorphologist in this regard is to be aware of the different topographic potentials of rock masses (1) where *foliation* exists as a part of the rock fabric and could be exploited by exogenic agencies and (2) where rock bodies are *massive* (lack segregated constituents). Since *migmatites* may exhibit metamorphic foliation or even ghost bedding structure, their potential topographic influence falls within the foliated rocks even though many petrologists regard them as igneous. With this in mind our current focus is on fabric effects of massive igneous rocks.

As rightly noted by Charles Twidale in his exhaustive study of structural landforms (1971), rocks of granitic composition are among the most widely distributed. We may safely presume that Twidale was referring to continents, and granite on continents is also the most abundant of crystalline rocks of intrusive igneous origin. Where massive, all such rocks share a number of fabric characters acquired at some depth below the earth's surface. Intrusive igneous rocks only appear at that surface to make a tectonic contribution to landscape where there has been a notable combination of uplift and erosion.

Since most if not all massive shield rocks are igneous, they may influence topography through their over-all external shape as well as internal fabric (Fig. 3.20). In general, igneous rocks are classified in accordance with whether their margins dominantly cut across foliation of surrounding countryrock or dominantly extend parallel to that foliation. Respectively, these types are termed *discordant* and *concordant*. The principal categories include *sills*, *dikes*, *laccoliths*, *batholiths*, *lopoliths*, and *stocks* (see Glossary for definitions). For the most part, the grain sizes found associated with each type depend on the size of the intrusion (larger masses being usually coarsest), and textures are usually presumed to relate to rates of heat loss and crystallization. Coarse textures are usually taken to indicate relatively

slow crystallization, but recent studies of *ophiolites* (Maxwell, 1970) suggest that coarse texture need not indicate great depth. Also, many ancient continental crusts that are coarse grained were surely also relatively thin when the intrusions were emplaced. Ergo, one may conclude that crystallization occurred at slight depth. Or, the crust is thinned.

Whether a given igneous body is topographically set off from surrounding countryrock depends largely on the relative resistance of each rock type to surficial processes (Fig. 3.21). The number of possible combinations of conditions is very great indeed, and a variety of examples are included in later chapters dealing with exogenic environmental effects. Of more immediate concern is the internal fabric of massive plutons which may find topographic expression. Granular relations (texture) and fractures are the two main elements of this fabric.

Textural relations of plutons are of potential geomorphic importance in two major aspects. Granular interfaces and cleavages (like larger fractures) are potential partings and possible avenues for fluid movement by groundwater or weathering solu-

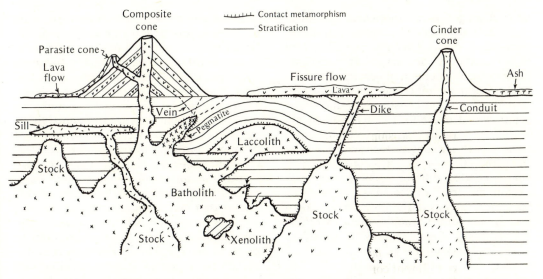

Fig. 3.20. Schematic cross section of various types of intrusive and extrusive igneous rock bodies and their interrelationships with one another and countryrock.

Fig. 3.21. Subdued topographic expression (right foreground) of a basic igneous dike cutting across granite (bouldery surface) on the Sweetwater Uplift, central Wyoming. (W. N. Gilliland.)

tions. Also, granular mineralogic variations may elicit a variety of responses to different kinds of surficial agencies, whether they be of weathering or erosion. Compositionally, with the exception of certain comparatively rare monomineralic rocks, most plutons have mineral constituents that resist solution and other forms of chemical activity and can be termed *inert* or *semi-insoluble*. Other constituents susceptible to the same effects rank as *metastable*. In a similar context, different common minerals exhibit an array of responses to thermal stresses as well as to shearing forces. Topographic variations brought about by differential weathering and erosion therefore begin at the granular level.

Twidale (1971) explores the effect of grain size on weathering susceptibility, but the results must be regarded as inconclu-sive. Systematic studies have yet to be made of rocks of identical compositions and different textures exposed to identical weathering conditions. The Palisades Intrusive along the Hudson River in New Jersey and New York offers a possible study subject. Though details are as yet lacking, it may be noted that an olivine-enriched zone within the general diabase mass is being subjected to intense granular disintegration, whereas the remainder of this relatively coarse-textured pluton is affected to a much lesser extent. One difficulty with studies of this type is that textural and compositional differences may parallel one another and simultaneously coincide with variations in fracture spacing. Thus, where susceptibility variations are apparent it may be difficult if not impossible to narrow the effect down to a specific cause. In the Palisades example,

one cannot be certain at this writing
whether the weathering variations are due
to grain size or compositional differences,
both, or other factors. Twidale (1971) cites
examples of coarse-grained "pink" granite
in the Mt. Lofty Ranges of Australia that
weather more readily than "gray" granite
in the same area, presumably because of
grain size, though other varients are obvious. The same writer refers to an example
in Dartmoor, England where coarse-grained
granites weathered less under frost action
than did fine-grained types (see Waters,
1964).

Though it is true that granites are the
most abundant type of plutonic rock on the
continents, they do not occur in oceanic
provinces.* Via plate tectonics, other igneous rock types find their way into the continental crust. Most oceanic rock types on
the lands occur as sills, dikes, plugs, vent
fillings or accumulations, linear flows, or
fissure flows. The extrusive volcanic rocks
will be considered at length elsewhere, but
most exhibit diabasic, basaltic, or andesitic
textures. As noted by Maxwell (1970),
ophiolites may represent large masses of
basic igneous material extruded under the
sea since they do not appear to be roofed
over by pre-existing rocks. According to
Maxwell, these bodies consistently display
crude stratification; the upper layer texture
is essentially that of *basalt*, commonly with
pillow structure (cf. Fig. 3.90); an upper
median texture is that of *diabase;* a lower
median texture is that of *gabbro;* and the
basal rock mass is *peridotite* (Fig. 3.22).
Though there is some uncertainty as to
whether, in ophiolites, we are dealing with
a strictly endogenic phenomenon, all of the
related lithologies are locally important in
the surface topography of the lithosphere
including that of shields. Ophiolites presently seem to be encountered near accreting
plate margins on sea floors where the contained lithologies may enter into various
metamorphic alteration phases in the presence of sea water. They may comprise the

Fig. 3.22. Intrusive basic igneous rock textures
commonly related to the ophiolitic series, from the
Monteregian Hills (Mount St. Hilaire), east of Montreal. Coarsest (lower left) is gabbro, intermediate
(diabasic) texture is of a contact zone with siliceous
sediment (above), and black, alphanitic rock (to
lower right) is basalt; scale, inches and centimeters.

closest thing we have to a view of the mantle, and for some that would compare with
a glimpse of paradise.

In the same context as ophiolites, *granodiorite* and *quartz monzonite* must be cited
as near kin, and closely related to granite in
composition and texture. Both occur rather
frequently on continents. Individual land
areas underlain by such acidic rocks do not
tend to be as large as comparable land areas
over granite (stocks rather than batholiths), but, in erosional contrast to other
rocks, the smaller intrusives often affect

* An exception is the Seychelles Islands in the Indian Ocean. Another is one of the Canary Islands off
Morocco. Both are probably continental fragments.

topography. The same remarks also apply to *andesite, syenite*, and *diorite*.

We have neither the space nor the inclination to survey all known igneous rock fabrics. Our interest lies in the topographic impact of the more common varieties. Since each may be differentially emphasized or suppressed by one or another of the subaerial geomorphic systems, specific examples are best deferred to detailed consideration of exogenic environments in later chapters. Eventual concerns as to specific fabric behaviors under a variety of conditions lead us to list the more important types with brief lithologic resumés. Readers wishing to pursue the fabric aspects of mas-

sive shield rocks apart from details of igneous lithology should turn to page 136.

GRANITE. A coarsely crystalline, light-colored, igneous rock with 5–30% quartz, feldspar (of several types), in which the ratio of potash feldspar exceeds calci-alkalic feldspar 5:3, and possibly mica (either biotite or muscovite), plus accessory minerals of which the most important geomorphically is probably magnetite (Fig. 3.23). Found in the largest plutons such as batholiths (> 40 sq mi area) and stocks (< 40 sq mi area) and occasionally in the centers of smaller intrusive masses. *Rhyolite* (Fig. 3.24) is the fine-grained equivalent.

GRANODIORITE. A coarsely crystalline,

Fig. 3.23. Direct photographic print of light passed through a thin section of Rockville Granite from St. Cloud, Minnesota (X3.4). This porphyritic, biotite granite has microcline (striated grains), oligoclase, and orthoclase as feldspars. Under white light (right), normally dark grains appear light; under polarized light (left), additional grains in positions of extinction also appear light.

Fig. 3.24. Fracture-bounded sawed slabs of aphanitic igneous rock. Large piece (right) is a rhyolite from the King Ranch, west Texas. Smaller piece is basalt from the Wachung flows, New Jersey. Both were sawed from columns produced by columnar jointing during cooling; scale, inches and centimeters.

light-colored igneous rock with more than 5% quartz, in which the ratio of alkali feldspar to lime-alkali feldspar lies between 1:1 and 5:3, with common major accessory minerals of biotite, amphibole, and pyroxene. Commonly somewhat darker in color than granite but found in similar intrusive associations. More sodic feldspar than monzonite.

QUARTZ MONZONITE. A coarsely crystalline, light-colored, igneous rock with more than 5% quartz, in which the ratio of alkali feldspar to lime-alkali feldspar lies between 5:3 and 1:1. Major accessory minerals and color is much like granodiorite and both these and granite commonly occur in coarse pegmatite facies (Fig. 3.25). More calcic feldspar than granodiorite.

SYENITE. A coarsely crystalline, light-colored, igneous rock characterized by alkali feldspar and containing little or no quartz with common dark accessories of pyroxene, amphibole, and biotite. Rare compared to granite; commonly in small stocks and dikes (Fig. 3.26).

DIORITE. A coarsely crystalline, gray to dark-gray igneous rock with less than 5% quartz and characterized by plagioclase feldspar with a calci-alkalic to alkalic ratio in excess of 5:3. Biotite, hornblend, diopside, and augite are common abundant dark accessories (Fig. 3.27). Usually found in stocks and thick sills and dikes. *Andesite* is the fine-grained equivalent.

GABBRO. A coarsely crystalline, dark-colored igneous rock consisting of plagioclase feldspar more calcic than andesine, usually labradorite. Occasionally abundant accessory minerals include pyroxene, hornblende, and olivine. Commonly found in lopoliths, stocks, dikes, sills, and ophiolitic

Fig. 3.25. Sawed slab of quartz monzonite pegmatite from the Reading Prong of the New England, Precambrian crystalline belt, northwest New Jersey. Large dark crystal is hornblende. Mottled medium-gray areas are feldspar. Dark-gray areas are quartz; scale, inches and centimeters.

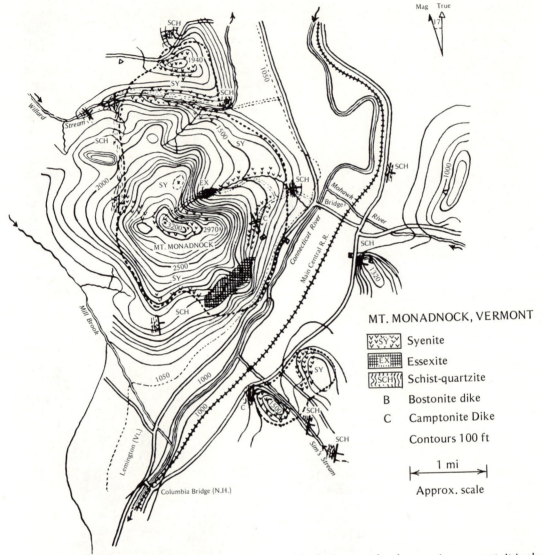

Fig. 3.26. Mount Monadnock, Vermont, is W. M. Davis's classic example of an erosion remnant. It is also a hill developed on an intrusive mass of syenite surrounded mainly by schistose quartzite. (After J. E. Wolff, 1929.)

masses (Fig. 3.28). *Diabase* and *Basalt* are, respectively, the medium- and fine-grained counterparts.

For more comprehensive treatments of these and other less common plutonic rocks, the reader is encouraged to refer to any of several texts and reference volumes including those by Wahlstrom (1950), Shand (1949), Turner and Verhoogen (1960), and Jackson (1970).

The second major fabric aspect of plutonic rocks in addition to texture is that of fractures; a *fracture* as we define it being any break in a rock. Closed breaks lacking evidence of movement are *joints* (Fig. 3.29).* Open breaks are *fissures* (Fig.

---

* In modern structural geology, the term *joint* is regarded as archaic. The slang use of the same word for a low-class bar or a marihuana cigarette may have something to do with this view. Even so, the only apparent alternative, "closed fracture," seems excessively cumbersome.

Fig. 3.27. Direct photographic print of light passed through a thin section of Salem Diorite from Essex County, Massachusetts (X3.4). Under white light (right) normally dark grains (biotite, magnetite, pyroxene) appear light, and plagioclase feldspars appear dark. Under polarized light (left), additional grains in positions of extinction appear light.

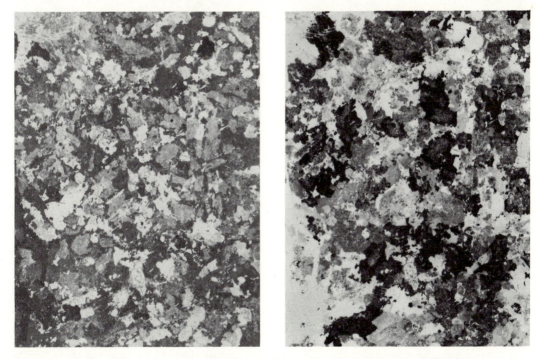

Fig. 3.28. Direct photographic print of light passed through a thin section of Salem Gabbro, a coarse-textured, dark-colored rock composed mainly of plagioclase feldspar, hornblende, and biotite (X3.4). The main feldspar is labradorite. Under white light (right), normally dark grains (biotite, accessory opaques) appear light; feldspars appear dark. Under polarized light (left), additional grains in positions of extinction appear light.

Fig. 3.29. Banded granite gneiss on the Altiplano [elevation 13,500 ft (4,000 m)] above Cuzco, Peru at the Inca fortress of Sachsawaman. Glacial scour has exploited foliation which trends toward viewer but has hardly modified closed fractures which trend essentially transversely. Though in the core of a mountain system (Andes), these rocks do not differ significantly from similar types in shield areas.

3.30). And breaks across which there has been relative movement along the plane of parting are *faults*. Plutonic igneous rocks exhibit examples of all of these, but they are not confined to such rocks (Fig. 3.31). Fractures are induced in rock bodies by stresses and once formed they constitute breaks in the homogeneity of the parent mass. Fractures are weaknesses along which other agencies may attack the integrity of the rock body once it enters some other environment.

More than in most other rocks, high-pressure and high-temperature stresses are applied to plutonic rocks at depth. The initial stresses usually come during the solidification process due to cooling, shrinkage, and lithostatic pressure, when the magma is accommodating to a reduced volume. Fractures induced then remain closed for a time, under most circumstances, because of lithostatic pressure. The breaks are there,

however, as zones along which granular texture has been disturbed by stress. Additional stresses are induced in deeply buried rocks by the process of exhumation. This uncovering, via the combined effects of erosion and uplift, reduces the lithostatic load on a given rock volume. It may permit pre-existing fractures to open into fissures or perhaps only until they are visible as partings or would permit more easy penetration by fluids. In addition, of course, plutonic bodies may be stressed and thereby fractured by diastrophic forces from without, following their solidification.

In view of the foregoing, it is not surprising to find that plutonic rock bodies are traversed by a bewildering assortment of fractures (Fig. 3.32). Some of these breaks nearly parallel present land surfaces, whereas others intersect the same surface at high angles. Variations in fracture attitude and associated differences in spacing have rather obvious geomorphic potentials. And it should be noted that field studies suggest that closeness of fracture spacing closely correlates with formative stress intensity. Thus, fractures normally exhibit their closest spacing adjacent to fault zones and fold axes and progressively greater spacing away from these zones.

Fig. 3.30. Fissure in basaltic rock near the south coast of Iceland. Water from the background falls passes through the opening but is apparently unrelated to its development. (W. N. Gilliland.)

Work by Cloos (1936) and Balk (1937), among others, demonstrates that at least some of the fractures found in batholiths can be related to stresses developed during intrusion (presumably after a degree of solidification had been achieved) and that other breaks relate to the regional tectonic framework. Observations of the writer and an associate are pertinent in this regard (Garner and Gibbons, 1969). Dikes of diabase in the Gettysburg, Pennsylvania area display large-scale, curved polygonal fractures that bound columns (10–12 ft in diameter) oriented vertically (Fig. 3.33). Moreover, the curved fracture surfaces exhibit a secondary polygonal fracture system (Fig. 3.34) which we interpret as an indication that notable additional shrinkage of the intrusive occurs after the development of a primary fracture system. There are additional implications, and the problem is under study.

Like many other far less homogeneous rock bodies, plutons commonly display at least two sets of high-angle fractures, often disposed nearly at right angles to one another. These breaks intersect somewhat curvilinear, nearly flat-lying fractures to form *orthogonal blocks* (cf. Twidale, 1971) or, in the presence of a third high-angle fracture set, polygonal blocks (Fig. 3.35). In the Palisades sill, polygonal fractures form nearly vertical columns, essentially normal to both the bounding wallrock of the intrusive and the presumed surface at the time of the intrusion. At least some of the high-angle fracture sets in intrusive bodies may well relate to stress fields developed outside the pluton, but the more nearly horizontal set of fractures probably relates to cooling-stress configurations within the pluton or is *sheeting* (horizontal fracturing) due to unloading (Fig. 6.9).

In summarizing the geomorphic significance of plutonic rocks in shields the following seems important. The physical weakness or strength of a given rock or fabric type can only be gauged in terms of the nature of the conditions to which it is later subject. Thus, since a major factor in chemical reaction rates is surface area,

Fig. 3.31. Faulting affecting the dominantly igneous terrain of Iceland. The Thingrellic Rift Valley, north-east, is a probable extension of mid-oceanic rifting and spreading. Faults are downthrown to left about 125 ft (40 m). (S. Thorarinsson.)

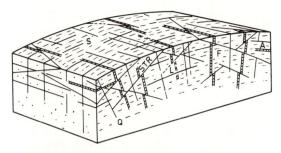

Fig. 3.32. Cloos's diagram showing the main kinds of fracture in a batholith. Q, cross joints; S, longitu-dinal joints; L, flat-lying fractures; STR, planes of stretching; A, aplite dikes; F, linear flow structures (cf. Twidale, 1971).

fracture spacing is an index to weathering susceptibility in a notably chemical set-ting.* But these fractures may localize igneous emmanations if the subsequent environment is endogenic, and the mineral-ized fracture zone may well be more re-sistant to later weathering or erosion than the remainder of the rock. Moreover, frac-tures at the textural level in plutons may be presumed to function somewhat as they do in other rock types with respect to fluids. A fine-grained rock may be more *porous* than a coarse-grained one, but it may well be less *permeable*, and this could influence solu-tioning and weathering. In dry environ-ments, pore character influence on fluids may be irrelevant.

* Fracture spacing is also important where frost action is a significant weathering mechanism.

Fabrics of Foliated Rocks. Foliated rocks are intimately associated with massive rocks in shield settings. The foliation may merely relate to original bedding partings in metasediments (or the rare patch of unmetamorphosed sedimentary rock), or it may appear as a consequence of mineral alignments, bedding partings, or flow structures in intensely metamorphosed or metasomatized rocks of either igneous or sedimentary parentage. Maxwell (1962) regards slaty (flow) cleavage as a form of foliation. Both flow and fracture cleavage show strong preferential development in folded terrain, the latter in relatively brittle rock types. Cleavage, unlike the other forms of foliation, is often distributed in relation to structural position. The resulting foliated rock bodies lack the granular and miner-

alogic homogeneity of plutons for the most part, and the fabric is notably aligned over large areas all apart from fracture patterns. Some Precambrian granite is interlayered *lit-par-lit* with megacrystalline schist and gneiss. Many similarly appearing rocks are actually migmatites. The possibilities of topographic expression of these lithic variations by differential weathering and erosion are obvious. The linear terrain texture that can result is generally not to be found in association with large plutons.

Actually, of course, intrusive igneous rock bodies of various shapes and dimensions penetrate or even surround one another as well as associated masses of metamorphic rock in shields. In the cases of metasediments, the lithic trends often match the trends one would expect in the

Fig. 3.33. Large-scale polygonal fracturing cutting diabase in the Devils Den area, Gettysburg Battlefield, Pennsylvania. The diabase is part of a Triassic dike and sill complex. Surfaces of the primary fractures (left foreground) are seen to carry secondary fractures (cf. Fig. 3.34).

Fig. 3.34. Continued shrinkage of the diabase intrusives, Gettysburg, Pennsylvania, after development of primary fractures is indicated by these polygonal fractures which are normal to the faces of the larger breaks; camera lens cover is 2 inches in diameter.

original sedimentary rock formations; they actually extend for miles in one general direction. Because the rocks are usually highly deformed, it is unusual to encounter tabular rock bodies paralleling the ground surface in shields. The foliation commonly intersects the surface at a high angle. A notable exception to this is in the Precambrian Roraima Series of the Venezuelan-Brazilian Shield (Fig. 3.36). But many tabular masses of rock in shields are so inclined that if they react in a resistant fashion to surficial agencies they form elongate uplands whereas, if nonresistant, they form similarly elongate lowlands (Fig. 3.37).

The principal foliated rock types encountered in shield areas include the usual metamorphic equivalents of the common sedimentary rocks plus a few types not developed to any extent elsewhere (e.g. *iron formations*), meta-igneous rocks (e.g. *greenschists*), and occasional masses of layered material including sedimentary and extrusive igneous rock types that are more or less deformed but essentially unmetamorphosed. The unmetamorphosed, foliated rock bodies are often locally mineralized, however, and are usually very thoroughly cemented, not uncommonly with silica.

Again, with no attempt to be all-inclusive in treatment, we will list and briefly describe textural and mineralogic relations of the more common foliated shield rock types. Later we will be concerned with their behavior under a variety of surficial conditions. Readers wishing to pursue the lithic

Fig. 3.35. Orthogonal and polygonal blocks of rock developed in granite exposed near the headwaters of the Arkansas River, Colorado. The more exposed pieces show rounding, probably related to weathering within a soil profile since removed.

Fig. 3.36. Nearly flat-lying Precambrian strata of the Roraima Series exposed in a high mesa, southern Venezuela, give rise to Angel Falls (the world's highest). A stream tributary to the Rio Caroni flows through V-shaped and notch-like clefts incised into elevated planation surfaces before plunging some 3,300 ft (1,000 m). The crest of this and many similar elevations in equatorial realms remained humid during several lower-level Pleistocene arid intervals in the same region.

Fig. 3.37.   A reach of the Rio Caroni, shield region, southern Venezuela, flowing left (north) through an elongate ridge of resistant Precambrian rock (foreground) after traversing a savanna lowland. Note the quadrate patches of jungle on the savanna and similarly rectangular flow pattern of the drainage network, one element of which parallels the foreground ridge.

fabric of shields apart from details of metamorphic lithology can pick up the thread of our discourse on page 146.

SLATE. The lowest-grade metamorphic rock with smooth regular foliation due to cleavage development (Fig. 3.38). Almost as fine grained as the parent argillite but probably with an incipient development of muscovite and chlorite which give the partings a sheen in which individual grains are not visible to the naked eye. Slate splits readily into slabs in which the dimensions vary with the spacing of cross fractures. Slate is generally somewhat less readily abraded than shale.

PHYLLITE. Strongly foliated metamorphic rock with a grain size intermediate between slate and schist. Mica grains are not megascopically visible but give a flaky character to the rock. Phyllite is commonly

dominated by white micas, chlorite, talc, or graphite. It splits readily into small fragments and is often as readily abraded as shale.

SCHIST (common varieties include mica, chlorite, garnet, sillamanite, and staurolite). Schist is a coarse-grained metamorphic rock with strong foliation caused by subparallel orientation of such minerals as mica, chlorite, talc, and graphite which are tabular and flaky (Fig. 3.39). Commonly there is some amphibole which is prismatic in habit. Quartz is often present, and grain foliation is commonly interrupted by more equidimensional grains such as quartz and garnet (*porphyroblasts*). The mechanical bonding between individual foliations is commonly weak because of the platy mineral cleavage.

GNEISS. This is a coarse-grained meta-

Fig. 3.38. Slate quarry near Bangor, Pennsylvania, developed in Ordovician age rocks. Sawing (left face) is done to take advantage of slaty cleavage (far face), and smooth slabs express both partings in spoil heap beyond.

morphic rock with more or less distinct banding provided by contrasting mineral segregations or orientations. The most common varieties of granitic derivation are quartzo-feldspathic with hornblende or biotite. Varieties of probable sedimentary derivation may display mineral suites of the common schist grades. Dominance of equidimensional grains helps explain the strong physical bonds between individual layers of gneiss and its consequent massiveness and coherence in the field (Fig. 3.40).

QUARTZITE. This is a metamorphic rock, consisting dominantly of quartz, which displays massive to weakly schistose fabric (Fig. 3.41). Though mainly of sedimentary derivation (sandstone, chert), original partings may be obscured. Mechanical bonding between apparent layers is usually strong but is dependent on original argillaceous content as reflected in the abundance of micaceous minerals.

MARBLE. Marble consists of metamorphosed calcite or dolomite rocks with a usually coarse, equigranular crystalline texture. Quartz is commonly present. Original stratification partings (which can usually be related to thin clay seams) have usually been altered to phyllite or schist. Metamorphic foliation results mainly from micaceous minerals, and bonds between apparent layers range from weak to strong.

STRATIFIED EXTRUSIVE ROCKS. A common lithic component of shields are extrusive, more or less layered igneous rocks, most of which were originally mafic in composition. Tabular, intrusive rocks of similar character often have a similar aspect (Fig. 3.42). Derivative foliated rock bodies range from basaltic to greenschist to garnet-sillimanite schist, depending upon metamorphic intensity. General coherency decreases with an increasing abundance of platy mineral facies.

sive and metasomatic masses of magnetite, hematite, and related oxides. Many of the latter intrusives contribute notably to shield landscapes (Fig. 3.43).

For a more comprehensive lithologic treatment of the foliated rocks (both sedimentary and metamorphic) the reader is referred to Kern Jackson's *Textbook of Lithology* (1970). Sedimentary rocks, of course, occur in shields but make only minor contributions to landscape there. To this point in our discussion we have focused on the more common textural and mineralogic attributes of foliated rocks in shields. Now we must consider the occurrence of fractures in foliated metamorphic rocks.

Like their igneous counterparts, most metamorphic rocks are traversed by a variety of fractures. These breaks are in ad-

Fig. 3.39. Garnet schist from the Adirondack Mountains, northern New York. The garnet porphyroblasts (dark grains) interrupt the schistosity, related mainly to muscovite and phlogopite micas in this case; scale, centimeters.

IRON FORMATIONS. Rocks with a volume and iron content not equaled elsewhere occur in several of the world's shield regions. The bulk of the stratified, sedimentary iron there is in the form of the carbonate (siderite) but it is also present as granular silicates and oxides, usually with bedded chert or quartzite. The other main shield occurrences are in the form of intru-

Fig. 3.40. Sawed slab of banded Precambrian gneiss, Lake Superior region. The light-colored bands are mainly potash feldspar and quartz; scale, inches and centimeters.

Fig. 3.41. Split slab of schistose quartzite, the Reading Prong of the New England, Precambrian crystalline belt, northwestern New Jersey. Schistosity is sufficiently poor to permit use in building stone facings; scale, inches and centimeters.

dition to the foliation and other potential or actual partings that relate to metamorphic textural conditions. Because of the frequent

intersection of fractures with foliation, metamorphic rocks are generally far less capable of yielding large detached blocks of rock (monoliths) than are plutonic rock bodies. Of the various metamorphic rock types discussed here, gneiss is by far the most structurally coherent over large areas and in large volumes. Its widespread use as a building stone reflects this property. Foliation in combination with other primary partings in most other metamorphic rocks comprises potential slippage surfaces under conditions of unbalanced stress as well as potential routes for fluid movement. Not only is it unusual to find a schist used as a load-bearing building stone in a really large structure, but schist combined in terrain with plutons and gneiss seems relatively susceptible to most forms of differential erosion (Fig. 3.39).

In fractured and foliated rocks of shields and most other terrains of similar country-rock, the dimensions of the clasts freed by weathering tend to depend upon the spacings between partings. These spacings are usually greatest in gneiss and quartzite and

Fig. 3.42. Mesas of Precambrian (Keewatin) diabase near the village of Nipigon, Ontario, north shore of Lake Superior. These flat-topped landforms reflect the tabular configuration of the igneous sill from which they are eroded.

Fig. 3.43. East slope and crest of the "iron mountain" Cerro Bolivar, southern Venezuela. This elongate inselberg rises about 1,000 ft (300 m) above the alluviated savanna some 50 mi (70 km) south of the Orinoco River; its deeply leached hematitic crest indicates continued chemical decomposition under plant cover at present and in times past when the surrounding lowlands were arid. Many lower, rounded inselbergs in this same Precambrian shield region have no appreciable weathered interval and little plant cover.

least in slate, phyllite, and schist and thus directly affect talus configurations (Fig. 3.44) and countryrock fracture, porosity, and permeability. One result of these relations is that monolithic blocks of rock freed by weathering may be important elements of landscapes developed on gneiss, quartzite, and many types of plutons in shields, since many monoliths are too large for most erosional mechanisms to remove (Fig. 3.45). Of course, glaciers and gravity can move almost anything. But the problem of monoliths can hardly be said to exist in a terrain developed in slate, schist, phyllite, and other foliated rocks of similarly weak coherence.

The Canadian Shield has been recently and intensively glaciated so that most such monoliths as it may once have exhibited are now *glacial erratics* elsewhere. Such differential erosion as may be expressed in the terrain of that shield is also largely a product of glacial erosion as modified by later events. Shield regions in lower latitudes have escaped this glacial stripping, however, and monoliths are commonly seen (Fig. 3.46). Where bedrock is immune to *exfoliation* phenomena, or where it has not been subjected to such rounding processes, angular blocks bounded largely by fracture surfaces are the usual result (cf. Fig. 3.45). Under other circumstances, removal of edges, corners, and faces in some degree produces less angular monolithic forms which Twidale (1971) wishes to term *boulders* (Fig. 3.46), but which many

Fig. 3.44. North side of the Delaware Watergap, New Jersey, showing disposition of quartzite talus beneath the escarpment of Silurian, Shawangunk Conglomerate from which it came. The large talus masses are so porous, cool air trapped within in winter pours out at the slope base much of each summer (cf. Fig. 2.7).

writers call *tors*. The rounding mechanisms are discussed in some detail in Chapters 5, 6, and 7.

Tectonic Landscape
Elements in Buried Cratons

Cratons comprise portions of older crystalline interiors of continents, and they are often overlain by a relatively thin veneer of more or less stratified sedimentary rock. In North America the sedimentary interval averages little more than 2,500 ft (800 m) thick and at several places (i.e. the Ozark Uplift in Missouri, the Black Hills Uplift of South Dakota, and major uplifts of the Colorado, Wyoming, and New Mexico foreland) the crystalline basement crops out at the surface. Vertical tectonic movements of sufficient magnitude to cause crystalline basement outcropping are relatively isolated in cratons, but occasionally such movements occur in parts well away from margins of oceanic tectonic plates. The structural styles, sediment thicknesses, and related topography of major cratonic uplifts, particularly in the western United States, resemble those of orogenic belts and are discussed in a later section. Crustal movements that affect the stratified cratonic rocks during lateral displacements of the continents by sea-floor spreading are largely limited to gentle uplift, subsidence, tilting, warping, and, more rarely, arching, or doming—all usually very gradually.

On occasion, the stratified rocks of cra-

Fig. 3.45. Angular blocks of dolerite and granite isolated by fracturing along Machakandana River, 1 mi southeast of Kumargaon, India, on the shield. The rapids are developed over the dolerite dike, and the open fractures clearly influence flow directions. (K. L. V. Ramana Rao.)

Fig. 3.46. Low, rounded granite domes locally developed on the Champua-Keonjhar Upland just west of Sirispal, peninsular India. Note the perched rounded boulders (tors), possibly the result of exhumation of core stones from a former weathered interval. (K. L. V. Ramana Rao.)

tons respond to epeirogenic stresses by fracturing and faulting. But the layers are seldom really folded by such activity apart from local drag along faults. The isolated, older basement rock exposures in cratons bring into play the same topographic elements of lithology and structure found in shields; these have already been discussed. However, the vast majority of cratonic landscapes and those of immediate concern here are developed on stratified rocks where the average inclinations range from about 0.5° to 1° over wide areas. In many places the rocks are actually flat-lying. Rarely, stratal inclinations amount to several degrees.

Stratified rocks in cratons can serve as a geometric reference system by which to gauge crustal deformation, a reference system that is generally absent over vast shield regions. Since many cratonic sedimentary rocks accumulated in sea water are horizontally layered, the amount of post-depositional uplift and deformation is usually determinable if not immediately apparent. Of course, nonmarine sedimentary deposits are not as useful in this regard, but the water-deposited ones have some attitude reference value. The over-all impression one gets of cratonic sedimentary rock structural configurations is that of broad, shallow elevations and depressions. What we see is presumably a muted and subdued reflection of much more abrupt basement uplifts and subsidences which are almost surely related to closely spaced faulting. As noted by Gibbons (personal communication, 1972), rock flow is hardly conceivable at the depths involved in most cratonic structures.

The isolated to interconnected depressions (*basins*) such as those in Michigan, Oklahoma, and Illinois in North America (Fig. 3.47), the Tindouf Basin in North Africa, the Paris Basin in Europe, and the Amazon Basin in South America are usually encircled, or separated from one another, by basement rock elevations. Admittedly, some of these depressions open toward the continental margins (i.e. Gulf Coast Embayment). The large basement structures of course have expression in the associated stratified rocks. The steepest dips usually are encountered on the flanks of the basins or uplifts, and the lowest inclinations are normally found in basin mid-portions and the crests of uplifts.

Though the precise relation of cratonic structures to plate tectonic phenomena is

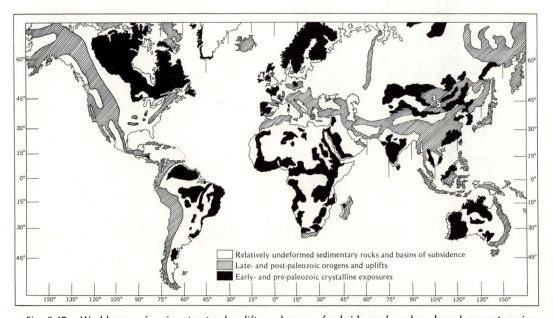

Fig. 3.47. World map of major structural uplifts and areas of subsidence based on broad age categories.

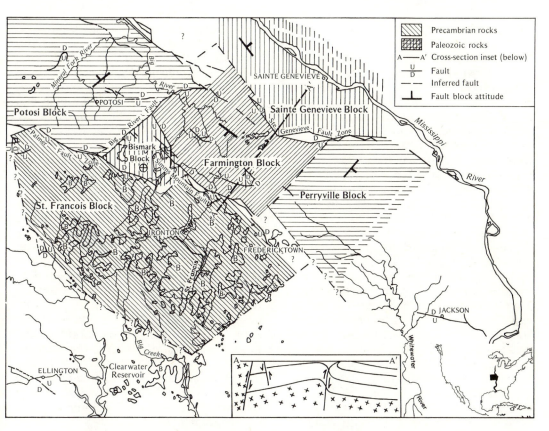

Fig. 3.48.   Upthrust block faulting, eastern Ozark region, map and cross section (inset). The area shown in southeast Missouri is regionally located (map inset). (Data, J. F. Gibbons, unpublished.)

uncertain in many cases, it is nevertheless true that most major structural features of cratons show periodic re-accentuation by crustal shifts over periods of scores to hundreds of millions of years. Studies of features such as the Ozark Uplift, Nashville Dome, and Cincinnati Arch in North America (cf. Fig. 3.47) all show repeated, periodic thinning of sedimentary rock formations across these highs beginning as far back as the Cambrian Period. Adjacent depressions similarly show intermittent subsidence until at least the end of the Paleozoic Era and even periodically since (Stanley and Wayne, 1971). Thus it is hardly surprising that many of the broad structural highs on cratons even now stand

topographically somewhat above adjacent structural depressions.

A recent detailed structural analysis of the Ozark Uplift by Gibbons (in press) demonstrates that this regional high owes at least a part of its expression to up-thrusting in which high-angle basement faults are modified into low-angle shear zones in the more surficial sedimentary rocks (Fig. 3.48). The deformational style resembles that recorded in the southern and middle Rocky Mountain Front Ranges by Foose *et al.* (1961), Berg (1962), Wise (1963), and Prucha *et al.* (1965). Such a tectonic style might develop as the continental crust was locally subjected to a convection vortex (*mantle plume*) in the subjacent mantle.*

---

* Such mantle plumes (fixed hot spots), the Tuzo Wilson idea cited by Dietz and Holden (1970), have been used to explain the development and differing ages of the Hawaiian islands.

Alternatively, such tectonism might relate to sea-floor spreading centers or triple plate junction activity under the continental crust. Since the continents presumably have been moved several times and in several directions during the period of deformation, it is difficult to see how inland continental structures could continue to relate to any particular plate tectonics mechanism. Perhaps they do not.

A survey of the literature dealing with cratonic uplifts and basins (Weller, 1960; Kummel, 1961) and some experience of the writer in the area (Garner, 1967) indicate that there are definite sedimentary rock lithologic differences between many of the regional highs and lows. The uplifts generally exhibit far more carbonate rocks and less argillite rocks than do the depressions. The sedimentologic reasons for this are several, but they include a tendency of carbonate (reef) facies to develop in shallow water (on uplifts), the common removal of argillites by winnowing from shoal areas, and the deposition of these compactable materials in adjacent depressions. The carbonates are relatively incompressible rock masses, and differential compaction of these two facies alone would tend to accentuate structural relief differences once the latter were established. The rock types just mentioned and several others are commonly associated in cratonic sedimentary covers, and it now seems appropriate to consider their fabrics and potential topographic role.

Fabrics of Sedimentary Rocks in Cratons. The lithologies and structural configurations of sedimentary rocks in cratons differ so markedly from those in older geosynclines and currently sedimented areas that a rather unique structuro-topographic association results in each instance. Accordingly, each lithologic suite will be considered separately and where it seems most appropriate. The several rather unique properties of cratonic sedimentary rock veneers include (1) a dominance of Paleozoic and Mesozoic rock ages with correspondingly heightened induration and cementation; (2) a dominantly tabular external configuration to most of the rock formations; (3) a dominantly near-horizontal bedding relationship so that major local lithologic variations are encountered in vertical sequence by any agency acting in that direction; (4) local lithic sequences are accessible "edge on" largely in areas of high relief (which in cratons are mostly erosional in origin); (5) many rock formations in cratons have great lateral lithologic persistence so that in the absence of erosional relief, lithic influences on landscapes only tend to emerge regionally with facies changes (e.g. the Ordovician St. Peter Sandstone of the central United States is estimated to cover some 35,000 sq mi); and (6) there are occasional extensive accumulations of nonmarine rock (e.g. the Tertiary formations of the High Plains). The principal sedimentary materials on cratons include both lithified and nonlithified *alluvium* plus *limestone, dolomite, sandstone, siltstone, mudstone,* and *chert.* Siltstone and mudstone are here collectively termed *argillites.* The lithologic descriptions that follow include a number of comments about the geomorphic role of each type.

Mudstone in cratons is usually compacted and is fissile (*shale*). In composition, particles less than 1/256 mm in diameter are dominant, and, in the older rock sequences, it is not uncommonly black, pyritic, and locally siliceous. Mudstone is of geomorphic significance in that it is the only common rock found on cratons that is notably subject to compaction under the lithostatic loads normally developed there by burial. Thin layers of mudstone are responsible for many if not most bedding plane partings of other sedimentary rocks. The most common minerals in mudstone are finely divided quartz and clay. The most frequently encountered clay minerals are *gibbsite, montmorillonite, kaolinite,* and *illite.* They are so seldom considered in beginning mineralogy courses they deserve some emphasis here. Jackson (1970) points out that gibbsite is the main constituent of the mineraloid *bauxite* and the soil type *laterite,* that it contains little or no silica, and that it is the product of complete humid tropical weathering and leaching where all cations including silica are removed. He further notes that kaolinite with its sheet-like atomic structure retains some silica and is the most common product of humid temperate weathering (cf. Fig. 5.24), whereas montmorillonite is even more incompletely leached (probably in an arid or semiarid environment), and its sheet structure retains a high cation-exchange capacity and variable water content. Illite has a layered structure, also, but it has a low cation-exchange capacity and is nonexpanding—if a weathered product, illite would require a parent alkaline (dry climate) soil. But, as Jackson (1970) notes, this clay may be produced in sea water as a diagenetic alteration product of other clay minerals.

Shale formations play several significant geomorphic roles in buried cratons in addition to that related to compaction. Most shale intervals are so impervious that they are rarely pervasively cemented. They may function as seals in artesian systems, as supports for *perched water tables*, and as infiltration barriers to meteoric water. Shale containing appreciable expanding lattice clay (such as montmorillonite) may swell greatly and contract notably where there is considerable periodic moisture variation. Such activity has several direct consequences on weathering processes, soil formation, and mass wasting. For example, in south Coastal Ecuador, the Oligocene Dos Bocas Shale (which contains expanding lattice clay) is oxidized and leached locally to depths in excess of 150 ft because of alternate swelling and heaving and shrinking and cracking. In that instance, iron sulfides and the calcium carbonate in foraminifera have reacted to produce secondary selenite gypsum along shrinkage fractures. In the same context it should be noted that the argillites (depending upon cementation) are the mechanically weakest common sedimentary rocks in cratons, with the result that they will not usually support steep slopes; their removal often jeopardizes the integrity of superjacent strata. Therefore,

where steep slopes are developed in essentially flat-lying rocks that include various rock types interstratified with shale, the shale will tend to be expressed as a low slope or bench. Shale commonly recedes on a slope face beneath tougher rock ledges to form overhangs and is rarely found to cap an upland (Fig. 3.49). If weathered mechanically, most shale yields thin flakes or irregularly polygonal chips. But in surficial chemical situations, the various clay minerals range from inert to metastable, and most types will disaggregate granularly under the right circumstances.

Siltstone is a clastic rock comprising particles from 1/16 to 1/256 mm in diameter; it usually includes quartz, other mineral flakes, and often a variety of lithic fragments which are commonly angular in shape. Because of its larger grain size, siltstone is usually more permeable than shale, and it may be thoroughly cemented by iron oxide, silica, or calcium carbonate. Clay mineral contents of siltstone are too low to permit consistent expression of the sheet-like clay mineral structure as fissility. Presumably because of low clay content, greater grain strengths, and more extensive cementation, fossils found in siltstone rarely show compactive deformation. The most common form of contemporary silt accumulation is as wind-blown dust (*loess*), and it seems probable that many ancient siltstone formations are of similar origin with particles derived from areas of sparse or nil vegetal cover. Significantly, many siltstones are characterized by abundant metastable rock and mineral chips— the latter often includes calcite, dolomite, and various feldspars. Some fluvial reworking is always possible and in some instances appears probable.

Marine siltstone beds in cratons are rarely more than a few inches thick, and, not uncommonly, they display admixtures of sand and clay. More thorough cementation usually makes siltstone beds more resistant than associated shale, and this, in combination with low porosity, may make siltstone better able to resist weathering than coarser-grained rocks. As noted by

Fig. 3.49. Benches developed on shale formations with intervening cliffs and bluffs in the Grand Canyon, Arizona, seen from the north rim. The more resistant lithologies forming the escarpments are mainly sandstone and limestone.

many authors, the sizes of blocks freed by weathering of sedimentary rocks is largely determined by bed thickness and the spacing and orientation of existing fracture sets. This applies to siltstone, and notably smooth-faced polygonal blocks of small dimensions often result (Fig. 3.50).

Sandstone is a clastic rock composed of fragments 1/16 to 2 mm in diameter, and, if we restrict the discussion to non-biogenic forms, in cratons most of the grains are quartz. Exceptions to the quartzose composition of cratonic sandstone are generally restricted to rocks accumulated directly on basement or around the more isolated cratonic uplifts where feldspar is also present in bedrock. There, first-cycle sandstone is commonly feldspathic, even to the point of forming *arkose*. Lower Cambrian sandstone "granite wash" is an example. Middle to upper Cambrian sandstone deposits and many Ordovician sandstones in central United States are noted for their pure quartzose content. But they and most other sandstone bodies depend upon some degree of cementation for durability.

Uncemented sandstone tends to be mechanically weak (*friable*). The variously cemented sandstones tend to be as chemically durable as their cements. Where quartz grains are cemented by silica the resulting rock (orthoquartzite) may rival its metamorphic counterpart in toughness. Where the cement is calcium carbonate, water can be expected to remove it little by little, and such a sandstone is little more resistant than is limestone to the same treatment (Fig. 3.51). Abrasively, the calcareous quartz sandstone normally exceeds the erosional resistance of a limestone.

Most of the clastic cratonic sediments contain some iron. Therefore, where leached by meteoric water, a calcareous sandstone will develop a more or less thick crust of quartz grains weakly bonded by a residue of iron oxide, and it may be difficult to take a fresh sample. Some such rock masses actually become cavernous.

Sandstone formations exhibit strata ranging from a few inches to many feet in thickness. Many lower and middle Paleozoic sandstone formations exhibit beds several tens of feet thick, and a basal Pennsylvanian sandstone deposit of the Winslow Formation in the Ozark region commonly crops out as a single bed 80–100 ft (25–

30 m) thick. Where jointed, such rock layers are clearly capable of yielding huge monolithic blocks (Fig. 3.52). Similar sandstone formations are found in the Mesozoic of west-central United States and in the Nubian sequence of North Africa. As one might expect, sandstone (particularly that cemented by iron or silica) commonly forms a resistant cap to hills or uplands and forms pronounced bluffs where erosionally incised in cratons. Though fracture spacing tends to be influenced by local stress concentrations, it is usually greatest in thick beds and least in thin beds.

Like shale, sandstone formations play several significant geomorphic roles in cra-

Fig. 3.50. Polygonal fracturing in a siliceous mudstone, southeast of Batesville, Arkansas. At least three sets of joints are involved plus bedding partings. The unit here is late Mississippian in age and occurs just above the Pitkin Limestone of northwest Arkansas and below a unit of similar lithology at Leslie, Arkansas.

tons in addition to those already mentioned. In one or another of its weathering- or erosional-resistance aspects, sandstone constitutes the toughest rock found in cratonic settings. Because of its commonly great permeability, it often serves as an aquifer. Sandstone is apparently not often susceptible to load compaction in cratons, though solution on grain contacts may cause intergrowths and reduced porosity. It, like siltstone and limestone, is most susceptible to erosion where it is inter-stratified with shale formations and most resistant where it rests on massive rocks of other types.

Carbonate rocks in cratons are mostly *limestone* ($CaCO_3$) and *dolomite* [$CaMg(CO_3)_2$] which constitute end-members for a group of rocks in which calcium and magnesium are mutually replaceable. Pure *magnesite* ($MgCO_3$) rocks are rare. Some silicification usually accompanies the process of dolomitization with the resulting reduced

solubility of dolomite formations as compared to those of limestone. With this in mind, the following remarks are largely intended to relate to limestone where primary structures are generally well preserved. Limestone is seemingly more abundant in late Paleozoic-to-Recent deposits than its counterpart.

Cratonic carbonate deposits are probably of organic derivation in most instances, and the configurations of the carbonate bodies tend to reflect parent organic influences to some degree (Fig. 3.53; cf. Fig. 2.38). Early Paleozoic limestone and dolomite formations in cratons are dominantly tabular, dense, fine-grained rocks in which matts of lime-secreting (blue-green) algae probably played a strong role during accumulation (Fig. 3.53A; 3.54). Notable carbonate cratonic facies variations began to appear (Silurian) with the development of shallow water faunas living in growth concentra-

Fig. 3.51. Differential solutional weathering in the lower Cretaceous Lakota Formation southwest of Colorado Springs, Colorado. The rock is a medium-grained, ferrugenous sandstone, and present aridity suggests some of the feature may be relict from former more moist conditions.

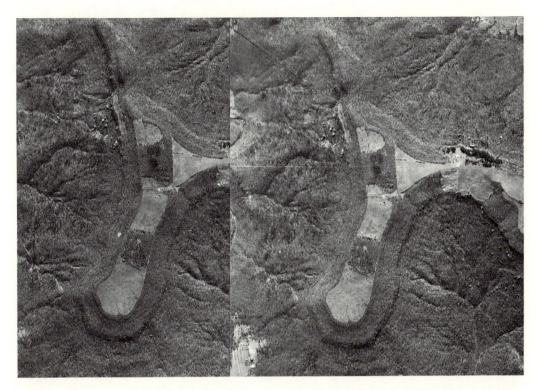

Fig. 3.52. Aerial stereo photos of huge orthogonal blocks of sandstone [±20 × 30 × 60 ft (6 × 10 × 25 m)] moving away from an escarpment and downslope, the Boston Mountains, northern Arkansas.

tion sites (reefs) with intervening deeper water areas of detrital accumulation. In the middle Paleozoic, during periods (or places) of warm water, strictly marine epicontinental sedimentation equates with widespread limestone accumulation where both reef and inter-reef facies are carbonates (Fig. 3.53B; 3.54). Times of terrestrial sand and mud sedimentation restricted organic carbonate reefs to localized high-relief accumulations between inter-reef facies of shale or sandstone (Fig. 3.53C; 3.54). Late Paleozoic and Mesozoic epicontinental seas generally correspond with the development of extensive barrier reef carbonate deposits and restricted seaways with associated evaporites (Fig. 3.53D).

Carbonates are, of course, best known geomorphically for the solutional imprint they may give to landscapes.* In cratons

the extent and manner in which carbonates affect topography necessarily varies with the form of carbonate facies complexes. The widespread tabular carbonates of North America and elsewhere are obviously susceptible to solution where exposed at the surface during humid times. Limestone is also a comparatively weak rock mechanically because of the softness of calcite compared with the most abundant detrital sediment (quartz). But it should also be noted that many carbonate deposits, and perhaps the majority, exhibit little inter-granular permeability. Solutions, including groundwater, penetrate limestone along fracture systems more commonly than between grains. Fractures are commonly observed to have re-healed by secondary calcite deposition. And, in the absence of water, limestone may fragment mechanically and part

* For a discussion of solutional topography see Chapter 5.

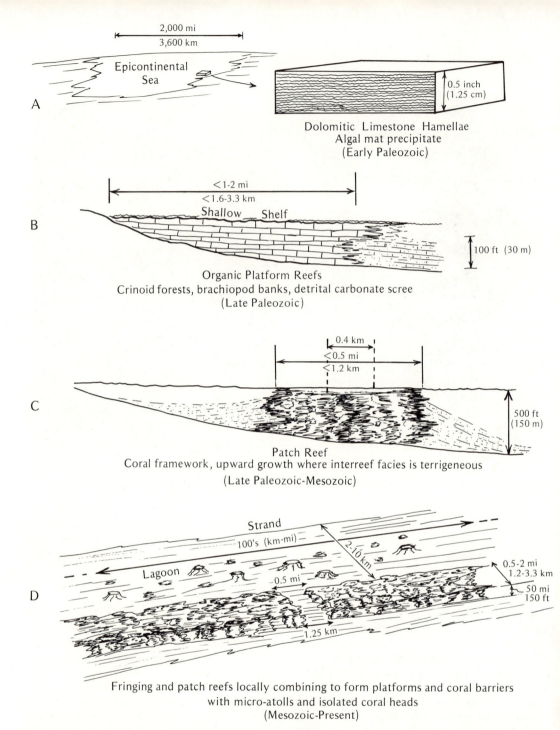

A

2,000 mi
3,600 km

Epicontinental
Sea

0.5 inch
(1.25 cm)

Dolomitic Limestone Hamellae
Algal mat precipitate
(Early Paleozoic)

B

<1-2 mi
<1.6-3.3 km

Shallow Shelf

100 ft (30 m)

Organic Platform Reefs
Crinoid forests, brachiopod banks, detrital carbonate scree
(Late Paleozoic)

C

0.4 km
<0.5 mi
<1.2 km

500 ft
(150 m)

Patch Reef
Coral framework, upward growth where interreef facies is terrigeneous
(Late Paleozoic-Mesozoic)

D

Strand
100's (km-mi)

2-10 km

Lagoon

0.5 mi

0.5-2 mi
1.2-3.3 km

50 mi
150 ft

1.25 km

Fringing and patch reefs locally combining to form platforms and coral barriers
with micro-atolls and isolated coral heads
(Mesozoic-Present)

Fig. 3.53. Schematic depictions of carbonate depositional configuration changes through time. (A) Early Paleozoic dolomitic limestone lamellae precipitated on floors of very shallow epicontinental seas, probably as a metabolic biochemical by-product of algal mat growth in quiet water; no relief; (B) middle and late Paleozoic platform carbonate deposits on shallow shelf areas devoid of terrigenous deposits—both reef (framework) and inter-reef (detrital) facies are present as carbonates with growth areas spreading laterally rather than growing upward. Continues to present, locally; (C) patch reef configurations of late Paleozoic to Recent, developed in areas of terrigenous sediment accumulation with main carbonate development in upward growing reef and flank detritus. Dimensions for an Eocene example with associated orbitoidal foraminifera in Ecuador; (D) Late Paleozoic and Mesozoic fringing and patch reefs locally merge to form coralline platform and barrier configurations as in Australia's Great Barrier Reef; several Mesozoic examples had associated evaporites in back-reef areas.

along bedding planes or joints. Limestone beds range from thin seams to massive units many feet in thickness. Dense reef cores, lacking fractures and almost totally impermeable, may be extremely durable features in the absence of effective abrasion.

In layered, nearly horizontal rock sequences within many cratons, limestone formations often make pronounced bluffs, particularly where overlain by such impervious rock as shale (Fig. 3.49). Occasionally, a carbonate deposit will be found capping a hill or upland. In the latter case, some signs of solutioning will commonly be evidenced. Buried limestone formations may fare better, but their bases commonly are sites of springs on hillslopes. Massive, reef core rocks may show some desiccation fracturing that has healed, but they commonly lack joints and are nonstratified. Inter-reef and reef flank detrital facies of carbonates are commonly both jointed and stratified. Variable partings distributions like the foregoing will certainly localize solutioning. Some of the most profound geomorphologic influences of carbonate rocks comes, however, where relatively incompressible patch reef core carbonates are accentuated via compaction in relation to inter-reef argillite facies as mentioned in Chapter 2 (Fig. 3.54). In a similar relation, elongate barrier reef masses may contrast with back-reef evaporite deposits and fore-reef basinal argillite facies (Golden Lane Region of Mexico). Though the sediments of the Mexican example are Mesozoic, the related reef is both structurally and topographically high at present.

Chert and its dark-colored counterpart (*flint*) are the common ex-solutional sedimentary rock expressions of silica (SiO_2) encountered in cratons. Though nodular and related forms are of interest, it is stratified chert that is of greatest importance geomorphologically. Bedded chert is usually regarded as of primary depositional origin, and truly extensive deposits of such material in cratons are relatively uncommon. Some of the best known are associated with Ordovician and Mississippian deposits in

Fig. 3.54. Common, reef-associated carbonate lithologies. Upper left, Ordovician stromatolitic algae from the Kittatinny Formation near Hackettstown, New Jersey. Lower left, Mississippian, inter-reef detrital facies with Gervinellid incrustations over dominantly encrinal and bryozoan fragments. Upper right, Mississippian reef core lime muds of lithographic character, commonly found within masses of solitary and colonial coralline frameworks (lower right). Mississippian examples from Pitkin Limestone, southern Ozark region; scale, inches and centimeters.

North America. In the Mississippian Boone Formation of the Ozark region, both bedded and nodular varieties are present, and at least some of this silica has apparently replaced carbonate material (Fig. 3.55). In the same region, both Ordovician and Devonian sandstone formations have been cemented and locally replaced by chalcedonic silica; similar effects are observed in Mississippian siltstone beds.

Chert layers observed in cratons seldom are more than a few inches thick, though the writer recalls 2- to 3-ft thick beds of chert in the Mesozoic of Ecuador in a "geosynclinal" setting. Chert beds are usually closely and intensely fractured, and these

Fig. 3.55. Interbedded and nodular chert and limestone of the Mississippian Boone Formation, northern Arkansas. The chert (light-colored zones) is more intensely fractured than the limestone. Machete blade (center) is about 16 inches long.

fractures are ordinarily curved, presumably because dense silica tends to break conchoidally. In most cherty formations, the chert layers interstratify with other rock types, including limestone and shale and where limestone is the other dominant rock of a formation or there is an otherwise well-defined reef mass, chert is usually associated with the inter-reef detrital facies. Though they are of course very brittle, chert fragments are otherwise among the toughest clasts derived from sedimentary rocks, and they are an important part of the corrasive load of many streams.

The geomorphic significance of chert and strongly siliceous sediment lies chiefly in its relative insolubility under many subaerial conditions.* A consequent, unique contribution to landscapes is the chert felsenmeer developed over much of the central Ozark region (Garner, 1967), where solution of cherty carbonate formations has left a nearly insoluble, coarse chert residual (Fig. 3.56).

Conglomeratic Rocks and Alluvium comprise the last category of sedimentary deposit found on cratons that will be considered in some lithologic detail. The degree of

* Lustig (personal communication, 1973) notes the occurrence of chert associated with basalt-capped mountains in the Basin and Range area, where it indicates relict drainage systems.

induration ranges from nil in some recent accumulations to well cemented. Though some lacustrine deposits are also present, the emphasis here will be on nonmarine, fluvatile mudstone, sandstone, and conglomerate and their unconsolidated counterparts, all of which are commonly intermixed and/or interstratified. Some of these materials clearly accumulated within an appreciable initial dip to their layers, some are virtually unstratified, and some, like the Triassic, Shinarump Conglomerate in the western United States, comprise a nearly tabular mass with an upper surface that

may originally have been as nearly level as the lag gravel deposits that cap the hammada in the central Sahara (Fig. 6.30). Stokes (1950) interprets the Shinarump as a possible pediment veneer accumulation. Strong initial dips are clearly present in the fans and bajadas that comprise large portions of the alluvial basin-fill deposits of the southwestern United States. Lower depositional inclinations are indicated in the Chadron and Brule alluvial formations of the Black Hills region in which badlands topography is now developed (Fig. 6.16).

Widespread alluvial deposits are largely

Fig. 3.56. Rubble of chert exposed north of Mountain View, Arkansas, where it forms a thick, insoluble felsenmeer over the Mississippian Boone Formation. Color variations mainly reflect red, iron staining. Machete is about 30 inches long.

concentrated in western and southwestern North America and other similarly arid parts of the world. They are also encountered in regions with histories of glacial or desert activity as noted by Gignoux (1955). In most such regions these materials and their lithified counterparts have been erosionally sculpted in various ways. Apart from the aforementioned badlands, erosion of alluvial materials has produced the Book, Roan, and Pink cliffs in the western United States, part of the locally steep-sided hammada of North Africa, and similar features elsewhere. Extensive, thick, coarse clastic fluvatile deposits of Cretaceous and Paleozoic age in cratons are concentrated largely around localized uplifts, though thinner accumulations veneer basement rocks and

occur along some unconformaties in younger stratigraphic sequences.

Most unconsolidated types of allumium are in recent deposits. Some are found on uplands, and they will commonly not stand in slopes steeper than the repose angle for the material involved. Walls of *arroyos* cut into alluvial fans are often vertical, however. Also, Tertiary gravels of the American High Plains and similar deposits in central Australia and elsewhere have developed surficial armors of caliche, iron oxide, and silicate material. The cemented surface interval (*duricrust*) thereupon begins to function as a durable caprock more resistant to weathering than the remaining deposit. If breached by some mechanism, a bluff and bench usually results (Fig. 3.57),

Fig. 3.57.   The caliche-capped rim of MacBride Canyon north of Amarillo, the Texas Panhandle.

Fig. 3.58. A tableland developed where the Colorado Plateau is incised by the Colorado River in Arizona. Note erosional remnants on the upland (a structural plain) and bench-bluff valley walls reflecting variable lithologies.

and, where the encrustation process has affected a series of elevated terraces, such pseudo-layered lithology may influence an entire landscape (cf. Dury, 1966). Indurated, nonmarine sand and gravel can play the same geomorphic role as their marine counterparts, where they are interstratified with other rock types in cratonic sedimentary sequences.

Apart from local, pronounced uplifts, the tectonic influence on craton landscapes tends to be largely negative. It is the general ABSENCE of rock deformation expressed by the nearly horizontal attitudes of strata that exerts the most profound morphologic influence on much cratonic geomorphology. Vast areas that have (somehow) escaped intense dissection are characterized by gently rolling, nearly flat, or only slightly inclined land surfaces often nearly paralleling the bedding attitude. At least some of

these surfaces should be termed *structural plains,* but the only tectonism involved in many cases was that presumably required to elevate the area above sea level. Indeed, Monument Valley in Arizona (Fig. 6.48), and similar low-dip portions of the Colorado Plateau such as the "canyon lands" (Fig. 3.58) and the Ozark Uplift (Fig. 2.35) differ in the tectonic sense mainly in their various elevations above sea level, and any consequent limits these differences MAY set for incision by running water. All share nearly horizontal bedrock formations of varying susceptibility to surficial processes over wide areas. Where the stratigraphic sequence above basement has been breached, the differential weathering-erosion responses of various lithologic types alone will suffice to produce one or more changes in the inclination of a resulting hillslope (Fig. 3.49).

Erosionally isolated remnants of flat-lying sedimentary rock bodies form most of the even-topped *tablelands, buttes,* and *mesas* of the world (Figs. 3.42; 3.58; 6.48). Where more strongly tilted and etched into relief by differential erosion, the same layering produces *cuesta-scarp* topography encountered along many cratonic borderlands (Fig. 3.59).

Comments about some of the types of fractures in cratons were made with each discussion of the common rock types. It remains to point out that fractures apparently do not develop a regionally parallel arrangement unless the sponsoring stress in similarly regional. Sedimentary rocks in medial portions of cratons usually display only one or two fracture sets, and those are widely spaced for the most part. Along cratonic margins where relatively recent orogenic forces have been active, it is common to en-

counter three or even four sets of closely spaced intersecting fractures, and such a joint pattern may extend far into the craton. For example, a fracture set that coincides with the stress pattern of the foreland fold belt of the Ouachita Mountains in central Arkansas has been detected as far north as the southern Missouri border more than 200 mi (280 km) away (Gibbons, 1967). A structural petrofabric analysis of twin plane gliding in calcite (Konig and Chin, 1973) confirms the fracture evidence (*see also* Friedman, 1964).

The second principal comment that should be made about craton fractures concerns what many geologists regard as "first-order" or "dominant" structures. Such linear expressions of basement architecture are called *lineaments*. In buried cratons, the lineaments are encountered in sedimentary rocks so that their precise relation to

Fig. 3.59. Curvilinear cuestas and escarpments developed on the Alcova and Chugwater formations as seen from South Pass Road, the Wind River Mountains, Wyoming. (W. N. Gilliland.)

Fig. 3.60. A segment of a lineament expressed in the northwest Canadian Shield along McDonald Lake near Great Slave Lake. The fracture appears faulted at this point (viewed to SW) and displays a scarp up to 900 ft (275 m) high. It has been traced for more than 300 mi (480 km) and separates rocks of the Yellow-knife province (right) from those of the Churchill province (left) (cf. Fig. 3.18). (National Air Photo Library, Ottawa.)

buried basement structure remains some-what speculative. Presumably, what we see is a renewed expression of some ancient fracture or flexure. Spencer (1969) suggests that many represent strike-slip faults. In shield regions, we can see nearly straight fractures (some of which are clearly faults) cutting a variety of rock types, and such features are presumably developed at depth under lineaments in buried cratons (Fig. 3.60). Be that as it may, basement features do apparently affect topography through intervening sedimentary rock sequences at least 2,000 ft (600 m) thick. Only a single example will be cited from the Ozark region.

A series of more or less parallel fractures 5 to 25 mi (7–35 km) apart trend from northeastern Oklahoma through northwestern Arkansas into southern Missouri on a bearing of North 30° East. They vary from essentially straight to curvilinear, and the longest is in excess of 75 mi (115 km). As noted by R. H. Konig (personal communication, 1972), the very continuity of these fractures indicates basement involvement. In places these lineaments have the appearance of faults, but they have a variable throw parallel to their strike. Magnetic data show changes associated with these fractures, but it is uncertain whether these responses are to basement lithology or to magneto-telluric currents. There is no record of recent seismicity. The faculty and students of the University of Arkansas have had these fractures under study for more than a decade. Though there are clear differences of opinion, available evidence also suggests that the fractures may have localized groundwater with some consequent solutioning of associated Paleozoic carbonate rocks. Certainly there are many places where near-surface strata have differentially subsided along these lineaments (Fig. 3.61).

Orogenic Continental Settings

Portions of essentially every continent are being subjected to strong vertical displacements or have been so subjected in the past

as to express this tectonism in high relief. In the western United States, pronounced, localized uplifts on the craton have formed mountains from causes largely unknown. Some vertical movements appear to be perpetuated at present merely through the subsidence one can relate to sediment accumulation. If the accumulation areas are elongate, and subsidence is great, then the term *geosyncline* seems appropriately descriptive. Areas of appreciable but reduced subsidence extending into cratons are *embayments*, and continental margin areas of similar soft sediment accumulations are *shelves*. Occasionally, and partly in response to the sedimentary loading of some of these areas, one encounters a variety of high-relief piercement structures, some of which are expressed topographically. Also, any of the areas of sediment accumulation

Fig. 3.61. Steeply dipping, fractured, late Paleozoic sandstone beds near Yoe Lake, Fayetteville, Arkansas. Bedding is inclined about 50° to the left (south). Deformation apparently relates to lineaments trending northeast through the area which involve basement rocks and probable solutioning and subsidence.

(like the associated shields and buried cratons) may be caught up by plate tectonic mechanisms and modified into mountain systems.

Lithostatic Subsidence and Topography

Areas of recent and continuing subsidence and sediment accumulation are concentrated along the continental margins, and most surficial earth materials encountered there are too poorly lithified to be termed rock. There are those who cherish the expression "shovelrock" for the substance in question. Rarely is it coherent enough to be broken. Certainly it is true that the materials in the same areas that are deeply buried are the most thoroughly indurated. But they are usually not at the surface and therefore do not ordinarily affect the landscape. Such topography as there is, and there often isn't much, is dependent on unconsolidated materials. A good many of the materials in question are marine sediment, but there are intercalated nonmarine layers from place to place. The surfaces of most of the areas in question are close to sea level (some above, some below), and their slopes are low. The Mississippi Embayment, for example, has an average slope of $1/180°$ (0.5 ft/mi) which raises it to an elevation only slightly over 200 ft more than 400 mi inland, and the inclination of much of the Gulf Coastal Plain is little more than that. Continental shelves average about $0.5°–1°$, and the continental slopes developed on sediment rarely exceed $2°$. From these marginal accumulations of debris there extend the underwater aprons of deltas, such as that of the Mississippi River, and continental slope sediments that in some instances merge with those of abyssal marine plains.

As noted by Marshall Kay and the earlier students of geosynclines long ago, the crust of the earth must subside in order to accommodate great thicknesses of sediment. The areas in question may be receiving sediment because they are low, but in many instances loading certainly contributes to subsidence and low elevations. Some such subsidence may characterize most sedimented continental margins, but there are places where subsidence tectonism (if we may call it that) is extreme. Along the Gulf Coast of the United States more than 65,000 ft (20,000 m) of sediment has accumulated, most of it since the Mesozoic Era. But the surface relief there is low, as it is in less active subsidence areas of recent sedimentation. Surface materials along coastal plains are generally too weak to stand in steep slopes, and there are few mechanisms available to generate such slopes in any case. Such *depositional plains* are among the lowest relief land surfaces in the world.

The geomorphology of coasts will be considered in Chapter 9, but here we will briefly discuss certain tectonic landscape effects common to areas of appreciable sedimentation and subsidence, many of which are coastal. Several earth materials are known to flow and deform plastically at relatively low pressures and temperatures as, of course, most substances will at high pressures and temperatures. The materials most susceptible to deformation in sedimentary environments are unconsolidated, water-saturated sediment, and, where these substances shift under the force of gravity, soft-sediment structures result. Gravity sliding and slumping occurs at several different scales of magnitude. Soft sediment deformation resulting in contorted bedding is only barely an endogenic phenomenon, and it may hardly alter the surficial form of a sediment deposit. On occasion, however, as at the mouth of the Magdelena River in Colombia, bottom topography is continually and drastically altered as newly deposited sediment slumps and moves seaward. Alternatively, loading of a crustal area by a great weight of sediment may mobilize the least competent materials at depth with great structural and topographic consequences. The materials in question are principally salt, anhydrite, and shale (see section on deltas, Chapter 7).

In several parts of the world, and notably Afghanistan, Romania in central Europe, the Gulf Coast of North America and the Gulf of Mexico, and part of the Mediterranean Basin, salt has become mobilized following burial. The structural result is a variety of piercement or near-piercement structures including diapers, domes, and anticlines cored by salt (Fig. 3.62). The general consensus on the origin of such phenomena (cf. Barton, 1933; Atwater and Forman, 1959; Bornhauser, 1963) is that a buried salt layer is differentially loaded. Many workers think it was the Louann Salt of probable Permian age in the Gulf region. In any event, the least loaded salt areas sponsor upward protrusions which penetrate overlying sediments even as the latter perpetuate the process by continued accumulation and further loading [Fig. 3.63 (drawing)]. If piercement by the mobile salt mass keeps pace with burial then the mobile salt body may remain essentially at the surface and thus may not actually have to penetrate thick sediment layers. In other cases, actual piercement of rock layers occurs, and they are bent upwards by drag; and, in still other instances (either with or without piercement), the moving salt mass

remains covered by an interval of strata. The landform consequences of this type of structure are several.

Where salt fails to penetrate to the earth's surface, it may nevertheless cause notable doming of overlying rocks to form a rounded hill. Such tectonic landforms are common to the regions previously cited and especially Gulf Coastal North America. At times the salt penetrates to the land surface where (in most parts of the world) it is immediately attacked by solution to form a depression which is not uncommonly thereafter occupied by a lake (Fig. 3.64). More rarely, the salt emerges in the driest of surficial environments as it does in the Afghanistan and Persian deserts. There, the salt is known to flow out onto the land surface as salt "glaciers" with at least some of the attributes peculiar to that type of erosional mechanism [Fig. 3.63 (photo)]. Possibly such features should be discussed in another section of this text as extrusive landforms. One additional expression of salt-induced topography should be noted

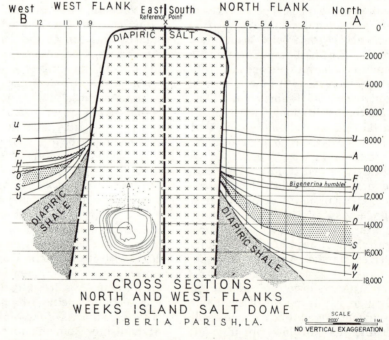

Fig. 3.62. Cross section of the north and west flanks of the Weeks Island salt dome situated about 150 mi (210 km) west of the Mississippi River Bird-Foot Delta, Louisiana. (After Atwater and Forman, 1959.)

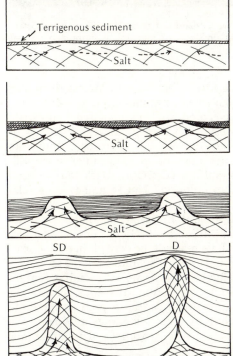

Fig. 3.63. Photo of Kuh-i-Namak salt glacier in Iran and schematic cross section of a salt dome (SD) and a diaper (D). Features are caused by salt flowage to places of least lithostatic load during post-salt deposition. The glacier represents a diaper which has penetrated to the surface in a desert mountain peak 4,300 ft high. (Aerofilms Ltd.)

Fig. 3.64.   Oblique aerial view of the surface expression of Avery Island salt dome, Louisiana. Near-surface solutioning has produced two lakes, in contrast to conditions in deserts where salt is more stable at the surface. (McLaughlin Air Service.)

where there is penetration of sea-floor sediments. In the Gulf of Mexico, for example, salt has formed a series of diaperic structures and topographic highs such as the Challenger Knoll (see Ballard and Feden, 1970; Kirkland and Gerhard, 1971).

Gypsum and shale may behave as mobile materials somewhat less readily than salt where there is simple lithostatic loading. The former are mobilized in orogenic areas where there are probably increased pore pressures or associated lithic stresses in addition to those induced by loading. Some shale is mobilized in salt piercement situations (cf. Fig. 3.63) but possibly only because salt was already moving. In any case, mobile masses of gypsum also produce structural and topographic effects similar to those of salt. Along the southern margins

of the Atlas Mountains in Algeria the writer has observed a number of piercement structures cored by vari-colored anhydrite masses which incorporate exotic blocks of igneous and metamorphic rock. Dome-like hills are one consequence, but of course calcium sulfate ($CaSO_4$) is less soluble than sodium chloride (NaCl) and is thus relatively stable where exposed to the Saharan environment. In somewhat analogous situations, Gilliland (1963) has described a mobile shale mass in the core of an anticline of the Central Rocky Mountains (Fig. 3.65), and Roach (1962) notes a similar feature in Louisiana. From this consideration of structure and topography produced in large measure by crustal loading, we turn to the complexities of orogeny and mountainous relief.

*Continental Accretion
and Orogenic Topography*

The usually linear portions of continents that presently display mountains seem to share at least one additional attribute. They apparently have all been affected by one or more of the several plate tectonic mechanisms—some of them relatively recently. Two separate geomorphic issues are involved here: the kinds of rocks incorporated and the types of stress to which the rocks were subject. Collectively, these factors establish a particular deformational style and consequent montane relief. We have already discussed the orogenic belts of the ocean floors that create dynamic relief

along the mid-ocean ridge and ocean trenches. Later in this chapter we will consider mountainous relief developed volcanically and by tectonic erosion as well as relief caused where continental interiors overlie sea-floor orogenic zones. Here, we will restrict our discussion to the broader tectonic aspects of orogenesis as apparently caused by collisions between continental blocks or encounters of such blocks with island arcs or oceanic trenches. The events in question eventually result in the broadening and thickening of continental crust—in short, *continental accretion* (Fig. 3.66). Chapters 4 to 8 deal with geomorphic mechanisms that occasionally produce residual (erosional) mountains. Chapter 10

Fig. 3.65.   Mountains of shale in the core of the Sanpete-Sevier-Valley anticline north of Glenwood, Utah. Protruding ridges are gypsum. Area has been structurally high since early Tertiary time due to shale flowage. (W. N. Gilliland.)

is devoted to the subaerial sculpture of orogenic belts, and Chapter 6 includes considerable discussion of the modes of sculpture of fault-block mountains.

One cannot become involved in the problems of orogenic geomorphology without becoming concerned with the bedrock fabrics that occur in or are even unique to orogenic zones. In their recent study, John Dewey and John Bird (1970) state, "Stratigraphic sequences of mountain belts (geosynclinal sequences) match those associated with present-day oceans, island arcs, and continental margins." With minor reservations, we agree. The reservations have to do with continental margins; the sediments around island arcs and on sea floors vary within much narrower limits.

In point of fact, we actually know very little about the comparatively complex sediment types along various continental margins around the world. In 1959, Drake,

Ewing, and Sutton discussed the sediments along the eastern North American seaboard and drew analogies between the materials on the shelf and slope and the miogeosyncline-eugeosyncline facies, respectively, of Marshall Kay's (1951) classic Paleozoic example, the Appalachian geosyncline (Fig. 3.67). There are rather widely accepted areas of correspondence. But there are also differences; for one thing, some Paleozoic miogeosynclines of North America contain more quartzose and less argillaceous sediment than present eastern shelf sequences exhibit. More to the point, continental margins differ greatly in probable long-term environmental norms (e.g. east versus west coasts), and these norms determine weathering modes and derivative sedimentary character. There are also obvious differences in *provenance* which range from shield and buried craton to more or less recent orogenic belts. Be that as it may, there

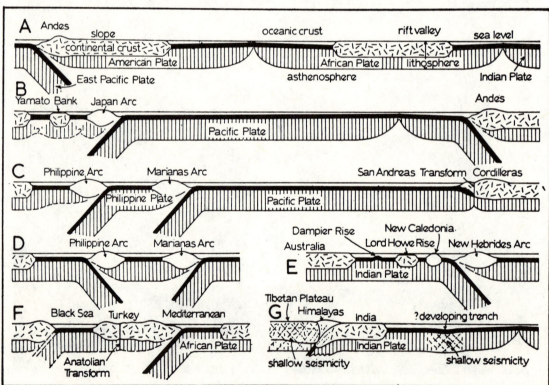

Fig. 3.66. Schematic sections showing crustal plate, ocean, continent, island-arc relationships. The (?) developing trench indicated in (G) is from Sykes (1970). (From Dewey and Bird, 1970.)

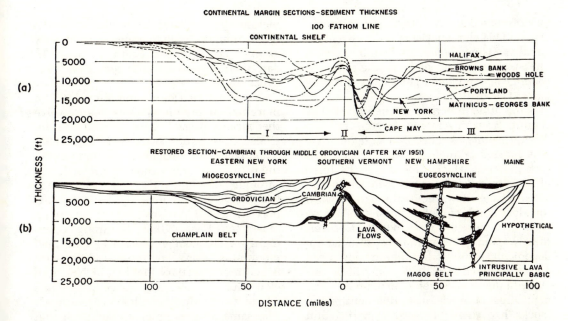

Fig. 3.67. Comparison of eastern North American continental shelf and restored geosyncline of the Paleozoic. (After Drake *et al.*, 1959.) (A) Present isopachous sections; (B) Appalachian geosyncline of Kay (1951).

are broad areas of lithologic correspondence in mountain systems, and these, rather than minor differences, will be emphasized.

The materials incorporated into marginal mountain systems by plate tectonics and continental accretion include (1) continental shelf sediments derived mainly from cratons and shields and, sometimes, from older orogens, and deposited along aseismic margins; (2) continental slope and ocean trench sediments accumulated along seismic margins; (3) material derived from developing orogens by particulate erosion and mass gravity sliding; (4) oceanic sediments with incorporated volcanic material (flows; guyots) translated into trenches during subduction; and (5) fragments of oceanic and continental type crust locally included during collisions between continents with irregular margins and/or intersecting island arc systems. Each of these materials can occur separately with the exception of crustal fragments (5), but typically they develop in combination with one another.

Marshall Kay noted in 1951 that materials accreted to the continental margins by orogeny often incorporated two principal lithic suites. He visualized them as accumulating in two essentially separate, parallel, sedimentary troughs (geosynclines), but we will here regard them as rock *facies*. The *miogeosynclinal facies* of Kay differs little from continental shelf sediments of type (1) in the preceding paragraph. Typically it includes ORTHOQUARTZITIC SANDSTONE, CARBONATE ROCKS, and minor SHALE. Subsidence is slight in the setting in question, the rock bodies are essentially tabular in shape and the principal lithic categories were described in the section dealing with cratons. The *eugeosynclinal facies* of Kay is said to comprise GRAYWACKE SANDSTONE,* BLACK SILICIC SHALE and CHERT, and VOLCANIC MATERIAL accumulated in elongate sedimentary masses, often of great thickness in areas of notable subsidence. These materials differ little from continental slope and ocean trench sediment of type

* For a lithologic description see next section.

(2) discussed in the previous paragraph. One notable characteristic of the eugeosynclinal sandstone bodies that commonly are interstratified with shale is their graded bedding and probable genesis as turbidity current deposits.

Some lithic materials frequently cited as coming from developing orogens are the detrital sediments called *flysch* and *molasse* plus masses of more or less consolidated rock material which slide under the force of gravity once they are sufficiently elevated. In our view, the distinction between this last kind of deformation and mass wasting is indeed a moot point. Moreover, the events in question may be initiated endogenically by uplift, but they are developed and terminated surficially and will be discussed later as mesogenic phenomena. The point must be made, also, that continental blocks like those of North America and Africa which are presumably being displaced by sea-floor spreading have trailing margins which are presently aseismic, have no associated trenches or volcanism. Hence, a typical eugeosynclinal lithofacies with its volcanic materials cannot now accumulate there. Neither can flysch or molasse facies (to the extent that they differ from eugeosynclinal rocks) unless the former are not actually dependent on a comparatively young mountain system with notable relict isostatic relief.

It would, in fact, appear that one set of lithic increments is added to the continental margin when the margin forms at a spreading zone with associated ridge-type faulting and volcanism. A second lithic set accumulates when the margin moves away from the spreading zone and becomes aseismic and maritime as opposed to continental. A third lithic set comes with the development of a marginal subduction trench and its volcanism and seismicity. A fourth set of lithic ingredients comes to continental margins with translation of oceanic sediments and volcanic materials into the trench by spreading and subduction. A fifth lithic set develops through plutonic assimilation in possible association with continental block or island arc collision. There igneous and metamorphic alteration products of the more common lithic increments previously assimilated in trenches begin to appear. To date there have been no recognizable beer cans.

According to Dewey and Bird (1970), marginal mountain systems that involve the various possible lithic ingredients include (1) a CORDILLERAN TYPE (Figs. 3.68; 3.69) where a trench develops along a continent margin and a mountain belt forms through dominantly thermal mechanisms with rising calc-alkaline and basaltic magmas (Andes Mountains) and (2) a COLLISION TYPE (Figs. 3.70; 3.71) through dominantly mechanical processes (a) with small mountains where continents collide with island arcs (Tertiary fold belt of Northern New Guinea) and (b) large mountains where two continents collide (Himalayas). The same authors go on to point out that cordilleran-type mountain belts have paired metamorphic belts (blueschist on the oceanic side and high temperature on the continental side), divergent thrusting, and synorogenic sediment transport from the high-temperature volcanic axis (cf. Figs. 3.69 and 3.78). They further maintain that collision-type mountains lack paired metamorphic belts but exhibit a single dominant direction of thrusting and synorogenic transport of sediment away from the side of the trench over the underthrust plate (Fig. 3.71).

In their study of continental margin sediments, Drake *et al.* (1959) conclude that "the major process necessary to convert the present continental margin [of eastern North America] into a mountain system is the one which thickens the crust under the outer, or eugeosynclinal trough. Since the miogeosyncline is already based on a crust of continental proportions, its deformation requires only a means of folding and thrusting the surficial sediments." It would now appear that the requirements they cite are met by the mechanisms of plate tectonics, particularly subduction and collision. From a geomorphologic standpoint, it is important to note that the initial orogenic relief related to continental accretion would be

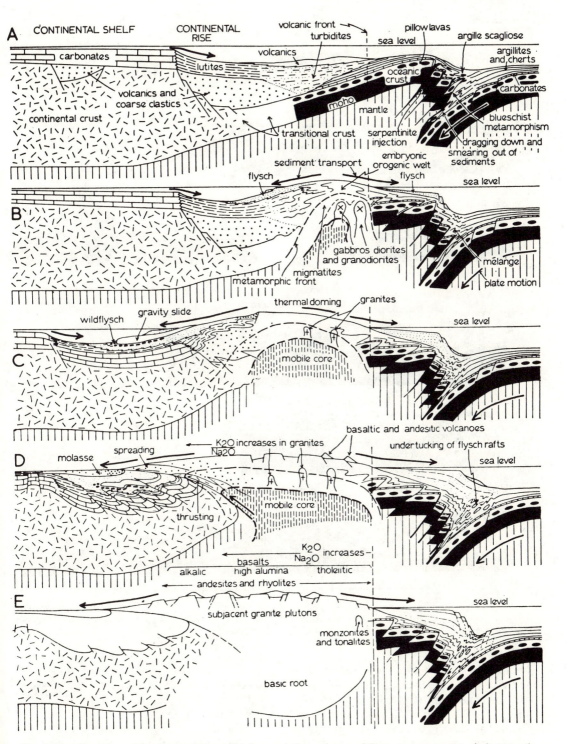

Fig. 3.68. Sequence of sections of a model for evolution of a cordilleran-type mountain belt (e.g., the Andes) developed by the underthrusting of a continent by an oceanic plate. (From Dewey and Bird, 1970.)

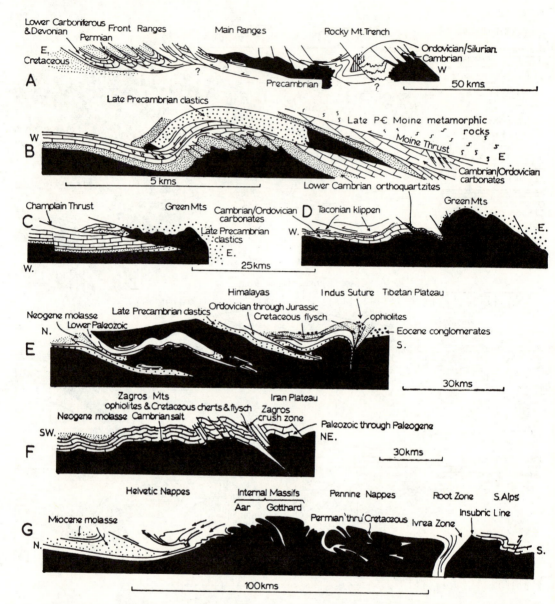

Fig. 3.69. Montane structural cross sections of (A) Rocky Mountains; (B) Scottish Highlands—northwestern orogenic margin; (C) and (D) western (Vermont) Appalachians; (E) Himalayas; (F) Zagros Mountains; (G) Swiss Alps, from various sources. (After Dewey and Bird, 1970.)

dynamic in character as sedimentary rocks and crust are foreshortened and plutonism with thermal doming is initiated (cf. Figs. 3.68B, 3.70C and 3.71C). Later, with crustal thickening, isostatic relief would become significant, and it would appear that ultimate balanced isostatic uplift cannot

occur until the dynamic compressional phase is essentially complete and crustal thickness is maximal (cf. Figs. 3.68D, E; 3.70D and 3.71D).

A second major geomorphic aspect of deformed orogens along continental margins relates to essentially surficial modifications of the rising rock mass. In part, of course, this involves classical subaerial mechanisms of erosion and deposition (cf. Chapter 10). It has recently become clear, however, that large-scale sliding of rock masses under the pull of gravity (tectonic denudation) is initially responsible for considerable montane sculpture. This phenomenon will be detailed in the next section. Moreover, in an orogen undergoing particle-by-particle erosion, compensative isostatic uplift will presumably be continual. But where large supraincumbent rock masses are quickly lost through sliding, uplift might appear to be sporadic. Also, each separate episode of crustal thickening to which an orogen is subject (probably at intervals of scores of millions of years) should renew uplift there. Such subsequent collisions would probably not affect the elevation of

a previously deformed belt already underlain by a thick crust.

A third major geomorphic aspect of continent margin orogens concerns the linear distribution of the several lithologic suites developed during the various orogenic phases, plus the larger structures that involve them. We will take up these matters now.

Rock Fabrics in Marginal Orogenic Belts. Materials incorporated into folded and faulted mountain belts include all three of the major groups of rocks. These three types have already been discussed in some detail with respect to shields and buried cratons, and it is only those aspects of bedrock fabric that are peculiar to orogens that will concern us here.

Sedimentary rocks related to orogens dif-

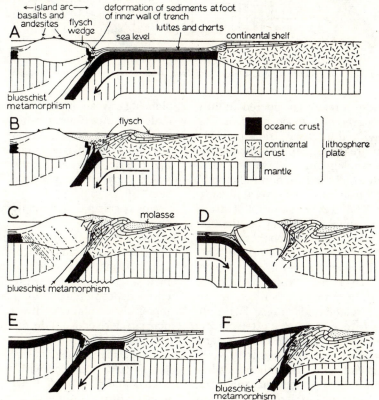

Fig. 3.70. Sequence of sections (A)–(D) depicting collision of a continental margin of Atlantic type with an island arc, followed by change in the direction of plate descent; (E) and (F) proposed mechanism for thrusting oceanic crust and mantle onto continental crust. (From Dewey and Bird, 1970.)

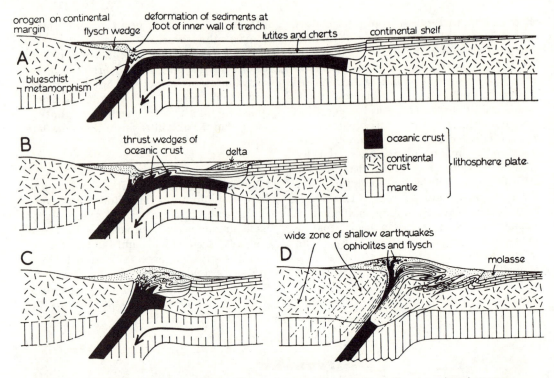

Fig. 3.71. Sequence of sections of the collision of two continents. (From Dewey and Bird, 1970.)

fer from those already considered mainly (1) in their tendency to be more or less tightly folded and intensely fractured, (2) in the often great sedimentary thicknesses of a single lithology, and (3) in the presence of a few lithofacies not normally found elsewhere. Consideration of several illustrations (Figs. 3.68 to 3.71) shows that stratified rocks accumulated along continental margins on shelves and in trenches and geosynclines (all of which are elongate) become laterally compressed during orogenesis and are frequently thrust outward to form a belt of imbricate shears, thrust sheets, and foreland folds (Fig. 3.72). Individual tabular rock bodies are often steeply inclined or even vertical and can be erosionally modified to form *hogbacks*, steep *cuestas*, and *flatirons* where there is some selective removal of weaker materials (cf. Fig. 2.43). Similar (high-dip) landforms are often encountered around high-relief cratonic uplifts (Figs. 3.73; 3.74). Argillite masses frequently function to localize shear

planes between blocks of more coherent material since they are impermeable and hence good pore pressure media. Erosionally breached fold belts occasionally yield the classic zig-zag mountain topography where they are etched by differential erosion (Fig. 3.75).

The principal sedimentary rock types present in orogenic belts, which do not occur in any abundance above basement on most cratons, include volcanic rocks, graywacke sandstone bodies, and siliceous sediments, all associated with the eugeosynclinal depositional realm. The terrigenous materials combine in flysch and molasse deposits derived from the developing orogen (cf. Figs. 2.43; 3.76).

Graywacke sandstone, according to Pettijohn (1957), is rock with a high detrital matrix and no cement, in which either feldspar or lithic fragments may dominate. It is noted by the same author to comprise a common constituent of *flysch* where it interstratifies with shale. Flysch often occurs in sedimentary masses tens of thousands of feet thick in geosynclinal settings, and at least some of these arenaceous beds (graded ones) are considered by many to be turbidite deposits. *Wildflysch* probably represents slumping of shelf facies into a trough in at least some facies. *Molasse* is noted by Pettijohn (1957) to comprise a combination of graywacke or subgraywacke sandstone and shale in a combination that is more arenaceous than flysch and in which there is notable conglomerate (cf. Figs. 3.77; 10.28).

There are almost as many definitions of graywacke sandstone, flysch, and molasse as there are instances of their use (cf.

Trümpy, 1960). In general we have tried for the consensus usage. And in the context of accretionary continental orogenesis, the flysch seems to accumulate mainly in the eugeosynclinal setting at the base of the continental slope and/or in a marginal trench where it may interstratify with pillow basalt, chert, and siliceous shale and incorporate thrust sheets and gravity slide masses to form *melange* facies (cf. Fig. 3.68). The molasse generally accumulates later than the flysch and in marginal orogenic troughs and foreland areas where it may be overthrust by flysch facies. Where both facies are well indurated, the molasse is normally erosionally more resistant than

Fig. 3.72.   Aerial view along the axis of the Saidmarreh (Kuh-i-Kailan) Anticline in Iran in the fold belt of the Zagros Mountains. The topography is clearly under "structural control." The foreground and background notches have been cut by transverse drainages (cf. Chapter 10, Fig. 10.31 and Fig. 2.43). (Aerofilms Ltd.)

Fig. 3.73. Aerial view of the Front Range of the Rocky Mountains from an altitude of 7,100 ft near Morrison, Colorado. The prominant foreground cuesta composed of upper Cretaceous, Dakota Sandstone steepens in the middle distance to form a pronounced hogback. Wedge-shaped masses of rock to the left are Pennsylvanian age Fountain Sandstone inclined upward toward a Precambrian crystalline highland. (T. S. Lovering, U. S. Geological Survey.)

Fig. 3.74. Flatirons developed from relatively resistant, massive sandstone on the east flank of the Navajo Uplift in southern Utah. Animals in foreground are goats. (W. N. Gilliland.)

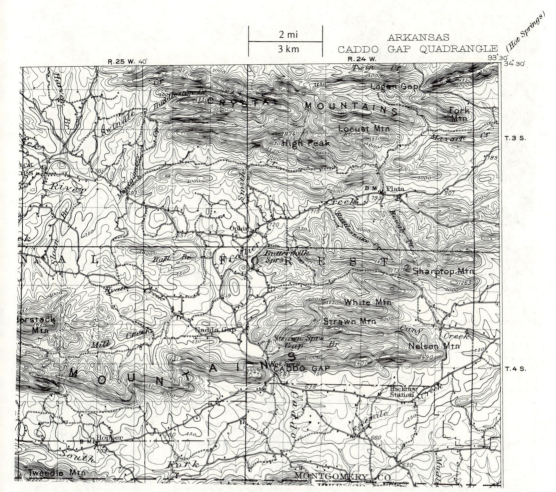

Fig. 3.75.   The northeast corner of the Caddo Gap 30′ Quadrangle Map showing zig-zag mountain topography developed by differential erosion of folded, steeply dipping sedimentary rocks in the Ouachita Mountains, Arkansas. Contour interval is 50 ft.

the flysch. Siliceous shale and chert are the erosionally most resistant eugeosynclinal sedimentary rocks and some deposits of each are very extensive—The upper Cretaceous Guayaquil Chert of Ecuador, for example, is 500–1,000 ft thick and crops out on a belt more than 100 mi long.

Of the various sedimentary rock suites encountered in orogenic belts along continental margins, relative to weathering and erosion, some comprise potentially resistant masses and some are comparatively weak. Chert (as on the craton) is generally tough, but graywacke sandstone contains meta-

stable rock and mineral fragments which are more chemically reactive than the quartz that dominates cratonic orthoquartzitic sandstone, and the graywacke nominally lacks cement. Shale and silt in orogenic zones have a tendency to be silicified and relatively resistant. Also, particularly in the axial portions of cordilleran-type orogens we encounter effects of heating, deformative stresses, and intrusions which bring about various degrees of metamorphism.

Metamorphic rocks in orogens may appear as lateral gradations of sedimentary

Fig. 3.76. Interbedded graywacke sandstone and shale of the Eocene Ancon Group (Seca and Socorro Formations) exposed in the sea cliff near Ancon Point, Ecuador; the rocks are cut by a high-angle reverse fault.

Fig. 3.77. Road-cut exposure of molasse exposed in the Mototan River valley, Estado Barinas, Venezuela, some 12 mi (19 km) downstream from the Mesa de Jajo (Fig. 4.28). Most of the material in view was apparently deposited under previously arid conditions.

facies or as distinct metamorphic zones such as the blueschist, high-temperature relations cited by Dewey and Bird (1970) that occupy essentially unique orogenic nitches (Fig. 3.78).

Blueschist metamorphic facies in cordilleran-type orogens are an alteration product of chert, argillite, alkalic basalt masses, plus carbonates which yields a high-pressure, low-temperature association of glaucophane, lawsonite, aragonite, and jadite.

High-temperature metamorphic facies in cordilleran-type orogens on the landward side of trenches and the blueschist zone comprise an association of andalusite, sillimanite, and cordierite mineral assemblages previously cited in metamorphic associations of shields and presumably of a somewhat similar origin.

When one considers the generally linear disposition of lithologies, structural zones, and metamorphic suites in orogenic belts developed along continental margins, it is hardly surprising that montane topography also tends to exhibit a notable linearity. There are departures from such patterns, of course, often caused by smaller plutons, though batholiths such as the Sierra Nevada in the western United States are distinctly linear in form. Foreland fold and thrust zones are generally marginal, least metamorphosed, and often quite heterogeneous lithologically. Axial zones of intrusion and metamorphism display generally increased integrity and resistance to subaerial processes. In addition, the axial zones tend to overlie the developing crustal root and hence undergo greatest ultimate uplift (Fig. 3.79). Apart from the precise changes wrought by surficial processes, therefore, the structural and lithologic essence of montane ridge-on-ridge topography exists prior to notable uplift.

Topography of Rifted Continental Rises

An obvious omission from our discussion of marginal continental orogenesis is the rifting that brings about block faulting and

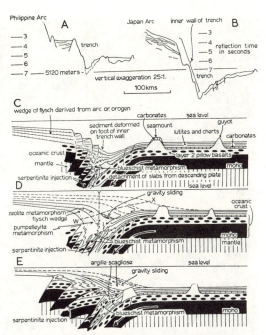

Fig. 3.78. Structures associated with trenches and trench metamorphic zones. (A) Manila Trench; (B) Japan Trench; (C)–(E) various relationships between structure, sediments, and metamorphism. (From Dewey and Bird, 1970.)

related mountainous topography. Beloussov (1969) has recently summarized the occurrence of such features, and numerous writers have noted that the probable prelude to sea-floor spreading in many instances is extension of mid-ocean ridge-type tectonic activity into the elevated portions of continents we call *continental rises* (cf. Fig. 3.5). Dewey and Bird (1970) illustrate the probable structural consequences as a series of tilted fault blocks (Fig. 3.80). Such a sequence of events may easily produce mountainous relief and should, in our estimate, be recognized as a form of orogeny but not one that is accretionary. It is, in fact quite the opposite. The present essentially buried basement topography of the eastern North American continental shelf, notably the Triassic fault block system, is one apparent result of continental rifting. It is of geomorphic interest, however, because counterparts exist elsewhere that are expressed as tectonic landforms.

Fig. 3.79. Exposure of the massive, crystalline rocks that comprise the Blue Ridge Mountains, Shenandoah National Park, Virginia, here mainly metamorphosed volcanics (greenschist facies).

Fault block mountains of the Germanic type have long been attributed to tensional stresses, and the Rhine Graben and Black Forest Horst comprise notable examples (Fig. 3.81), much modified by erosion of various kinds (Fig. 2.48). As noted by Wilson (1970), extensional rifting would be expected in an area of continental crust displaced athwart a mid-ocean ridge. This, of course, is the apparent deformational style of the Basin and Range physiographic province of the western United States (Fig. 3.83) and the rift valley systems of coastal Brazil (King, 1956), central Africa, and Israel. Moreover, portions of the mid-ocean ridge in the Indian Ocean can be traced into areas of similar structure and topography in the Gulf of Suez and the Red Sea (cf. Fig. 3.80). In that same region, an Ethiopian area known as the Afar Rift has been studied recently with great interest (Tazieff, 1970; Bonatti et al., 1971) with the conclusion drawn that rifting there is in progress and has produced a virtually unmodified tectonic landscape (Figs. 3.32; 3.91).

It should be noted that rifting associated with spreading centers appears to result in a series of ramp-like fault blocks all inclined in the same general direction on any given side of the spreading center as shown by Tanzieff (1970, p. 33) (cf. Figs. 3.80 and 3.82), whereas there is some suggestion that areas like the Basin and Range region involve alternate horsts and graben (Fig. 3.83).

Alternate horst and graben type orogenesis is also associated with continental margins now being rifted above a zone where a mid-ocean ridge extension is spreading. The well-known Los Angeles Basin fault system is one example. The related San Andreas transform fault has notably modified the Californian landscapes and life styles (cf. Fig. 2.14) and is continuing to do so. The San Andreas fracture has recently been said to reflect a shear between the Pacific and American plates. Similar movements have alternatively been related to transverse fractures of the East Pacific oceanic ridge though opinions differ as to whether such breaks penetrate the continental crust. Extensional stresses have already been noted in association with the margins of ocean trenches, and, along the west coast of South America where a trench and continent meet, the continental crust on the side near the trench has been intensely fractured and faulted. A series of horsts and graben have developed from southern Colombia to northern Peru, and several of the individual faults in Ecuador exhibit vertical displacements in excess of 8,000 ft (Fig. 10.10). Much of the relief so created during the Tertiary Period has been eliminated by sedimentation in the graben during the same period, but resistant Paleocene Azucar Sandstone that caps most of the horsts tends to stand higher than adjacent graben shale and marl (see also Figs. 3.85 and 3.86).

One additional orogenic deformational style should be noted in this section. There are close structural similarities between the southern Front Range Rocky Mountains and the Ozark Uplift even though the latter exhibits far less structural and topographic relief. Both regions exhibit thrust faults which appear to steepen downward (Fig. 3.84). In the Rocky Mountain Front Ranges, this *upthrusting* has created mountainous relief, possibly because the vertical movements occurred beneath an intracratonic basin which received several thousand

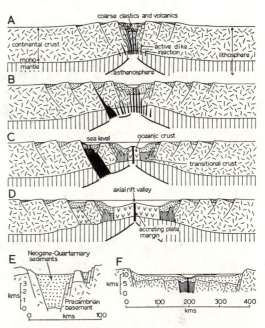

Fig. 3.80. Sections (A)–(D) illustrate a possible sequence of events in the rupture of a continent above a spreading center and early expansion of an ocean; (E) section across the Gulf of Suez (Picard, 1966); (F) section across the Red Sea (Girdler, 1966). (From Dewey and Bird, 1970.)

feet of sediment during the late Paleozoic and Mesozoic. The same high-angle rifting in Missouri affected a much thinner sedimentary section.

We are uncertain about the relation between upthrust orogenesis and particular plate tectonics mechanisms. But there are more than fifty mountain peaks in the state of Colorado that rise above 14,000 ft elevation, and the implications of crustal thickening seem clear. One cannot help but wonder if the postulated westward movement of North America which supposedly began about 200 million years ago may not have thrust the continental crust over a second sliver of the same composition or a massive sediment wedge lying to the west or possibly even both. Jurassic emplacement of the Sierra Nevada batholith shortly after the North American continent showed signs of movement in the Triassic could hardly be a fortuitous happening. Could it? Nor, for that matter, can the continental flexing that

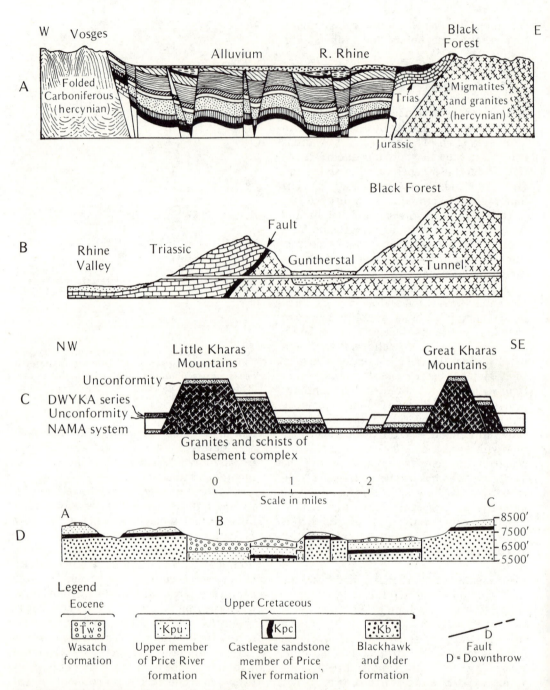

Fig. 3.81. Cross sections of several fault block mountain systems compiled from various sources. (A) The Rhine Graben north of Mulhausen (from Holmes, 1965)—length of section about 30 mi (42 km); (B) section across the boundary fault between the Black Forest (horst) and Rhine Graben revealed in a railway tunnel (after Hans Cloos); (C) horsts and graben of the Kharas Mountains, southwest of the Kalahari Desert, South West Africa (after C. M. Schwellnus)—length of section about 60 mi (84 km); (D) horsts and graben near Musinia, Sevier County, Utah, the Basin and Range Province (after Spieker and Baker).

Fig. 3.82.   En-echelon fractures that mark this seaward reach of the Afar Rift, the south end of the Red Sea, are believed to signify an emergent crustal spreading area. The resulting terrain is a series of inclined, ramp-like blocks. (Haroun Tazieff.)

caused the Upper Cretaceous epicontinental seaway be entirely divorced from the same Mesozoic chain of events. Perhaps an ocean trench forming under a continent looks like such a downwarp. These matters surely merit further study.

MESOGENIC TOPOGRAPHIC EFFECTS

A number of mechanisms that affect landscapes are initiated within the earth but attain completion at or above the surface of the lithosphere. Notable in this regard are some of the gravitational aspects of tectonism and volcanism. We will here consider these phenomena in that order and by so doing will be dealing with a group of attributes commonly associated with some

forms of orogenesis. Clearly, the effects are confined to neither the continents nor the ocean basins.

Gravity Tectonics and Topography

During our discussion of endogenic tectonism we made passing reference to sliding of portions of the lithosphere under the influence of gravity. Perhaps the first to cite this kind of structural event were geologists working in the Alps (see Fig. 3.69). Trümpy (1960) has summarized the alpine studies that concluded that many of the enormous recumbent folds (*nappes*) underlain by *décollement* zones there are in fact gravity slide masses which have moved from axial orogenic zones. It is evident that the sliding of portions of the crust requires

Fig. 3.83.   The crest of an almost buried range of hills associated with a probable horst just east of the Amargosa Mountains, Nevada. Block faulting is believed to characterize the Basin and Range Physiographic Provence.

initial relative uplift of the region from which the allochthonous masses of rock are to move (cf. Maxwell, 1959). Thus, the initial energy input into this type of tectonic system is that of orogenic uplift, and it is definitely of endogenic origin. The second increment of deformative force is, however, supplied by gravity acting essentially at the earth's surface and opposed by inertia and friction. The rock masses are variously transported through the atmosphere and/or hydrosphere in a realm that borders if it does not include the exogenic. It is these environmentally transitional events we here term *mesogenic*.

Lithic masses that have moved by gravity near the earth's surface all share the properties of low-temperature deformation. Many are tightly folded, fractured, sheared, and otherwise deformed. But they are rarely

metamorphosed. The only metamorphic fabric frequently encountered, and that on a local basis, is slaty cleavage. In a preliminary study of the Ouachita foreland thrust and fold belt and axial orogenic zone, Jackson (1967) was unable to detect even the common low-temperature zoeolite minerals that would indicate deformative temperatures of no more than about 300° C. It is the evidently shallow character of gravity deformation, and the low associated temperatures plus the lack of basement rock involvement, that has earned for it the designation "thin-skinned tectonics." The surficial gravity events are commonly associated with deeper-seated tectonic events tied to marginal continental accretion which occur with metamorphism at much higher temperatures. But the association of slaty cleavage with otherwise unmetamorphosed rocks in the Ouachita Mountains and in the Ordovician, Martinsburg Shale of the central Appalachian Mountains, may possibly be accounted for by abnormal pore pressures developed during gravity sliding (cf. Maxwell, 1962).

The materials involved in gravity sliding seem to range from hardly consolidated sediments to thoroughly indurated rock and from sheets of stratified material miles in length and breadth to isolated, more or less detached blocks. As noted by Dewey and Bird (1970), later gravity slides may incorporate plutonic and metamorphic materials previously formed at depth. Gravity tec-

tonics may be associated with either cordilleran-type or collision-type mountain systems, and the materials so transported tend to occur in areas marginal to the main orogen axis.

It seems evident that the removal of lithic material from a rising orogenic belt by the force of gravity is an important aspect of montane sculpture. Thus, it is not surprising to read that workers refer to areas from which slide masses originate as having undergone *tectonic denudation*. Depending upon the nature of the décollement zone as noted by Kehle (1970), sliding may be initiated on slopes as low as 0.1°. Therefore,

Fig. 3.84. East face of the Grand Tetons whose crest is more than 6,000 ft above Jackson Hole, Wyoming. As the eroded margin of an up-thrust block of crust, the Tetons have a western ramp-like flank of much less imposing nature. Foreground lodge-pole pine forests are developed on glacial moraine, fronted by two terrace levels of the Snake River.

tectonic denudation could begin to affect an orogen as soon as it becomes expressed as a positive structural-topographic element (see also Winterer, 1968). The writer has under study a portion of Guayas, Province, Ecuador, where there is an allochthonous mass of Paleocene estuarine, quartzose sandstone, and conglomerate (Azucar Group). This shallow-neritic to brackish-water deposit, now isoclinally folded and intensely sheared, is locally exposed over an area in excess of 10,000 sq mi in Ecuador on a series of horsts located on the coastal lowland (Fig. 3.85). Elements probably belonging to the same or a similar slide mass occur in adjacent Peru as far south as Talara. The allochthon(s) apparently slid from the axial zone of the western South

America, lower Eocene continental shelf (Andean miogeosyncline) into an island arc region lying to the west (a distance of as much as 150 mi or more). The early Eocene slide event probably marks initial denudation of the rising Andes orogen in that area, probably during a continent-island arc collision. The sliding may largely have occurred under water. This is one of the larger allochthons on record and certainly involves one of the greater displacement distances (see also Chapter 10).

The reader is by now aware that the distinction between tectonic denudation and mass wasting is neither sharp nor widely agreed upon. One could argue that the slide masses involved in gravity tectonics are larger and more coherent and move more

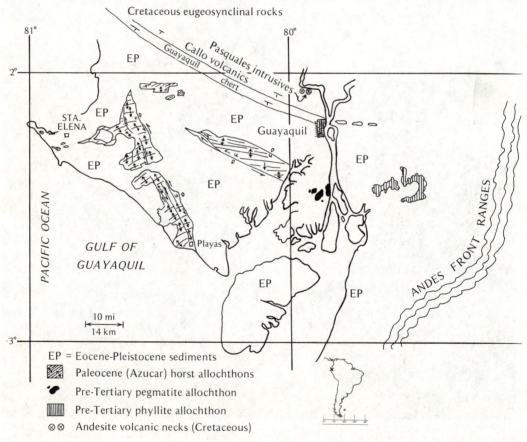

Fig. 3.85. Tectonic map, Guayas Province, Ecuador (location map inset), shows the major exposed elements of the Azucar allochthon and the fold and shear axes of these subsequently block-faulted units. (For further details see Figs. 10.15 and 10.16.)

Fig. 3.86. Flat-lying, Pleistocene Tablazo Limestone resting in angular unconformity on contorted and shattered wildflysch related to the Paleocene Azucar allochthon on the Sta. Elena Peninsula, southwest Ecuador.

slowly than masses affected by mass wasting. But the tectonic allochthons may move quite rapidly, whereas some forms of mass wasting are agonizingly slow. Kehle (1970) cites probable velocities of 1 km/10 years (100 m/year) for a slide on a salt bed to 10^{-4} km/million years if the décollement occurs in crystalline rocks. For these values, he assumes Newtonian behavior in which the equation relating the velocity to the controlling factors is

$$u = \{pg \sin \theta / \mu\}\{t(2d + t)\}$$

where p is the density of the material, g the gravitational constant, d the depth of burial to the top of the décollement zone, t the thickness of the décollement zone, θ the angle of dip of bedding, and μ the viscosity of the décollement zone.

Velocities may be even more rapid than Kehle suggests, for allochthonous slide masses such as that in the Ecuadorian example probably moved over an unconsolidated mud bottom in water. Indeed, velocity differences do not appear to provide a suitable criterion for differentiating between tectonic denudation and mass wasting. Coherence or size of the slide masses seem unlikely criteria either. The sliding masses bottomed by décollements are noted by several writers to have a bulldozer effect, ploughing up materials in their path and creating the chaotic deposit known as wildflysch which is in many ways similar to landslide debris (Fig. 3.86). In addition, as noted by Pierce (1957; 1960), in the Heart Mountain thrust, slide blocks may become detached (Fig. 3.87), thereby generating isolated lithic masses of considerable dura-

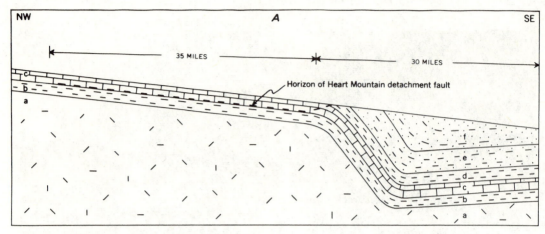

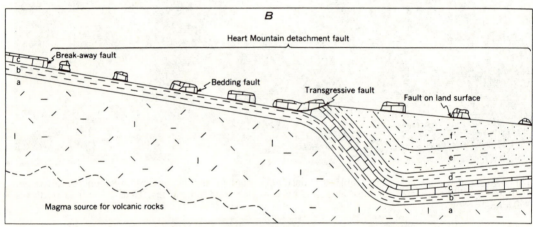

Fig. 3.87.   Cross sections illustrating the formation of the Heart Mountain detachment fault; (A) before faulting and (B) after last movement showing "break-away." a, Precambrian; b, Cambrian; c, Ordovician, Devonian, and Mississippian; d, Pennsylvanian, Permian, Triassic, and Jurassic; e, Cretaceous; f, Tertiary. (After Pierce, 1957.)

bility. Chief Mountain in Wyoming and the "rock" of Gibralter are among the more notable possible examples of detachment (Fig. 3.88).

It seems clear that gravity tectonics may or may not directly generate appreciable relief in the areas where the allochthonous masses come to rest. If slide blocks enter deeply subsiding troughs, they merely add to the sediment fill. However, the Helvetic Nappes north of the Swiss Alps seem to be allochthons of gravity tectonic origin which came to rest on a comparatively firm base-

ment, and they presently exhibit notable mountainous relief (cf. Fig. 3.69). It is also apparent from research to date that gravity tectonics has the significant geomorphic effect of reducing mountainous relief in much the same fashion as other erosional mechanisms. The materials transported tend to accumulate in zones marginal to the orogen so that the area of intense deformation is effectively broadened (cf. Winterer, 1968). However, it appears that most of the slide material is later subject to minimal isostatic uplift since it does not come to rest over the main zone of newly thickened crust.

Tectonic denudation may accelerate orogenic uplift by at least an order of magnitude during the early uplift phase. The allochthon previously cited from Ecuador averaged 7,500 ft thick prior to buttressing and contortion. If contiguous with the allochthonous mass of the same age in northern Peru, some 20,000 to 30,000 cu mi of rock that accumulated as"miogeosynclinal" sediment in Paleocene time had completed tectonic transport to the west of the Andean axis prior to the deposition of middle Eocene shale and limestone in the area in which the allochthon came to rest. There is evidence that similar allochthons also moved to the east from the same axial Andean region. A consideration of the geologic time scale for the Eocene (Chapter 1) suggests that some 5 to 7 million years are available for the western Ecuadorian tec-

tonic denudation events, but they may well have occurred within a few thousand years.

Volcanic Geomorphic Mechanisms

As a surficial expression of plutonism, volcanism shares with gravity tectonics the distinction of being initiated within the earth and culminating on the outside. The resulting topography incorporates materials with endogenic attributes (*lava*) as well as matter fragmented by the eruptive process (*pyroclastics*). Eruptions occur on the continents as well as in the ocean basins, but the vast majority of the continental eruptions are marginal (Fig. 3.89) and of oceanic lithologic makeup (basalt and andesite). The extrusive materials usually gain access to the surface of the lithosphere via more or less distinct vents or along essentially contiguous fissures, and each class

Fig. 3.88. Gibralter, a probable remnant of a detachment allochthon which slid under the pull of gravity into its present location, possibly from an easterly source. It consists of a mass of Lower Jurassic limestone above Upper Jurassic shale which, in turn, overlies Eocene Flysch. The steep eastern sea cliff is 1,400 ft (425 m) high. (*The Times, London*).

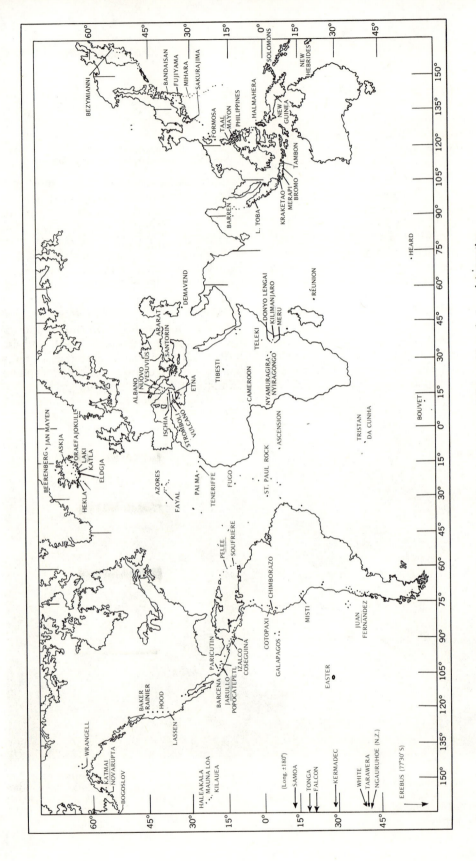

Fig. 3.89.   Distribution of the major active volcanoes of the world, exclusive of Antarctica.

of eruptive orifice tends to generate a distinctive group of landforms modified by the media (air, water) in which the eruption occurs (cf. Holmes, 1965; Green and Short, 1971; Longwell *et al.*, 1969).

Subaqueous Volcanic Topography

Essentially all subaqueous volcanism is also submarine, and in earlier sections of this chapter there is passing reference made to such activity. Examinations of bottom cores and deep sea drilling operations suggest that extrusive volcanism on the deep ocean floors is uncommon away from the mid-ocean ridges and rather well-defined vent eruptions. It is rare for a sea-bottom core to include flow basalt layers. Flow basalt emitted from fissures on the ocean floors appears to be common in the mid-ocean ridge zones, and bottom photos often illustrate its pillow structure (Fig. 3.90). As previously noted, Menard (1969) attributes a good bit of the mid-ocean ridge's elevation to this type of activity plus vent eruptions. Extensive flows are recorded in what appear to be subaerial portions of the mid-

Fig. 3.90. Ellipsoidal "pillow" masses of submarine lava on the flank of the South Pacific—Antarctic ocean ridge at a depth of 9,184 ft (2,800 m), longitude 145° W, Latitude 56° S. Similar extrusive basalt masses are common to sea-floor spreading areas of subaqueous eruption. (Smithsonian Oceanographic Sorting Center.)

ocean ridges in Iceland and the Afar Triangle (Fig. 3.82). If anything, such emissions tend to have a leveling effect on topography.

Submarine vent eruptions are, of course, not limited to the mid-ocean ridge. Sea-floor spreading also displaces vent deposits (Vogt and Conolly, 1971), and it is uncertain how the vent remains in association with the magma chamber unless the latter is included within the moving plate. Geomorphologically, such vent eruptions are of two types; those that develop entirely under water and those that penetrate through to the atmosphere. The majority of the flat-topped under-sea mounts (*guyots*) are believed to be volcanos (cf. Hess, 1946). For some decades after their discovery, their flat tops were ascribed to a combination of erosional truncation (presumably by waves) and sedimentary leveling by burial. Almost certainly there are guyots with level upper surfaces due to one or both of these phenomena. However, Tazieff (1970) illustrates two volcanoes in the newly emergent Afar Rift that exhibit essentially flat tops, apparently as a direct consequence of the underwater eruptive process (Fig. 3.91). A slight concavity is present on their upper surfaces but not the usual volcano crater. And it may well be shown that many if not most guyots are shaped in this manner.

On occasion, continued eruptions in the vicinity of a single vent will build the eruptive mass above sea level. Maintenance of a vent above a magma chamber during plate movements is, at least, puzzling. Sub-

Fig. 3.91. Mount Asmara, a flat-topped volcano in the Afar Rift on the Red Sea, is composed of volcanic glass shards of a kind formed during underwater volcanic explosions. It resembles the many flat-topped, under-sea mounts (guyots) and is probably an emergent example, suggesting that the configurations of such features are eruptive in origin rather than erosional or depositional. (Haroun Tazieff.)

aerial volcanic topography is one expected consequence, and this usually makes islands (cf. Hawaii). Of equal interest, however, are those volcanic masses that penetrate close enough to the ocean surface to permit sessile marine organisms to become established in organic reef communities which then form *fringing reefs* and *atolls* (Fig. 9.43). Vent deposits are noted by Cotton (1969) to include both pyroclastic material and lava. As one would expect, the vent eruptions that include lava are more resistant to wave erosion than are those composed entirely of ash and cinders. Good conformance to this relation was recently demonstrated during the eruption of the volcano Surtsey, near Iceland where one ash-cinder vent was quickly destroyed by wave action, whereas the pyroclastic/lava composite vent deposits formed a more durable island.

Subaerial Volcanic Topography

Volcanic topography developed under atmospheric conditions and particularly that which affects the continents is largely concentrated where mid-ocean ridge and ocean trench tectonism appears to infringe upon continental crustal blocks. Of course, one cannot always be certain as to which oceanic mechanism was involved in ancient examples, but the types of materials erupted are suggestive as noted by Lipman *et al.* (1971). Much of the basaltic volcanism in the Mediterranean region and in the African Rift Valley system is believed to relate to oceanic type rifting. The basaltic volcanism of the Basin and Range province of the southwestern United States was probably of similar genre. Arcuate strings of volcanic peaks in the oceans can clearly be related to ocean-trench, island-arc configurations, and Andes Mountain volcanism is evidently trench related.

Lavas extruded in trench settings tend to be andesitic in composition. The world's highest volcano, Mt. Chimborazo (Fig. 8.2),

and the world's highest active volcano, Mt. Cotapaxi (Fig. 3.92), are only two of about a dozen major vent accumulations in the Ecuadorian Andes (Fig. 3.93). The 10,000- to 12,000-ft-thick piles of lava and pyroclastic material there rest on a contorted and metamorphosed mass of rock of nearly 10,000 ft elevation within the Andean orogen bordering the Peru-Chili trench. Some of the most violent volcanic eruptions on record (Krakatoa, Mt. Pelee) were situated in island arc zones along trenches. This type of volcanism clearly contributes notable relief to orogenic belts and of course influences the bedrock fabric (cf. Lipman *et al.* 1971). The expulsion of lava may also lead to *Caldera Subsidence* with associated fault topography and ring-dike complexes (cf. Chapman and Chapman, 1940). Pyroclastic materials are well known for the ease with which they decompose to form fertile soils. Lavas exhibit their own upper surface flow topography (Fig. 3.94) and internal features such as *lava caves*. Flow features naturally tend to develop in the depressions that flows tend to follow. Such lava is generally more resistant to both weathering and erosion than is pyroclastic debris with the exception of welded tuff (*ignimbrite*), and differential erosion of volcanic deposits often generates distinctive landforms (cf. Fig. 3.95).

Classically, the volcanos with the steepest sides are presumed to be *cinder cones* (Fig. 2.41); *composite cones*, previously discussed (Figs. 3.20; 3.93), are of intermediate declivity, and *shield volcanos*, such as Mauna Loa in Hawaii, are composed largely of lava and are broad and of gentle slope (Fig. 3.96). Actually, *spatter cones* built from rapidly congealing lava blobs may have nearly vertical side slopes or even overhangs and are thus often considerably steeper than pyroclastic debris slopes. Some composite types are also exceedingly steep.

As previously noted, fluid extrusive volcanic material tends to accumulate in low places with a consequent leveling effect upon topography. The extreme expression of such leveling is the *lava plateau*, usually

Fig. 3.92. Cotopaxi, the world's highest active volcano whose ice-capped crest rises 19,339 ft (5,896 m) above the adjacent Pacific Ocean in Ecuador less than 1° from the Equator.

Fig. 3.93. The twin volcanic peaks of Illinesa rising more than 4,000 ft above the Altiplano south of Quito, Ecuador.

Fig. 3.94.   Lava-flow topography developed in association with the Craters of the Moon, southern Idaho. In the foreground is the rough, blocky surface of *Aa* type lava and beyond is the smooth, billowy surface of *pahoehoe* type lava backed by pyroclastic topography.

formed by fissure lava eruptions. One of the largest is in Washington and Oregon. There, the Columbia Plateau has an estimated average lava thickness of 0.5 mile and an area of about 200,000 square miles. Similar basalt plateaus include the Antrim in north Ireland (Fig. 3.97) and the Deccan "traprock" plateaus of India and Africa. We are admittedly uncertain as to the tectonic regimen that leads to such extensive outpourings of oceanic-type igneous material on the continents. Most of the deposits cited are close to the continental margins. Furthermore, in the North American example cited, radiometric ages of the basalt indicate a late Miocene eruption period. It is

probably no accident that the Columbia Plateau basalt, like the Basin and Range block faulting, occurred when western North America was presumably encountering the mid-ocean ridge in the Pacific (the East Pacific Rise).

A major bedrock fabric attribute of certain lava flows is columner jointing produced by intersecting cooling fractures. These 3-, 4-, and 5-sided polygonal fractures control the size of blocks yielded via weathering and erosion (Fig. 3.98). They typify thicker (>3 m) flow tongues and sheets. Where the vertical fractures are intersected by more or less flat-lying, closely spaced partings, rather equidimensional blocks may result. Widely spaced fractures cutting the polygonal pattern at a high angle often produce a series of post-like blocks. A very coarse talus commonly results. Columner fracturing is probably more common in basalt than in acidic rocks but is known in both types of material (Fig. 3.99). Consideration of flow lines in a rhyolite (Fig. 3.24) shows that the polygonal fractures are more or less independent of flow structure.

Fig. 3.95. A volcanic neck south of Mexican Hat, Arizona. Note the two near-vertical dikes cutting the essentially flat-lying strata in the foreground escarpment. (W. N. Gilliland.)

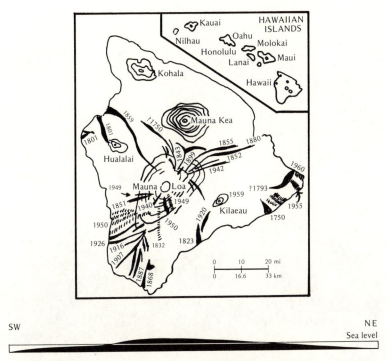

Fig. 3.96. Map and natural-scale cross section of the island of Hawaii and Mauna Loa, the type example of a shield volcano. The map shows lava flows from 1750 to 1960. Length of the section is 21 mi (33.6 km). (After Holmes, 1965.)

HOW IT ADDS UP

The stance taken in this chapter is consistent with the geomorphic philosophy developed historically in Chapter 1. We can only trust that these consistencies are not foolish for, as has been noted, foolish consistencies are the hobgoblins of little minds. The earth's rocks are made and moved in fashions largely mysterious, but fashions that more and more seem to be systematic. Some of the more important events appear to be taking place in the mantle, beyond our grasp, and were it not for the meanderings of geophysicists, beyond our reach as well. Other happenings of import occur beneath the oceans and are expressed as topography to the acoustic "eyes" of oceanographers. The tectonics of plates is a fascinating jigsaw puzzle with many of the nobler attri-

butes of a floating crap game. Like the game, plates form only to disappear and reform elsewhere. It may all eventually prove to be merely a charade, but if so, at least it is well organized.

Not only is the plate tectonic scheme dynamically and aesthetically satisfying, it is useful to geomorphologists whose aim is to interpret landforms. The continents and ocean floors have long been known to differ in kind. With the new influx of data we are finding out how much they differ and suggestions as to why this is so. Submarine geomorphology clearly has its own special problems. But these problems cannot be ignored by geomorphologists whose interests lie elsewhere, for solutions to plate tectonic enigmas relate to the development of topog-

Fig. 3.97. The Antrim (Thulean) Lava Plateau, Northern Ireland, seen looking eastward from above the Giant's Causeway (cf. Fig. 3.98). In the sea cliff are seen three tiers of basalt columns related to three separate flows. Subjacent lavas are not columnar. Note the cuspate erosional form of this sea coast, discussed in Chapter 9. (Aerofilms Ltd.)

Fig. 3.98.   Polygonal fracturing displayed by the upper ends of basalt columns in the Giants Causeway near the shore, County Antrim, Northern Ireland. The white rind is, in part, salt left by evaporation in the slight depressions (some here contain water) which characterize many of the columns on top (cf. Fig. 3.97).

raphy and rocks of the lands. Few problems of continental geomorphology entirely escape the implications of plate tectonic theory. Actual effects of sea-floor spreading appear to penetrate the very continental heartlands, at some time in the past, if not currently.

Tectonism enters our geomorphic constructs both statically and dynamically. And one of the unanticipated by-products of plate tectonic findings is a better understanding of the continental distribution of structural activity and deformative styles, lithologies, and relief as these relate to geomorphology. It no longer seems peculiar that one mountain system should be seismic and another aseismic or that one craton is

immobile whereas another appears to be breaking up. Plate tectonics wouldn't have it any other way.

We have talked about the way rocks are made within the earth and how they move toward or to the surface. And as rocks leave their formative environments, new stresses develop which will leave them changed. The transition when they are "kicked out-of-doors" is seen to vary from abrupt to gradual. But the ultimate contrasts are strong, for like the aged, the earth may have snow on the roof but there are "fires" within. We have scrutinized the "fires." Now we must consider the snow and whatever goes with it.

Fig. 3.99. Columnar fracturing developed in rhyolite in the wall of Yellowstone River Canyon, Wyoming. The scree slope below includes polygonal blocks from the igneous mass plus material from the subjacent conglomerate.

REFERENCES

Atwater, G. I. and Forman, M. J. (1959) "Nature of Growth of Louisiana Salt Domes and Its Effect on Petroleum Production," *Amer. Assoc. Pet. Geologists Bull.*, v. 43, pp. 2592–2622.

Balk, Robert (1937) "Structural Behavior of Igneous Rocks," *Geol. Soc. Amer. Mem.* 5.

Ballard, J. A. and Feden, R. H. (1970) "Diaperic Structures on the Campeche Shelf and Slope, Western Gulf of Mexico,"

Geol. Soc. Amer. Bull., v. 81, pp. 505–512.

Barton, D. C. (1933) "Mechanics of Formation of Salt Domes with Special Reference to Gulf Coast Domes of Texas and Louisiana," *Amer. Assoc. Pet. Geologists Bull.*, v. 17, pp. 1025–1088.

Beloussov, V. V. (1969) "Continental Rifts" [in] *The Earth's Crust and Upper Mantle*, P. J. Hart, ed., Wm. Byrd Press, Richmond, Va., esp. pp. 539–544.

Benioff, H. (1949) "Seismic Evidence for the Fault Origin of Oceanic Deeps," *Geol. Soc. Amer. Bull.*, v. 60, pp. 1837–1856.

Berg, R. R. (1962) "Mountain Flank Thrusting in Rocky Mountain Foreland, Wyoming and Colorado," *Amer. Assoc. Pet. Geologists Bull.*, v. 46, pp. 2019–2032.

Bird, J. M. and Dewey, J. F. (1970) "Lithosphere Plate-Continental Margin Tectonics and the Evolution of the Appalachian Orogen," *Geol. Soc. Amer. Bull.*, v. 81, pp. 1031–1060.

Blackett, P. M. S. (1961) "Comparison of Ancient Climates with the Ancient Latitudes Deduced from Rock Magnetic Measurements," *Roy. Soc. Proc. v.* A 256, pp. 1–30.

Bonatti, E., Emiliani, C., Ostland, G. and Rydell, H. (1971) "Final Desiccation of the Afar Rift, Ethiopia," *Science*, v. 172, pp. 468–469.

Bornhauser, Max (1969) "Geology of Day Dome (Madison County, Texas)—A Study of Salt Emplacement," *Amer. Assoc. Pet. Geologists Bull.*, v. 53, pp. 1411–1420.

Bowen, N. L. (1928) *Evolution of Igneous Rocks* (reprinted 1956 Dover, N.Y.), 332 pp.

Chapman, R. and Chapman, C. (1940) "Cauldron Subsidence at Ascutney Mountain, Vermont," *Geol. Soc. Amer. Bull.*, v. 51, pp. 191–212.

Clark, T. H. and Stearn, C. W. (1968) *The Geological Evolution of North America*, 2nd ed., The Ronald Press Co., New York, 570 pp.

Cloos, Hans (1936) *Einführung in die Geologie*, Gebrüder Borntraeger, Berlin, 356 p.

Coats, R. R. (1962) "Magma Type and Crustal Structure in the Aleutian Arc," *Amer. Geophys. Union Mono.* 6, pp. 92–109.

Cotton, C. A. (1969) "The Pedestals of Oceanic Volcanic Islands," *Geol. Soc. Amer. Bull.*, v. 80, pp. 749–760.

Deffeyes, K. S. (1970) "The Axial Valley: A Steady-State Feature of the Terrain" [in] *Megatectonics of Continents and Oceans*, H. Johnson and B. L. Smith, eds., Rutgers Univ. Press, New Brunswick, N.J., esp. pp. 194–222.

Dewey, J. F. and Bird, J. M. (1970) "Mountain Belts and the New Global Tectonics," *Jour. Geophys. Res.* v. 75, pp. 2625–2647.

Dietz, R. S. (1961) "Continent and Ocean Basin Evolution by Spreading of the Sea Floor, *Nature*, v. 190, pp. 854–857.

Dietz, R. S. (1970) "Continents and Ocean Basins" [in] *Megatectonics of Continents and Oceans*, H. Johnson and B. L. Smith, eds., Rutgers Univ. Press, New Brunswick, N. J., esp. pp. 24–46.

Dietz, R. S. and Holden, J. C. (1970) "The Breakup of Pangaea," *Scientific American*, v. 223, pp. 30–41.

Drake, C. L., Ewing, M. and Sutton, G. H. (1959) "Continental Margins and Geosynclines: The East Coast of North America North of Cape Hatteras" [in] *Physics and Chemistry of the Earth*, L. H. Ahern, F. Press, K. Rankama, and S. K. Runcorn, eds., Pergamon Press, London, esp. pp. 110–198.

Dury, G. H. (1966) "Duricrusted Residuals on the Barrier and Cobar Pediplains of New South Wales," *Geol. Soc. Australia Jour.*, v. 13, pp. 299–307.

Eardley, A. J. (1951) *Structural Geology of North America*, Harper and Bros., N.Y., esp. p. 183.

Fisher, R. L. and Hess, H. H. (1963) "Trenches" [in] *The Sea*, John Wiley and Sons (Interscience), New York, v. 3, pp. 411–436.

Foose, R. M., Wise, D. U. and Garbarini, G. S. (1961) "Structural Geology of the Beartooth Mountains, Montana and Wyoming, *Geol. Soc. Amer. Bull.*, v. 72, pp. 1143–1172.

Friedman, Melvin (1964) "Petrofabric Techniques for the Determination of Principal Stress Directions in Rocks" [in] *State of Stress in Earth's Crust* (Int. Cong., Santa Monica, Calif.) Elsevier Press, New York, pp. 450–552.

Garner, H. F. and Gibbons, J. F. (in press) "Fractures in Gettysburg Diabase," *New Jersey Acad. Sci. Bull.*,

Garner, H. F. (1967) "Batesville-Moorefield Stratigraphy and Sedimentation in Arkansas," *Geol. Soc. Amer. Bull.*, v. 78, pp. 1233–1246.

Gibbons, J. F. (1967) "Shear and Tension Fracture Patterns in Northwest and Northcentral Arkansas," *Ark. Acad. Sci. Proc.*, v. ix, pp. 32–37.

Gibbons, J. F. (in press) "Tectonics of the Eastern Ozarks," *Geol. Soc. Amer. Bull.*

Gignoux, Maurice (1955) *Stratigraphic Geol-*

ogy, W. H. Freeman, San Francisco, esp. pp. 1–5.

Gilliland, W. N. (1963) "Sanpete-Sevier Valley Anticline of Central Utah, *Geol. Soc. Amer. Bull.,* v. 74, pp. 115–124.

Gilliland, W. N. (1973) "Zonal Rotation and Global Tectonics," *Amer. Assoc. Pet. Geologists Bull.,* v. 57, pp. 210–214.

Grabau, A. (1904) "On the Classification of Sedimentary Rocks, *American Geologist,* v. 33, pp. 228–247.

Green, J. and Short, N. (1971) *Volcanic Landforms and Surface Features,* Springer-Verlag, New York, 519 pp.

Hart, P. J., ed. (1969) *The Earth's Crust and Upper Mantle,* Nat. Res. Counc. Pub. 1708, Wm. Byrd Press, Richmond, Va., 735 pp.

Hammond, A. L. (1971) "Plate Tectonics: The Geophysics of the Earth's Surface," *Science,* v. 173, pp. 40–41.

Heezen, B. C. and Fox, P. J. (1966) "Mid-Oceanic Ridge" [in] *The Encyclopedia of Oceanography,* R. W. Fairbridge, ed., Van Nostrand Reinhold, New York, esp. pp. 506–517.

Heezen, B. C. and Tharp, Marie (1965) "Tectonic Fabric of the Atlantic and Indian Oceans and Continental Drift," *Roy. Soc. London Trans.,* v. A 258, 90 p.

Heezen, B. C. and Tharp, Marie (1967) *Map of the Indian Ocean,* Nat. Geogr. Soc.

Hess, H. H. (1946) "Drowned Ancient Islands of the Pacific Basin," *Amer. Jour. Sci.,* v. 244, pp. 772–791.

Hess, H. H. (1962) "History of Ocean Basins," [in] *Geol. Soc. Amer. Buddington Volume,* J. and L. Engle, eds., pp. 599–620.

Holmes, Arthur (1931) "Radioactivity and Earth Movements," *Geol. Soc. Glasgow, Trans.,* v. 18, pp. 559–606.

Holmes, Arthur (1965) *Principles of Physical Geology,* The Ronald Press, New York, esp. pp. 287–339, 923–952, and 960–991.

Hopkins, W. (1839) "Researches in Physical Geology," *Roy. Soc. London, Phil. Trans.,* pp. 381–385.

Howell, B. F. (1970) "Coriolis Force and the New Global Tectonics," *Jour. Geophys. Res.,* v. 75, pp. 2769–2772.

Isacks, B., Oliver, J. and Sykes, R. L. (1968) "Seismology and the New Global Tectonics, *Jour. Geophys. Res.,* v. 73, pp. 5855–5900.

Hunkins, K., Mathieu, G., Teeter, S. and Gill, A. (1970) "Floor of the Arctic Ocean in Photographs," *Arctic,* v. 23, pp. 175–189.

Jackson, K. C. (1967) "Temperature Indicators of Arkansas Ouachita Deformation, *Geol. Soc. Amer.* (Abstract), p. 19, South-Central Meeting, program.

Jackson, K. C. (1970) *Textbook of Lithology,* McGraw-Hill Book Co., New York, esp. Chapters 5–7.

Johnson, H. and Smith, B. L., eds. (1970) *The Megatectonics of Continents and Oceans,* Rutgers Univ. Press, New Brunswick, N.J., 282 p.

Kay, Marshall, (1951) "North American Geosynclines," *Geol. Soc. Amer. Mem.* 48.

Kehle, R. O. (1970) "Analysis of Gravity Sliding and Orogenic Translation," *Geol. Soc. Amer. Bull.,* v. 81, pp. 1641–1664.

Kennedy, C. (1959) "The Origin of Continents, Mountain Ranges and Ocean Basins," *Amer. Scientist,* v. 47, pp. 491–504.

King, L. C. (1956) "Rift Valleys of Brazil," *Geol. Soc. South Africa Trans.,* v. 59, pp. 199–214.

King, L. C. (1958) "Basic Paleogeography of Gondwanaland during the Late Paleozoic and Mesozoic Eras," *Quart. Jour. Geol. Soc. London,* v. CXIV, pp. 47–70.

Kirkland, D. W. and Gerhard, J. E. (1971) "Jurassic Salt, Central Gulf of Mexico, and its Temporal Relation to Circum-Gulf Evaporites," *Amer. Assoc. Pet. Geologists Bull.,* v. 55, pp. 680–686.

Knopoff, L. (1969) "Continental Drift and Convection" [in] *The Earth's Crust and Upper Mantle,* P. J. Hart, ed., Wm. Byrd Press, Richmond, Va., esp. pp. 683–689.

Knopoff, Leon (1970) "Developments in Seismology and Georheology [in] *Megatectonics of Continents and Oceans,* H. Johnson and B. L. Smith, eds., Rutgers Univ. Press, New Brunswick, N.J., esp. pp. 113–146.

Konig, R. H. and Chin, A. A. (1973) "Stress Inferred from Calcite Twin Lamellae in Relation to Regional Structure of Northwest Arkansas," *Geol. Soc. Amer. Bull.,* pp. 3731–3736.

Kuenen, P. H. (1936) "The Negative Isostatic Anomalies in the East Indies (with Experiments)," *Leidsche Geol. Mededeel.* v. 8, pp. 169–214.

Kummel, Bernhard (1961) *History of the Earth,* W. H. Freeman, San Francisco, 610 pp.

Lipman, P. W., Prostka, H. J. and Christiansen, R. L. (1971) "Evolving Subduction Zones in the Western United States, as Interpreted from Igneous Rocks, *Science*, v. 174, pp. 821–825.

Longwell, C. R., Flint, R. F. and Saunders, J. E. (1969) *Physical Geology*, John Wiley and Sons, New York, esp. pp. 453–483.

Maxwell, J. C. (1959) "Turbidite, Tectonic and Gravity Transport, Northern Apennine Mountains, Italy, *Amer. Assoc. Pet. Geologists Bull.*, v. 43, pp. 2701–2719.

Maxwell, J. C. (1962) "Origin of Slaty and Fracture Cleavage in the Delaware Water Gap Area, New Jersey and Pennsylvania" [in] *Geol. Soc. Amer. Buddington Volume*, J. and L. Engle, eds., pp. 281–311.

Maxwell, J. C. (1970) "The Mediterranean, Ophiolites and Continental Drift" [in] *The Megatectonics of Continents and Oceans*, H. Johnson and B. L. Smith, eds., Rutgers Univ. Press, New Brunswick, N.J., pp. 167–193.

Menard, H. W. (1969) "The Deep Ocean Floor," *Scientific American*, v. 221, pp. 126–142.

Meyerhoff, A. A. (1968) "Arthur Holmes: Originator of Spreading Ocean Floor Hypothesis," *Jour. Geophys. Res.*, v. 20, pp. 6563–6565.

Meyerhoff, A. A. and Meyerhoff, H. A. (1972) "The New Global Tectonics: Major Inconsistencies," *Amer. Assoc. Pet. Geologists Bull.*, v. 56, pp. 269–336.

Pettijohn, F. J. (1957) *Sedimentary Rocks*, Harper and Bros., New York, 718 p.

Pierce, W. G. (1957) "Heart Mountain and South Fork Detachment Thrusts of Wyoming," *Amer. Assoc. Pet. Geologists Bull.*, v. 41, pp. 591–626.

Pierce, W. G. (1960) "The 'Break-Away' Point of the Heart Mountain Detachment Fault in Northwestern Wyoming, *U. S. Geol. Surv. Prof. Pap.* 400-B, pp. 106–108.

Prucha, J. J., Grahm, J. W. and Nickelsen, D. W. (1965) "Basement Controlled Deformation in the Wyoming Provence of the Rocky Mountains Foreland," *Amer. Assoc. Pet. Geologists Bull.*, v. 49, pp. 966–992.

Roach, C. B. (1962) "Intrusive Shale Dome in South Thornwall Field, Jefferson Davis and Cameron Parishes, Louisiana," *Amer. Assoc. Pet. Geologists Bull.*, v. 46, pp. 2121–2132.

Runcorn, S. K. (1962) "Paleomagnetic Evidence for Continental Drift and its Geophysical Cause" [in] *Continental Drift, Internat. Geophys. Series*, v. 3, pp. 1–40.

Schuchert, Charles (1910) "Paleogeography of North America," *Geol. Soc. Amer. Bull.*, v. 20, pp. 427–606.

Schuchert, Charles (1923) "Sites and Nature of North American Geosynclines, *Geol. Soc. Amer. Bull.*, v. 34, pp. 151–230.

Shand, S. J. (1949) *Eruptive Rocks*, John Wiley and Sons, New York, 488 p.

Sleep, N. H. and Biehler, Shawn (1970) "Topography and Tectonics at the Intersections of Fracture Zones with Central Rifts," *Jour. Geophys. Res.*, v. 75, pp. 2748–2752.

Spencer, E. W. (1969) *Introduction to Structure of the Earth*, McGraw-Hill Book Co., New York, esp. p. 123.

Stanley, K. O. and Wayne, W. J. (1971) "Epeirogenesis and Plio-Pleistocene Fluvial Sediment Deposition in the Northern High Plains," *Geol. Soc. Amer.* (Abstract), v. 3, pp. 716–717.

Stokes, W. L. (1950) "Pediment Concept Applied to Shinarump and Similar Conglomerates," *Geol. Soc. Amer. Bull.*, v. 61, pp. 91–98.

Tazieff, Haroun (1970) "The Afar Triangle," *Scientific American*, v. 222, pp. 32–40.

Turner, F. J. and Verhoogen, Jean (1960) *Igneous and Metamorphic Petrology*, McGraw-Hill Book Co., New York, 694 pp.

Toit, A. L. du (1937) *Our Wandering Continents*, Oliver and Boyd, Edinburgh, 336 pp.

Trumpy, Rudolf (1960) "Paleotectonic Evolution of the Central and Western Alps," *Geol. Soc. Amer. Bull.*, v. 71, pp. 843–908.

Twidale, C. R. (1971) *Structural Landforms*, The M.I.T. Press, Cambridge, Mass., 249 pp.

Vine, F. J. and Matthews, P. M. (1963) "Magnetic Anomalies over Ocean Ridges, *Nature*, v. 199, pp. 947.

Vogt, P. R. and Conolly, J. R. (1971) "Tasmanid Guyots, the Age of the Tasman Basin, and Motion between the Australian Plate and the Mantle," *Geol. Soc. Amer. Bull.*, v. 82, pp. 2577–2584.

Wahlstrom, E. E. (1950) *Igneous Minerals and Rocks*, John Wiley and Sons, London, 3rd ed., 367 pp.

Waters, R. S. (1964) "The Pleistocene Legacy to the Geology of Dartmoor" [in] *Dartmoor Essays*, I. G. Simmons, ed., Torquay, pp. 73–96.

Wegener, Alfred (1937) *La Gènese des Continents et des Oceans*, Nizet et Bastard, Paris, 236 pp. (A translation of the 1936 German edition of the original 1915 work).

Weller, J. M. (1960) *Stratigraphic Principles and Practice*, Harper and Bros., New York, 725 pp.

Wilson, J. T. (1965) "A New Class of Faults and their Bearing on Continental Drift," *Nature*, v. 207, pp. 343–347.

Wilson, J. T. (1968) "Static or Mobile Earth: The Current Scientific Revolution," *Amer. Philos. Soc. Proc.*, v. 112, pp. 309–320.

Wilson, J. T. (1970) "Some Possible Effects if North America Has Overridden Part of the East Pacific Rise," *Geol. Soc. Amer.* (Abstracts), pp. 722–723.

Winterer, E. L. (1968) "Tectonic Erosion in the Roberts Mountains, Nevada," *Jour. Geol.* v. 76, pp. 347–357.

Wise, D. U. (1963) "Keystone Faulting and Gravity Sliding Driven by Basement Uplift of Owl Creek Mountains, Wyoming," *Amer. Assoc. Pet. Geologists Bull.*, v. 47, pp. 586–598.

Woollard, G. P. (1966) "Regional Isostatic Relations in the United States" [in] *The Earth Beneath the Continents*, J. S. Steinhart and T. J. Smith, eds., Geophys. Mono. 10, A. G. U., Washington, D.C. pp. 557–594.

Woollard, G. P. (1969) "Regional Variations in Gravity" [in] *The Earth's Crust and Upper Mantle*, P. J. Hart, ed., Wm. Byrd Press, Richmond, Va., esp. pp. 320–341.

4 Surficial geomorphic systems

EXOGENIC ENVIRONMENTAL PHENOMENA

Our Chapter 3 journey "into" the earth gave us a glimpse of a body that belies an apparent solidarity where its crust shakes, leaks molten rock, and otherwise adjusts to stresses at a tempo often seemingly measured by a cosmic drummer who has yet to take a second stroke. Internal rhythms give way to external ones, and it is the surface of the earth and conditions there that now command our attention. Earth surface environments are created through interaction of the CRUST with several phenomena including mainly AIR, WATER, ICE, REGOLITH, ORGANISMS, and EXTRA-TERRESTRIAL OBJECTS plus GRAVITY and several kinds of RADIATION. In addition to environments generated through the "external" contacts between these phenomena, some of the same features mix with or otherwise penetrate others and so give rise to special "internal" conditions as when plant roots penetrate a soil or water contains sediment. Also, it should be understood that several of these phenomena are direct agencies for geomorphic change (e.g. EOLIAN, FLUVIAL, or GLACIAL activity) and others rather independently comprise environments (e.g. AQUEOUS, CRYERGIC). In short, depending upon one's perspective, the phenomena listed above can be regarded as AGENCIES or as ENVIRONMENTS or as ENVIRONMENTAL COMPONENTS. Where, in combination, they interact to generate landforms, they comprise SURFICIAL GEOMORPHIC SYSTEMS (*geomorphic = morphogenetic*). Almost any kind of surficial event can be geomorphically significant, but those most directly effective are expressed along the upper crustal surface where landforms develop. In short, the interface is "where the action is."

In general, each geomorphic system comes into existence, functions, and eventually ceases to exist with respect to conditions developed in some larger, often more or less enveloping context. These controlling conditions comprise geomorphic environments of which several distinct orders of magnitude are discernible. For extra-terrestrial phenomena impinging most directly on Earth, the sponsoring conditions include the Solar System, its mass, angular momentum and radiation, and similar attributes of our Galaxy and the Universe. Causal ruminations about universe origins range between the physical and metaphysical and give us pause. It is with mixed relief and disappointment that we discover few signs of land formation there. Closer to home we encounter the atmosphere whose genesis (probably continuing) is generally attributed to lithic degassing. Contained by the earth's gravitational attraction between the lithosphere and the void and vacuum of outer space, "The air," as noted by Leonardo Da Vinci, "moves like a river and carries the clouds with it." Water, similarly sponsored, is likewise implicated in the genesis of landforms and is more nearly enveloped by tangible matter in the form of the crust and atmosphere. The hydrosphere clearly tends to concentrate along the *interface* between earth and air where ice, regolith, and organisms also mainly occur.

If an environment endures as long as another entity with which there is a shared INTERFACE, any effects achieved through the relationship come about through the attributes of each as imposed on the interface BUT NOT BY THE APPEARANCE OR DISAPPEARANCE OF EITHER. These are the general relations between the earth as a whole and outer space. Effects along the mutual interface are largely matters of "action" and "reaction," of "force" or "event intensity" versus "resistance" in a sequence of "effectively" infinite duration from an "Earth-time" standpoint. As a rule, in such cases, some form of dynamic equilibrium develops along the interface. To a lesser degree, the same conditions exist between the deep ocean floors and the water covering them and between the air over the land and that land. However, these comparisons suffer because a third variable (the crust) has entered the scene. The air-water interface shifts. And of course when this happens the crust alternately shares interfaces with two different media.

LAND SURFACES, which are our immediate concern here, are only intermittently occupied by air, water, ice, regolith, and organisms. These phenomena can achieve geomorphic effects via the INTENSITY of their attributes WHERE they occupy a crustal interface, effects that are tempered by the DURATION of the contact. IN ADDITION, MOST OF THE SAME PHENOMENA CAN ALSO ACHIEVE PARTICULAR EFFECTS BY PERIODICALLY APPEARING AND DISAPPEARING FROM THE INTERFACE. Each geomor-

phic system exists in time to the extent that the environment inducing it dominates a portion of the crustal surface. Since a land area is involved, when that systemic dominance ceases, one or more other systems can take over. Thus, in any region undergoing morphogenesis, there are always at least two POTENTIAL environmental endmembers, each with its associated geomorphic system. If the environment changes, the system changes. Therefore, geomorphic effects achieved by environment change are matters of (1) environment-system KINDS, (2) environment-system SEQUENCE, and (3) environment-system change frequency (DURATION). Thus, as noted at the end of Chapter 1, a study of geomorphic effects is a study of an environmental space-time continuum. This statement applies equally to endogenetic, mesogenic, and exogenic phenomena, all of which interact at the earth's surface.

There is reason to believe that the geomorphic significance of extra-terrestrial objects involved in collisions with the Earth was formerly much greater than it is at present. There is also ample evidence (Fig. 4.1) that the Earth still occasionally endures meteorite impact. Because the atmosphere acts as a shield that frictionally incinerates most incoming objects, the crustal surface does not support an environment typified by impacts as it would seem the surfaces of the Moon and possibly other planets do. Tidal effects in water bodies caused by extra-terrestrial objects are important along most open-ocean coasts. Tidal forces acting on the "solid" Earth may be reflected in some aspects of tectonism, but generally the latter "body tides" appear too diffuse to be reflected in individual landforms. On Earth, meteorite falls rank as low-probability events and tidal attractions as high-probability events. Morphogenesis is achieved through the interplay of both frequent and infrequent happenings, but it is obvious that an event's probability does not necessarily indicate its impact. The other surficial phenomena listed in paragraph one, unlike meteorite impacts, do pro-

Fig. 4.1.   Aerial view looking southwest of the crater (Arizona) which is believed to have been formed by a meteorite impact. The Earth's environment is not characterized by such events, but it may once have been as it is clear the Moon was also. (John S. Shelton.)

tractedly characterize Earth surface conditions, and we will consider these features next.

Relative to the crust, air, water, ice, regolith, and organisms comprise comparatively dynamic environmental ingredients. But these comparisons are relative. As detailed in the previous chapter, rock fabrics in epeirogenic regions tend to be more acted upon than acting. In contrast, the crust becomes a dynamic geomorphic agency along lithosphere plate margins and their extensions where orogenesis prevails, and THERE tectonism may dominate the landscape. Similarly, surficial agencies commonly act on the crust in some combination, usually with one or another dominating, but from time to time one acts alone. Over the lands the all-encompassing condition through time is the atmosphere. Water, organisms, and "the rest" come and go, but the air does not.

True, the atmosphere is not always of the same geomorphic effectiveness, but it has great environmental "staying power." Consequently, the atmosphere tends to rule the other geomorphic agencies and periodically appears instrumental in displacing or replacing them. Once again we emphasize that the significant geomorphic events and effects in question are concentrated along the upper crustal surface.

Surficial environmental agencies tend to display both enduring and short-lived aspects, the latter betraying a constant state of energy flux by such phenomena as wind gusts, clouds, tides, rain squalls, currents, etc. Such attributes are imposed on the crust in varying degrees and in occasionally catastrophic form (storms, floods) that briefly but drastically reshape the Earth's surface and give rise to landforms. In more gradual or otherwise less spectacular displays, some of the same phenomena combine in the hour-to-hour and day-to-day conditions that often seem to accomplish little geomorphically and that, in the case of the atmosphere, we term *weather*. It seems clear that any surficial geomorphic effects achieved slowly over a long period must therefore come about (1) in spite of short-term agency variations, (2) because such varients are generally forced to function within rather well-defined limits, or (3) because extreme agency effects are more influential than mean effects in the genesis of landforms. As we shall see, there is reason to believe that all three conditions occasionally prevail. Nevertheless, particular landforms or landscapes like particular makes of automobile prove by their very similarities and numbers that they were formed under controlled conditions. This, in turn, implies particular processes acting upon particular materials under particular conditions. The key word here is "control." For the automobile there is a certain assembly line in a special factory. For the landforms analogous relations must periodically develop, and we will now consider these relations and their controls.

There are a good many indications of the nature of environmental controls over surficial geomorphic effects. For example, the behavior of some agencies is sustained at a particular energy level and in a particular direction (i.e. Trade Winds). In other instances, repeated agency events occur at a particular level of energy release and in a special topographic configuration (i.e. flood incidence) or transpire over a rather closely delimited land area with little apparent regard for topography (i.e. storm tracks). Alternatively, there are numerous cases where a normally variable agency has its sphere of influence curtailed or altered or its intensity muted by agency interaction. Thus, standing water, ice, soil, and plants all inhibit the atmosphere's access to bedrock in a variety of manners, places, and degrees. At the same time, water, ice, soil, and plants are to some extent induced by and modified by the very atmosphere whose effects they inhibit.

Much can be gained from special study of both short-lived agency behaviors and geomorphic effects and the more enduring environmental conditions and related geomorphic effects that are achieved slowly. Studies of earth surface conditions from place to place disclose that each surficial environment displays some relatively active agencies and some that are comparatively passive. In another environment the degree of activity of the same agencies may be reversed. Thus, wind is an effective agent of erosion where the ground is bare but is erosionally ineffective where there is continuous plant cover.

Consideration of the kinds of exogenic geomorphic effects cited in previous paragraphs demonstrates the complexity of subaerial morphogenesis. Yet even the many variables occur in particular patterns which we take to be of special significance where they directly involve the upper surface of the crust. (1) An agency (SUCH AS ICE) that can rather permanently occupy a rock surface being shaped (*geomorphic interface*) can evidently suppress and thus control the geomorphic effects of an agency that cannot simultaneously occupy the same interface (i.e. air, water), even though all agencies involved are somewhat interdependent. (2) Agencies that can tolerate the greatest range of physical conditions along a geomorphic interface will exert the most effective control there, since this tolerance allows the agency in question to persist through time and (other things remaining the same) an effect that lasts longest will be greatest. (3) A set of surficial conditions that maintains a given agency along a geomorphic interface to the essential exclusion of other agencies should achieve certain unique landscape effects. (4) As noted at the end of Chapter 1, *"where environments are interchangeable, controls must be interchangeable."* Thus, when a geomorphic environment changes, the former interface conditions will give way to another interface condition sustained by the new environment; i.e. geomorphic controls will have changed. (5) Where a given interface condition becomes established, actual landform modification comes about in accordance with the new geomorphic system's evolution toward a stable condition (as noted in Chapter 1, there tends to be disequilibrium caused by relicts followed by eventual elimination of the relicts and attainment of dynamic equilibrium). In this chapter we will examine the more enduring attributes of surficial environments, particularly as expressed on the geomorphic interface. Because of its apparent long-term dominance over other factors, we will begin with the atmosphere and gradually add other features.

GENERAL ATMOSPHERIC PARAMETERS

Earth's gaseous envelope is probably more pervasively in contact with the land than any other exogenic agency. True, the continents are sporadically wetted and penetrated by water, even to the point of local saturation. Similarly, organisms get into the act as does ice in various places, and as noted previously, both can influence atmospheric effects. But the air touches surfaces and penetrates voids in rock which even liquids cannot readily enter. Air moves, and in doing so moves other things; it joins in chemical reactions, conducts heat and moisture, sustains organisms, and intercepts extra-terrestrial matter and radiation, among other things. Before we embark on a detailed consideration of the atmosphere's movement and structure, we must consider its composition and its related "internal" thermal and physical properties that also influence the geomorphic responses invoked in the crust.

The atmosphere is a complex mixture of gaseous, liquid, and solid materials, with an average composition that is relatively constant; its admixtures vary notably from place to place and time to time. Illustrated first (Fig. 4.2) is a table of gaseous constituents. Among the gases, nitrogen (78%), oxygen (21%), carbon dioxide

Nitrogen, 78.09%	Helium, 0.000524%
Oxygen, 20.95%	Krypton, 0.0001%
Argon, 0.93%	Hydrogen, 0.00005%
Carbon dioxide, 0.03%	Xenon, 0.000009%
Neon, 0.0018%	Ozone, 0.000005%

Fig. 4.2. The chief constituent gases of the atmosphere—dry air, by volume, in per cent. (National Aeronautics and Space Administration.)

(0.03%), and ozone (0.000005%) are probably most important geomorphologically.* Only the first three of these are abundant near the earth's surface, and nitrogen tends to be inert chemically. There is an essentially uniform distribution of gases from the surface to altitudes of about 15 mi and this relation plus other standard meteorological atmospheric stratification aspects is illustrated also (Fig. 4.3). A surface pressure of 14.7 lb/sq in (29.92 mm Hg) is standard for air at sea level as is a coefficient of viscosity of 0.0002 at 20°C as compared to 0.01 for water at the same temperature. From the standpoint of geomorphic environments it is most significant that the atmosphere acquires its water vapor, most of its heat, and most of its solids at or near the earth's surface.**As noted by Rumney (1970), water vapor and dust tend to remain concentrated close to the surface. The average vertical distribution of water vapor (Fig. 4.4) and heat (Fig. 4.3) is of special concern to environmental studies and is particularly relevant to landform development problems insofar as the crust only intersects the *Troposphere*. The geomorphically significant major vertical changes in heat and moisture are largely of that zone and both constituents normally decrease upward in the part affecting the crust.

The main source of atmospheric heat is solar radiation, of course, and the unequal distribution of this heat with respect to the earth's poles and equator adds a latitudinal thermal gradient to the vertical one already noted.† It is estimated by Cole (1970) that the equatorial zone receives about three times as much heat as the polar regions—

* Since ozone acts as an ultraviolet ray filter in the upper atmosphere it makes terrestrial life possible.
**The point should be made that though the source of the heat is solar the atmosphere mainly warms from the surface outward.

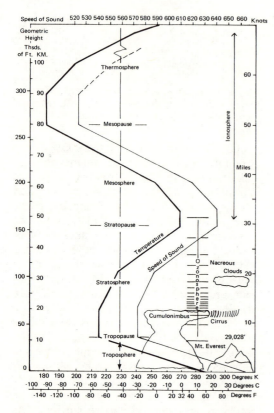

Fig. 4.3. Vertical profile of the atmosphere showing the four-layered structure based upon temperature change with height. Geomorphic problems develop in immediate contact with the Troposphere. (After U. S. Navy *Aerographer*.)

321 cal/cm²/year versus 133 cal/cm²/year. The vertical and lateral heat variations are undoubtedly of great geologic duration and place rather stringent limits on very-long-term geographic variations in heat and light. These long-term variations are due to such phenomena as equinox precession and the solar insolation cycles noted in Chapter 1. Unequal heating is also responsible for convective heat transfer down the various thermal gradients in accordance with the laws of thermodynamics. Thus, global air movements for thermal reasons are assured, all apart from movements induced by rotation and other causes.

It seems proper to mention at this point the hemispheric asymmetry of atmosphere heat distribution. The basic present imbalance can be traced to the fact that the *thermal equator* (zone of greatest heat input) is situated to the north of the geographic equator in relation to the corresponding heat deficiency of the Southern Hemisphere reflected by the Antarctic ice sheet. Solar radiation naturally penetrates through to the surface most effectively where cloud cover is consistently least—the low-latitude deserts. Most such deserts are in the Northern Hemisphere where the bulk of the land is located. This to a great extent accounts for the northern position of the thermal equator just as the presence of a landmass at the South Pole accounts in large measure for the larger ice mass of that hemisphere. We will return to this subject in a later section of this chapter.

It is not our intent here to generate a meteorologic or climatologic tome. For greater detail in these areas of study the reader is referred to texts by Geiger (1965), Rumney (1968), Trewartha (1968), and Cole (1970). Rumney in particular, notes the necessity of dealing with the "Geosystem" and the interdependence of various surficial agencies in problems of heat and moisture distribution as well as surficial environments in general. The data so far presented are therefore merely intended to highlight the chemical and physical characteristics of the atmosphere capable of contributing to lasting patterns on the geomorphic interface. Next we must consider

| Height | | Volume (%) |
(mi)	(km)	
5	8.3	0.05
4.5	7.5	0.08
4	6.7	0.13
3.5	5.8	0.19
3	5.0	0.29
2.5	4.1	0.37
2	3.3	0.46
1.5	2.5	0.63
1	1.6	0.80
0.5	0.8	1.06
Surface	Surface	1.30

Fig. 4.4. Vertical distribution of water vapor in per cent of volume within the atmosphere in the middle latitudes. (Based in part on Rumney, 1970.)

air movements and over-all resulting atmospheric configurations.

Atmospheric Circulation

Convective movements of the air related to solar heat have superimposed on them a system of air circulation related to earth's rotation around its axis. The most important of these air streams are concentrated in the lowest atmosphere layer—the Troposphere (Fig. 4.3). Rumney (1970) cites the main units of atmospheric circulation as (1) the Equatorial Trough—convergence zone of the Trade Winds, (2) Subtropical Oceanic Highs, (3) Subpolar Lows, and (4) Polar Highs (Fig. 4.5). Relations between these units are determined by upper level air (jet) streams, seasonal effects, and hemispheric inequalities in distribution of land, water, heat, light, and ice.

Air movements have both vertical and horizontal components with the latter about 1,000 times greater in magnitude than the former. The global wind systems shown here (Fig. 4.5) represent air transfer between the various major units of atmospheric circulation and, as noted by Rumney (1970), reflect a gravity response to air density variations. The same writer summarizes the general role of heat in atmospheric movements as follows

> *In brief, the incoming heat of global radiation taken up at the surface of the tropical seas is chiefly converted by evaporation into the latent heat of water vapor. Raised aloft by the turbulence set in motion through heating from below, the moisture-laden air is brought to condensation point. This releases heat which augments the continued upward motions of turbulence, water vapor thus providing the essential fuel in the overall dynamic process. Turbulence in the lower latitudes is thus of first importance in maintaining the continuous operations of the general circulation.*

Air movement conditions of greatest geomorphic importance include, (1) calms, notably the Doldrum Belt on the equator where vertical air movements reduce or eliminate the usual vertical atmospheric stratification and also the Subtropical Highs centering at about Latitude 30° where dry evaporating air descends to become a major

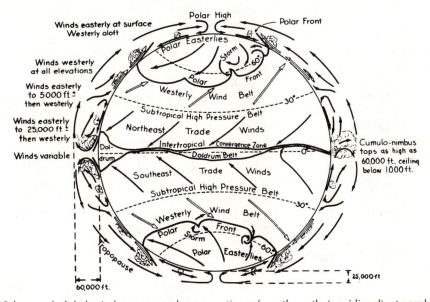

Fig. 4.5. Scheme of global wind systems and cross section of north-south (meridional) atmospheric circulation. (From U. S. Navy *Meteorology for Naval Aviators.*)

factor in the generation of Horse Latitude Deserts; (2) Trade Wind belts where there are continuous winds from only one direction much of each year; (3) zones of Prevailing Westerlies where large-scale eddy diffusion moved by the jet streams results in eastward migrating high and low pressure areas with rotating winds; (4) monsoonal winds relating to seasonal shifts in wind movements influenced by continent and ocean heating and cooling; and (5) the jet streams which appear to fix mid-latitude storm tracks. All air in motion is, of course, subject to the effects of the earth's rotation (*Coriolis and centrifugal forces*), plus friction.

Global air movements are of geomorphic importance largely to the extent that they influence land environment conditions. The latter comes principally in the form of varying patterns of moisture distribution along the crustal interface (hydrography). Atmospheric influence over land moisture mainly takes the form of (1) actual transfer of moisture by air masses until precipitation can be triggered in some manner and (2) control of cold and warm ocean currents that comprise evaporation sources of differing quality. A major factor governing the amount of moisture a given land will receive is how much water does associated air contain. A second related factor concerns available mechanisms for triggering precipitation. Thus, many deserts (e.g. the Sahara) have above them air masses that contain considerable moisture but lack precipitation triggering mechanisms on any consistent basis. Conversely, some humid lands receive comparatively great amounts of moisture from air masses containing relatively little absolute moisture because precipitation triggering mechanisms exist.

The atmosphere clearly acquires its moisture at the earth's surface, mainly via evaporation and transpiration (Fig. 4.9), but the transfer of water from the sea to the land and its effects there are of vastly greater concern to the geomorphologist than any effects land-derived moisture has

when it reaches the sea as precipitation. Consequently, it is atmosphere-hydrosphere interaction as it may be expressed on the lands that must be considered now.

CIRCULATORY INTERACTION— ATMOSPHERE-HYDROSPHERE

The portions of the hydrosphere that concern us here are the larger bodies of standing water, principally oceans and seas. These are the main primary evaporation sources of atmosphere moisture, and the location and other attributes of such water bodies determine their potential value as water sources for associated lands. We are not as concerned with the details of oceanic water movements as are oceanographers to whom subsurface and bottom characters may rank with those at the surface. The atmosphere mainly interacts with surface waters and such major temperature, salinity, and movement variants as are expressed along this interface.*

Fluid movements within larger bodies of standing water tend to fall in two categories which nevertheless interact to produce an over-all result. First, there are the mainly horizontal currents driven largely by the wind as modified by rotational (geostropic) and tidal forces. These may be underlain by deepwater counter currents. Second, there are essentially vertical shifts in water masses caused by density variations; the latter in oceans may be induced by temperature or salinity. The oceans are, however, the general thermal reverse of the atmosphere in that the waters closest to the lithosphere surface are generally coldest. Naturally, because of the distribution of solar energy, ocean water tends to receive most of its heat near the equator. Increased evaporation in the same zone tends to heighten salinity and related density, but this effect is countered somewhat in polar regions where freezing creates sea ice with less salt than normal sea water and asso-

* For oceanic salinity makeup, see Rumney (1970).

ciated cold ocean water with higher than average salinity and density.

Whether of normal or abnormal salinity, temperature, or density, surface circulation of the oceans is mainly at the mercy of global wind patterns. Circumglobal water movements parallel to lines of latitude are possible only in the Southern Hemisphere. There, prevailing westerly winds generate a major ocean current system (West Wind Drift) which has, with some justice, been termed "the motor of the oceans." Circling the Antarctic continent (Fig. 4.6) the frigid waters of the West Wind Drift are deflected to the left by Coriolis force and give rise to several major northward-trending currents where this water is intercepted by the south-ward jutting portions of the several south-ern continents (Fig. 4.6). These cold, northward-moving waters are expressed at the ocean surface as the Peru (Humboldt), Benguela, and West Australian currents, and they comprise the main evaporating water surfaces immediately to the west of South America, Africa, and Australia essen-tially up to the Equator.

Circulation of the southern oceans is taken up by the Southeast Trade Winds as modified by the Subtropical Highs which combine to form counter-clockwise current *gyres* in each of the more or less separate southern water bodies (Fig. 4.5A). Cold waters approaching the Equator from boreal regions are deflected to the left by rotational forces and move westward along the south-ern margin of the Doldrum Belt, gradually mixing with warmer waters and being heated by the Sun to comprise the southern equatorial currents. These warm currents are again deflected to the left to move south-ward along the eastern portions of the southern land masses. The warm East Aus-tralian, Brazil, and Mozambique-Agulhas currents comprise the main evaporative water surfaces to the southeast of South America, Africa, and Australia in the paths of the Southeast Trade Winds.

Circulation of the northern ocean surface waters is seen to be essentially the reverse of these south of the Equator (Fig. 4.5 A). Coriolis force in the Northern Hemisphere deflects moving bodies to the right so that the Northeast Trades combine with Sub-tropical Highs to produce clockwise ocean

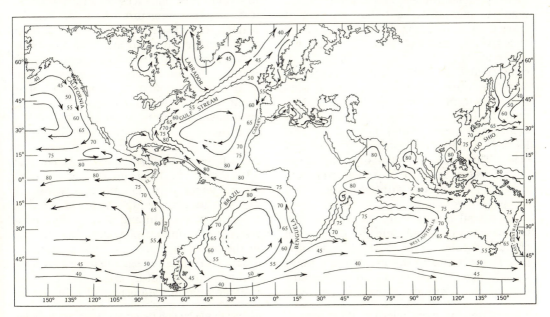

Fig. 4.6. Major ocean surface current directions and temperatures (°F). Wavy arrows indicate major warm currents transferring heat toward the poles.

current gyres. Since these gyres derive their warm water near the equator, the northward flowing currents along the eastern continental shelves are also warm. These include principally the Gulf Stream and Kuro (Japan) currents, and it has been suggested that the former is strengthened by a portion of the Benguela current that is deflected northward across the Equator, thereby transmitting Southern Hemisphere energy to the north. However that may be, the northward moving water masses mix with Arctic waters and move southward as the cold California and Canary currents to the west of northern continents.

In addition to the well-defined warm and cold surface currents that play a role in global heat distribution and place water bodies of rather predictable temperature on particular sides of continents, some comparatively deep ocean circulations and minor surface movements should be noted. Water chilled in polar areas tends to sink because of increased density and move equatorward where it locally returns to the ocean surface (*upwelling*) to augment areas of cold water which arrived there by more surficial routes. Since water has a much higher latent heat than air (e.g. it takes more calories to raise the temperature of a given quantity of water than air) the water tends to determine the temperature of associated air rather than the reverse. In addition, warm air is an equatorial phenomenon in the same sense that cold air is typically boreal since the gas is so readily heated and cooled. Water, however, retains its heat if warm and is difficult to heat if cold. Thus, we occasionally encounter warm water at high latitudes and cold water at low latitudes with resultant effects on land environments.

In a final commentary on hydrosphere circulation we should note the occurrence of the most important local eddy currents that flow counter to the mainstream flows already noted. These occur on the equator (Equatorial Counter Current) and are

rather "tucked in" at the distal ends of several major currents. The cold Labrador Current thus opposes the Gulf Stream along the eastern North America seaboard just as the cold Falkland Current opposes the warm Brazil Current along the Argentine coast. Conversely, a warm offshoot of the equatorial system (El Niño) counters the Peru Current west of South America. Under the present global circulatory system these minor flows are unimpressive. Under altered global conditions of the past they may have strengthened or possibly even countered some of the effects of their opposite numbers.

Atmospheric Moisture Transfer

The amount of moisture picked up by a given evaporating air mass leaving a surface (ocean, lake) is a function of both the air and water temperatures along the interface as noted by Rumney (1970). The amount of heat required to vaporize water decreases as the water temperature increases. As a consequence, air masses leaving cold water surfaces generally contain less absolute moisture than air masses leaving warm water surfaces.* The amount of moisture such an air mass can contain is also a function of its temperature, i.e., the warmer the air the more moisture it can hold without reaching saturation (dew point). These air-water temperature interrelationships govern most of the world's land moisture conditions either directly or indirectly, at least in areas with relatively open access to oceans. Thus, on the equator adjacent to the Doldrum Belt where both air and water are commonly very warm, air masses are frequently saturated AT SEA LEVEL and also can and do contain a relatively great amount of water. Any change in the air mass causing a reduction in the amount of water it can hold (e.g. a temperature drop) causes precipitation, often in large amounts. Consequent rainfall dis-

---

* Lustig (personal communication, 1973) notes that dense fog in the Namib Desert is a main precipitation form and is an immediate consequence of cold adjacent currents. Similar precipitation forms (mist-*Garua*) characterize the Peru-Chile deserts for similar reasons.

tribution on adjacent lands begins at sea level on most coasts and usually continues inland and upward to the highest elevations (the highest land near the Equator—in Ecuador—receives abundant precipitation in some form from sea level to elevations in excess of 21,000 ft).

Where cold oceanic water penetrates equatorial zones, evaporation is inhibited, air moving from the cold oceans is usually undersaturated, and some movement of the air masses inland and upward with consequent cooling (adiabatic) is usually required before precipitation can occur. Low-elevation deserts are the result, and in areas of cold-water upwelling or boreal currents the associated land moisture tends to be proportional to sea temperature. Thus, areal and vertical extents of these deserts increase as related ocean temperatures diminish. As one would expect from the previous discussion of ocean current circulation, low-elevation deserts presently occupy continental west coasts and have inland and elevational extents governed largely by topography and latitude. They are generally lowest and narrowest near the Equator and broaden and extend to higher elevations to about 35° North and South of the equator. Coastal mountains (e.g. in South America) tend to confine aridity to a narrow geographic zone, whereas lower coasts (e.g. southwest Africa, western Australia) permit oceanically induced dry conditions to extend inland where they can contribute to the aridity of existing subtropical deserts.

The Earth's coldest oceans (near the poles) yield minimal moisture to associated lands and that largely in the form of snow. In subpolar climes along cold current coasts there is usually sufficient air moisture to permit eventual saturation. All that is required is some precipitation trigger mechanism such as a cyclonic storm, freezing temperatures, or air movement upslope with resultant cooling through adiabatic expansion (orographic rainfall). Where near-equatorial warm-water temperatures penetrate higher latitudes along major currents (most east continental coasts), air masses leaving oceans achieve saturation

at or near sea level. Adjacent lands are consequently moist with minimal triggering of precipitation even if global air circulation (i.e. Subtropical Highs) tends to induce land dryness as in Argentina and southeast Africa. Where air masses leaving these cooling but still relatively warm currents are slightly undersaturated, corresponding land dryness may coincide with the beach zone and go unnoticed. General oceanic temperature reduction during glacial episodes almost certainly intensifies and extends low-elevation dry atmosphere conditions and related land moisture. Garner (1965) terms the commonly undersaturated air overlying cool to cold seas the *adiabatic gap*. Its effects may extend far inland. Generally such air must be cooled (usually by rising) before it can yield any moisture to the land. *Temperature inversions* (to be discussed shortly) may prevent the condensation event. The adiabatic gap and other related atmospheric moisture zonations commonly encountered within 35° either side of the equator are illustrated here (Fig. 4.7).

Figure 4.7 is an admittedly simplified expression of vertical atmosphere moisture distribution. Yet there are land environment patterns corresponding to each of the general atmospheric conditions cited. The intermediate atmospheric zone of near-saturated air may, in fact, often be satu-

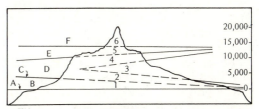

Fig. 4.7. Continental atmospheric moisture zonation and geomorphic climate belts diagrammed in east-west section with elevations in feet. (A) Sea level—evaporation surface; (B) adiabatic gap—undersaturated air; (C) minimum precipitation elevation; (D) near-saturated air—triggered precipitation; (E) moisture-depleted air; (F) frost zone. Corresponding geomorphic environments are (1) Coastal Deserts, (2) Semiarid Steppes, (3) Humid Uplands, (4) Semiarid Alpine Transition, (5) Desert Plateaus, (6) Glacial Peaks. Here, the sea to the left is assumed to be cold, the sea to the right warm.

rated air that exists in the absence of a precipitation trigger. Actual causes of precipitation vary with land-sea relations and latitudes. Among the more common causes are (1) diurnal convective heating, (2) monsoonal circulation, (3) migrating cyclonic storms, and (4) orographic effects. Obviously, more than one of these effects may combine to account for the amount and distribution of precipitation on a given land area. The monsoon is seasonally analogous to the diurnal heating phenomenon. Diurnal atmospheric heating of areally differing intensities is common to many deserts that are topographically diverse. Where combined with orographic influences the local convective circulation often causes precipitation to be concentrated in uplands. In land-water relations, diurnal heating initiates sea breezes. In undersaturated air these winds cause little consequent precipitation as along the west coast of South America. Elsewhere, in areas of near-saturated air, diurnal winds and local convective heating causes daily rainfall in the Doldrum Belt most of the year and seasonally at higher latitudes.

In regions characterized by migrating cyclonic storms the distribution of precipitation on land in the storm tracks tends to be governed by the location of evaporation sources as modified by topography. Generally, lands closest to water and uplands are most moist. Lands far from seas and lowlands are driest. Lands relatively upwind (prevailing) from water bodies are drier than those where land and water lie adjacently athwart the prevailing wind, and such land is in turn drier than land in the path of prevailing wind blowing over water. Portions of North America relative to the Gulf of Mexico are cases in point relative to the above conditions. However, since the winds in migrating cyclonic storms rotate (Fig. 4.8), the amount of precipitation on windward versus leeward land slopes varies little with respect to prevailing wind direc-

tion. Other factors become more significant. The role of the so-called topographic rain shadow has been much exaggerated with relation to prevailing winds where cyclonic storms are common. The prevailing Westerlies thus cause pressure cells to move eastward across much of North America. East slopes of many of the western mountains are dry, as much because of the lack of a close evaporation source to the east, south, or north as because Pacific Ocean moisture was dropped orographically farther west.* The eastern Appalachian slopes occupy essentially the same relative prevailing wind position as the eastern slopes of the western mountains but receive abundant precipitation via reverse cyclonic storm circulation of moisture from the Atlantic Ocean (i.e. against the prevailing wind). Similarly, western South America is largely dry because diurnal southwesterly winds move undersaturated air from a cold sea. The strong diurnal convection abetted by orographic effects simply does not have sufficient air moisture to generate notable precipitation. The fact that the Southeast Trades are blocked by the Andes, and the area fringes a Subtropical High, becomes mainly incidental. If the sea were warm, the land would almost certainly be humid.

Lands in the paths of Trade Winds differ in one major aspect from those where there are frequent cyclonic storms. Precipitation in Trade Wind belts is usually triggered orographically or by convective heating as noted by Rumney (1970). Orographic precipitation naturally tends to fall on the windward side, and the lee slopes of such lands fall in a true rain shadow and are exceedingly dry—presumably as long as the trade wind directions remain constant. Several islands in the Hawaiian Archipelago show excellent examples of this relationship (cf. Fig. 3.96).** Precipitation events at intermediate elevations over the Earth apparently strip moisture from rising air masses. This

_____

* Because of their height and mass, aridity on the east slopes of the Sierra Nevada is mainly an orographic effect.
**Because of the Northeast Trade Wind, the northeast side of Oahu receives 200–400 inches of rainfall whereas the lower southwest side receives less than 20 inches.

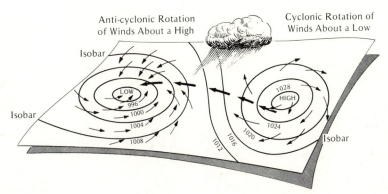

Fig. 4.8. Relationship between high and low pressure air circulations in the Northern Hemisphere. South of the Equator, wind directions are reversed. (From U. S. Air Force *Weather for Air Crews.*)

occurs so consistently at most latitudes that air much above 10,000 ft elevation is notably depleted in moisture as previously illustrated (Fig. 4.7). At such elevations where the air temperature is consistently above freezing the associated land surfaces are extremely dry. At somewhat higher altitudes, lower temperatures make snow and frost the common forms of precipitation and permit the resulting moisture to accumulate in snowfields and glaciers. It seems pertinent to remark at this point that the foregoing remarks about atmospheric moisture zonation and related land environments apply to a global zone some 70° wide centering on the Equator. It presently includes essentially all of Africa, most of South America and Australia, and the southern portions of North America, Asia, and all of India.

Approximately 35° North and South of the Equator an additional atmospheric aspect emerges through interaction of air and ocean temperature relations. Sea temperatures there may be cold, the more so in paths of currents of boreal origin. Air temperatures are also much cooler, on the average, and those over land areas drop faster in winter than those over water. The total amount of moisture such cool air can hold is less than the amount for warmer air in lower latitudes. Though oceanic evaporation is reduced, the relative humidity of the air masses nears saturation at much lower temperatures. The result is atmospheric

zones with near-saturated air in winter from near sea level to very high elevations between 35° and 40°. There is a resulting land moisture distribution that resembles the equatorial Doldrum Belt effects. This accounts for the high-latitude rain forests in Chile and along the Oregon-Washington coasts of the western United States.

As noted by Thornwait (1948), the amount of precipitation a land receives is not an adequate measure of the effective moisture there. Precipitation amounts, alone, tell us virtually nothing about what the moisture may accomplish geomorphically. Many other atmospheric, crustal, and organic factors are involved, notably *transpiration, evaporation, cloud cover, infiltration,* and *wind conditions* (Fig. 4.9). These, in turn, are affected by such properties as plant cover, air and soil temperatures, and such soil properties as permeability. It is evident that the annual precipitation of a region comprises only one clue to the effective moisture (Fig. 4.10), though admittedly it is an important clue. At latitudes above 40°, absolute precipitation amounts are commonly low, but reduced air and ground temperatures commonly result in moist land conditions up to latitudes where water is continually frozen. From many geomorphic standpoints (i.e. chemical activity, fluviation) perennially frozen water is virtually equivalent to no water. The Antarctic continent is the only land presently characterized by ice. In general, as noted

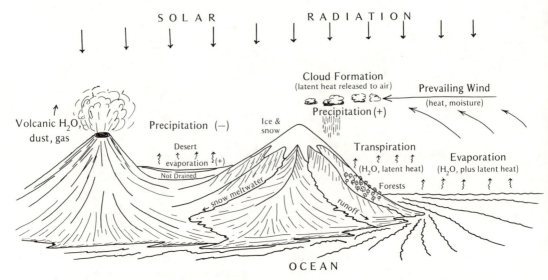

Fig. 4.9. Water and heat transfer related to several of the more common surficial environments. The expression "water cycle" is actually only appropriate to externally drained humid or glacial-cryergic regions.

by Rumney (1968), regions receiving the least precipitation show the greatest annual precipitation variation (i.e. from year to year) as shown here (Fig. 4.11). Clearly,

in such places geomorphic events that are dependent upon water will be similarly sporadic.

An additional atmospheric aspect that

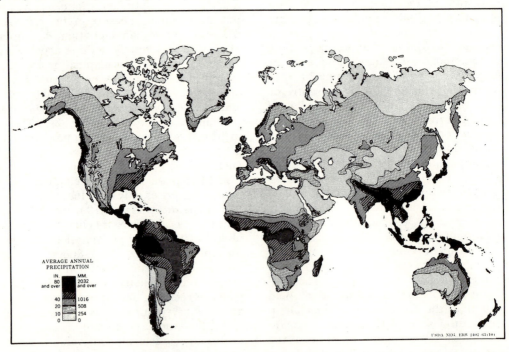

Fig. 4.10. Mean annual precipitation by regions. To gauge the value of plants as indices of moisture distribution, compare with Fig. 4.14. (U. S. Department of Agriculture map.)

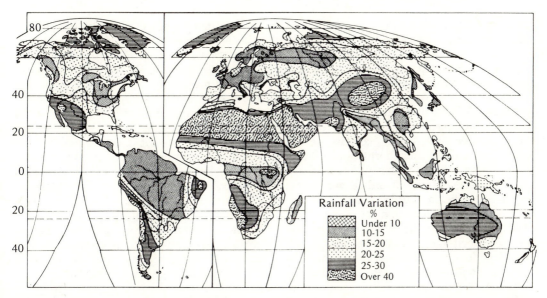

Fig. 4.11. Normal annual rainfall variability. Note that the greatest variability occurs in regions that receive the least moisture. (From Rumney, 1970.)

should be mentioned here because of its effects on land moisture concerns the phenomenon known as *temperature inversion*. This condition of the atmosphere involves a reversal in the order of upward temperature change so that warm air masses overlie cooler ones. The reversals of greatest concern to us occur in the troposphere, of course, though there are others at higher elevations (Fig. 4.3). Moisture condensation is often associated with temperature inversions, and the boundary between near-saturated air and relatively dry air may be exceedingly sharp. This abruptness is often reflected by a related distribution of land moisture as noted by Garner (1967). Among the more important inversions are those developed in trade wind belts and those that occur where cold ocean waters flow past continental west coasts in subtropical climes (see Neiburger *et al.*, 1961). Since the inversion tends to place a ceiling on the upward movement of low-level air and thus prevent its cooling and moisture condensation, the phenomenon contributes greatly to the development of the low elevation deserts previously discussed.

We have now briefly considered the main internal circulations of the atmosphere and related activities along the air-ocean interface. We have also reviewed mechanisms of moisture transfer and the principal relations between these mechanisms and land moisture patterns of long standing. None of these discussions are intended to imply that present environmental configurations are permanent. Far from it. But if the atmosphere and hydrosphere are subject to long-term fluctuations as the geologic record indicates, at least their present patterns are an established variant developed from prior ones, and so on. Our next objective is to consider the extent to which detectable land environments are of geomorphic significance.

CLIMATES, GEOMORPHIC ENVIRONMENTS, AND INTERFACES

Climate is usually defined as the average weather of a region, and this definition attests to an essentially meteorological basis. In their *Principles of Climatology*, Neuberger and Cahir (1969) refuse to define climate, but the definition just given at least delineates ONE KIND of climate, a meteorologic kind. We will consider others as our discussion progresses. A "meteorologic" climate area would therefore be the geographic expanse over which a given range of atmosphere conditions have been recorded—long-term averages of precipitation, temperature, wind velocity, cloud cover, and the like (cf. Thornthwait, 1948). In attempts to more precisely delineate climate types, geographers and climatologists often add criteria such as annual or seasonal variations in temperature, frost-free days, evaporation, and the manner and form of precipitation. Compilations of such data are depicted on the climatic maps of geographic atlases, and they are useful for a variety of purposes, particularly those involving immediate land utilization (agriculture, habitation, industry, recreation, or air travel).

Classifications of climate types have grown increasingly complex as this or that atmospheric feature was detected or chosen for emphasis. The ultimate in classification complexity is probably the Köppen System with its ten or so major climate types and more than twenty subtypes plus combinations of these (Fig. 4.12). As noted by Thornthwait (1948), Köppen regarded plants as meteorologic instruments which can be read to deduce the climate. Thornthwait, on the contrary, sees plants as a physical mechanism for evapo-transpiration and regards older climate classes as "vegetal regions climatically determined." In some measure we concur with Thornthwait's evaluation, but as we shall see, our concerns differ significantly. In essence, Thornthwait is advocating an "instrumental" kind of climatology, and certainly plants do serve as evapo-transpiration mechanisms IN ONE OF THEIR ROLES along the crustal interface (Fig. 4.13). Plants may also reflect conditions above the

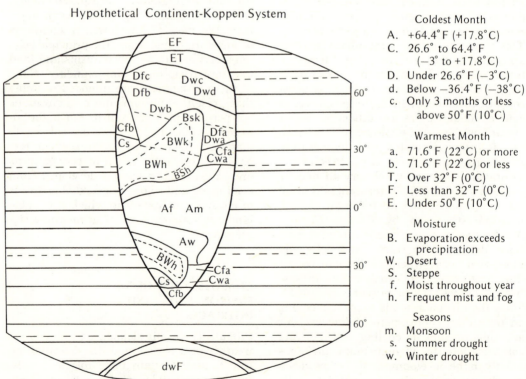

Fig. 4.12. Climate dispositions on a hypothetical continent centered at latitude 30° N, according to the Köppen system of climate classification.

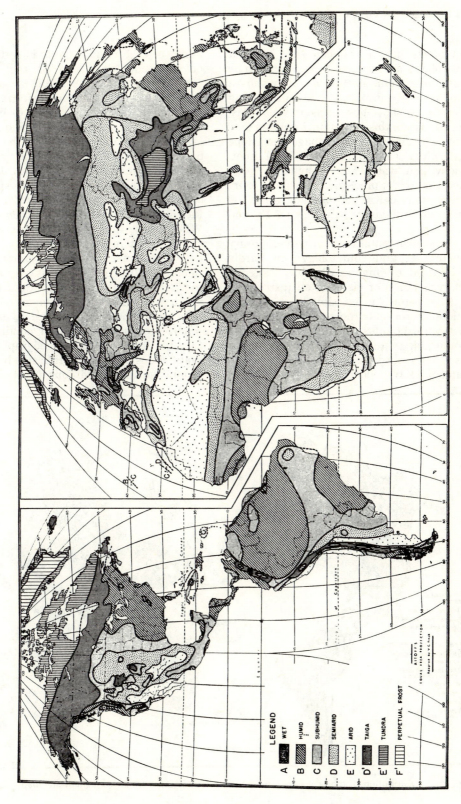

LEGEND
A ▨ WET
B ▥ HUMID
C ▦ SUBHUMID
D ⠿ SEMIARID
E ☐ ARID
D' ▤ TAIGA
E' ▥ TUNDRA
F' ▦ PERPETUAL FROST

Fig. 4.13. Geographic distribution of climates, defined meteorologically for the most part. (From *Climate and Man, Yearbook of Agriculture, 1941.*)

ground surface, as Köppen suggests, but most plants are sustained from beneath by their root systems so that they may subsist from a fossil groundwater condition long after the air above has ceased to yield moisture to sustain this condition, or plants may survive by underground seepage downslope along a dry watercourse into a land otherwise without water. Thus, Köppen's vegetal regions may not precisely correspond to instrumentally delineated climatic areas because of ambiguity (cf. Figs. 4.13 and 4.14).

The instrumental basis of contemporary meteorology and climatology can hardly be doubted by anyone who has visited a weather station or watched an instrument package being carried aloft by a weather balloon or rocket-borne satellite. However, on several counts, we must question the geomorphic value of climate categories based on such data. (1) Meteorologic climates largely reflect atmospheric condi-

tions (often far above ground) rather than environments developed at the air-rock (geomorphic) interface. This is not to say that conditions developed aloft are not expressed on the Earth's surface, but they may not be, or they may only be expressed in some degree determined at the surface by other factors as noted at the beginning of this chapter. (2) Basic climatic data, collected instrumentally, are often sparse or even absent over much of the Earth's surface where geomorphic problems exist. Where these are absent, other criteria of environment must be sought, and, where data are sparse, records of unrecognized microclimatic conditions may be misleading. (3) Rarely do climatic data span as much as a century of instrument recording. As a result, meteorologic climate data may reflect "recently" developed conditions, a fact hardly depicted on geographer's climatic maps or charts. Yet basic changes in ground surface conditions involving such

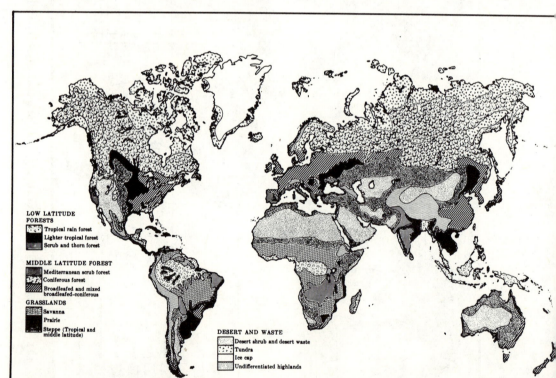

Fig. 4.14. Distribution of major vegetation categories. (U. S. Department of Agriculture map.)

things as water, ice, soil, or organisms, or near-surface configurations of a similar nature, may require hundreds or even thousands of years to develop. Once established, an environment may require tens or even hundreds of thousands of years to produce a particular geomorphic effect. (4) Climatic criteria like Thornthwait's evaporation-precipitation ratios do not fossilize any better than the figures upon which they are based. This intentionally facetious statement serves to re-emphasize a point made in Chapter 1. Analysis of a contemporary geomorphic setting necessitates the separation of relict from present conditions and effects. Present geomorphic environments are often only expressed on the upper crustal surface in intimate combination with geomorphic relicts. Thus, for the geomorphologist wishing to know environmental conditions on a land area, plants are of greater value as climatologic and surface environment indicators *à la* Köppen than they are as evapo-transpiration mechanisms of atmospheric climates *à la* Thornthwait (Fig. 4.14). The same may be generally said of water, ice, and regolith, though as previously noted all may act in several environmental roles. Thus, the evapo-transpiration role of plants is important in vegetated regions where it may influence such things as groundwater and runoff amounts or soil formation.

In essence, therefore, the environments that concern the geomorphologist occur on or near the crustal surface and include that crust as well as other geomorphic relicts in many cases. The relicts are as much a part of the formative environment as the meteorologic conditions expressed on the same interface. Thus a geomorphic environment involves more than just climate as conceived by meteorologists and geographers (cf. Garner, 1968a). Therefore, there is no necessary correspondence between meteorologic climate areas and the *geomorphic areas* that comprise the land surfaces occupied by geomorphic environments. Contemporary geomorphic identification and treat-

ment of surficial environments, therefore, often lacks consistency and fails to detect or isolate causal relationships. For example, King (1953), Schumm and Lichty (1963), Lustig (1965), and Leopold *et al.* (1966) designate as "SEMIARID" several geomorphic study areas with annual precipitation records that range from less than 2 inches to nearly 15 inches, with plant cover that varies from nil to forested, with ground surfaces that range from bare bedrock to alluvium to zoned profile soils, and with stream flows that range from ephemeral to intermittent. Moreover, the terms ARID and SEMIARID are used interchangeably by some of these workers, and the areas in question include pronounced orographic and seasonal precipitation variations in some cases but not others (cf. Garner, 1967a). We hasten to note that the stated basis for making these climate designations is in accord with the traditional meteorologic climate categories of geographers such as Bartholomew (1950) and include essentially the same range of conditions. However, it hardly seems surprising that a unified view of the geomorphic effects of "semiaridity" fails to emerge from a comparison of the studies cited.* Similar treatment can be cited for other climate types in landform studies. The lack of unanimity in results is particularly understandable if variations in ground surface conditions cause unique geomorphic effects.

In the introduction to this chapter it was suggested that the inception of an enduring effect on the land surface capable of modifying the action of other agencies on that surface is a probable prerequisite for the development of particular subaerial geomorphic effects. An experiment that explored the validity of this notion was carried out by H. H. Bennett in 1939. The findings of that experiment provide the rationale behind our identification of specific geomorphic environments and related geomorphic systems and areas.

H. H. Bennett described an erosion ex-

---

* We should observe that climatic categories are usually based on mean annual precipitation rather than frequency distribution of the total which governs geomorphic effects (cf. Fig. 4.11).

periment designed to establish the role of plant cover relative to fluvial erosion. Three erosion test plots were selected in the same region, with identical soil types, which received the same annual precipitation (36.46 inches), had the same areas, and had identical slopes (12°). The only significant variables involved ground cover. One plot was forested, one was grassland, and the third was barren. Sediment removed from the plots by overland flow was caught and measured with the results recorded here (Fig. 4.15). Bennett's findings clearly demonstrate the significance of plant cover variations in control of erosion. His data further show that a first-order increase in erosion coincides with a shift from essentially continuous plant cover to discontinuous plant cover without particular regard to floral taxa.

From the standpoint of geomorphic interpretation of natural geomorphic areas, Bennett's findings could be misinterpreted. As geomorphologists we require analogs that most closely approximate natural conditions. Unless evaporation rates were exceptionally high (or some similar moisture-detracting factor were involved), the amount of rainfall cited by Bennett in time would induce vegetal growth on a naturally barren surface and initiate forest growth over a grassland. Stated somewhat differently, naturally barren and grassy regions ordinarily receive less rainfall than Ben-

nett's erosion plots. This is dramatically shown on the second part of our diagram (Fig. 4.15). With no attempt to account for or express the effects of drier ground conditions, depressed groundwater tables, spotty vegetal occurrences, differing regolith, and the higher evaporation rates usually encountered in naturally dry environments, our illustration also shows (with figures in quotation marks) the general magnitudes of erosion one might expect to encounter in naturally grassy and barren regions. The rainfall amounts cited are common to present steppe and desert regions. In "natural" settings, other factors remaining constant, a barren area would erosionally yield approximately 475 times as much sediment as a grassland to a given volume of runoff.

Two other aspects of Bennett's data should be mentioned here. First, there is abundant evidence that forested regions do yield both solids and solutes to fluvial erosion processes. Thus, Bennett's figures that show zero material being removed from a woodland are suspect as a generalization. The apparent reason for the discrepancy is an absence of channels on the forested test plot. The vegetation prevented overland flow elsewhere. This serves to emphasize that fluvial erosion of particulate solids is mainly restricted to watercourses that normally occupy less than 1% of the land surface in vegetated areas. Second, it should be noted that Bennett's test plots were covered by regolith rather than bare bedrock, and the sediment volumes eroded are only compatible for nonvegetated natural land areas with a cover of soil.

The principal geomorphic environments of the Earth expressed by groundcover can be delineated largely on the basis of Bennett's study. He clearly arrives at no conclusions for ice-covered regions nor for extensive tracts of bare bedrock, though both conditions are widely encountered. He does establish the significance of continuous plant cover, subject largely to erosional modifications caused by watercourses. And he demonstrates the effects of runoff on barren regolith. Grassland would appear to be the most continuous plant cover that na-

Plant Cover	Runoff (percent rainfall)	Tons eroded/acre	H₂O
Woods	0.12	0.00	
Grass	6.5	0.04	36.46 inches
Barren	48.8	69.0	
"Woods"	"0.12"	"0.00"	36.46"
"Grass"	"6.5"	"0.02"+	15"
"Barren"	"48.8"	"9.5"	5"

Fig. 4.15. Erosion rates on test plots in relation to groundcover under 36.46 inches annual precipitation (after Bennett, 1939); predicted relations under similar groundcover types in actual correlative climates (in quotes).

turally survives in relatively low moisture lands, but Bennett's data show little distinction between grassland and forest in a fluvial erosion context. Overland flow is more nearly possible on the grassland than in the forest, but major sources for eroded material in both cases are watercourses. For all the immediate limitations they impose on solid sediment load acquisition by water or wind, overland flow of water, or sediment dispersal and accumulation, the sparse *Xerophytic* plants of deserts might as well not be there. Each of these plant categories comprise surface environmental indicators in the Köppen sense; they constitute evapo-transpiration mechanisms in the manner noted by Thornthwait and additionally function geomorphically in several other capacities involving erosion, weathering, and soil formation.

GEOMORPHIC SYSTEM COMPONENTS

Exogenic conditions capable of generating landforms and deposits clearly incorporate a variety of factors. Most important are (1) the more enduring meteorologic attributes of the atmosphere (a) in an environmental sense (precipitation, heat transfer, winds, etc.) where they induce such things as surface moisture, plants, ice, weathering phenomena, soils, and groundwater and (b) in an agency sense for such things as eolian erosion and deposition; (2) geomorphic interface conditions involving some combination of the phenomena listed at the start of the chapter (air, water, ice, regolith, organisms) that individually or collectively lead to eolian, fluvial, glacial, lacustrine, biotic, and pedologic activity, among others; and (3) geomorphic interface conditions contributed by crustal characteristics considered in Chapter 3 involving some combination of lithology, structure, and tectonic activity, the latter expressed as elevation, relief, earth movements, or some limited combination of these. Consideration of the three categories of phenomena just listed make it evident that just as environment involves more than climate,

morphogenesis involves more than environment.

Along the geomorphic interface there are environments, agencies, processes, and products all interacting through time. Together, these associations comprise the geomorphic systems briefly noted in Chapter 1, of which there are several kinds. We dealt with tectonic geomorphic systems in Chapter 3. Surficial systems selected for detailed consideration in several following chapters are here designated HUMID, SEMIARID, ARID, GLACIAL, CRYERGIC, and COASTAL (Fig. 4.16). Some of these same expressions have previously been applied to kinds of meteorologic climates, and there are degrees of correspondence as well as differences (cf. Figs. 4.13, 4.14, and 4.16). But each geomorphic system is initially recognized and defined by interface conditions. Recognition and definition would include observations on (1) ground cover or lack of it, (2) apparent weathering modes, (3) roles of wind, water, and ice, (4) forms of erosion and deposition, (5) runoff continuity in time and space, (6) drainage forms, (7) groundwater conditions, (8) mass wasting conditions, (9) hillslope forms, (10) regolith configurations, and (11) relief trends.

As an example of a geomorphic system, a humid system is here taken to include events and effects in an area characterized by continuous vegetal cover on land surfaces, dominantly chemical weathering modes, developing profile soils, few eolian effects, elevated groundwater table evidenced by seeps, springs, and swamps, canalized runoff flowing perennially out of the environment (usually to regional base level) through some type of drainage net, downslope movement of weathered debris usually lubricated by water and possibly clay, erosional exploitation of drainage lines by at least solution and usually corrasion so that channels and related valley bottoms are lowered faster than divides and relief increases as long as flow lines are above erosional base level. The technical bases for this synopsis will be presented in Chapter 5.

Such meteorologic data as may exist for

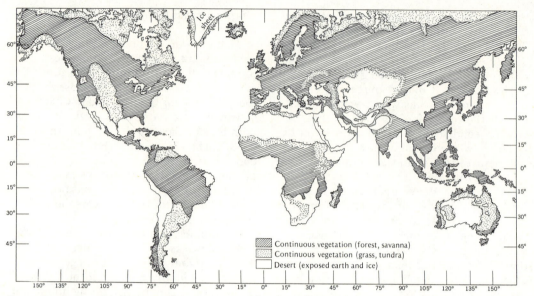

Fig. 4.16. Major global geomorphic areas, generalized, exclusive of the Antarctic. Depicted are barren lands, those with the two major forms of continuous vegetation (grassland and forest), and the Greenland Ice Sheet.

a given geomorphic system tend to indicate the present style and intensity by which agencies function. Thus for the humid geomorphic system just outlined, the basic geomorphic attributes persist under a variety of meteorologic conditions ranging from temperate to tropical. Therein, annual precipitation varies from 30 to 200+ inches and may or may not include snow; cloud covers, growing seasons, drouth incidences, floral assemblages, soil types, runoff amounts, mean annual temperatures, and other factors also vary, and, obviously, a wide variety of evaporation/precipitation relations exist that add to a "plus" on the moisture side of the budget insofar as plant growth, stream flow continuity, and groundwater are concerned. In short (as crudely indicated by a comparison of Figs. 4.13 and 4.16) a number of meteorologic climate categories are included within a single geomorphic system, in this instance, a humid one. The more extreme environmental variants (e.g. tropical) can be shown to contribute certain unique geomorphic features (e.g. laterite), but even the unique landform attributes are of a type (e.g. laterite is a soil),

and each attribute tends to fall within the over-all genetic pattern of the system (e.g. laterite is merely one kind of humid soil).

Criteria for individual geomorphic systems to be discussed later are taken both from regions where systems appear to have endured to the extent that relics of other systems have been eliminated and from other examples of lesser systemic duration where relics are at least identifiable. In Chapters 7, 8, and 10 we consider landscape effects and geomorphic systems that notably alternate in time and space. There, we will consider problems related to environmental displacement and environmental sequence. For example, How does landscape in a semiarid geomorphic system that was previously arid differ from a landscape in the same system that was previously humid? The previously noted existence of an ever-present danger in such discourse is worth repeating here; namely, the possible confusion of relics of former geomorphic systems with features presently forming which might thereby cause us to misconstrue either present geomorphic conditions, those of the past, or both.

One set of clues available to the geomorphologist as to the character of probable geomorphic relicts in a region may come from its global setting. Thus, we are unlikely to encounter recent glacial materials at low elevations in the tropics or relatively modern effects of intense chemical activity in boreal realms. Our previous discussion of temperature distributions and moisture transfer mechanisms generally indicates major geomorphic environmental types relative to latitude and elevation. The same discussion also indicates probable moisture variations with respect to land-water distributions and air-water temperature relations. Since mainly long-term atmospheric fabrics induce particular conditions on the geomorphic interface, the topography of that interface affects the manner in which geomorphic systems become established and are displaced. Topography, therefore, indicates indirectly the probable distribution of relict landforms and deposits. We must now consider topographic influences over geomorphic environments.

OROGRAPHY, SLOPE, AND GEOMORPHIC ADJUSTMENTS

As noted in Chapter 3, relief and elevation of the emergent portions of the continental rises appear to be a function of crustal thickness, with thicker crustal areas standing relatively high and thinner areas being low. Major topographic variations on each continent corresponding to such crustal thickness variations relate to elevated zones of recent orogenesis (<½ billion years old) versus older "erosional" and younger "depositional" lowlands. Records of Quaternary climate changes alone suffice to establish that protracted changes in conditions along the geomorphic interface occur with a periodicity much shorter than most tectonic events (cf. Chapter 1). There "briefer," but nonetheless protracted changes in geomorphic systems gradually modify the major tectonic components of landscape. And, largely because of the topographic positions

they occupy with respect to the atmosphere, high-elevation, high-relief areas often experience different geomorphic histories than low-elevation, low-relief areas. This section will consider these topographic-environmental relations.

When one takes into account the vertical and areal distribution of geomorphically distinctive interface conditions on the land surfaces, they appear to reflect a rather systematic interaction between the aforementioned zones of atmospheric moisture, global wind belts, and pressure systems, ocean current, and surface-temperature patterns and the just-noted major topographic elements. The long-term vertically disposed moisture and temperature zones of the atmosphere tend to parallel the earth's surface (cf. Fig. 4.6). Hence, the same horizontal zones tend to either parallel or include relatively low-relief land surfaces which occur at restricted elevation ranges (e.g. plains, plateaus). The atmospheric zones naturally tend to intersect land surfaces that are inclined at high angles, particularly those surfaces that extend to high elevations (e.g. mountain fronts).

Contemporary, pronounced wind and ocean-current temperature conditions may cause the major atmospheric moisture and temperature zones to deviate from the horizontal. Thus the contact (a temperature inversion) between the adiabatic gap and the zone of triggered precipitation along the western Andes Mountain front is inclined downward from south to north in Peru and Ecuador (Fig. 4.17). Changes in one or more of the controlling atmospheric or oceanic factors can clearly cause the zones to move upward or downward or cause their boundary zones to shift laterally along the geomorphic interface. If our previous analysis of atmospheric patterns and fabrics is correct, each geomorphic area (Fig. 4.13) reflects the present contact of a particular atmospheric pattern with the crust. As noted by the writer (1965), it is not necessary for the atmospheric patterns to remain geographically static for related geomorphic systems to induce a particular geomorphic

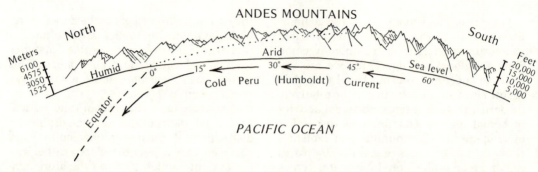

Fig. 4.17. Schematic portrayal of the arid-humid climatic zonation along the western Andes Mountain Front. Evidence presented in Chapter 10 suggests this zone boundary has periodically shifted north and south, probably in response to sea surface water temperature changes. Numbers along coast are degrees of latitude (S) where corresponding surface water temperatures (°F) starting at the Equator are, respectively, 70°, 65°, 58°, 50°, 38°.

response on the earth's surface. Possible relations are illustrated here (Fig. 4.18).

A second set of patterned atmospheric conditions are expressed on the crust essentially parallel to the equator. Related geomorphic systems are actually extensions of similar systems induced by the vertical atmospheric zonation. Thus, geomorphic systems also develop in response to the aerially disposed atmospheric conditions. An example is the Doldrum Belt on the equator. In essence, on land this belt generates humid geomorphic systems at virtually all elevations up to about 15,000 ft at latitude zero. Affected areas merge at higher latitudes with those occupied by similar systems developed at more restricted elevations. Similarly, the low latitude arid geomorphic systems partly reflect associated Subtropical Highs, and these too tend to merge with high- and low-elevation arid geomorphic systems at higher latitudes (Fig. 4.16). Related geomorphic areas are presumably subject to lateral adjustments as global wind and pressure systems are altered. Studies of the margins of such geomorphic areas suggest that their basic geographic shifts are toward or away from the poles and equator, subject to modification by local topography.

The regional slope of a land area would appear to be a major factor in the extent and distribution of its environments. This is largely a matter of geometry as illustrated here (Fig. 4.19). If an atmospheric zone capable of inducing a particular geomorphic system is 1,000 ft (300 m) thick and horizontal, the land area affected will be inversely proportional to the land slope. Such a zone would create a geomorphic area 1,000 ft (300 m) wide on a vertical cliff or one roughly 10 mi (16 km) wide on a 1° slope. But most plains and plateaus are much more nearly flat, and the same thickness of atmospheric zone would control a geomorphic area 125 mi (180 km) wide trending roughly north-south across the North American High Plains. Significantly, the major vegetation zones there presently exhibit the same general north-south trend.

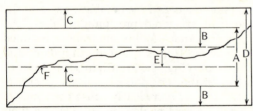

Fig. 4.18. High-level arid atmospheric zonation and gradation datum to which montane plateaus may be related. (A) Mean position of moisture-depleted atmospheric zone (cf. Fig. 4.6); (B) maximum zonal depression; (C) maximum zonal elevation; (D) vertical scope of long-term zonal oscillation; (E) residual, permanently, or dominantly arid geomorphic zone of protracted planation and aggradation; (F) regional slope change keyed to long-term climate zone boundary position and X_c, an erosional relation discussed in this context in Chapter 10.

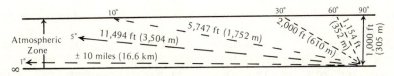

Fig. 4.19. Cross section showing land slope amounts—and hence related areas—affected by an atmospheric zone capable of inducing a particular land environment. The depicted zone is 1,000 ft (305 m) thick, but natural examples are capable of changing thickness and also of shifting up and down with corresponding climatic results.

These plains slope eastward an average of 8 ft/mi (1.5 m/km) or about 1/12°. Lower slopes would show greater affected areas, of course, and such a zone with its base at sea level could therefore control morphogenesis over the entire North American southern and eastern coastal plains plus extensive regions farther inland.

Geomorphic systems induced by atmospheric conditions that do not generally parallel the earth's surface are apparently somewhat less sensitive to topography for, as previously noted, the Doldrum Belt affects continental lowlands and mountains alike. Even there, the high-elevation atmospheric FROST ZONE (Fig. 4.6) provides an upper limit to humid morphogenesis. Similarly, in subtropical deserts where there is notable local relief, orographic effects with diurnal winds moving upslope tend to induce more precipitation on uplands than lowlands. Not uncommonly, therefore, humid or semiarid geomorphic systems oc-cupy relatively isolated uplands, while lowlands experience some form of arid morphogenesis. Whether for strictly orographic reasons, or because of horizontal atmospheric moisture-temperature zonation, or both, high-elevation, high-relief land areas commonly exhibit three or four major geomorphic systems disposed roughly parallel to one another and to sea level (Fig. 4.20).

It was previously noted that conditions such as temperature inversion may cause boundaries between distinct atmospheric conditions to be very abrupt. This abruptness shows up in vertical profiles through the whole atmosphere (Fig. 4.3) as well as in moisture and temperature data recorded within the troposphere (Fig. 4.4). Similarly abrupt boundaries are evidenced by conditions developed along the geomorphic interface, particularly by plant and snow cover. The SNOW LINE, as the expression implies, is a singularly abrupt transition which displays only limited seasonal fluctuation and

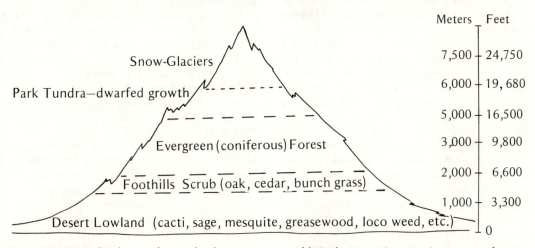

Fig. 4.20. Generalized groundcover development on a mid-latitude mountain range in an area of atmospheric zonation.

corresponds to vertically disposed atmospheric conditions of restricted nature. The writer (1967) describes an unusually abrupt contact between the atmospheric ADIABATIC GAP and ZONE OF TRIGGERED PRECIPITATION in coastal Ecuador (Figs. 4.6; 4.17) which is directly reflected by plant types and ground cover; adjacent land in contact with the lower zone exhibits xerophytic plants in an arid geomorphic system, whereas the higher zone induces continuous deciduous forest in a humid geomorphic system.

Tree lines are noted especially because of their frequent abruptness at higher altitudes and latitudes. On upper mountain slopes the tree line tends to be of uniform elevation on constant slopes of similar exposure. On a regional basis, however, montane tree lines vary in height considerably where local topography and exposure also vary, and it seems evident that microclimatic influences modify any atmospheric zonation effects to some extent. Geiger (1965) has summarized in detail the micro-environmental conditions existing along a forest border that contribute to tree growth limits. For our purposes of the moment, it is sufficient to note that a complex interplay of such factors as seeding, wind, shading, root exposure, precipitation inducement, heat retention, lighting, and convection all contribute to a given position of a natural forest border. In the foregoing forest-border context, two other tree line types are notably abrupt. One is the forest-steppe border typified by the present western limit of tree growth along the eastern edge of the North American High Plains. The transition is dramatic and first appears to the traveler going east as a dark "stain" on the prairie horizon (Fig. 4.21). Similar conditions exist along the borders of steppe regions on the several continents. Forest-steppe tree lines are not usually straight or necessarily absolute, their configurations being due to subtle topographic variations which affect relative exposure, etc. There may be outlying groves or there may not be, depending to some extent (as discussed in Chapter 7) upon which direction the most recent forest border shift was—toward the open areas or away from them.

A third type of tree line is that developed as the polar limit of tree growth. Although somewhat analogous to alpine tree lines, on exposed high-latitude lowlands of low-relief, topographic factors are clearly subordinate to global atmospheric environmental patterns. In northern Canada, for example, the coniferous growth becomes distinctly lower for a few score yards and then ends abruptly at the tundra. North of this boundary there are occasional expanses of bleached, barkless tree trunks virtually devoid of branches presumably relating to times when growth extended farther. The Indians are said to refer to this phenomenon as the "land of the little sticks." Elsewhere, north of the continuous dense forest, one occasionally encounters trees in two states. In one state, isolated trees have somehow grown a few feet above ground level in exposed positions and are still alive but have no branches or growth signs on their exposed (northern) sides. Alternatively, shallow (1- to 2-ft) depressions contain dwarfed trees growing up to surrounding ground level but no higher. It seems clear that a biologic-environmental threshold is being reflected by the general forest border which delimits both geomorphic systems and some general global atmospheric pattern roughly centered on the poles.

Boundary conditions between vegetal types that provide continuous ground cover under low-moisture situations (grassland) and discontinuous ground cover (sage, bunch grass) are less obvious than tree lines but no less important geomorphically. Viewed along the western or southwestern margins of the North American High Plains, the east edge of the Kazakstan Steppes of Russia, or along the southern Sahara, there is an irregular but definite grassland border which appears to be strongly influenced by subtle variations in such factors as topography, regional slope, soil conditions, ground water distribution, prevailing wind directions, directional exposure, and blowing sand (Fig. 4.21). As along the tree lines, desert borders reflect

Fig. 4.21. Groundcover transitions. *Below*: the tree line where grassland meets forest border on the High Plains of central Oklahoma in open, nearly level terrain. *Above*: semi-desert sage giving way to sandy barren waste, northeastern Arizona. Clumps of vegetation are situated on low mounds of silty sand, a relation postulated for the development of prairie mounds (Chapter 7).

complex relationships, but the near-threshold state of vegetal cover is amply demonstrated by the effects of unusually severe drouths or fires just as a near-threshold state for soils in similar settings is occasionally and tragically demonstrated through man's "in-depth" destruction of plant cover.

RECENT GEOMORPHIC SYSTEM DISPLACEMENTS

In the preceding discussions of geomorphic systems and the atmospheric patterns that apparently induce them, it is evident that a shift in the position of the patterns will cause displacements in related systems.

Within the troposphere two basic types of environmental patterns were distinguished, a vertical moisture-temperature zonation essentially paralleling the earth's surface (Fig. 4.6) and latitudinally disposed global zonation related to general atmospheric circulation (Fig. 4.5). The two types of patterns are clearly somewhat interdependent causally, but they do not necessarily react simultaneously or in like degree to a given influence. The onset and disappearance of continental glaciers has certainly been a major factor in the relatively recent past, presumably involving shifts in latitudinal zones with respect to the poles and equator. Related general cooling and warming of the oceans may lag considerably behind glaciation and deglaciation, yet ocean temperature changes may notably affect disposition of atmosphere zones roughly paralleling the earth's surface.

The writer (1965, 1967a) discussed adjustments in the vertically disposed atmospheric zones, and Butzer (1957), Büdel (1953, 1958), and the writer (1966, 1967, 1968) have explored the nature of latitudinal zone displacements, mainly with respect to glacial and interglacial episodes. Damuth and Fairbridge (1970) consider the effects of intensifying various atmosphere circulations at the expense of others during the Ice Ages. Garner (1959a, 1967, 1968b) makes repeated reference to land environment shifts that can be related to altered ocean current patterns and temperatures, atmospheric temperature inversions, and possible accentuation of hemispheric environmental inequalities. Büdel focuses primarily on latitudinal zone shifts in which he presumes glacial episodes to be pluvial and interglacial times to be arid in low latitudes. Flint (1971) questions the notion that glacial times correlate globally with increased precipitation, noting that reduced temperatures would generally reduce evaporation. The problems involved are, of course, immensely complex, and it is not always possible to separate cause from effects. Flint offers an excellent recent summary of Quaternary environmental condi-

tions on the various continents. It seems nonetheless advisable to list a number of the major paleogeomorphic relationships that must be accounted for in any over-all geomorphic-environmental theory.

(1) Pleistocene continental glaciers were dominantly high-latitude, land localized phenomena with bordering frost environments only a few kilometers wide (cf. Flint, 1971); the extent of alpine glaciation in low latitudes appears to have been vastly overestimated (cf. Bowman, 1916; Garner, 1959a), and Flint concludes that temperature changes in middle latitudes were stronger than in the tropics.

(2) Presently there are both high- and low-budget ice sheets (the Antarctic example being low and the Greenland example displaying both conditions) but it is probably wrong to conclude that glaciers grow, persist, and disappear under constant environmental conditions, either atmospherically or oceanically (see Chapter 8).

(3) In general, long-term environmental records show geographic diversity at any one time, in configurations roughly corresponding to the main atmospheric patterns. It must be emphasized that the correspondence is crude, suggesting that the patterns have fluctuated with respect to some long-term mean set of positions rather than that the patterns are fixed in either time or space.

(4) Geographic shifts of diverse conditions along the geomorphic interface means that the "direction" of these shifts will not be uniform with respect to geomorphic systems, since a given system may move into one region while moving out of another area. This observation is compatible with data that show increased effective moisture and extension of upland vegetation downslope during glacial stages in presently arid southwestern North America (cf. Morrison, 1968), whereas dry conditions with sparse vegetation invaded many humid tropical lowlands during the same episodes (cf. Bigarella, and de Andrade, 1965; Garner, 1958, 1966a; Damuth and Fairbridge, 1970). In effect, the lands, apart from high

mountains, seem to become dry from the bottom up and moist from the top down.

(5) Some east-west environmental contrasts on continents can be anticipated because of ocean surface current temperature differences, and these contrasts are intensified where topographic relief exhibits notable north-south alignment. Differing geomorphic histories were suggested for Andean South America east and west of the mountains by this writer (1959), and this has been recently re-affirmed by findings of Vanzolini and Williams (1970) and Vuilleumier (1971). Similar east-west environmental asymmetry (and hence non-synchroneity of like geomorphic events) is indicated for North America where portions of Alaska experienced arid morphogenesis while glaciers were maximal at much lower latitudes farther east on the same continent as noted by Flint (1971).

(6) In his summary discussion, Flint (1971) notes that a lack of absolute radiometric dates prior to about 50,000 B.P. (C_{14} is limited to younger materials) prohibits a definite statement about synchroneity of Quaternary glacial and interglacial episodes, either trans-oceanically or between the northern and southern hemispheres. However, the obvious hemispheric inequalities in heat input and absorption plus unequal ice, land, and water amounts ensures that geomorphic system displacements north and south of the equator will also be non-symmetrical and, thus (in some degree), non-synchronous. One symptom of hemispheric non-synchroneity of environmental happenings would be sea level changes out of phase with glacial cooling such as Friedman and Sanders record (1970). The Southern Hemisphere would clearly have a "head start" into any forthcoming glacial period and lags behind the Northern Hemisphere in adjustment to what appears to be the beginning of an interglacial episode, at least in terms of glacier ice bulk and heat equator location. In a recent study, Mercer (1972) notes that glacial events in Chile correlate rather well with Eastern North American glacial events

but appear to differ with New Zealand sequences and do not show up at all in Antarctic paleo-temperature curves based on oxygen isotope ratios.

For an in-depth consideration of the nuances involved in the preceding environmental considerations, the reader may wish to consider Charlesworth's two-volume work on the Quaternary (1957) and more recent related studies by Wright and Frey (1965) and Flint (1971). In addition, there are literally hundreds of articles dealing with geomorphic and climatologic syntheses, one category of which (deserts) is indicated by Lustig's bibliographic inventory (1967). In short, the foregoing paleo-geomorphic-environmental evaluation is only minimally referenced here but is thoroughly documented in the literature. Let us now consider actual instances of geomorphic system displacement.

Actual cases of geomorphic system displacement are documented on several continents. In middle and low latitudes there is evidence for changes from continuous to discontinuous plant cover and vice versa. As noted by Davis (1969), from 4,000 B.P. to 7,000 B.P., Arkansas and probably areas to the north, south, and west comprised part of a desert with sparse plant cover and a fauna that included peccary and armadillo. The present forest boundary is 200 to 300 mi west of the former eastern limit of arid morphogenesis, and the nearest present combination of arid morphogenesis, discontinuous plant cover, and desert fauna lies some 300 to 400 mi farther to the south and west (Fig. 4.22). As previously noted in Chapter 1, barren conditions with free-blowing sand relate to the same geomorphic area shifts and are presently represented by relict dune sand in western Nebraska, Kansas, and Oklahoma and the loess blankets extending east from these dunes (Fig. 2.28). In general, the north-south trending vegetal zones appear to have shifted alternately east and west. Details of this morphogenesis will be taken up in later chapters. It should be noted, however, that specific meteorologic adjustments behind

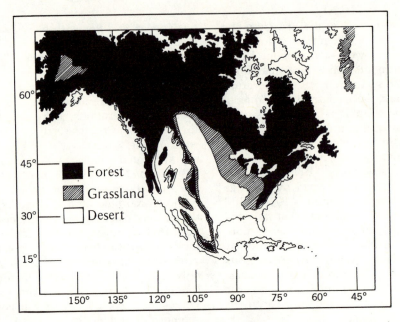

Fig. 4.22. Generalized paleoclimatic map of North America during the Altithermal interval.

the geomorphic shifts remain uncertain, but could involve a northward shift in storm tracks combined with the temperature increases widely postulated for the interval (Altithermal) in question. Related studies in the region (cf. Lugn, 1960; Quinn, 1956; 1961; Garner, 1966b) indicate that there have been a series of geomorphic alternations in this region during the Quaternary.

Geomorphic system displacements similar to those just noted in North America are also indicated for the barren-vegetated transition zones in Australia, Asia, parts of South America, and on either side of the equator in Africa. Garner (1959b) indicates the apparent distances of geomorphic zone shift on the several continents (Fig. 4.23) and further documented these adjustments (1966a, 1967, 1968) by pedologic and drainage configurations. Grove (1958, 1970), Prescott and White (1960), and Grove and Warren (1968) also note dune and vegetal configurations along the southern Sahara and northern Kalahari where conditions of plant cover have changed repeatedly. Curiously, along the southern Sahara, grassland and some forests have shifted northward to invade dune and re-

lated barren lands over distances ranging from about 150 mi in the west to 500 mi in the east. Recent geomorphic zone displacements in South Africa ALSO appear to have been northward (cf. Grove, 1971), with barren Kalahari Desert conditions now extending 200 to 300 mi farther into the Zambezi River and Okavango Delta regions that formerly (Fig. 4.24).

Most African geomorphic patterns closely parallel latitudinal climatic zonations with some deviations relative to topography and in coastal regions, presumably because of marine influences. The parallelism suggests that adjustments in global atmospheric circulation patterns may have induced geomorphic area displacements, and in Africa a notable hemispheric asymmetry is suggested. The record is meager but indicates that vegetal zones did not merely move toward and away from the equator. Rather, the entire geomorphic complex seems to have shifted alternately northward and southward across lines of latitude with the most recent shift, noted by Garner (1967) and Grove (1970), being northward (Fig. 4.25). Butzer et al. (1972) confirm an African equatorial lacustrine record relating to

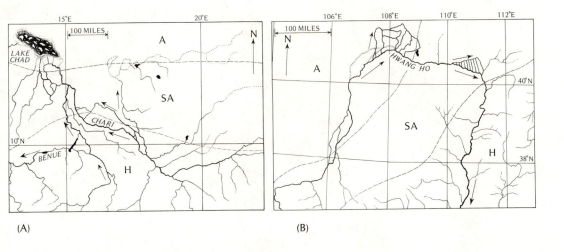

(A)

(B)

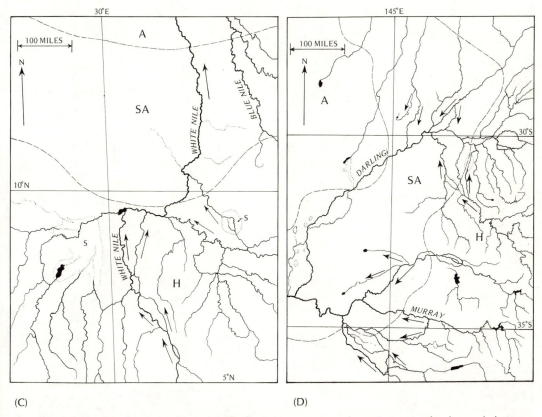

(C)

(D)

Fig. 4.23. Evidence of geomorphic zone displacements on several continents in the form of drainage dispositions, particularly where runoff flows from humid (h) through semiarid (sa) to arid (a) regions. (A) The Chari River, North Africa; (B) the Hwang Ho (Yellow River), China; (C) the White Nile River, North Africa; and (D) the Darling-Murray drainage system, Australia. The former extents of deserts tend to coincide with the areas of network channel systems. (For a more thorough discussion of drainage network development, see Chapter 7.)

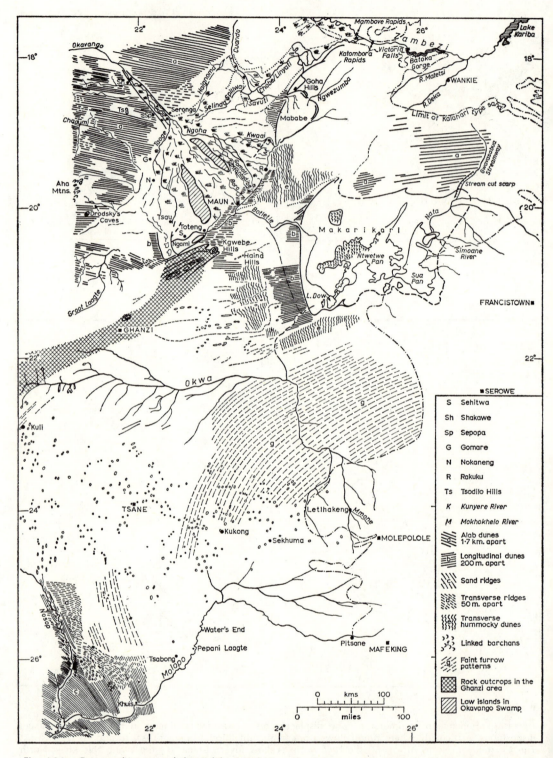

Fig. 4.24. Geomorphic map of the Kalahari and Ngamiland. Note particularly the Zambezi River in relation to the Okavango Inland "Delta" area in which the channel networks compare favorably with those occurring elsewhere where there has been climate change (cf. Fig. 4.23). (From Grove, 1971.)

late Quaternary Glacial conditions that may be, "delayed, multiple-factor or inverse . . ."; the point being that lacustrine maxima do not precisely match glacial times, are sometimes interglacial, and nonglacial factors may be involved.

Geomorphic system distributions and recent changes in these in South America are rendered more complex than those of Africa by several factors. The north-south trending Andes Mountains is a major influence that tends to accentuate the east-west continental environment differences one would ordinarily anticipate for oceanic and atmospheric reasons in any case. Also, there are suggestions of pronounced atmospheric temperature inversion influences on moisture distribution, though these affect parts of Africa also. The resulting geomorphic patterns are complicated by topographic features which intersect various atmospheric moisture zones in some instances and fail to do so in others. A number of recent geomorphic shifts are indicated, however (cf. Bigarella and Garner, 1974). West of the Andes, the boundary between vegetated and barren geomorphic systems appears to have sporadically shifted north and south along the mountain front and coastal lowlands.* Since the boundary is inclined upward from north to south (Fig. 4.17), crestal peaks in the Peruvian Cordillera Occidental have sporadically been affected (cf. Garner, 1959a), whereas coastal areas there have remained arid. In Ecuador, the coastal lowland has been alternately arid and humid with the most recent apparent shift being a southward displacement of humid morphogenesis. Radiometric dating is lacking for these geomorphic adjustments just as it is for many similar events in South America. Since the Ecuadorian land conditions correspond to surface temperature conditions in the adjacent Pacific Ocean, causes for the adjustments may involve changes in the positions and/or temperatures of marine currents, notably the cold Peru Current and warm equatorial counter current (El Niño).

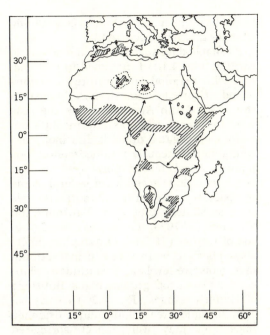

Fig. 4.25. Major geomorphic zone shifts, Africa (generally involving expansion of humid, vegetated areas shown by parallel lines), mainly since the Altithermal but involving tendencies initiated at the end of the Wisconsin Glacial Stage.

East of the Andes Mountains, geomorphic studies by Garner (1958, 1959a, 1959b, 1966a, 1967, 1968), Bigarella *et al.* (1965), Bigarella and Ab Saber (1964), Bigarella and Mousinho (1966), Mousinho de Meis (1971), and others have recently been supported by faunal and floral researches by Haffer (1969), Vanzolini and Williams (1970), Vuilleumier (1971), and by sedimentologic findings of Damuth and Fairbridge (1970). The clear consensus here is that much of the present area of the Doldrum Belt became arid during glacial episodes—e.g. plant cover has several times been widely discontinuous where there are now tropical rain forests—the most recent interval of widespread aridity being roughly correlative with the Wisconsin Gracial Stage (see Bigarella and Garner, 1974). The paleogeomorphic map of South America prepared by the latter authors is

---

* For an in-depth discussion of climatic orogenesis, see Chapter 10.

helpful in visualizing the geomorphic changes that have occurred.

Available data suggest that the adiabatic gap (cf. Fig. 4.18) functioned in conjunction with regional temperature inversions to induce low-elevation dryness in eastern equatorial South America, while the seas were most cold, whereas intermediate elevations on isolated uplands and mountain fronts simultaneously retained continuous plant cover and humid morphogenesis. The latter conditions account for several of the persistently moist floral and faunal *refuges* cited by Haffer (1969) and Vanzolini and Williams (1970)—the *humid climatic nuclei* of Garner (1959b). Orographic effects also appear to be involved in maintaining high moisture levels and continuous plant cover on some topographic highs. But some of Haffer's refuges (Fig. 4.26) clearly occupy lowland areas along major drainage lines and may well reflect spring-fed swamps and/or continued sub-alluvial

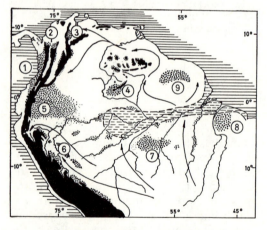

Fig. 4.26. Presumed forest refuges for various organisms requiring humid conditions during dry climatic episodes of the Pleistocene, central and northern South America. Arrows indicate northward-advancing nonforest faunas of central Brazil. Shown are: (1) Chocó refuge; (2) Nechí refuge; (3) Catatumbo refuge; (4) Imerí refuge; (5) Napo refuge; (6) East Peruvian refuges; (7) Madeira-Tapajós refuge; (8) Belém refuge; (9) Guiana refuge—hatchured area indicates extremely problematic inter-glacial Amazonian embayment due to sea level rise above that of the present. Black areas are at elevations above 1,000 m and remained humid except at extreme elevations. (After Haffer, 1969.)

seepage of groundwater from more moist uplands during episodes of low-elevation aridity. Certainly, the tremendous genetic diversity of tropical fresh-water fishes from this region (like that of birds and reptiles) is understandable as a product of species differentiation during isolation. An analogous case of fish isolation and evolution under desert conditions has been described by Brown (1971) in western North America which appears to have become arid while parts of equatorial lands were becoming humid.

As was the case in Africa, there is some suggestion of South American, latitudinally disposed geomorphic zones being shifted north and south. Present steppe-desert associations in Venezuela and Argentina at least in part seem to relate to Subtropical High Pressure belts (cf. Fig. 4.5). Northern morphogenesis is presently occurring under some form of plant cover (glassland or forest) essentially everywhere; though several of the previously cited reports document previously barren conditions extending from the north coastal ranges of Venezuela at least into most of northern Brazil and the Guianas, humid climatic nuclei persisted in isolated uplands. Work by Bigarella and others (1965) noted that forests in southern Brazil extend far southward into formerly barren regions. The distribution of relict arid alluvial deposits in the adjacent Andean Cordillera Oriental is, however, notably asymmetric (cf. Garner, 1968). The southern valley alluvium is less extensive than that to the north but is less completely incised by runoff (cf. Figs. 4.27 and 4.28). Presumably, therefore, the most recent vegetated geomorphic zone shift in South America was possibly somewhat southward at intermediate elevations (2,000–9,000 ft) in the sierra but may have been generally northward at low elevations just as it appears to have been in Africa.

The writer (1967a) attempted an analysis of geomorphic zone shifts, both with respect to the actual causal mechanisms environmentally and resultant changes in groundcover and with respect to environmental space-time considerations and prob-

Fig. 4.27. Alluvial terrace being developed through incision of an alluvial valley fill along the Urubamba River, eastern Peru. Incision depths range up to 150 ft (50 m) in fills up to 600 ft (190 m) thick (cf. Fig. 4.28).

able relict distributions. This and several subsequent studies (some cited above) tend to support several generalizations. ˙

(1) Geomorphic systems do not commonly spring into existence from nothing, so to speak. They usually represent lateral geographic expansions of formerly more restricted environmental conditions or, conversely, isolated remnants of previously widespread environments.

(2) The high-elevation zone of barren morphogenesis related to moisture-depleted air is geomorphically important only on the higher mountain systems, and its lateral geographic expansion is limited by the general absence of land surfaces at requisite elevations.

(3) Remnants of formerly extensive geomorphic areas commonly act as points of origin (*nuclei*) from which subsequent expansions may start. These nuclei are also usually biologic refuges for both plants and animals. In addition, geomorphic nuclei are places where a given type of morphogenesis

has endured longest and, thus, where dynamic equilibrium (*climax*) landforms and deposits are most likely to be encountered (Fig. 4.29).

(4) Moist, vegetated geomorphic nuclei, apart from those induced by fossil groundwater bodies, tend to occur where geographic, atmospheric, oceanic, and topographic factors combine most favorably. Like many of Haffer's refuges, vegetated geomorphic nuclei seem to occur on intermediate elevation uplands, near to and either cross- or downwind from warm water bodies with related precipitation triggering of great consistency—usually combining some form of orographic, convectional, and cyclonic storm precipitation. Portions of the southern Appalachian Mountains constitute a probable example (Fig. 4.22).

(5) Barren, dry geomorphic nuclei, apart from those high on mountain ranges, are commonly situated on very low lands, near the probable long-term locus of cold currents if coastal, or far from warmer

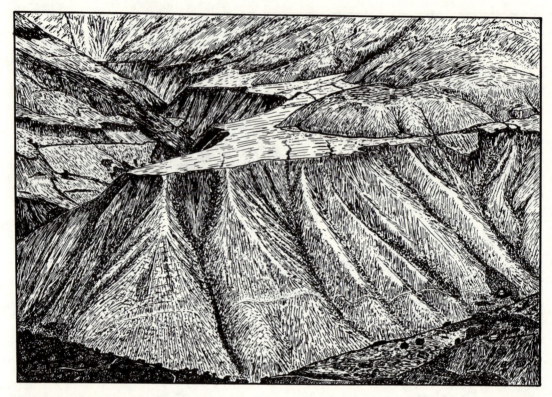

Fig. 4.28. The Mesa de Jajo dissected valley fill in the Rio Motatan Valley, Sierra de Merida, Estado Barinas, Venezuela, the northeast Andean flank—view from main highway looking east. Estimated alluvial fill thickness is 2,500–3,000 ft (800–900 m). House (lower right) at the end of the trail provides scale. See also Fig. 4.27 for comparative alluviation and subsequent incision depths in the southern Andes Mountains.

water bodies, more probably upwind and in closed depressions or in the lee of highlands. The deserts of southwestern United States and northern Mexico as well as the central Sahara qualify as inland representatives, whereas the Atacama in Peru and Chile and the Namib Desert of South West Africa conform to the coastal arid geomorphic nuclei.

(6) Just as the locations of geomorphic nuclei indicate the probable locations of climax landforms and deposits, so these nuclei in relation to regional land slope and geomorphic zone shift direction become a space-time framework for estimation of the duration of the geomorphic system beyond the limits of the nucleus (cf. Garner, 1967a, 1968, Fig. 4.29). Generally, regions farthest from a given type of nucleus show evidence of the most brief exposure to related geo-

morphic effects (possibly none), whereas those closer reflect this proximity in greater relict development (Fig. 5.52).

USE OF THE GEOMORPHIC SPACE-TIME CONCEPT

Our initial intention in this chapter has been to identify the main surficial landform-making systems and to relate them to the various environmental frameworks that appear to be causal. The frameworks, it turns out, are neither rigid nor even permanent. Rather, they remind one of a tent, being both collapsible and portable in most cases. The detailed attributes of the major geomorphic environment-systems will be developed in the next and also in later chapters. Here we have only attempted to de-

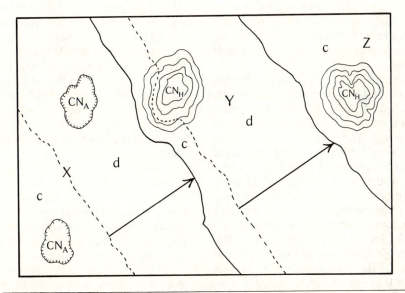

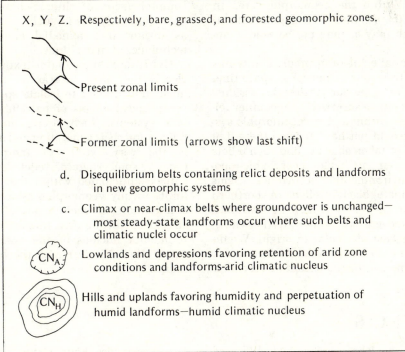

X, Y, Z. Respectively, bare, grassed, and forested geomorphic zones.

Present zonal limits

Former zonal limits (arrows show last shift)

d. Disequilibrium belts containing relict deposits and landforms in new geomorphic systems

c. Climax or near-climax belts where groundcover is unchanged—most steady-state landforms occur where such belts and climatic nuclei occur

CN_A Lowlands and depressions favoring retention of arid zone conditions and landforms-arid climatic nucleus

CN_H Hills and uplands favoring humidity and perpetuation of humid landforms—humid climatic nucleus

Fig. 4.29. Diagram of geomorphic zone relationships and displacements in relation to topography, time, and climatic nuclei. Nuclei are, logically, biologic refuges for organisms of correlative type and, simultaneously, sites for enduring landform genesis under a particular condition.

limit the recognizable geomorphic system categories, their usual ranges in time and space along crustal interfaces, their apparent controls, and some of their broader interactions.

One consequence of our synthesis is that geomorphic systems are seen to occupy areas within which related landform-deposit changes may exist in several stages of development. The nature of the changes tends to be relatively uniform in kind within a given system, but only relatively, since rate and type of change were earlier shown (Chapter 1) to be a function of time which (in the geomorphic zone context) may be indicated spatially by geographic-topographic location (cf. Fig. 5.51). The distribution of landform deposit types and stages within the geomorphic system that is forming them is therefore usually non-random. Within the geomorphic area in question, there is thus a developmental pattern which may or may not be readily apparent.

A potential element of predictability has emerged from our geomorphic space-time discourse to guide our further discussions. We would not be surprised to encounter relict features within a given geomorphic system that could not have formed there, but we would be taken aback to find such relicts beyond their probable geographic-topographic-environmental occurrence. Thus we anticipate glacial relicts in northern Europe or in North American lowlands, since we recognize that the causal environments are basically polar in origin. We do not expect such relicts in the southern parts of the same continents except at very high elevations and would be astonished to find widespread evidence for recent low-elevation glacial activity near the equator. Similarly, arid relicts in a humid area or humid relicts in arid regions should not astonish us, unless they were found in an area supposed to be a geomorphic-environmental nucleus of the opposite type. Thus, an arid relict in a presumed humid geomorphic nucleus would indicate that the nucleus was incorrectly identified and/or located.

In point of fact, it is the geographic distribution of geomorphic relicts that determines the former geographic ranges of causal geomorphic systems cited here, and certainly many of these former ranges are very imperfectly known. Searching out such paleo-environmental patterns through the study of landforms and deposits is a worthy objective for geomorphologists. An eventual comprehension of climate-change mechanisms probably lies in such studies as well as insight into remedial environmental techniques of many kinds.

Our future course in this text is therefore clearly indicated. Insofar as possible we must first attempt to isolate specific landform-deposit effects of particular geomorphic systems. Such effects are potential relicts, and their generation and destruction is the very "stuff" of morphogenesis. Clearly, some stage of relict development can be anticipated within the geographic limits of any geomorphic system, but the culmination of such a set of processes should be encountered in the actual environmental nucleus of the system if one exists. We will start with the humid geomorphic system.

REFERENCES

Bartholomew, John (1950) *The Advanced Atlas of Modern Geography*, McGraw-Hill Book Co., New York, 115 p.

Bennett, H. H. (1939) *Soil Conservation*, McGraw-Hill Book Co., New York, esp. pp. 145–148.

Bigarella, J. J. and Ab Saber, A. N. (1964) "Paläogeographische und päläokimatische Aspekte des Känozoikums in Südbrasilien," *Zeits. für Geomorphology*, bd. 8, pp. 286–312.

Bigarella, J. J. and de Andrade, G. O. (1965) "Contributions to the Study of the Brazilian Quaternary," *Geol. Soc. Amer. Spec. Pap.* 84, pp. 433–451.

Bigarella, J. J. and Mousinho, M. R. (1966)

"Slope Development in Southeastern and Southern Brazil," *Zeits. für Geomorphology*, bd. 10, pp. 150–160.

Bigarella, J. J. and Garner, H. F. (1974) "Preliminary Paleogeomorphic Map of South America during the Quaternary Period," *Paleogeography and Pleistocene Periglacial Phenomena Nanka, Moscow*, 16 mscp pp., 1 map.

Bowman, Isaiah (1916) *The Andes of Southern Peru*, Henry Holt Co., New York, 336 pp.

Brown, J. H. (1971) "The Desert Pupfish, *Scientific American*, v. 225, pp. 104–110.

Büdel, J. (1953) "The 'Periglacial' Morphologic Effects of the Pleistocene Climate over the Entire World" (Engl. Transl.), *Int. Geol. Rev.* 1959, v. 1, pp. 1–16.

Büdel, J. (1959) "Periodische und episochische Solifluction im Rahmen der klimatischen Solifluktconstypen," *Erdkunde*, v. 13, pp. 297–314 (Engl. Summary).

Butzer, K. W. (1951) "The Recent Climatic Fluctuation in Lower Latitudes and General Circulation of the Pleistocene," *Geogr. Ann. Stockh.*, v. 39, pp. 91–111.

Butzer, K. W., Isaac, G. L., Richardson, J. L. and Washbourn-Kamau, C. (1972) "Radiocarbon Dating of East African Lake Levels," *Science*, v. 175, pp. 1069–1076.

Charlesworth, J. K. (1957) *The Quaternary Era*, Edward Arnold, London, 2 vols.

Cole, F. W. (1970) *Introduction to Meteorology*, John Wiley and Sons, New York, 388 pp.

Damuth, J. E. and Fairbridge, R. W. (1970) "Equatorial Atlantic Deep-Sea Arkosic Sands and Ice-Age Aridity in Tropical South America," *Geol. Soc. Amer. Bull.*, v. 81, pp. 189–206.

Davis, L. C. (1969) "The Biostratigraphy of Peccary Cave, Newton County, Arkansas," *Ark. Acad. Sci. Proc.*, v. 23, pp. 192–196.

Flint, R. F. (1971) *Glacial and Quaternary Geology*, John Wiley and Sons, New York, 892 pp.

Friedman, G. M. and Sanders, J. E. (1970) "Coincidence of High Sea Level with Cold Climate and Low Sea Level with Warm Climate Evidence from Carbonate Rocks," *Geol. Soc. Amer. Bull.*, v. 81, pp. 2457–2458.

Garner, H. F. (1958) "Climatic Significance of Anastomosing Channel Patterns Typified by Rio Caroni, Venezuela," *Geol. Soc. Amer. Bull.* (Abstract), v. 69, pp. 1568–1569.

Garner, H. F. (1959) "Stratigraphic-Sedimentary Significance of Contemporary Climate and Relief in Four Regions of the Andes Mountains," *Geol. Soc. Amer. Bull.*, v. 70, pp. 1329–1368.

Garner, H. F. (1959b) "Interpretation of Globally Distributed Anastomosing Channel Drainages," *Geol. Soc. Amer. Bull.* (Abstract), v. 70, p. 1607.

Garner, H. F. (1959c) "Recent Climate Change Patterns—A Mode of Control," *Geol. Soc. Amer. Bull.* (Abstract), v. 70, p. 1607.

Garner, H. F. (1965) "Base-Level Control of Erosion Surfaces," *Ark. Acad. Sci. Proc.*, v. 19, pp. 98–104.

Garner, H. F. (1966a) "Derangement of the Rio Caroni, Venezuela," *Rev. Geomorphology Dyn.*, v. 16, pp. 54–83.

Garner, H. F. (1966b) "Groundwater Aquifer Patterns and Valley Alluviation along Mountain Fork Creek, Crawford County, Arkansas," *Ark. Acad. Sci. Proc.*, v. 20, pp. 95–103.

Garner, H. F. (1967a) "Geomorphic analogs and Climax Morphogenesis," *Ark. Acad. Sci. Proc.*, v. 21, pp. 64–76.

Garner, H. F. (1967b) "Rivers in the Making," *Scientific American*, v. 216, pp. 83–94.

Garner, H. F. (1968a) "Climatic Geomorphology" [in] *The Encyclopedia of Geomorphology*, R. W. Fairbridge, ed., Reinhold Book Co., New York, esp. pp. 129–130.

Garner, H. F. (1968b) "Tropical Weathering and Relief" [in] *The Encyclopedia of Geomorphology*, R. W. Fairbridge, ed., Reinhold Book Co., New York, esp. pp. 1161–1172.

Geiger, R. (1965) *The Climate Near the Ground*, Harvard Univ. Press, Cambridge, Mass., 611 pp.

Grove, A. T. (1958) "The Ancient Erg of Hausaland, and Similar Formations on the South Side of the Sahara," *Geogr. Jour.*, v. 124, pp. 526–533.

Grove, A. T. (1971) *Landforms and Climate Change in the Kalahari and Ngamiland*, *Geogr. Jour.*, v. 135, pp. 191–212.

Grove, A. T. and Warren, A. (1968) "Quaternary Landforms and Climate on the South Side of the Sahara," *Geogr. Jour.*, v. 134, pp. 194–208.

Haffer, J. (1969) "Speciation in Amazon

Forest Birds," *Science*, v. 165, pp. 131–137.

King, L. C. (1953) "Cannons of Landscape Evolution," *Geol. Soc. Amer. Bull.*, v. 67, pp. 721–752.

Leopold, L. B. Emmett, W. W. and Myrick, R. M. (1966) "Channel and Hillslope Processes in a Semiarid Area, New Mexico, *U. S. Geol. Surv. prof. pap.* 352-G, pp. 189–253.

Lugn, A. L. (1960) "The Origin and Sources of Loess in the Great Plains in North America," *Rept. Int. Geol. Cong. XXI, Copenhagen part XXI*, pp. 223–235.

Lustig, L. K. (1965) "Clastic Sedimentation in Deep Springs Valley California," *U. S. Geol. Survey, Prof. Pap.* 352-F, pp. 131–192.

Lustig, L. K. (1967) "Appraisal of Research on Geomorphology and Surface Hydrology of Desert Environments," [in] *Deserts of the World*, Univ. of Arizona Press, Tucson, pp. 93–283.

Mercer, J. H. (1972) "Chilean Glacial Chronology 20,000 to 11,000 Carbon-14 Years Ago: Some Global Comparisons," *Science*, v. 176, pp. 1118–1120.

Morrison, R. B. (1968) "Pluvial Lakes" [in] *The Encyclopedia of Geomorphology*, R. W. Fairbridge, ed., Reinhold Book Co., New York, esp. pp. 873–883.

Mousinho de Meis, M. R. (1971) "Upper Quaternary Process Changes of the Middle Amazon Area," *Geol. Soc. Amer. Bull.*, v. 82, pp. 1073–1078.

Neiburger, Morris, Johnson, D. S. and Chien, Chen-Wu (1961) "The Inversion over the Eastern North Pacific Ocean" [in] *Studies of the Structure of the Atmosphere over the Eastern Pacific Ocean in Summer*, v. 1, Univ. of California Press, Berkeley, pp. 1–94.

Neuberger, Hans and Cahir, John (1969) *Principles of Climatology*, Holt, Rinehart and Winston, New York, 178 p.

Prescott, J. R. U. and White, H. P. (1960) "Sand Formations in the Niger Valley between Niamey and Bourem," *Geogr. Jour.* v. 126, pp. 200–203.

Quinn, J. H. (1956) "Plateau Surfaces of the Ozarks," *Ark. Acad. Sci. Proc.*, v. XI, pp. 36–43.

Quinn, J. H. (1961) "Prairie Mounds of Arkansas," *Ark. Archaeological Soc., Newsletter*, v. 2, pp. 1–8.

Rumney, George R. (1970) *The Geosystem*, Wm. C. Brown Co., Dubuque, Ia., 135 pp.

Schumm, S. A. and Lichty, R. W. (1963) "Channel Widening and Floodplain Construction along Cimarron River in Southwestern Kansas," *U. S. Geol. Surv. prof. pap.* 352-D, pp. 67–88.

Thornthwait, C. W. (1948) "An Approach toward a Rational Classification of Climate," *Geogr. Rev.*, v. 38, pp. 55–94.

Trewartha, G. G. (1968) *An Introduction to Climate*, McGraw-Hill Book Co., New York.

Vanzolini, P. E. and Williams, E. E. (1970) "South American Anoles: The Geographic Differentiation and Evolution of the Anolis Chrysolepsis Species' Group (Sauria, Iquanidae), Pt. I, text; Part II graphs & maps, *Arq. Zool. Estado Sao Paulo*, v. 19, pp. 1–298.

Vuilleumier, B. J. (1971) "Pleistocene Change in the Fauna and Flora of South America, *Science*, v. 173, pp. 771–780.

Wright, H. E. Jr., and Frey, D. G., eds. (1965) "The Quaternary of the United States, *Internat. Assoc. Quatern. Research, 7th Cong.*, Princeton Univ. Press, Princeton, N. J., 922 pp.

5 Humid geomorphic systems and landforms

In this chapter we will be mainly concerned with the moist, vegetated setting as a regimen for landform genesis. The principal mechanisms available therein for shaping terrain include differential weathering, groundwater solutioning, fluvial activity along surficial watercourses, and mass wasting. The character of most of this activity is strongly influenced or even governed by a groundcover of vegetation so we will consider that relation first. Then we will examine weathering, both as a mechanism for differentially generating relief and as a preparatory stage for erosional removal of rock debris. We will also comment on mass wasting as a debris transfer phenomenon related to and shaping hillslopes, and particularly slopes developed in conjunction with running water. More or less simultaneously we will scrutinize fluvial erosion, both as a means of generating relief and as a mechanism for transporting surplus water and rock waste out of the system. Finally, we will discuss the over-all effects of agency interaction within the humid geomorphic system, in terms of landforms and deposits, and beyond the systemic boundaries where expelled water and sediment cause morphogenesis elsewhere.

THE VEGETATED SETTING

Consider the places on earth where plants grow in abundance. What attributes do these regions have in common? The logic behind this question is comparatively simple. The essentially "invariable" ingredients of a geomorphic system must govern its ultimate effects. Incidental or passing factors may impose a temporary aspect on a landscape but should eventually become suppressed by the over-all evolutionary tendency of the system. To be valid, the evolutionary tendency should be somewhat expressed, in spite of special geologic settings or unique environmental sequences, and should become more pronounced with the passage of time. We will, therefore, now summarize the "invariable" attributes of humid geomorphic systems as recorded by researchers throughout the world. A detailed analysis of each will follow. In humid geomorphic systems:

1. There is a CONTINUOUS GROUNDCOVER

OF LIVING PLANTS (grasses, shrubs, trees) occupying the interface between ground and air over 95% of the land area on the average, atmospheric effects being therefore largely imposed on the plant cover rather than directly on the lithosphere. The actual taxonomic makeup of the vegetation, its density and phyletic variability differ considerably from one humid region to another (Fig. 5.1).

2. There is A VARIETY OF MOISTURE-LOVING ANIMALS representing almost every phylum, most important of which are bacteria and a suite of crawling and burrowing insects, worms, reptiles, mammals, and arachnids which exhibit close dependency on the plant community and the subjacent weathered interval.

3. There is HUMIC MATERIAL OCCUPYING THE GROUND SURFACE and generally mixing with the mineral matter of that surface in a vertical interval from less than a centimeter to more than a meter in thickness; usually dark-gray, brown, or black in color, the material ranges from newly fallen leaves, twigs, and branches to thoroughly decomposed vegetal debris.

4. There is an INTERVAL OF UNCONSOLIDATED ROCK DEBRIS in which the bulk of the vegetation is rooted and some of which consists of materials that are the chemical decomposition products of mineral alteration —either from subjacent bedrock, from that upslope, or from associated transported regolith. This *soil* of dominantly chemical origin may range from a fraction of a centimeter to many meters in thickness; it has its highest organic content near the surface, it tends to be distinctly layered in many thick examples developed on nearly horizontal surfaces, and it usually includes only minor mechanically fragmented constituents.

5. More water enters vegetated areas as precipitation than is utilized there so that A WATER SURPLUS MOVES DOWNSLOPE OUT OF THE VEGETATED SETTING into adjacent lands or water bodies and carries with it

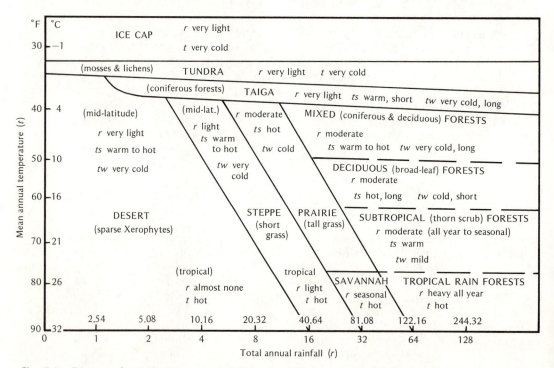

Fig. 5.1. Diagram showing major types of groundcover in relation to rainfall and temperature. Key: r, character of rainfall; t, temperature; t_s, summer temperature; t_w, winter temperature. (After Jackson, 1970.)

products of weathering, erosion, and biologic activity. Water within the humid geomorphic system is OVERTLY manifested for much or all of each year by moist ground surfaces, standing water areas, distinct watercourses which on the surface aggregate less than 1% of the environmental area, and water-saturated pore space in the ground from the surface or a near-surface *watertable* to the bottom of the *zone of fracture* within what is normally termed the *groundwater body* (Fig. 5.2). Abundant, more or less permanent water in appreciable amounts is further SUBVERTLY indicated by each of the invariable attributes of humid geomorphic systems just cited including plant growth and associated organic activity, soil chemistry, and rock solutioning.

6. NATURAL BEDROCK OUTCROPS ARE SCATTERED AND RATHER TEMPORARY, aggregate less than 3% of the average humid land area, commonly lack plant cover, and are concentrated along watercourses. Through direct contact with atmospheric conditions, such rock exposures indicate the degree to which chemical versus mechanical weathering dominates in the absence of regolith or plants—solid materials freed from rock exposures tend to enter the soil-making events occurring farther downslope.

7. SOME LANDSLOPES EXHIBIT DISTINCT INCLINATIONS TOWARD WATERCOURSES, the latter being either located at the surface or expressed there by partly subterranean solutional voids. These slopes range from the low banks of shallow watercourses to regional land surfaces extending to major drainage divides, and there is evidence, on the inclined areas, of downslope movement of weathered material.

The vegetated settings appear to have no major attributes of either landform or deposit that appear to be invariably associated other than those just listed. Bedrock types, relief amounts and configurations, regolith kinds or distributions (apart from zoned soils), and stream load characters and amounts all vary greatly from one humid land to another. Some humid areas are

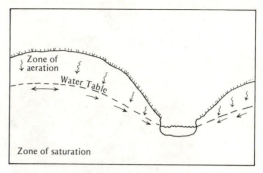

Fig. 5.2. Cross section showing water movement in the ground in relation to the groundwater body and surface runoff in a humid region.

nearly flat plains; some are mountainous; some are alluviated, and some are not; some yield large amounts of solid sediment, whereas others yield virtually none. The implication seems clear that these highly variable factors either change as the humid geomorphic system evolves, or they are introduced independently by agencies of other geomorphic systems.

We would not wish to have our use of the term "invariable" misunderstood with regard to attributes of geomorphic systems listed previously or in sections to come. For example, continuous plant cover DOES invariably typify humid lands. But there are breaks. In artificial settings, such breaks are common, but, in "natural" humid settings, breaks in plant cover are generally limited to slopes too steep to retain soil or provide extensive root attachment (Fig. 5.3), banks of watercourses (Fig. 5.4), landslide scars, erosional or structural escarpments, "new land" (e.g. volcanic terrain, cf. Fig. 3.94), and poisoned or sterile surfaces —for example pyritic shale may react with rainwater to form acidic ground locally toxic to plants as seen in the Mississippian age Fayetteville Formation and similar Paleozoic formations along the southern margin of the Ozark Uplift (Fig. 5.5). The point should also be made that only the plant cover breaks along watercourses tend to persist, presumably because of continuing erosion there. The other types tend to heal with the passage of time as slopes adjust

Fig. 5.3. Locally barren, solutionally over-steepened slopes of tropical karst developed in various non-carbonate bedrocks, the eastern Andes Mountains, Peru, along the Rio Urubamba, an Amazon tributary.

and soils thicken and become more extensive.

The preceding paragraph is intended to re-emphasize a point students of geology quickly begin to realize, namely, that there are exceptions to almost everything. The issues needing resolution in most scientific discussions relate to dominant versus sub-ordinate causes and effects rather than to unique solutions. Generalizations, including those previously made herein about the humid geomorphic system, represent efforts to establish over-all, long-term, and, therefore, controlling characters as opposed to localized, brief, and, consequently, subordinate effects.

In our attempts to understand humid geomorphic systems, it is essential to initially consider the actual physical conditions that prevail there, it being a tacit assumption that the "invariable" attributes of the humid system just cited are an essentially direct consequence of these physical conditions. We shall consider secondly the geomorphic features (e.g. landforms and deposits) that appear to owe their genesis, either initially or with the passage of time, to the humid geomorphic system. From time to time we will comment on the effects of special geologic settings or particular antecedent non-humid environments. In the course of these discussions we will also consider the effects on and by humid systems, and coexisting geomorphic systems

of other types, particularly through the effects of wind and running water.

Humid environments are presently widespread; they are now known to have undergone sporadic displacement, reduction, and accentuation down through geologic time, but they have probably always been expressed "somewhere" on the earth's surface since a hydrosphere came into existence. The earliest moist regions lacked vegetation, of course. Because conditions in humid lands and along rivers flowing therefrom are among the most habitable by man, records of geomorphic effects there are among the most extensive available. As a consequence the literature abounds with accounts of phenomena PRESUMABLY peculiar to humid regions. Recent recognition

Fig. 5.4.   Possibly the densest form of vegetal cover, a tropical rain forest, in the valley of the Rio Javita, southwest Ecuador. On the outside of a stream bend, an outcrop of Paleocene, San Eduardo Limestone is visible (light central area), apparently maintained through undercutting by channel gravels.

Fig. 5.5. An expanse of sterile, pyritiferous, black shale of the late Mississippian Fayetteville Formation exposed southwest of Fayetteville, Arkansas. Acids in the loose shale chips inhibit plant growth and soil formation in this humid and otherwise densely vegetated region.

conditions are relics from other types of prior conditions. We therefore require a set of geomorphic standards for humid conditions to serve as a basis for comparative evaluations of humid and non-humid effects.

Phenomena, to serve as a comparative standard for humid morphogenesis, should be encountered in regions that have been most intensively subject to humid conditions for the longest periods. In other words, in considering humid physical conditions, we should include regions that have somehow escaped the widespread effects of environmental alternation. This escape may involve extended exposure to only humid conditions, or alternatively, humid effects may merely have dominated in time and intensity over opposing conditions. Not surprisingly, therefore, many of the geomorphic features cited hereafter as "typically humid" were recorded in settings that various researchers believe were humid for very extended periods (e.g. probable humid "climatic" nuclei noted in Chapter 4). In a few well-known instances, recent endogenic or mesogenic events (i.e. volcanism) have projected new land directly into humid geomorphic systems without prior exposure to other surficial environments (cf. Fig. 9.4). Once the essential landform-deposit effects of humid morphogenesis have been discussed, we can consider more fully the time-dependent changes of these geomorphic features and evaluate various non-humid associations. A similar procedure directed toward other types of subaerial geomorphic systems in later chapters should ultimately permit us to identify and to evaluate relics of essentially any system whether they are presently associated with their parent condition or not.

that global environmental successions include non-humid episodes make it evident that many things encountered under humid

HUMID GEOMORPHIC PARAMETERS

PRECIPITATION AND TEMPERATURE VARIANTS

A comparison of world precipitation distribution (Fig. 4.10) with the present distri-

bution of vegetal cover (Fig. 4.14) is sufficient to establish that vegetated lands generally receive the most precipitation (cf. Fig. 5.1). Leopold *et al.* (1964) illustrate variations in process intensity in geomorphic regions "climatically" defined in terms of rainfall-temperature relations (Fig. 5.6). The obviousness of many of the vegetation-moisture relationships compels us to point out that the absolute amount of precipitation in a region is often less important geomorphically than is the amount of water remaining after evaporative, transpirative, hydrative, and similar consumption—water remaining that can become involved in erosional and related processes that directly generate landforms and deposits. Thus, in the erosional experiment of H. H. Bennett (1939) described in Chapter 4 (Fig. 4.15), the precipitation water available for surficial erosion in the form of runoff amounted to 0.12% in a forested setting and 6.5% in a grassland. Translated into water quantities representative of those falling on natural grasslands and forests, and assuming that Bennett's percentages would be proportionally valid, it is interesting to note that a deciduous hardwood forest in southwestern Missouri receiving 30 inches (75 cm) of precipitation annually should yield about 0.036 inches (1 mm) in the form of runoff. A similar forest in northcentral Arkansas at about the same latitude receiving 45 inches (205.2 cm) of precipitation would yield approximately 0.054 inches (1.4 mm) as runoff. And a grassland in central Oklahoma at the same latitude receiving 15 inches of precipitation would yield 0.98 inches (25 mm) as runoff. These figures may be assumed to be in error, because terrain and climatic factors differ from those in the original study. Yet the right orders of runoff magnitude are probably indicated for a general location in southcentral North America on terrain developed in Paleozoic sandstone, shale, and limestone generally dipping less than two degrees. And it should be emphasized that the erosional effects of such runoff would be concentrated along watercourses.

In Chapter 4, we generalized that regions with essentially continuous vegetation receive from less than 30 to more than 200 inches of annual precipitation (75–500 cm). Actually, if one regards the short prairie grasses (not bunch grass) as the continuous vegetal type with the lowest moisture requirements, the lower precipitation limit is near 20 inches (50 cm). But temperature is clearly a major determining factor. Northern coniferous forests (Taiga) flourish where annual precipitation is only 12–15 inches (30–40 cm), but summer forest-floor temperatures there rarely exceed 60°F, and forest floors are often continuously moist when they are not frozen. Temperate and tropical forests tend to have correspondingly higher amounts of precipitation and more elevated temperatures with the highest tropical rainfall near 200 inches (1000 cm) falling where the mean annual temperature approaches 75°F. The precise amounts of moisture used by the various types of vegetal growth are not known, mainly because the amount of water losses to pedologic (soil-making) processes and transpiration are very uncertain (cf. Thornthwait, 1948). Runoff can, of course, be measured, and, if temperature is known, evaporation can be calculated very accurately but the unknowns diminish the significance of the figures arrived at in terms of over-all moisture distribution within humid systems. One thing is certain. There is sufficient moisture present to grow the vegetation we observe and to yield some runoff.

Functioning of the humid geomorphic system is not only influenced by the amount of water entering the system but also by the manner of its introduction and distribution in both time and space. From a long-term standpoint, a shift from a barren condition to one that will support grassland or forest clearly involves a gradual increase in the effective moisture of a region. However, the abrupt character of regional vegetal borders noted in Chapter 4 suggests that threshold effects, perhaps in degree of soil development, soil moisture levels, attainment of the water table by root systems, or similar achievements govern the actual appearance of major phytic aspects of humid geomorphic systems (Fig. 5.7).

With respect to the seasonal and individual meteorologic events that tend to govern the intensity of geomorphic change, the form and rate of precipitation becomes very important. Opinions differ somewhat, but

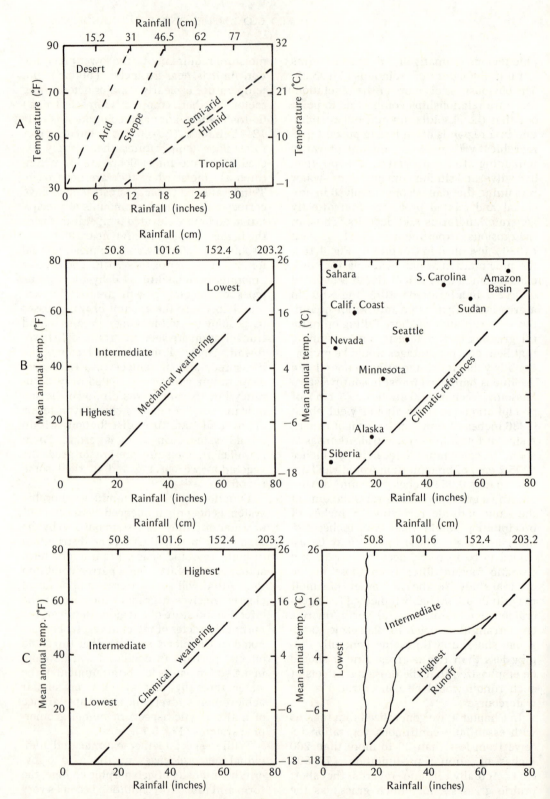

Fig. 5.6. Rainfall-temperature expressions in climatic types (A) and geomorphic processes, weathering, and runoff (B) and (C). (After Leopold, Wolman, and Miller, 1964.)

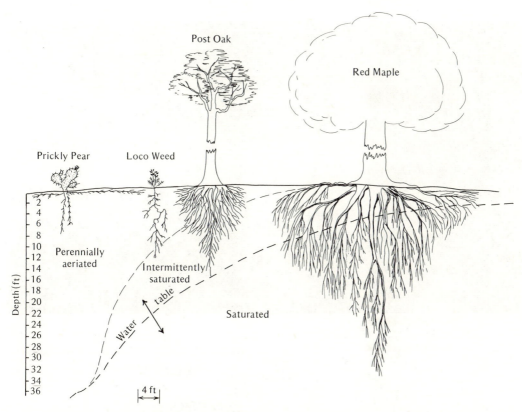

Fig. 5.7. Diagram of selected botanic thresholds related to root system configurations and groundwater distributions in arid and humid climates. The prickly pear and loco weed are not dependent upon the zone of saturation, and at least the former is apparently adversely affected by sustained ground moisture. Post oaks in the Ozarks have woody leaves and can survive severe drouths in spite of ball-like root configurations but may die from the slightest ground disturbance that alters deep seepage. Deciduous hardwoods like the maple depicted rely on the zone of saturation and die when the water table is excessively depressed. (Above-ground dimensions are not to scale.)

the consensus seems to be that regional precipitation events related to weather fronts account for the bulk of the precipitation in vegetated regions, whereas localized, often more violent storms related to differential convective heating usually are subordinate. The opposite relationship may hold for barren regions and some grasslands, though this is disputed by Russell (1936). The local storms (thunderstorms of many temperate and tropical regions) rarely last for more than a few minutes but may be of very high intensity (Fig. 5.8). Three experienced by the writer resemble the many on record in meteorologic archives. Two storms near the eastern margin of the North American High Plains yielded 8 inches (19.5 cm)

of rainfall in 20 minutes in 1957 and 11 inches (28 cm) of rainfall in 45 minutes in 1961. Each included some hail. These "gully-washers," as they are termed in the southern Ozarks, live up to their names in terms of erosional effects produced along existing watercourses and on bare ground. A third thunderhead in 1947 dumped 6 inches (15.2 cm) of rain in 15 minutes on a surveying plane table under which the writer was crouched southwest of Colorado Springs. Other things got wet also. The local character of precipitation in dry regions renders precipitation records there suspect (cf. Reich, 1963).

So-called "flash floods" are a commonplace result of local convective storms in

Fig. 5.8.   A typical tropical downpour over the selva terrain of the North Coastal Range of Venezuela (above) and isopluvial maps of the Americas, Eurasia, Africa, and Australia (below). The maps show maximum precipitation for a 1-hour interval within any 2-year period. (After Reich, 1963.)

drier regions, where parched watercourses suddenly and briefly fill to overflowing. A rather distinct erosional environment is developed in regions of this nature. The storms in question often affect a zone only a mile or so wide and a few miles long, and commonly only part of a small drainage basin or a single reach of a watercourse. Many writers have commented on the fact that a very high per cent of a thunderstorm's precipitation may flow away over the land surface, even where the ground was previously dry and porous, simply because the water falls too fast to soak into the surface or to be taken up by plants (Fig. 5.9).

In the Colorado storm just noted, after about 5 minutes, runoff 1 to 2 inches deep was developed on essentially all ground surfaces in an area of gulleyed granite having some 6 to 8 ft of local relief with a surficial scree of 0.5- to 1-inch diameter angular rubble and a discontinuous plant association of scrub oak lacking an appreciable understory of growth excepting an occasional patch of grass or prickly pear cactus. Runoff from this storm carrying sand and gravel eroded a gully 3 ft deep and 2 ft wide in calcareous, Cretaceous shale where it was locally canalized by a road siding and deposited a notable sediment fan just downslope where the water could spread out. One year previously a storm in the same area filled a gully about 20 ft (6 m) wide and of similar depth with boulder gravel over a distance of more than 100 ft (30 m) in about 20 minutes. During the 1957 Ozark storm, four dumptruck loads of 0.75-inch angular limestone aggregate (±16 cu yd) awaiting a cement surface in the writer's driveway were displaced almost entirely by runoff onto a neighbor's newly sodded lawn 100 ft away down a 8° slope. Neither of us were pleased.

Regional storms, excepting those occasionally induced by hurricanes, rarely cause as intense rainfall as thunderstorms, and such regional storms are mostly associated with frontal passages or monsoonal conditions. Rainfalls related to frontal passages often affect thousands of square miles of land area including many smaller drainage basins and even nearly all of larger drainage systems on occasion. Regional precipitation events commonly last for several hours or even for a few days though aggregate accumulations of moisture associated with frontal passages rarely exceed 2 to 3 inches (5 to 7.5 cm). Because this precipitation normally falls more slowly than in local storms, there may be a period of several minutes to even a few hours before signs of ground saturation appear, depending of course on local groundwater conditions (cf. Fig. 5.9). Naturally the direct force of all types of precipitation in humid lands tends to be dissipated against the vegetation rather than on the ground. Actual erosion due to raindrop impact is insignificant. Away from habitation, early water movements on the ground take the form of infiltration, sub-vegetal percolation, and seepage, and, in associated water courses, peak runoff responses to a given precipitation event may not occur for hours or even days.

Rainfall forms in monsoonal lands merit brief special comment. In such regions heavy rains aggregating several feet of precipitation have been known to fall uninter-

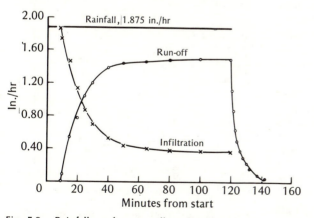

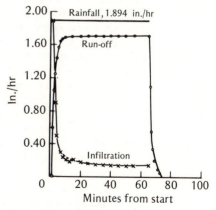

Fig. 5.9.   Rainfall, surface runoff, and infiltration (A) on forested land (55-year-old pine) and (B) on bare and abandoned land, the Tallahatchie River Basin, Mississippi. (After Sherman and Musgrave, 1942.)

ruptedly over a period of days or even weeks. In some Doldrum Belt areas, the monsoonal rains may be expressed as daily rains usually developing each afternoon during the wet season. Since the monsoonal areas also usually experience a pronounced seasonality of precipitation and endure severe drouths, monsoonal rains initially tend to fall on parched surfaces underlain by depressed water tables. As the seasons change, soil moisture levels and groundwater table elevations usually fluctuate markedly, and discharge levels of watercourses tend to reach a peak at the end of the rainy season. It is not unusual for smaller tributary channels in humid regions to be dry a good bit of the time, since the groundwater body upon which they draw is limited. But in monsoonal regions comparatively large rivers range from bank-full or even overflowing to essentially dry or with only isolated pools and sub-alluvial seepage between rainy and dry seasons.

In the context of precipitation forms in vegetated lands, we would be remiss not to note the annual snowfall accumulations of north-temperate, subpolar, and mountainous areas, much of which melts during a few weeks of each Spring. Because subjacent ground is frozen at this time of year such snow meltwater cannot appreciably soak into the surface, and the per cent of winter precipitation that develops into runoff is very great. In heavy snowfall areas such as New England and northern New York, amounts of water released by Spring snow melting have necessitated extensive reservoir flood control systems (e.g. along the Hudson River). Relatively moist mountain peaks in the western United States and parts of Asia often discharge snow meltwater into lowland deserts each Spring to provide the bulk of the discharge of main streams with upland tributaries.

From the foregoing it is evident that precipitation of vegetated lands tends to be spread out over an appreciable portion of each year, whereas a desert may receive its entire annual precipitation from a single shower. When one considers this in conjunction with the quantities of annual pre-

cipitation involved (100, 200, and 500 cm being common amounts recorded in heavily vegetated places and 5, 10, and 25 cm being near maximum for many barren lands), it is tempting to conclude that running water is of only minor significance in deserts compared to humid regions and morphogenesis of dry lands is thus so slow as to be insignificant. Some researchers have been unable to resist the temptation, so it seems important to point out that appearances can be misleading.

The per cent runoff and erosion rates cited by Bennett (1939) and discussed in Chapter 4 are generally corroborated by erosion rates, as cited by Langbein and Schumm (1958) (see Fig. 5.10). Both cite effects due to plant cover variation expressed in sediment yield rates. In point of fact, if we accept Bennett's runoff percentages as being of the right general magnitude, an essentially barren desert receiving an annual average of 5 inches (12.75 cm) of precipitation could be subject to as much

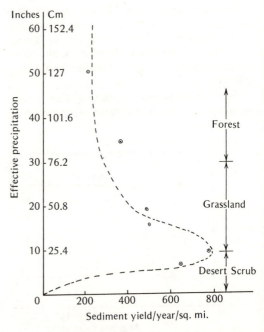

Fig. 5.10. Sediment yield variation by runoff in relation to various types of plant cover. (From Langbein and Schumm, 1958.)

as ten times the amount of erosional energy in the form of runoff (2.44 inches or 6 cm) as would a humid vegetated area receiving 200 inches (1000 cm) of rainfall with a runoff of only 0.12% or 0.24 inches (0.6 cm). Langbein and Schumm's figures indicate only a fourfold difference, but it is of similar nature. It is freely admitted that essentially all the numerical values for runoff just cited are probably in error, because of differences in evaporation rate and regolith character between naturally barren and vegetated lands, if for no other reasons. But the issue at hand concerns orders of magnitude rather than precise numerical values of fluvial erosion intensity. And the point seems to be made. There is as much or more potential fluvial energy introduced by precipitation into barren lands as runoff each year than there is entering most vegetated areas. Where the runoff goes in each setting and what it accomplishes will be taken up in later sections.

The relationships indicated by Leopold *et al.* (1964) between geomorphic process types, intensities, and precipitation-temperature relations, and others also illustrated here (Fig. 5.6) are far from invariable. There is an element of validity if we assume constancy of geologic factors (e.g. bedrock weathering susceptibility, terrain permeability) and environmental sequence. However, the sequences in which environments develop and their durations will be shown to vary and thereby vastly affect process intensity. The precipitation/temperature relations to major vegetal variants has been indicated (Fig. 5.1), but temperature rather independently enters the picture in a variety of threshold phenomena. For example, the decomposition of humic materials is notably retarded by the low mean annual temperatures of north-temperate and subarctic lands (cf. Fig. 5.18). Peaty, or otherwise humic-rich soils are a notable consequence. Also, the 32° *isotherm* delineates areas where freeze-thaw phenomena do and do not occur with corresponding direct consequences on biologic, weathering, and erosional processes. Present mean annual precipitation and temperature relations of vegetated lands (cf. Figs. 4.10, 4.13, 4.14, and 5.1) indicate major sectors in which potential variants could occur.

VEGETAL COVER

For reasons previously outlined, we are not concerned merely with the occurrence of plants on the earth's surface but rather with the special differences that appear to exist between areas where plants provide an essentially continuous "cover" for the earth and those places where plant cover varies from localized to virtually nonexistent. The first illustration of this chapter (Fig. 5.1) lists the major plant categories that appear to satisfy the requirements of the "continuous cover" category. The alternative circumstance will be treated in Chapter 6.

The obvious physical differences between the major plant cover categories previously cited provide some insight into geomorphic variants that occur within regions otherwise classed as humid. Among other characteristics, floral groups differ in density, canopy height above the ground surface, depth of root systems, need for water (and hence actual occurrence of major kinds of plants indicate water amounts available), homogeneity of expression, seasonality, and taxonomic variability. And perhaps the best way to determine the significance of floral assemblage variability is to consider briefly the apparent geomorphic potential of plants where they are abundant.

As a general condition, in humid lands plants are understood (1) to absorb the energy of falling precipitation to the extent of essentially eliminating rain splash erosion; (2) to frictionally resist surface water flow effectively everywhere except along water courses—usually to the point where actual overland flow elsewhere is prevented; (3) to aid infiltration of water by loosening ground in growth areas, generating root passages, and providing a favorable habitat for burrowers; (4) to use water in a variety of plant "metabolic" processes associated with photosynthesis and to transpire large amounts back to the atmosphere (cf. Thornthwait, 1948); (5) to affect groundwater solvency and hence regolith and bedrock leaching through additions of organic acids and carbon dioxide; (6) to influence soil textures and fertility by growth, additions of humic material, and nitrogen fixation; (7) to shield weathered debris from erosion by wind and running water; (8) to resist down-slope movements of regolith under the force of gravity through root anchors extending to bedrock (Fig. 5.11); (9) to inhibit circulation and otherwise contribute to sediment entrapment in swamps, along lake margins, and elsewhere where there is an appreciable infall of wind-blown dust; (10) to mechanically exploit weaknesses in soil grains or bedrock by root extension; (11) to shade or otherwise insulate the ground surface from short-term air temperature extremes, both heat and cold; and (12) to indicate by the major phytic categories present the amount of water actually available at and near the ground surface.

From the list of plant attributes just cited, it is readily apparent that plant cover constitutes a humid geomorphic parameter the value of which is difficult to overestimate. The researching geomorphologist may have first gleaned some insight into a region's environment from a climatic description, map, or chart, but most such

Fig. 5.11.   Aspens and pines near the north rim of the Grand Canyon, the Kaibab Plateau. The tree growth form expresses mass-wasting modes. Young aspens lack a taproot to bedrock and tilt downslope because of downslope regolith movement (left, foreground), but mature trees attain a bedrock anchor and straighten with a resulting trunk curvature (right, fore- and background).

sources are seen to comprise crude indications compared with an on-the-spot examination of groundcover. In contrast to deserts where plants commonly grow in isolation or locally bunched, it is the very continuity of humid plant associations that is most impressive. But such continuity carries with it the implication of a favorable growth surface. True, a plant will grow in a crack in an otherwise bare rock (Fig. 5.12), but no farmer in his right mind would sow his seed on bare rock. So, where plant cover is dense and continuous, the observer can generally be sure that the growth surface is some type of rather finely divided granular substance, though not necessarily one developed under plant cover.

Recent sediments, whether marine, fluvatile, lacustrine, eolian, glacial, or pyroclastic, are commonly poorly indurated and are known to provide adequate plant rooting surfaces without notable prior weathering (cf. Fig. 5.4). Tertiary, Mesozoic, and older rocks commonly must be decomposed to some degree by weathering before plant cover can become continuous across them. Conversely, without the presence of some mechanism on a bare rock surface that will serve to trap fine-grained weathered material (inter-granular cracks, plant roots, loose clasts, etc.), that material can hardly withstand erosion by wind or water to remain, accumulate, and anchor more vegetation. Thus, the basic interplay between establishment of both plant and soil cover over bedrock surfaces is evident. It probably makes little ultimate difference whether the initial bit of cover on bedrock is organic or inorganic. Development of "continuous" plant cover and subjacent soil in place on indurated bedrock may require scores or even hundreds of years, and the areal extent of such relations as opposed to bedrock exposures can be taken as one measure of the duration of related humid conditions. On the other hand, development of continuous plant cover on relict granular materials following the onset of humid conditions may occur in only a few years, the rate being dependent upon re-seeding or other propagation mechanisms available as noted

Fig. 5.12.   The root system of a sycamore tree has been flattened during growth in a joint developed in the Devonian Centerville Reef (a calcareous siltstone) near Stroudsburg, Pennsylvania. Such root developments are an important mechanical weathering mechanism.

by Marks and Bormann (1972) and Gómez-Pompa *et al.* (1972).

Experiments on erosion like that of Bennett (1939) which was described in Chapter 4 and studies like those of Langbein and Schumm (1958) on amounts of stream sediment with respect to moist and dry regions make it evident that plant cover anchors products of rock weathering, thus making such material susceptible to further decomposition. But even where the groundcover is essentially continuous, various plant taxa tend to be only locally developed or concentrated. Variations in plant assemblages often show strong responses to changes in drainage effectiveness, soil types, or bedrock kinds. Thus, swamp floras hardly characterize well-drained uplands. In the hardwood forests of northern Arkansas, the redbud and dogwood trees tend to be found in natural concentrations having the form of belts following the contours of hills and low mountains. This occurrences correspond with the disposition of calcareous sandstone and limestone formations, most of which are essentially flat-lying. A rooting surface with a high calcium carbonate content appears to be required. Similar bedrock-floral or soil-floral associations are recorded elsewhere. And even though the entire land surface is vegetated in the cases in question, notable differences in morphogenesis from place to place is possible where the phytic variants correspond to or influence other factors. Thus, one group of plants—phreatophytes—have been found to greatly accentuate transpirative water losses along watercourses in which they grow, with consequent discharge volume reduction and effects on sediment transport.

Like the other major elements of a geomorphic system, plant assemblages have been shown to evolve as related environments change or endure—the changes being both a cause and an effect of the evolution. This parallels the plantcover-soil relations previously discussed where each contributes to the development and existence of the other. Plant environment changes are probably best dealt with in relation to the botanic concept of a *climax flora* —i.e. a group of plants showing essentially complete adaptation to a particular surficial environment. In essence, two essentially distinct types of cases are involved. The general case concerns the changes in plant assemblages that occur when one geomorphic system (e.g. humid) replaces another (e.g. arid). The special case involves floral changes in response to adjustments within a particular type of geomorphic system as the system evolves.

The special case will be considered first since it tends to parallel botanic thinking with respect to the term CLIMAX. Thus when a lake depression fills with sediment (as most such depressions seem to do under humid conditions), the subaqueous lake flora gradually gives way to marsh or swamp plant associations of mosses and sedges as the water depths decrease and photic conditions change on the bottom. Water-loving trees such as willow or cyprus may invade the boggy area, but, if sediment infilling continues, the flora adjusted to poorly drained land may give way to a prairie grassland and/or forest plant assemblage. A botanic consensus would probably apply the term climax to one of the final and presumably most stable plant associations. But viewing each lacustrine stage as representing a distinct geomorphic environment capable of having a plant association in complete adjustment, one might as properly speak of a climax lacustrine flora or a climax marsh flora as to make reference to a climax grassland or climax woodland plant assemblage. For in an environmental continuum there is no real end or beginning that one can insist is thus more or less important because of "when" it occurs.

The evolution of lacustrine depressions under humid conditions was noted in theory long ago by Davis (1899). The associated floral sequence may be preserved in spore and pollen form for paleobotanic study and is, furthermore, a landform-deposit adjustment that is time-indicative during humid conditions. This is particularly significant since most lacustrine depressions are of non-humid origin (e.g. volcanic, eolian, glacial), though their filling and drainage may require more than a single exposure to humid conditions. The topographic and botanic adjustments in question are often paralleled by regional phytic adjustments relating to the general case of a climax flora.

In the case of the general floral climax which is of immediate geomorphic concern we are interested in the degree plants are adjusted to the prevailing surface environment as opposed to some environmental alternative. The fact that substitutions can and do occur is well documented, for example, by palynologic studies of North American mid-continental glacial sequences which show many alternations between prairie grassland floras and either deciduous hardwood or coniferous forests (cf. Frenzel, 1968; Wells, 1970; and Gruger,

1972). As noted by Flint (1971), there are vast areas of North America where the vegetation can probably be regarded as reflecting present environments (Fig. 5.13), particularly the broad tracts of boreal coniferous forest, temperate mixed forest, and prairie grassland. These would be CLIMAX FLORAS. There are, however, areas of mixed forest assemblages probably reflective of non-climax relations (cf. Davis, 1969), particularly where northern coniferous spruce-pine assemblages and deciduous hardwoods merge along the United States-Canadian border (cf. Fig. 5.1).

In southcentral United States, the few virgin forest remnants include virtually no boreal tree associations over wide areas of the Mississippi Valley, and a CLIMAX FOREST FLORA adjusted to humid temperate conditions was almost certainly being approached in the Ohio-Kentucky-Tennessee region when North America was being discovered again—by Columbus. However, in the southwestern portion of the same region —where, as noted in Chapter 4, there are landform-deposit records of former aridity —the temperate deciduous oak-maple hardwood forests were noted by the writer (1967) to include relict patches of thorn scrub and prickly pear cactus that presently cling to rocky, very well drained ground. Thus, the western Louisiana, Arkansas, Missouri region displays a *climatic floral blend* as opposed to the *climax flora* to the east.

The incidence of desert *xerophytes* increases westward from Central Arkansas and southwestern Missouri into Oklahoma and Kansas, respectively. A similar floral association was also noted by the writer (1967) in southern coastal Ecuador in a vegetated region adjoining the Peruvian Desert. The desert areas have only sparse xerophytic plants. In Ecuador, barren areas give way to forests where cacti and thorn scrub are mixed with a partially deciduous tropical mahogany-guayacan hardwood floral assemblage. This low-elevation deciduous forest area merges to the north, east, and generally upslope with a broad-leafed evergreen rain-forest flora which at

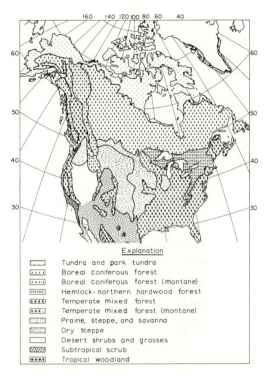

Explanation

	Tundra and park tundra
	Boreal coniferous forest
	Boreal coniferous forest (montane)
	Hemlock-northern hardwood forest
	Temperate mixed forest
	Temperate mixed forest (montane)
	Prairie, steppe, and savanna
	Dry steppe
	Desert shrubs and grasses
	Subtropical scrub
	Tropical woodland

Fig. 5.13. Generalized natural vegetation zones in North America at present. (From Flint, 1971.)

elevations above 1,000 m may be essentially at climax. Geomorphic evidence in southern Ecuador also suggests former aridity (Garner, 1959), particularly at lower elevations, and the blended arid-humid character of the plant cover tends to support this view.

In general, the geomorphologist would appear to be on firm ground if he considers an apparent floral blend to be a botanic suite in a state of environmental transition. The classical savannah flora consisting of grassland with patch forests is a case in point. There have been efforts to explain savannahs as products of tropical agricultural practices. However, the small forest areas on savannahs differ botanically in no great measure from more continuous rain forest elsewhere, and the grasslands resemble those of adjacent steppes and prairies. Moreau (1963) presents ornithologic evidence that suggests that present African forests concentrated on isolated moist uplands (Fig. 5.14) have formerly been continuous across lower, drier, intervening regions presently with savannah cover, a notion Flint (1971) found appealing also. In this scheme the savannah vegetal cover should reflect incomplete reforestation of a grassland, following drier climatic episodes, a notion voiced by Cole (1960) for portions of South America. She sees an evolutionary transition between savannah and rainforest botanic suites with the latter developing from the former as humid conditions persist. The present wide extent of savannah plant communities may possibly be accounted for by the former extensive steppe conditions and the natural reforestation problems confronting the plant types that make up heavy tropical rain forests noted by Gómez-Pompa *et al.* (1972).

The weight of arguments and data so far presented would seem to support the idea that savannah vegetal assemblages reflect a floral blend rather than a climax situation, presumably indicative of an incomplete

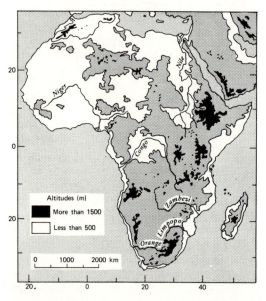

Fig. 5.14. Generalized relief map of Africa. During dry episodes, forests were probably restricted to moist uplands, and present savannah vegetation over much of the area between 500 and 1500 m elevation probably represents incomplete re-forestation. (After Moreau, 1963.)

recovery of forest groundcover in some instances or an incomplete loss of such cover in others. If this analysis is correct, many of the patch forests on savannahs would actually represent relict woodland ecosystems or refuges in some cases. A study of such situations might well reveal some association of hydrologic, pedologic, or geologic factors which favor the local perpetuation of the patch forests. There is, however, no particular reason to expect notably different topography to develop in the differently vegetated portions of savannahs unless the plants there are found to represent an equilibrium response to a persistent condition and unless distinct types of morphogenesis are found to relate to forest as opposed to grassland. Neither relation is definitely known to hold.

It has long been recognized that topographically diverse regions with the highest relief tend to be most botanically diverse. On a single mountain flank in the northern Venezuelan Andes one can move from a nearly barren coastal desert with a plant population of cactus and other xerophytes upslope through a thorn-scrub cover into a tropical rainforest in a vertical distance of less than 350 ft (100 m). Elsewhere on the same mountain system the mid-elevation rainforests usually give way to either dry upland thorn-scrub and xerophytes as in southern Peru or to moist high elevation coniferous or eucalyptus forests and tundra associations as in Ecuador. Indeed, for a variety of reasons involving atmospheric-topographic interaction as outlined in Chapter 4, uplands (in both humid and dry regions) tend to receive the most moisture (cf. Fig. 5.14).

Isolated hills and mountains rising above tropical savannahs are often heavily forested as in the African situation just discussed (cf. Fig. 3.43). Similarly, highlands in deserts often have upper slopes that receive notably more precipitation than adjacent lowlands, as noted by Lustig (1965), who cites a case where mountain tops (±3,400 m elevation) in a desert area of California receive some 13 inches (33 cm) of precipitation as opposed to only 3 inches

Fig. 5.15. Juniper-piñon tree growth on the flanks of an eroded Tertiary volcano, southwestern Colorado, attests to the increased precipitation at higher elevations in deserts. The adjacent lowland plateau supports only sparse sage cover. Similar moisture relations typify mountains in desert regions.

(7.5 cm) in adjacent lowlands (elevation ±1,500 m) (Fig. 5.15). The writer (1967) noted that hills rising 100 to 150 ft (±50 m) above the south coastal desert in Ecuador and some 5 to 10 mi (8 to 17 km) inland exhibit hardwood forest "caps" with trees draped in Spanish moss in contrast to the nearly barren adjacent lowland. Hills and low mountains slightly farther inland have rainforests on crests rising over 1,000 ft (±310 m) above sea level. In point of fact, therefore, as defined previously, a vegetated upland in a desert would be in a different geomorphic system than the adjacent, essentially barren lowland. We will develop the significance of this idea more fully in our consideration of arid geomorphic systems (Chapter 6). It is well to note here, however, that the upland vegetation not only indicates the localization of precipitation but also defines areas where surplus runoff can be expected to originate.*

In a somewhat contrasting situation, dry micro-climates develop in the north-south trending portions of river valleys draining the eastern Andes Mountains. There, in the path of the moist Southeast Trade Winds, the eastward facing slopes and crests of ridges are humid and forested, whereas, even now, many of the deep valley floors and lower valley sides are dry and sparsely vegetated through local rain shadows.**

WEATHERING MODES AND EFFECTS

It has long been recognized that rock weathering in moist environments tends to be dominantly chemical in nature. The effectiveness of chemical activity (i.e. its completeness) is usually stated in terms of *weathering maturity*, an immature weathered residue being incompletely decomposed and a mature residue consisting

---

* Lower upland temperatures may, in some instances, help induce higher effective moisture through reduced evaporation.
** For further discussion of such effects see Chapter 10.

mainly of chemical equilibrium by-products which are sensibly insoluble. Certainly, where there is abundant ground moisture and temperatures are above 32° F (0° C) much of the time, chemical changes in bedrock supersede comparatively minor mechanical effects achieved by tree roots and other organisms. Ice wedging certainly achieves some mechanical breakdown of exposed bedrock in some humid environments, even under regolith with a thickness less than the maximum depth of frost penetration. Since there is a general tendency for gradual thickening and lateral extension of weathered material with consequent bedrock burial, mechanical weathering effects should diminish with time of exposure to humid, vegetated conditions.

From the foregoing comments it seems clear that under humid conditions the character of weathered residuals, relief related to weathering, and debris removed by erosion will be dominated by chemical decomposition effects which tend to generate solutes, clays, and other generally fine-grained, semi-insoluble substances as noted by Garner (1959) and Jackson (1970). Holmes (1965) observes that the weathered products in chemically active settings show increases in negative ions at the expense of positive ones, mainly through oxidation (O^{2-}) and hydration $(OH)^-$ and hence greater stability (resistance to further chemical change). Minerals most resistant to chemical weathering occur in inverse relation to Bowen's reaction series (Chapter 3). Quartz and muscovite are most stable, followed by the feldspars which together with the first two generally compose granitic rocks. More easily altered are biotite, hornblende, augite, and olivine which are rich in iron and magnesium and dominate the oceanic (basic) igneous rock suites.

The chemical weathering effects that dominate under humid, vegetated conditions and induce rock decomposition, thereby simultaneously generate rock waste residues and potential sediment plus local relief. The weathered residues called *soil* are of immediate concern to us. And for sake of clarity we must define our terms. In his recent text, Flint (1971) defines *soil* as the part of the regolith that can support rooted plants as opposed to a common usage of engineers that applies the same term to the entire interval of surficial unconsolidated material which we will here term *regolith*. Flint's usage is close to that of some geologists and pedologists, but it does not satisfy our needs since, as already noted, plants can be supported in essentially any granular material including volcanic ash. We therefore favor a definition of *soil* close to that in the "A. G. I. Glossery of Geologic Terms," in which the genetic role of weathering and plants is emphasized. The term *soil* as used here applies to a surficial interval of weathered material dominantly formed by biochemical activity under plant cover.*

Soil may develop directly from subjacent transported detritus such as glacial drift, alluvium, lacustrine, marine, or pyroclastic sediments, or from consolidated bedrock essentially in place (*in situ*). It should be pointed out that many soils contain transported material and have grown in thickness through accretion of such material from above. Eolian additives are common as in the loessal soils of Israel noted by Dan and Yaalon (1966) and have been recorded by many researchers in central Europe and the central United States as well as in the eolianite soils of Bermuda. Elsewhere, soil wash additives may produce a clay-rich "gley" facies as noted by Flint (1971).

Usually there are volume changes in the materials affected by chemical weathering,**and, since the effects are most accentuated on corners and edges (of both mineral grains and bedrock outcrops), a general rounding and softening of angular profiles and forms is a usual result.

---

* This deviation from previous usage seems warrented since the weathered relicts of the several geomorphic environments must be differentiated, both in the field and in discussions that follow.
**Studies of Appalachian piedmont soils (saprolite) show little volume loss from parent material but notable density loss.

As we will use the term soil, it is admittedly close to the term *saprolite* as discussed by Fairbridge (1968). It should also be noted here that the gross structural and textural characters of soil tend to be destroyed by mass movement which in vegetated lands initially takes the form of slow downslope *creep*. The product of such movement occurs on and along the lower parts of slopes and is the rather heterogeneous mass of detritus termed *colluvium*. In regions that appear to have long histories of essentially uninterrupted humid conditions, soil and colluvium tend to mantle the entire bedrock surface, and the colluvium commonly extends to and constitutes the banks of watercourses. An example is the "saprolite"-colluvial relations described by Cady (1950) in the southern Appalachian piedmont region.

SOILS AND WEATHERING PROFILES

An excellent general discussion of weathering and soil is given by Flint (1971), and these are the matters that now concern us. In his diagram (Fig. 5.16) he notes the distinction that is often made between the portion of the earth's regolith or bedrock being subjected to weathering (*zone of weathering*) and that part that has actually been altered into soil (*soil zone*). The two zones often fail to coincide. This is understandable, if we consider the fact that the depth to which surficial effects penetrate and which establish the zone of weathering depends upon such factors as terrain permeability, groundwater table location, land slope, subsurface drainage, and regolith and bedrock lithology.

Since the existence of the weathering zone depends only on the establishment of a set of conditions, it may therefore be delineated rather early in a given humid episode and then adjust only gradually as related factors evolve. For example, as a watercourse is deepened, the related water-table may subside slightly under an adjacent hillslope and upland, thereby permitting a greater depth of oxidation and solutioning. In contrast, the soil-making (*pedologic*) processes initiate an alteration of regolith and/or bedrock beginning at the surface and gradually penetrating downward, mainly starting with the advent of vegetal cover capable of holding and protecting the pedologic residues (Fig. 5.17). In essence, a soil represents an accumulation of substances that are least destructable or have been least subject to such destructive processes as operate at the ground-vegetation interface and within the

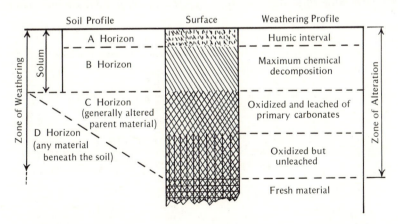

Fig. 5.16. The relationship between the weathering profile (zone of alteration) and the corresponding soil profile in a humid region. The zone of weathering as conceived here could extend much deeper and would include the theoretical downward limit of alteration as opposed to the actual altered zone. (After Flint, 1971.)

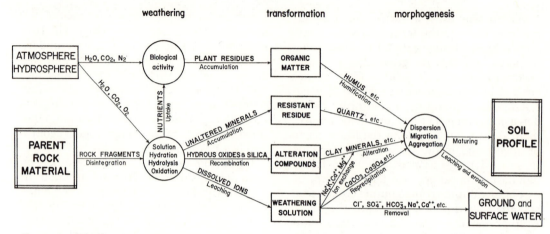

Fig. 5.17. Scheme showing reactions and processes of soil profile development from parent earth materials involving three main groups of processes—chemical decomposition, biologic activity, and profile morphogenesis, all occurring more or less simultaneously. Different combinations, variable rates, and accentuation of some processes cause different soil results. (After Yaalon, 1960.)

subjacent zone of weathering. The soil zone will therefore tend to thicken as rock and vegetal material accumulates and will thereby more nearly coincide with the weathering zone—given time. Initial coincidence of these zones is clearly impossible, and ultimate matching is unlikely unless the causitive humid conditions persist sufficiently to allow the base of the soil zone to "catch up" with the base of the weathering zone. Such a "catching up" would be a theoretical if not an actual impossibility in a region where weathering zone depths are steadily increasing in response to continually increasing relief.

In humid lands the pedologic processes fall into two broad categories, (1) those where the net long-term transfer of weathered materials is from the surface downward by water, countered by only minor upward transfer of matter through evaporation and organic activity, and (2) those where the downward transfer of soil materials is countered by sufficiently strong upward forces (capillary rise, evaporation) as to maintain a near-surface concentration

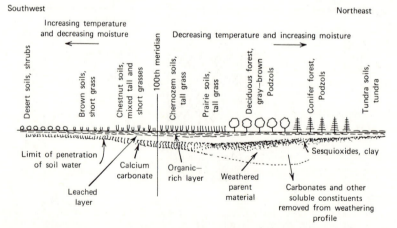

Fig. 5.18. Regional relationships of weathering profiles and soils to climate and vegetation in North America from the southwestern United States to Arctic Canada. (After Hunt, 1967.)

of relatively soluble substances, particularly carbonates. The first situation generally accompanies forest groundcover or areas of tall (more humid) grasses. The second equates with terrain characterized by short grasses or "dry" steppes sometimes termed semiarid. These relations are well illustrated by Hunt (1967) for North America (Fig. 5.18).

The principal groups of soils encountered in the world's humid regions differ mainly in the degree to which cations such as calcium and sodium have been removed by leaching and in the relative amount of humic material which has developed as a function of plant growth versus decay (Fig. 5.19). In general, the decay of vegetal matter shows a very high temperature sensitivity, and the most highly organic soils are in cold regions. Leaching of cations, however, may reflect such factors as drainage, temperature, amounts of moisture, and character of solvents. One broad soil category appears to reflect the relative insolubility of iron, aluminium, and silica under comparatively cool, moist conditions. The resulting residues are enriched in the oxides of the metals and hence are called *pedalfer soils*.

I. Pedalfer Soils

A. Immature, frost-shattered, stony tundra soils with surficial peaty accumulations reflect minimal chemical decomposition and organic decay in a cold, moist environment vegetated by mosses and lichens.

B. Peaty, gray* soils (*podzols*) with a bleached zone in the A horizon** reflect low-grade chemical activity and reduced organic decay in boreal coniferous forest settings.

C. Gray-brown forest soils (podzols) with reduced humic material in A horizons and proportionally thickened B horizons containing brown, red, and yellow oxides of iron reflect accentuated chemical decomposition in warm, temperate, decideous forest regions (Fig. 5.20).

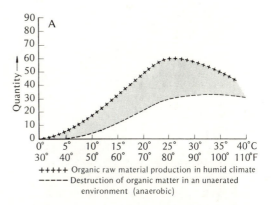

+++++ Organic raw material production in humid climate
- - - - Destruction of organic matter in an unaerated environment (anaerobic)

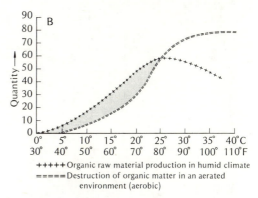

+++++ Organic raw material production in humid climate
===== Destruction of organic matter in an aerated environment (aerobic)

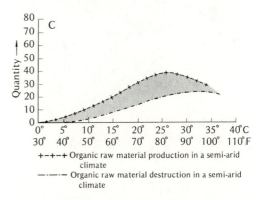

+-+-+ Organic raw material production in a semi-arid climate
—·—·— Organic raw material destruction in a semi-arid climate

Fig. 5.19. Organic decay rates versus production of organic matter under humid (anaerobic) conditions (A), humid (aerobic) conditions (B), and semiarid conditions (C). (After Senstius, 1958.)

---

* Records of color are highly subjective unless made with reference to some standard color-symbol system such as the Munsell soil colors (cf. Ruhe *et al.*, 1961; Goddard *et al.*, 1948).
** *See* Fig. 5.16 for soil horizon locations.

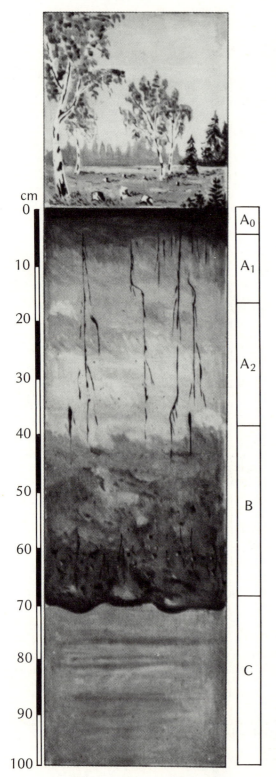

cm
0 — A₀
10 — A₁
20
30 — A₂
40
50 — B
60
70
80 — C
90
100

A second broad soil category appears to reflect the notably reduced effectiveness of solutioning and increased evaporation in grassland environments which results in soil residues enriched in carbonates called *pedocal soils.*

II. Pedocal Soils

A. Black prairie soils (*chernozems*) exhibiting relatively thick, dark-colored **A** horizons (black, gray) and secondary carbonate accumulations in **B** horizons reflect formation in tall grass prairie regions (Fig. 5.21).

B. Brown (chestnut) prairie soils with reduced humic material and much secondary carbonate reflect incomplete leaching and accentuated evaporation in dry (short grass) steppes; poorly drained areas may exhibit surficial carbonate impregnation (*hardpan**) which in low areas is rendered impervious by accreted clay thereby causing water to stand for a period after a shower.

A third broad soil category appears to reflect hypersolutioning and accentuated organic decay capable of destroying organic debris almost as it forms and of removing in solution essentially all cations including silica; this results in residues of complex hydrous oxides of iron and aluminum, here called *tropical soils.*

III. Tropical Soils

A. Red-brown lateritic soils with reduced **A** horizon, minor humic and clay content, and a thick **B** horizon of hydrous iron oxides developed under evergreen

Fig. 5.20. Development of gray-brown forest soil (podzol) on the Yoldia argillite deposit in the vicinity of Leningrad, Russia. Soil horizons are **A₀**, zone of dead vegetal litter and black humis, pH = 4–5; **A₁**, gray-black, humic earth, pH = 5; **A₂**, bleached (white) and leached eluvial horizon which has lost nearly all of its iron and other soluble cations in which the groundmass consists of finely divided silica with a faint horizontal structure, pH = ±5; **B**, zone of illuviated red and tan iron oxides and hydoxides derived from **A₁** and **A₂** plus some bases from the same level precipitated in concretionary structures, sometimes cemented into a tough layer (Ortstein), pH = 6–7; **C**, bedrock, a noncalcareous sandy argillite, pH = 6–7. (From Erhart, 1939.)

* A slightly different use of the term hardpan by Holmes (1965) is noteworthy (cf. **Fig. 5.19** caption).

tropical forests and savannahs subject to monsoon or otherwise seasonal drouth conditions.

B. Laterite and bauxite soils consisting of concentrated hydrous iron or aluminum oxides generally lacking humic content or a distinct A horizon develop under continuously humid tropical rainforests or tall grasslands (Fig. 5.22).

The soil categories previously listed clearly represent somewhat arbitrarily selected types from a broad spectrum of chemical weathering residues developed in the presence of continuous plant cover. Yet they are the most widely recognized examples. The types of clay formed by pedologic processes express various degrees of cation removal and were cited in Chapter 3. It seems evident that some of the soil types cited represent rather extreme forms of chemical maturity or immaturity, whereas others are more transitional. In point of fact we have been hovering around the fringes of a concept borrowed from botanists by Senstius (1958)—the notion of CLIMAX forms of rock weathering, or, in other words, the idea of a weathering residue that is in essential chemical equilibrium with its formative environment (see also Chapter 1, pp. 27–29).

The climax weathering concept would be rather superfluous if we were merely confronted with differing degrees of soil development expressed by varying thicknesses of the horizons, etc. Instead, we periodically

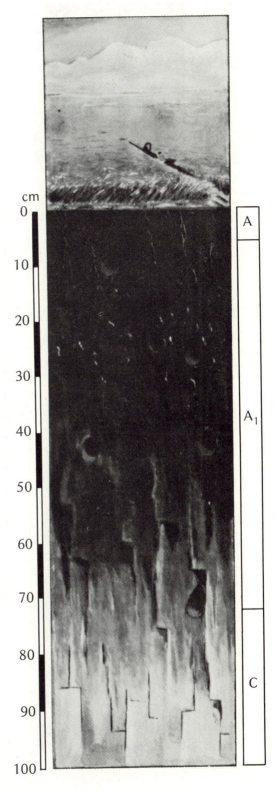

Fig. 5.21. Development of black prairie soil (chernozem) from calcareous loess on the Lessigiana steppe in the vicinity of Saratov, Russia. Soil horizons are **A**, root zone for grasses. Each small root carries one or more small, clotted granular masses of $CaCO_3$ pierced by multitudes of fine pores. Faint effervescence in dilute HCl, pH = 7.6; A_1, uniformily black zone of granular structure like **A**, with very fine white filiments of $CaCO_3$ in a pseudo-rootlet configuration (loess kinchen) or lining earthworm borings and precipitated around granular masses of earthworm excrement. The passage into unaltered parent material is gradual and indistinct with much $CaCO_3$ and a faint effervescence with dilute HCl, pH = 7.8. Humic content 9–10%; **C**, parent material, gray-yellow, strongly calcareous loess. (From Erhart, 1939.)

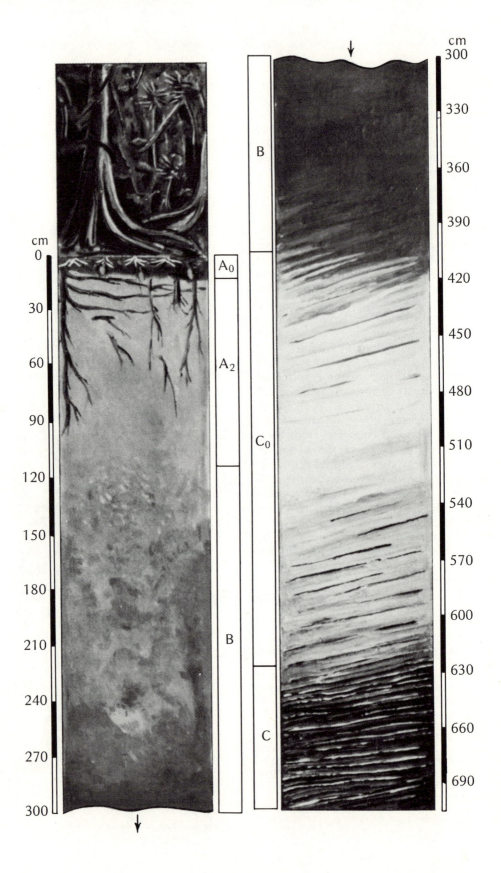

encounter soils that have essentially always, or at least for an effectively infinite period of time, been exposed to only one intensity and kind of humid pedologic process. These are climax soils. And where one variety of pedologic process has been replaced by a second variety of pedologic process, we may find relict soil structures as well as developing ones and we thus need the concept of an environmentally *blended* (degraded or secondary) *soil*.

Not all types of humid soil-making conditions are likely to alternate without such drastic reshuffling of geography as could be accomplished only through plate tectonics. On a relatively short-term basis, we would thus not expect a tundra soil to be transposed through a short-term climate change into a setting capable of generating a laterite or vice versa. However, we have abundant evidence that grassland pedologic processes periodically invaded laterite regions (Fig. 5.23), and vice versa, both at intermediate and low latitudes. Elsewhere, in the middle latitudes, forest and prairie soils have been affected by pedologic effects due to vertical accretion of loess from deserts situated elsewhere (Fig. 5.24). And of course there are cases where desert weathering effects have both replaced pedologic conditions and been displaced by them (cf. Fig. 1.16).

Most of the soil categories listed previously were cited because they appear to represent comparatively extreme soil-making effects with respect to moisture and temperature. The very recency and intensity of climate changes in some regions (as noted in Chapter 1) confronts us with the possibility that some of the examples cited are climax soils, whereas others may be blended

soils, albeit blends of widespread occurrence and considerable stability (cf. Figs. 5.23 and 5.24). We regard the laterite and bauxite soils as humid, tropical climax types which differ mainly in the original metaliferous content of their parent materials. This estimate is based not only on their extreme chemical maturity with respect to hot, humid environments, but also on the common occurrence of such soils in tropical rain-forest settings and especially those areas designated humid floral or faunal refuges by various authors (cf. Chapter 4); they are elsewhere termed humid climatic nuclei in South America, Africa, Australia, and southeast Asia (Garner, 1967).

The red-brown lateritic soils, however, are commonly associated with the savannah type grassland-patch-forest vegetal suites. Not only are the associated floral assemblages phytic blends suggestive of environmental change, but related geomorphic evidence in near-equatorial parts of South America, Australia, and Africa suggest recent geomorphic system alternations. The soils in many tropical savannahs are what one would expect if rainforest laterization were imposed on prairie grassland soils, and the converse of this. The relations are discussed in detail by Senstius (1958) and illustrated here (Fig. 5.23), but of course all are blends and not climax soils. Many of the buried soils of North American mid-continental glacial sequences exhibit characters acquired in more than one environment as noted by Holmes (1965), Ruhe (1969), and Flint (1971). There, both temperate-forest and prairie-grassland soils display characters apparently induced by opposing pedologic conditions (cf. Fig. 5.20).

Fig. 5.22. Development of primary lateritic forest soil under virgin rain-forest cover in the region of Moramanga, Madagascar. Soil zones are A_0, thin layer of vegetal debris in the process of rapid decomposition and with little actual humus in the usual sense of that term, pH = 5; A_2, leached, mottled, yellow and red lateritic "argillite" without apparent humus, 5% iron hydroxide, 23% aluminum hydroxide, 36% clay, pH + 5.6 (in climax examples no clay would be present); **B**, lateritic argillite ranging from yellowish red through deep red to purplish red, 16% iron hydroxide, 28% aluminum hydroxide, and 48% clay, pH = 6.2; C_0, zone of altered bedrock where feldspars are altered to hydrous clay deficient in silica, and the iron-rich micas are oxidized and appear as reddish layers—iron hydroxides 1.5%, aluminum hydroxides 12%, insoluble (i.e. undissolved) residue 82%, pH = 7.4; **C**, bedrock consisting of banded, biotite gneiss. (From Erhart, 1939.)

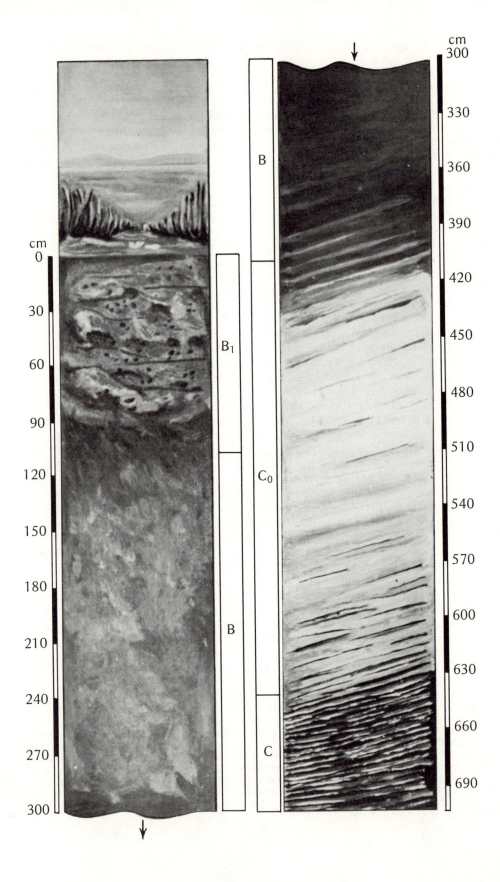

Soil Development

It is usual in discussions of soil development to cite the general equation of Jenny (1941)

$$s = (cl, o, r, p, t, \ldots)$$

where s is soil, cl climate, o organisms, r topography (relief), p parent material, and t time. It is well to note that the apparent intention here is merely to imply that the "character" of a soil will vary with these other factors. Actually, it can be demonstrated that a variety of parent materials will yield a single type of soil under the same climatic conditions. As noted by Holmes (1965), for example, "The black soil of the Russian steppes . . . is equally well developed from such different parent rocks as granite, basalt, loess and boulder clay." The implication seems clear that parent material does not always vary SIGNIFICANTLY. Conversely, as noted by Holmes ". . . a single rock type, like granite, gives grey soils in temperate regions (*podzol*), black soils in the steppes (*chernosem*), and reddish soils in tropical regions of seasonal rainfall (*lateritic earths*)." It would appear, therefore, that parent material only tends to govern basic soil type in extreme cases (laterite, bauxite) where climax weathering residues reflect an original difference—in the cases cited the differences are in metals, e.g. iron versus aluminum. Alternatively, where the parent material is itself a product of intense chemical decomposition—i.e. some types of sedimentary rock such as shale—little soil can form in a moist environment since the rock is already chemically stable.

Because the pedologic organisms (plant and animal) clearly vary with the climate, and it can be shown that the same generally holds true for erosional relief, Jenny's equa-

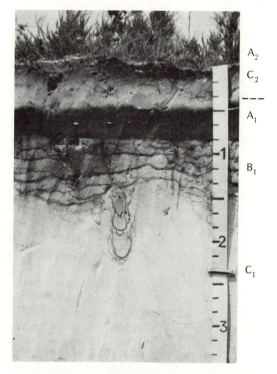

Fig. 5.24.   Soil profile developed on sandy alluvium near Endhaven, Netherlands, with secondary layer and soil probably developed through vertical accretion of loessal material. Vertical scale in feet. Zones include A_2, thin, slightly humic zone developed under heather on C_2, a 6-inch interval of gray, loessal material; B, not notably developed. A_1, paleosol of compact, ferro-humic concentrates plus some clay particles brought down from above by water; B_1, argillaceous, mottled layer with concretionary iron hydroxides, locally banded and extended downward into C (sandy alluvium) along permeable zones. (Roy W. Simonson, U. S. Department of Agriculture.)

tion could be rephrased without notable loss in validity as

$$s = (cl, t)$$

where s is soil, cl the climate as expressed by plant cover and related organisms, and

Fig. 5.23.   Expression of a degraded lateritic soil as a ferruginous duricrust developed on gneiss near Tompoketsa, Ankazobe Region, Madagascar. Zones are B_1 [zone A_0 is to all intents missing because of loss of forest cover and an arid episode according to Erhart (1939), since only a little organic matter occurs in cracks within the crust due to present growth of sparse grass] compact, ferruginous lateritic duricrust; effectively a rock consisting of 52% aluminum hydroxide and 46% hydroxides of iron; B, red lateritic clays, essentially the same as for the B horizon in Fig. 5.22; C_0, zone of altered bedrock corresponding to the same level in Fig. 5.22; C, bedrock consisting of banded, biotite gneiss. (From Erhart, 1939.)

t time. Even this statement leaves much to be desired, but clearly not all factors cited by Jenny are equally important. Richmond (1962) has shown, in fact, that major soil variations tend to parallel plant and precipitation variations on a single mountain mass in the western United States. There he notes a mechanically weathered residue in a desert at a mountain base grading upward through brown prairie soil under short grass, black prairie soil under tall grass, a gray forest soil, and, in the high alpine zone, under tundra vegetation, an immature stony soil. The vertical interval involved is noted by Flint (1971) to approach 3,000 ft (~1,000 m) with precipitation ranging from 10 to 31 inches (~250 to ~800 mm).

Actual changes in the parent rock material resulting in soil under newly established humid conditions are probably initiated under a moss-lichen plant cover accompanied by fungal and bacterial activity. The initial chemical reaction is often presumed to be *oxidation,* but it is almost certainly accompanied by some degree of *solution* and *hydration* (cf. Fig. 5.17). In any event, the deepest soil zone (Fig. 5.16) is usually oxidized, but where original carbonates exist, they are commonly not thoroughly leached; typically the **C** horizon displays the brown-yellow and red hues of iron oxides. In a buried **C** horizon developed on pre-Cambrian granite in the St. Francis Mountains of southern Missouri (Fig. 5.25), the first altered zone above fresh bedrock displays hydrated and chloritized muscovite as well as oxidized magnetite but essentially unaltered potash feldspar and quartz; 15 to 20 cm higher, in the same weathered zone, the micas are thoroughly chloritized, quartz shows some rounding and pitting, and approximately 50% of the feldspar is kaolinized. Another 15 to 20 cm higher, iron has been altered to limonite, there is complete alteration of feldspar to white kaolin, and many of the quartz grains show what appears to be intensive solutional rounding and pitting. The latter level may constitute a transition to a **B** soil horizon, but this is uncertain since higher ma-

Fig. 5.25. Three expressions of granite associated with a basal Cambrian weathered zone developed in the St. Francis Mountains, southeastern Missouri. Dark fragment (top) is red granite having no visible alteration of feldspar. Center piece has about 50% of feldspar kaolinized (white areas). All feldspar in bottom piece is kaolinized (dark areas are chloritized micas). See text for details; scale, inches and centimeters.

terial is missing. The precise nature of the conditions under which these weathering changes occurred is, of course, not known, but the chemical nature of the changes is unmistakable.

In a humid vegetated setting with abundant moisture, the downward percolation of groundwater tends to remove as soluble bicarbonates such metals as calcium, magnesium, and iron and to mechanically transport downward grains of clay until they are caught and concentrated in the finer pore spaces near the top of the **C** horizon, thereby forming the clay-rich **B** soil horizon. Since the **B** horizon clay concentrations tend to be relatively impervious, further leaching below it is inhibited. In and above the **B** horizon, leaching continues, and the **A–B** zonation matures chemically which, in general, means a reduction in the amount of mineral matter available as plant nutrients. Apart from its humic content which was previously noted to relate to rate of organic decay, the **A** soil horizon tends to consist of the most durable substances in the parent material. Because of the nearly ubiquitous occurrence of quartz, most soil **A** horizons include some grains of this material, often loosely bonded with some of the relatively insoluble hydroxides of iron. In the humid tropics which appear to favor laterization, even the silica may be removed in solution, and, above laterite, there may be little or no accumulation of material that would merit the designation **A** horizon. Of course, as noted by Holmes (1965), chemical solutioning of limestone will produce no soil in any moist setting unless the parent rock contains relatively insoluble impurities such as quartz, clay, or iron. If present, the iron and other material may form reddish soil residues (*terra rossa*) like those of eastern Tennessee and on the Karst Plateau of the Adriatic side of Yugoslavia. Both of these residual soil regions show geomorphic characters of humid climatic nuclei as does the Nigerian upland discussed in this same context by Tricart (1956a).

It seems clear that readily soluble monomineralic rocks may generally yield little true soil, and, in Tricart's hot, humid example, he describes the quartzite that comprises the main countryrock in a region drained by the upper Senegal River. There the bedrock "resembles nothing so much as a lump of sugar which has been immersed for a moment in hot coffee." The silica is being pervasively dissolved, and, since its main appreciable impurity is minor iron, only a thin ferruginous encrustation passes for soil. A study of related watercourses (cf.

Figs. 1.8, 5.40, and 5.41) discloses that water in them only transports solutes (Tricart, 1956a).

PRIMARY STRUCTURES IN SOILS

It has long been known that certain primary characters of regolith or bedrock may persist to influence the characters of derivative soil. The influence may take the form of soil topography expressed at the ground surface or it may be expressed by internal aspects. In Chapter 2 the distinction was made between depositional and erosional landforms, and the original internal structures clearly may differ. The same is true of associated soil topography. Downslope movements of soil tend to destroy primary structures so their presence is evidence of *in situ* pedologic processes.

In depositional landforms, the primary influence on the surface expression naturally relates to original deposit form (sand dune, mega-ripple, ground moraine, etc.) as modified by subsequent geomorphic events. As previously noted, most depositional landforms are composed of unconsolidated, granular material at some angle of repose. If the parent material is generally homogeneous, the volume changes that accompany the humid pedologic processes may hardly alter the original surface topography. However, altered water saturation and changed clay contents as the humid condition persists may cause changed repose angles. Since unconsolidated rock material is often highly permeable, the weathering zone may penetrate rapidly and to great depth. Given similar relief and drainage, an unconsolidated deposit will usually be more deeply weathered than associated bedrock.

Erosional landforms are at least partly identifiable on the basis of a consolidated bedrock composition, in most cases, so that soil topography necessarily relates to the relative effectiveness of weathering and erosion with respect to bedrock variation. Nor is there so sharp a line between weathering and erosion as one might wish. As the Chinese are fond of noting: "A journey of a thousand miles begins with but a single step." Solutioning is common in rock weathering and contributes to pedologic changes, but it is also a powerful form of erosion capable of extensive landscape modifications. A terrain developed on a single bedrock lithology may well reflect the lithic uniformity in related soil topography, e.g., a weathered level surface remains level even after soil forms unless erosional relief also develops. More commonly, even where there is textural uniformity, there is rock structure. Fracture systems are as often exploited by weathering as by erosion to generate relief on the soil surface. Also, if there are a variety of bedrock types present which vary in weathering susceptibility, the types resistant to chemical decomposition may stand topographically apart from the others simply because of weathering (Fig. 5.26), even though the processes were initiated on an even surface.

Internal structures of depositional landforms may also be detectable in soil intervals, usually mainly in the C horizon which is least altered and disturbed by organisms. In detrital deposits such as alluvium and glacial drift that are being weathered, the

Fig. 5.26. Cross section of a laterite weathering profile with a hard ferruginous (ferricrete) crust (1), passing down into a ferruginous, pisolite or concretionary zone (2), then a "spotted zone" (3) of residual clay, a "bleached zone" (4) of weathered, but partially preserved rock, and (5) the fresh crystalline rocks of the Precambrian basement. An ancient quartz dike (6) penetrates the entire sequence and thus proves that the overlying soil was formed in place. A former stream filling (Tertiary) with fluvial gravels (7) was also involved in the weathering process and subsequent duricrust development in this example from western Australia. (After Walther, 1924.)

individual clasts survive in varying degrees, in part as a function of time (cf. Flint, 1971). Chemical decomposition begins at the clast surface and penetrates inward to form thin shells of oxidation and hydration called *weathering rinds*. This is analogous to the *spheroidal weathering* phenomenon, to be discussed shortly, which also affects bedrock. Blackwelder (1931) found that the ratio of weathered to fresh granite clasts provided a basis for distinguishing between two glacial tills in the Sierra Nevada. Where weathered clasts consist of minerals that are metastable (e.g. feldspar) and semi-insoluble (quartz), the complete dissolution of the parent clast usually involves alteration and/or removal of the weaker substances leaving only a concentrated mass of stable residue. In the Wisconsin Age ground moraine near Ames, Iowa, for example, exposures show many patches of quartz granules imbedded in clay; these are all that remain to represent former granitic cobbles and boulders. In instances where clasts being weathered consist of materials of nearly uniform resistance to decomposition, the entire clast may be represented only by a local color change. The writer well recalls a saprolite developed from a Cretaceous, andesitic, volcanic agglomerate just north of Guayaquil, Equador. In this soil, formerly angular clasts were clearly visible, reddish and yellowish "ghosts" which had no more substance than the clay and lateritic materials comprising the remainder of the accumulation.

Soil C horizons often also show expressions of bedrock or altered bedrock structures and rarely the effects penetrate to the ground surface (Fig. 5.26). Bedrock partings provide easy avenues of attack for weathering processes, and the individual blocks of rock are increasingly isolated from one another by weathering rinds in the process generally termed *spheroidal weathering*. Since rates of chemical reaction are proportional to reacting surface area, the corners and edges of joint blocks are destroyed more rapidly than the faces, and progressively more rounded residual fragments are generated (Fig. 5.27). These re-

Fig. 5.27. Spheroidal weathering developed in a wind gap at Hook Mountain, northern New Jersey. The rock is Triassic basalt. Lens cover (right center), 2 inches in diameter.

sidual clasts have been termed *core stones* by Linton (1955). A classic study of their development by Ruxton and Berry (1957) also illustrates the value of bedrock structure in gauging the *in situ* genesis of a soil from bedrock (Fig. 5.28). It is also evident that a transported clast in a weathered detrital deposit might be mistaken for a core stone, a test for this possibly being the usual lithologic identity of a core stone with sub-

jacent bedrock and a related downward continuity of C horizon ghost structures into bedrock.

Many of the aspects of the Hong Kong setting disclosed by the Ruxton and Berry study (1957) suggest that a near climax soil has been developed under a persistent humid environment. In spite of local, exhumed core stones, most of the terrain is thoroughly decomposed under heavy vege-

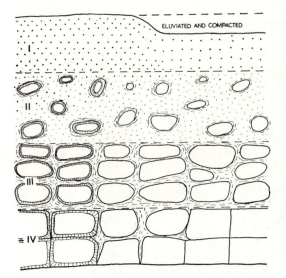

Fig. 5.28. Zones (I–IV) of a mature weathering profile on granite developed on the island of Hong Kong showing separation and definition of *core stones* under a humid environment. Iron staining inside core stones is shown only to left. (After Ruxton and Berry, 1957.)

tation in a moist tropical setting. Maturity of the weathering profiles in Hong Kong and elsewhere are perhaps best gauged by studies like that of Walstrom (1948). The latter researcher records the degree of decomposition of various minerals in a soil developed on the Pikes Peak Granite in Colorado where it was deeply decomposed during Pennsylvanian time (Fig. 5.29).

It has already been noted that soil formation appears to gradually reduce the extent of bedrock exposures in a humid, vegetated region. It should also be pointed out that the changes that occur in rock exposed to the air differ notably from those in the same material if buried. Exposed rock surfaces undergo extremes of diurnal air temperature and moisture changes, rapid wetting and drying via evaporation, and weathering conditions that differ markedly from the rather continually moist, even-temperatured activities along a soil-rock interface.* Since the weathering of bare rock surfaces hardly typifies humid geomorphic systems,

it will be discussed in relation to environments where it is more common (Chapter 6). Nevertheless, it must be recognized that similar weathering effects do occur to a minor extent in vegetated lands where there are exposures of bedrock. The vast majority of bedrock exposures in humid lands develop naturally as a result of erosion, and many if not most weathered fragments yielded therefrom become involved in pedologic processes occurring downslope. Some such weathered materials, of course, enter directly into watercourses.

As a summary comment on humid weathering it would be well to understand that a regolith under vegetation outside of a boreal tundra environment is almost certainly a relict left by some prior nonhumid subaerial condition if it consists of appreciable mechanically weathered material (cf. Fig. 1.16). The only apparent alterna-

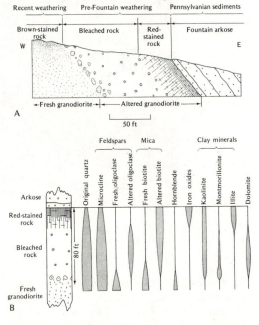

Fig. 5.29. Pre-Fountain weathered zone developed on the subjacent granodiorite on Flagstaff Mountain, Arizona shown in cross section (A) and in terms of the vertical persistence of minerals within the weathered zone (B). (After Wahlstrom, 1948.)

* Repetitions of evaporation sequences and moisture transfer may be more important than temperature changes.

tive explanation to such an occurrence is that the mechanical increment is an accreted additive (i.e. loess) from some geographically extraneous environment. In time, pedologic processes of the humid geomorphic system would normally tend to destroy such a relict or additive. It follows that only material normally generated by humid, chemical weathering processes should ideally be expected to dominate the lithic character of derivative sedimentary detritus (cf. Garner, 1959).

FLUVIAL EROSION AND MASS WASTING

Because the emphasis of this chapter falls on characteristic effects of humid geomorphic systems, it would be inappropriate to attempt to separate the erosional effects induced by gravity from those caused especially through running water WHICH MAKES THE GRAVITY EVENTS POSSIBLE. We say this last, just for the record, for in the absence of terrestrial relief and slope, gravity can accomplish little beyond compaction. Gravity's role in tectonic denudation was outlined in Chapter 3. But in humid, vegetated settings, tectonic events must rank as random happenings, not without their geomorphic impact, but certainly not especially typifying a mode of relief genesis where there is abundant running water. Of the seven "invariable" attributes of humid, vegetated settings listed at the start of this chapter, the only topographic criterion cited was that "some landslopes exhibit distinct orientations toward watercourses." A genetic relationship is therefore indicated between landslopes and running water in humid lands. This was perhaps first implied in the suggestion of Agricola, and later Hutton and Playfair, all of whom wished to make rivers erosionally responsible for the valleys in which they flow, an idea no longer seriously opposed as a generalization.

If running water is the principal cause of landslopes in most humid settings, and mass wasting is inappreciable in the absence of such slopes, it follows that the two processes must function in such close conjunction as to be effectively interdependent. Davis (1902) essentially recognized this interdependence in his pioneer work on the "normal" erosion cycle. More recent developments in geomorphic thought lead us to take a somewhat different approach to the problem than Davis did and with rather different over-all results. For example, there is the problem of where does it all begin? All, in this case, refers to the average humid geomorphic system.

Davis (1902) and later Penck (1924) and King (1953) dated their theoretical erosional episodes from the uplift of a block of continental crust above sea level. There is a problem already. For, if plate tectonics concepts previously outlined in Chapter 3 are valid, continents as high standing crustal areas are "old," have thickened, and hence have gradually become relatively more elevated above the ocean floors in some areas, probably more or less as the ocean basins filled with water; and "new" land is pragmatically less of a problem than are greater and lesser amounts of "old" land. In other words, through most of geologic time, landscapes have formed and been destroyed WHILE continents fluctuated somewhat in their degree of emergence.

For geomorphologists, it might have been simpler if lands had only submerged and emerged. Then, for openers, we would merely have to contend with the nuances of global environmental patterns outlined in Chapter 4 which indicate that a newly emergent land on the equator, or about 40° of latitude north or south, would probably be humid as it emerged, whereas land emerging in between would probably exhibit at least one arid side. Admittedly, with newly emergent terrain, one must also contend with problems common to newly formed depositional surfaces. The topographic complexity of such areas is hardly cause for concern, for depositional areas are often so featureless as to delight anyone who wishes to initiate an erosional episode on an even surface. However, there is the problem that some marine sediments, particularly terrigenous clastic detritus lacking cement, may remain unconsolidated for long periods, whereas others, notably calcareous and other exsolutional sediments, seem to lithify relatively quickly. This is important geomorphically, since there is reason to believe that initial water flow patterns developed on unconsolidated materials may differ significantly from those formed on bedrock.

In any event, the development of fluvial erosion phenomena on a regional surface previously entirely under water is probably rather rare. More complete emergences of already-existing lands is probably normal. This brings a second set of factors into play in the evolution of drainage phenomena, namely, the geographic, topographic, and meteorologic (e.g. hydrographic) factors that govern the distribution of humid, vegetated conditions. In a good many instances, newly emergent coastal regions, whether humid or arid, would receive runoff from moist land areas already in existence at higher elevations. One can also anticipate the downslope extension of drainage lines.* Later fluvial activity on newly emergent coastal land surfaces would certainly take place over some slopes erosionally related to these drainage line extensions. This leads us to conclude that it is probably not possible to arrive at a realistic scheme for the virginal development of humid fluvial erosion systems through study of newly emergent coasts which would be modified by upslope runoff before direct precipitation and runoff could be expressed. Perhaps it is just as well that with recent de-glaciation and the resulting rise(s) in sea level, newly emergent sea floors are in short supply. It is interesting to note that the newly submergent lands (mostly continental shelves) are abundantly notched and grooved by what appear to be former drainage line extensions, some of which match up with existing subaerial drainages and some of which do not (Fig. 7.35).

Ideally, we would like to trace the development of humid fluvial erosion effects "from scratch," so to speak. An erosionally virgin land surface would do nicely—one formed depositionally and, therefore, rather featureless. Such low-relief depositional landforms are created intermittently by volcanism, desert aggradation, lacustrine and marine sedimentation, glaciation and, of course, artificially created by man, the latter usually by construction, cultivation, or the draining of water bodies. With minor variations, each depositional type resembles the others in granular surface composition and form so that similar erosional results could be anticipated from humid gradational activity acting upon them. Each type of surface tends to occupy a particular topographic-geographic setting, however, and its erosional history could differ for this reason. It would in any case, of course, be most inappropriate to discuss humid fluvial effects on any surface that did not shortly acquire vegetal cover and soil. It is apparent, however, that prior to the establishment of plant cover, even under the most favorable of plant propagation circumstances, there would be some period of fluvial activity on bare ground.

DRAINAGE SYSTEM DEVELOPMENT

As noted at the beginning of this chapter, humid, vegetated regions are characterized by, "a water surplus," one which moves downslope into adjacent lands or water bodies. In brief, humid lands drain. Gravity is clearly instrumental. And because of friction, there is an interplay between the draining water and the surface on which it is flowing. Landforms and deposits result. Drainage configurations that show the most

---

* As a consequence, the downstream portions of such drainages would be their youngest parts.

complete conformance to bedrock structural trends are generally believed to have
formed essentially in contact with bedrock
(cf. Figs. 2.19 and 2.20). Drainage patterns
essentially independent of subjacent bedrock litho-structural trends are usually
thought to have developed on some covermass of unconsolidated, homogeneously
granular material such as might form
from the depositional mechanisms just
mentioned. The flow direction taken will
naturally be determined by initial slope,
and this slope-directed flow is what we will
discuss first. In order to simplify the discussion, we will consider the changes that
occur on an inclined, erodable surface of
low relief, both with respect to various proposed erosional models and in relation to
actual erosional observations.

Inclined Granular Surfaces

One of the pioneer quantitative studies of
drainage system development is that of Horton (1945) in which runoff behavior on an
erodable surface was observed. We should
note here that essentially no naturally
formed surfaces are entirely featureless,
and hydrologic observations of various researchers confirm the effects of these irregularities on flow directions of initial
(*consequent*) runoff. However, where relief
is low and runoff is abundant, initial flow
tends to spread out into a sheet of water.
After some flow and accumulation of runoff
volume for a critical distance (designated
x_c by Horton) a series of shallow parallel
grooves (*rills*) begin to appear which are
oriented essentially down the regional
slope. These rills are erosional in origin and
mark the initial step in the development of
a drainage system (Fig. 5.30). Horton
(1945) notes that, "A rilled surface presents a striated appearance in plan and a
finely serrated appearance in cross section"
(cf. Fig. 5.31, cc′bb′, ee′).

The energy required to initiate rill erosion
can be calculated. The amount of energy
expended in frictional resistance per foot of
slope length on a strip 1 ft wide running

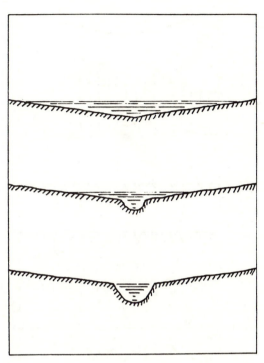

Fig. 5.30. Cross sections of successive stages of rill-
channel development. (After Horton, 1945.)

down the slope, per unit of time, for steady
flow is noted by Horton to be equal to the
energy of the volume of water passing over
a unit of area per unit of time. If we let
x = distance from a divide or watershed
line, measured on and along the slope (not
horizontally); δ_x = depth of overland flow
at x, in inches; w_1 = weight per cubic foot
of water in runoff, including solids in suspension; α = slope angle; v = velocity of
overland flow at x, in feet/second; we may
compute the product of the weight, fall, and
velocity, or

$$e = w \frac{\delta_x}{12} v \sin \alpha$$

It is of further genetic significance to note
that the energy e equals the force times the
distance moved. Hence the force exerted
parallel to the soil surface per unit of slope
length and width is

$$F_1 = \frac{e}{v} = w_1 \frac{\delta_x}{12} \sin \alpha$$

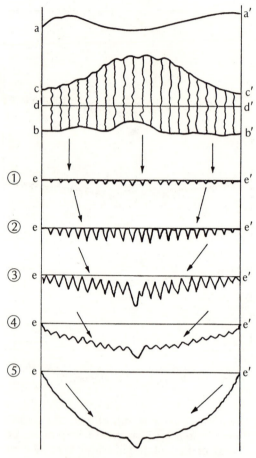

Fig. 5.31. Plan and cross-sectional views of a developing rill system (sections e-e' taken along line d-d') in which cross-grading of rills and gullies cause valley development. (After Horton, 1945.)

fied erosionally before the establishment of plant cover is highly problematical. On a terrestrial surface already impregnated with seeds and spores, the rilling process might hardly get started before plants appear. On a newly emergent sea or lake floor or a new volcanic terrain, considerable rill modification could clearly be expected before a continuous plant cover is developed. In any event, rills seem to form where water depths are greatest in areas of sheet flow. These places, in turn, tend to coincide with primary surface irregularities, but the writer has observed rill systems developed by wave back wash on beaches of fine sand which were sensibly very even-surfaced locally though broadly cusped otherwise (cf. Fig. 5.31 and Fig. 5.32). As noted by Horton (1945), rills appear to be modified into a *drainage net* by deepening and as a direct consequence of the overtopping and breaking down of ridges between rills during heavy overland flow periods (cf. Figs. 5.33; 9.13). As illustrated, there are a number of cogent arguments that can be made to show that the diversions are consistently from higher rills to lower rills. However, the reasons for diversions are not totally related to either over-topping a ridge by water or undercutting it by erosion as Horton indicates. Many rills are so small initially that diversions in them relate to obstructions formed by unusually large clasts of the material being eroded or even blockages by newly sprouted plants.

Studies of the incidence of drainage line deflection to the right or left with respect to slope in a drainage net suggests that it is essentially a random process (Fig. 5.34). It may, in fact, be the random diversion of rills or even gullies that develops the initial tributary pattern of a given *drainage basin* (Fig. 5.35) as well as the ultimate juncture of trunk streams from adjacent drainage basins (Fig. 5.36).* In this or some effectively similar manner, the system of gullies and channels form, deepen, and unite on a granular surface to comprise a *dendritic*

Known as DuBoys formula, this equation is a rational expression which will serve as a basis for determining eroding force/square foot of soil surface. It is of little value in assessing available erosion energy where plant cover has become established because frictional resistance is offered by the plant material essentially everywhere but in watercourses. Neither does it take into account the erosive force of falling rain considered in Chapter 6.

The extent to which rills would be modi-

---

* Drainage basin junctures with each other would only relate to rills indirectly through developed gully and ravine systems.

Fig. 5.32. Development of rills into drainage nets on the very fine sand beach near Cape May Point, New Jersey, by sheet runoff which is part of wave backwash. The sheetflow initially develops a micro-rill system which merges to form a rhombic rill pattern (in very fine sand under water less than 1 inch deep) (foreground). Downslope, the rhombic rills randomly diverge and intersect while deepening to form an initially dendritic and secondarily anastomotic drainage net (background). The rhombic pattern tends to be replicated on ever-increasing scales within the anastomotic flow area. The coin is a 25-cent piece. (See also Fig. 9.13 for potential rill effects on beach cusp formation.)

drainage system of the type so common to humid regions (Fig. 5.37). There are no widely agreed upon dimensions for the components, but rills may be only a fraction of an inch deep and wide, gullies range from a few feet deep and wide to perhaps a score of feet in both dimensions. Needless to say, valleys are bigger, and one definition of a gorge is a valley so deep it scares you to look over the edge.

In his major study of drainage develop-

ment, Horton (1945) notes that, during rill evolution on a granular cultivated surface of erosion, there is a tendency for angular bends in rill systems to develop into rounded meanders because of lateral thrust of current on the outsides of curves, particularly on lower land slopes (Fig. 5.37). The writer observed a rill system develop in New York State on the floor of Lake Sacandaga in sand and silt exposed when lake levels were artificially reduced in order

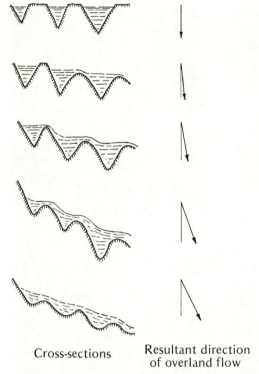

Cross-sections   Resultant direction
of overland flow

Fig. 5.33.   Cross sections showing successive stages of rill obliteration by overtopping of micro-divides. (After Horton, 1945.)

to accommodate runoff from winter snows. The rills formed after a shower much as Horton describes, but small ones ±0.25 inch deep and somewhat narrower quickly developed strong meanders with amplitudes of 0.75 to 1 inch in the finer-grained sediment. Though repeatedly obliterated by incoming waves from the lake over a distance of 1 to 2 ft, the micro-meanders immediately reformed, again and again. Regardless of the precise combination of runoff volume, sediment type, and amount and slope required to produce meanders (we will consider this in some detail in Chapter 7), it seems clear that a meandering flow pattern may appear at the onset of humid conditions where unconsolidated granular material is being eroded.

Drainage and Slope Development

Humid fluvial erosion is characterized by the removal of detritus from the flow lines by the force of rather rapidly moving water. This type of erosion has been observed to generate slopes inclined toward the watercourses. Prior to the establishment of vegetal cover on such slopes, they too are subject to direct fluvial erosion. Davis (1902) long maintained that landslope angles were slowly lowered during river degradation, in contrast to King (1953) who claimed that slope angles once established tended to be maintained (cf. Fig. 1.4). Horton's erosional observations tend to support King as do those of Schumm (1956).* The latter worker records both parallel and declining slope retreat in clay pit erosion where vegetationless surfaces were artificially induced. Schumm also notes that slopes in his study were maintained by creep of the material composing them.

Once a land above a watercourse is vege-

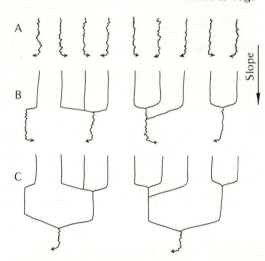

Fig. 5.34.   Plan view diagram of consequent runoff developing rill systems on an inclined granular surface with random diversions. (A) Initial rill development and diversions; (B) primary drainage net and secondary diversions; (C) secondary drainage net and tertiary diversions—rill overtopping may obliterate all the foregoing in development of the eventual drainage configuration.

* As noted in Chapter 6, backwearing in excess of incision is only possible where incision is inhibited. When one considers the normal vectoral components of gravitational force, this is hardly surprising.

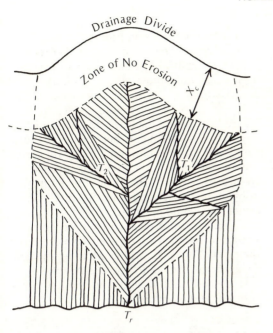

Fig. 5.35. Lines of flow after cross-grading of first pair of tributary areas (T_1 and T_2) in an incipient drainage net with trunk stream T_r. (After Horton, 1945.)

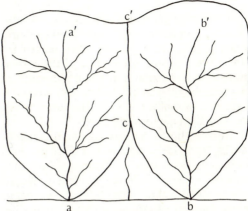

Fig. 5.36. Final development of two adjacent drainage basins on newly exposed land under humid conditions. (After Horton, 1945.)

tated, the humid erosional system has been observed to change. Direct erosion by runoff becomes essentially confined to watercourses where either regolith or bedrock are exposed. Channels are normally being deepened by the removal of rock material and, if being incised into unconsolidated material, the surfaces contiguous with the channel tend to assume a slope which is the stable repose angle for the material in question. Deepening of the flow line automatically oversteepens adjacent slopes, beginning at their bases, thereby triggering a mass wasting response, usually in the form of slope creep, which is gradually propagated upward until the entire slope surface is in adjustment (Fig. 5.38). As a consequence, the gully or valley being formed through the combined effects of fluvial erosion and mass wasting widens more or less as it deepens.

Horton's x_c parameter mentioned at the beginning of this section delineates a zone of nil surface erosion by runoff on either side of each drainage divide, even prior to

acquisition of vegetal cover (Fig. 5.39). A barren divide area would, however, be subjected to rain-splash erosion. Horton (1945) also notes that the divide configurations are little changed once the essential drainage pattern is established. The reason for this is readily seen once it is realized that the main forces acting on a sharp, vegetated divide would be those related to groundwater solutioning and such increments of slope creep shear as are expressed in valley widening and divide lowering. The hillslope proper is being subjected to both solutioning and to creep along a maximum slope. The drain-

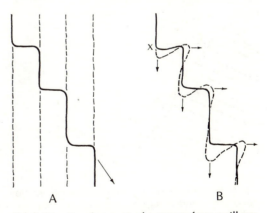

Fig. 5.37. Development of a stream from a rill system by cross-grading. (A) Plan view showing initial angular bends; (B) rounded bends by thrust of water where flow is on granular, unconsolidated material. (After Horton, 1945.)

age line itself is undergoing erosion by solutioning, undercutting, and bank caving (Fig. 5.38) and by direct force of stream flow. The morphologic results of these relationships, as Horton notes, is a reduction of drainage divides at a rate determined by the rate of channel incision as the latter is expressed upslope through creep.

During the fluvial erosion of unconsolidated material which is our present concern, any deepening of drainage lines at a rate in excess of divide reduction would oversteepen valley sides and cause earth slips. Horton (1945) notes that such slips are rare, but this may merely reflect the comparative rarity of truly large valleys eroded in unconsolidated material (cf. Fig. 4.28). Along such valleys the valley-wall materials may be so rapidly undercut that

it is difficult for vegetation to maintain a root purchase. Once consolidated bedrock is added to the drainage development picture, as it eventually must where drainage lines are cut through existing covermasses, the rules of slope development must change. There are many bedrock materials fully capable of standing in both vertical faces and even in overhanging cliffs. In cases where vegetated divides are composed of indurated bedrock, they would no longer be subject to lowering by creep, and valley widening would no longer necessarily accompany channel deepening. Deep gorges with nearly sheer walls are normal consequences (Fig. 5.3), and slope creep in such areas is restricted to those surfaces near divides with inclinations low enough to retain soil and plant cover.

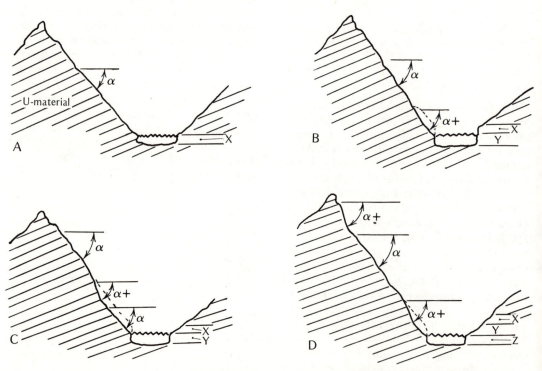

Fig. 5.38.   Series cross sections of channel incision, valley deepening, and valley widening along a watercourse in a humid region underlain by comparatively nonresistant regolith or bedrock. (A) Valley wall in stable repose at angle α followed by incision of channel to depth X. (B) mass-wasting response to incision depth X is accelerated creep accompanied by local steepening of valley wall to angle α^+, followed by incision of channel by increment Y. (C) and (D) steepened slope segments propagate upward, and valley wall is progressively regraded as additional increments of incision are developed.

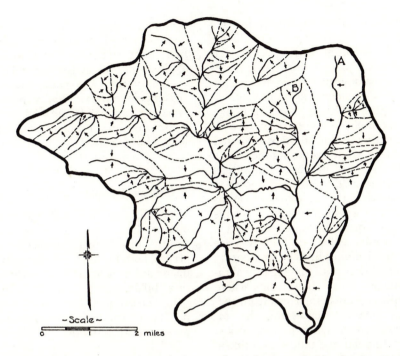

Fig. 5.39.   Drainage basin of Pennypack Creek above Valley Falls, Pennsylvania, showing areas from which surface runoff is derived. (After Horton, 1945.)

It would appear that drainage system development under humid conditions is in accordance with the processes so far elucidated and has the over-all topographic result of an increase in land relief and a consequent increase of the area in slopes. Horton (1945) comments that the ultimate surface of erosion within a main drainage basin boundary is not "usually as close to being a plane as was the original surface area from which it has been derived." He further states that under ideal conditions it is ". . . closely similar to a segment of a parabaloid cut by a plane which is not parallel with the axis of the parabaloid." The parabaloidal surface is ripped with ridges which represent the divides between the streams. The amount of relief generated by this type of fluvial process would seem to be limited only by the amount of land elevation extending to the maximum upward limit of humid conditions (13,500–14,000 ft), and the lower limits of fluvial incision (locally scores of feet below sea level on several

rivers including the Amazon and Mississippi).

The humid, fluvial processes expressed by watercourse incision so far discussed are clearly time dependent. Depending upon the degree of consolidation of the material being eroded, deepening of a watercourse takes time. Other factors being the same, a deep valley should represent a longer interval of incision than a shallow one. However, it will be noted that age would not necessarily be expressed by different forms of cross-valley profile as was suggested by Davis (1902); cross-profile would rather tend to continue in a general V-shape since there is no mechanism indicated by current research which would generate a valley flat as long as humid conditions persist without interruption.

Certain other time-indicative trends appear to correlate with drainage system development. In drainages initiated on a granular covermass, the watercourse incision would be characterized by an uncovering of bedrock, or, as we shall term it here, *de-aggradation.** A second mode of fluvial incision would be initiated with the abrasive down-wearing of bedrock once such material is exposed along the bottom of a watercourse. This *corrasional* erosion stage is only just beginning along many drainage lines. For example, a comparison of bedrock outcrop dimensions on maps prepared of northwestern Arkansas during the 1960's with geologic maps of the same areas (Winslow Quadrangle, 1904) disclosed a notable increase of rock exposure areas through stream gravel veneers. It seems clear that the corrasion of channels in a region will persist as long as abrasives are supplied to the drainage lines by weathering and mass wasting or until pre-existing abrasives are flushed away. In settings where insoluble minerals are being freed from bedrock during chemical weathering, the corrasional erosion mode may persist almost indefinitely. In polar and temperate climes, and elsewhere where drainages draw their sediment loads from settings where mechanical weathering prevails or has occurred in the past, corrasional erosion typifies stream incision. Perhaps most of the world's drainage lines are characterized by this type of erosion. In the humid tropics, however, chemical decomposition apparently tends to eliminate abrasive sediments. Complete success in this is apparently limited to *climatic nuclei*. In our discussion of humid weathering, we noted that Tricart (1956a) recorded a case of entirely solutional erosion along the upper Senegal River (Fig. 5.40). In that instance a *corrosional* erosion mode endured long enough to permit fluvial incision in quartzite to depths of several tens of feet along two sets of joints which intersect at nearly right angles. Tricart notes that the solutional erosion is not only documented by the absence of solid bed load, but also is proved by the fact that the stream bends persist at nearly 90° angles following a long period of incision (Fig. 5.41). Only very protractedly

humid conditions such as one encounters in a *climatic nucleus* would be expected to exhibit a CLIMAX EROSIONAL RELATIONSHIP of this kind, for as seen in the discussions on weathering and plant cover, it seems reasonable to regard dominantly solutional erosion as an expression of fluvial dynamic equilibrium in a humid, vegetated environment characterized by chemical activity.

Drainage Density

In our discussion of drainage system development it is apparent that the combined lengths of all watercourses tends to increase, both by downslope extension and by headward erosion as well as through tributary bifurcation.** The *drainage density* (D_d) of a drainage basin is simply the channel length per unit area. This may be expressed as an equation

$$D_d = \frac{\Sigma L}{A}$$

where ΣL is the total length of streams, A is the area, and both are in units of the same system.

The quantitative expression of drainage density was first worked out by Horton (1932) to more precisely express the degree to which drainage is established in a basin. Its use should convey a more exact idea of the extent of fluvial incision than such expressions as well drained or poorly drained. El-Ashry (1971) has suggested specific numbers of streams to represent different degrees of drainage density (Fig. 5.42). The accuracy with which drainage density calculations can be made from a map is limited by map scale, and maps of sufficiently large scale should be used to show all watercourses present. On some maps, drainage lines that do not carry water constantly (*perennial*) are not indicated. Indeed, as noted also by Horton (1945), headwaters reaches of perennial streams often only carry water part of each year (*intermittent*

---

* To appreciate the value of this term, cf. DEGRADATION and DENUDATION in the Glossery.
** Drainage density may not correlate well with water or sediment discharge (cf. Lustig, 1965a).

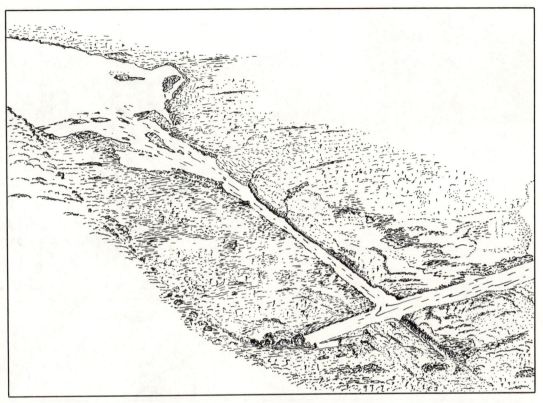

Fig. 5.40. Rapids and right-angles bends on the middle Bafing River, western Mali, associated with the Senegal drainage system. Fractured Ordovician sandstone is truncated by erosion but directs flow routes. The sandstone pavement has scattered boulders, but the stream carries little solid load. (See also Figs. 1.8 and 5.41.) (Based on an aerial photo and data provided by J. Tricart.)

flow) or even only during and just after rainfall (*ephemeral flow*), though neither of the latter are considered to typify flow continuity in humid lands. Temporal flow continuity is, of course, an expression of the relation between the water table, the groundwater body, and watercourse elevations. In headwaters areas, the volume of groundwater above the drainage lines is minimal for a given area, and dry seasons often depress the water table below stream channel bottoms there with consequent reduced flow continuity (Fig. 5.43). In downstream areas of humid lands, the water table rarely is depressed below flow lines, and, therefore, surface return of groundwater supplements flow between precipitation events to produce perennial runoff.

The density of drainage lines in a given basin is limited by Horton's x_c value which establishes the minimum terrain catchment area required before erosion can occur. In general, the more tributaries a stream has, the higher is the drainage density. This is expressed by the *stream order* of the drainage basin. The stream order system devised by Horton (1945) is widely used to make quantitative comparisons and analyses of drainage systems. However, Abrahams (1972) questions basin comparisons that disregard environmental history (cf. Melton, 1958). In Horton's system, the trunk stream is the highest order. Stated somewhat differently, the smallest unbranched tributaries would be first-order, tributaries or streams of the second order receive branches of tributaries of the first order only; a third-order stream may receive one

Fig. 5.41. A reach of the upper Senegal River (Souma Falls), near Kindia, Guinea, where the channel drops over an escarpment in quartzite (left, foreground), makes a right-angle turn to the left, and follows a narrow solutional notch. (Based on a photo by J. Tricart.)

or more tributaries of the second order but may also receive first-order tributaries. Using this system, main streams are assigned the highest order number. Horton's stream order system is useful for comparing drainage development in drainage basins of comparable size, but as he notes, the stream order tends to increase with drainage basin size. It should also be observed that stream

Numerical Scale for Grading Drainage Density

		Watercourses/sq. mi.
Low density = Coarse texture		= Less than 20
Medium density = Medium texture		= 40 to 50
High density = Fine texture		= More than 80
Very high density = Ultrafine (badland) texture		= More than 200

Fig. 5.42. Rating system for drainage density based on the number of watercourses per unit area. (After El-Ashry, 1971.)

order is entirely independent of the idea of drainage patterns. Thus, with identical lengths and numbers of streams a drainage pattern could be dendritic, rectangular, radial, or some other geometric form (cf. Figs. 2.19; 2.20).

The development of a drainage net in a humid geomorphic system would appear to logically reflect the tendency of such natural systems to evolve toward a least-work, equal-area energy expenditure relationship. The latter attributes (initially cited in Chapter 1) are those assigned by Howard (1965) to the dynamic equilibrium condition in natural systems. It is therefore of interest to relate the apparent ultimate attributes of humid drainage nets and topography in the equilibrium context. In both Horton's theoretical example (1945) and in regions believed to be humid climatic nuclei, we find a *ridge-ravine, serrate,* or *feral* topography essentially identical to that termed *selva* in the tropics in which there is a maximum land area in slopes (Fig. 5.44). This is essentially the same landscape that Davis (1902) termed "mature," but it is lacking in flat-bottomed valleys and meandering streams. Hack (1960) describes essentially identical terrain in the southern Appalachian Piedmont in an area he regards as being completely in equilibrium with humid weathering and erosion. All of the same areas exhibit near maximum drainage densities and are entirely sheathed in vegetation and soil, excepting for watercourses and rare, isolated outcrops. The bottoms of valleys in such regions are occupied by channels of watercourses so that there is abundant evidence of vertical stream incision but essentially none of lateral corrasion (Fig. 5.45).

Hack (1960) assesses his examples of ridge-ravine terrain as being landscapes in a state of dynamic equilibrium with respect to a humid, vegetated environment. This

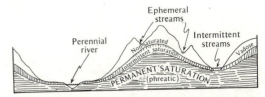

Fig. 5.43. Cross section of the usual runoff flow continuities in a humid region where upland streams are commonly ephemeral, those of intermediate elevation are dry much of each year, and main streams flow constantly. This is also the order in which dryness is imposed following a climate change to aridity as discussed in Chapter 7. Corresponding groundwater zones are also indicated.

seems reasonable, though there seems to be no particular need to invoke the concept of "grade" as Hack does in making his assessment. Rather, the terrain configuration with a maximum area in slopes and a similar maximum density of drainage lines is a logical geometric reflection of hydrologic processes designed to drain surplus water out of the environment with a minimum expenditure of energy in either flow or sediment transport. It thus seems appropriate

Fig. 5.44. Selva topography in the north coastal range of Venezuela under heavy tropical rain forest. The land is entirely in slopes, ridges are very sharp, and outcrops are rare. Crest peak of Naiguata is over 10,000 ft (3,050 m) in a situation some 7° north of the Equator.

Fig. 5.45. Mountain valley on the heavily vegetated, dominately humid north slope of the Sierra del Norte coast range, Venezuela. Weathered valley sides incline directly to channel banks, and bare rock channel floor slopes about 15° away from the viewer. Heavy rains in this and similar regions occasionally cause weathered vegetated tracts to tear loose in very large gravity slides (cf. Figs. 2.29 and 5.44).

here to designate this type of selva landscape *humid climax topography*.

It would appear that an environment's terrain, soil, vegetal cover, and erosional modes need not all achieve a climax form simultaneously, particularly since any interruption of the processes involved need not affect all agencies and products within a system in like manner or degree. However, in situations where essentially all agencies and products in a setting have climax expressions, we note that chemical weathering processes are generating fine-grained insoluble residues and solutes, materials moved by running water with the least possible energy drain. The hillslope angles that mainly control mass-wasting rates (creep) have been shown to develop in response to watercourse incision. And since the weathered residues are only appreciably available to runoff along watercourses, the streams are erosionally "fed on demand," so to speak. In effect, as illus-

trated previously (Fig. 5.38), in the humid climax setting, it requires an erosional event by the stream before mass wasting can provide additional sediment. Thus, in the humid geomorphic system that has attained a climax expression, there is far more water than sediment in the watercourses.

If we view Gilbert's law of stream capacity (cited in Chapter 1) in light of erosional mechanisms operative in humid climax topography, it will be seen that there is apparently no problem of sediment transport in a related watercourse. Many streams in such regions carry little solid material. As a consequence, it will be noted that in most humid climatic nuclei, many of the topographic attributes commonly associated with watercourses elsewhere are missing. Included in the list would be valley flats (floodplains), meanders developed after valley incision, terraces, and channel alluvium. Rarely is there sufficient solid sediment in transit in such areas to form extensive point bars on the inside of stream bends, and it is doubtful if such a stream could maintain a delta even on a near-zero energy coast such as that along northwest Florida. Indeed, one would have to conclude that drainage systems with the foregoing attributes not common to humid climatic nuclei are probably displaying effects related to some sort of prior non-humid condition.

Litho-Structural Drainage Effects

We have considered drainage development on a granular, unconsolidated surface and must now devote our attention to effects of bedrock on drainage development. Naturally, there is little recorded on the subject in terms of actual observations simply because indurated bedrock responds to erosion very slowly. Persons doing stream-table experiments almost always avoid using bedrock for a flow surface. Nevertheless, there are many drainage expressions that can clearly be related to either bedrock lithology or structure. It has already been pointed out that close drainage conformance to bedrock characters usually implies drainage development in absence of an appreciable granular covermass over bedrock. Under these circumstances, both runoff and groundwater are free to exploit fracture systems, soluble rocks, and similar "weaknesses" during the development of flow lines (Fig. 5.39).

Relations between bedrock and drainage are usually outlined in elementary geologic texts in two main aspects: (1) angular geometric flow patterns indicative of rock structure and (2) solutional landforms. Some of the flow patterns most commonly attributed to structural influences and previously illustrated (Figs. 2.19 and 2.20) actually deal with arrangements between drainage nets rather than between individual watercourses. Thus, in *radial* drainage, the nets usually originate on some local upland of tectonic origin, commonly a volcano, and radiate outward from a common center. *Rectangular* patterns commonly reflect bedrock fracture sets intersecting at high angles. Higgins (1961) suggests that such fractures can be expressed upward through a covermass and thereby influence drainage patterns (cf. Garner, 1966). *Trellis* patterns are usually developed where individual drainage nets exploit fractured rocks of variable erosional resistance exposed by denudation of folded strata. And *annular* drainage patterns are usually encountered where structural domes, anticlines, or basins have strata of variable lithic character being attacked by linear runoff.

In most of the foregoing examples of structurally controlled drainage, there is the implication that tributaries have been extended by the headward erosion of tributaries along zones of bedrock weakness. But there have been rather elaborate mechanisms devised to explain erosional modes of streams in deformed and fractured rocks (cf. Davis, 1899). To begin with, if we assume initial fluvial erosion occurring on a lithified surface, there are virtually scores of primary lithic features capable of directing and localizing runoff and erosion. In volcanic terrain, for example, there are the original surface crenulations and allied irregularities of lava flows, cooling fractures, *lava caves*, contacts between lava and pyroclastic material, or pre-flow soil (Fig. 3.94). Water-deposited sedimentary rocks frequently display ripple marks, mud cracks, scour depressions, compactional irregularities, organic reef mounds, and lateral textural and compositional variations, to name only a few of the kinds of features capable of influencing runoff character. Consequent runoff patterns developed on consolidated bedrock can hardly develop in the rill-gully-valley sequence discussed by Horton (1932) since on bedrock initial flow routes would be determined more by terrain irregularities than by surface erodability. In point of fact, most consolidated rock types become water saturated only through prolonged immersion. Water briefly impressed on rock surfaces (as by precipitation) is largely repelled as runoff, and Horton's x_c value would be zero in many cases. That is, a belt of nil erosion dependent upon appreciable terrain permeability next to divide areas of drainages developed on bedrock would hardly be possible prior to development of a weathered zone.

Modification of drainage patterns through the effects of underground solutioning is common under certain special, humid, vegetated conditions. In temperate latitudes, only limestone, dolomite, and rock salt are readily soluble bedrock types of widespread occurrence. Solutional landscapes (*karst topography*) are extensive in carbonate bedrock areas of Yugoslavia, eastern Kentucky, and Tennessee and also in such tropical settings as Barbados, Puerto Rico, and Jamaica. In these examples, solutioning seems to be primarily

localized by fractures and therein surface drainage frequently goes underground at *sinks*, there to either join the groundwater body in slow percolation or occasionally to flow essentially unimpeded in underground cavern stream systems. In non-humid settings, such sinks would be relict landforms (Fig. 5.46). As already noted, limestone may fail to produce soil during chemical weathering if it is lacking in impurities, and erosion in such regions is mainly corrosional along associated watercourses.

Caverns produced by percolating groundwater are noted by Bretz (1942) to display both *vadose* and *phreatic* features (cf. Fig. 5.43). The general consensus would appear to be that dripstone features are formed mainly by carbon dioxide loss through agi-

Fig. 5.46. A sink adjacent to the Pecos River south-southeast of Roswell, New Mexico—view to northwest. Though clearly a solutional feature, in its present environment this sink is a relict. (John S. Shelton.)

tation and evaporation in the vadose zone. *Stalactites, stalagmites,* and similar features grow slowly and apparently signify episodes when water percolated through existing cavern systems (Fig. 5.47). The cavern openings themselves, however, are generally believed to form in large part within the phreatic zone during bedrock saturation by groundwater. If so, cavern formation largely relates to times of elevated groundwater tables (e.g. relatively humid but prior to deep stream incision). Dripstone formation in this theoretical context would correlate with times of depressed water tables (e.g. relatively dry climates or in humid regions with deeply incised drainages). It should be emphasized that incision depths increase under humid conditions with the result that cavern formation would presumably tend to occur as a stage of fluvial erosion, possibly rather early.

Under tropical conditions, rock in addition to carbonate types and salt may be notably soluble. Tropical karsts include terrain developed on both igneous and metamorphic rocks. Holms (1965) illustrates solutionally enlarged fractures (*lapis*) in olivine basalt on the island of Oahu, Hawaii, and Tricart (1956b) describes extensive solutional landforms developed in metamorphic rocks in Belo Horizonte, Brazil as shown here (Fig. 5.48). Where protracted stream incision has developed deep gorges with rock walls in the humid tropics (cf. Fig. 5.3), there can be little doubt that the dominant form of sculpting on hillslopes and ridge crests is by solutioning in a wide variety of rock types. This contention is strongly supported by numerous slopes far too steep to retain soil which lack any soil scree or talus at their bases. In similar settings developed on volcanic terrain, valleys commonly display amphitheater-like heads as noted by Jones (1938); such features are common in Hawaii and Ecuador (Fig. 5.49).

In Chapter 3 it was noted that the penetration of salt piercement masses to the surface in humid regions may lead to formation of solutional depressions—often

Fig. 5.47. A dripstone column and assorted complex stalactites in Luray Caverns, Virginia.

occupied by lakes. And, of course, the presence of solutional features in relatively dry areas may signify formerly moist conditions. The Carlesbad Caverns in New Mexico is an outstanding example of limestone solutioning now in a desert. Frye and Schoff (1942) described sinks in Kansas, that induce centripetal drainage and are believed to be due to solutioning of salt at depth (Fig. 5.50). Elsewhere, Peabody (1954) has discussed *travertine* deposits and cave formations related to former surface drainage lines in a portion of South Africa that presently ranks as one of the driest regions on earth. There are numerous similar solutional phenomena on record, many no longer situated in humid lands. At least some of the latter presumably relate to formerly more moist, vegetated conditions.

Fig. 5.48. Solutionally enlarged joints in a banded gneiss near Belo Horizonte, Brazil. In tropical settings as humid as this, the silicate minerals dissolve readily, and karst topography may develop from various lithologies (cf. Fig. 5.3). (J. Tricart.)

Trans-Environmental Rivers

We will draw our discussion of humid geomorphic systems to a close with a brief consideration of the downslope effects of humid runoff in other environments. A *transenvironmental river* is defined here as a stream from a humid region that flows downslope into a different kind of land environment rather than directly into an ocean or sea. When such a river leaves the humid, vegetated setting, it tends to impose certain attributes of the humid geomorphic system upon the downslope geomorphic system. In most instances, the downslope environment is that of a desert, and the transmitted attributes (e.g. perennial flow) are altered, both by accentuated evaporation and by reduced surface return of

groundwater. It has been noted that most streams in humid regions maintain their sediment carrying capacity by downslope runoff volume increases, largely via runoff from tributary streams, but in part from contiguous groundwater bodies. Once a humid river enters a desert, both runoff and groundwater sources disappear. Transenvironmental rivers in deserts are often joined by normally dry washes and such runoff and sediment as these may add ephemerally (cf. Butzer and Hansen, 1968). However, the long-term environmental consequence of a desert on a humid stream is a runoff volume reduction and a correspondingly reduced ability to transport sediment.

Effects of downslope environments upon streams from vegetated regions vary considerably, even among deserts. These effects seem to depend mainly upon (1) the volume of water and sediment leaving the humid environment, (2) the intensity of downslope environmental effects, particularly in terms of runoff volume changes, and (3) the environmental history of the transition zone. North America has relatively few major trans-environmental rivers at present. Notable examples that penetrate deserts include the Colorado River, the Snake and Columbia rivers, and the Rio Grande. All of these rivers enter their dry reaches with sufficient runoff volume to be able to penetrate through to the ocean. Prior to dam construction, the first three of these were effectively corrading their channels in their middle and lower reaches. Present natural and artificial volume losses along the Colorado River cause increased salinity and aggradation along lower reaches and in the delta area, though over-all sediment

and water volume have diminished. As noted by the writer (1966), Africa and Australia have a number of trans-environmental rivers (Fig. 4.24). In Africa, the Senegal, Niger, and Nile rivers all reach oceans in spite of obvious volume reductions in a desert. So do the Hwang Ho and Darling-Murray drainage systems in Asia and Australia, respectively (Fig. 4.24 B, D). The Chari (Shari) River in north Africa is a notable example of a drainage that dries up in a desert, and, of course, the desiccation leads to deposition of the sediment being transported and a loss of trunk channel identity (Fig. 4.24 A). Essentially all the trans-environmental drainage examples cited outside of North America display deranged reaches. The *drainage derangement* in question is typified by the aggradational blockage of trunk drainage lines, *distributary* channels with junctures having acute angles pointing upstream, and a variety of

Fig. 5.49.   An amphitheater valley head developed in pyroclastic material at an elevation of about 10,000 ft (3,000 m) in the Ecuadorian Andes Mountains. Escarpment retreat, possibly of a special kind, is involved.

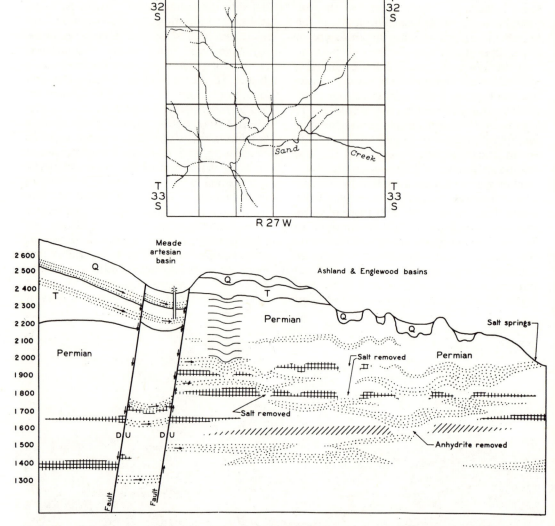

Fig. 5.50. Cross section and drainage map of the Meade Basin and "Jones Sink," Kansas. Arrows in section indicate supposed course of circulating groundwater. Note especially the centripetal drainage pattern. (After Frye and Schoff, 1942.)

related alluviated flow and impoundment configurations. In one of the more obvious cases recorded by Forbes (1932), some of the deranged reaches continue to affect present-day runoff patterns (cf. Fig. 5.51).

Present trans-environmental drainages now being aggraded as a consequence of volume loss show braided flow patterns and related features. These features appear after a period of exposure to desert conditions of evaporation and infiltration; the period in each case being expressed as flow distance within the desert. In the same regions, vegetally immobilized dunes and related alluvium appear to signify formerly reduced humid, vegetated areas and probably related runoff reductions in many associated trans-environmental streams (cf.

Grove, 1958). This relict desert regolith in the upstream humid areas currently augments stream loads from those places and thereby lends substance to fluvial deposits being made further downstream.

Trans-environmental rivers are clearly able to transmit fluvial geomorphic effects from humid, vegetated environments downslope into adjacent land environments. The geographic extent of such effects and whether they are characterized by erosion or deposition is apparently dependent in large measure upon the amount of water the humid environment yields. Many factors are involved, yet the intensity of precipitation and the actual area of catchment that is vegetated and correspondingly moist would appear to be major considerations. In the case of at least one such drainage (the Darling-Murray system in Australia) there is a continuing debate (cf. Langford-Smith, 1960; Butler, 1960; 1961; Cotton, 1963; Dury, 1963; Schumm, 1965) as to whether the streams in question aggraded their courses to form the Riverine Plain during

pluvial times or under accentuated aridity. Though the issue cannot be said to be resolved, it is evident that the ultimate locus of deposition for runoff leaving a humid upland can and will shift up- and downslope along main drainage reaches as related arid and humid areas and their respective abilities to absorb and supply runoff vary reciprocally (Fig. 5.52); see also Chapter. 6. It should also be noted that the interwoven flow patterns formed as a result of desiccational aggradation of a trans-environmental river (cf. Figs. 5.50; 5.51; 5.52) are morphologically similar to runoff patterns developed where humid environments replace arid or glacial ones as discussed in Chapters 7 and 8.

Our discussion of trans-environmental rivers would be incomplete if we failed to note the several unique features of many drainages on the north continental slopes of Eurasia and North America. A number of rivers including the Ob, the Yenisei, and the Lena in Eurasia and the Mackenzie in North America originate in moist, vegetated

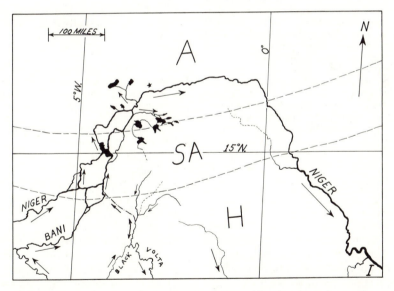

Fig. 5.51.   Drainage derangement along the upper Niger River, Nigeria, along the southern margin of the Sahara Desert. In addition to the network channel system caused by alluviation under formerly more southerly arid conditions, a chaotic flow pattern involving several lakes results where sand dunes blocked the main watercourse during times of low discharge during the Pleistocene. Some flow along the Black Volta also enters the Niger channel network. Former desert conditions were at least 150–200 mi farther south.

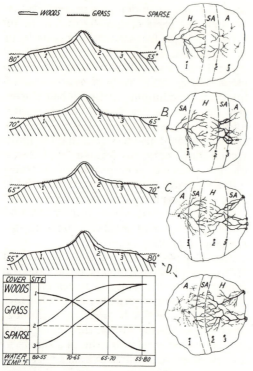

Fig. 5.52. Effects of a climate change extending over thousands of years upon vegetation and drainage of a hypothetical landmass just north or south of the Equator. The region is shown in plan views (right) and in corresponding sections (left). Land environment shifts are related to changes in ocean surface water temperatures adjusted to periodic glaciation. The drainage configurations adjust in relation to the areal extents of the humid and arid regions and correspondingly altered runoff volumes and continuities. In sequence, (A), the land near the warm sea receives ample moisture and forests begin at the strand; the opposite ocean is cold so

that air masses are undersaturated, rise, and cool before precipitation can occur; there is a coastal desert; (B)–(D) the situations become reciprocal climatically, as the ocean conditions reverse. In the bottom graph, three randomly located areas (1–3) record differing climatic and geomorphic histories during the *same* time span.

areas and flow into perennially frozen seas. Most of these rivers exhibit great downstream reaches which are completely frozen over throughout the Winter and early Spring. Water flowing in these channels is sealed in like that in the pipes of the usual municipal water system, since it is under a hydraulic head. Where water pressures in the channels exceed the strength of the ice seal, hydraulically induced explosions can result (*ice bursts*). Moreover, the fluvial erosion alternates semiannually due to the shift to open channel flow each Spring when the ice goes out (Fig. 5.52). The headwaters areas of these rivers are in notably warmer environments than their mouths, with the consequence that the latter are the last to thaw and loose their ice cover at the Winter's end. Also, in the Spring, ice jams not uncommonly dam flow routes and cause extensive flooding. Because essentially all the rivers in question derive their sediment loads from some glacial debris and regoliths otherwise generated with a high incidence of freeze-thaw activity, the detritus tends to include abundant mechanically fragmented solids (see Chapter 8, especially Fig. 8.73).

SUMMARY

The humid geomorphic system has been presented insofar as possible as an independent mode of landform-deposit genesis. Our primary concerns have been with the unique geomorphic effects that characterize lands where moist, vegetated conditions have long endured. We find the humid system dominated or controlled by several attributes, most notable of which are (1) a water surplus in excess of all environmental requirements within the humid realm which must therefore drain away and which in so doing removes sediment and generates watercourses, hillslopes, and, ultimately, relief; (2) plant cover which largely restricts erosion by overland flow to watercourses and otherwise protects weathered residues from erosion; (3) soil and related colluvium as an essentially residual, dominantly biochemical accumulation of rock and organic substances, which dominates the character of derivative fluvatile sediments and curtails direct interaction between the atmosphere and bedrock.

The climatic nuclei in which humid geomorphic systems appear to have had most sustained activity are characterized by land surfaces mostly in slopes, high relief, V-shaped or narrow slot-like valley cross-profiles with channels occupying essentially the entire bottom of the notch, relatively thick weathering profiles including soils that are chemically mature decomposition products for the setting and related bedrock, climax floral associations thoroughly adapted to moist conditions, and stream sediments dominated by solutes and semi-insolubles. Examples of humid climatic nuclei were cited in Chapter 4, but they are taken to include the elevated portions of the Guiana Highlands and intermediate elevation portions of the southern Appalachian Mountains (and Piedmont) which either escaped or has eliminated effects of recently terminated high-elevation frost weathering.

Students of geology as well as geomorphology will be quick to note that our discussion of fluvial activity has so far omitted any thorough consideration of such topics as the hydraulic geometry of stream channels, valley cross-profile variations, such

depositional and erosional landforms as point bars, terraces, meanders, and floodplains, plus a variety of hillslope characteristics and processes. These omissions are intentional since it has seemed most reasonable to exclude from the foregoing discussions those landscape attributes apparently created under non-humid subaerial conditions. Thus, if the drainage development concepts of Horton and the hydraulic findings of Gilbert (1880) are correct, excluded phenomena would include those dependent upon extensive alluvial valley deposits. Landforms developed under humid conditions because of the presence of such alluvium could not rank as "typically humid" features, but only as humid modifications of relics left by some prior condition. Each surficial environment that involves a major change in groundcover appears capable of generating its own suite of landforms and deposits upon which humid geomorphic systems can be imposed and for which humid systems can be antecedent. The next chapter will focus upon the unique landscape effects of arid environments. The reader should keep in mind, therefore, that the several geomorphic phenomena including fluvial activity will be considered again under arid conditions in Chapter 6 and again in later chapters dealing with other geomorphic systems.

REFERENCES

Abrahams, A. D. (1972) "Environmental Constraints on the Substitution of Space for Time in the Study of Natural Channel Networks," *Geol. Soc. Amer. Bull.*, v. 83, pp. 1523–1530.

Bennett, H. H. (1939) *Soil Conservation*, McGraw-Hill Book Co., New York, 993 pp.

Blackwelder, Eliot (1931) "Pleistocene Glaciation in the Sierra Nevada and Basin Ranges," *Geol. Soc. Amer. Bull.*, v. 42, pp. 865–922.

Bretz, J. H. (1942) "Vadose and Phreatic Features of Limestone Caverns," *Jour. Geol.*, v. 50, pp. 675–811.

Butler, B. E. (1960) "Riverine Deposition during Arid Phases, *Australian Jour. Sci.*, v. 22, pp. 451–452.

Butler, B. E. (1961) "Groundsurfaces and the History of the Riverine Plain," *Australian Jour. Sci.*, v. 24, pp. 39–40.

Cady, J. G. (1950) "Rock Weathering and Soil Formation in the North Carolina Piedmont Region," *Soil Science Soc. Proc.*, v. 15, pp. 337–342.

Cole, M. M. (1960) "The Origin of Savannah

Vegetation in Eastern Brazil," *Geogr. Jour.*, v. CXXVI, pp. 241–273.

Cotton, C. A. (1963) "Did the Murrumbidgee Aggradations Take Place in Glacial Ages?," *Australian Jour. Sci.*, v. 26, p. 54.

Dan, J. and Yaalon, D. H. (1966) "Trends of Soil Development with Time in the Mediterranean Environment of Israel, *Nat. and Univ. Inst. Agri. Contrib.*, no. 1040-E, 8 pp.

Davis, M. B. (1969) "Palynology and Environmental History during the Quaternary Period," *Amer. Scientist*, v. 57, pp. 317–332.

Davis, W. M. (1899) "The Geographical Cycle," *Geogr. Jour.*, v. XIV, pp. 481–504.

Davis, W. M. (1902) "Base-level, Grade, and Peneplain," *Jour. Geol.*, v. X, pp. 77–111.

Dury, G. H. (1963) "Prior Stream Deposition," *Australian Jour. Sci.*, v. 25, pp. 315–316.

El-Ashry, M. T. (1971) "Quantitative Method for Grading Drainage Density," *Geol. Soc. Amer. Bull.*, v. 82, pp. 1703–1706.

Erhart, H. (1939) "Altération des roches et mode de formation de principaux types de sol" [in] Pub. 12, Bur. d'Études Géol. et Min. Coloniales.

Fairbridge, R. W. (1968) "Regolith and Saprolite" [in] *The Encyclopedia of Geomorphology*, R. W. Fairbridge, ed., Reinhold Book Corp., New York, esp. pp. 933–936.

Flint, R. F. (1971) *Glacial and Quaternary Geology*, John Wiley and Sons, New York, esp. pp. 287–301, 500–511.

Forbes, R. H. (1932) "The Desiccation Problem in West Africa: The Capture of the Sourou by the Black Volta," *Geogr. Rev.*, v. 22, pp. 97–106.

Frenzel, B. (1968) "Pleistocene Vegetation of Northern Eurasia," *Science*, v. 161, pp. 637–649.

Frye, J. C. and Schoff, S. L. (1942) "Deep-Seated Solution in the Meade Basin and Vicinity," *Kansas and Oklahoma, Amer. Geophys. Union Trans., Hydrology repts. and papers*, pp. 35–39.

Garner, H. F. (1959) "Stratigraphic-Sedimentary Significance of Contemporary Climate and Relief in Four Regions of the Andes Mountains," *Geol. Soc. Amer. Bull.*, v. 70, pp. 1327–1368.

Garner, H. F. (1966) "Derangement of the Rio Caroni, Venezuela," *Révue de Géomorphologie Dynamique*, no. 2, pp. 54–83.

Garner, H. F. (1967) "Geomorphic Analogs and Climax Morphogenesis," *Ark. Acad. Sci. Proc.*, v. 21, pp. 64–76.

Goddard, E. N., Trask, P. D., DeFord, R. K., Rove, O. N., Singewald, J. T., Jr., and Overbeck, R. M. (1948), *Rock Color Chart*, Geol. Soc. Amer., New York.

Gómez-Pompa, A., Vázquez-Yanes, C., and Guevara, S. (1972) "The Tropical Rainforest: A Nonrenewable Resource," *Science*, v. 177, pp. 762–765.

Grove, A. T. (1958) "The Ancient Erg of Hausaland, and Similar Formations on the South Side of the Sahara," *Geogr. Jour.*, v. CXXIV, pp. 526–533.

Grüger, E. (1972) "Pollen and Seed Studies of Wisconsinan Vegetation in Illinois, U. S. A.," *Geol. Soc. America. Bull.*, v. 83, pp. 2715–2734.

Hack, J. T. (1960) "Interpretation of Erosional Topography in Humid Temperate Regions," *Amer. Jour. Sci., Bradley Volume*, v. 258-A, pp. 80–97.

Higgins, C. G. (1961) "Effect of Buried Topography on Drainage in Sonoma County, California" (Abstract), *Geol. Soc. Amer. Bull.*, v. 72, p. 73.

Holms, Arthur (1965) *Principles of Physical Geology*, Ronald Press, New York, esp. pp. 386–406.

Horton, R. E. (1932) "Drainage Basin Characteristics," *Amer. Geophys. Union, Trans.*, pp. 350–361.

Horton, R. E. (1945) "Erosional Development of Streams and Their Drainage Basins," *Geol. Soc. Amer. Bull.*, v. 56, pp. 275–370.

Howard, A. D. (1967) "Drainage Analysis in Geologic Interpretation: A Summation," *Amer. Assoc. Pet. Geologists Bull.*, v. 51, pp. 2246–2259.

Hunt, C. B. (1967) *Physiography of the United States*, W. H. Freeman, San Francisco, esp. p. 93.

Jackson, K. C. (1970) *Textbook of Lithology*, McGraw-Hill Book Co., New York, esp. pp. 84–90; 95–113.

Jenny, Hans (1941) *Factors of Soil Formation*, McGraw-Hill Book Co., New York.

Jones, S. B. (1938) Geomorphology of the Hawaiian Islands: A Review," *Geomorph. Jour.*, v. 1, pp. 55–61.

King, L. C. (1953) "Canons of Landscape Evolution," *Geol. Soc. Amer. Bull.*, v. 64, pp. 721–752.

Langbein, W. B. and Schumm, S. A. (1958)

"Yield of Sediment in Relation to Mean Annual Precipitation," *Amer. Geophys. Union Trans.*, v. 39, pp. 1076–1084.

Langford-Smith, T. (1960) "The Dead River Systems of the Murrumbidgee," *Geogr. Rev.*, v. 50, pp. 368–389.

Leopold, L. B., Wolman, M. G., and Miller, J. P. (1964) *Fluvial Processes in Geomorphology*, W. H. Freeman, San Francisco, 522 pp.

Linton, D. L. (1955) "The Problem of Tors," *Geogr. Jour.*, v. 12, pp. 470–487.

Lustig, L. K. (1965) "Clastic Sedimentation in Deep Springs Valley, California," *U. S. Geol. Surv. Prof. Pap.* 352-F, pp. 131–192.

Lustig, L. K. (1965a) "Sediment Yield of the Castaic Watershed, Western L.A. County, Calif." *U. S. Geol. Surv. Prof. Pap.* 422-F, 23 pp.

Marks, P. L. and Bormann, F. H. (1971) "Revegetation following Forest Cutting: Mechanisms for Return to Steady-State Nutrient Cycling," *Science*, v. 176, pp. 914–915.

Melton, M. A. (1958) "Geometric Properties of Mature Drainage Systems and their Representation in an E_4 Phase Space," *Jour. Geol.*, v. 66, pp. 35–64.

Moreau, R. E. (1963) "Vicissitudes of the African Biomes in the Late Pleistocene," *Zool. Soc. London Proc.*, v. 141, pp. 395–421.

Peabody, F. E. (1954) "Travertines and Cave Deposits of the Kaap Escarpment of South Africa, and the Type Locality of Australopithecus africanus Dart," *Geol. Soc. Amer. Bull.*, v. 65, pp. 671–706.

Penck, Walther (1953) "Morphological Analysis of Land Forms, (translated from the 1924 version by H. Czech and K. C. Boswell), Macmillan and Co., London, 430 pp.

Reich, B. M. (1963) "Short-Duration Rainfall Intensity Estimates and Other Design Aids for Regions of Sparse Data," *Jour. Hydrol.*, v. 1, pp. 3–28.

Richmond, G. M. (1962) "Quaternary Stratig-raphy of the La Sal Mountains, Utah," *U. S. Geol. Surv. Prof. Pap.* 324, 135 pp.

Ruhe, R. V., Cady, J. G., and Gomez, R. S. (1961) "Paleosols of Burmuda," *Geol. Soc. Amer. Bull.*, v. 72, pp. 1121–1142.

Ruhe, R. V. (1969) *Quaternary Landscapes in Iowa*, Iowa State Univ. Press, Ames, 255 pp.

Russell, R. J. (1936) "The Desert-Rainfall Factor in Denudation," *Int. Geol. Cong. XVI, Rept.*, Washington, D.C., pp. 753–763.

Ruxton, B. P. and Berry, L. R. (1957) "Weathering of Granite and Associated Erosional Features in Hong Kong," *Geol. Soc. Amer. Bull.*, v. 68, pp. 1263–1292.

Schumm, S. A. (1956) "Evolution of Drainage Systems and Slopes in Badlands at Perth Amboy, New Jersey," *Geol. Soc. Amer. Bull.*, v. 67, pp. 597–646.

Schumm, S. A. (1965) "Project Summary: Geomorphology of the Prior Rivers of the Murrumbidgee River System" [in] *Report of Water Resources Research*, July 1964–June 30 1965, U. S. Geol. Surv.

Senstius, M. W. (1958) "Climax Forms of Rock Weathering," *Amer. Scientist*, v. 46, pp. 355–367.

Thornthwait, C. W. (1948) "An Approach to a Rational Classification of Climate," *Geogr. Rev.*, v. 38, pp. 55–94.

Tricart, Jean (1956a) "Types de fleuves et systèmes morphogénetiques en Afrique Occidentale," *Soc. Geogr. Comm. Trav. hist. scient. Bull.*, pp. 303–432.

Tricart, Jean (1956b) O Karst das Vinzinhanças Setentrionais de Belo Horizonte," *Brasileiro Gergrafia Revista*, v. XVIII, pp. 451–470.

Wahlstrom, E. E. (1948) "Pre-Fountain and Recent Weathering on Flagstaff Mountain near Boulder, Colorado," *Geol. Soc. Amer. Bull.*, v. 59, pp. 1173–1190.

Wells, P. V. (1970) "Postglacial Vegetational History of the Great Plains," *Science*, v. 167, pp. 1574–1582.

6 Arid geomorphic systems and landforms

In this chapter we are mainly concerned with the dry, sparsely vegetated geomorphic setting to which Twenhofel (1932) assigns about one-fifth of the Earth's surface (11,500,000 sq mi). Areas of discontinuous plant cover exceed Twenhofel's estimate. A list of the principal mechanisms available for shaping terrain in such a setting would look very much like the list for a humid geomorphic area at the beginning of Chapter 5. It would include, again, differential weathering, groundwater solutioning, fluvial activity, and mass wasting. But for the arid realm we would have to add effects by the wind, both in erosion and deposition, and this addition expresses a major difference between the two environments.* Not that wind by itself grossly changes the landscape (though on occasion it may), but rather that essentially all desert agencies are strongly influenced and now and then even governed by the almost complete exposure of bedrock or regolith at the ground surface that makes wind effects possible.

Unlike the situation in humid lands, where water is obviously a potent landscape factor, the only desert agency for which continual activity is readily observed is the wind. Yet it is widely conceded that much landform genesis in deserts is by running water. Gullied terrain there is far from uncommon, and there is little reason to believe its development differs greatly from that already described in vegetated areas. However, *gully systems* that unite into drainage nets with readily identifiable downstream channel continuations are much less common in dry regions. For reasons to be outlined later in this chapter, some such channel continuations can be considered to be humid relicts (cf. Cotton, 1968). All of these relations makes one suspect something different is going on in deserts. So we will first direct our attention to the comparatively few, detailed, first-hand accounts of desert precipitation and runoff effects. Then we shall consider other types of evidence of water occurrence, both above and below ground as well as some of the more obvious effects of isolated plants. Rock decomposition will then command our attention together with its effects on free faces of hillslopes, individual clasts, and the regolith as a whole. We will then consider the interrelated activities of mass wasting and running water over essentially barren land surfaces, both in terms of observable relations and in theory. Effects of the wind will be

---

* Wind access to the crust under humidity is usually local, temporary, and either artificial or coastal.

examined in some detail, and ultimately we will consider the over-all terrain effects of the arid geomorphic system in relation to apparent long-term environmental consequences in arid climatic nuclei.

THE DESERT SETTING

We now direct our attention to those characteristics that essentially all desert regions have in common. These "invariable" attributes, briefly summarized here, will then be considered in detail.

1. THERE IS AN EXTREMELY DISCONTINUOUS DISTRIBUTION OF LIVING PLANTS. Thus, the expression "plant cover" is inappropriate. Most of these plants (cf. Fig. 5.1) are classed as *xerophytes*, are moisture-conserving types with thick-skinned leaves or spines, and tend to grow in isolation or isolated clumps with possible concentrations along dry watercourses and more extensively as caps of higher hills in areas of notable relief (Fig. 5.15).

2. MECHANICALLY FRAGMENTED ROCK MATERIAL IS PRESENT IN AMOUNTS RANGING FROM PATCHES TO ALMOST CONTINUOUS BLANKET DEPOSITS OVER BEDROCK. In regions of notable relief, the largest amounts of rock debris tend to be at the bases of slopes and in depressions. Clast dimensions generally decrease away from uplands or escarpments. Surface layers are often extremely coarse and may be impregnated by a mineral cement; deposits as a whole tend to be dominated by angular and subangular clasts, considerable areas of little or no sorting, poorly defined and discontinuous stratification, and virtually no soil (as defined in Chapter 5).

3. THERE ARE A VARIETY OF ANIMALS (REPRESENTING ALMOST EVERY PHYLUM) DISPLAYING A VARIETY OF EXTREME FORMS OF ADAPTATION TO LOW-MOISTURE CONDITIONS. In general, there may be less OBVIOUS over-all survival dependence upon the plant community because the plants are rather less edible and shed less humic debris. But, in general, also, the lower plant density is reflected by a reduced animal population compared to that in humid lands.

4. THE WIND HAS ESSENTIALLY UNLIMITED ACCESS TO THE GROUND SURFACE. Thus, relatively fine-grained materials (whether organic or inorganic) are subjected to *eolian erosion* and transportation as soon as they are exposed. Debris finer than sand size (<1/16 mm) tends to be removed from the environment entirely. Sand tends to be shifted about and locally concentrated (dunes).

5. Less water generally enters areas of discontinuous vegetal cover than can be utilized there: WATER IN THE FORM OF RUNOFF ESCAPES DESERT ENVIRONMENTS ONLY UNDER UNUSUAL CONDITIONS AND VERY RARELY. Water within the arid geomorphic systems is mainly manifested during and shortly after precipitation events within the area of precipitation and for relatively short distances downslope. The general paucity of moisture is further reflected by the rarity of permanent springs, lakes, or flowing streams and by the usually depressed character of the watertable or the actual absence of an appreciable groundwater body.

6. NATURAL BEDROCK OUTCROPS ARE COMMON, LOCALLY COMPRISE A HIGH PER CENT OF THE GROUND SURFACE, AND ARE PROBABLY MOST EXTENSIVE IN ESCARPMENTS AND UPLANDS. There they frequently exhibit angular outlines, extensive fracturing, and downslope debris accumulations which appear mechanically fragmented.

7. LANDSLOPES FREQUENTLY CANNOT BE RELATED TO EXISTING DRAINAGE LINES AND COMMONLY APPEAR TO REFER TO AREAS OF EOLIAN EROSION AND/OR DEPRESSIONS OF UNCERTAIN ORIGIN. Where watercourses exist, landslopes may locally incline toward them, but such watercourses and related

slopes may be absent and are perhaps most common on elevated areas.

8. Signs of downslope movement of weathered debris exist on most inclined surfaces; BUT, THERE APPEAR TO BE RATHER EXCEPTIONALLY EXTENSIVE AREAS OF LOW INCLINATION AND LOW RELIEF WHERE MASS WASTING IS INHIBITED BY LOW SLOPE.

9. Because of the general absence of plant cover the EFFECTS OF A NUMBER OF SURFICIAL PHENOMENA BESIDES THE WIND ARE IMPOSED DIRECTLY ON THE LAND SURFACE IN THEIR MOST EXTREME FORMS. These phenomena include air temperature changes, rainfall, runoff, and solar radiation.

The desert geomorphic setting appears to have few or no attributes that are "invariably" associated other than those just cited. There is, perhaps some tendency for deserts to include plains. But some deserts are notably mountainous. Some are extensively alluviated, whereas others are much less so. Similar types of regolith are found in some humid lands. Most deserts yield rather large amounts of wind-blown dust. Again we have the question, What forms there and what does not? As in the case of vegetated settings, the implication again seems to be that the highly variable factors encountered in deserts either change with the passage of time, as the desert endures, or some features have been introduced into desert settings independently by agencies of other geomorphic systems. In any event, none of the nine "invariable" desert relationships cited previously are shared in precisely the same form and degree by other geomorphic systems, particularly vegetated ones. Because (as noted in Chapter 5) it is the "invariable" attributes of a geomorphic system that can be expected to exert special influences and thereby generate unique landforms, the differences between arid and humid types should prove to be as significant as Davis suggested (1932).

In our attempt to comprehend the desert geomorphic system, we must consider the actually existing physical conditions of the arid environment, just as we previously did for the humid condition. Our attention first goes to precipitation and temperature whose interaction establishes the initial occurrence and availability of water. Then rock weathering and its usual lithic consequences on various rock types will be noted followed by a consideration of what happens to weathered debris in deserts under the influences of gravity, running water, and wind. Finally, we will appraise the over-all apparent topographic effects of all these agencies in both time and space and the ultimate genesis of landscape under desert conditions.

ARID GEOMORPHIC PARAMETERS

The main factors that determine the nature of desert topography include the EARTH'S CRUST; the related ATMOSPHERE; the HEAT BUDGET as determined mainly by solar radiation (which in many deserts is of unusual intensity and duration), absorption, reflection, and refraction (see Chapter 4); WATER, whose existence and behavior resembles that of a droplet on a hot stove lid; a few ANIMALS; a very few PLANTS. In a sense, all these are GEOMORPHIC parameters. But in the following section we will restrict that designation to interrelationships involving heat and water, the activity of which tends to establish the character of most of the other features.

PRECIPITATION-TEMPERATURE EFFECTS

A consideration of the world's desert regions discloses certain relationships between precipitation and temperature that appear crucial to the genesis of landforms. We have already noted that plants tend to be sparsely distributed in deserts. Yet plant occurrence is strongly indicative of the distribution and abundance of precipitation within arid regions. One is therefore impressed with the comparatively uniform distribution of plants across desert plains, always sparse, possibly in bunches or clumps but with bare areas between; often there are virtually no plants. On low-relief surfaces, plants such as sage which grow in ovate or circular clumps may act as traps for drifting sand. Their root systems do not reach the groundwater body (cf. Fig. 5.7). Only in desert oases do we encounter plant communities dependent on groundwater. Studies of watercourse plant concentrations in deserts almost invariably disclose a groundwater seepage condition, and, in the drier deserts, such seepage must be related to a relict (fossil) groundwater body left by some prior and more moist environment or to underground flow from a more humid extraneous upland (Fig. 6.1).

In deserts that exhibit high relief, the occurrence of vegetation is often far less uniform than in plains regions. Crests of topographic highs are frequently more heavily vegetated than lowlands, and some uplands are densely vegetated, even though adjacent lowland watercourses show little plant growth (Fig. 5.15). Vegetal relations indicate that precipitation is relatively uniform in desert plains regions and uneven in areas of appreciable relief. There can be little question that such precipitation and groundcover variation should drastically affect landform genesis (cf. Davis, 1905; Lustig, 1967).

As previously noted, a region within a desert that exhibits continuous plant cover, to all intents and purposes, supports a humid geomorphic system and would, in all probability, display some of the landform-process associations of such a system. Vegetated uplands in deserts must be regarded as humid geomorphic areas in any broader consideration of desert landscape development. Accompanying remarks about desert precipitation-temperature relations exclude associated humid-vegetated tracts of land even though the latter appear to comprise

Fig. 6.1   Irregular dark areas of middle background are date palms of Tahrit oasis in the southwestern Algerian Sahara profiled against the light-colored sands of the Grand Erg Occidental.

an almost invariable element of arid morphogenesis in certain types of high-relief terrain. Also, in attempting to define the precise landscape effects of arid geomorphic systems (just as in our previous discussion of humid geomorphic systems), we will be guided by first-hand accounts of processes and their effects, supplemented by landform relationships observed in regions where geographic, topographic, atmospheric, and oceanographic factors suggest it has been arid for very long periods of time (ARID CLIMATIC NUCLEI).

Chapter 5 contains a number of comments about the nature of precipitation in deserts and particularly its sporadic character, its normally localized and often high-intensity occurrence, and its usually brief duration. Lustig (1967) correctly emphasizes the paucity of accurate hydrographic data from deserts related to the precipitation form and type. The tendency for precipitation to concentrate on desert uplands was just mentioned. But an additional feature of such precipitation should be emphasized. The water falls directly on the ground surface over a very high per cent of desert lands. Schumm and Lusby (1963) note that in moderately arid portions of western Colorado (Badger Wash), plant cover ranges from 8 to 15% of the area. Vast tracts of the central Sahara are 98% barren. No CONTINUOUS blanket of vegetation acts as a cushioning layer to inhibit lateral water movement along the land surface or to lessen raindrop impact. The detailed consequences are well worth considering.

Let us calculate the energy directed against a single acre of bare ground exposed to a 15-minute rain shower during which 2 inches (5.8 cm) of water falls. For simplicity of calculation and clarity, all figures cited will be in the English system; let us also assume that the event occurs at sea level where the normal acceleration of gravity would be 32 ft/sec/sec. One acre of ground includes some 43,560 sq ft of area, and each square foot would ultimately receive a weight of water equal to a layer 2 inches deep (1/6 cu ft) which we will assume to be falling at terminal velocity.

Since a cubic foot of water weighs 64.3 lb, and each square foot would receive the impact of 1/6 cu ft, the weight falling on each square foot would be 10.7 lb. Multiplying the weight times the area

$$10.7 \times 43,560 = 466,092 \ \text{lb/acre}$$

The force exerted should equal the mass times the velocity

$$F = MV$$

substituting

$$F = 466,092 \times 32$$

$$F = 14,914,944 \ \text{ft/lb}$$

The force indicated would be that if the entire weight fell in a single second. Since the storm in question is of 15 minutes duration (900 seconds), it will be seen that the acre area would be subjected to

$$\frac{14,914,944}{900} = 16,572 \ \text{ft/lb/sec}$$

Therefore, each and every second of a 15-minute rain storm on a barren surface the ground would be subjected to the impact 16,572 ft lb of force per acre.

The foregoing example is intended to emphasize the consequences of the absence of plant cover with respect to precipitation and runoff. ESSENTIALLY NONE OF THIS FORCE WOULD HAVE ACCOMPLISHED ANYTHING GEOMORPHICALLY ON A VEGETATED SURFACE. Physical evidence of the force involved, and its consequences on bare ground, is frequently apparent around construction sites where bare ground is adjacent to newly constructed walls. In such areas, *rain-splash erosion* can readily be observed and leaves its record in the form of clay, silt, and fine sand particles adhering to the walls from 18–24 inches (45–60 cm) above the ground surface. In exceptional instances rock material is lifted as much as 30 inches (76 cm) by raindrop impact erosion. It is abundantly evident that not all of the force potentially available in this form will actually be erosionally effective in deserts, since some drops fall on large clasts or bare rock surfaces or even on areas of standing or flowing water. Some may lack terminal velocity. Nevertheless, the significance of rain-splash erosion and its role in the entrainment of sediment in desert runoff would be difficult to overestimate. In a humid region we might meaningfully calculate the potential energy of the 466,092 lb of water/acre with an eye on potential erosional accomplishments during the runoff journey to sea level. But, as we shall see, that journey in a desert may never occur.

The example of precipitative erosive force cited above is hardly an extreme case in terms of intensity. First-hand accounts of the effects of more intense storms are comparatively rare. Some are cited in Chapter 5, and others are recorded by several authors including McGee (1897) and Savegear and Peel (in Grove, 1960). All of these writers record the erosive effects of raindrop impact on bare ground. Once the water falls, however, and sometimes even "as it falls" the role of temperature becomes evident. Streams of rain descending from isolated convectional clouds in arid, southern Colorado can often be observed to terminate downward in an apparent recurved misty halo and, because of high air temperatures, the falling water often fails to reach the ground, evaporating back upwards faster than it condenses and falls.

Air temperatures in deserts may be very high or very low. In the desert wastes around the Antarctic ice sheet, mean annual temperatures hover around 32° F (0° C). The region is dry because most of the water present remains in a solid state, and it is barren because of freezing temperatures also. Lower-latitude deserts are much given to diurnal temperature extremes. Extreme air temperatures are impressive, but temperatures near the ground surface are probably more significant geomorphically. Because of the general absence of cloud cover most of each year, desert heat radiates back into space very rapidly once the sun sets. Temperature contrasts between day and night are rigorous. The writer will not soon forget arising at 3 A.M. in the Algerian Sahara in early September. The bucket of water for washing

presented a thin film of ice, and a heavy wool sweater and leather jacket were comfortable while riding in an open jeep. By 9:30 A.M., the air temperature was nearing 85° F; by 11:30 A.M., a pocket thermometer registered over 105°, and at 2:00 P.M. the same thermometer registered 127° F. What passed for a local pub in a nearby oasis cooled its beer by wrapping the bottles in wet sacking and laying them in the sun. Evaporation occurs at an almost unbelievable rate under such circumstances, and effective ground moisture levels are fantastically low. The foregoing account amounts to a record of a daily temperature variation of at least 95° F. The beer was delicious.

Air temperatures are on record in excess of 135° F in low-latitude deserts in the summertime. Stone (1968) cites 136.4° F at Tripoli and 134° F in Death Valley. One estimate in the Egyptian desert places the average ANNUAL PRECIPITATION at about 4.5 inches (11.5 cm) and the DAILY EVAPORATION RATE at nearly 5 inches (12.75 cm). One can say with some certainty that there is a water shortage. But air temperatures are only a part of the story. The ground may be much hotter.

Hadley (1972) notes that xerophytic herbaceous plants can tolerate dry air temperatures in excess of 125° F (55° C), that prickly pear cacti (*Opuntia* sp.) can survive up to nearly 145° F (63° C), and that desert lichens on rock can withstand temperatures in excess of 158° F (70° C). Sun temperatures on bare rock commonly exceed 170° F in Horse Latitude deserts during the day. Grove (1960) records that at dawn the surface sand temperature in a part of Tibisti in the central Sahara was 67° F and it was concurrently 90° F at a depth of 1 ft. Heat generated at ground level by direct solar radiation is probably generally greater in low-latitude deserts than nearer the equator in humid regions, where the atmosphere usually has greater opacity because of haze and clouds. For example, Branner (1896) records a number of bare rock and ground temperatures in arid parts of Brazil as well as a few readings taken in the air. He notes, "In 1876, in the dry interior of Pernambuco near Aguas Bellas, I made some observations to ascertain the difference between the temperature in the sun and in the shade. The weather was sultry, and at the hottest part of the day—between 2 and 2:30 P.M. —the thermometer in the sun, and covered with one thickness of black woolen cloth, registered, respectively, 40, 48, and 50 degrees higher in the sun than in the shade on

Place	Maximum	Minimum	Variation
Manaus, Amazonas	153	71	72
Pará	145(?)	70	75
Oeiras, Piauhy	149	—	—
Ceará*	115	66	49
Pernambuco	134	75	59
Bonito, Pernambuco*	146	64	82
Aguas Bellas*	145	62	83
Piranhas, Alagôas	150	72	78
Bahia	145	61	84
Rio de Janeiro*	117	70	47
Rio de Janeiro	117	46	71
Rio de Janeiro	117	65	52
Rio de Janeiro*	152	70(?)	82(?)
Rio de Janeiro*	120	56	54
Barbacena*	139	36	103
Queluz, Minas	136	46	90
Pirapóra, Minas	148	45	103
São Paulo	141	47	94
São Paulo	129	47(?)	82(?)
Itú, S. Paulo	139	52	87
Tatuhy, S. Paulo	146	47	94
Campos do Jordão, S. Paulo	127	50	77
Curitiba (20 ks. E of)*	145	—	—
Paraná	150	24	126
Rio Grande do Sul	135	62	73
Tombador, Matto Grosso*	144	68	76
Upper Paraguay	152	45	107

* Maximum reading in sun from direct observations.

Fig. 6.2. Rock surface temperature variations (°F) in Brazil. (After Branner, 1898.)

the three days on which the observations were made" (cf. Fig. 6.2). He goes on to comment on rock temperatures where it is seen that several exceed 150° F, further noting that, "Melloni has shown that radiation reduces the temperature of some bodies 8 degrees centigrade below the temperature of the air (*Amer. Jour. Sci.*, 1849, vol. viii, p. 47), and that at the ground the temperature is 5 or 6 degrees centigrade lower than it is 4 or 5 feet above it . . . ." Branner (1896) notes, however, that the temperature for rocks openly exposed to the Brazilian sun is often between 140 and 150° F for 3 or 4 hours during the day.

The temperatures just cited for arid regions have significance not only for their relation to processes such as rock weathering and organic decay but also for their immediate effects upon exposed water. Rain falling on such heated surfaces initially turns to vapor which differs from steam mainly in its somewhat lower temperature. Evaporation tends to diminish during a rainstorm as ground surfaces cool, and air moisture contents are temporarily elevated. But water losses back to the air are considerable, and, as noted by McGee (1897) in his discussion of the Sonoran desert of the southwestern United States and northern Mexico, by "reason of heat and aridity . . . the vegetation is too scant, stunted, and scattered to protect the surface from storms; the meager flora forms little or no humus, and thus there is no soil and little of that chemic action initiated by vegetal growth and decay."

WEATHERING AND MASS WASTING

McGee's remark is a fitting prelude to a discussion of rock decomposition in deserts and the correlative movement by gravity of the waste so produced. We should take up these matters here since some landforms result and since erosional processes are in a large measure directed at unconsolidated surface detritus. The character of that detritus is mainly governed by the weathering and mass wasting processes. The same material also influences the nature of associated erosional events.

Weathering in deserts can generally be considered in two major aspects. One aspect concerns changes where rock is in direct contact with the atmosphere. The second aspect concerns rock decomposition and modification beneath the ground surface in both bedrock and regolith. It will be recalled that rock exposures were earlier noted to be most abundant in deserts along escarpment faces and uplands, whereas lowlands display some tendency to be alluviated. McGee (1897) describes extensive "stripped plains" where vast areas of bare rock are exposed (Fig. 6.3), and, as will be seen later, there is some reason to believe that each of the described conditions represents stages in desert geomorphic evolution. In any event, we will now consider both types of rock weathering more or less independently.

Weathered rock fragments in deserts commonly appear to be freshly and mechanically shattered (Fig. 6.4). To some extent, this relation is independent of original granular texture and composition, and it may be largely a matter of primary parting spacing in some cases (cf. Schumm and Chorley, 1966). This potentially alters topographic effects of bedrock lithic variations in deserts. Because of the general absence of plants and water in sufficient abundance to induce significant frost wedging, effectively mechanical disruption of rock debris in deserts must relate to other processes. The possibility that stresses are set up in rocks through extreme heating and cooling (and related expansion and contraction) with resultant fracturing has been questioned by Blackwelder (1925) though, of course, long-term repetition effects are almost impossible to gauge experimentally. It has long been known that different minerals have different coefficients of expansion as do different rocks, and certainly heat absorption and radiation is strongly affected by color. Moreover, rock surfaces are subject to wetting and drying as well as heating and cooling which rock interiors escape or are affected by to only a minor degree.

Fig. 6.3.   Inclined layers of Devonian rocks exposed in a crudely planar region near the Oued Saoura in western Algeria. Only thin ferruginous layers and concretions within the argillaceous series yields rubble to form a discontinuous reg on an otherwise stripped bedrock surface.

Fig. 6.4. Desert pavement (gibber) developed in the Amargosa Desert northwest of Las Vegas, Nevada. Many clasts are ventifacts and are additionally etched, possibly by solution from dew.

Branner (1896) cites Forbes's observations (1849) that temperature changes do not penetrate rocks far, as shown by temperature variations of 15° F at a depth of 3 ft (91.5 cm), 10° F at 6 ft (183 cm), 5° F at 12 ft (366 cm), and 1.5° F at 24 ft (732 cm); he comments that the depth to which a given temperature penetrates varies with the square root of the period of exposure. It seems clear that rock material at the surface would expand and contract more than that it would at depth. For example, Branner notes that gneiss expands 1 part in from 187,560 to 228,060 parts for every degree Fahrenheit, and, taking 1 part in 200,000 as average, he calculates the linear expansion of a gneiss surface 300 ft long, relative to depth (Fig. 6.5). The effects recorded may, in part, account for the observed tendency of outer layers or shells of rock to separate from inner ones (*exfoliation*). This phenomenon (Fig. 6.6) and the separation of individual mineral grains in a rock from one another (*granular disintegration*) constitute two of the main expressions of the weathering of bare rock surfaces. Both exfoliation and granular disintegration are known from a variety of climatic settings including humid and arid, and it is uncertain if one or the other or both characterize a particular environment. There are certain indications, however, that chemical activity is also involved and that we may be observing stages of rock decomposition rather than merely different kinds of weathering activity.

Two apparently related expressions of chemical activity are known as *desert varnish* and *case hardening*. Both appear to involve the transfer of mineral matter to the rock surface (or nearly to it) by water that has penetrated from dew or other precipitation and is then drawn back out by either capillary forces or evaporation or some combination of these. The varnish appears as a patina or thin film mainly composed of iron or manganese oxides and silica on finer-

Depth (ft)	Temperature (°F)	Expansion (inches)
Surface	103	1.854
3	15	0.27
6	10	0.18
9	7(?)	0.126
12	5	0.09
15	4(?)	0.072

Fig. 6.5. Linear expansion for a surface of gneiss 300 ft long due to temperature changes with respect to depth (Brazil). (After Branner, 1898.)

grained rock materials such as chert (cf. Fairbridge, 1968a). The latter writer notes that the varnish is a relatively tough residue which may be polished to a high gloss by the

Fig. 6.6. Exfoliation of granite boulders at Trona, California, near the southern end of Panamint Valley. The nearest clast is slightly more than 6 ft (2 m) in diameter. The area is north of the Mojave Desert where Oberlander cites exhumed core stones of similar composition.

wind which also tends to remove softer halides (Fig. 6.7). Common colors are tan, reddish brown, and black. There is some difference of opinion as to the precise conditions favoring varnish formation. Records in the Mojave Desert of western North America indicate that a varnish layer will form in as little as 25 years, but since the coating makes the surface increasingly impervious the process probably slows down both because of this and, in some instances, because the supply of mineral matter from which the varnish forms is being depleted. In this context it should be noted that varnish on essentially isolated clasts on the ground surface must largely be drawn from within the clast, whereas a fragment closely bounded by others in an alluvial matrix could theoretically derive dissolved mineral matter in much larger amounts from much greater rock mass. Many *ventifacts* display desert varnish, but a highly abrasive eolian environment probably would tend to destroy such a coating.

Fig. 6.7. Pebbles of chert with a coating of tan-brown desert varnish. The fragment (upper left) has been broken to disclose its cream-colored interior and thus the extreme thinness of the varnish.

Case hardening is a form of rock induration, of special interest to geomorphologists, which is related to desert varnish and a variety of similar phenomena. Typically, case hardening affects comparatively coarse-textured materials that have already undergone some lithification. As noted by Fairbridge (1968b), the three main chemical compounds involved are calcium carbonate ($CaCO_3$), iron oxide (Fe_2O_3), and silica (SiO_2). As in the case of desert varnish, these compounds appear to be dissolved by penetrating water and returned to the surface of the material by capillary action and evaporation. The latter causes precipitation of the mineral matter at or near the rock surface with the resultant formation of a toughened outer crust which may later tend to peel away (Fig. 6.8). In some measure, therefore, case hardening contributes to the weathering phenomenon known as *exfoliation* (Fig. 6.6). Exfoliation should not be confused with a similar phenomenon termed *sheeting* (cf. Gentilli, 1968) which appears to be due to horizontal or concentric fracturing in otherwise massive intrusive igneous rocks due to internal lithic zonations or unloading stresses (Fig. 6.9). If we limit exfoliation to the spalling or peeling phenomena due to weathering, then it is only known to affect the outer few inches of a rock surface, whereas sheeting may produce layers many feet thick.

The outer crust on a clast or outcrop formed by case hardening tends to generate two distinct weathering environments related to the same bare rock surface. In coarse-grained rocks (igneous, sedimentary, or metamorphic), material beneath the outer crust tends to decay granularly, possibly since it is more often moist. Not infrequently, a coherent appearing boulder exposed to the air will be found to consist of a rather tough outer crust and a thoroughly disintegrated inner mass of granular material. Moreover, since the case-hardening phenomenon appears to require a degree of prior coherency in a rock surface, once the outer crust of a case-hardened surface is shattered and the loose inner grains are exposed, case hardening probably cannot be

Fig. 6.8. A case-hardened tombstone composed of Triassic arkose, Bloomfield, New Jersey. The tough outer shell has yielded to the elements, and the rotted interior is undergoing granular disintegration.

Fig. 6.9.   Spectacular sheeting developed in granite in a cirque headqall, Little Shuteye Pass, Sierra Nevada, California. Individual layers average 1–3 ft thick. (N. King Huber, U. S. Geological Survey.)

Fig. 6.10.   Granite undergoing granular disintegration to form grus in the Sierra Nevada Mountains near Sequoia National Park. Unlike the grains shown in Fig. 6.31, the feldspars in this instance are kaolinized, suggesting greater humidity, and the exposed rounded bedrock knobs occupy the positions of core stones in a weathering profile. See geologic pick at lower left.

readily resumed. It would appear that the amount of cementing mineral matter required to re-bond already loose grains is not usually present in a single, isolated, decayed clast—tombstones frequently decay in this manner (Fig. 6.8).

A more extreme form of surface incrustation and induration is commonly encountered in deserts where alluvial formations are bonded from the surface to some depth by precipitated mineral matter termed *duricrust* by Woolnough (1927). The three main adhesive substances consist of Fair-

bridge's aforementioned (1968b) *calcrete* ($CaCO_3$), *ferricrete* (Fe_2O_3), and *silcrete* (SiO_2), plus a fourth variety, cited by Ramana Rao and Vaidjanadhan (1970) and by Ramana Rao (1972), in India where manganese oxides appear to be the bonding material.* Each of the three main types is known locally under a variety of names. Duricrusts appear to form mainly on surfaces of low relief and low inclination, perhaps implying a greater possibility (with inhibited drainage) that the fragments being bonded remain immersed in cement-

* Goudie (1973) discusses duricrusts in some detail and mentions aluminum-rich varieties as well as various modes of genesis.

ing fluids. In any event, the same basic requirements for desert varnish and case hardening appear to extend to duricrusts. The consensus from a number of writers is clearly that the mineral residue forms where evaporation at the surface concentrates substances mainly brought up from depth in solution. The mineral-impregnated crusts in India have been the source for strongly bonded angular clasts, and reworked duricrusts (e.g. *canga*) are not uncommon (cf. Ramana Rao, 1972).

The precise environments of origin of surface mineral encrustations remain uncertain, though it seems clear that the solutions carrying the mineral matter must evaporate on the surfaces and neither be diluted nor flow away. Thus, one of the drier surficial environments would appear to be favored. Even so, a duricrust, once formed, is an extremely durable feature, silcrete and ferricrete both being very well bonded by relatively inert substances. Some duricrusts probably reflect repeated episodes of formation. And it may well be that particular environmental stages are more important in their genesis than others—for example, much mineral matter probably moves toward the ground surface when groundwater bodies are being destroyed by evaporation following the onset of arid conditions.

Granular disintegration has already been mentioned as a main form of rock weathering and as a phenomenon commonly following destruction of a case-hardened rock surface. Whether ALL granular disintegration follows case hardening is uncertain. Furthermore, it has been observed proceeding in a variety of environments, some moist, some dry (cf. Figs. 3.21, 6.6, and 6.10). The main point to be made here is that granular disintegration appears to require the presence of some moisture, and, in deserts, moisture rarely remains on exposed rock or mineral faces. Apart from the stresses already discussed in relation to differential expansion in areas of thermal contrast, intergranular forces related to salt crystallization, hydration of micas, or kaolinization of feldspars appear sufficient to account for much of the observed grain

separation. Water from precipitation events presumably penetrates micro-partings and along mineral grain contacts where incipient clay mineral formation or hydration can proceed briefly until capillary forces and evaporation again eliminate the moisture. Forces of crystallization of clays and other minerals along grain boundaries are probably severe (cf. Gentilli, 1968). More to the point, as soon as a grain is freed from the rock and exposed to the air in a desert, any surface alteration is polished away by the wind, and essentially unaltered, "fresh" grains remain. In the case of granite, the *gruss* that forms thusly from quartz and felspar granules appears to have been formed "mechanically," whereas some if not most of the processes involved were distinctly chemical in nature (cf. Fig. 6.31).

Fragments freed by weathering in deserts respond to gravity just as they do elsewhere (see Roy and Hussey, 1952). Under arid conditions, however, some later events in talus accumulations or alluvial cones or fans and pediment veneers are notably chemical in character mainly where mineral encrustations are forming. Elsewhere, the debris falls loose to be further acted upon by wind, running water, and gravity (cf. Fig. 6.11). But little clay or water exists to lubricate downslope movement of weathered debris. It will be noted that this condition contrasts with humid, vegetated settings where weathered debris tends to enter into chemical, soil-making events downslope from bedrock exposures.

In view of the foregoing discussion it would appear advisable to speak of desert weathering as being EFFECTIVELY MECHANICAL in most instances. This remark is not intended to discount purely mechanical rock fragmentation in deserts by thermal fracturing, root wedging, crystallization pressures, etc. But rock breakdown in deserts is apparently far more chemical in character than was formerly realized. The lithic products, however, clearly reflect the "effectively mechanical" consequences of the chemical events. Weathered residues in deserts tend to mirror their bedrock sources mineralogically and therefore usually in-

Fig. 6.11. Inselbergs being buried in their own weathered debris, eastern Algerian Sahara Desert. Most larger rock fragments depicted are ventifacted.

clude both metastable and semi-insoluble materials in a notably "immature" rock and mineral assemblage. Gravel deposits and talus at the bases of free faces are commonly feldspathic in granitic terrain, include limestone, dolomite, and even gypsum fragments where appropriate sedimentary rocks are exposed in arid sites, and incorporate iron micas, amphiboles, and even olivine fragments in desert regions where basic igneous rocks are found cropping out. Oberlander (1972) cites evidence that some granitic boulders in the Mojave Desert are relict core stones from prior soil profiles—their size, distribution, and composition would therefore not reflect desert weathering or mass wasting.

To this point in our discussion of rock weathering in deserts we have been emphasizing surficial phenomena, mainly because the surface is where such moisture as develops has its brief tenure. A "groundwater body" is rarely thought to enter the rock decomposition picture in deserts other than retrospectively as noted by Oberlander (1972), yet there is one contemporary relationship where it has been argued to be important. Rather recently, Mabbutt (1966) suggested that the base of the alluvial regolith in deserts is the site for weathering related to concentration of moisture at the bases of hillslopes. This is similar to the earlier suggestion of Ruxton (1958) that such moisture tends to collect at the notch (*piedmont angle*) between the hillslope and the pediment, thereby aiding chemical rock decomposition there. The same notion has more recently been reaffirmed by Ramana Rao (1972) who illustrates the break in slope very well on pediments in India (Fig. 6.12). Twidale (1962) terms the same phenomenon "soil moisture weathering" and attributes to it the steepened sides of inselbergs in south Australia. It should be em-

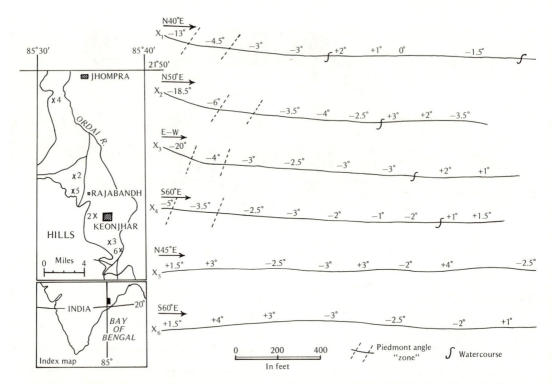

Fig: 6.12. Profiles showing inclination of pediment surfaces along the Ordai River, northeast India. Note development of "piedmont angle" in some instances. (K. L. V. Ramana Rao.)

phasized that even this "chemical" environment is localized and comparatively temporary although it may be very significant geomorphically.

In closing our discussion of weathering and mass wasting in arid, sparsely vegetated environments it seems important to emphasize certain relationships. As noted by Gignoux (1955) and the writer (1959), the weathered residue produced in deserts that have a variety of bedrock types tends to be relatively coarse, particularly since eolian activity tends to remove fine-grained substances. As elsewhere, absolute block dimensions freed by arid weathering are limited by bedrock parting spacings (cf. Chapter 3). Naturally, rocks undergoing

granular decomposition yield little coarse debris. Where the rocks yield large clasts, the parent bedrock exposures tend to exhibit angular profiles. Bare rock surfaces in deserts are differentially abraded by several agencies including the wind, but such effects must, of course, diminish to the extent that outcrops tend to become buried or are too high to undergo sand abrasion. The coarsest-grained materials in deserts tend to be found close to escarpments, and the finest usually are encountered out in the midst of any broad depressions, though there are exceptions to these relations dictated in part by climatic history, by bedrock lithology, and by hydraulic factors which we will next consider.

DESERT FLUVIAL ACTIVITY

The geomorphic effects of running water in deserts naturally occur in conjunction with other phenomena brought about by such agencies as gravity and the wind. Yet, as will be seen, these happenings are rather less well coordinated here than they are in humid, vegetated regions. It will be recalled that many geomorphic events in moist areas appeared notably systematic. Precipitation there makes plants grow which, in turn, tends to maintain localized water flow and cause fluvial incision. Incision by running water generates slopes along which mass wasting can occur and simultaneously influences the rate of such wasting by determining slope angle to some degree. Many of the basic humid agency interactions are rather well defined. In deserts, however, fluvial events are often isolated "happenings." The arid geomorphic "system," in a fluvial context, is more difficult to identify than its humid counterpart; it is, perhaps, less obviously systematic. Moreover, wind is erosionally most effective on dry surfaces and can generate slopes totally independent of prior water flow lines, both by erosion and by deposition. Running water may be geomorphically effective. Where it rains. And when. And thereafter for perhaps a short distance downslope. If there is a slope. As we shall see, the effects of water in deserts also appear to differ from those in humid lands.

It is apparent that the geomorphic effects in an arid, sparsely vegetated setting are in some measure dependent upon antecedent happenings. Again, as in our earlier humid geomorphic discussion, a question to be raised and faced immediately is, "In what type of structuro-topographic setting do arid geomorphic events usually occur?" Is "usually" the right word? Tectonically speaking, there are various possibilities. As was noted in Chapter 5, there is reason to believe that newly emergent land areas in intermediate latitudes may well tend to be arid, with the possibility that higher areas of the same land may become humid if uplift continues. In such a hypothetical case, the desert ero-

sional and depositional events would be initiated on a marine surface of deposition which could well be rather featureless. Our paleogeomorphic discussion (Chapter 11) takes up this problem, but, as already noted, newly emergent sea floors are scarce at present. Other types of surface that might be modified by desert events would include those formed volcanically, glacially, and by humid conditions. The latter two situations will occupy us in Chapters 7 and 8. Yet it appears that it would be unrealistic of us to consider arid geomorphic systems in any particular tectonic setting, whether dynamic high relief or cratonic plain. Regardless of suspected morphologic tendencies, all present desert regions exhibit localized uplands as well as lowlands, and it appears that arid fluvial activity should be considered in varied physical settings.

Immediately following pages are devoted to arid geomorphic events in diverse topographic settings, some with uplands—even mountains—and some lowlands that are essentially planar. For the sake of simplicity, and for the moment to set aside any detailed consideration of non-arid surficial geomorphic effects, uplands and related surface roughness will be regarded as structural (i.e. block faulted) or volcanic, and all surfaces are considered erodable and at least locally granular.

X_c-E_c-D_c—THE ARID FLUVIAL EVENT

At this point it appears best to turn to first-hand accounts of rainfall and runoff effects on various desert terrains. The symbols in the heading are, however, intended to emphasize certain distinctions between the humid geomorphic setting that was the focus of Horton's drainage study (1945) and what we must now consider. In this regard, it will be recalled that in Chapter 5 we mentioned a distance, X_c, first cited by Horton (1945) as the critical distance of overland flow for water that must occur before erosion can occur. This distance is in some measure dependent upon terrain permeability (cf. Schumm and Lusby, 1963), but we must note that precipitation in deserts does not generate a continuous ground-cover of plants, and the mere falling of rain on loose, bare earth is sufficient to initiate erosion on granular material. Thus, the distance X_c in deserts may be zero in some cases,* not only at the beginning of an erosional cycle, but also throughout. Fluvial activity in deserts also involves two other critical time-distance relationships expressed in our heading as E_c and D_c. We will consider the meanings of these latter symbols in the context of first-hand accounts of desert fluvial events.

Accounts of Desert Water and Mud

A study of arid fluvial activity which must rank as a classic of its kind is W. J. McGee's 1897 account of rainfall and runoff in desert areas of southwestern United States and northern Mexico. He distinguishes between the behavior of water in terrain of some local relief and that elsewhere on roughly planar surfaces. He notes the tendency of water flowing over an erodable plain to collect in distinct flow lines on granular desert surfaces in some instances, but states ". . . since the streams formed in this way at once begin to scour and overload themselves and thus check their own flow, this tendency is soon counteracted and the water is distributed again; so that the ultimate tendency is toward movement in a more or less uniform film or sheet," which he terms a *"sheetflood."*

McGee goes on to observe that minor precipitation amounts are largely absorbed (cf. Schumm, 1956), whereas larger runoff amounts are potent geologic agents, and ". . . during the great storms occurring from decade to decade or from century to century whole plains are flooded; yet so dry are air and earth that the deluge is absorbed within a few miles or scores of miles." He also comments that in deserts (the Sonoran) waters generated by precipitation "never reach the sea [hence] the territory is complete in itself as a geologic provence; the storm waters gather detritus in the mountains and transport it into the valleys, but their agency is limited to shifting the rock matter from one point to another in the same vicinity, and thus degradation and aggradation go hand in hand and gradation is completed within the district." These remarks appear to be generally applicable to the desert areas of the world as a whole, since they reflect a general consensus on the topic by many researchers.

McGee observes that in deserts the "typical valley is waterless save during storms, and its ephemeral waterways [tributaries] are multitudinous in the bounding mountains and few or none in its flattened interior." He also remarks on the pronounced tendencies of desert uplands to receive the most precipitation and notes that "The chief source of stream water is the sporadic storm, especially the thunder-gust or cloudburst, which fills old channels and gouges out new ones, though the flow may last but a few minutes, and seldom continues more than a few hours." Again: "Outside the sierras the typical channel is at first a rugged or flat-bottomed barranca [steep-walled gully] cut in the country rock; it soon diminishes in depth and increases in width and becomes lined with boulder beds; still further down stream it changes into a broad, steep-banked arroya cut in alluvium and burdened with gravel beds or sand sheets; and it finally ends in an alluvial fan, usually of imperceptible slopes miles in length and furlongs in width." In Death Valley, California, Beaumont and Oberlander (1971) observed braided runoff on a fan apex channel as well as in a canyon above the fan head. Their hydraulic findings will be given more attention later.

Speaking of running water in deserts, King (1953; 1968) states that broad, laminar sheet flow, non-erosive in quality, has actually been observed on pediments in Africa, as has deeper, highly erosive sheet-

---

* X_c in the sense of erosion occurring BECAUSE OF runoff would always have a value above zero—but rainsplash erosion would already have occurred.

flood in which turbulent linear flow in sheet form is dominant. He adds that,". . . wave trains have been observed to develop upon the sheets of water, and these may have almost catastrophic erosive effect." The latter phenomenon was earlier mentioned by McGee (1897) without, however, any specific comment on erosive effects. Other researchers express rather different ideas of the effects of desert runoff. Speaking of the Tibesti region in the Libyan Sahara, Grove (1960) notes that the volcanic highlands form hydrographic centers from which runoff originates to water the much more arid lowland plains. He reports: "As we saw for ourselves in a heavy shower at Debasser, the rain runs quickly off the compact silty and gravelly surface of the plateaus. After ten minutes of heavy rain on that occasion, the water had collected into streams that plunged over the cliffed sides of the *enneri* [wadi] and from our vantage point overlooking the oasis from 600 ft, we saw a tongue of water creep down the rush lined channel and spread over oasis gardens to the consternation of the villagers."

In a personal comment included in Grove's 1960 report, Savigear notes details of stream-floods and sheetfloods in Agades near the southern edge of the Sahara. There he notes a stream-flood some 6 ft wide and 2 ft deep flowing like a mountain torrent after a rain, highly turbulent because of suspended sediment and including large, visible, moving clasts. Savigear also observed water sheets on a nearby sandy plain, all of which soaked into the ground within 15 minutes; prior to this it flowed in a braided, turbulent pattern. Of great interest are his comments on the effects of about 1.5 cm of rain falling in 15 minutes where he states

I was able to observe the result of rain falling on a surface of coarse sand on which there were occasional large waste fragments and tufts of vegetation. For the first three to four minutes the water was entirely absorbed. After that it began to accumulate and then to flow in the direction of the surface slope braiding, anastomosing and causing corrasion as it moved

linearly between the larger waste fragments or vegetation clumps. Raindrop impact was clearly of some importance in putting grains into suspension and in forcing others to saltate. Since all the pediment surfaces I saw in west Africa were mantled with waste fragments of varying sizes and/ or vegetation clumps I find it difficult, after this experience, to appreciate how the movement of water on such surfaces can be non-erosive [cf. King, 1953] since larger fragments of rock waste and vegetation will always tend to concentrate it into channels where corrasion will occur.

In the same study by Grove (1960), Mr. R. F. Peel raised the question of water soaking into plateau surfaces and reappearing at escarpment bases as seepage to augment weathering at the piedmont notch previously suggested by Ruxton (1958) and Mabbutt (1966) as noted here in the section on weathering. Grove stated he observed no seepage at escarpment bases but that sand mantling them would permit it to occur unobserved. He also concurs with Savigear that water on pediments moves mainly in rills with threads of high velocity in sheet flow and observes that this movement mode is reflected on pediment surfaces which air photographs show to be covered with shallow channels of streams and rills.

The several preceding paragraphs present actual observations of desert rainfall and runoff of water. Our discussion is incomplete, however, if we fail to cite the records that show changes in the erosive and transportative medium (water) during desert fluvial events. McGee (1897), for example, in speaking of a desert flood not far north of Nogales on the Mexican-American border states

The water was thick with mud, slimy with foam, and loaded with twigs, dead leaflets and other flotsam; it was seen up and down the road several hundred yards in either direction or fully a half mile in all, covering the entire surface on both sides of the road save a few islands protected by exceptionally large mesquite clumps at their upper ends. The torrent advanced at race-horse speed at first, but slowing rapidly, died out in irregular lobes not more than a quarter of a mile below the road; yet, though so broad and tumultuous, it was nowhere more than about 18 inches and generally only 8 to 12 inches in depth, the diminution in depth in the direction of flow being less rapid than the diminution in velocity. The front of the flood was commonly a low, lobate wall of water 6 to 12 inches high, sloping backward where the flow was obstructed by shrubbery, but in the open curling over and breaking in a belt of foam like the surf on a beach; and it was evident that most of the water first touching the earth as the wave advanced was immediately absorbed and as quickly replaced by the on-coming torrent rushing over previously wetted ground. Within the flood, transverse waves arose constantly, forming breakers with such frequency as to churn the mud laden torrent into mud-tinted foam; and even when breakers were formed it was evident that the viscid mass rolled rather than slid down the diminishing slope, with diminishing vigor despite the constant renewal from the rear. Such were the conspicuous features of the sheetflood—a thick film of muddy slime rolling viscously over a gently sloping plain; and this film was a transformed stream still roaring through a rugged barranca only a few miles away.

In essence, and with great vividness, McGee has given us a description of flowing water being converted by sediment acquisition and water loss into a *mudflow (lahar)*. It is an invaluable description because of its great detail and the fact that such phenomena are rarely closely observed. Mudflows of the type just cited are not to be confused with mass-wasting phenomena where water-saturated masses of regolith or soil in humid regions break loose on hillslopes and move down until arrested at a slope base by some obstruction and/or friction. Crandell (1968) maintains that mudflows have water contents ranging from 10 to 60%. In the general mass-wasting context, this may be true. However, it is clear that during desert fluvial events the moving media involved range from 100% water, before sediment is picked up, through varying lesser amounts until a muddy consistency is achieved, and then down to an ultimate

near 0% water when the solidified mudflow desiccates on a desert surface (Fig. 6.13).

Considered as the essentially inevitable fluvial consequence in desert regions where there is appreciable surficial debris with argillaceous content, the mudflow and its desiccated depositional resultant comprise first-rank geomorphic phenomena. Crandell (1968) observes that "Mudflows possess a remarkable ability to transport very large masses of rock; this ability is in large part due to a relatively high specific gravity which ranges from less than 2 to at least 2.4." The same writer notes that some mudflows no more than 4 ft thick have transported blocks having dimensions of $9 \times 11 \times 16$ ft, and thicker mudflows have carried rock masses measuring at least $20 \times 30 \times 40$ ft. He goes on to note that, "Mudflows seem to have limited erosional ability; materials such as grass, loose volcanic ash and forest litter have been overriden by mudflows without being removed." Crandell adds descriptions of mudflow deposits, the most significant features of which include general lobate form, poor size sorting, lack of stratification, vertical size gradation upward from coarse to fine, abundant voids in the matrix, and content of angular to subangular stones, some of which are very large (Fig. 6.14) (cf. Stokes, 1950; Dury, 1970).

Lustig (1965) cites a number of reports on the character and behavior of mudflows, some of which imply erosive abilities. In general, however, the implication is that it is the more fluid (non-viscous) portion of the arid fluvial event that is erosive (cf. Wooley, 1946). This erosive phase would therefore correlate with the upslope reaches of runoff prior to the acquisition of a full sediment load. This notion is supported by Lustig who states, "The views suggest that debris is stripped from the upper reaches of each area and is transported to and deposited on the lower slopes. This is precisely the relationship observed on most alluvial fans in the Basin and Range province today. The catchment areas up to the divides, and the apex regions of the fans, are source areas for mudflows; deposition occurs below

Fig. 6.13. A mudflow (black), in Big Pine Creek Canyon, the east side of the Sierra Nevada Mountains, photographed mid-morning. The actual mudflow deposition occurred some time after the previous midnight, and a later stage of water flow has cut through the entire deposit (which at one stage occupied the entire channel at the point shown). (L. K. Lustig.)

midfan. Clearly different spatial relations could occur under other conditions."

Erosion: Transportation? Deposition

Consideration of the foregoing detailed descriptions of fluvial activity in a dry, sparsely vegetated region allows us to more accurately identify specific geomorphic aspects of erosion, movement of sediment, and sediment accumulation in a water-mud medium. The X_c value previously discussed gives us the critical distance that runoff must move before volume or other factors initiating erosion become effective. It is evident, however, that factors that limit the availability of weathered debris to watercourses in humid lands are essentially inoperative in deserts. Generally unprotected by groundcover, weathered rock material is clearly available to desert storm runoff in essentially unlimited amounts. Whereas streams in vegetated lands may often be carrying only a fraction of the potential sediment load that could be moved if the material were available (gauged by existing hydraulic energy), it would appear that no appreciable vegetal limitations on sediment load exist for desert runoff.

With due respect to occasional barren bedrock surfaces where little loose material exists to be eroded, after the flow distance X_c, desert runoff can acquire a sediment load up to the maximum amount and clast size allowed by such factors as water viscosity, slope, surface roughness, and volume of runoff. Bennett's research clearly documents the sediment *capacity* relationship (Chapter 5). On the other hand, the maximum particle size (*competence*)

Fig. 6.14. Alluvial section exposed in an arroyo cut into a fan, the east side of Death Valley, California. Crude stratification is evidenced in some places, but sorting is generally poor and compatible with that of other mudflow deposits (cf. Fig. 7.11); hammer for scale.

moved in desert runoff is dependent on these and additional factors such as fluid density and source rock character (Lustig, 1965). Lustig observes that maximum particle size is usually equated with flow velocity as stated in terms of particle diameter (cf. Menard, 1950; Fahnestock, 1961), but that because of very large particle sizes encountered in such sites as alluvial fans, and the fact that the transport medium itself is rarely observed, any velocity power law is best avoided. Lustig (1965) alternatively suggests using tractive force (τ) as a gauge of competency, expressed as

$$\tau = \tau dS$$

where τ is the specific weight of the transporting medium, d is the depth of flow, and S is the slope of the energy gradient. The tractive force thus defined pertains to the shear exerted on the upper layer of the bed material.

Lustig (1965) goes on to note that, ". . . at best, only an approximation to the true competence can be sought in the field and that hence more sophisticated formulae should be avoided" (cf. Rouse, 1950, p. 122). This is so because values needed for density, velocity, and drag coefficients required by such formulae would be mere guesses in most instances. Actual measurements made during observed desert storm runoff should certainly reduce the guesswork in specific settings (cf. Beaumont and Oberlander, 1971), and the latter workers were able to obtain reasonable values for both Manning's Roughness Coefficient and Sternberg's Critical Velocity in a case involving channeled flow (see Chapter 7). However, as already noted, in a desert water-to-mud metamorphosis such factors as fluid density and viscosity would change continually downslope from X_c. Therefore, one is not surprised to discover, as did Lustig (1965), that calculations of tractive force on desert surfaces based on slope and maximum clast size do not correlate well with distance from source. In short, very large clasts are moved relatively great distances over very low slopes (without, we might add, the size reduction, shape changes, and sorting per unit distance one would expect with running water). We are in agreement with Lustig when he concludes that, ". . . transport by density flows, attended by bouyancy and momentum effects, provides the most reasonable answer to the observed anomaly."

On the basis of the foregoing discussion, it seems clear that after a period of rapid sediment load acquisition, desert runoff tends to become fully loaded—it is carrying all the sediment its energy will permit. As generally defined, fluvial erosion is evidently occurring during this critical interval and over a correlative flow distance contiguous with and downslope from X_c and here designated E_c. What occurs downslope from E_c is somewhat less clear, but the changes in hydraulic slope profile of ephemeral streams recorded by Cherkauer (1972) and pediments noted by Ramana Rao (1972) and others (cf. Figs. 6.12; 6.15) may possibly correlate with the shift from low-density to high-density transport within E_c, or, less probably, with the attainment of maximum load at the downslope end of E_c.

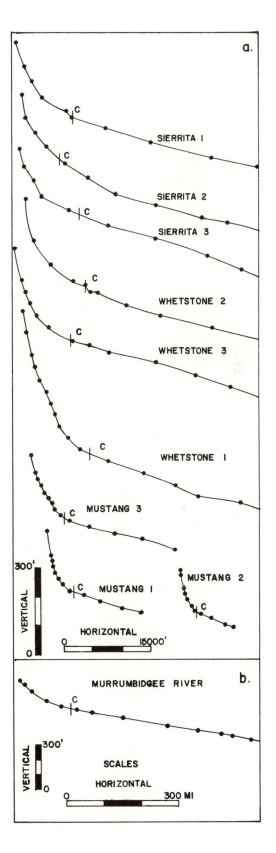

Fig. 6.15. Longitudinal profiles of nine streams (ephemeral) in southeastern Arizona and the Murrumbidgee River of New South Wales, Australia. Segments change concavity (C), presumably in relation to changes in hydraulic regime as flow occurs. (From Cherkauer, 1972.)

Cherkauer (1972) studied concavity variations of ephemeral stream profiles developed on both granitic and sedimentary rock. He states that, ". . . the observed concavity change . . . is related to drainage area, used to approximate discharge." Designating the upslope portions of his profiles "concave" and the lower "straight," (Fig. 6.15), Cherkauer observes that the profile differences relate to the longitudinal variation of drainage area (A), bed-material size (D_{mm}), and channel width: depth ratio (W/D) which, respectively, approximate discharge, sediment load, and roughness, factors that have been found to control stream profiles in other regions but that are ordinarily unavailable for ephemeral streams. Through inclusion of basin relief (R) and lithology, he derives the following statistically significant equations expressing gradient (S)

Sedimentary Concave
$$S = \frac{(6.6 \times 10^{-5})R^{.80}D_{mm}^{.20}(W/D)^{.14}}{A^{.40}}$$

Sedimentary Straight
$$S = \frac{(1.4 \times 10^{-3})R^{.38}D_{mm}^{.26}}{A^{.19}}$$

Granitic Concave
$$S = \frac{(7.6 \times 10^{-6})R^{1.02}D_{mm}^{.28}(W/D)^{.22}}{A^{.51}}$$

Granitic Straight
$$S = \frac{(1.6 \times 10^{-3})R^{.42}D_{mm}^{.062}}{A^{.21}}$$

Cherkauer notes that the width:depth ratio was statistically insignificant in the straight segments but that the remaining variables exceed a significance level of 90%.

The foregoing expressions of slope gradient necessarily apply to geomorphic effects achieved during the E_c portion of the arid fluvial event, in the case cited for streams draining fault-block mountains in Arizona and presently somewhat incised into the mountains and into downslope alluvial deposits. Under different, and presumably prior conditions, E_c must have terminated within the profile areas studied in order to generate the basin fill deposits. An ephemeral stream profile associated with pediments studied by Lanford-Smith and Dury

(1964) lacked the profile break observed by Cherkauer (1972), did not in any real sense have a catchment "basin," and had gradients sensibly matching those of associated pediments.

To return for a moment to the arid fluvial event, in theory, downslope from E_c there may be an interval of time (and a unit of distance traveled) where runoff is merely engaged in sediment transport with neither a net gain or loss as has been suggested for some reaches of humid streams and which has been termed a "graded" condition. Cherkauer's findings (1972), certain related hydraulic factors, and Lustig's observations (1965) make such a likelihood, to borrow a term, "unlikely." We will elaborate.

From the descriptions of McGee (1897) it appears without much question that overland flow of water in deserts begins to suffer volume losses almost immediately—even within areas of rainfall and more so while traversing parched ground beyond such areas. These volume losses would normally result in a reduced sediment-carrying capacity and deposition and would, in most instances, do so over some critical interval of water flow, here designated D_c—the depositional interval that terminates the arid fluvial "event." It would seem from descriptions quoted previously, however, that an alternative terminal phase develops where acquisition of relatively fine-grained sediment by the runoff gradually increases its viscosity and otherwise alters its physical character and sediment-carrying capacity, even as the water is disappearing. In the highly viscous, dense medium of the mud-flow, therefore, debris transport may continue until ultimately brought to a halt by mud rigidity. Thus, D_c, the final depositional episode, may occur from a medium of either water or mud or some transitional medium between them. And it may occur on essentially any slope from nearly vertical to horizontal. (See Schumm and Chorley, 1966, photo 14.)

As a result of the three phases of fluviation recorded in deserts, the deposits made by the more fluid media may be rather well sorted, shaped, and even graded (cf. open sedimentary systems of Krynine, 1955). Deposits within solidified mudflows may, however, display minimal shaping and size reduction and neither sorting nor grading, and, in fact, may comprise the epitome of Krynine's *closed sedimentary system*. Everything is deposited. The three parameters cited in our heading are therefore seen to represent the three major identifiable phases of the desert fluvial event: X_c, pre-erosion water accumulation and flow; E_c, erosion and load acquisition to capacity; D_c, deposition. We must now attempt to locate these phenomena with respect to the overall desert realm and in relation to specific landforms and deposits.

INCISIONAL REALMS

Our awareness that arid geomorphic systems can bring about landscape effects distinct from those of other systems could cause us to be too zealous in our search for unique processes and their products. In point of fact, a brief comparison of the immediately foregoing paragraphs and the section in Chapter 5 dealing with drainage system development discloses marked similarities. In the earlier chapter it was noted that initial fluvial events in a newly humid region would probably occur on non-vegetated surfaces. The fact that this surface in a humid region would later acquire groundcover does not alter the situation—given similar erodable, at least partly granular terrain, there is no reason to expect that overland flow WHICH HAS NOT YET ACQUIRED A FULL SEDIMENT LOAD will behave in a significantly different manner in a desert than runoff described by Horton (1945) in a humid drainage system development. This is to say, in normally erodable (granular?) areas of deserts, one can anticipate the development of the rill-gully-drainage net configuration and increased surface relief (cf. Figs. 5.30 to 5.36). They have been observed there by many workers.

Some qualifications to the above relation-

ship should be added. First, because desert precipitation tends to be local, the drainage-divide-basin framework of reference used in humid areas may be inappropriate simply because any fluvial erosion is tied to the rainfall area and downslope areas contiguous thereto. And the desert rainfall area frequently does not extend to a divide, whether major or minor. McGee (1897) cites instances where rain appears to have fallen almost exclusively on a plain, albeit one with some slight slope. In such cases, Horton's X_c distance would presumably be measured from the upslope limit of the rainfall area or from some ill-defined region within the area. Clearly, prior and subsequent rains need not coincide geographically with any given rainfall event, though, as we have already intimated, they may tend to. Because of the apparently random occurrence pattern of rainfall ON LOW-RELIEF DESERT AREAS, as indicated by vegetal distribution, it would seem that the depositional phases of such events would tend to be randomly superimposed upon erosional phases (areally) and, in the absence of any sensible localization of one type, the effects would tend to cancel one another out; thus, no drainage net could form. In brief, THE PLAIN WOULD TEND TO REMAIN A PLAIN. To the extent that this apparent genetic relationship holds valid, a plain should be an equilibrium landform in a desert. More on this in the section on climax arid landforms.

A second possible qualification on the development of drainage nets in deserts concerns the erosional versus depositional fluvial phases just discussed. Runoff that is fully loaded cannot continue to erode and incise flow routes by picking up additional rock material without a change in its energy budget. A fully loaded stream of water can only erode further after depositing something. Thus, if there are regions in deserts where the erosional (incisional) phases of rainfall and related fluvial events tend to be repeatedly superimposed on granular surfaces, there is no apparent reason why at least incipient drainage nets should not form. To the extent that downslope depositional phases also tend to geographically coincide, these drainage nets should give way to phenomena characteristic of desert hydrographic dissipation and sediment accumulation. This last is essentially the relationship just quoted from Lustig (1965) in which he describes rainfall catchment areas and alluvial fans (Fig. 6.16); here the catchment areas are those places where fluvially generated relief would be anticipated in the contexts just discussed.

A third possible qualification on the development of drainage nets in deserts involves bedrock and regolith relations. As noted in Chapter 5, there appears to be some doubt that the typical dendritic drainage net would form directly on well indurated bedrock, though superposition of the pattern would be possible.* *Badland topography* in deserts is almost invariably developed from material that is comparatively fine grained, often with a high argillaceous content (Fig. 6.17). Such topography merely reflects intense dissection of easily erodable material by drainage nets of the type discussed by Horton (1945) and studied by Strahler (1952), Schumm (1956), and many others. More significantly, with respect to the typical arid fluvial event, badland topography reflects the occurrence of the requisite terrain material in a (catchment) area where runoff is not yet fully loaded and where the area is sufficiently elevated so that the runoff can incise linearly. Drainage densities in such nets tend to reflect terrain permeability, among other things (see Chapter 5), but most such nets terminate downslope in either full-scale or miniature pediments or fans (Fig. 6.18). Where requisite lithologies occupy divide areas, dissection downslope may be limited either by deposition or resistant lithologies or some combination of these. Where, as in the Black Hills of South

* Note section on inselbergs, p. 374.

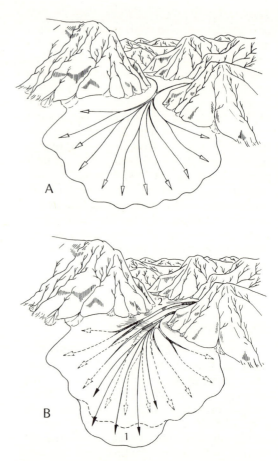

Fig. 6.16. Formation of alluvial fans according to the climatic hypothesis. (A) During more arid episodes when the D_c portion of the arid fluvial event falls within the upland drainage net or just downslope. The desiccating stream floods aggrade and spill out of braided flow routes (auto-diversion), building the fan upward and, to a degree, outward before drying up; (B) during more moist episodes the D_c phase of the fluvial mechanism is displaced toward the fan toe or even out of the changing environment area. Runoff leaving the more vegetated upland carries less load, and fan trenching occurs as more permanent watercourses are established and deepened. If only slightly more abundant runoff develops, the fan toe (1) may be extended while fanhead trenching occurs. (See also Fig. 6.23.) (From Lustig, 1965.)

Fig. 6.17. Badland topography developed mainly by fluvial processes acting on fine-grained (argillaceous) materials. *Above:* texture developed in the Triassic, Chinle Shale north of Hankesville, Utah. *Below:* terrain in Badlands National Monument, western South Dakota.

Fig. 6.18. Miniature pediments developed from argillaceous rocks of Badland National Monument, western South Dakota. Simple dendritic drainage nets are developed at the pediment heads in the dissected interfleuves; hammer for scale.

Dakota, the most erodable lithologies occur as nearly flatlying formations, the development of badland topography is somewhat restricted to those formations.

In addition to an obvious variety of bedrock lithologies, runoff in deserts often also encounters highly resistant surface mineral incrustations in the form of the duricrust already discussed, which may lend increased erosional stability to plains regions and the alluvium commonly encountered there (cf. Langford-Smith and Dury, 1965) (Fig. 6.19). There seems to be little question that underloaded runoff traversing bed-

rock surfaces in deserts tends to exploit certain lithologies and structures thereon, just as it would in humid lands, and thus develop drainage nets reflecting these factors until such time (and distance) as erosion is replaced by deposition or otherwise inhibited.

A fourth apparent qualification on the development of drainage nets in deserts involves the uneven distribution of precipitation related to irregular topography. It was previously noted that desert rainfall tends to be orographic in character. That is to say, uplands (and few regions do not have some) tend to get the most rainfall; relatively lower areas tend to be drier. Grove (1960) and others have noted the relation for the central Sahara; Davis (1903), Lustig (1965), and others have recorded similar relations in the western United States; and other writers have recorded evidence of similar effects on other continents. Available data indicate that the elevated expanse of terrain need be neither areally extensive nor exceptionally high. Ayres Rock, for example, rises in splendid isolation on an Australian desert plain (Fig. 6.20), but it seems to catch appreciably more precipitation than the surrounding lowland.

It seems probable that the increased moisture on uplands in deserts is due mainly to air masses being forced to rise because of high land. But there may be some tendency, either seasonally or diurnally, for uplands in deserts to focus rising air masses for reasons not yet fully explored. Differences in heat absorption and

radiation between alluvium and bedrock may also be involved, since these two types of material, respectively, tend to comprise lowlands and uplands in many deserts; sand may substitute for bedrock. In any event, in arid regions of notable relief, precipitation is apparently concentrated on elevated areas, in effect because a moisture-yielding portion of the atmosphere contacts those areas (cf. Fig. 4.7). This increased moisture is reflected by the concentrations of vegetal cover in such areas (cf. Fig. 5.15). By inference, concentrated precipitation is also indicated by the fact that drainage lines consistently originate in elevated parts of deserts, regardless of the geologic origins of the elevated area. In fact, apart from drainage lines related to trans-environmental rivers (cf. Chapter 5), recognizable drainage nets are almost entirely concentrated in desert uplands or their flanks (cf. Davis, 1905). There are obvious exceptions. And with respect to hydraulic theory re-

Fig. 6.19.   Canga, a duricrust (ferricrete) typically developed over laterite in peninsular India, in more arid regions. Reworked laterite and ferricrete clasts are both incorporated. (K. L. V. Ramana Rao.)

Fig. 6.20. Ayres Rock, an inselberg situated some 200 mi (320 km) west-south-west of Alice Springs, central Australia. The monolithic mass of nearly vertical Precambrian strata rises 1,100 ft (335 m) above the surrounding pediplain and is about 2.25 mi (3.6 km) long and 1.25 mi (2 km) wide. (See also Fig. 1.12.) (Australian Information Service, New York.)

lated to desert fluvial observations, any incisional flow line system must reflect a genesis by linear runoff that is not fully loaded with sediment. There are definite TENDENCIES for the occurrence of drainage nets in deserts. And these are what the geomorphologist must live by.

DEPOSITION—WHEN, WHERE, AND HOW

It has already been observed that vegetation tends to localize runoff, and the more nearly plants cover an upland the more nearly it

qualifies as a humid geomorphic system in which running water should induce the development of "PERMANENT" drainage nets. Actually, of course, the more dense the upland vegetation, the less available is surficial weathered debris (cf. Bennett's experiment, Chapter 4) and the farther a given volume of runoff from such a source can be expected to flow without becoming fully loaded. The actual "area" that is most moist also influences derivative runoff volume and hence affects E_c (distance of effective erosion) as well as D_c (locus and area of deposition) (cf. Cherkauer, 1972). It would therefore appear that the relation between vegetal density and area on an upland in a desert (as a reflection of a long-term precipitation locus) will ultimately determine (1) the distance storm runoff will flow and tend to develop a drainage net before becoming fully loaded, (2) the amount of sediment acquired per unit distance, (3) something of the character of the initial sediment load since weathering could be expected to vary with changed plant cover and moisture amounts, (4) the location of the general transition zone between fluvial incision and deposition—for one particular set of conditions this is well expressed in McGee's description (p. 328) of the desert stream line that becomes progressively shallower and wider before disappearing in a deposit, and (5) the ultimate distance to which water will move within or without the desert—if within, to more or less immediately deposit the previously eroded detritus.

These considerations tend to modify the previous assessments of the influences of slope upon deposition and the formation of such features of alluvial cones and fans. For some decades, and consistent with general Davisian fluvial concepts, it has been traditional to associate changes in hydraulic aspect, and at least some desert deposition, with breaks in slope. Miniature fans in quarry depressions and gravel pits can clearly be associated with breaks in slope. Since the entire environment in question is often humid, and the ground is saturated with water, issues of evaporation or excess infiltration do not seem to be involved in deposition. Moreover, there is usually a distinct break in slope between the catchment drainage lines and the fan surface, whereas, in full-size fans, this is often lacking (cf. Fig. 6.21). Differences of opinion exist as to the conditions that govern fan deposition versus fan trenching [see Lustig, (1965); Hooke, (1965) and Bull (1968)]. Yet, from the foregoing descriptions of desert fluvial activity, it seems clear that deposition can occur on slopes and without a notable slope break as noted long ago by Johnson (1901). Contrary to what is implied by Bull (1964; 1968), however, available data demonstrate no necessary causal relation between tectonic activity and catchment area incision or related downslope deposition (cf. Lustig, 1965; Cherkauer, 1972).

The point should be made here that precipitation events in desert regions, particularly those in relatively moist uplands, through infiltration can locally generate a temporary (and probably perched) groundwater body. Therefore, for some minutes or hours after a storm and its initial erosive runoff surge, surface return of water that has seeped underground in upland areas commonly causes minor surface flow in extant downslope channels. This flow is commonly clear, can be observed to shift minor sandy bedload, and ultimately seeps into downslope porous surfaces and disappears. Such water may flow considerable distances considering its small actual volume, mainly because much of the distance traversed is over surfaces previously moistened by the initial runoff surge, and it is often observed briefly on alluvial fans and pediments.

It also seems important at this point to note the relationship that must exist (but that has been little studied) between the rate of production of weathered debris in deserts and the frequency with which runoff develops to move such detritus. This relationship, as much dependent upon lithology and weathering mode and susceptibility as it is upon orography and meteorology, can strongly influence the quantity of sediment the average storm runoff of a region will contain. Schumm and Chorley (1966) note variations in talus production with lithology in the Colorado Plateau and conclude that coarse, durable talus diminishes scarp retreat rates. It has already been noted that such weathered material as is present in deserts is more or less completely available to runoff of suitable competency because of the absence of plant cover. Clearly, however, rocks resistant to breakdown or those yielding readily decomposable fragments may have their surfaces completely stripped of loose detritus even

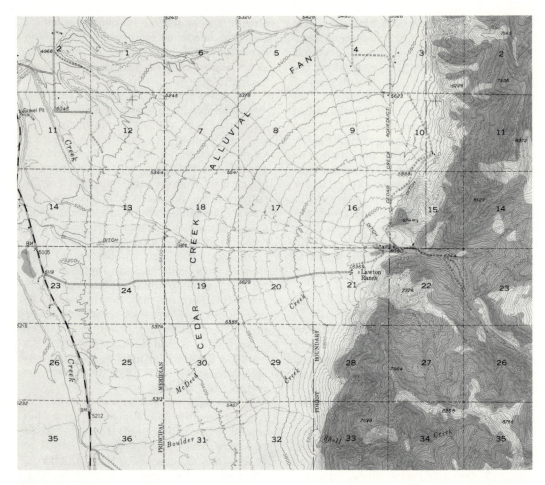

Fig. 6.21. Topographic map of the Cedar Creek Alluvial Fan near Ennis, Montana (N4515-W11130/15 1949). Note particularly that the contour spacing (and hence slope) does not change at the mountain base (fan head) but continues with the same declevity back into the feeder valley for more than a mile. (Contour interval 40 ft—numbered squares are sections measuring 1 mi on a side.)

where precipitation incidence is extremely low. Alternatively, mixed lithologic suites which include readily decomposed as well as durable bedrock fragments may yield material at a rate sufficient to mantle slopes, clog gullies and ravines, and even obliterate rills (cf. Schumm, 1962; Schumm and Lusby, 1963) in spite of precipitation and runoff at relatively frequent intervals. Other factors being the same, storm runoff in areas having the latter characteristics would always acquire loads quickly, whereas those in settings of the first type

might move considerable distances before becoming fully loaded.

A most interesting case of balance between runoff frequency and detritus supply was recounted by Lustig (personal communication, 1973) for the Kuiseb River in the central Namib Desert. The Kuiseb's channel separates the sand dune area of the southern Namib from the gravel plains to the north. The sands from the south accumulate in the channel in time but never manage to cross this barrier. Water flowing down the length of channel to the sea, about once every ten years, is apparently sufficient to remove the accumulated dune sand.*

It seems evident that depositional foci of loaded versus underloaded runoff could differ notably. From runoff loaded to capacity, accumulation of sediment could begin within, or almost immediately downslope from the precipitation catchment area. On the other hand, deposition from underloaded runoff might not occur until some expanse of erodable material (E_c) had been traversed beyond the precipitation area. In instances where runoff is fully loaded as it leaves an area of precipitation, one could anticipate alluvial encroachment upon upland drainage nets and at least some aggradation of existing channels and gullies. Alluviated valley floors of the net would exhibit essentially the same declivity as alluvial accumulations beyond the upland margins (e.g. fans). (See Fig. 6.21). As observed by several researchers, including Bull (1968), extension of underloaded runoff onto existing, non-channeled alluvial deposits will necessarily induce incision. Continuing debates involve the relative importance of such factors as slope, rainfall frequency and intensity, uplift and subsidence, load caliber, vegetation density, etc., in this regard. Presumed results include

fan-head trenching, pediment gullying, and drainage net extension across lowland areas.

In light of the data assembled here, there can be little doubt that marked diminution in sediment availability along drainage lines would be required before linear incision and drainage line extension out of a desert could be effected. There seems to be little question that this would require increased effective precipitation and continuous plant cover (cf. Chapter 5). The per cent of a desert "upland?" region that would have to be modified in this manner before exhoric drainage could be achieved is uncertain and would certainly vary with each setting. More to the point, available records show that most deserts during their existence undergo periodic fluctuations in upland precipitation, vegetation, and, hence, runoff without actually ceasing to experience arid, sparsely vegetated conditions across most of their lowland extents. This relation, then, would appear to be the "normal" high-relief desert environment—one where elevated areas are relatively moist and lowlands comparatively dry. It should logically follow that the lower a desert region's relief is, the more uniform should be its environment and geomorphic regimen.

PEDIMENTS AND PLANATION

"The time has come," as the Walrus said in Lewis Carroll's poem "to speak of many things," including how bedrock beveling occurs in deserts where "observed" running water appears to either incise or deposit on the land surface. Thus, there are some basic questions. It will be recalled that the arid fluvial event as presently understood tends to cut drainage lines in upslope regions and

deposit previously eroded material not far downslope. Where, then, do the truncated bedrock surfaces that comprise a major attribute of pediments originate? How, in fact, do escarpments retreat? Why do some regions exhibit only upland precipitation catchment areas that have been gullied into incipient drainage basins but with downslope extensions that appear instead as allu-

---

* At several points we have had occasion to note that deserts are GENERALLY undrained exteriorly. Obviously there are exceptions related to rare rains falling close to coasts and equally rare moist years or intense storms that lead to drainage link-up and exterior flow as in western Australia in the early 1950's and in the Kalahari Molopo drainage in the early 1960's. The generalization is nonetheless valid, as such, and desert landscapes are hardly sculpted with respect to exterior drainage systems.

vial fans and clearly reflect aqueous deposition, or, alternatively, that assume the form of pediments with a bedrock substrate that shows erosional features and a meager or even absent alluvial cover? Is there, possibly, some unobservable desert chronologic relationship that establishes the genetic scheme between their various features, their formative environment(s), and certain processes? In short, what's going on?

THE ROLE OF WATER

Let us consider the facts, insofar as they are available to us, and some of the possibilities: An examination of the first-hand accounts of desert fluvial events discloses that essentially all of the erosive activity clearly evidenced is essentially linear in form. This not only applies to upland catchment areas where rills and gullies concentrate runoff, it also appears to apply to downslope planar surfaces where water can spread out. In the excellent account of McGee (p. 329), it is the linear flow lines in sheetfloods that are seen to scour, and this view is echoed by King and Grove, though King has elsewhere maintained that it is the planar form of runoff that shapes pediments (1953). In terms of actual observation, the data on erosion do not seem to support this latter view. Runoff may mold itself to the form of a surface, but it can also modify the shape of the surface. Yet, it seems reasonable to ask: "Does water spread out in order to erode a plain because of some hydraulic factor, or is the fluid merely accommodating itself to a low-relief surface?" As can be shown, there is some reason to believe the latter is the dominant relationship.

If running water is to account for planar landforms (including pediments) in deserts, and if the erosionally most effective form of runoff is linear, what is the genetic interrelationship? Several suggestions have been made. McGee, for example (1897), describes local gullying beneath a sheetflood, with small fans of debris being formed at the downslope ends of the depressions and the gullies themselves being refilled almost as soon as they are formed by later runoff on a more or less random basis. Tator (1949, 1952) has alternatively argued that streams emerging from mountain valleys become loaded to capacity by cutting linearly down through the 2- to 3-ft alluvial veneer commonly mantling a pediment surface and then by aggrading these linear incisions in the waning stages of runoff. Later storms must therefore incise new channels elsewhere since older flow lines are blocked, and bedrock is thereby presumably beveled laterally by randomly located linear incisions, the depth of which is controlled at any one area by the alluvial veneer which must be removed before bedrock erosion can occur.

Twidale (1968) notes that many workers now consider that a complex of processes—weathering, rill wash, sheet wash, and sheet flow—are responsible for the molding of pediments. The same author notes the general absence of marked meandering of flow lines and also questions the significance of laminar flow over ground surfaces which, as previously noted, are often rather rough. In point of fact, it seems clear that the amount of water necessary to completely innundate a surface is dependent upon the relief of that surface. It seems reasonable to suspect that the sheetflooding McGee describes would be rare over a rough pediment or fan surface. Runoff will clearly be channeled where channels exist until or unless they are aggraded, just as runoff not fully loaded but in the presence of suitable debris will erode, but if fully loaded, and losing energy, must deposit. However, none of the foregoing discourse really answers the question why pediments and not fans or basin fills.

BASE LEVELS AND SCARP RETREAT

An intensive survey of the literature of pediment formation (cf. Bryan, 1923, 1935; Pallister, 1956; Dury, 1966; Hadley, 1967;

Denny, 1967; Lustig, 1967, Dury, 1970; Oberlander, 1972) suggests that certain relationships have been greatly emphasized, whereas others may have received too little attention or have been overlooked almost entirely. For example, the planar landform we call a pediment, sloping as it usually does in a gently concave upward form, must be erosionally graded to a particular base level. This seems evident. Yet little use has been made of this fact. To what level is a pediment graded and why? Pediment longitudinal profiles that are concave upward (Fig. 6.22) and the longitudinal profiles of streams (Fig. 2.13) are essentially identical in form. This is probably no accident. But it should be emphasized that a humid stream is (presumably) usually sloped with some respect to regional base level, whereas a pediment slopes to a local base level (cf. Langford-Smith and Dury, 1964). Something in the pedimentation process clearly requires the establishment of a local erosional datum. A consideration of humid fluvial processes (Chapter 5) discloses that landslopes in that environment are oriented with respect to a constantly lowering erosional datum established by and inclined with respect to incising channel bottoms that are above regional base level (Fig. 5.38).

Examination of Fig. 5.38 discloses that watercourse enlargement under humid conditions involves simultaneous incision and slope back-wearing. SO LONG AS THE TWO INCREMENTS OF SHAPE CHANGE COINCIDE IN TIME AND AMOUNT, IT IS HARDLY POSSIBLE TO DEVELOP TWO NOTABLY CONTRASTING SLOPE CONFIGURATIONS (I.E. SCARPS VER-

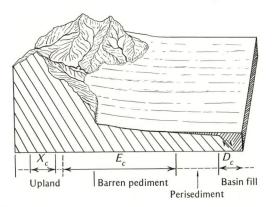

Fig. 6.23. Probable relationships under arid conditions between landscape and deposit elements and correlative phases of related fluvial events focused by orographic effects upon the upland. Depicted are X_c (zone of nil runoff erosion within upland precipitation locus), E_c (zone of eroding but underloaded runoff), and D_c (zone of fully loaded, depositing runoff). Transitions between fluvial phases are zonal with respect to any extended period but may be abrupt during any given fluvial event.

SUS FLATS). Though the downstream end of a major drainage system may remain essentially fixed for a period, the remainder of the drainage basin will continue to be governed topographically by the just mentioned slope development in two directions. This should continue as long as drainage lines are above regional base level or are not influenced by local, erosionally resistant *knickpoints*. Even where a local knickpoint in a humid area inhibits channel deepening, any backwearing of slopes under vegetal cover, creep, and chemical weathering would be extremely slow without slope steepening related to incision. AN ALTERNATIVE BASE-LEVEL RELATIONSHIP CLEARLY DEVELOPS IN THE DESERT GEOMORPHIC SETTING because of the X_c, E_c, D_c phases of arid fluvial events. Diagrammed schematically (Fig. 6.23), the desert erosion-deposition relations disclose a very important geometric relationship. The upslope margin of D_c (i.e. the highest part of the area over which desiccation-induced deposition occurs) becomes the erosional datum for any drainage area existing farther upslope whose erosion is causing the deposit. Stated somewhat differently, A DRAINAGE NET SUPPLY-

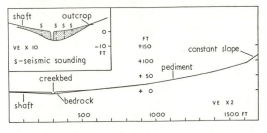

Fig. 6.22. Pediment profile surveyed transverse to Middle Pinnacle Creek, Australia, showing alluviated channel (inset). (After Dury, 1966.)

ING WATER AND DEBRIS TO A PEDIMENT OR
AN ALLUVIAL FAN OR BASIN FILL IN A DES-
ERT CANNOT INCISE ITSELF BELOW THE
LONG-TERM DOWNSLOPE LEVEL OF INITIAL
DEPOSITION (cf. Lobeck, 1939, p. 244). Since
rainfall intensity will vary from storm to
storm and year to year even under a particular desert environment, the depositional datum is more probably a zone than
a horizon and could at any given moment
display evidence of erosion as well as deposition. As noted long ago by Davis (1905),
erosion surfaces controlled by such a depositional datum need not relate to regional
base level (cf. Garner, 1959; Cotton, 1960;
Garner, 1963). This would apply equally
well to deserts near sea level and to those
on plateaus near the crests of major mountain systems. In the latter instances, the
high-elevation zone of moisture-depleted
air (cf. Chapter 4) would presumably delimit the arid fluvial effects including D_c as
suggested by Garner (1965).

It would appear that the area affected by
a fluvial event in a desert includes a "built-
in" erosional datum to or toward which
higher related fluvial slopes should incline.
Therefore, a drainage net developed in a
desert runoff area where flow is generally
under-loaded and incising (E_c) has ONLY
one of the two major attributes of the average humid drainage net. That is to say,
HILLSLOPES ALONG DESERT DRAINAGE LINES
ARE SUBJECT TO BACK-WEARING (IN THIS
INSTANCE UNDER LITTLE OR NO PLANT
COVER), BUT THE DRAINAGE NET CANNOT
BE DEEPENED SO LONG AS THE DEPOSITIONAL
DATUM IS ESTABLISHED AND MAINTAINED
(Fig. 6.24)! Hillslope recession above the
datum keyed on existing scarps and drainage configurations should initiate pedimentation at that level (cf. Schumm, 1956).
Related pediment slopes should conform to
the usual concave-upward profiles of normal drainage nets (cf. Figs. 1.2 and 2.10).
Presumably, back-wearing of the related
hillslopes would involve sheetwash, mass
wasting, and weathering, the latter possibly
both subaerial and suballuvial. But as long
as the runoff through the developing pediment is dominantly under-loaded, little

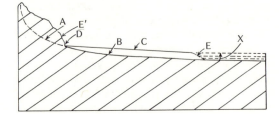

A. Longitudinal profile of upland drainage net
B. Extension of A pedimented on fixed base level
C. Extension of A pedimented after X-rise in base level
D. Piedmont angle (sharpness varies with pediment slope)
E. Scarp immobilized by burial
E.′ Recessional scarp

Fig. 6.24. Diagram of pedimentation keyed on an
upland drainage net with fixed or rising base level.

more than a thin (in-transit) alluvial
veneer should develop. And there might be
essentially no detrital cover at all.

Viewed in relation to the automatic base
level of erosion associated with the average
arid fluvial event, planation in a desert is
seen to involve several alternative situations, dependent to a large extent on special
relict relationships. The block-faulted tectonic situation, presumably common to
the Basin and Range province of North
America, is seen as a special condition,
since the fluvial incision of highlands there
might have several alternative changing
downward limits. Initially, a graben floor
might comprise the only downward limit on
incision. Later, basin fill materials accumulated in graben or related structural depressions could inhibit downward incision of
marginal drainages, but it seems clear that
the structural breaks in slope (Fig. 6.25)
need not coincide with the upslope limit of
D_c. Thus (dependent upon such factors as
catchment area locations, sizes, sediment
availabilities, and runoff volumes), depositional loci defined by D_c could occur within
a structural upland as well as in adjacent
structural lowlands (Fig. 6.26).

The arid volcanic situation is again
rather different. A scarp cannot recede until
it forms and many eruptive events generate
nearly featureless depositional terrain (cf.
Figs. 3.96 and 9.31). A volcano's slopes are
clearly a potential target for scarp recession. But until runoff originating on the

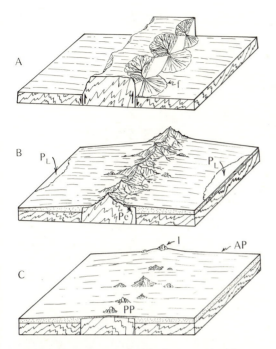

Fig. 6.25. Three-stage block diagram of arid geo-morphic events in an area of fault-block mountains. (See Fig. 6.26.) (A) Development of a horst and preliminary incision of its most moist portions (crest) with development of drainage nets and basins upslope from fans (f); (B) establishment of basin fills, local base levels of erosion, and mar-ginal pedimentation of horst (Pe), possible playa development (Pl); (C) reduction of horst to pedi-plain (pp) surmounted by inselbergs (I) and sur-rounded by an aggradation plain (Ap).

volcanic upland channels its flanks and/or adjacent lowlands, or they are rifted by faults, it seems unlikely that extensive planation could occur. In some respects, erosionally developed coastal escarpments resemble those formed by regional drainage systems. At issue here, again, is the effec-tiveness of regional base level as an ero-sional datum—it being assumed that a coastal scarp in a desert will recede appre-ciably as long as it remains essentially bar-ren. One distinction must be made between the fault-block, volcanic, and coastal es-carpment relations, however. That is, whether deposits downslope from the reced-ing scarps have an elevation that (in the long run) is rising or remaining essentially fixed. Each situation should define a dis-

tinct planation morphology (see Twidale, 1968).

In a closed desert basin, fluvial effects dictate a steadily rising base level of erosion keyed to the upper surface of the alluvial basin fill deposits as these are added to, unless this addition is more than counter-acted by (1) structural subsidence of the basin and/or (2) eolian deflation and re-moval of material from the basin occurs faster than it is added. In the absence of either of the latter relations, therefore, the basin alluvium must eventually encroach on pedimented upland areas as *perisedi-ment* (cf. Figs. 6.24; 6.25). Thus, such sediment would tend to bury pediments dis-tally, even as they form proximally adja-cent to receding scarps. Since the base level of planation determined by deposition could be rising constantly, only the unburied por-tion of the pediment developed on a drain-age net should display upward concavity (cf. Figs. 2.9; 6.24), whereas the aggraded surface could be essentially planar. For sur-faces developed on an essentially fixed planation datum with volcanic, structural, fluvial, or marine escarpments not subse-quently related to drainage nets, essentially flat or only slightly inclined pediments could form (cf. Lustig, 1967).

The sediment-pediment relationship is illustrated in several desert basin situations of the southwestern United States. A vol-cano, if high enough to influence moisture distribution, could presumably induce its own unique erosional and depositional rela-tionship in a desert (cf. Grove, 1960). Coastal escarpment situations are rather different. In those settings, the capacity of an adjacent ocean basin for sediment can be considered to be effectively infinite. Ex-cepting an actual strand shift, the desert erosional-depositional pattern might or might not intersect the strand with any con-sistency, and pediments developed by re-cession of sea cliffs should not be notably alluviated (cf. Figs. 6.23 and 6.24). An ad-ditional aspect of pedimentation therefore lies in the set of co-requisites that seem to be required for a pediment to form. These co-requisites include (1) an erosional

Fig. 6.26. Examples of various erosional stages and changing depositional loci during reduction of desert mountains. *Top;* Panamint Range, California, with alluvial fans developed at the foot of a moderately incised fault block; *center,* Ibex Mountains, California, where broad pediments embay the range around outlying bedrock remnants; *bottom,* Cima Dome, California, a broad pedimented surface surmounted by isolated erosional residuals now too small to induce appreciable precipitation. (Eliot Blackwelder.)

datum (established by deposition in many cases) at the distal end of the forming pediment and (2) inhibition of vertical fluvial scour of the pediment surface by the fact that the enlargement of the pediment probably tends to destroy its runoff source. There is little question that reduction of uplands by hillslope recession should gradually diminish localized upland precipitation concentrations and thereby reduce over-all runoff volumes from highland catchment areas as they become smaller and lower. Davis (1905) also suggested this. In time, the E_c dimension of the arid fluvial event should proportionally diminish along with the D_c dimension, since deposition is not possible without erosion, and neither is fluvially possible without slope.

It seems worth while to conclude our discussion of pediments and planation by contrasting the three recorded arid fluvial erosion styles that relate to these landforms. One style involves water in channeled upland terrain where the flow is clearly linear but does not incise (indeed it cannot) below the erosional base level established by deposition just downslope. Related drainage nets can only be deepened and extended if the D_c phase of the arid fluvial event shifts downslope or out of the environment. A second fluvial style relates to water on sparsely vegetated surfaces of fans or pediments in sheet or braided form within the D_c zone. Here it does not incise the flow area consistently in any one locale because of heavy sediment loads and constantly shifting flow and precipitation patterns in which randomly disposed erosional effects are nullified by equally random deposition, and the final phase of each fluvial event is always depositional.* A third fluvial style is that observed by Grove (1960) where sheetflow or braided runoff traversing a plain or plateau flows off an escarpment to a lower surface with consequent fluvial sapping and related erosion of the escarpment. Such sapping, like that occurring at the base of waterfalls, must be added to the other recorded hillslope recession mechanisms (cf. Schumm, 1956) where barren surfaces under sheetwash and mass wasting undergo some erosion essentially "everywhere" with a resultant perpetuation of escarpment attitude during recession.

WIND, SAND, AND SCULPTURE

Deserts have long been known for their biting, sand-filled winds, chill by night, and scorching by day. Actually, apart from gusts of occasional two- or three-day wind storms, desert nights tend to be distinctly tranquil. Also, as a point of fact, the concept of a desert as a sand-buried waste was probably born in a novel or a Hollywood movie. According to Holm (1968) an average of 25 to 35% of the earth's desert lands are actually mantled by sand deposits; these include North America (2%), Sahara (11%), and Arabia (50%). Indeed of the two major aspects of deserts usually cited, "stony" and "sandy," the former easily outbulks the latter. Both are strongly influenced by wind activity, and before we depart overmuch from considering wind frequency and intensity, it should be noted that desert winds tend to be distinctly diurnal, rising with the sun, moderately strong by mid-morning, gusting by noon, and fading again with the sun, the heat of which appears to be convectively responsible for much of the air movement. Daily, near-surface wind gusts frequently exceed 15–20 mph (24–32 kph). Twenhofel (1932) notes that in the Libyan Desert where sand grains average 0.5 to 2 mm, sand moves at 13 mph (21 kph) and the air is sand charged at 23 mph (37 kph).**

* Of course, many fans have been trenched by a D_c shift downslope and subsequent runoff thereon is confined in relict channels.
** For a general geologic discussion of Africa and its deserts, see Furon (1960) and Butzer (1968).

MOVING ROCK

Air moving in deserts naturally tends to transport rock fragments that are small enough to move. In general, eolian access to such materials is limited mainly by surficial rock fragments too large to be moved since plants are so sparse. Clasts most readily moved by air (which is about 1000 times less viscous than water) are clay size (1/256 mm), silt size (1/256–1/6 mm), and sand size (1/16–2 mm), though the writer recalls an occasional tornadic gust [ca. 50–60 mph (80–96 kph)] blowing 3/4-inch, angular flint chips from a low-slope roof in northwest Arkansas (cf. Fig. 6.27). Such wind velocities are uncommon in deserts, but desert winds readily shift sand and remove unconsolidated silt and clay to other environments, clay to uncertain and often distant destinations where water or wind can trap the detritus, and silt mainly to vegetated regions immediately downwind. Thus, wind in deserts is analogous to the trans-environmental rivers discussed in Chapter 5; wind is the only agency known to move consistently beyond the environmental limits of the average desert. Unlike those rivers that must respond to gravity and move downslope, wind frequently moves along or up slopes, though it is a rare locale that lacks a prevailing wind direction.

Prevailing winds of a given desert region can, of course, vary in direction on a long-term basis. Such variations may be determined by the distribution of eolian deposits such as *loess* and by the orientations of *deflation depressions* as well as by studies of cross-stratification attitudes in dune de-

Wind velocity (m/sec)	M.P.H.	Maximum clast size moving (mm)
4.5 - 6.7	±10	0.25
6.7 - 8.4		0.5
8.4 - 9.8		0.75
0.8 - 11.4		1.0
11.4 - 13.0	±30	1.5

Fig. 6.27. Relationships between wind velocity and maximum moving clast dimension. (After Twenhofel, 1932.)

posits and orientations of particular types of dunes. Grove (1960) notes the relation between wind direction and *yardangs;* "Streamlined pinnacles with steep faces to the . . . wind and gentle slopes to the leeward side." Clues to wind direction may also be provided by ventifact orientations, but these are less reliable wind-direction indicators because they may be undercut and thus move, and their facets are cut by sand carried by near-surface air streams that can be seen to undulate laterally. Bagnold (1935) was a pioneer in the study of wind-blown sand.

Geomorphic expressions of wind activity take the form of eolian deposits, mainly loess accumulations outside of deserts and dune sand plus erosional features within the desert proper. As noted in Chapter 5, loess is often an accretionary additive to soil, and it occasionally accumulates to appreciable thicknesses where trapped by vegetation over long intervals of time (see also Chapter 7). Since the general process is depositional in effect, loess contributions to landscape morphology are usually subtle and mainly pedologic (cf. Fig. 5.20).

ARMORS AND CRUSTS

Eolian erosion and deposition effects within deserts are striking but must be viewed in proper perspective. Early assessments of desert landscapes tended to strongly emphasize wind effects. Taken without reservation, this evaluation appears to have been in error on several counts.

(1) Careful study shows that unconsolidated deposits of mixed clast size subject to eolian erosion quickly tend to loose fine-grained sediment fractions by *deflation*, thereby acquiring a surficial, coarse-grained residue of clasts too large to be moved by the wind. The coarse residue is variously termed *reg, lag gravel, gibber gravel*, or, if bonded by some adhesive material, *desert pavement* (Fig. 6.28) and it armors the ground against further deflation. Furthermore, experiments in the Mid-East have shown that the gravel armor, where

Fig. 6.28. Desert pavement developed in the central portion of Death Valley, California. Areas between large clasts are partly masked by sand in transit between adjacent dune areas.

breached to expose finer subjacent material, reforms in two or three years, in part by creep and sheet wash. Thus, following fluvial events, fine-grained detritus left exposed by water tends to be removed, and scour depressions gradually are filled.

(2) Surficially precipitated mineral matter on desert plains (*duricrust*) tends to bond surface clasts to one another, and thus, it further protects finer material from the wind. This process therefore tends to counteract the abrasive reduction of surface clasts by wind-blown sand expressed by wind-faceted rocks. The latter *ventifacts* are common among the pebble-to-cobble–

size gibber gravels and may be both polished and faceted (Fig. 6.29). Originally, clast materials in gibber gravels may come from fractured bedrock or from duricrust that has been broken up and re-worked (cf. Dury, 1970). Both types of material can be shaped by the wind. In the Algerian Sahara, as winds rise in the morning, blowing sand in streamers is readily visible moving 2 to 3 inches above the ground in anastomotic paths between wind-polished cobbles on the duricrusted plateau (*hammada*) (Fig. 6.30). Conjunctive formation of lag gravel and duricrust on desert plains would tend to establish a least-work eolian relation

Fig. 6.29. A wind-faceted cobble from the Sahara. This ventifact is basalt with small amygdules (white patches) and has a pitted as well as an etched exterior with at least three major faceted surfaces; scale, inches and centimeters.

there, plus inhibiting acquisition of sediment by sheetflow. Such factors must figure in the establishment of any eventual dynamic equilibrium relationship in the arid geomorphic system. See section on arid weathering.

(3) Another factor limiting desert eolian effects relates to a discrepancy between weathered debris, clast magnitude, and eolian competency. Many clasts freed by desert weathering are too large to be moved by the wind. Granites commonly yield granules of quartz and felspar (2–4 mm) and similar sizes of other, less durable substances originate from other coarse-grained igneous rocks (Fig. 6.31). Most are initially

too coarse for wind to move. Fine-grained igneous rocks like basalt tend to yield coarse fragments rather than grains. The same relations tend to hold for coarse-textured or well-indurated sedimentary and metamorphic rocks. Minor clay is slowly generated by chemical activity. Only the argillaceous rocks and friable sandstones readily break down into fragments the wind can easily move. Clearly, deflation of primary weathered debris is severely curtailed by the foregoing relations. Moreover, these statements are confirmed by the size and composition of most dune sands which are dominantly quartz of medium-to-fine sand size, whereas weathered residues in deserts of mechanical origin tend to be poly-mineralic. Abrasion influences in this respect are considered in the fourth point.

(4) The next limiting factor on eolian effects in deserts concerns the scope of wind abrasion and hardness effects. Much of the chemical maturity of desert dune sands can be accounted for by the abrasive elimination of minerals softer than quartz. Much, that is, but probably not all. The size, sorting, and composition of most sand dunes documents one or more episodes of non-eolian (fluvile?) movement, size reduction, shaping, and related selective elimination of fragile minerals, probably augmented by one or more intervals of chemical weathering. Sand studies such as that of Kuenen (1960) and Krinsley and Smalley (1972) document both types of effects. Remember, however, that we are speaking of sand, a particle size usually moved only by *saltation;* sand rarely moves out of deserts and is capable of abrading and shaping only what the grains hit. The abrasion is clearly a matter of differential hardness so that substances softer than quartz (<7) are polished and worn. In general this abrasion shapes both the grains and the surfaces along which the sand normally bounces (plains), and entire rock outcrops may be polished (Fig. 6.32), plus surfaces encountered during bounces (large clasts, scarps).

Larger, wind-shaped clasts have been discussed. Storm winds in deserts like the

Fig. 6.30.   The Hammada (a desert plain in the Algerian Sahara) dominates the even horizon of this scene at the oasis of Beni-Abbès. The foreground slope is the side of the dry watercourse called Oued Saoura. The apparent light color of the clasts on the ground surface is illusory and is due to reflected light from wind-polished surfaces.

Sahara are known to raise sand hundreds of feet into the air and even to blow grains out into the eastern Atlantic Ocean. But bouncing sand grains in average desert wind velocities rarely rise more than eye level— 5–6 ft (1.5–2 m) and therefore only the lower portions of outcrops exposed in high scarps are readily abraded by saltating sand grains (cf. Bagnold, 1941). French experiences with wooden AND STEEL telegraph posts severed about 1 m above the Saharan plains surface establish the "height" locus of saltating sand which must limit the elevation to which eolian abrasion can be most effective. Therefore, intricate sculpting of high-relief desert terrain (regardless of the origin of the relief) is most probably a product of differential weathering and eolian deflation rather than abrasion (cf. Fig. 2.55).

DUNES

It has already been pointed out that the materials composing most sand dunes show signs of processing by agencies of erosion and transport other than the wind and evidences of weathering other than mechanical. Yet the gross shapes of dune formations are widely agreed to be imposed by the wind (Fig. 6.33). Technically, this makes most dunes monogenetic landforms, though their compositional dependence upon other conditions is evident. Migrating sand bodies are known to cross plains (Fig. 6.34), but detailed mapping by the French in Africa and similar work in Arabia indicate that the largest permanently located dune sand concentrations are situated on relative uplands. This is consistent with the known fact that convectively induced winds frequently

Fig. 6.31. Angular fragments of feldspar and quartz (grus) produced by the mechanical breakdown of granite in the St. Francis Mountains, southern Missouri. The accumulations of grus there are in part relict from former more arid times.

move upslope. As much or more sand is presumably fed into such ergs than blows away. This is logical only if particular long-term air current patterns are maintaining the sand concentrations (cf. Folk, 1971).

It is worth reflecting upon these documented atmosphere-landform (duneform) relationships when considering the feasibility of inducing and maintaining particular climates and genetically related landscapes or erosional datums by adjustments in certain major areal or vertical fabrics of the atmosphere (cf. Chapter 4).

Dunes are of several shapes and sizes (cf. Holm, 1953; 1960).* Occasional dunes are clay, but most are sand. Several terminologies exist to describe the main variations

in dune form. One, by Brosset (1939), which has been rather widely employed, lists *elb* (pl. *alab*): symmetrical ridges of sand parallel with the dominant wind (cf. Fig. 6.41); *silk* (pl. *slouk*): longitudinal dunes—also termed *seif*—slightly oblique to wind direction that are smaller, less regular, and less continuous than alab, with Y junctures between segments having apexes pointing downwind (Fig. 6.35); and *akle* (pl. *akeile*): crescentic tranverse dunes = ? *barchane dunes* (Fig. 6.34) or possibly parts of star (*oghurd*) dunes (Fig. 6.37). Folk (1971) attributes the oghurd dunes to essentially fixed rising atmospheric vortexes (cf. Figs. 6.36 and 6.37) in contrast to elongate dunes he attributes to helicoidal circulation of horizontally moving air streams (Fig. 6.38). Wind ripple marks on dunes are oriented in what is often assumed to be right angles to the wind on the basis of the steep leeward and gentle windward slope relationship. Twenhofel (1932) states that ripples average 2–4 inches (5.8–11.6 cm) in wavelength and 1/8–1/4 inch (3–6 mm) in amplitude (see Fig. 6.39).

It should be noted that the morphologic distinctions between elb and silk dunes are often not sharp. Also, as noted by Lustig (1967), the assumed relations between wind direction and dune form fail to explain why some dunes should be transversely aligned to prevailing winds, and others are more or less parallel to the same winds. Terms such as lee and windward have altered relationships for such dunes (cf. Figs. 6.33, 6.37, and 6.38). One problem is that we do not know how long it would take for a series of dune forms to become reoriented after a major change in wind direction. In the case of generally non-migratory dunes through which sand is passed it may require a very long time, and some of the dune forms on record (akle?) may merely reflect one longitudinal dune form evolving into another in response to a change in the prevailing wind direction. As Folk (1971) observes, however, there is lit-

* Giant "sand mountains" on the order of 1,000 ft (300 m) high occur in Arabia's Rub al-Khali and in the southern Namib.

Fig. 6.32.   Wind- and sand-polished Ordovician limestone on the Hammada near Oued Saoura, western Algeria.

tle question that the wind patterns concentrate the sand, which in turn perpetuates the wind patterns *ad infinitum*. Presumably the *ad infinitum* ends when a major shift in long-term wind pattern or direction occurs.

Some dunes such as barchanes and seif types may indicate limited sand supply, whereas others may be presumed to reflect prevailing wind direction during accumulation (cf. Figs. 6.34; 6.36). Actual sand seas (*ergs*) (Fig. 6.40) are relatively isolated masses of arenaceous material which presumably reflect some fortuitous combination of favorable sand source, topography, and long-term regional air currents; the latter is presumably initially influenced by bedrock topography. Maps of ancient ergs and studies of same (cf. Grove and Warren, 1968) would appear to document major adjustments in regional atmospheric circulation (Fig. 6.41). It should be kept in mind, however, that most of the ergs of the earth

appear to be very old, mainly because of the size of the sand bodies involved, grain shape, size, sorting, and mineralogic maturity.* Only polycyclic sedimentation can account for these attributes (cf. Kuenen, 1959, 1960; Krinsley and Smalley, 1972). Compositionally, they almost qualify as uni-

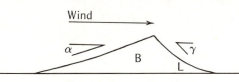

Dune Sand Relations
  Windward slopes (α), av $8°-10°$
  Lee slopes (γ), av $30°$ (range, $28°-40°$)
  B. Main body sands (Libyan Desert), $1/8 - 1$ mm
  L. Lee sands (Libyan Desert), $1/16 - 2$ mm

Fig. 6.33.   Cross section illustrating prevailing wind direction in relation to windward and leeward dune slopes. Star dunes developed in regions of rising air may not exhibit pronounced slope asymmetry (cf. Fig. 6.37). Data from several sources.

* It has been suggested that lowered sea level in the Pleistocene may have exposed sand sources not now existing in continental shelf areas. Potential bedrock sources such as the Nubian Sandstone are obvious on continents.

Fig. 6.34. Aerial view of a group of barchan sand dunes situated 1 mi southeast of Laguna, New Mexico. Such dunes indicate a wind direction from the convex side (right to left in this instance) and are usually taken to suggest a limited sand supply. (John S. Shelton and Robert C. Frampton.)

versally steady-state landforms. Morphologically, several may constitute landscapes in chaos, having recently gone through an eolian "shredder" (in the form of a wind direction shift that created a wave-form degeneration from which they have not yet recovered (cf. Figs. 6.37; 6.40). It should be noted that gypsum sands, such as comprise White Sands National Monument in the United States are compositionally unusual.

DEFLATION

The lifting of any fragment from a land surface by the wind qualifies as *deflation* so we have, of course, already touched upon the phenomenon. There is some indication that locally rising air currents induced by rather uneven topography favor the process. Much unevenness, however, relates to crustal movement and fluvial effects, and on plains some is apparently a by-product of defla-

Fig. 6.35.   Longitudinal (*seif*) dunes in the Simpson Desert, central Australia. Wind direction is toward the observer. (Robert F. Folk.)

Fig. 6.36. Oghurd (star) dunes in the Irrarene dune field of east-central Algeria. For vertical aspect see Fig. 6.37. (B. McCasland.)

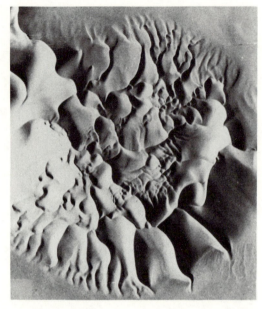

A Grease

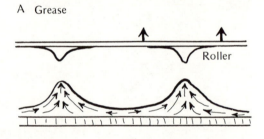

Roller

B Desert

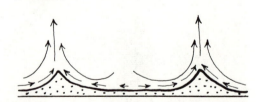

Fig. 6.37. Air photo of oghurd (star) dune from an unknown locality in the Sahara Desert and diagram of inferred air flow lines determined by experimental rolling on grease. Star dunes may be due to locally fixed rising air currents. (From Folk, 1971.)

tion. And naturally, materials of minimum particle size are most readily picked up. Deflation is therefore most effective in those desert regions where suitable air currents gain access to the finest-grained unconsolidated materials.

Sharply defined depressions created by eolian deflation in alluvium or bedrock are termed *blow-outs*. Deflation depressions should thus, in part, coincide with those areas of bedrock composed of such material as friable sandstone, shale, or possibly phyllite in which marginal coherency is diminished by sand grain impacts. Other-

wise deflation may affect poorly consolidated surficial materials which include areas or pockets of sand, silt, or clay. Included here are alluvial and lacustrine de-

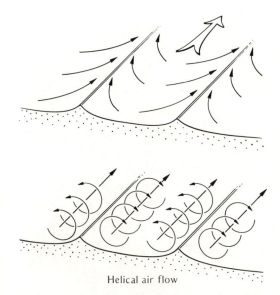

Helical air flow

Fig. 6.38. Diagram of air movements and helical air flow inferred for longitudinal dune areas (after Folk, 1971). Conclusions based on experimental grease flow versus air flow mechanics. Cf. Fig. 6.36 for dune forms created. Analogous types of water movement may induce meandering (see Fig. 7.30).

posits. Alluvium with concentrations of fine-grained material is perhaps most common along dry watercourses in deserts (Fig. 6.11), whereas lacustrine clay and silt is usually encountered in structural basins (*playas*) and locally on plains (*pans*). It has already been noted that desert fluvial processes tend to shift finer-grained sediment into depressions away from scarps and uplands. In the absence of coarse rock fragments which can move into the depression and armor its surface, the deflation process, like dune building, probably is somewhat self-perpetuating, particularly on plains. Thus, a slight surface irregularity, perhaps caused by fluvial action, can induce an eddy current in the air which removes fine-grained material and thereby increases the relief of the surface irregularity and the intensity of the deflating eddy, *ad infinitum* (Fig. 6.42). Turbulence developed downwind from the eddy may cause loss of energy and deposition of the deflated material (cf. Fig. 6.54).

Fig. 6.39. Ripple marks developed on a dune backface and trough (foreground) near the branching ridge of an oghurd (star) dune in the Grand Erg Occidental, southwestern Algeria.

Fig. 6.40. Dunes of the Grand Erg Occidental, western Algeria, make it clear why the expression for sand sea (erg) is used for this, one of the world's largest sand masses.

It should be noted that wind velocities across plains are probably higher than those in irregular terrain and therefore should produce stronger eddies where disturbed. In contrast to this relation, which may generate some pans, winds in structural basins (*bolsons*) are more commonly due to uneven heating and convection. Thus, one often sees *dust devils* (which have clearly picked up clay and silt) spinning across playas. Winds related to periodic storms or seasonal conditions (sirocco winds of North Africa) of course may also deflate dry lake beds. It seems pertinent to observe, however, that playas often consist of desiccated argillites plus evaporite minerals, and both types of material may be very coherent in the absence of saltating grains of sand.

Deflation of desert surfaces naturally lowers them, and resulting depressions stand (or, as it were, recede) in contrast to concentrations of sand accumulated by wind on some uplands. Wind, therefore, has a limited potential to increase desert relief in rather erratic opposition to previously described arid fluvial processes which appear to reduce relief by upland erosion and lowland deposition. Consideration of regions in which there appears to be a record of long-term aridity (in the next section) should help determine which relief tendencies are usually dominant. Suffice it to observe here that slopes and depressions developed by eolian activity may not conform with those formed fluvially, either in magnitude, direction, or position. As a consequence, desert landscapes may exhibit slope configurations that are not consistently oriented toward drainage nets or drainage basins.

Fig. 6.41. Aerial photo mosaic of relict elb dunes under partial cultivation about 100 mi east of Kano, Nigeria, which demonstrates the formerly greater southern extent of Saharan aridity. (Directorate of Overseas Surveys.)

ARID MORPHOGENESIS

We have now considered desert process associations in some detail, both in terms of the physical setting and the overt manifestations of wind, water, gravity, and rock. Some effort has been made to emphasize actual observations and measurements of these phenomena in regions where, as Lustig (1967) laments, there is a serious shortage of information. Some paragraphs have been devoted to the nature of groundcover, precipitation, weathering, and the behavior of surface water, the latter displaying variable runoff attributes in different terrains as well as assuming distinct guises with travel time and distance. Now it seems appropriate to consider the long-term interaction of these and related phenomena (e.g. organisms) as components of the arid geomorphic system insofar as particular landforms are created.

Since many presently arid regions have been subjected to a multiplicity of environmental changes in the recent past (just as have the humid regions already discussed), some of the landforms and deposits encountered therein may safely be assumed to be relicts, left there by non-arid conditions. Other features may reflect transitions between environments and the related disequilibrium conditions that appear to accompany (or follow) climate change. In addition, and pinpointing the thrust of this section, at least some parts of certain desert landscapes should reflect prolonged arid conditions. In the latter areas, climax landforms should provide a basis for determining the unique geomorphic effects of aridity and, thus, criteria for gauging attainment of these effects in varying degrees.

CLIMAX ARID LANDFORMS

Landforms in regions that have been arid for a very long time should be dynamic equilibrium expressions of the arid geomorphic system. Such *climax landforms* should be essentially restricted to *arid climatic nuclei*—areas where relief, elevation, land-water relations and meteorological factors combine "happily," in the sense that they tend to perpetuate dry, sparsely vegetated conditions. Reference to pertinent sections of Chapter 4 discloses that aridity is favored by land-surface locations of three distinct types: (1) those areas generally above 12,000–14,000 ft elevation (3,650–4,260 m), particularly between about 10° and 30° either side of the Equator but below the snow line; (2) those lands near sea level (some above, some below), particularly adjacent to cold ocean surface currents or rather inland and up-wind from potential evaporation sources; and (3) lands some 10° and 30° either side of the Equator at low and intermediate elevations where geostropic winds approaching the equator descend, become warm, and evaporate (Horse Latitude Deserts).

Deserts that include regions with the foregoing attributes include plateau portions of the Colorado, Peru-Bolivian, and Tibetian deserts in type (1); inland and relatively low portions of the Saharan, Arabian, Thar-Iranian, and Australian deserts, particularly to the lee of mountain ranges, which combine properties of types (2) and (3) but mainly type (3); Sonoran, Californian, Atacama-Peruvian, Kalahari-Namib, and Patagonian deserts, particularly in their lower elevations near cold seas, which are mainly type (2) but which exhibit some characters of type (3). For the related discussion that follows, a map should be help-

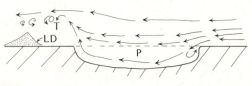

→ Air currents
P. Pan; probable deflation
T. Turbulence due to intersecting air currents
LD. Lunette dune deposits, localized by T and P

Fig. 6.42. Diagram of probable air stream and eddy currents developed over a pan during deflation and formation of a leeward lunette dune.

ful (Fig. 6.43). From the figure, it is evident that only portions of the presently arid areas cited actually have the "ideal" combination of geographic, meteorologic, and topographic attributes that would appear to favor prolonged or dominant aridity. Desert areas excluded from immediate consideration tend to be marginal to more humid lands, or unusually high, or in paths of trans-environmental rivers, or so otherwise situated as to have experienced non-arid conditions in all probability. This is not to say the areas excluded have no arid climax landforms. They may. But conditions appear not to favor it. And the game we play is one of probabilities. So it should be interesting to note the landscape attributes of the arid nuclei selected with respect to our previously delineated arid geomorphic process associations.

In a sense we are going to make process-landform comparisons, but only in a sense. Actually, in working with desert landscapes, we can for the most part only describe a landform and then refer to available process descriptions in an effort to identify the agency or phase that could have formed it.

The technique has its limitations. But in the absence of generally observable fluvial activity or significant quantitative data, we are forced to compromise. Of the deserts cited, most should have been shaped under subaerial environments dominated by arid mechanisms. None are known to have been glaciated. Some could show fluvial effects originating elsewhere, especially via trans-environmental runoff from uplands or outlying humid areas. And, of course, any could be tectonically or volcanically altered.

An assessment of arid climax landforms should also be tempered by the following considerations. Arid fluvial processes clearly differ from humid ones in being bimodal; within deserts, fluvial erosion and deposition appear to be essentially balanced during each precipitation-runoff event, whereas runoff effects in humid lands are distinctly erosional (cf. Chapter 5). Deposition of any permanence occurs elsewhere. A comparatively simple ridge-ravine terrain is generally agreed to be the humid geomorphic result. But the arid fluvial mechanism has the apparent potential of generating at least two types of terrain morphology:

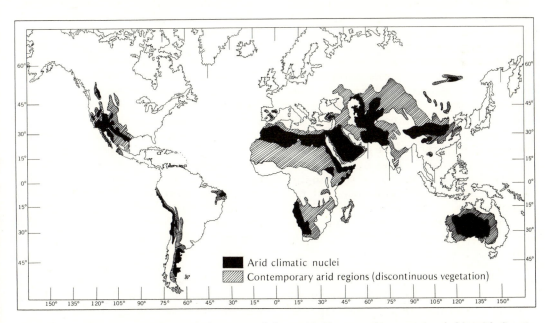

Fig. 6.43. Contemporary arid geomorphic areas of discontinuous vegetation versus probable arid climatic nuclei where dry conditions have persisted under essentially all other environmental patterns.

one essentially erosional; one essentially depositional. The transition between them is an unknown factor geomorphically. Moreover, eolian effects can obviously contribute additional morphologic dimensions, probably at least two. Crustal movements, volcanism, possible trans-environmental rivers, and similar factors obviously can add additional geomorphic ramifications to any desert landscape regardless of a location that is "ideal" otherwise. Thus, desert topography is potentially very complex, even where other subaerial environmental effects are not anticipated, and we should expect considerable landscape variety in an arid climatic nucleus in comparison with the humid climatic nuclei previously discussed.

Desert Plains and Uplands: Theory and Process

In his sweeping inventory of research in desert geomorphology and surface hydrology, Lustig (1967) comments on the principal types of landforms encountered in most of the previously cited desert areas. Heading Lustig's list are the planar landforms he terms "plains and flats," and he observes that these and local "depressions" are the predominant features of deserts in general but are most widely developed in the Kalahari, Sahara, Arabian, and Australian deserts (cf. Fig. 6.43). Less extensive, but nonetheless imposing areas of low relief and slight inclination are cited in the Somali-Chalbi portion of the greater Sahara and in the Namib, Thar, Atacama, and Patagonian deserts. Nor does it seem to this writer to be without significance that the high-elevation deserts of Colorado, Tibet, and Peru are termed "plateaus." What then, is the meaning of planar landforms in arid climatic nuclei. What kind of plains are they? And what does the obvious absence of the feature to which the plains attest (relief) signify? Let us consider these questions first in terms of theory, second in relation to observed arid process associations,

and third in relation to actual arid climatic nucleus terrain morphology.

Desert erosion theory goes back through King's (1953) statement that arid and humid climatic effects do not differ in kind, to Davis's prior assertion (1932) that they do differ. Davis's initial statement on desert landforms is his most comprehensive (1905). Therein, he states: "There is no novelty in the idea that a mountainous region of interior drainage may be reduced to a plain by the double process of wearing down the ranges and filling up the basins, and that the plain thus formed, consisting partly of worn-down rock and partly of built-up waste, will not stand in any definite relation to the general base-level of the ocean surface. . . ." With extensive reference to similar ideas voiced by Passarge (1904a, 1904b), Davis concludes that vast plains would compise the end morphology of his arid erosion cycle and that on such planar surfaces the occasional shifting of rock debris by water would readily counteract any eolian tendency to create erosional relief. In terms of terminal desert landscape morphology, Penck (1924) and King (1953) depart little from Davis, both describing an ultimate planar landscape possibly surmounted by isolated erosion remnants.

Most deserts have uplands, and, as Davis (1905) was quick to note, these uplands vary considerably in their form, size, height, and relative degree of isolation from one another. To our knowledge, no one has ever seriously suggested entirely divorcing the genesis of desert uplands from the erosion of the plains that develop coincidentally. The overwhelming weight of opinion and evidence (cf. King, 1953; Quinn, 1957) relates the creation of isolated uplands in deserts to escarpment retreat, though as we shall see in Chapter 7, there are other possible mechanisms. The scarp retreat genesis presupposes the initial existence of escarpments or other forms of local relief in an environment where erosional relief apparently tends to decrease with time. In an arid climatic nucleus such relief is unlikely to relate to other climatic effects apart from trans-environmental rivers, but one can expect to encounter scarps originating along the banks of such streams, at sea coasts, or through crustal movements or volcanism. In any case, the area of uplands is generally presumed to become smaller as that of lowlands increases. In this context, a plains area with remnant uplands in a desert would be very old; only a plain formed there, and now without uplands, would be older. Scarps, subjacent to such a plain, that have not yet receded would be very young. Let us now consider these possibilities in arid climatic nuclei and related portions of some of the same deserts.

It is clear that earlier theories of arid erosion indicate that plains should form as a major topographic consequence. Added to this we have the actual reported domination of such landforms in arid climatic nuclei. One cannot help but be tempted by such a boundless vista of circular reasoning relating each of these phenomena to the other, alternately. Instead, let us consider the observed arid process associations. Foremost among these is the widely acknowledged—and observed—fact that a depositional phase is an integral part of each arid fluvial event. This should make depositional surfaces (which may be of surpassing evenness) virtually inevitable elements of every desert landscape—elements with an extent and perfection that must be generally time dependent. Added to this must be the previously mentioned spreading of runoff on flattened surfaces with a consequent dissipation of vertically directed erosive energy. Also we must consider accounts of the notably random character of desert precipitation which must surely be reinforced by the equally random incision and deposition recorded within areas where sheetflow occurs; both of the latter clearly tend to maintain plains as plains. In fact, considering the total arid fluvial event, there is an interval (E_c) where linear erosion can incise bedrock. There is the area of deposition (D_c) that establishes a lower limit on this incision. Finally, upslope from E_c, there is the area delineated by Horton's distance X_c that

falls within the radius of any desert precipitation locus and within which fluvial erosion should be unusual. Within such a precipitation locus, the downslope margin of an area with a radius X_c should define the minimal area to which fluvially assisted scarp retreat should be effective (Fig. 6.44). Inselbergs with such dimensions should be durable features, since water on them would only erode solutionally, if at all, and their further destruction would mainly relate to weathering and eolian effects. This brings us to a consideration of several regions of probable prolonged aridity.

Desert Plains and Uplands: Actuality

What are the actual terrain characteristics within arid climatic nuclei? Let us begin with a North American example. Most attributes of the North American arid climatic nucleus suggest very prolonged aridity. Much of the terrain is low, some below sea level. All of it is downwind from a cold sea, and most falls in the lee of a major physiographic barrier (Sierra Nevada Mountains) and upwind from the only other major evaporating water body (Gulf of Mexico). But there are problems, three in particular. First, local relief is high, frequently in excess of 6,000 ft (2,000 m) and so are maximum elevations [12,000–14,000 ft (3,500–4,000 m)]. Not only are the uplands

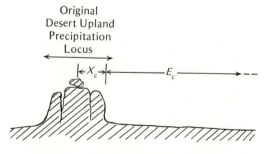

Fig. 6.44. Cross section depicting the relationship (theoretical) between the X_c zone of nil erosion, orographically localized upland precipitation, and the development of an inselberg, all in a desert environment characterized by local fluvial events rather than systematic regional drainage.

extensive and rather closely spaced, their tops catch considerable moisture and at various times past appear to have been even more moist, even to the point of generating notable lakes in some basins. Second, there is a considerable and still-growing body of evidence that attests to the creation of much of the Basin and Range province relief by block-faulting and local volcanism during the late Tertiary (cf. Chapter 3). Movement along some fractures continues. Therefore, over much of this region we are dealing with an arid geomorphic system modifying and being modified by a tectonic suite of landforms. Third, the region is traversed by several trans-environmental rivers, the most notable being the Colorado and its tributaries, almost all of which are very deeply incised in response to prolonged runoff from direct source area precipitation and snowfield meltwater (see Fig. 1.7).

From the previous paragraph, it seems evident that some difficulty attends any effort to discuss only arid geomorphic effects in the North American arid climatic nucleus. However, if we exclude those regions that display the rather obvious effects of rivers, lakes, structural movements, and volcanism, a few areas of considerable interest remain, two in particular. In the Sonoran Desert of northern Mexico and Arizona and the Mojave Desert of California, in areas of rather homogeneous metamorphic and granitic rocks, there is a tectonically stable area of extensive plains, locally rising to broad low domes (Fig. 6.45); and some of these plains are surmounted by what appear to be erosional remnants (i.e. *inselbergs*) (cf. Figs. 1.12 and 6.46). These inselbergs plus the vastly more extensive plains comprise essentially the entire terrain; and, as noted by McGee (1897), some portions of the plains are bare bedrock, whereas other areas are alluviated. In such surroundings, it is rather easy to visualize complete elimination of all upland remnants and development of a completely planar landscape.

A second region that falls within the North American climatic nucleus and seems to have escaped extraneous influences apart from arid conditions for some time is that of Monument Valley in Utah. The plains of this region are relatively less extensive, and bedrock stratal texture provided by essentially flat-lying sedimentary rocks is evident on many of the lowland surfaces as well on the steep sides of the many pinnacles, buttes, and mesas (Figs. 6.47 and 6.48). Some of the isolated uplands are fringed by extensive pediments with notable upward concavity, whereas

Fig. 6.45. Cima Dome, the produce of extensive planation in the Mojave Desert, southern California, can be considered a near-endpoint in the processes of desert morphogenesis. Note the isolated inselbergs.

Fig. 6.46. Wase Rock, an inselberg in Nigeria displaying the common sugar-loaf configuration of such erosion remnants. Isolated huts in surrounding fields and along roads show scale. (Aerofilms Ltd.)

others are nearly flat. As in the Sonoran region, the lowlands are only locally veneered by alluvium for reasons that, in neither case, are entirely clear. What is clear, is that planation is much less nearly complete in the Monument Valley area than it is in the Sonoran region.

Neither of the North American areas discussed shows any indication that uplands catch appreciably more precipitation than surrounding lowlands, at least not enough more to give rise to stream channels, and

there is no sign of recent faulting or volcanism. In these two regions, we may therefore have some topographic expression of very prolonged aridity, a notion that can be tested by comparing their terrains with those in other arid climatic nuclei. With certain exceptions, much of the remainder of the North American arid nucleus displays the topography one would expect in recently formed fault-block mountains with desert lowlands and sporadically moist crests. The attributes, in addition to high relief, include fault scarps, incompletely dissected and hardly pedimented bedrock uplands, and incompletely alluviated structural basins containing readily identifiable alluvial cones, fans, and bajadas, plus playas (cf. Figs. 6.23; 6.24; 6.25). If one assumes that desert plains form by progressive reduction of uplands, modification of most of the Basin and Range region by aridity is far from complete.

In extending our search for climax arid landform analogs, there are many regions to which we could refer (Fig. 6.43), but the writer is inclined to seek out those to which his own first-hand experience extends. Let us consider the Saharan climatic nucleus where there are areas of flat-lying strata, folded strata, and a bit of volcanism. None of the major rock deformation appears to be more recent than Mesozoic so that, apart from volcanoes, the relief is erosional in origin. Compared with the previous examples, we are dealing with a complex landscape, but, as the reader was warned, this can happen in deserts. Plains of exceptionally low inclination are extensive (Fig. 6.49), but they are broadly undulatory, rising (or falling) on the order of 400 ft (120 m) in a distance of some 40 mi (70 km). This amounts to an average slope of about ⅑ of a degree. Most are littered by ventifacted gibber gravels in regions of blowing

Fig. 6.47. Vista of Monument Valley, northeastern Arizona, showing isolated erosion remnants (inselbergs) variously designated needles, buttes, and mesas. Subjacent pediments with little concavity are developed where shale has been removed beneath a more resistant sandstone caprock. (W. N. Gilliland.)

Fig. 6.48. Inselbergs developed by planation in Monument Valley, Arizona. Because both shale and capping sandstone disentegrate granularly there is little coarse talus (cf. Fig. 6.47).

sand which characterize the most arid interior. They also display a crude anastomatic swale pattern extending generally parallel to slope on the broad, regional relief undulations (Fig. 6.50). The plain surfaces often parallel nearly flat-lying sediments in areas of northern Algeria and Tunisia, and in that area they are here and there surmounted by small, isolated, butte-like inselbergs resembling those in Monument Valley but much more scattered. Larger, high-standing structural blocks are rare but do exist near the Atlas Mountain region and along the Mediterranean coast.

The interior plains (*Hammada*) are locally duricrusted and are garnished by occasionally isolated sand dunes (cf. Fig.

6.49) and regionally buried by sand seas (*ergs*) (Fig. 6.39). The Hammada also is seen to truncate areas of folded Paleozoic rock in southwestern Algeria. The same strata also rise as much as 800 ft (250 m) above the adjacent plains in the vicinity of Beni-Abbes to form erosional remnants (Fig. 6.51).* Notable breaks in relief apart from volcanoes farther east include (1) the banks of shallowly incised, dry washes (*oueds*) which in Algeria trend generally north-south and appear to head in the Atlas mountains in some cases (e.g. Oued Saoura); (2) rather steep-sided but shallow, quasi-lacustrine depressions (*pans*); and (3) the aforementioned erosional remnants that form steep, cuesta-like, curvi-

* For detailed cartographic coverage of the region the reader is referred to published maps (scale 1:200,000) of the 19th International Geologic Congress, particularly the Beni-Abbés-Ougarta special printing.

Fig. 6.49. The Hammada, a lag-gravel-strewn desert plain in the Algerian Sahara, here surmounted by a comparatively isolated oghurd dune. Essentially all clasts on this surface are ventifacts.

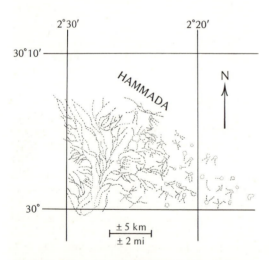

Fig. 6.50. Map view of the anastomotic swale pattern and localized closed depressions on the Hammada, southeastern Algeria, near the edge of the Grand Erg Occidental. Taken from the Beni-Abbés special sheet of the 1:200,000 scale map of the Sahara (XIXth International Geological Congress.)

linear uplands from Cambro-Ordovician quartzite and superjacent sedimentary rocks.

The oueds of Algeria are incised into the Hammada as much as 75–100 ft (25–30 m) and consequently are rather younger than the elevated plain although clearly pre-Recent. Pans are also locally inset into the Hammada and therefore also younger than that surface, but they could be changing at present. The oueds have at least locally widened by scarp retreat and incorporate at least one terrace level in addition to the main oued floor. The latter surface is locally alluviated or sand strewn, and oases along the oueds prove that there is at least some sub-alluvial seepage of water (cf. 6.1). The retreat of some of the steep scarps that border oueds has exposed truncated rock strata which may or may not have an alluvial veneer. Some of the more argillaceous rock formations, particularly of Silurian and Devonian ages, can generate little coarse debris through weathering, and where these strata are inclined 10° to 15° they have been differentially eroded by wind and sheetwash to produce crudely planar areas with low-relief wash-board surfaces (Fig. 6.52). In a few of these same formations the most resistant clasts produced are ferruginously cemented fossils, and in places essentially all of the gibber gravels consist of these. In general, the most durable gibber materials in the region apart from lower Paleozoic quartzite clasts appear to be fragments of re-worked duricrust.

The few bedrock uplands in southeast Algeria consist mainly of rocks like quartzite, which is of course a very durable rock type, but identical lithologies have also been beveled by erosion elsewhere in the same region. Most of the erosion remnants in this area lack the "sugar-loaf" form which presumably characterizes inselbergs, and the majority are elongate and are mantled on crest and flanks by coarse, angular talus

which locally constitutes a *felsenmeer* and which is evidently resistant to both eolian erosion and sheet wash (cf. Fig. 6.51). Their erosional reduction must be indeed slow. Juncture of the uplands with the Hammada is abrupt, and there is no present indication, vegetally or in the form of pronounced radial drainage centered thereon, that the uplands receive more moisture than the plains. Small, shallow gullies are associated with the flanks of some of the uplands, and past conditions may have differed.

In central portions of the Saharan climatic nucleus, volcanism has entered the picture and accounts for the major uplands. This is true of the Tibisti region, discussed by Grove (1960), where there is relief of several thousand feet generated volcanically. That author notes abundant evidence for formerly increased precipitation in the form of low-level watercourses originating at higher elevations. Apart from the Nile, the central Sahara is notably affected by

Fig. 6.51. Djebel Dir el Ougsa, linear erosion remnants of cuesta form developed from inclined Cambro-Ordovician quartzites southwest of Beni-Abbès, the western Algeria Sahara. View is to the northwest.

Fig. 6.52. A washboard surface of inclined Devonian rocks exposed in the floor of the Oued Saoura surmounted by talus from exposed benches related to the western Algerian Hammada.

trans-environmental rivers mainly in the Chad (cf. Garner, 1967). Both Grove (1960) and Cotton (1968) note ancient channelways therein, well within the arid nucleus, which must signify former fluvial incision by water originating on either isolated uplands or beyond the desert limits.

We do not pretend to encompass all the details of desert morphology of the Sahara or any other similarly arid region here. However, the basic climatic landforms of the Saharan arid nucleus are identifiable. (1) Plains are vast, perhaps more so than in the Sonora Desert, broadly undulatory, surmounted by only very scattered erosional remnants and separated from more localized lower levels by low, steep scarps. Bedrock structures are widely beveled. (2) Scarps on the few isolated uplands reflect bedrock structure and also some previous relief-making mechanism subsequent to

which there has been extensive hillslope recession. Low-level scarps relate to fluvial incision originating generally beyond the arid nucleus at some prior time and presumably from some more moist environment. Duricrust on the Hammada often forms the crests of subjacent escarpments. (3) Dunes, mainly occupying upland plains, far exceed in bulk and variety of form anything encountered elsewhere except possibly in Arabia and constitute major aspects of the landscape. Apart from the extent of dune development, essentially all landforms listed have counterparts in the arid climatic nuclei previously discussed. Examination of a third arid climatic nucleus should tend to confirm or deny the consistency of occurrence of such features under prolonged arid conditions.

The arid climatic nucleus of South Africa (cf. Fig. 6.43) displays far more plant growth than either of the regions previously considered, but only along the northern portions of the Kalahari can continuous vegetal "cover" be said to exist. It is a somewhat cooler desert than the Sahara, and there are some local indications of former moisture levels both lower and higher than at present. However, to the extent that landforms previously cited from other arid nuclei are reflective of prolonged aridity, the Kalahari has all the apparent qualifications. In his synopsis of the Kalahari, Lustig (1967) recognizes a separate coastal desert under the name Namib, in which he makes note of several inselbergs. He also makes the significant comment: "The pans of the Kalahari constitute one of the relatively few departures from featureless plains of the region. . . ." One of the most recent studies of the Kalahari by Grove (1971) is most useful and informative. In it he echos the previous comment on Kalahari terrain by noting that relief in Botswana ". . . is extraordinarily low. . . ." The same writer notes that there are drainage lines that relate to former accentuated precipitation in surrounding upland "plateaus" and "highlands" and probable former trans-environmental drainage from the Zambezi region that left anastomotic channel systems in the fashion previously described by the writer (1967).

Grove speaks of the "old rock floor" developed on Precambrian, lower Paleozoic, and early Tertiary rocks that include igneous, metamorphic, and sedimentary types, some of the latter being continental deposits of the Karroo System. This rock floor which constitutes the main bedrock substrate of the Kalahari plains is only locally exposed through widespread sands, and the most extensive exposures of bedrock relate to isolated hills along a NE-SW trending fault scarp associated with current seismic instability. Grove (1969) also notes that surface alluvium on the plains has thick *calcretes* which tend to stabilize planar topography by induration of water terraces and scarps.

Kalahari dune systems tend to be of the parallel variety in relation to prevailing winds of the present (cf. Fig. 6.36). Grove records an aleb dune system over a 6,000 sq mi (16,000 sq km) area in which the dunes 50 ft (15 m) high are equivalent to a uniform sand sheet 25 ft (8 m) thick. In addition, he describes "sand ridges" 15 to 25 ft (5 to 8 m) high in southwest Botswana equivalent in sand volume to a former uniform sheet 1 to 2 m thick. Grove establishes prevailing wind direction from the northwest during dune formation by orientation of lee dunes adjacent to pans (Fig. 6.53) in a NW-SE direction. He notes the aleb dunes influence (are parallel to) drainage channels in the northwest, whereas the "sand ridges" appear to be younger than drainage lines and locally penetrate valley floors.

The Kalahari, like the Sahara, also has pans. In his discussion of pans and playas, Lustig (1967) notes that the North American playas and similar features elsewhere are products of Basin-and-Range–type topography and attendant climatic variability. He further observes that the features termed "desert pans" (or just *pans*) are flat-floored, shallow depressions usually surrounded by plains of low relief, generally containing only a few feet of argillites and little or no evaporitic material. Their origin is uncertain. At times they contain minor amounts of water. They are also acted upon by wind. And they are clearly a feature common to arid climatic nuclei.* Lustig adds that pan distribution conforms well with ancient drainage paths, now locally sediment choked and subject to local ponding of water following equally local rainfall. In his Kalahari study (1969), Grove failed to encounter the coincidence of pans and drainage lines, recording instead belts of pans along a height of land, each a kilometer or more in diameter, up to 15 m deep, and most bordered on the SW, S, and SE sides by one or two *lunette dunes* which are concave to the north (Fig. 6.54). The form, origin, and location of pans therefore remains generally unaccounted for, in spite

---

* But pans are not confined to such nuclei.

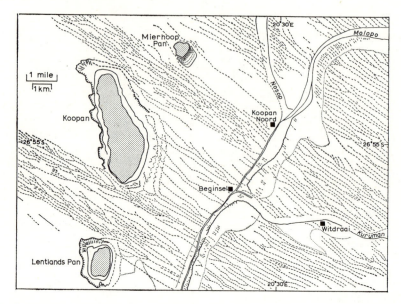

Fig. 6.53.   Sand ridges and pans near the confluence of the Nosop and Molopo rivers, the Kalahari Desert. (From Grove, 1971.)

of evident eolian involvement, and further study is clearly warranted.

Exclusive of the Namib area, the Kalahari region is perhaps the most nearly planar region discussed and one of the most featureless plains on record for its size. Inselbergs appear to be somewhat more numerous in the Namib region and Lustig (1967) notes the need for precise petrographic data in the assessment of these features. In this regard he cites the occurrence of four isolated inselbergs in the central Namib desert within an area of 5 sq mi, all "casually" granitic in composition and each about 100 ft in height. Of them Lustig says

One consists entirely of large blocks of rock that are several to perhaps 10 feet in diameter. A second is similar in appearance, but upon inspection the blocks may be seen to surround an unbroken bedrock core. The third is relatively smooth and exhibits only scattered patches of coarse sand and fine gravel upon its surface. The last resembles nothing so closely as a glaciated bedrock knob; not a trace of weathered debris can be found. Moreover, this last inselberg contains an extraordinary dendritic network of drainage channels with smooth, semicircular cross sections,
that are incised into bedrock to a depth of 2 to 3 feet and contain small plunge pools along their course. And this in a region where the mean annual precipitation is presently less than 2 inches.

The same author concludes, by pointing out, ". . . many topographic surveys in western United States have indicated that the slopes surrounding inselbergs range from about 2° to 7°," but in other arid areas the surfaces surrounding such features may be "perfectly horizontal," and in the Tibisti region (previously mentioned) he cites pinnacle-like examples surrounded by depressions (see section on pediments and planation).

CLIMAX ARID LANDFORMS— AN ASSESSMENT

Though we have only considered topography associated with three of the world's arid climatic nuclei, a pattern emerges which is largely echoed by descriptive literature relating to arid nuclei in Arabia, Australia, and elsewhere (cf. Chapman, 1971). With

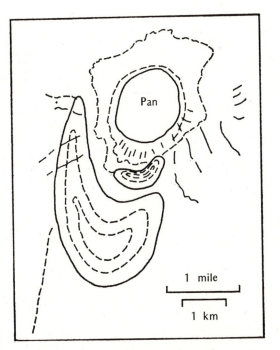

Fig. 6.54. A pan in the Kalahari Desert of South Africa near Tsane showing lunette dunes which are probably relict features formed by formerly northerly winds. Note morphologic similarities with Carolina Bays (Fig. 7.56). (From Grove, 1971.)

minor deviations we seem to be dealing with plains, a few inselbergs and pans, and sand in amounts that vary considerably. None of these features are restricted to the arid nuclei or even to the deserts which generally surround them, but the plains are more extensive and uninterrupted by breaks in slope in the nuclei than anywhere else. As a point of fact, the vast majority of slope breaks in the arid nuclei seem to relate to dry watercourses in which runoff evidently relates to former moist conditions, either on uplands or beyond the desert environmental limits. On this basis, it is therefore taken to be genetically significant that the arid fluvial event includes depositional phases (D_c) which can provide a planation datum by limiting incision. It is also probably genetically significant that process interaction of wind and water appears to tend to perpetuate planar landforms once the latter develop. Even if pans are eolian in origin, and

taking dunes into account, the dominance of fluvatile leveling over wind effects in deserts is amply documented. Consequently, plains are here taken to constitute one kind of climax arid landform, approximating, as they appear to, a dynamic equilibrium adjustment to the arid environment.

The extent and perfection of plains in deserts is presumably a function of time, and even in arid nuclei it is evident that they rarely escape modification by outside agencies, whether water, wind, or tectonism. If our assessment of plains is correct, in the extreme case (resembling the Kalahari) water simply spreads out, dissipates its energy, soaks in, evaporates, and, in evaporating, contributes to a duricrust which further indurates the plain. If this is correct, the inselbergs in deserts simply reflect the incompleteness of the planation process much as has been postulated previously by several authors—or it may not be so simple (cf. Chapter 7). If earlier assessments of eolian activity are correct, generation of sand suitable for dune accumulation is a geologically complex process and requires considerable time. Also requiring considerable time is the eolian concentration of the sand into large, essentially fixed bodies, probably reflecting certain air movement patterns. The apparent immobility of the free-blowing sand seas in North Africa and Arabia is suggestive of an adjustment of their related environments, and these ergs may also be climax arid landforms.

The occurrence of pans in arid climatic nuclei is of uncertain significance, mainly because we are so unsure of their precise origins. This writer tends to believe that they are essentially eolian depressions developed over the most readily deflatable lithologies, whether alluvial or bedrock. But even if this is true, pans also probably reflect competitive interaction between two agencies (wind and water) which are semi-independent of one another and hence never quite in balance. Water certainly tends to fill pan depressions with coarse debris as Davis (1904) suggests, but some of the more readily deflated rock types yield little debris suitable for gibber gravel. And on a nearly flat plain, wash probably does not move gravel very effectively. It is a problem. But then, so are plains and inselbergs and sand seas. So are anastomotic swales on hammadas, depressions surrounding inselbergs, and apparent pediments that lack upward concavity. The broad features of desert morphogenesis are emerging, but many of the details remain obscure. Even our use of terms such as "plain" and "flat" are not genetically diagnostic. On close inspection, most of the extensive low-relief areas are found to be compound landforms which incorporate areas of appreciable alluvium (*aggradation plain*) and other expanses where beveling of rock structures is a dominant relationship (*pediplain*). Much work remains to be done.

Our remarks in this chapter have been directed primarily at the conditions that prevail in arid environments, the associations of related processes, and the landforms that exist where these conditions and processes have apparently functioned for a very long time. We noted that influences of various types of bedrock lithology and structure were expressed primarily in upland erosional remnants, since the plains show little detailed relief related to substrate. But the issues of antecedent terrain were restricted to a brief consideration of block faulting and volcanism plus the modification of an upland initially incised by running water and subsequently planated following the creation of a depositional-erosional datum. In his pioneer work on the subject, Davis (1905) focused primarily on erosional and depositional modification of Basin-and-Range type terrain produced by high-angle faulting. His original synthesis of long-term landscape evolution of this special tectonic setting under arid conditions requires little modification to be in accord with conclusions just presented (Fig. 6.24). He proposed also that, ". . . the scanty rainfall of an arid region will be decreased as its initial highlands, which originally acted as rain provokers, are worn down . . . ," a position that more recent findings would tend to support fully.

SUMMARY

One could hardly justify a chapter of this length and detail merely to document arid geomorphic changes which occur in a special type of tectonic topography of limited geographic occurrence. Such documentation was not, however, our sole purpose. Rather, we wished to establish the process-landform tendencies under the arid geomorphic system. This was necessary since those tendencies appear to have been imposed on a variety of antecedent terrains developed under many different kinds of conditions which will be considered in following chapters.

One arid geomorphic tendency that is especially evident, and for which there is an essentially complete affirmative consensus is, THE GRADUAL REDUCTION OF RELIEF THROUGH INTERACTION OF RUNNING WATER, GRAVITY, AND THE ATMOSPHERE WITHIN THE ARID ENVIRONMENT. There is no reason to believe that the origin of the relief makes any great difference in the eventual result; material removed from elevated areas accumulates in lower areas in a time-dependent relationship which we should be able to gauge by study of the deposits and through quantification of scarp retreat rates.

A second apparent arid geomorphic tendency which is significant in some regions but less widely agreed upon involves the extent of increased relief by wind action. SAND ACCUMULATION IN DUNES CERTAINLY INCREASES LOCAL RELIEF OF A SPECIAL TYPE, and, to the extent that dune concentrations center on uplands, regional relief may also be affected. It is rather less certain that major depressions are generated by eolian activity or that even minor features (pans) which may be related to wind effects express any long-term trend.

A third tendency expressed in desert landscape genesis relates to the absence of a water cycle and resulting FIXATION OF (MINERAL MATTER) AS EVAPORATIVE ENCRUSTATIONS ON ROCK SURFACES. Whether derived by leaching of subjacent bedrock or through atmospheric and hydrologic importation, affected landforms achieve increased coherency by the form of induration which in magnitude appears to be time dependent as long as the surfaces in question remain pervious.

A fourth tendency expressed in deserts by the essentially closed fluvial sedimentary systems is THE CREATION OF DEPOSITIONAL LANDFORMS OF SEVERAL KINDS IN WHICH THE COMPOSITION REFLECTS LOCALLY DERIVED SOURCE LITHOLOGIES SITUATED NOT FAR UP SLOPE and a general but imperfect tendency for grain size in the deposits to diminish with distance from adjacent scarps and uplands (cf. Bosworth, 1922). A related grain-size trend (involving decreased particle dimensions from the base of a major deposit toward its top as distances toward source uplands increase with scarp retreat) may be obscured by eolian deflation of fine fractions and formation of gibber surfaces. Depositional patterns related to trans-environmental drainage lines that bring in detritus from a distance may be superimposed on local desert patterns, a matter to be considered in some detail in Chapter 7.

A fifth tendency for desert regions noted by Giguoux (1955) and Garner (1959) is that they tend to loose sub-sand–size rock particles through wind action with A CONSEQUENT RELATIVE INCREASE OF RESIDUAL DETRITAL CLAST SIZE. Viewed in conjunction with planar surfaces, duricrust, and dune formation, an over-all "ideal" dynamic equilibrium relation as previously defined can readily be visualized where a "least work and equal area energy expenditure" relation is encountered on a desert land in that (1) sand particles that are fine enough to wash away, in dunes readily absorb precipitation but are too large to blow away and (2) surfaces on which water might flow without soaking in are so even as to cause spreading and are often littered with clasts generally too large to shift by shallow laminar water flow. Some of the plains in arid climatic nuclei comprise probable steady-state features of great potential age.

The landscape trends just listed for arid geomorphic systems and those cited in Chapter 5 for humid geomorphic systems will be discussed in Chapter 7 insofar as they alternate in time and space. Once again we will grapple with environmental parameters—the possibly more subtle conditions that relate to transitions between aridity and humidity, and back. Fluvial activity, mass wasting, work of the wind, and weathering will once again receive a nod—in a rather more dynamic context. And instead of concern with climatic nuclei where there should be minimal change, we will turn to the far more vast regions situated geographically between them where we have every reason to expect changes and the effects of these, sometimes many changes and many effects. The possibilities are indeed intriguing. And unlike situations pertaining to deserts, data of most kinds are devastatingly abundant. It will be our task to try to understand them.

REFERENCES

Bagnold, R. A. (1941) *The Physics of Blown Sand and Desert Dunes*, Wm. Morrow and Co., New York, 265 pp.

Beaumont, P. and Oberlander, T. M. (1971) "Observations on Stream Discharge and Competence at Mosaic Canyon, Death Val-ley, California," *Geol. Soc. Amer. Bull.*, v. 82, pp. 1695–1698.

Blackwelder, Eliot (1925) "Exfoliation as a Phase of Rock Weathering," *Geol. Jour.*, v. 33, pp. 793–806.

Bosworth, T. O. (1922) *Geology of the Tertiary*

and Quaternary Periods in the Northwest Part of Peru, Macmillan and Co., London, 434 p.

Branner, J. C. (1896) "Decomposition of Rocks in Brazil," *Geol. Soc. Amer. Bull.,* v. 7, pp. 255–314.

Bryan, Kirk (1923) "Erosion and Sedimentation in the Papago Country, Arizona," *U. S. Geol. Surv. Bull.,* v. 730, pp. 19–90.

Bryan, Kirk (1935) "The Formation of Pediments," *15th Int. Geol. Cong. C. R.,* v. 2, pp. 765–775.

Brosset, D. (1939) "Essai sur les Ergs du Sahara Occidental," *Bull. de l'IFAN,* v. 1, pp. 657–690.

Bull, W. B. (1964) "Geomorphology of Segmented Alluvial Fans in Western Fresno County, California," *U. S. Geol. Surv. Prof. Pap.* 352-E, pp. 89–129.

Bull, W. B. (1968) "Alluvial Fan, Cone" [in] *Encyclopedia of Geomorphology,* R. W. Fairbridge, ed., Reinhold Book Corp., N. Y., pp. 7–10.

Butzer, K. (1968) *Desert and River in Nubia,* Univ. of Wisconsin Press, Madison, 562 pp.

Cherkauer, D. S. (1972) "Longitudinal Profiles of Ephemeral Streams in Southeastern Arizona," *Geol. Soc. Amer. Bull.,* v. 83, pp. 353–366.

Cotton, C. A. (1960) "The Origin and History of Central Andean Relief: Divergent Views," *Geogr. Jour.,* v. 126, pp. 476–478.

Cotton, C. A. (1968) "Relict Landforms" [in] *Encyclopedia of Geomorphology,* R. W. Fairbridge, ed., Reinhold Book Corp., New York, pp. 936–940.

Crandell, D. R. (1968) "Mudflow" [in] *Encyclopedia of Geomorphology,* R. W. Fairbridge, ed., Reinhold Book Corp., New York, pp. 763–764.

Davis, W. M. (1903) "The Mountain Ranges of the Great Basin, *Mus. Comp. Zoo. Bull.,* v. XLII, pp. 129–171.

Davis, W. M. (1905) "The Geographical Cycle in an Arid Climate," *Jour. Geol.,* v. 13, pp. 381–407.

Denny, C. S. (1967) "Fans and Pediments," *Amer. Jour. Sci.,* v. 265, pp. 81–105.

Dury, G. H. (1966) "Duricrusted Residuals on the Barrier and Cobar Pediplains of New South Wales," *Geol. Soc. Aust. Jour.,* v. 13, pp. 299–307.

Dury, G. H. (1970) "Morphometry of Gibber Gravel at Mt. Sturt, New South Wales," *Geol. Soc. Aust. Jour.,* v. 16, pp. 655–666.

Fahnestock, R. K. (1961) "Competence of a Glacial Stream," *U. S. Geol. Surv. Prof. Pap.* 424-B, pp. 211–213.

Fairbridge, R. W., ed. (1968a) "Desert Varnish (Patina)" [in] *Encyclopedia of Geomorphology,* Reinhold Book Corp., New York, pp. 279–280.

Fairbridge, R. W., ed. (1968b) "Induration" [in] *Encyclopedia of Geomorphology,* Reinhold Book Corp., New York, pp. 552–556.

Folk, R. L. (1971) "Genesis of Longitudinal and Oghurd Dunes Elucidated by Rolling upon Grease," *Geol. Soc. Amer. Bull.,* v. 82, pp. 3461–3468.

Forbes, J. D. (1849) "Observations on the Temperature of the Ground," *Roy. Soc. Edinb. Trans.,* v. 14, pp. 189–236.

Furon, R. (1960) *Geology of Africa,* Payot, Paris.

Garner, H. F. (1959) "Stratigraphic-Sedimentary Significance of Contemporary Climate and Relief in Four Regions of the Andes Mountains," *Geol. Soc. Amer. Bull.,* v. 70, pp. 1327–1368.

Garner, H. F. (1965) "Base-Level Control of Erosion Surfaces," *Ark. Acad. Sci. Proc.,* v. 19, pp. 98–104.

Garner, H. F. (1967) "Rivers in the Making," *Scientific American,* v. 216, pp. 84–94.

Gentilli, J. (1968) "Exfoliation" [in] *Encyclopedia of Geomorphology,* R. W. Fairbridge, ed., Reinhold Book Corp., New York, pp. 336–339.

Gignoux, Maurice (1955) *Stratigraphic Geology,* W. H. Freeman, San Francisco, esp. pp. 1–5.

Goudie, Andrew (1973) *Duricrusts,* Clarendon Press, Oxford, 174 pp.

Grove, A. T. (1960) "Geomorphology of the Tibesti Region with Special Reference to Western Tibesti," *Geogr. Jour.,* v. 126, pp. 18–31.

Grove, A. T. (1971) "Landforms and Climate Change in the Kalahari and Ngamiland," *Geogr. Jour.,* v. 135, pp. 191–212.

Grove, A. T. and Warren, A. (1968) "Quaternary Landforms and Climate on the South Side of the Sahara," *Geogr. Jour.* v. 134, pp. 191–208.

Hadley, R. F. (1967) "Pediments and Pediment-Forming Processes," *Geol. Ed. Jour.,* v. 15, pp. 83–89.

Holm, D. A. (1953) "Dome-Shaped Dunes of Central Neja, Saudi Arabia," *19th Int. Geol. Cong., C. R.,* v. 7, pp. 107–112.

Holm, D. A. (1960) "Desert Geomorphology in

the Arabian Peninsula," *Science*, v. 132, pp. 1369–1397.

Holm, D. A. (1968) "Sand Dunes" [in] *Encyclopedia of Geomorphology*, R. W. Fairbridge, ed., Reinhold Book Corp., New York, pp. 973–979.

Hooke, R. LeB. (1965) *Alluvial Fans*, Calif. Inst. Tech. Ph.D. Thesis (on file), Pasadena, Calif., 192 pp.

Horton, R. E. (1945) "Erosional Development of Streams and Their Drainage Basins," *Geol. Soc. Amer. Bull.*, v. 56, pp. 275–370.

Johnson, W. D. (1901) "The High Plains and Their Utilization," *U. S. Geol. Surv. Ann. Rept.*, v. 4, pp. 601–741.

King, L. C. (1953) "Canons of Landscape Evolution," *Geol. Soc. Amer. Bull.*, v. 64, pp. 721–752.

King, L. C. (1968) "Pediplanation" [in] *Encyclopedia of Geomorphology*, R. W. Fairbridge, ed., Reinhold Book Corp., New York, pp. 818–820.

Krinsley, D. H. and Smalley, I. J. (1972) "Sand," *Amer. Scientist*, v. 60, pp. 286–291.

Krynine, P. D. (1955) "Systems of Sedimentation and Their Phases," *Amer. Assoc. Pet. Geologists Bull.* (Abstract), v. 39, pp. 133–134.

Kuenen, Ph. H. (1959) "Sand—Its Origin, Transportation, Abrasion, and Accumulation," *Geol. Soc. So. Africa, Trans. Proc.*, v. 62, 33 p.

Kuenen, Ph. H. (1960) "Sand," *Scientific American*, v. 202, pp. 21–34.

Langford-Smith, T. and Dury, G. H. (1964) "A Pediment Survey at Middle Pinnacle, Near Broken Hill, New South Wales, *Geol. Soc. Aust. Jour.*, v. 11, pp. 79–88.

Langford-Smith, T. and Dury, G. H. (1965) "Distribution, Character, and Attitude of the Duricrust in the Northwest of New South Wales and the Adjacent Areas of Queensland," *Amer. Jour. Sci.*, v. 263, pp. 170–190.

Lustig, L. K. (1965) "Clastic Sedimentation in Deep Springs Valley, California," *U. S. Geol. Surv. Prof. Pap.* 352-F, pp. 131–192.

Lustig, L. K. (1967) *Inventory of Research on Geomorphology and Surface Hydrology of Desert Environments*, Series v. IV, Office of Arid Lands Research, Univ. Arizona, Tucson, 189 pp.

Mabbutt, J. A. (1966) "Mantle-Controlled Planation of Pediments," *Amer. Jour. Sci.*, v. 264, pp. 78–91.

Menard, H. W. (1950) "Sediment Movement in Relation to Current Velocity," *Jour. Sed. Pet.*, v. 20, pp. 148–160.

McGee, W. J. (1897) "Sheetflood Erosion," *Geol. Soc. Amer. Bull.*, v. 8, pp. 87–112.

Oberlander, T. M. (1972) "Morphogenesis of Granitic Boulder Slopes in the Mojave Desert, California," *Jour. Geol.*, v. 80, pp. 1–20.

Pallister, J. W. (1956) "Slope Form and Erosion Surfaces in Uganda," *Geol. Mag.*, v. 93, pp. 465–472.

Passarge, S. (1904a) *Die Kalahari*, Reimer, Berlin, 2 vols.

Passarge, S. (1904b) "Rumpffläche und Inselberge," *Zeits. der Deut. Geol. Gesel*, v. 56, pp. 193–209.

Penck, Walther (1924) *Morphological Analysis of Landforms*, Macmillan and Co., London, 430 pp. (trans. 1953).

Quinn, J. H. (1957) "Paired River Terraces and Pleistocene Glaciation," *Jour. Geol.*, v. 65, pp. 149–166.

Ramana Rao, K. L. V. and Vaidyanadhan, R. (1970) "Recognition, Correlation and Age of Erosion Surfaces in the Iron Ore Ranges of Bihar and Orissa," *Geol. Soc. India, Jour.*, v. 11, pp. 178–181.

Ramana Rao, K. L. V. (1972) *Studies on Planar Surfaces, Laterites and Pediments in the Keonjhar Region, Eastern India*, Andhra Univ. Ph.D. Thesis (on file), Andhra, India, 150 pp. 14 Plates, 2 Maps.

Rouse, Hunter (1950) *Engineering Hydraulics*, John Wiley and Sons, New York, 1039 pp.

Roy, C. J. and Hussey, K. M. (1952) "Mass-Wasting on Table Mountain, Fremont County, Colorado," *Amer. Jour. Sci.* v. 250, pp. 35–45.

Ruxton, B. P. (1958) "Weathering and Subsurface Erosion in Granite at the Piedmont Angle, Balos, Sudan," *Geol. Mag.*, v. 95, pp. 353–377.

Schumm, S. A. (1956) "Evolution of Drainage Systems and Slopes in Badlands at Perth Amboy, New Jersey," *Geol. Soc. Amer. Bull.*, v. 67, pp. 597–646.

Schumm, S. A. (1962) "Erosion on Miniature Pediments in Badlands National Monument, South Dakota," *Geol. Soc. Amer. Bull.*, v. 73, pp. 719–724.

Schumm, S. A. and Chorley, R. J. (1966) "Talus Weathering and Scarp Recession in the Colorado Plateaus," *Geomorph. Ann.*, v. 10, pp. 11–36.

Schumm, S. A. and Lusby, G. C. (1963) "Seasonal Variation of Infiltration Capacity and Runoff on Hillslopes in Western Colorado," *Jour. of Geophys. Res.*, v. 68, pp. 3655–3666.

Stokes, W. L. (1950) "Pediment Concept Applied to Shinarump and Similar Conglomerates," *Geol. Soc. Amer. Bull.*, v. 61, pp. 91–98.

Stone, Richard (1968) "Deserts and Desert Landforms" [in] *Encyclopedia of Geomorphology*, R. W. Fairbridge, ed., Reinhold Book Corp., New York, pp. 271–279.

Strahler, A. N. (1952) "Dynamic Basis of Geomorphology," *Geol. Soc. Amer. Bull.*, v. 63, pp. 423–458.

Tator, B. A. (1949) "Valley-Widening Processes in the Colorado Rockies," *Geol. Soc. Amer. Bull.*, v. 60, pp. 1171–1184.

Tator, B. A. (1952–53) "Pediment Characteristics and Terminology," *Assoc. Amer.*

Geogr. Ann., v. 42, pp. 295–317; v. 43, pp. 39–53.

Twenhofel, W. H. et al. (1932) *Treatise on Sedimentation*, Williams and Mathews, Baltimore, esp. pp. 642–795.

Twidale, C. R. (1962) "Steepened Margins of Inselbergs from North-Western Eyre Peninsula, South Australia," *Zeits. für Geomorph.*, v. 6, pp. 51–69.

Twidale, C. R. (1968) "Pediments" [in] *Encyclopedia of Geomorphology*, R. W. Fairbridge, ed., Reinhold Book Corp., New York, pp. 817–818.

Wooley, R. R. (1946) "Cloudburst Floods in Utah, 1850–1938," *U. S. Geol. Surv. Water-Supply Pap.* 994, 127.

Woolnough, W. G. (1927) "The Chemical Criteria of Peneplanation, (II) The Duricrust of Australia," *Roy. Soc. N.S. Wales, Jour.*, v. 61, pp. 1–53.

7 Alternating arid-humid geomorphic systems

THE ARID-HUMID ENVIRONMENTAL CONTINUUM

Geomorphic effects on the earth's surface may be thought of as occurring in relation to an essentially endless sequence of conditions, hereafter termed an *environmental continuum*. The conditions involved in the sequence differ with geologic and geographic setting as well as with long-term fabrics of the atmosphere and oceans and changes in such interrelated patterns as may affect the over-all hydrography. Thus, at relatively high latitudes or altitudes, glacial conditions may comprise a part of the environmental continuum, whereas at low latitudes and elevations, where this does not appear to be possible, the principal variants involve effective moisture. Usually (outside of climatic nuclei), at least two major kinds of surficial environment will be found to alternate with one another over a particular land area. As noted in Chapter 4, there is often some tendency for one of the environmental end-members to dominate over the other. The surficial conditions of immediate concern to us here are the arid and humid geomorphic systems where they periodically replace one another.

It should also be pointed out that effects of crustal movements, volcanism, and even marine transgressions may be imposed upon subaerial conditions more or less ran-domly. The former (endogenic) conditions, described in Chapter 3, generally provide the background for subaerial environmental events or may rather randomly disrupt and modify surficial conditions. And at least some of the surficial environments appear to be notably and regularly cyclic (cf. Fig. 1.26).

Also, from discussions in Chapter 4, it will be recalled that surficial environmental adjustment in time is believed to express a correlative adjustment in space. Thus, in the space-time continuum of environments that we here contend with, there are extremes of moisture and dryness, of plant growth and barrenness, of shade and sun, of rocks that dissolve and rocks that rupture.

Since we are attempting to deal with an environmental continuum having two cyclically displaced end-members, it is theoretically possible to begin our discussion any place since, in covering the main eventualities, we would eventually come "full circle." Actually, there appear to be a few "most logical" starting places, determined in part by what we know about the effects of the arid and humid geomorphic systems. Our knowledge leads us to begin by considering the geomorphic events that accompany the conversion of a humid geomorphic system into an arid one. And, at the outset, it is evident that the possibilities include an infinite series of landscapes grading in form between the ones that develop in humid or in arid climatic nuclei where there is little or no climate change. It is clearly impossible to treat in any detail humid terrains exposed to aridity for, shall we say, 50 years, versus others exposed 100, 200, 300, 400, 500, 1,000, 5,000 and so on up to perhaps several million years. Nor is it presently within our abilities to discriminate among many of these categories. Rather, we will select geomorphic case studies that highlight the transitions involved in terms of major stages of adjustment in landforms and deposits.

In considering the nature of climate change, one immediately must be concerned with mechanisms and rates of change in environmental parameters. For example, if a desert region becomes humid, in a hypothetical shift from an annual rainfall of 4 inches (11.0 cm) to one of 45 inches (113.3 cm), as would be required by forested dunes on the New Jersey coast, or in southeastern Louisiana, is the shift abrupt or does a given area experience a gradual rainfall increase with transitional annual precipitation rates of, say, 10 inches (25.4 cm) at one stage and 20 inches (50.8 cm) at a later stage? Would changing precipitation rates also be reflected in, say, temperature or evaporation rates? Clues are available in depositional records of climate change and in the geographic disposition of environments. Continental sedimentary rock sequences convey a sense of abrupt climate change where, as in the Kalahari Desert of South Africa, a knife-edge contact separates a layer of wind-blown sand from one of chemically formed soil or lacustrine clay (cf. Peabody, 1954). The stratigraphic abruptness is belied by zones of environmental transition which presently lie geographically between such extremes as humid and arid. In high-relief terrain, the arid-humid transition expressed meteorologically can occur over a distance of only a few hundred feet (cf. Garner, 1959). On a plain, the same changes may occur over a zone scores or even hundreds of miles wide.

Yet, as noted in Chapter 4, the contact between forest and grassland or barren versus vegetated tracts on the same transition zones may be singularly sharp.

A considerable body of data assembled in Chapter 4 indicates that environmental zones and related geomorphic systems (expressed by variations in groundcover) reflect atmospheric fabrics. It was further suggested that changes in environment reflect adjustments in that fabric closely correlated with oceanic conditions, and though relatively small vertical atmospheric zone shifts may be involved, they may be expressed by large horizontal increments of groundcover change over low-inclination planar terrain (see Fig. 4.19).

Implications of the foregoing relations in terms of climate change seem to center mainly on the physical conditions that favor glacial, humid, and arid conditions. In the present context, this can be narrowed down to humidity and aridity—the latter particularly as favored by low elevations and cold seas. Each continent has its climatic nuclei. More significantly, the hydrographic conditions that would appear to favor the intensification of humidity also appear to favor simultaneously the weakening of aridity in regions between such nuclei. To this extent, one condition does not "push" another condition out of the way in the manner of a bulldozer pushing a pile of rubble or glacial ice advancing on a periglacial realm. Rather, it is simply that as a region in a transition zone between an arid nucleus and a humid one becomes "less wet" and vegetated it also becomes "more dry" and barren.

The Pleistocene paleoclimatic record suggests that climate zone boundaries tend to migrate away from a climatic nucleus where the condition is being favored hydrographically over wider and wider areas (cf. Figs. 4.7 and 4.29). A climatic nucleus would mainly be exposed to effects characteristic of its opposite number (i.e. opposing extreme) on those rare instances when conditions of the opposite number are most pervasively developed. Where low-level aridity has been intensified, sparsely vegetated conditions appear to invade uplands, and, in an extreme case, leave their briefest and weakest related geomorphic effects along the fringes and lower reaches of humid nuclei. Where intermediate-high elevation humid conditions become more intense, the opposite relations seem to prevail (see Fig. 4.23). In essence, these are the situations recorded on most continents, and they will serve to anchor our initial discussion of changing climates. In most conditions of terrain and elevation, one can anticipate a gradual adjustment in the meteorologic parameters that define the transitional environments. Yet, as noted in Chapter 4, the major geomorphic systems (defined in terms of groundcover) depend upon the attainment of certain threshold conditions, not only meteorologically, but also, in some cases, pedologically, as related to groundwater and plant-propagation mechanisms. The fact that at least some environmental changes occur on a threshold basis expressed geomorphically may account for sharp breaks in depositional records of climate changes previously noted (Peabody, 1954).

As the climate changes, the interaction of various agencies within a given geomorphic system becomes most apparent. In the type of change now in question, involving a shift from humidity to aridity, we are concerned with a progressive decrease in precipitation and either a relative or an actual increase in evaporation. The situation is paralleled by the many drouths recorded in Europe, Asia, Australia, Africa,* and North America. During drouths, several progressive changes are known to occur more or less simultaneously. The groundwater table becomes more and more depressed as does the average amount of soil moisture, the effects being most intense on hilltops where watertables are normally at greatest depth even during humid conditions. Old and diseased trees begin to die as do certain types of shrubs. The depressed watertable is also reflected in dry wells and in flow continuity changes along drainage lines. Even deep wells frequently show increased salinities. Beginning with head-water tributaries and progressively extending to lower elevations, stream flow changes from perennial to intermittent to ephemeral (cf. Fig. 5.43). During severe drouths, such deeply incised trunk drainages as the Arkansas or Rio Grande or Niger rivers usually continue to transmit some seepage if not actual surface flow. With a shift to actual aridity, even surface flow would presumably terminate on any but an ephemeral basis (Fig. 7.1).

In our discussion of temporal runoff-continuity changes during a shift from humidity to aridity, we have also documented the adjustment from exterior to interior drainage. That is to say, when main trunk drainages cease to transmit runoff, even intermittently, a region has lost its *exhoric* drainage. A MINIMUM RUNOFF-VOLUME THRESHOLD has been crossed. The system of oriented slopes and channels extending to or beyond the limits of the humid environment for purposes of containing and draining away excess water is no longer needed. More on this later. Other things are known to happen. For instance, most trees depend in part upon the groundwater body for moisture. Depending upon the rate of the meteorologic adjustments taking place, a minimal-moisture level for tree growth will be reached after which most trees will die. (Thousands of trees were lost in the West Texas drouth of the 1960's.) Those along water courses are apparently the last to go, and some few may survive in oases. Many types of grass have a tolerance for extended periods of dryness. It should perhaps be noted, too, that brush, grass, and forest fires have a high incidence under drouth conditions and presumably during climate change conditions of the type in question.

For the moment it can be assumed that some of the grasses originally present in a region would persist after the TREE-LOSS THRESHOLD was crossed. Certain other kinds of changes have also been found to occur. Soluble mineral matter present in near-surface portions of the regolith and soil tends to be drawn toward the surface during evaporation of water. In a shift to

* Even at press time, the southern Sahara border is drouth striken again.

Fig. 7.1. Stream-cut gorge in the northern Atlas Mountains, Algeria, which are presently almost without runoff except for snow meltwater. The stream in the gorge bottom is ephemeral, and formerly more humid conditions are indicated.

aridity, this may possibly begin to modify pedalfer soils into pedocals. The extent of this mineralization is surely open to question, since a thoroughly leached humid soil or regolith might actually contain little soluble material (cf. Chapter 5). Deeper bedrock sources for solutes might exist in some settings. Certainly, leaching of shallower soil horizons would decrease in intensity with the passage of time and increased aridity and, ultimately, should cease except during duricrust formation. With due respect for the moisture requirements of the various types of plant cover, it may safely be assumed that grasses would ultimately go the way of trees in a complete shift from humidity to aridity. It is uncertain whether the sequence involved parallels the general relationship, tall grass to short grass to bunch grass, observed in many plains regions as one crosses these to approach desert areas. If this grass hierarchy reflects the sequence of ground uncovering, then we are dealing with a gradual botanic adjustment. It nevertheless seems evident that a minimal GRASS MOISTURE THRESHOLD is also crossed and that actual loss of groundcover would be accelerated in many instances by fire, burrowing animals, eolian action, and probably some forms of runoff (see Bartholomew, 1970).

HUMID-TO-ARID SEMIARIDITY

A moment's consideration of the humid-to-arid climate change problem is sufficient to establish that some of the transitional physical parameters fall within the limits of the condition generally termed *semiaridity*. Semiaridity is recognized as a climate in its own right and is regarded by some workers as having a special geomorphic significance (cf. King, 1953; Leopold *et al.*, 1966). Why then do we cite "humid-to-arid semiaridity?" Why not just "semiaridity?" The reasons are comparatively simple. First, we are concerned here with semiaridity as a transitional condition between humidity and aridity. It was pointed out in Chapter 1 that an environment's effects are modified by the relicts upon which it acts. If this is true, "geomorphic" semiaridity following humidity should differ from "geomorphic" semiaridity following aridity (see Garner, 1959). A second point in this regard is that semiaridity may comprise a climatic end-member (at least in terms of physical parameters) in alternation with humidity, aridity, or some other surficial condition. However, even though it is possible to delineate a group of physical attributes that qualify in many definitions of semiaridity [e.g., 10–20 inches (25.4–50.8 cm) of precipitation; grass groundcover; intermittent stream flow], in terms of its probable geomorphic effects, the semiarid condition appears to be much more closely allied with classic "humidity" than with "aridity."

If the foregoing evaluation of semiaridity is correct, and we will consider the evidence shortly, semiaridity plays one set of environmental roles as a transitional condition. Alternatively, semiaridity may qualify as a sort of humid geomorphic "subspecies," where it acts as an environmental end-member, one where its effects would be particularly notable in contrast with such conditions as aridity but where its geomorphic effects in alternation with humidity are uncertain and probably not extreme.

Setting of the Stage— Humid Geomorphic Relicts

Our initial concern with terrains relating to the humid-to-arid environmental transition has to do with the kinds of relict humid landforms and deposits that will be subjected to the new conditions. The main humid geomorphic end products were outlined in Chapter 5. They include (1) a terrain more or less thoroughly incised by running water with hillslopes contiguous to drainage nets and probable regional drainage systems; (2) a regolith consisting in part or entirely of soil with a texture that may range from fine-grained chemical residues or accreted detritus (clay, silt, sand, humic debris) to large rock masses isolated by concentric weathering (*core stones*); (3) comparatively small amounts of debris undergoing fluvial transport in watercourses; (4) possible lacustrine deposits; and (5) possible solutional features. It is these substances and terrain configurations that will presumably be modified (a) by the transition to aridity and subsequently (b) by more or less enduring aridity.

We will make no more precise assumptions about the character of the humid terrain becoming arid than those just cited. That is to say, the relief present is typical of humid regions and is dominently fluvial (as opposed to structural examples treated in Chapter 6), but its amount could clearly vary from a slightly channeled plain to a thoroughly dissected high-relief region essentially all in slopes except for the bottoms of watercourses. Similar variation is clearly possible for thickness and maturity of weathered zones. As was pointed out in Chapter 5, humid geomorphic systems tend to accentuate relief by fluvial means, and, for portions of the crust above regional base level, the amount of relief generated is basically time dependent. In this context, therefore, other factors being the same, the highest fluvial relief and deepest weathering should occur within and near humid climatic nuclei (Figs. 5.44 and 5.45); moderate incision and weathering depths for valleys keyed to local precipitation should be encountered roughly medially disposed between arid and humid nuclei (Fig. 2.35); and minimal incision depths and weathering maturity should be found in regions adjacent to arid nuclei (Fig. 6.46). Certainly, the amount of terrain modification required to adjust each of these terrain and regolith examples to an arid geomorphic system differs considerably.

Humid-to-Arid Disequilibrium

A precise chronologic correlation between the attainment of the various geomorphic thresholds during the humid-to-arid transition has not yet been established. Many variables are involved, and these include the long-term rate of depression of the watertable that will at one stage kill forests and at another terminate external drainage on a regular basis but that may not closely correlate with the actual elimination of vegetal groundcover by other, independent conditions. The coexistence of vegetation and intermittent watercourses that seasonally gives rise to external drainage from grassland regions like the North American High Plains and the Serangeti Plains of Africa leaves us to puzzle over whether external drainage or groundcover would be the last typically humid agencies to go. We do not really know.

There are several humid-to-arid transition relations about which we can be reasonably sure. One is that desiccation of main watercourses would cause deposition of any sediment being carried (cf. Garner, 1963), and much of this material should retain the fluvial attributes of open sedimentary system deposits in terms of sorting, rounding, grading, and chemical maturity as outlined in Chapter 6 (cf. Hubert, 1960). A second, almost inevitable relation accompanies the exposure of the humid weathering profile to the effects of wind, water, and mass wasting following loss of groundcover. When an agency of erosion is supplied with a large amount of rock material having ideal transport dimensions, accelerated erosion can be anticipated as noted by Bowman (1916). Excepting possible corestones, rock and mineral debris in humid soil profiles is dimensionally ideal for both water and wind transport. Farmers who plowed and planted hillslopes to corn in southern Iowa in the 1920's and 1930's saw soil A horizons up to 2.5 ft thick and portions of B horizons 1 to 2 ft thick stripped off 5°–10° slopes by fluvial agencies in as little as two decades. The vegetation there inhibited wind erosion, but in an incipient desert no such inhibition exists. As noted by the writer (1959), a loss of vegetal cover in a region

> . . . which has been humid for a long period would not only expose the humid soils to the ephemeral severity of arid runoff but, probably more significantly, would expose a vast amount of fine clastic material to eolian deflation. Under these circumstances early aridity would be characterized by the removal of the fine humid soils similar to the Dust Bowl activity in the United States during the 1930's and periodically since then.

The foregoing description is essentially that of an episode of environmental disequilibrium brought about when relicts from one condition (humid) are placed in an alternative condition (arid) as outlined in Chapter 1 (Fig. 1.20). The results of an active humid-to-arid climate change have never been observed. But, as noted by the writer (1959), they have been approximated by man aided by a drouth in the Dust Bowl of the 1930's. Descriptions of some of those events should be instructive in the comprehension of the significance and character of climate changes. The dust storm ("Black Duster") of May 12, 1934 is notable, though similar events are on record since about 1900. Dust was stirred up by high winds in Kansas, Oklahoma, Texas, Colorado, New Mexico, and adjacent states. A vast cloud of blinding, choking, pulverized rock moved eastward down the prevailing winds, ultimately to be deposited, but initially causing nearly total darkness at noon in Iowa, Illinois, Indiana, Pennsylvania, and most of the eastern seaboard, engulfing New York City skyscrapers and affecting ships 300–500 mi (420–700 km) out in the Atlantic Ocean (Fig. 7.2).

Estimates based on local measurements of the effects of this dust storm indicate that some 100 tons/sq mi of loess were deposited over the eastern United States and western North Atlantic or about 300,000,000 tons of soil lost from the deflated area. It has been further estimated that for each ton of dust deflated and carried away, 2 to 3 tons of coarse silt and sand were drifted about in and near the source region. One of the most carefully studied dust storms occurred on February 7, 1937 where the dust landed on snow, amounting to 34.2 tons/sq mi near Ames, Iowa, 14.9 tons/sq mi at Marguette, Michigan, and 10.0 tons/sq mi in southern New Hampshire. A similar storm deposited 80–100 tons/sq mi in south Germany in 1859.

It is difficult if not impossible to imagine similar eolian events occurring several times each year and being repeated year after year following the onset of actual aridity. Yet there can be no real doubt that this is precisely what would happen after a climate change from humid to arid until such fine-grained material as could blow away were either completely stripped or protected by some layer of lag gravel. Under these circumstances, it is possible to visualize the

Fig. 7.2. A "black duster" induced by turbulence associated with the passage of a cold front as seen at Manteer, Kansas, in April, 1935. Though related to drouth and marginal-land cultivation, in this instance, similar storms involved with soil deflation probably accompany climate changes from humidity to aridity. (National Oceanic and Atmospheric Administration.)

complete destruction of a soil interval down to bedrock over a vast region in a matter of a few decades. Such theorizing must be tempered, however, by the fact that deflation would tend to focus along the margin of a receding grassland in some cases. Nevertheless, what is being discussed would by most definitions of the term be designated CATASTROPHIC EROSION even as it is being cited as a normal aspect of the environmental transition from humidity to aridity. A geologically insignificant period of time would apparently be required.

The foregoing account of accelerated erosion takes on added significance in the light of studies such as that of Oberlander (1972) in the Mojave Desert of California. He describes boulder-littered slopes that can be traced laterally along a geomorphic surface to a chemically decomposed soil profile developed on quartz monzonite and granite (cf. Chapter 3 for descriptions) and buried under basalt flows radiometrically dated at 8 million years. Oberlander concludes that present granite weathering in arid conditions is by exfoliation and states, "The existing boulders were originally isolated as corestones by subsurface chemical weathering and have been exposed by the stripping of a thick weathered mantle [regolith] formed during pre-Quaternary periods of greater moisture availability" (see Fig. 6.6). Rather similar features develop in other manners (Fig. 7.3). Oberlander notes several alternative and related modes of genesis for domed residuals (Fig. 7.4) and concludes

Fig. 7.3. Concretions produced by differential weathering and erosion of an Eocene limestone near Kharga Oasis some 200 mi west of the Nile River, Egypt. Like core stones illustrated (Figs. 3.35, 3.46, and 6.6), the concretions are resistant masses probably freed initially within a weathering profile, and, in any case, exposed in part by deflation. (Tad Nichols.)

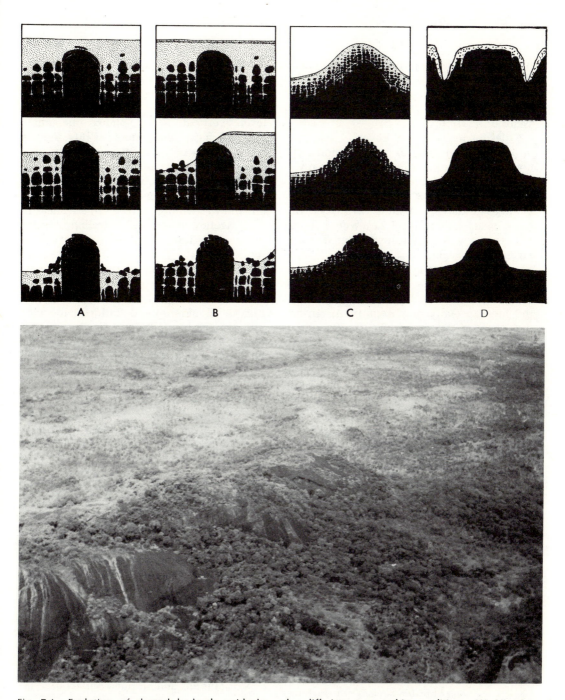

Fig. 7.4. Evolution of domed bedrock residuals under differing geomorphic conditions. (A) Nigerian bornhardts isolated by differential subsurface weathering and exposed by surface lowering (Thomas). Subsurface decay continues so relation is perpetuated. (B) West Australian inselberg isolated by differential subsurface weathering under humidity and exposed by subsequent retreat of duricrusted escarpment under aridity (Mabbutt). Weathering front stagnant under aridity. (C) Domed inselberg of Mojave Desert isolated as soil-covered hill and later stripped of soil, decayed rock, and residual boulders. Dome surface was downward weathering limit (chemical) in Tertiary (Oberlander). (D) Domed inselberg of southern Venezuela, isolated by stream incision and shaped by a chemical weathering front under humid conditions, later denuded of soil and reduced by scarp retreat under aridity, now acquiring plant cover under tropical humidity (see photo).

that the gradual onset of aridity terminated subsurface weathering and accelerated surface erosion, bringing about massive exposure of core stones and a joint-controlled "weathering front."

A partial loss of plant cover is also apparently documented in some regions by mound-like accumulations of wind-blown soil constituents. Called *prairie mounds* by Quinn (1961; 1968), many can be attributed to the clumped growth style of desert vegetation (Fig. 4.21) where roots and stems catch eolian dust deflated from barren inter-vegetated areas (Fig. 7.5).* The mounds are known from California and Oregon (*mima mounds*), the Gulf Coastal Plain (*pimples*), northern Mexico, and are especially numerous in Arkansas and eastern Oklahoma. There, they are found in groups—"fields"—on relatively level surfaces blanketed by comparatively fine-grained, commonly loessal material which had been partly weathered into soil under plant cover prior to mound construction. In fluvially dissected regions, such as northwest Arkansas, the prairie mounds are encountered on plateau remnants and on

valley flats above the present flood level. Averaging 30 to 60 ft in diameter and 2 to 4 ft in height, they tend to be circular to ovate in plan, and they appear on high-level, vertical aerial photos as small, grouped, light-colored dots (cf. Fig. 7.6). A complete loss of plant cover would presumably destroy the mounds.

There is little question that exposed humid soils would be affected by both runoff and the wind. And the effectiveness of each agency would be controlled in some measure by the actual composition of the relict regolith. Incompletely decomposed relict bedrock fragments of concretionary masses of mineral matter precipitated in soil zones during semiarid weathering (cf. Fig. 7.3) could inhibit deflation by the wind and comprise runoff loads. Gardner's study (1972) suggests that duricrusts form much later under aridity, and it should be kept in mind that the armoring effects of lag gravel would be much reduced on steep slopes left by humid erosion where the clasts could be more readily shifted by sheet wash or gravity than is the case on desert plains.** In any event, it would appear that some relict humid soil constituents should comprise initial arid fluvial deposits. IT IS FURTHER EVIDENT THAT THE MECHANICAL WEATHERING OF BEDROCK UNDER ARID CONDITIONS CANNOT BEGIN UNTIL THE BEDROCK IS EXPOSED through stripping of relict regolith (mainly soil). Through accelerated interaction of wind and water, this stripping is assumed to take very little time, after which the arid weathering processes, described in Chapter 6, could begin to provide rock material to the arid geomorphic system.

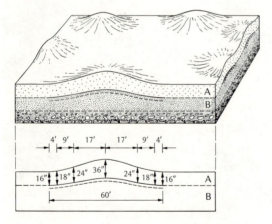

Fig. 7.5.   A group of prairie mounds near Highway 112 on the University of Arkansas Experimental Farm north of Fayetteville, Arkansas. **A** and **B** are soil horizons developed on loess. The dashed line below **B** is the base of the loess on a relict alluvial deposit of probable arid derivation. (From Quinn, 1961.)

Streamfloods and Stripped Surfaces

A landscape consisting in some substantial degree of the ridge-ravine topography generated under humid geomorphic conditions tends to exhibit a somewhat rounded and

* There are many other theories of mound origin, but this appears to be most widely applicable.
** The generally complete destruction of relict soil in deserts is attested to by the rarity of preserved soil remnants in deserts.

Fig. 7.6.   The prairie mounds (Mima Mounds) of southwestern Washington viewed in an oblique, low-level aerial photograph. (A. M. Ritchie.)

subdued expression of the same relief configuration on the bedrock surface below the weathered interval (Fig. 7.7). In general, weathering depths are greatest on hill and ridge crests where seasonal, vertical water table fluctuations are greatest, leaching and subsurface drainage are accelerated, and alteration processes can penetrate most deeply. Alternatively, lower valley sides and floors usually are less deeply weathered where water table positions are more stable, and unaltered bedrock is commonly at the surface in channel banks and bottoms (cf. Figs. 5.2, 5.44, and 7.7). Stripping of regolith should, therefore, alter slope amounts but, generally, not alter slope directional configurations developed along drainage nets. However, loss of plant cover should increase the per cent of runoff in general accordance with Bennett's findings (1939) outlined in Chapter 4 (see Fig. 4.15). This should induce a runoff increase from less than 1%, in some cases, to nearly 50%.* Sediment loads would increase correspondingly as loose debris is exposed.

With the loss of groundcover it seems evident that a RAINDROP EROSION THRESHOLD is attained. This threshold, in combination with sheetwash effects, could account for much of Bennett's accelerated fluvial erosion record (1939). The turbulent discharge of storm runoff through channels in deserts has been termed *streamfloods* by several

* The actual magnitude of increase would clearly depend upon such factors as terrain permeability and evaporation rate. During a winter rain, evaporation might be nil. On bare rock nearly 100% of the precipitation could run off.

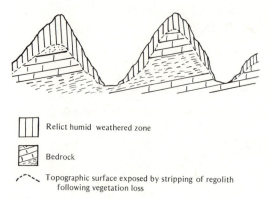

▯▯ Relict humid weathered zone

▱ Bedrock

⌁⌁ Topographic surface exposed by stripping of regolith
following vegetation loss

Fig. 7.7. Cross section of relations between weathered regolith and bedrock in terrain exposed to humid conditions for a protracted period. Loss of vegetation, either artificially or through a shift to aridity, would expose bedrock to sheetwash and deflation. Depending upon regolith constituents, bedrock could be stripped completely bare by the adjustment.

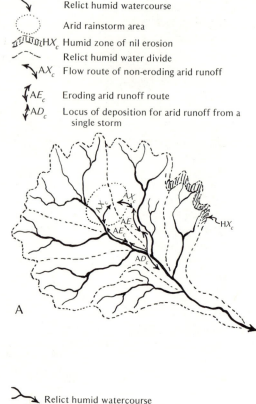

↘ Relict humid watercourse

◌ Arid rainstorm area

⌇⌇HX_c Humid zone of nil erosion

⌁⌁ Relict humid water divide

↘AX_c Flow route of non-eroding arid runoff

↟AE_c Eroding arid runoff route

↟AD_c Locus of deposition for arid runoff from a single storm

A

↘ Relict humid watercourse

⣿ Accumulating alluvium

⌁⌁ Relict humid divide

⋯ Upslope limit of aggradation locus

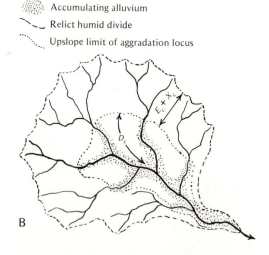

B

Fig. 7.8. Relict humid drainage nets depicting (A) main hydrologic effects related to a single arid rainstorm showing one possible distribution of X_c, E_c, and D_c, and (B) long-term depositional locus within relict humid drainage net where desert rainfall is orographically localized on former water divides.

writers, including Davis (1905). This denotes a degree of confinement presumably missing for *sheetfloods*, and, since the runoff would be localized by relict slopes, we may safely assume that incision would be possible as long as the runoff is not fully loaded. In effect, therefore, the change from a vegetated to a barren condition on an incised terrain is a change from (a) Horton's humid drainage net (1945) with an upper "nil-erosion zone" delimited by X_c and a downslope system of watercourses, where fluvial incision can occur (see Fig. 5.35), to (b) the arid fluvial event occurring "somewhere" on a drainage net, where X_c is followed downslope by an erosional interval (E_c) during which there is excess fluvial kinetic energy, followed by a depositional interval (D_c) during which energy available to erode is zero and energy available to transport diminishes, ultimately, to zero (see Chapter 6).

This résumé of fluvial metamorphosis during humid-to-arid climate change points up the fact that the hydraulic phenomena of the arid phase do not necessarily match the fluvial relics (drainage nets) left by the humid phase (Fig. 7.8). In the first place, humid drainage nets are, in large part, a response to regional precipitation phenomena, so that humid divide areas (whether major or minor) receive some moisture each year. Arid precipitation may focus on divides for orographic reasons, in some cases, or it may be essentially random over terrain that does not attain great elevation but has some relief. Second, with the advent of aridity, the D_c zone must eventually fall somewhere within a relict drainage net. Thus, detritus washed from inter-stream areas must accumulate in downstream reaches of the same relict drainage net. Third, since the arid fluvial event is keyed on a local area of rainfall, in most instances, the X_c distance need not be measured from the divide of a relict drainage net. Rather, X_c must relate to the rainfall area proper (Fig. 7.8A). The erosional phase E_c and the depositional phase D_c are somewhat less random in location than is the rainfall, since both erosion and deposition would be influenced by relict slopes (cf. Fig. 7.8A). Presumably, under arid conditions, downstream portions of relict humid drainage nets should tend to become alluviated (Fig. 7.8B). And, unlike alluvium emplaced by a through-flowing perennial humid stream, the arid alluvium could as readily be derived from valley sidewalls as from upvalley and should, in most instances, not move great distances.

In a 1968 discussion of quantitative geomorphology, Strahler considers the effects of "Upsets of Steady State" on drainage basin dynamics with an example of a ". . . forested land surface [that] is denuded of its vegetative cover and intensely cultivated." The climate-change situation just discussed approximates Strahler's example, even though there is no particular reason for the antecedent terrain to be a humid climax (steady-state) association. Strahler describes the changes (summarized in the first paragraph of this section) in terms of an abrupt increase in *Horton number* (Q_rK) in which Q_r is the runoff intensity expressing volume rate of flow/unit area of surface, which is simply the excess of rainfall intensity over infiltration capacity; K is an *erosion proportionality factor* expressing mass rate of removal/unit area divided

by force/unit area; Q_r has the dimensions of velocity LT^{-1}, where L is stream length from gauge to point on divide measured in miles; and K has the dimensions $L^{-1}T$, the inverse of velocity expressed in seconds/foot.

It is curious that Strahler, who was not considering climate change and therefore presumably did not anticipate a rainfall reduction or increased evaporation, does in fact suggest that aggradation of the main drainage net channel would ensue and that upstream areas would be intensely dissected into badlands (cf. Fig. 7.9), with the result, "That a new steady state of erosion is achieved at a much higher level of intensity" and "Steepening of mainstream channel gradients through aggradation is also a characteristic result of the transformation." The anticipated effects closely parallel those indicated for humid-to-arid climate change (cf. Figs. 7.8A-B and 7.9). The reason for aggradation in the latter climatic situation seems clear enough, but we are uncertain as to why the runoff in Strahler's model should not, in general, increase downslope in volume as well as load

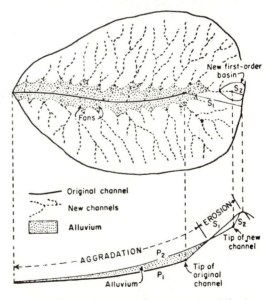

Fig. 7.9. Drainage density transformation following a disruption of a steady-state drainage situation by artificial deforestation and cultivation. (After Strahler, 1958.)

(cf. Gilbert's Law of Stream Capacity, Chapter 1, p. 10) and thus at least maintain velocity and no more than sporadically and locally agrade. Whatever the correct fluvial mechanism may be, it should be reflected in the deposits.

One cannot observe the pattern of arid alluvium emplacement and related patterns of runoff, erosion, deposition, and precipitation. Individual desert pluvial events are too scattered, too random in location and intensity, and, hence, too unpredictable for present surveillance systems to record. Probably the best one can hope for in this phase of geomorphic analysis is a study of the resulting deposits that would distinguish between (1) open sedimentary system channel deposits made as a humid stream reworks a valley fill or dries up with the onset of aridity and (2) closed sedimentary system deposits formed as desert runoff loses transportive power, desiccates, and, as water or mud, stops. There can be little doubt that many studies have dealt with these two kinds of deposits in areas of arid-humid climatic alternation (cf. Miller, 1958; Miller and Wendorf, 1958; Miller *et al*, 1963). But such studies must rely upon subtle sedimentologic distinctions, often of uncertain significance. Few studies indeed set out with the stated purpose of making this differentiation. What would be helpful in this matter, would be a field relationship in the requisite environmental setting which would be analogous to injecting a radioactive tracer into an animal's bloodstream in order to ascertain where a substance goes and how. Unfortunately, and understandably, there are indeed few rivers that carry sediment that is unequivocally distinguishable from bedrock or regolith otherwise traversed. Probably even fewer occupy the proper climatic location. The writer only knows of one. Fortunately, it has been studied (Doria-Medina, 1962; Garner, 1966).

Because of its geomorphic value as an analog and its uniqueness in meeting the foregoing restrictions, a synopsis of the study just cited seems worth while here. Mountain Fork Creek is the trunk stream of a drainage net tributary through Lees Creek to the Arkansas River along the Arkansas-Oklahoma border (Fig. 7.10). Its study was undertaken because the region displays signs of alternating aridity and humidity (cf. Figs. 1.22, 2.2, and 2.3). This drainage also happens to lie in the southern Ozark upland about midway between North America's southern Appalachian humid climatic nucleus and its main arid climatic nucleus in southwestern United States and northern Mexico (see Fig. 4.16). This presently humid, vegetated Ozark area catches rather more precipitation than the lower plains to the northwest, west, and south, and it is partially dissected by drainage nets incised into a series of plateau levels (cf. Quinn, 1965). However, the prime value of Mountain Fork Creek in diagnosing modes of arid versus humid alluviation lies in one unique attribute. Near its headwaters, the main channel is incised some 40 to 50 ft into a small *inlier* of gray and white chert and cherty limestone referable to the Boone Formation of Mississippian age (Fig. 7.10). The inlier is the only bedrock source for fragmental white chert within the Mountain Fork Creek drainage basin (cf. Figs. 3.55 and 3.56).

The geomorphic significance of the white chert inlier cut by the channel of Mountain Fork Creek is that the unique white clasts provide a long-term sedimentologic tracer for deposits made by water moving down the valley along the stream channel. Any alluvium in the valley deposited by the stream since the inlier formed should contain chert. Alluvium derived directly from the valley sides (which consist mainly of Pennsylvanian age sandstone, siltstone, and shale with minor limestone) should be chert free as should the alluvium similarly originated that has moved only short distances down valley by runoff not associated with the Mountain Fork Creek channel (e.g. arid sheetwash or streamfloods of local character). Efforts to establish modes of alluvial emplacement along Mountain Fork Creek took several forms. A graduate student under the direction of the writer undertook a morphometric study of the valley

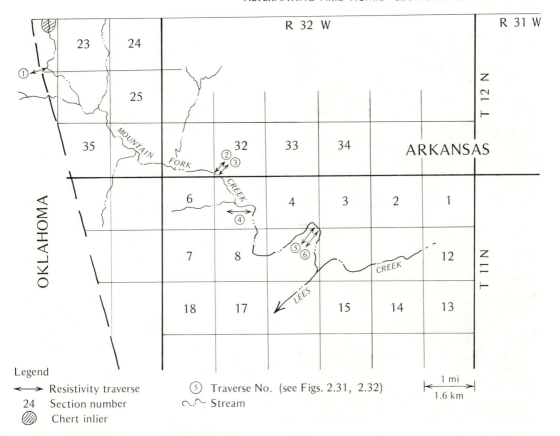

Fig. 7.10. Map of a portion of northwest Arkansas along the border with Oklahoma showing the drainage form of Mountain Fork Creek, resistivity meter traverse locations (cf. Figs. 2.31 and 2.32), and position of the parent tracer lithology outcrop (Boone chert) cited in the accompanying discussion.

gravel to determine if alluvium deposited by the stream differed in manners other than chert content from other materials seemingly present (Doria-Medina, 1962). The writer then conducted an earth resistivity meter survey of alluvial thicknesses along and transverse to the valley (Garner, 1966) and further recorded compositional variations in alluvium. In the latter study, the Arkansas Geological Survey provided data to cross-check compositional, thickness, and aquifer properties by drilling eleven boreholes to bedrock (cf. Fig. 2.32).

The several conclusions of the foregoing studies appear to have a bearing on modes of desert alluviation as well as on the behavior of humid stream flow on relict arid valley alluvium. We will defer consideration of the latter to the section of this chapter

dealing with the arid-to-humid transition. The study of Doria-Medina (1962) definitely indicated the presence of two distinct types of valley alluvium in terms of sedimentary parameters. Much of the alluvial veneer mantling lower valley walls and flats was a chert-free, poorly sorted, non-stratified, nearly impervious mixture of clay, silt, sand, and gravel dominated by angular to sub-angular, ferrugenous siltstone or sandstone clasts up to cobble size. Similar materials were found in outcrops in superjacent hillslopes or in exposures no more than 1.5–2 mi (±2–3 km) up valley (Fig. 7.11). Gravel encountered along the present stream channel and in a few other drainageways on the valley flat contained an evident chert admixture and was, in addition, much more distinctly sorted, and locally cross-

Fig. 7.11. Contrasting types of alluvium associated with Mountain Fork Creek, northwestern Arkansas. *Above*: stream-deposited gravel containing many white chert clasts from the Boone Formation which crops out near the headwaters. The matrix contains sand but little or no clay. *Below*: chert-free alluvium from the stream bank which includes much mud and silt in the matrix, probably deposited by a desert mudflow.

bedded, with many sub- to well-rounded clasts. In addition to the clear evidence of relatively far travel down valley (chert), the cherty deposits were notably pervious and contained water where encountered in bore holes. In all, the morphometry of the valley alluvium showed that the cherty gravels were better rounded and better sorted than non-cherty alluvium at the same place along the valley and rarely contained sub-sand–size material or metastable rock fragments indicative of mechanical weathering. The non-cherty alluvium was essentially just the opposite of the cherty

deposits in terms of rounding, sorting, fine-grained material and metastable rocks (limestone). The cherty materials were apparently deposited in an open sedimentary system, e.g. by running water in the stream. The non-cherty detritus has most of the attributes of closed sedimentary system deposition by a high-density medium, possibly desert mud flows in part. The Mountain Fork Creek gravel morphometry and conclusions as to origin just cited are in general agreement with the findings of Lustig (1965) in Deep Springs Valley, California.

In summary, the valley and drainage net of Mountain Fork Creek was apparently alluviated before conditions of present humidity by water that rarely moved rock debris more than 2 miles down valley from an outcrop and commonly began to deposit even as it left adjacent valley slopes. Since the region was not glaciated, and since it is several hundred miles from the nearest glacial margins, the rigorous environment indicated by metastable fragmented rocks in the non-cherty alluvium is believed to have been arid. Corroborative evidence to this effect has been presented heretofore (Chapter 1) and more will follow where appropriate. More to the point, the alluvium buried older channels to depths of as much as 20 ft (±7 m) and veneered valley flats and adjacent hillslopes an average of 2–3 ft (0.6–1.0 m). Some of the hillslopes resemble pediments (Figs. 2.31 and 2.32). The alluviation may relate to the Altithermal Interval—4,000–7,000 years B.P. (Fig. 1.24)—but it is not really extensive in terms of volume (cf. Fig. 2.32) and could hardly account erosionally for the average valley widths [500–600 ft (150–180 m) along straight reaches] by hillslope retreat. But our intention is not to elucidate the geomorphic history of the region. Rather, we merely wish to document the means by which relict humid valleys become alluviated with the onset of aridity. The extent of this alluviation being a time function makes it evident that the alluvial burial of a relict drainage net is not only possible; in rather special instances, it is probable (Fig. 7.12).

The extent to which a relict humid drainage net may be alluviated under a subsequent arid geomorphic system (indicated mainly by terrace levels, discussed in a later section), is perhaps not quite what we should be directing our attention to at this time. If our first assessment of arid processes (Chapter 6) is correct, THE DEPOSITIONAL LEVEL IMPOSED BY DESERT HYDRAULIC PARAMETERS IS DETERMINED BY WHERE THE WATER DRIES UP. Control over that level may be supplemented by a local structural basin, or, as is the case for many relict humid drainage nets, there may be no such basin. If water flowing down a drainage net (such as the Mississippi River System) begins to dry up and deposit (at St. Louis, for example) then the base level for bedrock planation will be St. Louis. The resulting deposit may adjust its slope to the Gulf of Mexico and, in the process, bury associated terrain in varying degrees up to, and even including, some major divides. According to Butler (1960, 1961) the Australian Riverine Plain is a good example of a river system that has apparently done this several times (Fig. 4.23). But that is almost beside the point. The point is that above the depositional datum, incision of watercourses will be inhibited, and hillslope retreat causing planation can occur. If Strahler's analysis of the hydraulic effects, previously discussed in this section, is correct, some increase in drainage density of non-aggraded surfaces could be expected (Fig. 7.9). Intensity of gullying would also probably have an effect on planation rate, since hillslope retreat is a surface-area phenomenon. The more finely dissected an upland becomes under arid conditions, the faster it should be destroyed by hillslope retreat once an incisional limit is established by down-valley aggradation (see Chapter 6).

An ancillary aspect of the relict drainage net phenomenon not already considered involves the relative scope of the humid-to-arid climate change with respect to drainage net size. There is considerable evidence that at least the southeastern portions of the Mississippi River Basin remained humid at various times during the Pleistocene when areas to the south and west became much drier. Thus, a question might be raised as to whether runoff from humid portions was materially altered by the arid relations. If Schumm *et al.* (1972) are correct, effects are appreciable when rivers like the Arkansas have dumped relict arid debris into the Mississippi valley, thereby steepening the latter's gradient. Several Mississippi drainage basin events, including sporadic glaciation of northern areas, cloud the issue. The Mississippi relationship differs from the drainage nets we have already discussed in the North American west and in Australia—where headwaters divide areas continue to receive notably more moisture than downstream areas, and the depositional-erosional foci are established by what amounts to a bi-climatic interplay (see Chapter 6). Our present concern involves drainage nets (like that of Mountain Fork Creek) where the relief and elevation are moderate, the basin size is relatively small, and, essentially, the entire drainage net has apparently become arid.

The Mountain Fork Creek deposit characters indicate that deposition within that basin mainly occurred from sporadic pre-

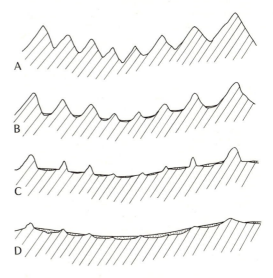

Fig. 7.12. Sequential cross sections (A)–(D) illustrating drainage basin planation and burial through reduction of secondary and tertiary divides.

cipitation, also within the basin. In general, alluvial thicknesses increase in a down-valley direction. Since a buried scarp cannot recede, downslope portions of relict drainage nets being buried by alluvium under desert conditions should have rather different erosional regimens than upslope dissected areas above the depositional datum. One would anticipate upstream cross-valley profiles showing more or less traditionally concave upward pediment forms (Fig. 2.31A, B) and downstream cross-valley profiles with more nearly abrupt side walls of greater height (Fig. 2.32B, C). Similar cross-valley profile relations could, however, be explained in other manners (cf. Garner, 1959). The matter needs more study. But, from our previous discussion, it seems clear that the effects of arid runoff on a relict humid drainage net are those of reduced relief, and, if protracted, the combined effects of planation and aggradation could render the region nearly featureless (cf. Fig. 7.12). In less extreme cases, drainage lines have been observed in varying degrees of burial or blockage by alluvium and wind-blown sand (Fig. 5.51), and divides may be obscured by deflation depressions and scarp retreat in a process of progressive *drainage derangement*, where the geometry of drainage nets is partially or completely obliterated.

The relicts of humid geomorphic systems encountered in deserts show by their degree of modification how long or how intensely they have been exposed to the new environment. As noted in Chapter 6, gullying can occur in deserts, subject to certain lithologic restrictions, but, if restricted to uplands, gullying (within a desert) may reflect micro-climatic, high-moisture conditions rather than relicts. As noted by Fenner (1948), lowland desert watercourses almost always show some alluviation. They are largely relict from climate changes or are due to trans-environmental runoff, now discontinued. Most such channels evidently tend to become buried (see Chapter 4). The fragility of humid soils in deserts is attested to by the fact that one simply does not find humid type soil exposed at the

surface in deserts, even in those regions shown by C_{14} dates and related data to have been humid only a few thousand years ago (cf. Dury and Langford-Smith, 1970). Mostly, soils are encountered in deserts as buried remnants under alluvial deposits or lava flows (see Cotton, 1968; Williams and Pollach, 1971; Oberlander, 1972). The over-all effects of aridity in other contexts have already been summarized (Chapter 6). We will now consider what happens when deserts become humid.

ARID-TO-HUMID SEMIARIDITY

We will begin this discussion of the transition conditions from aridity to humidity by emphasizing again that we are not considering semiaridity in the role of a climatic end-member. Also, as noted in the previous section, the transitional effects of humid-to-arid adjustments on humid relicts are essentially destructive (cf. Chapter 1). Most immediately and effectively, this is the case for vegetation and soils, and, in the long run, it appears to hold also for drainage systems, mainly by eolian or fluvial burial, blockage, slope re-orientation, and divide lowering. Davis (1905) termed the sum of these effects "drainage disintegration," and he attributed most changes of this type to work of the wind. All in all, most of the changes initiated under humid-to-arid semiaridity (i.e. soil loss, runoff desiccation) appear to be only intensified under subsequent aridity, so that the landform-deposit record tends to be that of the end-member under which the landscape is observed or its opposite extreme rather than the transition. More detailed work, particularly in paleo-pedology, may help us detect transitional environmental phenomena (cf. Williams and Pollach, 1971).

The type of semiaridity of concern here is, of course, initiated in association with a group of arid geomorphic relicts potentially much more diverse than those left by a humid geomorphic system. The environmental parameters of this type of semi-aridity may involve the same range of meteorologic characteristics as the other, but the situation is one of increasing relative moisture, not only in the rock and soil and on the ground surface, but also in the air. Again, it is the interaction of geomorphic agencies that apparently generates the main topographic effects. And just as drouths simulate some aspects of incipient aridity, so occasionally intensified rainfall like that in the Kalahari Desert in the 1960's, provides clues to the nature of a desert made humid. For the actual climate change, increased precipitation and/or relatively reduced evaporation is again assumed. Also, as noted in Chapter 4, there are several possible combinations of temperature and precipitation that could achieve essentially the same ultimate effect—the generation of a continuous plant groundcover.

The precise chronologic relationships between the attainment of various critical geomorphic thresholds is, again, uncertain. Plant propagation was considered in some detail in Chapter 5, and it is known that a desert regolith often contains many spores and seeds which spring to life with the slightest precipitation. Mostly these are grasses and low shrubs, though tree seeds may be present or can be blown in. Plant refuges (botanic relicts) may, of course, provide growth centers, especially for forests. On granular surfaces, plant-cover acquisition could be rapid. Pleistocene sequences suggest that grassland floras precede forests in many cases, as noted in Chapter 2, probably in rather close dependence upon rising water table levels. Complete plant cover would naturally end eolian activity of any significance as noted by Grove (1969). There can be little doubt that stream-flow continuities change (in both time and space) with the development of a *positive infiltration increment* (e.g. more water eventually soaks in than evaporates back to the surface). Humid soil development should also start at this point. Attain-

ment of external (exhoric) drainage should generally coincide with intermittent runoff (flow part of each year), depending somewhat on terrain capacities for impounding water. Thus, some closed structural depressions receiving intermittent or even perennial runoff (cf. Lake Chad) lose it to the atmosphere by evaporation rather than by filling and overflowing. Under most circumstances, high-volume perennial runoff should develop flow routes through to the oceans. The precise nature of the foregoing changes is clearly dependent upon the nature of arid geomorphic relicts, our next topic.

Setting the Stage—Arid Geomorphic Relicts

Perhaps foremost on the list of relicts from deserts are the xerophytic plants which may survive for a time in humid settings but which tend to be replaced by other species (Garner, 1967). The same situation generally holds for desert animals. The main arid landforms and deposits of more immediate concern include dune sand; talus; alluvium in several external configurations including fans and pediment veneers; pediments, inselbergs, and plains of several types; deflation depressions; and regional drainage nets (the nets in one of several possible stages of deterioration from exposure to arid morphogenesis).* We will consider each of these main categories of relicts in terms of their behavior during the arid-to-humid transition and to a degree thereafter during later phases of humidity.

Dune Sand in isolated masses under vegetal cover is difficult to identify unless of very great extent, since individual dune shapes tend to be softened in outline or otherwise obscured. Vegetally immobilized dune systems and ergs have been identified in various parts of the world. The ergs along the south margin of the Sahara and in Nebraska, both under grass cover, have

---

* Drainage nets restricted to uplands could reflect gullying under arid conditions.

already been noted. Otvos (1971) describes smaller dune areas under plant cover along the Gulf Coast of North America. Also, there is a small erg and more isolated dunes noted by Minard (1954, 1966) under pine forest not far inland from the New Jersey coast and another relict erg described by Connally *et al.* (1972) in northern New York under decideous forest.* Regardless of their precise origins (the New York example may relate to winds over proglacial barrens), the quartz grains which are the main dune constituent confine notable groundwater bodies, are chemically almost inert, and may well "just sit there" until laid bare by climate change once more unless incised by stream action (see Fig. 6.41).

Alluvium left by desert weathering tends to be rich in such metastable constituents as subjacent bedrock may contain, and these materials placed in a setting that is increasingly chemical must begin to decompose (Fig. 1.22). Along the margins of deserts in Tunesia, Ecuador, South Africa, Australia, and in other similar regions not long vegetated, the soils are thin and stony but very rich if irrigated. In the course of chemical weathering, zoned soils (cf. Chapter 5) should be produced, of course, but the geomorphic effects of such changes also extend to mass-wasting phenomena. Rock fragments from free faces of hillslopes and scarps under arid conditions tend to enter debris, scree, or talus slope relations of repose with angles stable for dry clasts accumulated in air. Mud flow deposits and even fluviatile gravels left where water evaporated in deserts also often come to rest on slopes, some of them steep. The arid-to-humid environmental transition, either through a rising watertable or by chemical weathering, often has the effect of injecting the relict arid regolith with two very effective lubricants—water and clay. Two thresholds affecting mass wasting would therefore be crossed with the onset of humidity, one relating to

the growing groundwater body and a second dependent on progressive chemical weathering. Either type of change is capable of altering the repose angles of the relict arid detritus. In general, a water-saturated, clay-impregnated mass of rock debris will stand less steeply than detritus accumulated in the absence of such substances (Fig. 7.13).

Relict desert regolith on steep slopes exposed to humid conditions within the past few thousand years is often very unstable. Undisturbed, the material in question may undergo accelerated creep (cf. Fig. 1.22). Undercut by a stream or artificial excavation, or saturated by heavy rains, the same material may break away in active solifluction, mudflows, debris slides, or even avalanches (cf. Garner, 1959; Hack and Goodlet, 1960).** The same relict arid debris, desiccated in a desert or during a drouth, may be essentially immobile. Geologists interested in quantifying mass-wasting phenomena should keep such circumstances in mind. It might also be well to point out that downslope movement rates for regolith depend in part on depth to bedrock, and this depth on a given slope depends

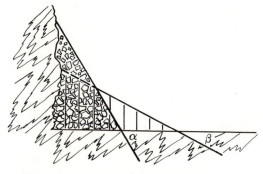

α Dry repose angle for mechanically weathered rock ± 40° max
β Wet repose angle for chemically decomposed regolith containing clay—? < 10°

Fig. 7.13. Cross section of regolith repose angle adjustment following a climate change from aridity to humidity (or an analogous one). In general, dry and/or cold environments that give way to warm and wet environments cause slope adjustments.

upon the relationship between bedrock weathering rates and downslope movement rates. A regolith developed in adjustment to one such set of limits under aridity would doubtless be forced to adjust to a new set of limits under humidity (see Fig. 1.22). The arid-to-humid transition would presumably be the time when such instabilities would be initiated, and some might continue well into a subsequent humid episode.

Terrain left by arid geomorphic systems (apart from specific influences of alluvium and dune sand or of high tectonic relief where upland moisture varies), which could be modified by more humid conditions, tends to fall into two general categories. First, and by far the most common, there is previously arid terrain that still retains the main topographic characters of an earlier humid geomorphic system. The essential slope elements of humid drainage forms still exist (trunk valley and tributary valleys), though former channels and even deeper portions of some valleys may be alluviated. Soil is gone or buried. Second, and much more rarely, there is a near-climax suite of arid landforms dominated by plains, inselbergs, and eolian configurations, in relation to which any relict humid drainage slopes are discontinuous or fragmentary. Considered in the order just cited, for the remainder of this chapter we are primarily concerned with these two terrain extremes, exposed to running water during the arid-to-humid transition.

Moving Water and Sediment

The subject at hand has probably received more intense scrutiny than any other single subject in geomorphology. There is some justification for this, but it might be well to place our effort here in proper perspective with regard to earlier chapters. In Chapter 5 we considered running water during the generation of drainage nets under humid conditions—water that if left in such a geomorphic system would tend to carry a minimum amount of sediment.

Such permanently moist regions probably comprise less than 5% of all land areas. In Chapter 6 we examined running water in deserts, often with a surplus of sediment and commonly unchanneled on plains, which tended to dry up. Here we will consider water moving through a relict drainage system under increasingly humid conditions containing an abundance of sediment left largely by a former arid geomorphic system. By far the majority of the world's rivers carry detritus created under non-humid conditions (cf. Dury, 1966). Much of the debris in northern North America, Europe, and Siberia is glacial. The bulk of the remainder was probably generated in deserts. (We will leave consideration of the glacial effects to Chapter 8.) Relict arid effects in humid drainage systems dominate the world's humid regions in the low latitudes.

In terms of a relict humid drainage net that has been partially alluviated under arid conditions, the movement of water and sediment during the arid-to-humid transition will necessarily be directed by relict slopes (see Fig. 7.8). The valley floors and former channels are commonly buried, and the hill slopes may exhibit a covering of rock waste, or they may be barren (cf. Fenner, 1948). But the essential inclinations of the former drainage net developed on bedrock remain to direct water flow, both from direct precipitation and water related to surface return from a steadily rising groundwater table. Some relict alluvial slopes may also influence water movements (cf. Figs. 2.2, and 2.3). This is the relation that prevails through much of the central and southern Ozarks and the adjacent hilly regions of Oklahoma, Texas, Louisiana, and Mississippi; much of hilly southern and southeastern Europe; the northern Mediterranean region beyond the reach of direct glacial effects; and vast hilly, areas of Africa, Asia, India, and Australia now characterized by grassland, savannah, or scrub forest. The emphasis here is on regions, displaying some evident fluvial dissection, that are nonetheless somewhat alluviated. The relict slope influences in plains regions are much less distinct.

As explained in an earlier section, we will use the term *deaggradation* to describe the initial erosional events that follow the development of perennial external drainage in an alluviated, incipiently humid land. This term is used to emphasize the "uncovering" of bedrock along flow routes. Such uncovering appears to be the essential long-term consequence of the initial fluvial events in question and not merely the erosion of bedrock and regolith alike that seems to be generally implied by the words *denudation* or *degradation*. The point here is that direct erosion of bedrock by runoff may be inhibited or even prevented for some time by a relict layer of rock waste, depending upon maximum depth of fluvial scour (cf. Chapter 6). Talus, scree, and similar relict arid material covering hillslopes of a relict humid drainage system were already noted as potentially subject to accelerated downslope movements under conditions of increased moisture and clay content, and this movement would be toward the positions of former as well as developing watercourses. That this post-arid mass wasting is often incomplete by the time plant cover is established is amply demonstrated by the relict arid colluvium encountered under vegetal cover and soil that has not yet reached perennial watercourses situated downslope (see (Fig. 1.22).

When slope inclinations and lithologies on either side of a valley are essentially equal, relict arid colluvium on a short slope above a slightly incised first-order relict humid tributary should be eliminated by mass wasting before similar relict arid colluvium on a longer slope above that tributary's deeper alluviated valley is eliminated (Fig. 7.14). In the illustration, it can be seen that the downstream reach of the first-order stream should continue to receive relict arid debris by mass wasting more than three times as long as the headwaters. A relationship similar to this is required to explain alluvial distributions in drainages of many recently humid regions,

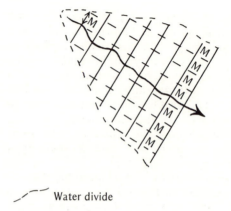

_ _ _ Water divide

$\updownarrow$M   Distance from tributary head to divide up valley side projected onto horizontal plane

Fig. 7.14. Plan view of the sequence of alluvium introduction in a valley head of a relict drainage net under humid conditions following a period of alluviation. On a purely areal basis, the watercourse "8" units downslope should receive about four times the amount of alluvium (in terms of either time or volume) as the upper "1" unit segment of the same stream. Distribution of mass wasting units (M) helps explain streams lacking alluvium (Fig. 7.15) where downstream reaches are alluviated.

for headwaters reaches of many drainage nets in the southern Ozarks, the Zagros Mountains of Iran, and the hills of southeastern Europe are swept free of alluvium even though downstream areas are not (see Fig. 7.15). First-order tributaries in the southern Ozarks in gullies less than 20 ft deep are often mere troughs cut into bare bedrock, whereas downstream areas of the same nets are extensively alluviated. In some instances, the paucity of alluvium in minor tributaries could relate to a more humid, upland setting and less effective prior mechanical weathering and alluviation. Even with this possibility in mind, the relation shown (Fig. 7.14) probably accounts for one of the most common transitional alluvial configurations on record and thus documents one aspect of the deaggradational process.

Alluvium flushed downstream through drainage nets following the onset of humidity clearly is not removed uniformly.

Fig. 7.15.   Only a single boulder remains where it was bypassed in this bare bedrock trough of a tributary of War Eagle River, Madison County, Arkansas. Downstream reaches of the same drainage, like most others in the region, are still extensively alluviated following an arid episode.

Regardless of precise relict drainage net geometry (e.g. dendritic, rectangular), RE-LICT arid alluvium in the drainage lines apparently does not move through RELICT humid straight reaches and bends with equal ease. We say this because, in many environmentally transitional drainage systems, masses of alluvium are often encountered at the lower end of a straight reach and the head of a bend, with a rather deep, slow-moving pool upstream from the alluvial "plug" and a riffle (rapids) beginning at the "plug" and extending around the bend between the point bar and cut bank (Fig. 7.16). It seems appro-priate and important at this time to point out that many studies, to date, of the behavior and effects of sediment and water in channels in humid regions have been predicated on the assumption that there is an essentially continuous supply of BOTH sediment and water. For quantities encountered in streams draining humid climatic nuclei, this would presumably be true. But relative to the onset of humid conditions following aridity, it is not true. The quantity of relict detritus left in a desert when a climate changes is finite. It can be consumed. Whether this occurs by weathering, erosion, or other means is, per-

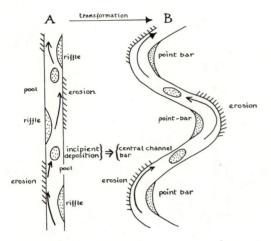

Fig. 7.16. Conformation of straight and meandering channels with respect to erosion, deposition, and channel morphology. Morphologic aspects of the channel are invariant under transformation. (From Tinkler, 1970.)

haps, irrelevant. The amount in a developing humid drainage system steadily becomes less.

There is another aspect to the foregoing problem. All too many conclusions about the size (*caliber*) of material being moved by a stream in a valley have been based on a study of clasts from a relict alluvium in the valley, as in the Pyrenees stream gravels studied by Tricart (1958). If the relicts are recognized, all may be well. Failure to recognize the relict relationship is common and probably goes back to the early notion that streams in humid regions NORMALLY alluviate their own valleys. Materials deposited in valleys under conditions where alluviation or reworking is possible usually show a dimensional relationship to the *competency* of runoff passing through the same drainage system (cf. Miller *et al.*, 1963). Actually, since much relict arid alluvium is deposited by laterally sponsored transient desert mudflows, the dimensions of the clasts along a particular reach of river valley may far exceed the competency of even peak flood runoff during a later humid episode. Such relict

arid clasts may comprise obstructions to the ready discharge of humid runoff. Some can apparently be tumbled to a place of lodgment by an occasional turbulent surge of water. But much *sedimentary bypassing* occurs, *à la* Pettijohn (1957), where relict fine-clastic sedimentary fractions are flushed out, and relatively (or completely) immobile increments remain to lodge in the channel. The lodgment of these immobile clasts is mainly a matter of original position, complemented by hydraulic geometry and channel floor roughness. Once the average discharge of a drainage system is established under a particular form of humidity, the materials in question could easily be confused with the *bed (traction) load* of a stream, but they need not, in fact, be undergoing regular transport (cf. Gage, 1953).*

On the bed of a stream that has just begun to transmit perennial runoff, bypassed gravel effectively armors that bed against corrasive incision if the channel is not free to shift laterally. As such, the clasts on the bottom are analogous to the eolian lag gravel in a desert. Quinn (1959) touches upon the subject. But unlike the eolian residues, the fluvial residue tends to be interlocked and wedged and to remain so unless disturbed—whether by high velocity discharge during flooding, changed current directions, or non-hydraulic means (bank caving) probably makes little difference. The writer recalls crossing the Rio Yaguachi in a recently arid portion of the Andes Mountain foothills in southern Ecuador. The channel at the ford was straight, averaged 2–3 ft (0.6–1 m) deep and some 150 ft (50 m) wide (Fig. 7.17). Water was clear, but flow was turbulent and estimated at 2–3 ft/sec. The entire channel bottom was paved with cobble- to boulder-size material, none of which was apparently moving. This last was ascertained when the horses and mules on which we were riding began to move across the ford. At first, there was only the sound of rushing water. Then, in faster flow areas, when

* Gage notes that large clasts in New Zealand rivers are solutionally rounded below and angular above (on the channel floor) thereby establishing their immobility at present.

Fig. 7.17. Sketch based on a photo of the Rio Yaguachi, Ecuador, a westward flowing stream in an alluviated, comparatively broad, open valley in the foothills of the Andean Sierra Occidental. Though the flow velocities are high, the cobble and boulder bed debris is hardly moving.

the animals bucked, plunged, and strove for footing and kicked loose the odd clast, an underwater avalanche followed. The clasts so dislodged could be seen and heard to roll, bounce, and hit others, apparently knocking them loose also, by impact. This type of chain reaction continued, apparently until each of the moving clasts lodged against others too large or well anchored to shift; a matter of 20–30 seconds and 75–100 feet of downstream movement. The mules didn't seem to mind. But the horses and some of the riders didn't like it at all.

Whether a fluvial pavement of relict arid clasts in a channel is essentially continuous, as in the Ecuadorian example, or occurs in the form of discontinuous "plugs" (as observed in the southern Ozarks, and elsewhere, and previously discussed), is probably largely a matter of original relict alluvial volume and coarseness. That same alluvial volume also determines the water discharge capacity of a

given relict flow route. The ability of a land-surface depression (channel, valley) to transmit runoff without overflowing is dependent upon its cross-sectional area and, to some extent, its slope. Alluviation of such a depression (known to occur in deserts) reduces this cross-sectional area and may, at least locally, change the slope and thus tend to cause overflowing (*flooding*).

A *flood* (unconfined runoff), in terms of an essentially fixed water-sediment ratio under humid conditions, becomes a matter of humid rainfall intensity, duration, and incidence, and several flood-frequency studies have proceeded along these lines. But in the context of an arid-to-humid transition, a flood is an expression of the extent to which a watercourse can contain peak runoff volumes. This tends to depend on the degree to which runoff has eliminated alluvium. A deep, V-shaped valley or canyon in a humid climatic nucleus can hardly flood, for all flow through it is confined. Floods are thus an expression of environmental transitions where watercourse discharge capacities have been impaired (Fig. 7.18). Floods are inevitable on desert plains, for there is almost no confinement. They appear to comprise the steady-state runoff form on such plains. Floods occur on the floors of alluviated valleys where channel enlargement has not yet matched peak discharges. And confinement of runoff should become progressively greater as humidity endures. Flood incidence (as an expression of disequilibrium following climate change) should therefore progressively decrease.

ARID-TO-HUMID DRAINAGE ADJUSTMENTS

The geomorphic adjustments that occur during an arid-to-humid climate change do not necessarily terminate when the humid condition (defined vegetally) is achieved. Indeed, how long it takes for an adjustment to occur in weathering, mass wasting, or drainage clearly depends upon the magnitude of the change required to induce equilibrium and upon the intensity or effectiveness of the agency involved. Thus, it should take longer to alluvially bury a deep valley than a shallow channel under similar arid conditions. It should require more time to thoroughly decompose a thick alluvial deposit than a thin veneer. The words so far left out are, of course, "everything being the same." And, again, some types of drainage change following aridity require more time and energy than others. Therefore, some of the effects already mentioned involving water and sediment are probably completed under semiarid conditions, whereas others persist for some time under the humid conditions that follow. The same is apparently true of many of the drainage phenomena to be scrutinized shortly.

In the foregoing context, it should be mentioned that climate changes need not involve a sequence of environments of equal intensity. Studies of possible causes of glaciation, for example, suggest that intensified glaciation does not occur because of a change in a single factor but rather because of changes in several interrelated factors in which the cumulative effects are additive—that is, they all express a similar tendency. In the opposite case involving the same factors, some may be substractive—that is, they act to nullify other effects—and thus generate an opposing trend, de-glaciation, with variations.* As a consequence, a series of arid-humid climatic alternations may similarly involve instances where each climate phase (humid, for example) is more intense and/or prolonged than that before. In terms of drainage adjustments during such climatic alternations, some drainage trends may persist through several moist episodes in spite of intervening dry periods. The opposing relationship involving aridity is also possible. These remarks apply to such phenomena as stream channel redevelopment, channel evolution, and channel maintainance during the arid-to-humid transition. Records of such drainage events being interrupted short of some final climax expression are frequently preserved in river terrace systems and deltaic deposits. Each of these relationships will now be examined in some detail.

Channel Redevelopment in Drainage Nets

Redevelopment of channels in a drainage net previously alluviated in a desert apparently does not occur with the simplistic neatness of a hypothetical rill system developing on a smooth, inclined, granular surface as discussed by Horton (1945) and illustrated in Chapter 5. The relict alluvial surface is granular, of course, and, depending upon the vagaries of prior arid eolian and fluvial agencies, should be generally if somewhat irregularly inclined down valley. Local reversals in valley slope

* See Chapter 8.

Fig. 7.18. The floodplain of the North Fork of the Kentucky River at Lothair and Hazard, Kentucky, is completely inundated in this view recorded during the catastrophic flood of March 12, 1963. Such discharges of water are far in excess of the amounts existing channels can transmit because the valleys have been alluviated. *(Courier-Journal* and *Louisville Times.)*

observed in deserts (which could be inherited) relate to such features as alluvial fans developed at the mouths of side valleys (cf. Figs. 2.2; 2.3), sand dunes (Fig. 5.51), talus, mudflow deposits, and deflation depressions plus structural escarpments or uplifts. Where relict water-flow depressions in headwaters areas are most narrow, runoff is observed to cover the entire gully or ravine bottoms, and many of these tend to be swept free of debris and converted into bare rock channels (see Fig. 7.15).

The coarsest fraction of the relict arid detritus flushed from a headwaters area may be presumed to be that first encountered in a downstream direction; this fraction has been reworked in an open sedimentary system and usually consists of a poorly rounded gravel lacking appreciable sub-sand–size material. Usually, the deposit in question is found as a discrete mass of alluvium lodged at the lower end of a straight reach of stream channel. Also, the deposit is normally somewhat vegetated by scrub and trees, such as willow and sycamore, and is therefore anchored by root systems and whatever valley floor irregularity, constriction, or other configuration as may have originally induced entrapment of the clasts. This *al-*

luvial *"plug"* functions as a dam which is then crossed by a shallow "spillway" channel occupied by a *riffle* or rapids, usually on the outside if on a bend (Fig. 7.19). If the "plug" is large, there tends to be an area of impounded water upstream (*pool*) that traps larger clasts moving down the channel and thus adds to the mass of the "plug." Since water moving past the "plug" often has lost some of its load, scour, often extending to bedrock, is common just downstream. If the relict alluvial veneer is thin, a considerable downstream reach may again be stripped of debris down to bedrock. The coarse fraction from this material then forms a second "plug" just below, and so on, downstream.

Since the quantities of relict debris are finite (cf. Fig. 7.14), additions to the "plugs" should gradually diminish as the humid condition endures. Also, it will be noted that a "plug" should form at a place of lodgment as soon as the volume of coarse debris required to block the channel cross section has been bypassed upstream. In regions where the relict alluvium is areally of comparatively uniform texture, the "plugging" should occur at relatively uniform intervals along the channel. Hydraulic explanations for riffle-pool spacings have been offered by several researchers

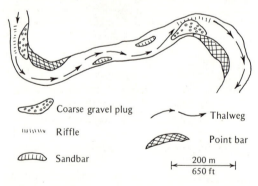

Coarse gravel plug
Riffle
Sandbar
Thalweg
Point bar

|← 200 m →|
650 ft

Fig. 7.19. Map view of a reach of channel showing the development of coarse gravel plugs in southern Ozark streams from relict alluvium which is being bypassed. Where fully stabilized, the reach occupied by sandbars may instead be occupied by a long, deep pool.

including Leopold *et al.* (1964, p. 297), Dury, (1967), and Tinkler (1970). These seem applicable where valley flats are appreciable, and flow is free to meander. However, *Thalweg sinuoisity* may influence alluvial "plug" disposition in the narrow valley bottoms being discussed here (see Fig. 7.24 and also Fig. 7.30 in the section on channel evolution and maintainance).

Where an alluvial "plug" has developed at the upstream end of a bend in a stream, a shallower, usually discontinuous flow route may exist on the inside of the alluvial mass, and there are also usually a group of discontinuous, anastomotic scour pockets situated between clumps of vegetation on the "plug." The latter are only under water at times of high discharge. A similar relation develops in some valleys alluviated by glacial outwash (cf. Chapter 8). Flow routes other than the main ones on the vegetated alluvial surface are often only indicated by the directions in which rooted reeds and grasses have been bent, just as the depths of flood waters are shown by the height at which flotsam catches in tree branches. Curiously, the latter indicators in the southern Ozarks show that many alluvial plugs consisting of gravel up to 8–10 inches (20–25 cm) in maximum diameter are under 7–10 ft (2–3 m) of very fast water several times each Spring. The coherency of the "plugs" is apparently considerable.

The tractive force required to move relict particles from an alluviated drainage net is presumably much the same as that already discussed for gullying alluvium in a desert (cf. Chapter 6). As noted by Sundborg (1956) and Morisawa (1968), much lower flow velocities are required to move sand than to move either silt or gravel, and, for grains larger than 0.5 mm, erosion velocity increases directly with grain size; for particles smaller than 0.5 mm, the smaller the particle the harder it is to move (Fig. 7.20).* The curves shown only suffice to outline very broadly the realms of erosion, transportation, and deposition of material

---

* In general, small particles have greater cohesion and larger ones greater inertia.

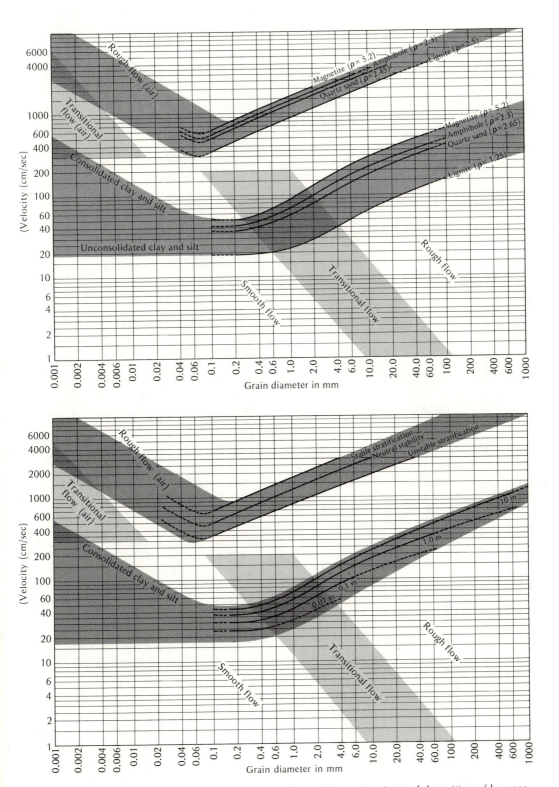

Fig. 7.20. Curves showing velocity/particle size relationships relative to erosion and deposition of homogeneous sedimentary material. (Redrawn from Sundborg, 1956.)

by a stream of water. For stream erosion of heterogeneous alluvial bodies, such as those formed in deserts, the individual particle size/velocity erosional relations should not apply. For example, saltating grains of one size (i.e. sand) can by impact cause entrainment of smaller particles (i.e. silt and clay) at a velocity below that required for the finer material alone, and undercutting of cobbles and boulders can cause such clasts to also move at a velocity below that required for such large fragments alone. This would accelerate the erosion of poorly sorted relict arid detritus by humid stream flow in its early stages of development. Later, when individual tracts of channel floor become occupied by clasts of one general size through sorting processes, the caliber of sediment being moved should more closely correlate with velocity.

Deposition of sediment within a stream channel or by overbank flow within the potential sweep of a migrating meander system can be considered temporary (cf. Garner, 1963) and is usually related to a slackening of current. Stokes' Law of settling velocity applies to most of the grains involved, if one assumes sphericity, since those particles that move after bypassing is completed tend to be small. As previously noted, many mechanically weathered clasts are angular. When larger clasts move it is generally as a part of the traction load, and, as already noted, their deposition is often a matter of lodgment to which Stokes' Law does not apply. For small grains, a particle drops out of suspension and is deposited when the upward velocity components provided by currents fall below the settling velocity for a given grain size. The latter velocity, for spherical particles, depends upon the resistance offered by the fluid, which in turn depends upon the surface area of the grain, $6\pi r$, viscosity of the fluid, μ, and the velocity of fall, V. The bouyant force tending to support the particle in the fluid is equal to the volume of the particle (a sphere), $\left(\frac{4}{3}\right)\pi r^2$, times the density of the fluid, d_2, times the force of gravity, g. For a suspended particle the

bouyant force is balanced by the force pulling the grain downward that equals the volume of the particle, $\left(\frac{4}{3}\right)\pi r^2$, times its density d_1, times the force of gravity, g. Equating upward and downward forces for a grain in suspension

$$6\pi r\mu V + \left(\frac{4}{3}\right)\pi r^3 d_2 g = \left(\frac{4}{3}\right)\pi r^3 d_1 g$$

$$6\mu V = \left(\frac{4}{3}\right)r^2 g(d_1 - d_2)$$

$$V = \frac{2}{9}\frac{gr^2(d_1 - d_2)}{\mu}$$

This mathematical relation shows that settling rates depend primarily upon grain size and density, since gravity, viscosity, and density of water are constant for a given time and point along a humid stream. This is, of course, not true for a desert mudflow, nor would it be true for many streams carrying relict detritus because of angular clast shapes, if for no other reason.

At the point along an alluviated drainage net where channel width is notably less than valley-flat width (? one-half), one begins to encounter masses of relict alluvium which appear to have escaped re-working by contemporary runoff. This relation varies from region to region, of course, and upon the nature of flow patterns—meandering, anastomotic, or other. Where consequent alluvial slopes have influenced the locations of channels, and there is no subsequent meandering, there may be signs of early anastomotic flow on the valley flat followed by later confinement to a single channel as was seen in the case of Mountain Fork Creek (previously discussed). In that drainage, the earlier anastomotic pattern and the present channel are both marked by the distribution of the lithic tracer (white chert) shown in cross section (Figs. 2.31; 2.32) and here in plan view (Fig. 7.21).

Whether a channel developed on alluvium in the relict depression of a former drainage net consists of crudely straight reaches and bends, or meanders, or has a braided flow pattern apparently depends upon a number of factors. A valley fill into which a channel has been cut may, of

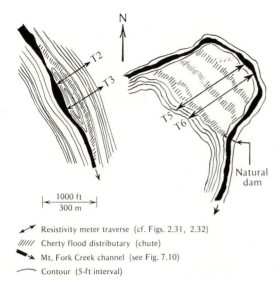

Resistivity meter traverse (cf. Figs. 2.31, 2.32)

Cherty flood distributary (chute)

Mt. Fork Creek channel (see Fig. 7.10)

Contour (5-ft interval)

Fig. 7.21. Maps of two segments of Mountain Fork Creek, northwest Arkansas (Fig. 7.10), depicting channel positions, flood overflow routes marked by cherty alluvium, and resistivity meter traverses designed to show alluvial thicknesses (cf. Figs. 2.31 and 2.32).

course, be periodically flooded and is widely termed a *floodplain*. The bulk of the material upon which valley flats have been developed may have been thoroughly reworked by swinging meanders or may represent detritus left entirely by other agencies—in the present context, some desert phenomenon. As noted by Leopold *et al.* (1964), overbank flow during floods may leave some veneer of fluvial debris behind on floodplains. And, as a channel is incised into a valley fill, its ability to confine runoff increases, and flooding incidence should diminish, finally, to the point where the valley flat is no longer flooded. At this stage, the elevated remnants of dissected alluvial fill are termed *alluvial terraces* (cf. Fig. 4.28). Such terraces differ from *rock-cut terraces* (Fig. 2.31), which have bedrock cores, and tend to be paired on either side of a valley (see Quinn, 1957; Howard *et al.*, 1968).

In beginning geology courses, the student soon becomes familiar with the sinuous bend of a river termed a *meander* and quickly learns that the velocity is com-

paratively uniform along a straight reach of a river as compared with that on a curve (Fig. 7.22). He or she usually also appreciates that the higher velocities of flow on the outside of a bend cause some lateral erosion, to generate a steep *cut bank*, whereas the slack water on the inside induces formation of a lunate alluvial deposit called a *point bar*. The lateral and downstream migration of meanders induced through this combined erosion and deposition of granular material produces a variety of valley flat features including *meander scrolls*—arcuate scars showing former channel locations; *sloughs* —slack water or boggy depressions formed by flood scour or meander cutoff; *oxbow lakes*—cutoff meander segments occupied by standing water; *natural levees*—channel banks, especially concave ones, raised above general flood plain level by deposition of coarse fraction flood sediment; and *backswamps*—marshy areas where slack water from overbank flow is ponded, and fine-grained sediment accumulates between natural levees and a terrace or valley side (see Fig. 7.23). A general description of

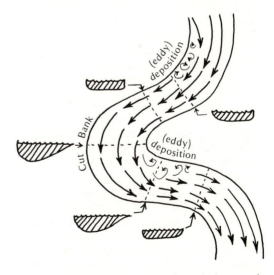

Fig. 7.22. Diagram illustrating channel shape and velocity variations along a meandering channel developed in alluvium. Velocities are roughly proportional to arrow lengths. Deposition tends to occur in slack-water eddies on the inner parts of bends and results in point bars.

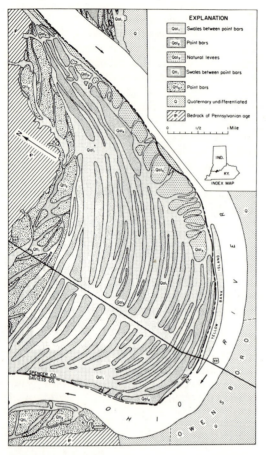

EXPLANATION

Qal₄	Swales between point bars
Qal₃	Point bars
Qal₂	Natural levees
Qal₁ₛ	Swales between point bars
Qal₁ₚ	Point bars
Q	Quaternary undifferentiated
P	Bedrock of Pennsylvanian age

0   1/2   1 Mile

INDEX MAP

Fig. 7.23. Surface expressions of point bar deposits on the inside of bends and natural levee accumulations along the Ohio River near Owensboro, Kentucky. (Paul Potter, 1967.)

the main valley flat features related to the phenomenon of meandering does not elucidate the basic causes of meandering. This we shall attempt to do now.

It was observed previously (cf. Chapters 1, 5) that the development of meandering flow patterns apparently requires the presence of a granular surface. That is not to insist that meandriform flow is not intrinsic to fluid movement but only that a granular medium may be required for meandering to be expressed. Consequent slopes on relict arid alluvium may possibly inhibit the process just as would a narrow rock-walled gully. But there is apparently

more to the phenomenon than this. An anonymous researcher for the United Kingdom Ministry of Technology (Anon., 1966) experimented with sediment injection rates on an outdoor sand bed 300 ft long and 30 ft wide. Below a critical sediment injection level, meanders would not form however long the experiment was continued. Above the critical injection level, meanders were induced in a straight channel where shoals formed on alternate sides of the channel and later developed into meanders as they worked downstream. At very high injection rates, braiding occurred. Clearly, in a natural setting, injection of sediment would depend upon a variety of factors including relict alluvial erodability (in part as indicated by clast sizes), bank caving, depth to bedrock, etc.

Similar experiments conducted by Schumm and Khan (1972), and Schumm *et al.* (1972) on variations of slope (equivalent to valley slope, not channel gradient) and sediment load detected threshold values for both at which channel patterns were altered notably. They noted that at very low slope and sediment load the channels remained straight but that at a discharge of 0.15 cu ft/sec a meandering-thalweg channel formed at slopes greater than 0.002% —probably the equivalent of the shoal stage previously noted (Fig. 7.24). Further increases in slope and load caused thalweg sinuosity to reach 1.25, and at slopes greater than 0.016 a braided channel formed. A truly meandering channel was not induced through slope variations and was only achieved finally with a sinuosity of 1.3 when a suspended-sediment load (3% kaolinite) was introduced (cf. Figs. 7.25; 7.26). *Sinuosity* of a channel is the ratio of thalweg length to valley length.

The two foregoing studies tend to complement one another. The first induced meandering and braiding WITH A CONSTANT LOAD CALIBER OF VARYING AMOUNTS, whereas the second VARIED LOAD CALIBER to cause meandering BUT HAD TO INCREASE SLOPE TO CAUSE BRAIDING. In terms of the arid-to-humid transition, it would appear

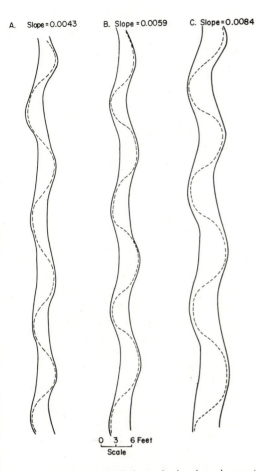

A. Slope = 0.0043   B. Slope = 0.0059   C. Slope = 0.0084

0   3   6 Feet
Scale

Fig. 7.24.   Diagrams of channels developed experimentally in which the deepest flow route (thalweg) meanders. Solid lines are boundaries of bank-full channels; dashed line is thalweg. Note that in spite of a thalweg sinuosity of 1.25, a straight line drawn down the channel center would touch neither bank. For cross sections of this channel type, see Fig. 7.25. (From Schumm and Khan, 1972.)

that increased clay and suspended loads would accompany prolonged humidity (Garner, 1959). But expulsion of relict alluvial deposits should generally reduce regional slopes and hence inhibit braiding, if slope is a major cause. Also, before plant cover is complete, sediment injection rates into watercourses should be very high, because of this, and also because heterogeneous sediment is being flushed from first-order tributaries. Winnowing out of relict arid fine-grained residual should pro-

vide initially accelerated injection of sediment, but this should diminish as bypassed gravels become relatively coarser. This relation could initially induce meandering or braiding and later inhibit maintenance of the same patterns.

Studies of the relations between braided, meandering, and straight flow patterns and slope/discharge values on actual rivers have been made by a number of researchers including Lane (1957), Leopold and Wolman (1957), and Ackers and Charlton (1971), and these also disclose threshold values (Fig. 7.27) that can be compared with the data of Schumm *et al.* (1972). The comparisons are puzzling, however, for in the comparatively brief span of the average environment, tectonic slope changes are usually nil, those slope changes due to erosion and deposition are slight and only doubtfully capable of causing

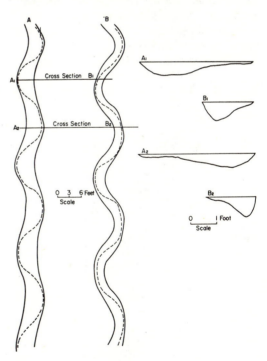

Cross Section

Cross Section

0   3   6 Feet
Scale

0   1 Foot
Scale

Fig. 7.25.   Maps showing a channel (A) before and (B) after introduction of suspended sediment load into an experimental channel (cf. Figs. 7.24 and 7.26). Cross sections show changes in channel dimensions and shape. Slope was 0.0064. (From Schumm and Khan, 1972.)

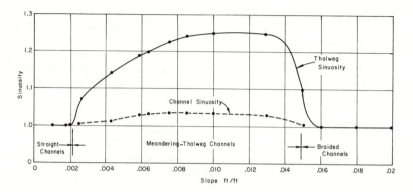

Fig. 7.26. The relation between slope of surface on which experimental channel was formed (valley slope) (Fig. 7.24) and thalweg sinuosity versus channel sinuosity. (From Schumm and Khan, 1972.)

some flow patterns (braided), and there is no apparent causal relation between relict alluvial clast dimensions and volumes and subsequent humid discharges. Sediment quantities, dimensions, and availabilities, all of which are interrelated and affect injection rates, are seen to vary greatly with climate phase and seemingly affect flow patterns significantly.

Adequate and meaningful quantification of the load-discharge relations is yet incomplete. The writer is reminded of a

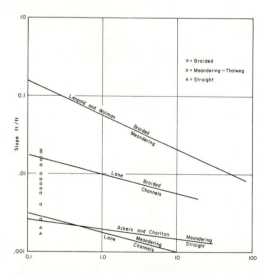

Fig. 7.27. Relation between slope and discharge and threshold slopes at each discharge as defined by Lane (1957), Leopold and Wolman (1957), and Ackers and Charlton (1971). (From Schumm and Khan, 1972.)

meandering stream observed on the presently semiarid interior plains of Venezuela (Llanos) that various data (Garner, 1966) indicate were recently arid. The stream in question (Rio Pao) is also braided (Fig. 7.28), the braiding occurring within the meandering channel. There are no signs of recent tectonism, and, from a consideration of the pattern relations, it would appear that the meanders were initiated on a steeper alluvial slope than that upon which the braided pattern is presently being maintained. Also, braiding followed meandering. Only variations in relative sediment and discharge amounts appear capable of accounting for the pattern changes recorded. The region has a record of both more and less moisture than at present (Garner, 1966).

Alluviated Channel Evolution and Maintenance

Apart from channel incision into bedrock of headwaters reaches of relict drainage nets following alluvial flushing, most initial major humid drainage adjustments involve channel modifications in alluviated, downstream reaches. The majority of the flow patterns involve meandering, though a few North American examples including the Platte River in Nebraska, and the Cimmeron, Arkansas, and Little Red rivers to the south are notably braided, particu-

Fig. 7.28.   Aerial view of a reach of the Rio Pao, a meandering, braided stream on the Venezuelan interior plains (Llanos), a combination pediplain and aggradational plain that is presently semiarid.

larly during times of low water. It will be noted at this point that a *braided* pattern of flow is caused when runoff deposits sedimentary obstructions WITHIN a channel. Most of our attention will go to meanders, however, and they have been most thoroughly studied.

Several European hydrologists including Prus-Chacinski (1954) and Leliavsky (1955) explain the growth of successive symmetrical bends in a river as a result of helical flow. Analogous movement of air is utilized by Folk (1971) to explain development of longitudinal dunes in deserts (cf. Figs. 6.37 and 6.38). If helical flow is instrumental in meander development, it is not yet clear how it is expressed in related channel widths, curvatures, and meander wavelengths. Perhaps the most lucid recent statement in terms of hydraulic

drag is that of Grorycki (1973). The early explanations of Matthes (1941) and others that meandering will occur in the presence of bank erosion are hardly adequate unless —in the light of experiments cited—certain injection thresholds are attained. Yet, as noted by Leopold *et al.* (1964), the Gulf Stream meanders without either sediment load or banks (Fig. 7.29).* And it should be noted that the water in the aforementioned experiments moved down channels of uniform shape, slope, roughness, and depth, yet it induced sinuosity in the erodable material by what can hardly be other than an internal velocity variation— probably in spiral form (Fig. 7.30) (see Gorycki, 1973).

Meanders are commonly described and compared in terms of wavelength and radius of curvature (cf. Fig. 7.29) which

---

* There are water salinity-density variations, however, which could possibly be construed as "load," i.e. dense water masses in contact with less dense masses.

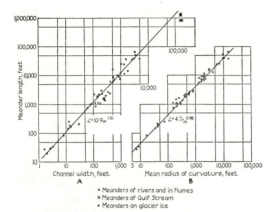

Fig. 7.29. Relation of meander length to width (A) and to radius of curvature in channels (B). (From Leopold, Wolman, and Miller, 1964.)

are, in turn, discussed in terms of discharges and sediment loads. According to Schumm (1972), "The width (w), depth (d), meander wavelength (l), gradient (s), shape (w/d), and sinuosity (P) of stable alluvial river channels are dependent on the volume of water moving through the channel (Q_w) and the type of sediment load conveyed through the channel (Q_s)."

The discharge relationship

$$Q_w \propto \frac{w, d, l}{s}$$

demonstrates that channel width, depth, and meander wavelength will INCREASE with an INCREASE in discharge but that channel gradient will DECREASE. Schumm points out that, for a constant discharge, a change in the average quantity of bed load (Q_s) which is sand size or larger moved by the stream is related to channel morphology as

$$Q_s \propto \frac{w, l, s}{d, P}$$

which shows that as the sand load or bed load INCREASES, channel width, meander wavelength, and gradient INCREASE, whereas channel depth and sinuosity DECREASE, and the width/depth ratio INCREASES.

In most natural settings, a change in discharge will be accompanied by a change in sediment load, but the channel response to such changes will be complex. Schumm (1972) cites an example where both discharge and percentage of bed load INCREASE—a common natural relationship. In this case, channel morphology will be altered as follows

$$Q_w^+ Q_s^+ \propto \frac{w^+ l^+ F^+}{P^-} s^+ d^+$$

where F is the width/depth ratio. Since the direction of change of each variable is indicated by a plus or minus superscript, the possible solutions are almost as numerous as the variables. Schumm observes that the width of channel, width/depth ratio, and meander wavelength would INCREASE but sinuosity would DECREASE. However, since bed load and discharge effects are opposed, depth and gradient effects would tend to be neutralized. At least this would be the case if it were not for the fact that REDUCED sinuosity should INCREASE the gradient. Also, INCREASES in both width and width/depth ratio should either DECREASE the depth or leave it unchanged. DECREASED discharge and percentage of bed load should, of course, have opposite effects to those shown in the last equation. After rereading the foregoing lines, the writer is perplexed. The logic seems acceptable, but the utility is marginal. There must be another way.

The idea in calculations like those just cited is to establish quantitative relationships between particular meander dimen-

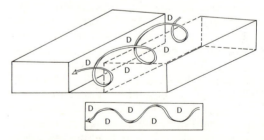

Fig. 7.30. Diagram approximating the character of helical flow of water in a channel (above) and the same in plan view (below) showing probable areas of deposition (D). Areas of erosion would be opposite. Helical flow may also be involved in air streams which move sand (cf. Fig. 6.38).

sions and forms and genetically equivalent amounts of water and sediment. If this can be done for a large number of contemporary examples, the data should aid in the evaluation and interpretation of other drainages—either modern channels or paleochannels—which exhibit similar parameters, but for which discharge and sediment data are lacking. Thus, a valley floodplain, a delta, or a riverine plain like that in Australia, may carry the imprints of several separate sets of meanders, each set indicating by its different dimensions formerly different hydraulic (load-discharge?) relationships. Dury *et al.* (1972) make use of these concepts in their study of the River Severn on Shrewsbury Plain in Wales and indicate that a drainage that shifts from former large meanders to present small meanders is *underfit*. The same authors observe that meander wavelength is commonly between 8 and 12 times the bedwidth which is compatible with Bagnold's finding (1960) that resistance to flow around an open channel bend is at a minimum where the radius of curvature lies between two and three bedwidths. Take the average meandering geometry, $2 \times$ radius $= \frac{1}{2}$ wavelength and $4 \times$ radius $= 1$ wavelength $= 8$ to 12 bedwidths. In this relationship, bedwidth and meander wavelength are related to dominant discharge in the form

$$w, l\ Q_w\ 0.5$$

where w, l are, respectively, bedwidth and meander wavelength, and Q_w is the dominant channel-forming discharge.

Several workers regard the channel-forming discharge as bank-full, and in many streams the dominant discharge is far less than this. In his interesting study of the Angabunga River in New Guinea, Speight (1965) discusses the several problems of estimating bank-full discharges, distinguishing between channel versus bank sediments, and ascertaining meander wavelength modes. The problem of measuring bank-full discharges which provide crucial data on flooding and channel formation is rendered more difficult by a lack

of widely accepted limits (Fig. 7.31) and by the fact that bank-full flow for one reach of a stream may be overbank flow for another reach. Stream discharge is not always the factor dominating meander form, and, in his comparison of several rivers in the western United States and Australia, Schumm (1969) notes causal meander-wavelength and load-caliber interrelations.

It was noted in Chapter 1 that earlier workers in geomorphology assumed that stream flow velocities decreased in a downstream direction since slope becomes less in that direction. This notion was used to explain alluviation of downstream reaches of rivers by Davis (1890) and others. Leo-

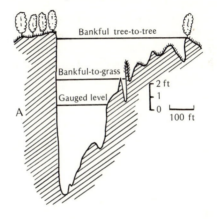

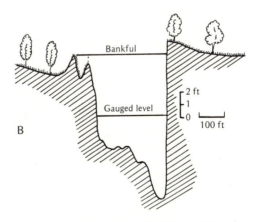

Fig. 7.31. Cross sections depicting relationships between gauging level and bank-full (ascertained on two bases) along the Anabunga River, Papua, New Guinea. (After Speight, 1965.)

pold and Maddock (1953) challenged this velocity notion. And, as recently pointed out by Carlston (1969) in his study of several North American rivers, there are a variety of mean velocity relationships along major rivers. Some rivers increase in velocity downstream, many maintain a constant velocity, and a few decrease in velocity. Carlston concludes that the most common relationship on long river segments is a constant flow rate with notable increases or decreases mostly occurring on smaller streams. His analysis also shows that very large rivers such as the Mississippi accommodate downriver increase in discharge through depth increases, whereas lesser rivers adjust to the same increase by becoming wider.

In their 1953 study, Leopold and Maddock point out that width (w), depth (d), and velocity (V) change in a downstream direction as simple power functions of mean annual discharge (Q) so that

$$W = aQ^b$$
$$d = cQ^f$$
$$V = kQ^m$$

where a, c, and k are the coefficients for width, depth, and velocity, respectively, and b, f, and m are the corresponding exponents. By definition, b, f, and m must add up to 1.00 and the logs of a, c, and k 0.00, or the products of the three coefficients must equal 1.00.

Leopold and Maddock's average value for exponents (1953) based on mean annual discharge are

$$b = 0.5$$
$$f = 0.4$$
$$m = 0.1$$

but Carlston (1969) restudied the ten river basins that provided Leopold and Maddock's data, and, using a least squares solution on a computer, arrived at means for the exponents of

$$b = 0.461$$
$$f = 0.383$$
$$m = 0.155$$

Carlston also echos Mackin's 1963 observation that flood discharge parameters are geomorphically more significant by far in effecting drainage line changes than is mean annual discharge.

Deaggradation: Fluvial Terraces, and Deltas

We have considered the character of arid relicts and the environmental adjustments that accompany the onset of humidity together with assorted thresholds attained and disequilibria induced. The thresholds and disequilibria of greatest geomorphic impact mainly relate to weathering, mass wasting, and fluvial erosion, though the role of ancillary controls such as the water table and plants is unmistakable. In these circumstances, the development of a water surplus, requiring drainage, more or less incidentally involves topographic depressions containing appreciable loose rock detritus. In effect, this involvement produces a much higher solid sediment/water ratio than that normally found in streams that do not flow on relict alluvium. Meandering and braiding are common. In the course of developing A LEAST WORK RELATIONSHIP with respect to the relict terrain and deposits, this water leaves some or much of the coarse alluvial fractions behind—at least initially—and, if Langbein and Leopold (1966) are correct, the same water develops meandering habits as a "most stable" drainage form on granular surfaces while making adjustments in channel width, depth, and related hydraulic geometry. Braided patterns are less common and presumably less stable. Smith (1971) relates braiding to the development and dissection of transverse bars.

Relict arid alluvium amounts to an energy drain on a developing humid fluvial system, hence the evident tendency to eliminate the relict. Elevated alluvial terraces along the margins of many valleys amply demonstrate that the deaggration process is often incompletely accomplished —often several times. Nevertheless, a migrating meander belt may completely rework the alluvium in a valley fill to the maximum depth of scour over a valley flat many miles in width. Much material is flushed seaward as a consequence, and, given time, a humid stream should be able to deepen its channel to bedrock while clearing out most of a relict valley fill. Several possible relationships between terraces and one or more valley fills are illustrated by Leopold *et al.* (1964), and this emphasizes the fact that rivers flowing on alluvium tend to exhibit two erosional styles. Some additional possibilities are shown here (Fig. 7.32). One erosional style amounts to essentially vertical incision, and the other involves lateral channel migration with little or no deepening beyond some mean scour depth. Actually, of course, it is not strictly speaking an either/or situation, since both erosive styles are removing sediment from the valley bottom, and each shares the vertical and lateral movements of the other to some degree.

In spite of the common traits exhibited, it appears nonetheless that some streams do tend to cut downward into alluvial fills to particular depths and, through meandering, gradually remove the fill to produce a

valley flat at a lower elevation, a flat that may be bounded laterally by terrace risers (Fig. 7.33). Alternatively, a notch may merely be locally deepened into an alluvial valley flat and, possibly, on down into bedrock with little apparent limit on vertical cutting. Early explanations of these relationships (cf. Davis, 1899) usually involved shifts in the base level to which the stream flowed, it being reasoned that a fixed base level would limit incision, a falling base level would permit incision, and a rising one would cause aggradation. The fixed base level presumably accounted for valley flats developed during lateral erosion, whereas the falling base level allowed valley flat incision (cf. Howard *et al.*, 1968).

The foregoing arguments have strong overtones of the graded stream concept discussed in Chapter 1 and thereby imply that lateral erosion is an indication of a graded state during which incision would not occur. A recent study of the Truckee River in Nevada helps put this notion in perspective. Here Born and Ritter (1970)

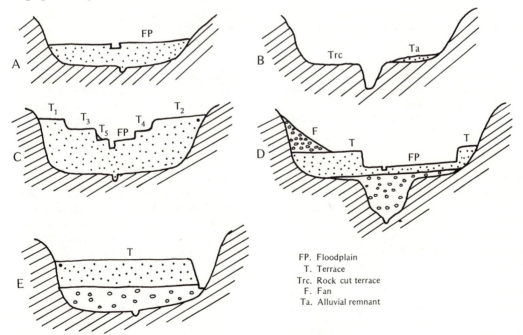

Fig. 7.32. Cross sections of selected valley fills and related erosional and depositional features. For alternative configurations, see Leopold, Wolman, and Miller (1964).

Fig. 7.33. The Urubamba River incising relict arid alluvial fill to produce stream terraces along this tributary to the Amazon drainage system at an elevation of about 9,000 ft (2,700 m). Incision depths into alluvium at this point are in excess of 100 ft (30 m).

recorded an artificially induced continuously falling base level for the stream over a period of 44 years, beginning in 1910. During the period in question, the stream formed at least six well-developed erosional terraces between 10 ft (3 m) and 35 ft (11 m) above the present stream, while the base level established by Pyramid Lake dropped 80 ft (24 m) (Fig. 7.34). It seems clear that the location and direction of movement of base level in this case permitted the over-all downcutting but was not instrumental in the several recorded shifts from lateral to vertical erosion. Hydrologic effects are apparently involved under a single type of environment, and, because of the brief geologic time, tectonic influences presumably can be disregarded.

Much of the apparent confusion in the literature about the genesis of fluvial terraces probably stems from uncertainty as to the origin of the associated alluvium. In the context of alternating arid and humid environments, the possibilities are rather more restricted. Mechanisms for extensive self-aggradation of river channels under HUMID CONDITIONS ALONE do not seem to exist. Minor open-system gravels may be deposited when a humid stream dries up with the onset of aridity, and more open- and closed-system detritus may be added to channels and valleys where runoff, either from local precipitation or from trans-environmental rivers, desiccates in a desert. Sediment can also be introduced into stream valleys by glacial outwash or can be impounded therein where their lower reaches are drowned by lake- or sea-level rises (cf. Saucier and

Fleetwood, 1970). A few valleys including those of the lower Mississippi and Rhone rivers appear to contain all three types of material.

We will consider the glacial landform situation in Chapter 8. But glacial climates coincide in time with other kinds of climates elsewhere and also affect sea levels. Evidence suggests that some regions were humid during glacial times (the American Southwest), whereas others were arid (parts of equatorial South America and Africa). The opposite climatic relation is apparently

also valid. High sea level is ostensibly correlated with interglacial times, and some interglacial rivers débouché into drowned lower reaches which are being subaqueously sedimented. Thus, such a river could develop a set of terraces in subaerial reaches that were previously alluviated and, with a drop in sea level due to glaciation, could cut a second terrace system in emergent estuarine deposits (cf. Russell, 1958). Depending upon the setting, therefore, some terrace systems could be glacial and some interglacial, but the in-

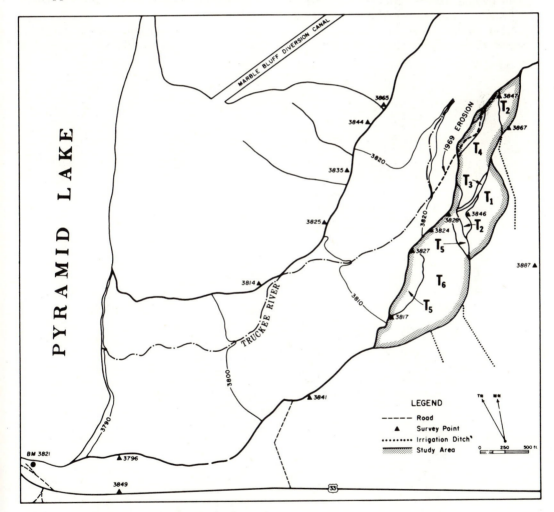

Fig. 7.34. Map of lower Truckee River region showing terrace sequence developed since lake level began to fall in 1900. Surveyed October, 1968. Dashed line 1969 erosion indicates the extent of valley widening during the 1969 water year. (From Born and Ritter, 1970.)

cisional phases that produce scarps would presumably always tend to occur during a humid phase when sediments could be flushed out, as noted by Saucier and Fleetwood (1970). Meanders or braids related to such incision could thus develop under a variety of discharge and sediment load relations as exhoric runoff (initiated during arid-to-humid semiaridity) builds to a humid climax and then diminishes to zero during subsequent humid-to-arid semiaridity. The interval so defined comprises a *humid erosional phase*, one that could be spent by one stream variously meandering or braiding through relict alluvium, while another watercourse in another setting merely deepens its channel in bedrock.

The morphologic changes in valley fills during their removal, which we see recorded in terraces and inset valley flats and/or channels, plus traces of meandering or braiding, signify simultaneous deposition in some lake or sea at the mouth of the same river. When sea levels are low during glacial stages, major river débouchérs are extended downslope to or near to the margin of the continental shelf which is naturally incised and often marginally notched (Fig. 7.35). Subaqueous slumping of the resulting deposits is probably the major cause of *turbidity (density) currents* which are widely viewed as depositionally responsible for graded terrigeneous (i.e. land-derived) deposits in

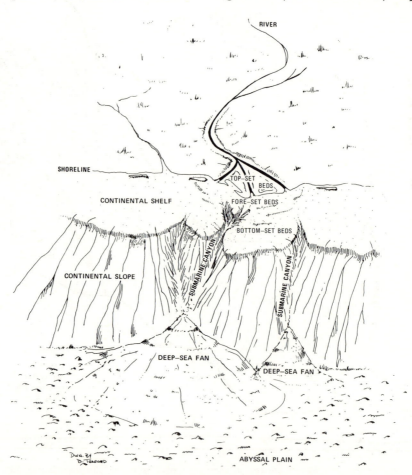

Fig. 7.35. Components of a post-glacial (Holocene) subaerial-submarine dynamic system on a continental margin. Note especially the notches in the continental shelf into which long-shore currents may feed sediment. (From Moore and Asquith, 1972.)

Fig. 7.36.   Satellite view of the Nile delta as seen from the Gemini IV spacecraft in which the triangular shape of the feature is clearly delineated by vegetation (dark, textured area). To compare this remote sensing record with a map see Fig. 7.39 (NASA.)

deep water areas and as erosionally responsible for the development of many submarine canyons (cf. Daly, 1936; Shepard, 1951, 1963). Other possible origins of such canyons are considered in Chapter 11. When sea levels are high, and shelf seas most extensive along continental margins, rivers empty onto shallow ramps of low inclination (1.5°–1°), and the sediment accumulates there unless removed by marine erosion. One possible consequence is a *delta* (Fig. 7.36).

Deltas have, in a sense, been taken for granted. They are landforms known to occur at the mouths of rivers, presumably as a response to the solid sediment deposited there. A site rather sheltered from the full force of open coast erosion seems favored. And one does not really expect to find a delta at the mouth of a drainage originating in a desert, since it is widely acknowledged that a continuous supply of sediment is required in order to counter its simultaneous removal by marine waves and currents. But the discovery that humid regions can become arid, and vice versa, places deltas in a rather new geomorphic light, with ramifications extending to such areas as stratigraphy and petroleum geology. Since a delta can only be maintained by a perennial humid river carrying an appreciable solid load, if too much of the source drainage area becomes arid, the delta will no longer be maintained. Through marine erosion, at least the delta's subaerial aspect should disappear, and presumably certain subaqueous elements would also be altered. Resumption of perennial stream flow in the presence of abundant sediment should renew delta formation. The latter relation generally characterizes the arid-to-humid transition and at least the early phases of humidity. Elimination of residual arid alluvial deposits from a drainage system by related accelerated erosion should materially diminish a river's ability to maintain a delta, even though its flow continues to be perennial

(see Fig. 4.39). This occurs since the volume of alluvium left by a given arid episode is finite, and, as noted in Chapter 5, prolonged humidity tends to reduce solid stream loads.

Modern subaerial deltas of several rivers, including those of the Mississippi and Orinoco, have been found to have formed in about the last 10,000 years or less (cf. Andel et al., 1954; Muller, 1959; Coleman, 1968). It is during this interval that much of the Orinoco drainage basin seems to have changed from arid to humid (Garner, 1966; 1967). During essentially the same interval, the Mississippi drainage basin changed from glacial to humid in northern areas and, in at least the southwestern portions, experienced a probable brief arid interlude (4,000–7,000 B.P.). Elaborate comparative chronologies have been attempted between terrace sequences and depositional units in deltas (cf. Doering, 1958: Fairbridge, 1968). However, it should be pointed out that a delta in an area of subsidence (Fig. 7.37) may preserve much of the erosional history of the related drainage basin, whereas one formed on a stable shelf may be destroyed almost completely down to wave base each time stream deposition falls below marine erosion capacity; see the discussion of the Carolina Capes, Chapter 9.

If the foregoing analysis is correct, apart from limitations on the intensity of marine erosion in accumulation sites, deltas form primarily at the mouths of humid

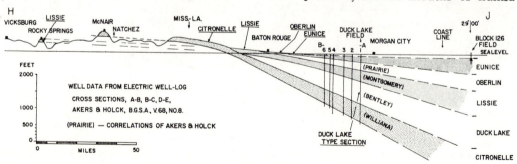

Fig. 7.37. Profile and section correlating fluvial terraces of the lower Mississippi valley below Vicksburg with the subsurface Quaternary formations of the Mississippi delta. Actual timing between terrace development, marine deposition, and climatic episodes remains somewhat uncertain. (From Doering, 1958.)

streams engaged in removing relict valley fills. In the present context, the alluvium is mainly formed in deserts, though glacial debris apparently serves equally well. Thus, perennial streams with properly sheltered débouchér areas that lack deltas are the logical consequence of the two or three situations already discussed. One category would include streams in which the drainage basins fall mainly in probable humid climatic nuclei and thus lack adequate solid loads. Probable examples include several rivers draining south through Alabama, Georgia, and Florida into the Gulf of Mexico. A second category includes streams in which the valley flats and alluvial fills are limited, and with channels where there has been much bypassing of coarse gravels, little meandering, and evident stabilization of bank and bottom materials. The Delaware and Hudson rivers, both of which drain glaciated terrain, are probable examples. A third category of streams with perennial flow that lack deltas includes those in which discharge volumes have either been diminished through the onset of drier conditions or have not yet increased sufficiently under present more humid conditions. The Rio Grande and several other Texas rivers are probable examples. As already noted, the Colorado River has had both its volume and load reduced artificially, and its delta is being destroyed (see Chapter 12). Finally, of course, a river and its drainage basin may be too small to maintain a delta with any combination of drainage basin hydraulic or climatic histories, given its existing débouchér erosion rates.

Sea coasts are, of course, environmental interfaces, and we will consider their landscape characters in this context in (Chapter 9). In a delta, however, we are dealing with a coastal phenomenon dominated by river hydraulics and deposition, a situation that is at least simulated in certain lakes and along some rivers (Fig. 7.38). When the fluvial dominance ends, as it invariably must with fluctuating environments, shorelines cease to build outward (prograde),

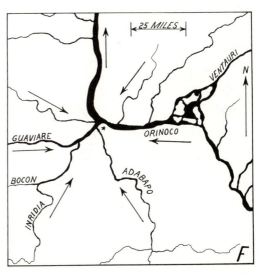

Fig. 7.38. Map showing the delta (inland) of the Ventauri River where it joins the Orinoco River in southern Venezuela (cf. Fig. 7.55). Note also the centripetal drainage patterns near the mouth of the Rio Guaviare (star) and in the upstream area of the Rio Adabapo.

and the delta morphology is altered if not destroyed. In essence, a delta is a depositional landform with certain erosional overtones. The gross deposit may often be lenticular, the more so in a subsiding area, but it generally conforms otherwise to the original bottom shape of the bay, gulf, or inlet in which the sediment is accumulating. The surface tends to be planar, and, where the distal limit of accumulation is established by marine erosion, the resulting plan form ranges from roughly triangular (delta) to semi-circular (Fig. 7.39).

A subaqueous delta, like that of the Amazon River, is not strictly speaking a "landform." And here we are concerned with the morphology of subaerial deltas. In such deltas we observe a phenomenon that is approximated in a variety of other settings and appears to result where running water carrying appreciable sediment is apparently unable to maintain a single flow route. On alluvial fans and some pediments, runoff that is drying up deposits in and blocks channels, thereby causing

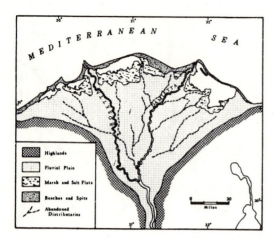

Fig. 7.39. Subaerial deltaic plain of the Nile River (cf. Fig. 7.36). (From Coleman, 1968.)

"autodiversion" and channel redevelopment elsewhere (Fig. 6.21). In braided streams, hydraulic inefficiencies during transport of an abundant sediment load produce temporary local bar deposition and multiple flow routes within a channel (Fig. 7.40). And in deltas, river flow is checked by standing water, and deposition of associated sediment forms a system of shifting *distributary channels (anabranches)*, all more or less simultaneously occupied by runoff (Fig. 7.41).

As noted by Bird (1962, 1968, 1971) in his study of the deltas of the Gippsland Lakes in Australia, depositional morphology is characterized by a tendency for natural levee formation through *vertical accretion* along anabranches during flooding as in upstream areas of alluvial rivers (cf. Russell, 1936; Fisk, 1944). *Horizontal accretion* along distributary channel margins tends to project the natural levees seaward as sediment jetties (Fig. 7.42), and this phenomenon accounts for the digitate distal margins of some deltas in quiet water areas including the "Birdfoot" Delta of the Mississippi River. The relatively coarse sediments of the levees (silt, sand?) tend to grade laterally to fine-grained intervening marsh deposits (clay?) (Fig. 7.43).

Coleman (1968) observes that a major feature of deltas is the shifting of distribu-

taries by breaches in levees (*crevasses*) during flooding. Since areas between levees are relatively low, a new group of radially branching distributaries forms downslope from a new crevasse (Fig. 7.44). As a given group of such anabranches prograges, their gradients flatten, and water velocity and sediment-carrying ability diminish. Cessation of sedimentation and abandonment may follow, in a cycle that Coleman says averages about 100 years for the Mississippi Delta. Each such event forms a lens of relatively coarse, detrital deposits bounded by finer, organically rich materials (see Fig. 7.43). The loading of deltaic sediment onto prodelta marine clay can cause the upward intrusion of the clay into a small domal landform called a *mudlump* (see Morgan, 1961). These appear analogous to subsidence structures discussed in Chapter 3.

Arid-Humid Deranged Drainages

Our discussion of humid drainage phenomena following an episode of aridity has been almost exclusively devoted to those situations in which major slope configurations of humid drainage nets survive arid morphogenesis. This is apparently the usual situation in regions between arid and humid climatic nuclei where effects of desert planation and aggradation are not extreme, and ravine-ridge topography —sometimes in subdued form—is the most persistent element of the landscape. We shall now consider those regions of the Earth where drainage net slope configurations have been essentially destroyed by dominant or prolonged aridity so that subsequent runoff is chaotic. This apparently happens rather close to, but not usually in, arid climatic nuclei.

In the section entitled "Streamfloods and Stripped Surfaces," it was observed that prolonged aridity might essentially destroy a humid drainage net. The associated processes involved are considered in Chapter 6. Their effects are aptly summarized in Davis's (1905) expression, "drainage dis-

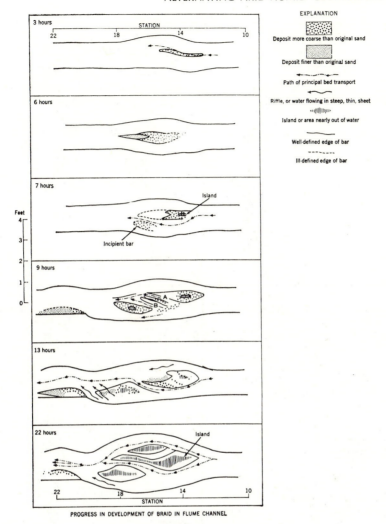

Fig. 7.40.   Successive plan views of central bar formation during development of a braided flow pattern in a laboratory flume. (From Leopold, Wolman, and Miller, 1964.)

integration," wherein channels, valleys, and divides undergo progressive loss of identity. Each of the latter features comprise an expression of surface relief relative to which runoff must ultimately seek the water's own lowest level. It should be noted at the outset that for running water to be directionally controlled by a landslope, the bulk mass of the water must be confined by the upper limit of the slope. Submerged land surface irregularities may induce water surface morphologies (or reflect them) as in the generation of *dunes* or *ripplemarks*. But it is the surface form of the water that is being affected in such instances, and not its basic direction as determined by regional slope and confining slopes.

Several types of forces affect the movement of water on the Earth's surface. *Cohesive forces* acting between water molecules tend to pull them together. *Adhesive forces* tend to pull water molecules toward other surfaces (*wetting*). Water on a oily surface tends to pull into globules or droplets when cohesive forces are stronger than

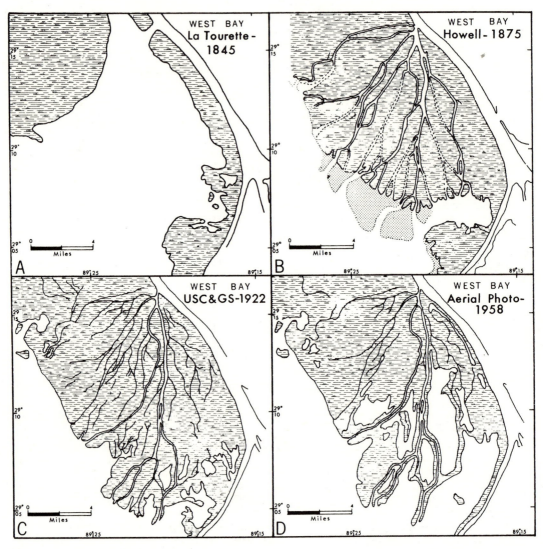

Fig. 7.41.   Historic maps of the West Bay subdelta, Lower Mississippi River Delta. See Fig. 7.44 for location. (From Coleman and Gagliano, 1964.)

adhesive forces. We say this surface resists wetting. *Gravity* is, of course, a third major force. On a perfectly planar, horizontal surface, a drop of water would tend to be pulled by gravity and adhesion into a flat molecular film, resisted, of course, by cohesion and friction.

Drainage in a humid region tends to be essentially a "container" problem. The containers are organized valleys or channels that are wettable and into which gravity pulls water. Once collected, the water tends to follow regional slopes within the local organized "containers" to or toward base level. If there is a "key" phrase here it is "within organized containers." And the dominant motive force is gravity opposed by friction.

Arid planation and aggradation tend to eliminate the local, "organized" slopes that constitute "containers" for prior humid runoff. As the land surface more and

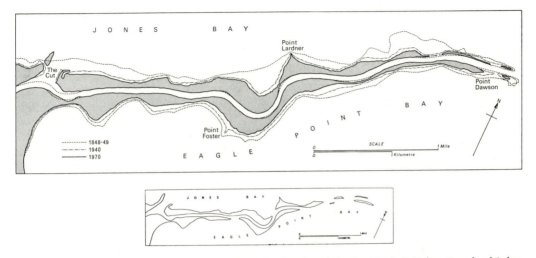

Fig. 7.42. Depiction of lateral accretion along a delta distributary in the Mitchell Delta, Gippsland Lakes, Australia, as it was in 1849 (above) and in 1959 (below). (From Bird, 1971.)

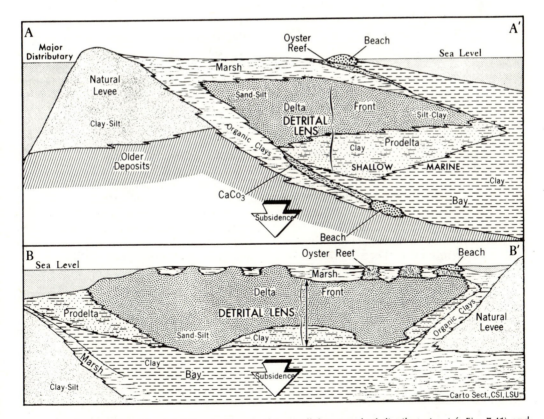

Fig. 7.43. Cross section of a subdelta. Section A–A' is parallel to trend of distributaries (cf. Fig. 7.41) and section B–B' is transverse. (After Coleman and Gagliano, 1964.)

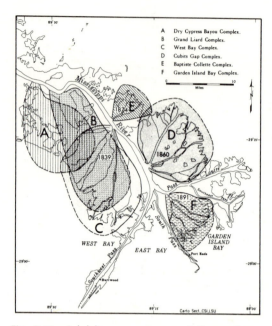

Fig. 7.44. Subdeltas or crevasses of the modern Bird-Foot Delta of the Mississippi River. Dates are year of crevass breakthrough. (After Coleman and Gagliano, 1964.)

more nearly approximates a plane, and the local "containers" disappear, the effectiveness of gravity as a localizer of runoff diminishes. Its effectiveness as a regional collector of water may also change as the regional slope changes. But as a surface becomes more nearly planar and horizonal, the forces of friction, cohesion, and adhesion become more important, the force of gravity becomes less important, and intrinsic hydraulic factors including flow forms (helical, turbulent, laminar) apparently can have greater influence. These major force variants are expressed by the hiararchy of drainage morphologies already discussed. (1) Runoff that is originally confined in essentially straight channels, particularly in bedrock, tends to remain straight and confined since downcutting under gravity and from friction are the major forces expressed erosionally. Weaker flow influences are expressed depositionally by such attributes as thalweg sinuosity developed in gravel which is probably the most evident expression of helical flow in a straight channel (Fig. 7.22). (2) Weakened confinement expressed by such materials as alluvial channel banks, with reduced discharge, permits internal flow patterns to be expressed as braids or meanders (Figs. 7.21; 7.40) (see Smith, 1971). It is as if the presence of increased bed load flattens the helical flow spiral (Fig. 7.30). Because there are instances where reaches of a single stream are alternately meandering and straight, one can hardly argue in favor of reduced gravitational force. Rather, the EFFECT of gravity is being altered, permitting other coexisting forces to be expressed. (3) Runoff that is not confined by banks and valley walls, except sporadically and locally, in association with gradient flattening by arid erosion and deposition, apparently marks the opposite extreme, where incision by gravity is only marginally effective and lateral forces are nearly uninhibited except by friction (cf. Keller, 1972).

On many desert plains, local confinement of runoff is virtually nil on any permanent basis, and only regional slope remains to consistently influence runoff direction. *Sheetfloods* are the order of the (rainy) day. The principal physical constraints on runoff in a climax or near-climax desert terrain are inselbergs, deflation depressions, dunes, the odd stretch of incompletely alluviated relict valley, pediments, pediplains, aggradational plains, pans, playas, and structural or volcanic irregularities. Few of these tend to oppose lateral water movement. Low-relief, low-inclination regions were seen to dominate long-term arid landscapes (Chapter 6), and subject to the aforementioned irregularities, runoff imposed in intermittent or perennial form on such terrain would tend to move down regional slopes otherwise subject to minimal confinement. Over low gradients, subject to accentuated lateral hydraulic oscillations and ·at least initially, with almost unlimited access to relict desert alluvium, a semi-planar, more or less chaotic type of flood runoff could be expected. Since a region's local relief essentially defines the volume of runoff that would be required to submerge a local area, the flatter the surface the less water would be needed. One may well ask where such drainage records exist. For they appear to require the more extreme forms of desert planation followed by one of the more intense aspects of humidity.

Once in a great while, a lateral shift in atmospheric fabric may impose intensely humid conditions over a land surface that mainly reflects arid planation and aggradation (see Chapter 4). This would create increasing and potentially great runoff amounts (for vegetated areas) over a relatively low-relief region. Alternately, runoff from a humid region may invade a desert terrain situated downslope. In this case it is not the differences in intensity of successive environments in one area that determines the extent of the flooding, but rather the degree of difference between adjacent arid and humid environments. Both types of morphogenesis appear to be on record.

Derangement Morphology
and Evolution

The patterns taken by runoff on a planar desert terrain during the arid-to-humid transition must initially be dictated by the landsurface morphology. We are speaking, then, of a nearly planar sheet of water having depth irregularities that reflect minor inherited terrain highs and lows, locally interrupted by patches of land representing more pronounced original elevations whose height exceeds the water depth (see Fig. 7.45). The runoff form illustrated is little different morphologically from the ephemeral *sheetfloods* described by McGee (1897) and discussed in detail in Chapter 6. Their occurrence is intermittent in northern Australia during the "Wet," in the broad, shallow depressions that pass for valleys of many minor streams on the North American High Plains, in the Kazakstanian Steppes of the U.S.S.R., and in the indefinite downstream reaches of the Shari River in central Africa and the

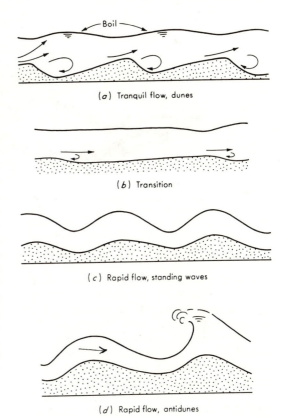

(*a*) Tranquil flow, dunes

(*b*) Transition

(*c*) Rapid flow, standing waves

(*d*) Rapid flow, antidunes

Fig. 7.45.   Dune and antidune development under tranquil- and rapid-flow conditions. (From Simons and Richardson, 1960.)

Okavango delta region in South Africa described by Grove (1969). Perennial examples are more rare but include that illustrated from southern Venezuela (Fig. 7.46) on the Rio Caroni (cf. Fig. 7.55 for location). Since these terrains are generally more irregular than the usual pediment, the lateral water continuities are less than for some sheetfloods, and, for this reason and the greater temporal flow continuities, the writer called this form of runoff a *quasi-sheetflood* (Garner, 1966). In known examples, its lateral limits are not eroded banks, but rather, the still, feather edge of a flood on a plain.

We have already discussed the effects of sheet runoff moving across a granular surface (Chapter 5), but the phenomena under discussion are at least an order of

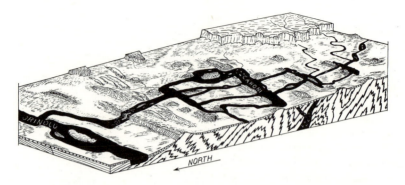

Fig. 7.46. Schematic block diagram of a portion of the Caroni River drainage area in the shield region of southern Venezuela in a view to the southeast. The planar savannah lowland with its anastomosing channel systems and residual alluvium demonstrates alternating aridity and humidity, whereas the high plateaus that have remained humid and under solutional erosion are hardly incised so that the world's highest waterfalls are a result (cf. Fig. 3.36).

magnitude larger and some relief (relict) already exists. Furthermore, we are talking about a terrain that is in the process of acquiring a vegetal cover. Plants will resist flow unless water depths are considerable, and they will frictionally resist erosion by reducing velocity as well as by binding sediment in root systems. Thus, in a quasi-sheetflood, there appears to be *laminar flow* in the marginal, shallow-water, vegetated flood areas characterized by very low flow velocities. As illustrated by Sundborg (1956) and Morisawa (1968), laminar flow involves low-velocity movement measurable in fractions of a millimeter/second where the water moves in parallel layers, each shearing past the other (Fig. 7.47a). Shearing is a form of

deformation resisted by the molecular adhesion of a substance's particles to one another and expressed by the term *viscosity*. Shearing stress is proportional to the resistance of the layers to movement as determined by viscosity and by the velocity differences from one layer to another. Thus, for Fig. 7.47a, at any point y distance above the bed, where the layer of water is flowing at velocity v, the shearing stress τ can be expressed as

$$\tau = \mu \, \frac{dv}{dy}$$

where μ is the viscosity, and dv/dy is the change in velocity from one layer to the next (velocity gradient). As noted by Morisawa (1968) laminar flow cannot support solid particles in suspension and is nonerosive, since the velocity with respect to bed or channel walls would be zero. Though uncommon in naturally channeled runoff, it probably characterizes flow in shallow, vegetated fringes of quasi-sheetfloods.

Higher velocity flow lines are observed in open river channels and in portions of quasi-sheetfloods where water is deeper and vegetation less (Fig. 7.48). There, flow is turbulent, and erosion occurs. Turbulent flow, characterized by chaotic movements, variable velocities, and secondary eddies superimposed on the main flow direction (Fig. 7.47b), can occur where velocity exceeds a critical value most com-

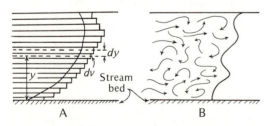

Fig. 7.47. Diagrams illustrating (A) laminar and (B) turbulent flow forms in relation to velocity. At a distance y above the bed in laminar flow, the velocity is a function of the velocity of the layer dv and the velocity difference between adjacent layers dy. A relatively smooth curve results, but in turbulent flow a smooth velocity profile cannot develop because of turbulent counter eddies.

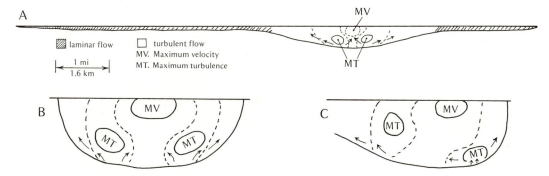

Fig. 7.48. Cross sections depicting velocity relations. (A) Quasi-sheetflood; (B) symmetric channel; and (C) asymmetric channel, indicating areas of maximum turbulence (MT), maximum velocity (MV), and laminar flow. (After Leighley, 1934, in part.)

monly expressed by the Reynolds number

$$N_R = \rho \, \frac{VR}{\mu}$$

where ρ is density, V is mean velocity, R is hydraulic radius, and μ is viscosity. The hydraulic radius of a channel is the cross section divided by the wetted perimeter (Fig. 7.49). In a quasi-sheetflood, the wetted perimeter is less hydraulically significant, since much of the inundated area is occupied by laminar flow, and the depression is, in any case, not a channel of erosional origin and often lacks discrete banks. Fortunately, as noted by Morisawa, the hydraulic radius in a channel is approximately equal to the water depth d, so that value may be substituted for R in the above equation. Flow is laminar for small values of Reynolds number and turbulent for higher ones. Morisawa (1968) states that the Reynolds number for streams is generally over 500 and varies from 300 to 600. Sundborg (1959) marks 500 as the boundary between laminar and turbulent flow.

Only one of the two main kinds of turbulent flow (*streaming flow* and *shooting flow*) are common to quasi-sheetfloods. This is the streaming flow found in most streams where there are no rapids, and the absence of shooting flow may relate to inadequate water depths of quasi-sheetfloods. Shooting flow, which occurs at high velocities, apparently develops as erosion

within quasi-sheetfloods, deepens flow routes to form discrete channels, and rapids appear (Fig. 7.50). Conditions for streaming versus shooting flow may be expressed by the Froude number

$$F = \frac{V}{\sqrt{gd}}$$

where V is velocity (celerity) of a small surface wave, g is the force of gravity, and d is the depth of the water. Froude numbers less than 1 express streaming flow, whereas those greater than 1 express shooting flow (cf. Sundborg, 1956).

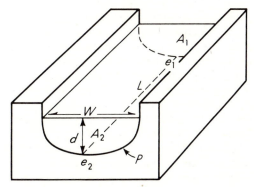

Fig. 7.49. Stream channel dimensions: W, actual water width; P, wetted perimeter (surface in contact with water in cross section); A, water cross section; d, depth, approximates the hydraulic radius R, the cross section, divided by wetted perimeter ($R = A/P$). Stream gradient S is the drop in elevation ($e_1 - e_2$) between two points on the channel bottom divided by the projected horizontal distance between them (L).

Fig. 7.50. Aerial view of developing channels within a quasi-sheetflood of the Rio Caroni, southern Venezuela. Foreground flow area shows large, low islands and many patches of vegetation. Five major distributaries may be discerned in runoff areas flowing to right (north) in a network about 20 mi (33 km) wide. For location see Fig. 7.55.

Flow conditions for streaming and shooting flow can usually be detected visually in a moving body of water in instances where measurement is difficult or impossible (see Fig. 7.51). The white water of rapids is, of course, a sign of high turbulence, and, at the head of a rapids, the water surface commonly lowers in response to the velocity increase. A velocity decrease may occur locally within a rapids due to channel bottom roughness commonly in the form of large, bypassed blocks of rock, so that shooting flow locally reverts to streaming flow, and the water rises into a standing wave (Fig. 7.45). When one is forced to traverse a large standing wave in a canoe, a right-angle (direct) approach is advised since an oblique attack frequently results in an inverted canoe which is soggy and difficult to paddle.

The erosive effectiveness of a turbulent reach of channel is indicated by the shearing force, r, which in turbulent flow can be expressed mathematically as

$$r = (\mu + \varepsilon)\, \frac{dv}{dy}$$

where the symbols are the same as for laminar shear just discussed, and ε is the eddy viscosity or intensity of turbulent mixing. Turbulent flow appears to increase in the quasi-sheetflood drainage forms being considered as channels are defined by deepening. This relation focuses our discussion on the fact that we are considering intermittent-to-perennial runoff in which the debris being eroded is not immediately redeposited as it would be in a desert but rather is being carried out of the drainage lines and the environment. As pointed out by Garner (1966, 1967), the deepening of flow lines within a quasi-sheetflood must eventually result in confinement of the run-off within a system of interwoven channels. Along the Rio Caroni in southern Venezuela, and a great number of drainages elsewhere in South America, Africa, and Australia, this increasing confinement is apparently occurring at present (Fig. 7.51). More precisely, in most cases, the canalization is essentially complete, though overflows are extensive during times of high discharge.

The networks of channels that apparently develop through incision of turbulent flow lines in quasi-sheetfloods have been termed *anastomosing channel systems* (Garner, 1958) following earlier usage for similar features by Bretz (1923) and Bretz *et al.* (1956). In plan view, the flow patterns in question resemble those of *braided streams*, but the distinction be-

tween the two types of flow networks is significant. An anastomosing channel system is AN EROSIONALLY DEVELOPED NETWORK OF CHANNELS in which the insular flow obstructions represent relict togographic highs and often consist of bedrock. A braided stream pattern is one DEVELOPED DEPOSITIONALLY WITHIN A CHANNEL in which the flow obstructions are sand and gravel deposited by the water moving around them. The anastomosing channel system differs from a quasi-sheetflood in that the latter lacks distinctly incised banks, and has relatively more water area and less land in the inundated region. Flow obstructions of an anastomosing channel system may be relict arid alluvium (cf. Fig. 7.52), and, in the case of the Rio Caroni in Venezuela, seasonal runoff changes convert some anastomosing channel distributaries into braided streams (Fig. 7.53).

Fig. 7.51. Aerial photograph, looking southeast, of an exceptionally well-incised reach of an anastomosing channel system along the Rio Caroni, southern Venezuela. The associated hills and elongate ridges of Precambrian crystalline rock range from 500–800 ft (150–200 m) high. (See cover.)

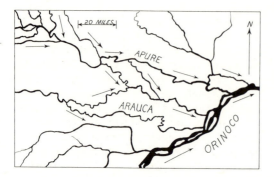

Fig. 7.52. Anastomosing channel systems involving the Apure and Arauca rivers, western Venezuela (cf. Fig. 7.55). Portions of the island-like drainage obstructions are alluvial as illustrated in Fig. 1.15 where rainforest growth has now replaced former arid aggradation.

In his studies of anastomosing channel systems, Bretz (1923) was quick to echo earlier ideas that a multiple channel discharge system is unstable. Because of evident differences in slope, discharge, and bed lithology within a network, erosion will be more effective in deepening some distributaries than others. Thus, favored routes of flow should gradually incorporate higher percentages of the total discharge of a given network system and ultimately evolve into the single-channel system we call a *river* (Fig. 7.54). The more effectively deepened channels "pirate" the runoff from the others. This *channel piracy* within networks like that of the Rio Caroni results in abandoned channels. In systems like the Apure-Arauca drainage on the Venezuelan Llanos (Fig .7.52), where the relict alluvial obstructions permit relatively rapid channel piracy, some of the abandoned flow routes (*caño viejo*) show up on old maps. Since such depressions continue to direct local precipitation, they show up in bedrock terrains as *barbed tributaries*. The latter relation holds for much of the lower Rio Caroni (cf. Fig. 7.54) and appears to characterize evolving anastomosing channel systems (see Garner, 1966).

The complete integration of relict arid terrain configurations into a well-drained landscape is clearly a complex process, and the fluvial disequilibrium expressed by

quasi-sheetfloods and anastomosing channel systems represents an extreme case. Undrained desert depressions presumably acquire lakes with the onset of humid conditions, as was the case in the North American southwest during the Pleistocene. Such depressions frequently become the focus of centripetal drainage patterns (Fig. 2.20), and similar patterns are common to regions where deranged drainage relates to arid-humid alternation (cf. Fig. 1.16). The lakes in the southern Venezuela region are now drained as are those of adjacent Brazil, and the centripetal drainages have been incorporated into the regional drainages (Fig. 7.55).

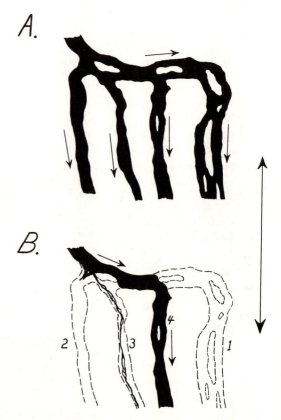

Fig. 7.53. Diagrammatic map of the effects of seasonal runoff changes in anastomosing channel systems such as the Caroni, southern Venezuela. (A) High water stage; (B) dry season runoff pattern with braiding. Numbers 1–4 indicate order of abandonment with a water-level drop and hence relative channel depth.

There is good reason to believe that a series of sharply outlined, ovate-to-elliptical, water-filled depressions that are widespread from Maryland to northern Florida along the Atlantic coastal plain are actually eolian deflation depressions and pans relict from former deserts there. Carbon-14 dates from accumulations (peat) in some of the features indicate development during the Pleistocene Epoch. Most of the depressions show similar distinct orientations. Generally termed the "Carolina Bays,"

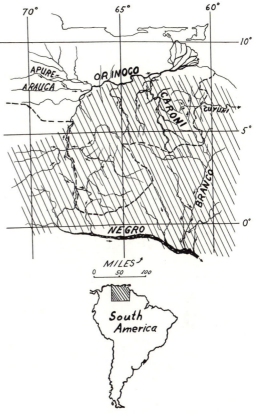

Fig. 7.55.   Regional location map of drainages discussed from northeastern South America showing indistinct divides between the Amazon and the Orinoco drainage basins, centripetal drainage patterns and anastomosing channel systems of the Rio Caroni (Figs. 7.45, 7.46, 7.50, and 7.51), the Apure-Arauca system (Fig. 7.51), the Orinoco (Fig. 7.58), and the Negro (Fig. 1.16).

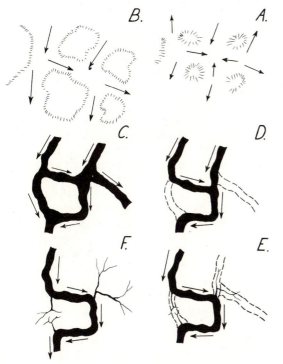

Fig. 7.54.   Development of channeled drainage (A) following extreme arid aggradation and planation (B)–(F) through prolonged humidity as typified by the Rio Caroni, Venezuela. Initial expression (A) is of ephemeral sheetfloods on low-relief alluviated surface. Humid runoff modifications of (A) are expressed by (B) intermittent to perennial transmission of runoff through quasi-sheetfloods which are essentially unconfined and only locally and linearly erosive, (C) anastomosing channel systems formed through deepening of sheetflood flow routes, (D) channel piracy within the network by selectively deepened channels, (E) further piracy, local drainage reversals with barbed tributaries and development of trunk channel, and (F) systematic, directed incision by units of an integrated drainage net.

the east coast phenomena have oriented counterparts in the Llano Estacado, in the High Plains of Texas and New Mexico, and in the Beni Basin Lakes of Africa (cf. Price, 1968). The present assemblage of bays on the Atlantic coast has the appearance of a series of superimposed pans, complete with lunette-shaped dunes on probable lee sides (Fig. 7.56). Price summarizes the more than fifteen theories of bay origin including genesis as spring basins, meteorite craters, solution depressions, submarine eddy scour, and eolian deflation. He concludes that the latter

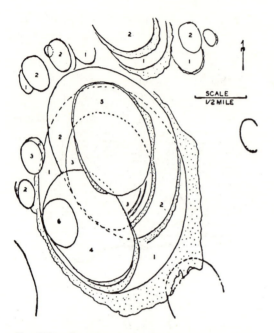

SCALE
1/2 MILE

Fig. 7.56. Superimposed ovate depressions (bays) in an interior longitudinal dune field belt, Cumberland County, North Carolina (dune-like beach ridges are stippled). Basin age succession 1–6 as numbered by Prouty (1952). The clear affinity (morphologic) for desert pan configurations can be seen in Figs. 6.53 and 6.54.

eolian genesis, earlier proposed by him (1951) and Prouty (1952), is compatible with all the known field evidence, noting particularly the bay orientations, and association with a relict longitudinal dune system, and localization on alluvial and deltaic plains composed of deflatable materials. The bays—or relict pans, for this is what they would appear to be—are presently forested around their margins, and many are nearly filled with sediment. Some have been flooded by the postglacial rise in sea level and affect coastal morphology (cf. Chapter 9).

Other indications of drainage disequilibrium, in regions like those in eastern South America where aridity has been replaced in large measure by humidity, include, of course, the relict alluvial deposits (Fig. 1.15) and the many knickpoints in streams and related watercourses (Fig. 7.57) (see also Fig. 3.36). Falls and rapids reflecting

knickpoints naturally destroy the relatively smooth, concave upward profile of the "normal" humid stream, not that such profiles are perfectly smooth or evenly curved in every detail. Many are not. In addition to such features, the obscurity of drainage divides is typical of regions where aridity has deranged the drainage. Strahler (1952) and others wish to use the drainage basin as the basic geomorphic unit. But, in regions of climate alternation, basins often lack individuality. In eastern South America, for example, the three main drainage basins are not really separate. Not only do the individual drainage nets of the Orinoco, Amazon, and Paraguay lack single trunk channels along some reaches (Fig. 7.58), headwaters streams along the vague divide that marks the Venezuelan-Brazilian border develop distributaries which supply runoff to both the Amazon and Orinoco rivers. A similar situation exists in the obscure watershed separating the Amazon and Paraguay rivers to the south. Along the southern edge of the Sahara (Fig. 4.23), the Chari River sends some of its runoff into Lake Chad in the Sahara and the rest by way of the Benue River and the Niger River to the South Atlantic Ocean. And, near the northern edge of the Kalahari Desert in South Africa, there is evidence that the Zambezi River may, in times past, have supplied water to both the Okavango inland delta and the Indian Ocean (see Grove, 1969) (Fig. 4.24). Of course, if a drainage can have multiple distributary channels, it can have more than one *débouchér*, even without a delta.

The arid-humid drainages that display signs of drainage derangement generally require relatively large volumes of runoff traversing nearly planar expanses of terrain. In major drainages, most derive at least some of their runoff from the elevated regions that show evidence of more continual humidity. The derangements on record are, therefore, due in part to accentuated precipitation in a former desert and in part to trans-environmental runoff into the same desert. Which condi-

Fig. 7.57. View of Caroni Falls where the drainage system of that name, in the form of an anastomosing channel system, drops over a 50–75 ft (15–20 m) escarpment some 6 mi (10 km) south of the Orinoco River on the northern edge of the Venezuelan Shield (see Fig. 7.55).

tion is most important varies from region to region. But the majority of the drainages, like the Rio Caroni, have dendritic tributary systems with anastomosing channel systems downslope. In effect, these drainages become longer in a downslope direction following the onset of humidity. This is, in fact, the only reasonable explanation to date for distributaries diverging downslope around bedrock obstructions. The divergence can hardly be the result of headward erosion. In the final analysis, however, we have yet to discern anything in the flow patterns that distinguishes between braided distributaries formed by desiccation of runoff entering a desert (Fig. 4.24) and anastomosing channel systems caused by climate change. The former would develop within a desert, of course, for accentuated evaporation and

infiltration is encountered there. The latter can form in a humid vegetated region and should display some bedrock in runoff obstructions together with possible relict alluvium. Probably only a detailed field investigation can ultimately make the distinction in some cases.

The occurrence of anastomosing channel systems is rather uneven in terms of continents, and not all continue to carry water in multiple channels, if they carry water at all. North America and Eurasia have very few deranged drainages that appear to relate to arid-humid conditions, possibly because the major planar arid regions are not downslope from the most distinctly humid lands. One relict channel network exists near the head of the Mississippi Embayment (Fig. 7.59), though the extent to which glacial meltwater may

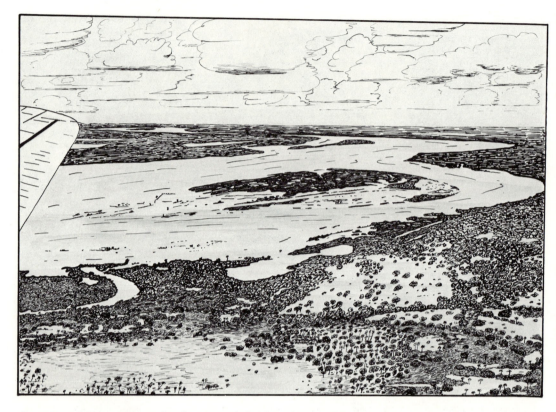

Fig. 7.58. Orinoco River at flood near Cuidad Bolivar, Venezuela, with discharge through multiple channels as well as in overbank flow. Inundated area is 25–30 mi (40–50 km) wide.

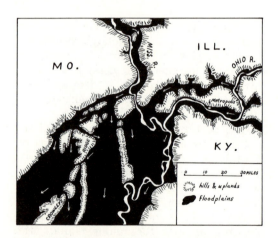

Fig. 7.59. "Hill islands" and a network of flood-plain-level drainageways in the upper Mississippi Embayment. Arrows show the inferred directions of flow by flood waters within a channel maze prior to development of present river channels.

have been involved there is uncertain. The Channeled Scabland in the state of Washington (Fig. 7.60), made famous largely through the researches of Bretz (1923), presently lies in a desert and was apparently alluviated prior to development. The huge channels there, plus enormous cataract scars and mega-ripples, among other features, suggest discharges of tremendous magnitude (cf. Figs. 8.54 and 8.55), possibly related to breaching of a glacial ice dam on Glacial Lake Missoula one or more times (see Chapter 8).

South America, Africa, and Australia all have extensive desert regions downslope from more humid ones, and active channel labyrinths are recorded from each (Fig. 4.24). More than half the major tributaries of the Amazon River display multichannel trunk stream reaches (Fig. 7.61).

Fig. 7.60. The anastomosing channel systems of the Washington Scablands, northwestern United States. Arrows show inferred runoff directions during the several apparent inundations of the network due to a series of ice-dam ruptures. (Adapted from Bretz *et al.*, 1956.)

And, in the central Sahara Desert south of Tibisti, there is a well-known relict channel labyrinth, shown on most physiographic maps (cf. Bartholomew, 1950; Cohen, 1973), which must have transmitted appreciable runoff at some time in the past. Perhaps superposition of requisite arid-humid climate extremes and related geomorphic systems is more common in the tropics. However that may prove to be, the floods and chaotic flow patterns that appear to be a consequence of such events can be taken to reflect the disequilibrium that follows the inception of a new humid environment even as the meandering and braided flow paterns of otherwise "normal" rivers do, to a lesser degree. Meandering, as such, may constitute an equilibrium flow pattern for water crossing an alluviated surface with certain sediment-discharge characters. But water crossing an alluviated surface IS NOT A CLIMAX HUMID DRAINAGE RELATIONSHIP, and the "least-work relation" seen in alluvial meanders by hydrologists is not as "least" as the same flow pattern "frozen" in bedrock by incision, and lacking the granular bed.

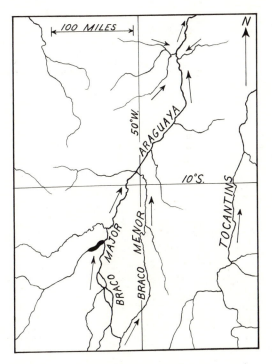

Fig. 7.61. Anastomosing channel systems and a centripetal drainage pattern on the Araguaya River, a northward flowing tributary of the Amazon, Brazil.

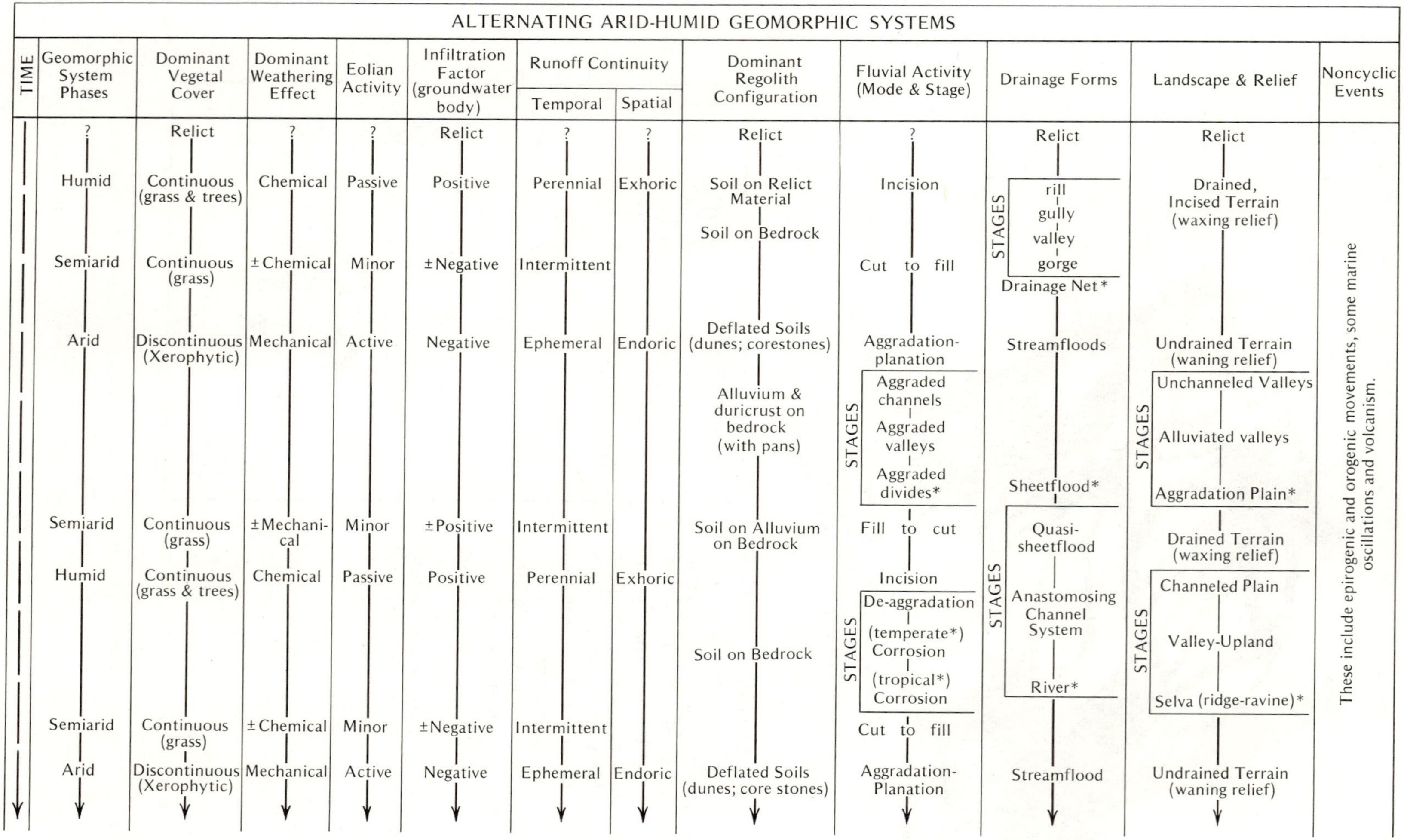

ALTERNATING ARID-HUMID GEOMORPHIC SYSTEMS

TIME	Geomorphic System Phases	Dominant Vegetal Cover	Dominant Weathering Effect	Eolian Activity	Infiltration Factor (groundwater body)	Runoff Continuity — Temporal	Runoff Continuity — Spatial	Dominant Regolith Configuration	Fluvial Activity (Mode & Stage)	Drainage Forms	Landscape & Relief	Noncyclic Events
	?	Relict	?	?	Relict	?	?	Relict	?	Relict	Relict	These include epirogenic and orogenic movements, some marine oscillations and volcanism.
	Humid	Continuous (grass & trees)	Chemical	Passive	Positive	Perennial	Exhoric	Soil on Relict Material	Incision	rill–gully–valley–gorge Drainage Net* (STAGES)	Drained, Incised Terrain (waxing relief)	
	Semiarid	Continuous (grass)	±Chemical	Minor	±Negative	Intermittent		Soil on Bedrock	Cut to fill	Streamfloods	Undrained Terrain (waning relief) — Unchanneled Valleys / Alluviated valleys / Aggradation Plain* (STAGES)	
	Arid	Discontinuous (Xerophytic)	Mechanical	Active	Negative	Ephemeral	Endoric	Deflated Soils (dunes; corestones); Alluvium & duricrust on bedrock (with pans)	Aggradation-planation / Aggraded channels / Aggraded valleys / Aggraded divides* (STAGES)	Sheetflood*		
	Semiarid	Continuous (grass)	±Mechanical	Minor	±Positive	Intermittent	Exhoric	Soil on Alluvium on Bedrock	Fill to cut / Incision	Quasi-sheetflood	Drained Terrain (waxing relief) — Channeled Plain / Valley-Upland / Selva (ridge-ravine)* (STAGES)	
	Humid	Continuous (grass & trees)	Chemical	Passive	Positive	Perennial		Soil on Bedrock	De-aggradation (temperate*) Corrosion / (tropical*) Corrosion (STAGES)	Anastomosing Channel System / River* (STAGES)		
	Semiarid	Continuous (grass)	±Chemical	Minor	±Negative	Intermittent			Cut to fill			
	Arid	Discontinuous (Xerophytic)	Mechanical	Active	Negative	Ephemeral	Endoric	Deflated Soils (dunes; core stones)	Aggradation-Planation	Streamflood	Undrained Terrain (waning relief)	

*steady-state (climax) expressions.

Fig. 7.62. A flow-chart portraying major geomorphic stages during the alternation of arid and humid geomorphic systems as part of a space-time continuum in a more slowly evolving geologic setting. Landforms, deposits, and agency conditions cited at any particular "time" level are anticipated results of correlative geomorphic environment. In actual settings, a given geomorphic phase may be interrupted at any stage of development, and climax expressions (asterisks) are only rarely or locally achieved.

Arid-Humid Geomorphic Analysis

In a sense we have now come full circle in this chapter, for we began by considering the humid-to-arid transition and have concluded with the arid-to-humid transition, elucidating during this discourse the major attributes of the two KINDS of semiaridity. But Thomas Wolf's remarks about the impossibility of going home again also apply here. For our environmental space-time continuum includes landscapes and geomorphic mechanisms that evolve, in addition to those discussed in Chapters 4, 5, and 6 (climatic nuclei) which have virtually ceased to evolve apart from negative allometric growth (cf. Chapter 1). And our evolving geomorphic systems (when repeated) never quite return to the starting point. Each humid environment is a bit different from the preceding one and accomplishes slightly different things while also accomplishing many of the same things. Thus, nature simulates but never duplicates, and the space-time environmental continuum, like a helical spiral, goes round and round without retracing its path.

To summarize the geomorphic implications of this chapter, and to aid in the geomorphic interpretation of related features in the field, it seems worth while to give a synopsis, in chart form, of important geomorphic developments (Fig. 7.62). What has been done, in essence, is to plot time against the principal associations of geomorphic processes and products one normally encounters in particular environments, in particular, sequences of environments, and for particular durations of environments, in this case alternating arid and humid ones. A friend of the writer's often says that quantitative geomorphologists do very well until they multiply by t (time), and in some measure the chart is designed to compensate for this. Many of the categories listed begin with question marks to indicate that the antecedent condition is either unknown or problematic. Others indicate the possible presence of some physical relict depending on the

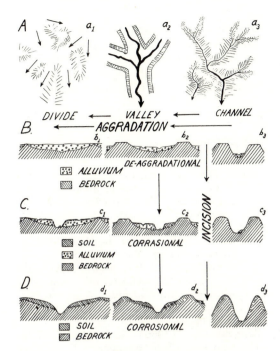

Fig. 7.63. Erosional relief development and destruction under alternating arid-humid conditions. Adjacent elements of (A), map view, and (B), cross section, reading from right to left are, respectively, channel, valley, and divide aggradation stages $(a_3b_3 - a_1b_1)$. Incisional stages, top to bottom, portray humid modifications of drainage beginning with (1) a slight channel aggradation $(a_3 - d_3)$, (2) a moderate valley aggradation $(a_2 - d_2)$, and (3) a severe divide aggradation $(a_1 - d_1)$. Dominantly arid relief relations (a_1b_1) contrast with dominantly humid relief configurations (a_3d_3). Note the implied changes in stream-load character as regolith is altered.

antecedent condition to be determined. Where a particular process or effect is known to change through time, its notable modes or stages are indicated along with its probable climax expression.

In this chart, it should be recognized that the stages merely represent "possible" and not "necessary" achievements for a given geomorphic system. Relative to climatic nuclei and regions between them, any stage of landscape development can provide the setting for subsequent evolution under a later geomorphic system. This is schematically indicated (Fig. 7.63) to emphasize the multiplicity of possibilities that can develop in our geomorphic continuum with variations in both space and time. Yet the final illustration of this chapter brings out the essential simplicity of gross terrain morphologies that develop with respect to only two environmental end-members. The third main global environment (glacial) will be the topic for Chapter 8, and there we will also consider how that environment interacts with others to produce landforms.

REFERENCES

Ackers, Peter and Charlton, F. G. (1971) "The Slope and Resistance of Small Meandering Channels," *Inst. Civil Engineers Proc.*, Supp. XV, Paper 73625.

Andel, T. van, Postma, H. and Kruit, C. (1954) "Recent Sediments of the Gulf of Paria," *Orimoco Shelf Exped. Rept.*, v. 1.; *K. Nederl, Akad. Wetensch., Verh.*, v. 20.

ANON. (1966) "Meandering of Channels in Alluvium" [in] *Hydraulics Research 1965* (United Kingdom Ministry of Technology), pp. 21–24.

Bagnold, R. A. (1960) "Some Aspects of the Shape of River Meanders," *U. S. Geol. Surv. Prof. Pap.*, 282-E.

Bartholomew, Bruce (1970) "Bare Zone between California Shrub and Grassland Communities: the Role of Animals," *Science*, v. 170, pp. 2110–2112.

Bartholomew, John (1950) *Bartholomew's Advanced Atlas of Modern Geography*, McGraw-Hill Book Co., New York, esp. pp. 81–84, 100–101, 106–107.

Bennett, H. H. (1939) *Soil Conservation*, McGraw-Hill Book Co., New York, esp. p. 147.

Bird, E. C. F. (1962) "The River Deltas of the Gippsland Lakes," *Roy. Soc. Victoria. Proc.*, v. 75, pp. 65–74.

Bird, E. C. F. (1968) "Delta Dynamics" [in] *Encyclopedia of Geomorphology*, R. W. Fairbridge, ed., Reinhold Book Corp., New York, esp. pp. 252–255.

Bird, E. C. F. (1971) "The Disappearing Mitchell Delta," *Roy. Soc. Vic. Proc.*, v. 84, Art. 16.

Born, S. M. and Ritter, D. F. (1970) "Modern Terrace Development near Pyramid Lake, Nevada and its Geologic Implications," *Geol. Soc. Amer. Bull.*, v. 81, pp. 1233–1242.

Bowman, Isaiah (1916) *The Andes of Southern Peru*, Henry Holt, New York, esp. p. 144.

Bretz, J. H. (1923) "The Channeled Scablands of the Columbia Plateau," *Jour. Geol.*, v. 31, pp. 617–649.

Bretz, J. H., Smith, H. T. U., and Neff, G. E. (1956) "Channeled Scabland of Washington; New Data and Interpretations," *Geol. Soc. Amer. Bull.*, v. 67, pp. 957–1050.

Butler, B. E. (1960) "Riverine Deposition during Arid Phases," *Australian Jour. Sci.*, v. 22, pp. 451–452.

Butler, B. E. (1961) "Groundsurfaces and the History of the Riverine Plain," *Australian Jour. Sci.*, v. 24, pp. 39–40.

Carlston, C. W. (1969) "Downstream Variations in the Hydraulic Geometry of Streams: Special Emphasis on Mean Velocity," *Amer. Jour. Sci.*, v. 267, pp. 499–509.

Cohen, S. B., ed. (1973) *Oxford World Atlas*, Oxford Univ. Press, New York, esp. pp. 22, 25.

Coleman, J. M. (1968) "Deltaic Evolution" [in] *Encyclopedia of Geomorphology*, R. W. Fairbridge, ed., Reinhold Book Corp., New York, esp. pp. 255–261.

Connally, G. G., Krinsley, D. H. and Sirkin, L. A. (1972) "Late Pleistocene Erg in the Upper Hudson Valley, New York," *Geol. Soc. Amer. Bull.*, v. 83, pp. 1537–1542.

Cotton, C. A. (1968) "Relict Landforms" [in] *Encyclopedia of Geomorphology*, R. W. Fairbridge, ed., Reinhold Book Corp., New York, esp. pp. 936–940.

Daly, R. A. (1936) "Origin of Submarine Canyons," *Amer. Jour. Sci.*, v. 31, pp. 401–420.

Davis, W. M. (1899) "The Geographical Cycle," *Geogr. Jour.*, v. 14, pp. 481–504.

Davis, W. M. (1905) "The Geographical Cycle in an Arid Climate," *Jour. Geol.*, v. 13, pp. 381–407.

Doering, J. A. (1958) "Citronelle Age Problem," *Amer. Assoc. Pet. Geologists Bull.*, v. 42, pp. 764–786.

Doria-Medina, J. H. (1962) *A Study of Chan-*

nel, *Bank and Bottom Gravels along Mountain Fork Creek, Crawford County, Arkansas*, Univ. Ark. M.S. Thesis (on file), Fayetteville, Arkansas.

Dury, G. H., ed. (1966) "The Concept of Grade" [in] *Essays in Geomorphology*, American Elsevier Pub. Co., New York, esp. pp. 211–234.

Dury, G. H. (1967) "Some Channel Characteristics of the Hawkesbury River, New South Wales," *Australian, Geogr. Studies*, v. 5, pp. 135–149.

Dury, G. H. and Langford-Smith, T. (1970) "A Pleistocene Aboriginal Camp Fire from Lake Yantara, Northwestern New South Wales," *Search*, v. 1, p. 73.

Dury, G. H., Sinker, C. A. and Pannett, D. J. (1972) "Climate Change and Arrested Meander Development on the River Severn" [in] *Area*, Inst. Brit. Geographers, v. 4, pp. 81–85.

Fairbridge, R. W., ed. (1968) "Terraces, Fluvial—Environmental Controls" [in] Encyclopedia of Geomorphology, Reinhold Book Corp., New York, esp. pp. 1124–1138.

Fenner, C. N. (1948) "Pleistocene Climate and Topography of the Arequipa Region, Peru," *Geol. Soc. Amer. Bull.*, v. 59, pp. 895–917.

Fisk, H. N. (1944) *Geological Investigation of the Alluvial Valley of the Lower Mississippi River*, Miss. River Comm., Vicksburg, Miss.

Folk, R. L. (1971) "Genesis of Longitudinal and Oghurd Dunes Elucidated by Rolling upon Grease," *Geol. Soc. Amer. Bull.*, v. 82, pp. 3461–3468.

Gage, Maxwell (1953) "Rounding and Transport of Large Boulders in Mountain Streams," *Jour. Sed. Pet.*, v. 23, pp. 60–61.

Gardener, L. R. (1972) "Origin of the Morman Mesa Caliche, Clark County, Nevada," *Geol. Soc. Amer. Bull.*, v. 83, pp. 143–156.

Garner, H. F. (1959) "Stratigraphic-Sedimentary Significance of Contemporary Climate and Relief in Four Regions of the Andes Mountains," *Geol. Soc. Amer. Bull.*, v. 70, pp. 1327–1368.

Garner, H. F. (1963) "The Fountain Formation, Colorado: A Discussion," *Geol. Soc. Amer. Bull.*, v. 74, pp. 1299–1302.

Garner, H. F. (1966) "Derangement of the Rio Caroni, Venezuela," *Revue de Géomorphologie Dynamique*, no. 2, pp. 50–83.

Garner, H. F. (1966) "Groundwater Aquifer

Patterns and Valley Alluviation along Mountain Fork Creek, Crawford County, Arkansas," *Ark. Acad. Sci. Proc.*, v. 20, pp. 95–103.

Garner, H. F. (1967) "Geomorphic Analogs and Climax Morphogenesis," *Ark. Acad. Sci. Proc.*, v. 21, pp. 64–76.

Gorycki, M. A. (1973) "Hydraulic Drag: A Meander-Initiating Mechanism," *Geol. Soc. Amer. Bull.*, v. 84, pp. 175–186.

Grove, A. T. (1969) "Landforms and Climate Change in the Kalahari and Ngamiland," *Geogr. Jour.*, v. 135, pp. 191–212.

Hack, J. T. and Goodlet, J. C. (1960) "Geomorphology and Frost Ecology of a Mountain Region in the Central Appalachians," *U. S. Geol. Surv. Prof. Pap. 347.*

Horton, R. E. (1945) "Erosional Development of Streams and Their Drainage Basins," *Geol. Soc. Amer. Bull.*, v. 56, pp. 275–370.

Howard, A. D., Fairbridge, R. W. and Quinn, J. H. (1968) "Terraces, Fluvial—Introduction" [in] *Encyclopedia of Geomorphology*, R. W. Fairbridge, ed., Reinhold Book Corp., New York, esp. pp. 1117–1124.

Hubert, J. F. (1960) "Petrology of the Fountain and Lyons Formations, Front Range, Colorado," *Col. Sch. Mines Quart. Jour.*, v. 55, 242 pp.

Keller, E. A. (1972) "Development of Alluvial Stream Channels: A Five-Stage Model," *Geol. Soc. Amer. Bull.*, v. 83, pp. 1531–1536.

King, L. C. (1953) "Canons of Landscape Evolution," *Geol. Soc. Amer. Bull.*, v. 64, pp. 721–752.

Lane, E. W. (1957) A Study of the Shape of Channels Formed by Natural Streams Flowing in Erodable Material," *M.R.D. Sediment Series*, No. 9, U. S. Army Engr. (Mo. River Corps), Omaha.

Langbein, W. B. and Leopold, L. B. (1966) "River Meanders—Theory of Minimum Variance," *U. S. Geol. Prof. Pap. 422-H.*

Leliavsky, S. (1955) *An Introduction to Fluvial Hydraulics*, Constable, London, 257 p., esp. pp. 111–141.

Leopold, L. B. and Maddock, Thomas (1953) "The Hydraulic Geometry of Stream Channels and Some Physiographic Implications," *U. S. Geol. Surv. Prof. Pap. 252*, 57 p.

Leopold, L. B. and Wolman, M. G. (1957) "River Channel Patterns: Braided, Meandering and Straight," *U. S. Geol. Surv. Prof. Pap.* 282-B.

Leopold, L. B., Wolman, M. G. and Miller, J. P. (1964) *Fluvial Processes in Geomorphology,* W. H. Freeman, San Francisco, esp. pp. 151–188, 433–474.

Leopold, L. B., Emmett, W. W. and Myrick, R. M. (1966) "Channel and Hillslope Processes in a Semiarid Area, New Mexico," *U. S. Geol. Surv. Prof. Pap.* 352-G, pp. 193–253.

Mackin, J. H. (1963) "Rational and Empirical Methods of Investigation in Geology" [in] *The Fabric of Geology,* C. C. Albritton, ed., Addison-Wesley Pub. Co., Reading, Mass., pp. 135–163.

Miller, J. P. (1958) "High Mountain Streams," *New Mex. Bur. Min. and Miner. Res. Mem.* 4, 51 pp.

Miller, J. P. and Wendorf, F. (1958) "Alluvial Chronology of the Tesuque Valley, New Mexico," *Jour. Geol.,* v. 66, pp. 177–194.

Miller, J. P., Montgomery, A. and Sutherland, P. K. (1963) "Geology of Part of the Southern Sangre de Cristo Range, New Mexico," *New Mex. Inst. Mining and Tech., Mem.* 11, 106 p.

Minard, J. P. (1964) *Geology of the Roosevelt Quadrangle, New Jersey,* U. S. Geol. Surv. Quadrangle Map, G.Q.-340.

Minard, J. P. (1966) "Sand Blasted Blocks on a Hill in the Coastal Plain of New Jersey," *U. S. Geol. Surv. Prof. Pap.* 550-B, pp. 87–90.

Morgan, J. P. (1961) "Mudlumps at the Mouth of the Mississippi River," *Louisiana Dept. Conserv. Geol. Bull.,* v. 35, pp. 1–116.

Morisawa, Marie (1968) *Streams: Their Dynamics and Morphology,* McGraw-Hill Book Co., New York, 175 pp., esp. pp. 29–60.

Muller, Jan (1959) "Palynology of Recent Orinoco Delta and Shelf Sediments," *Micropaleo.,* v. 5, pp. 1–32.

Oberlander, T. M. (1972) "Morphogenesis of Granitic Boulder Slopes in the Mojave Desert, California," *Jour. Geol.,* v. 80, pp. 1–20.

Otvos, E. G. (1971) "Relict Eolian Dunes and the Age of the 'Prairie' Coastwise Terrace, Southeastern Louisiana," *Geol. Soc. Amer. Bull.,* v. 82, pp. 1753–1758.

Peabody, F. E. (1954) "Travertines and Cave Deposits of the Kaap Escarpment of South Africa, and the Type Locality of *Austra-*

lopithecus africanus Dart," Geol. Soc. Amer. Bull., v. 65, pp. 671–706.

Pettijohn, F. J. (1957) *Sedimentary Rocks,* Harper and Bros., New York, esp. pp. 243–261.

Price, W. A. (1951) "Wind Caused Patterns," Sci. Newsletter, v. 327.

Price, W. A. (1968) "Carolina Bays" [in] *Encyclopedia of Geomorphology,* R. W. Fairbridge, ed., Reinhold Book Corp., New York, esp. pp. 102–109.

Prouty, W. F. (1952) "Carolina Bays and Their Origin," *Geol. Soc. Amer. Bull.,* v. 63, pp. 167–224.

Prus-Chacinski, T. M. (1954) "Patterns of Motion in Open-Channel Bends," *Assoc. Internat. d'Hydrologie,* v. 3, pp. 311–318.

Quinn, J. H. (1957) "Paired River Terraces and Pleistocene Glaciation," *Jour. Geol.,* v. 65, pp. 149–166.

Quinn, J. H. (1959) "Bed Load and Stream Entrenchment," *Geol. Soc. Amer. Bull.* (Abstract) v. 70, pp. 1659.

Quinn, J. H. (1961) "Prairie Mounds of Arkansas," *Ark. Archaeological Soc. Newsletter,* v. 2, pp. 1–8.

Quinn, J. H. (1965) "Monadnocks, Divides and Ozark Physiography," *Ark. Acad. Sci. Proc.* v. 19, pp. 90–97.

Quinn, J. H. (1968) "Prairie Mounds" [in] *Encyclopedia of Geomorphology,* R. W. Fairbridge, ed., Reinhold Book Corp., New York, esp. pp. 888–890.

Russell, R. J. (1936) "Lower Mississippi River Delta," *Louisiana Dept. Conserv. Geol. Bull.,* v. 8, pp. 8–199.

Russell, R. J. (1958) "Geological Geomorphology," *Geol. Soc. Amer. Bull.,* v. 69, pp. 1–22.

Saucier, R. T. and Fleetwood, A. R. (1970) "Origin and Chronologic Significance of Late Quaternary Terraces, Ouachita River, Arkansas and Louisiana," *Geol. Soc. Amer. Bull.,* v. 81, pp. 869–890.

Schumm, S. A. (1972) "Fluvial Paleochannels" [in] *Recognition of Ancient Sedimentary Environments,* J. K. Rigby and W. K. Hamblin, eds., S.E.P.M. Spec. Pub. 16, pp. 98–107.

Schumm, S. A. and Khan, H. R. (1972) "Experimental Study of River Patterns," *Geol. Soc. Amer. Bull.,* v. 83, pp. 1755–1770.

Schumm, S. A., Khan, H. R., Winkley, B. R. and Robbins, L. G. (1972) "Variability of River Patterns," *Natural Physical Sci.,* v. 237, pp. 75–76.

Shepard, F. P. (1951) "Mass Movements in Submarine Canyon Heads," *Amer. Geophys. Union Trans.*, v. 32, pp. 405–418.

Shepard, F. P. (1963) "Submarine Canyons" [in] *The Sea*, M. N. Hill, ed., John Wiley and Sons, New York. Interscience.

Smith, N. D. (1971) "Transverse Bars and Braiding in the Lower Platte River, Nebraska," *Geol. Soc. Amer. Bull.*, v. 82, pp. 3407–3420.

Speight, J. G. (1965) "Flow and Channel Characteristics of the Angabunga River, Papua," *Jour. Hydrol.*, v. 3, pp. 16–36.

Strahler, A. N. (1952) "Dynamic Basis of Geomorphology," *Geol. Soc. Amer. Bull.*, v. 63, pp. 923–938.

Strahler, A. N. (1968) "Quantitative Geomorphology" [in] *Encyclopedia of Geomorphology*, R. W. Fairbridge, ed., Reinhold Book Corp., New York, esp. pp. 898–912.

Sundborg, A. (1956) "The River Klarälven," *Geografiska Ann.*, v. 38, pp. 125–316.

Tinkler, K. J. (1970) "Pools, Riffles, and Meanders," *Geol. Soc. Amer. Bull.*, v. 81, pp. 547–552.

Tricart, Jean (1958) "Etudies sur quelques cailloutis fluviatiles actuels des Pyrénées Orientales et du Massif Central," *Zeits. für Geomorph.*, bd. 2, H. 4, pp. 278–304.

Williams, G. E. and Polach, H. A. (1971) "Radiocarbon Dating of Arid-Zone Calcareous Paleosols," *Geol. Soc. Amer. Bull.*, v. 82, pp. 3069–3086.

8 Glacial and cryergic geomorphic systems and landforms

Let us now consider the role of ice in the genesis of landforms. Ice takes a variety of forms and has a number of related agencies and conditions, but it appears geomorphically most significant in *glaciers*, in areas of perennially frozen ground (*permafrost*), and in associated effects brought about by freeze-thaw conditions. With respect to the earth's environmental continuum, with which we have been preoccupied in the three previous chapters, A GLACIAL REALM is any region where glaciers periodically tend to form and produce their effects. Most are either close to the poles or at high altitudes. Presently, a few such regions are in part occupied by glaciers, but some are not (Fig. 8.1).

Areas where glaciers now occur can be thought of as approximating *glacial climatic nuclei*, inasmuch as the "Present" is "deglaciated" if not actually "interglacial," and existing ice masses appear to occupy environmental niches that favor their endurance. Even so, some contemporary glaciers are evidently more permanent and "healthier" than others. Certain of the regions that have been glaciated have topography dominantly composed of glacial landforms, even though they are not under ice at present. Other areas, though recently glaciated, are much less pervasively affected.

Glacial geomorphic systems can be thought of as associated with one of the earth's major surficial environments. At relatively low latitudes, glaciation is known to alternate with humid and arid conditions, but at higher latitudes, the principal alternate environmental end-member with glaciation is that related to the tundra and permafrost realm or boreal deserts. We will consider the general effects of such alternations as our discussion proceeds here. An adequate treatment of the subject of glaciation requires initial consideration of the causative conditions—in both time and space. Also, our discussion should include an evaluation of ice as a geomorphic mechanism in its main modes and areas of activity, plus the principal topographic results. We will grapple with the problem of causal conditions first.

GLACIAL ENVIRONMENTS

Our understanding of the conditions under which glaciers presently exist is rather good, thanks to research stations permanently maintained at or near both poles by various nations, periodic expeditions across the Greenland Ice Sheet, and continuing studies of various valley glaciers, particularly in the Alps. This understanding, however, should not be confused with the notion that we truly comprehend the conditions that caused glaciers in other areas to form, grow, and ultimately waste away several times during the Pleistocene Epoch. In reality, it appears that glaciers simultaneously present us with at least two environmental problems; contemporary conditions for existing glaciers and causes of former glacial environment changes.

EXISTING GLACIAL REGIMES

Study of an existing glacial environment tends to work its way down to the question of whether a maintenance budget exists. In essence, it is essential to know if the meteorologic parameters being recorded (e.g. precipitation, temperature) relate to a condition that merely sustains the ice mass or, alternatively, causes it to grow or to shrink. Definitions of what is required for glacial growth, subsistence, and wastage depend upon such knowledge. However, difficulties encountered in obtaining accurate precipitation records in the 100-mph winds frequently encountered on ice sheets are compounded by vast uncertainties about several other matters. For example, what percentage of actual snowfall does the *annual firn increment* measured in bore holes actually represent? Also, there are problems of wastage rates expressed by *sublimation* where ice or snow enters a gaseous state in the atmosphere without passing through a liquid phase. By carefully measuring accumulation and flow rates, and making estimates of losses via phenomena such as sublimation and iceberg calving, we get some

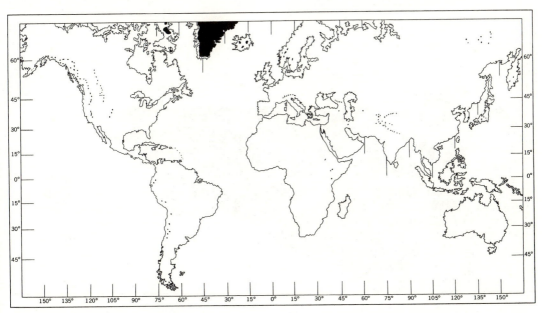

Fig. 8.1. Distribution of major existing areas of glacier ice, exclusive of Antarctica. Essentially all the black dots are in high mountains.

notions about glacier budget. In general, glaciers occur where mean annual temperatures are 32° F (0° C) or less, and much of the precipitation occurs as snow. Lower limits of favorable accumulation are indicated on mountain peaks by the *snow line* (Fig. 8.2). As noted by Flint (1971), the over-all economies of the Greenland and Antarctic ice sheets are uncertain, that of the latter possibly being positive (i.e. accumulation may exceed wastage). Losses of glacier mass by melting and evaporation (sublimation) are termed *ablation*.

On a global basis, glaciers presently appear to exist under what may be termed high-budget and low-budget conditions. The low-budget ice masses mainly involve areas like the central part of the Antarctic Ice Sheet and the northern part of the Greenland Ice Sheet where annual precipitation ranges from less than 2 inches (5 cm) in the former area to 6 inches (15 cm) in the latter. The low precipitation rates are counterbalanced by very low mean annual temperatures which curtail wastage by melting. Because of air dryness, evaporative losses may be high. High-budget ice masses include the more southern part of the Greenland Ice Sheet along coasts where precipitation up to 34 inches (85 cm) is balanced by much more intense summer wastage in this lower latitude setting. Flint (1971) notes that the Vatnajökull glacier in Iceland in an even warmer and more moist situation receives in excess of 120 inches (300 cm) of annual precipitation, some in the form of warm rain which accelerates ablation. It seems clear that high- or low-budget glaciers could display a positive, negative, or balanced economy. In general, the high-budget glacial economies appear to be in relatively low-latitude, low-altitude coastal situations. This has a bearing on the character of ancient glaciation extending to

Fig. 8.2.   Aerial view of Mount Chimborazo, the world's highest volcanic peak (middle background) with an elevation of 20,556 ft (6,267 m), in which the snow line is clearly delineated at about 15,500 ft (4,750 m) about 2° from the Equator.

rather low latitudes and upon possible causal mechanisms.

PAST GLACIAL FLUCTUATIONS

The environmental conditions that exist over and adjacent to an extant glacier may reflect a sustaining condition. But, at the same time, those conditions must also reflect a somewhat self-induced regimen. We say this, for a glacier exists as part of a geomorphic system in which feedback plays a significant role, and the "icebox" effect, once a glacier comes into existence, must invariably chill and thus alter the original sponsoring condition. Thus, an existing glacier is at once a form of groundcover and an erosional mechanism that is a product of a set of conditions that may not be wholly typified by the cold surroundings of the ice mass. As noted by Charlesworth (1957), Wright and Frey (1965), Flint (1971), and many others, glaciers appear to form, endure, and disappear in response to climate changes, changes that nevertheless relate to shifting oceanic conditions and altering orography and global temperature distributions as ice masses grow higher and spread out.

Unlike the situation involving plants, where the groundcover exists under a wide range of temperatures which nevertheless permit attainment of critical moisture levels, ice in any enduring form requires a temperature below freezing. Of course, this represents a temperature threshold. Yet, world-wide changes in temperature, as seem to be reflected by greatly depressed Pleistocene snow lines, must have direct effects on precipitation. And the automatic juxtaposition of terms like "cold glacial" and "hot interglacial" could be a dangerous practice. Addition of the terms "wet" and "dry" to either combination also involves certain assumptions that may be only locally valid, if at all. This is not to say that the oceans did not appear to cool off during glacial episodes. Just consider what an ice cube does for a highball. And there seems to have been oceanic warming in interglacial times. But the world-wide glacial-climatic arrangement may not be quite that neat (cf. Friedman and Saunders, 1970). One can account for the indicated temperature changes by various mechanisms including the solar radiation cycles discussed in Chapter 1.* But the relationship between mean global temperature and oceanic evaporation is probably direct; relations between sea surface temperatures and adjacent land moistures certainly often tend to be direct (cf. Chapter 4), and more than one researcher has argued that larger ice sheets may tend to starve themselves by chilling their surroundings and freezing over nearby evaporation surfaces (cf. Ewing and Donn, 1956).

In the final analysis, the actual conditions that lead to the formation of glacial episodes probably can be determined by the study of glacial economies and meteorology, glacial stratigraphy, and allied geomorphic records that show regional patterns of climate change and hydrographic effects, all coupled with a careful use of radiometric dating techniques. With mounting evidence that some regions become drier as others receive increased precipitation coincident with glacial episodes, terms like "glacio-pluvial" can have little more than local application, and it becomes more and more difficult to ascertain if a particular glacial adjustment correlates with a net global increase in precipitation or the opposite. Flint (1971) places great emphasis on variations in solar radiation, but he couples these effects with increased land elevation during the late Tertiary to account for Pleistocene glaciation. Ewing and Donn (1956) attempt to utilize variable sea ice conditions on the Arctic Ocean to account for glacial fluctuations, the writer (1959a) emphasizes Antarctic influences over ocean current patterns and east-west continental climatic asymmetry, and Damuth and Fairbridge (1970) advo-

---

* There are unresolved problems here, including the precise nature of the causal solar mechanism and the brevity of the solar data span.

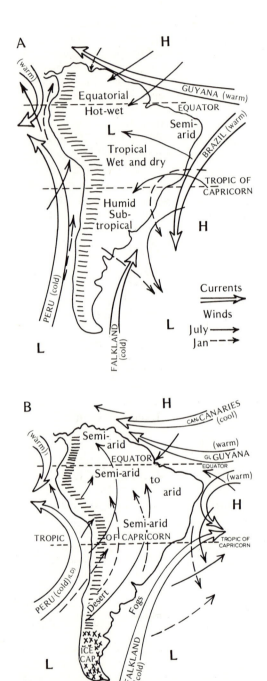

Fig. 8.3. Atmospheric and oceanic circulation patterns and climatic effects related to South America in (A) interglacial and (B) glacial episodes. (After Damuth and Fairbridge, 1970.)

cate altered atmospheric circulation to the same end (Fig. 8.3). There have been many other suggestions; some are very similar (cf. Figs. 8.4 and 8.5).

Atmospheric and oceanic interaction under the influence of variable solar radiation, coupled with the probable attainment of a polar position by Antarctica via continental drift during the late Tertiary, appears adequate to account for glacial episodes elsewhere—major fluctuations being explicable on the basis of solar variations superimposed on an array of interdependent phenomena. Existing high- and low-budget glacial conditions suggest that some glaciers may form simply because reduced temperatures permit related snow accumulation to exceed wastage. Others may develop because high-intensity precipitation induced by increased air-sea temperatures permits land moisture (ice) accumulation in excess of wastage enhanced by glacial self-refrigeration and in spite of relatively elevated equatorial temperatures. Added to such possible effects are the probable delays in cooling of the oceans after icesheets begin to form, plus the lag in time from the inception of an ice mass until its medial portions become so distant from precipitation sources, due to peripheral spreading, that starvation sets in. How it all works, in fact, remains very uncertain. But in their study of C_{14}-dated river terraces (Chapter 7), Saucier and Fleetwood (1970) suggest that glacial episodes are reflected in southcentral United States by initial moisture increases (stream incision) and then later aridity (valley aggradation) followed (during the next interglacial) by diminishing aridity (reduced alluviation) and, ultimately, increased humidity (renewed stream incision). The events indicated correlate rather well with the glacial self-starvation notion—provided, of course, that allowances are made for certain climatic biases. The latter becomes necessary in light of hemispheric heat asymmetry (cf. Chapter 4) and climatic nuclei which are effectively immune to these changes. Even so, data from equatorial realms such as those of Mousinho

(1971), whose studies in the Amazon Basin are for the same general time interval, show essentially the same moisture pattern and sequence. Findings of Grove and Goudie (1971) on lake development and fluvial activity in Africa are also in general accord.

A final observation appears to be in order before we leave our discussion of glacial conditions. It relates to the mobility of ice. Glacier ice, like water in humid regions, may escape the sponsoring environment. Like a trans-environmental river, a glacier may leave a cold, elevated source for more or less temperate climes—whether the elevation be expressed above sea level or by latitude. And to a degree not matched by a river, glaciers tend to take their own source environments along. That they are not entirely successful is of course reflected by the fact that they do not advance indefinitely. Yet the potential for ice penetrating great distances into inhospitable realms is indeed considerable. One has only to peer out through the leafy canopy of a New Zealand rain forest at the snout of an alpine glacier to grasp the possible environmental juxtapositions. And if the accumulation centers of a continental ice sheet, initially localized by land topography, are later governed primarily by a wandering hydrography focused on an equally vagrant ice orography, a past glacial border environment in central Iowa could differ notably from one in central Canada or from those present in Greenland or Antarctica. We will discuss this later in some detail.

GLACIAL GEOMORPHIC EFFECTS

The creation of landforms and deposits by glacial ice is intimately related to the manner of glacial movement and associated erosion, deposition, and wastage plus peripheral effects brought about by glacier presence. Once again we are concerned with a geomorphic system, like those already discussed, that may be presumed to be affected by antecedent topographic setting as well as by interaction of various glacial agencies and products. The topographic extremes usually cited for glaciers are *alpine* and *continental*, the latter usually having the connotation of a relatively low-relief, low-elevation setting as compared to the former.* The geomorphic distinctions that appear to exist between these two types of glacial terrains will be noted in due course. But the basic character of ice formation and movement in both settings appears to be similar, so this is a logical starting point for our discussion.

GLACIER MOVEMENT

The view taken here is that a glacier is a mass of ice and *firn* at least part of which is actively flowing. If the ice has flowed but is not flowing now, then it is a *relict glacier*, which, like a pond, has some attributes of a stream but not all. The movement of glacier ice is closely caught up in the sequence of physical state (phase) transformations that occur in the materials usually associated with a glacier. And both the movement and the phase transformations have an effect on mechanisms of erosion and deposition. Keeping things general for a moment, we can say that IN THE BEGINNING THERE MUST BE SNOW accumulated in excess of *ablation* to a depth where the strength (*elastic limit*) of the buried ice crystals is exceeded by the superincumbent load of frozen water. In effect, the load is compactive and generates an internal shearing stress within

* It is nonetheless true that at least parts of Antarctica and the area of the North American Cordilleran ice sheet were mountainous.

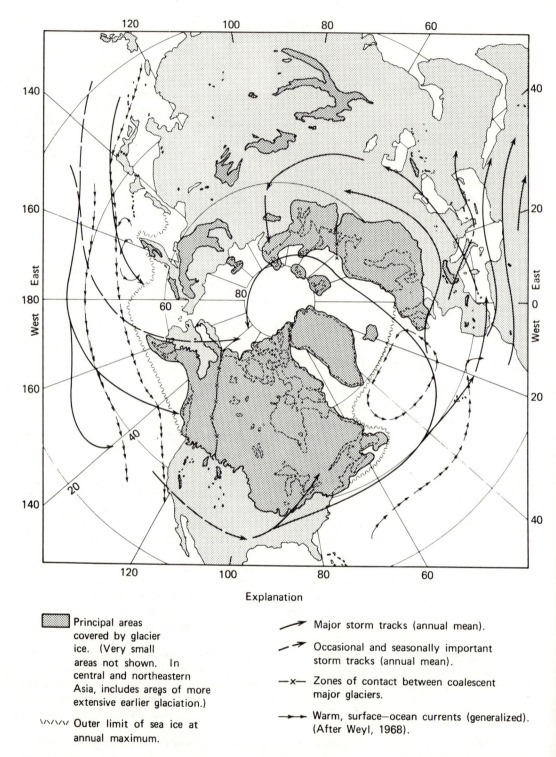

Explanation

Principal areas covered by glacier ice. (Very small areas not shown. In central and northeastern Asia, includes areas of more extensive earlier glaciation.)

\/\/\/ Outer limit of sea ice at annual maximum.

Major storm tracks (annual mean).

Occasional and seasonally important storm tracks (annual mean).

—x— Zones of contact between coalescent major glaciers.

Warm, surface—ocean currents (generalized). (After Weyl, 1968).

Fig. 8.4. Part of the Northern Hemisphere showing glaciers, selected ocean currents, sea ice, and storm tracks during (A) glacial episodes and (B) at present. (From Flint, 1971.)

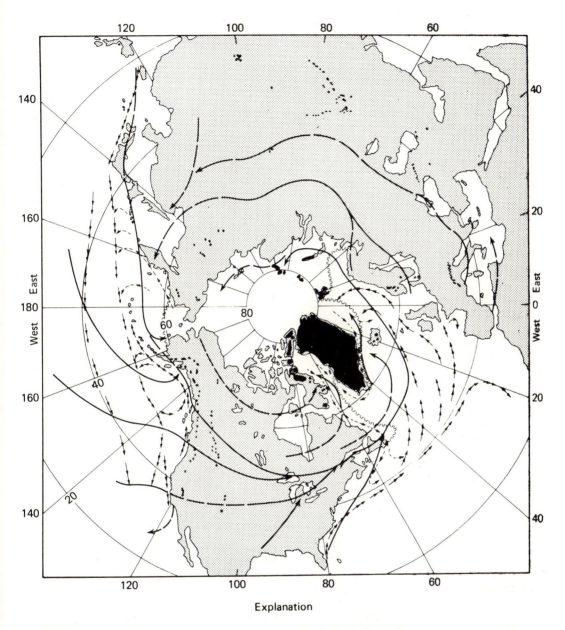

Explanation

■ Principal areas
now covered by
glacier ice. (Very
small areas, and
areas in southern Asia
and East Africa not shown).

ʌʌʌʌ Approximate present outer limit
of sea ice at annual maximum.

↗ Present—day major
storm tracks (annual mean).

↗ Present—day occasional and seasonally
important storm tracks (annual mean).

→ Warm surface—ocean currents (generalized).

--→ Cold surface—ocean currents (generalized).

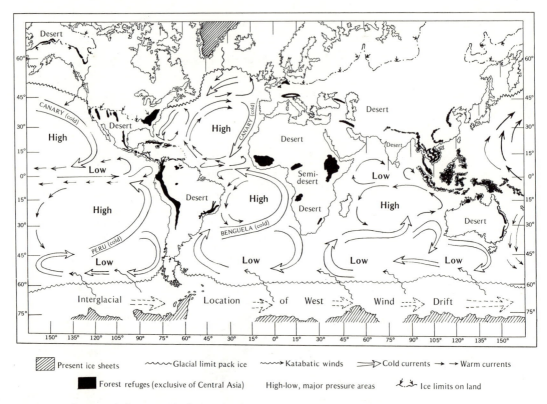

Fig. 8.5. Theoretical diagram of effects of intensified Antarctic glaciation and extension of sea ice athwart the prevailing westerlies. Elimination of West Wind Drift on a circumglobal basis should generally weaken and locally reverse currents to the north as the seas become colder (cf. Garner, 1959a).

and/or at the base of the ice mass. In the absence of confining terrain, or in the presence of land slope, or both, the load stress is unbalanced and can be released by shearing within the ice mass or other internal movement forms as well as by basal sliding. But what of the materials?

Snow accumulating on glaciers with dry densities as low as 0.05 and 95% porosity has been found by several researchers to gradually alter by recrystallization into the compact, granular material termed *firn* or *névé* with a density of some 0.5 and 50% porosity. Some of the most interesting studies of this transition have been made by the glaciologists of the Jungfraujoch Research Party in their laboratory in the Great Aletsch Glacier in conjunction with

the Swiss Snow and Avalanche Commission in Davos, Switzerland (cf. Hughs and Seligman, 1938; Perutz and Seligman, 1939).* They conclude that the transition from snow to firn is mainly by melting, sublimation, and refreezing and may occur in about a year in low-latitude alpine glaciers. The same transition may require many years in colder polar settings. The conversion of firn into glacier ice is apparently only a continuation of the same process, whereby air spaces in the firn become reduced to intercommunicating capillaries during growth of hexagonal ice crystals as recorded by Bader and Haefeli (1939) and illustrated here (Fig. 8.6). It has been found that a high per cent of firn crystals near the névé surface have

---

* There is a good bit of excellent, recent, low-temperature experimentation and ice-study research by the Japanese.

their hexagonal axes at right angles to the surface and parallel to the temperature gradient but that these crystal orientations later become random through individual grain movements. It should also be pointed out here that the firn appears to lack internal shear planes but may display seasonal or annual layering, and this layering may later reappear as blue bands of ice in a glacier tongue. Other blue bands may be due to former *fissures* or *crevasses* or to thrust planes caused by laminar motion between ice layers.

As noted in an earlier chapter, glacier ice is both a mineral and a metamorphic rock. Its average density is 0.9059, but, with the common air-bubble inclusions, it may fall below 0.90. The hexagonal crystals generally increase in size with time and distance of movement, but, as noted by Seligman (1949), crystal growth is inhibited by cold and stimulated by warmth —even in the absence of meltwater—and probably aided by certain types of movement. Views on the nature of glacial movement are well summed up by Seligman who states

> *At the turn of the century the many conflicting views simplified themselves into two main schools of thought, the one following Hess in what was called the Theory of Plastic Flow in which the unit of flow is a molecule, a group of molecules or a single crystal. The other school developed from principles perhaps best enunciated by Forbes in what was called the Theory of Laminar Flow. It might be safer to call the latter non-plastic flow, it being assumed that the ice became divided into small or large units owing to differential speed in different parts of the stream.*

The units were presumably separated by thrust planes. But it should be noted that examples of both plastic and laminar flow have been observed in glacier ice tongues (see also Paterson, 1969).

Though properly the concern of glaciologists, forms of ice movement and the physical states of glaciers and glacier ice must also be our concern since they have a direct bearing on modes of sediment load

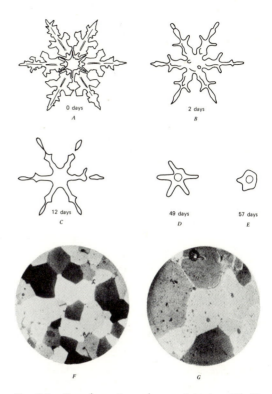

Fig. 8.6. Transformation of snow into ice. (A)–(E) Changes in snowflake after 2, 12, 49, and 57 days, respectively; (E) essentially granular ice; (F) ice made artificially from snow soaked in water, analogous to ice made from firn; (G) ice subjected to shear deformation for six days, analogous to glacier ice. Note evident increase in grain size of (G) over (F). (From Bader *et al.*, 1939.)

acquisition and transport as well as on mechanisms of abrasion and scour. Glacier ice movement down slopes has appropriately been termed *gravity flow*. The occurrence of *glacial surges* (very rapid ice advances for short periods) appears to document at least some *basal slippage* as one type of movement increment. Calculations of Strieff-Becker (1938b) on the Claridenfirn glacier in the Alps appear to support the occurrence of *extrusion flow*, since snowfall on the glacier surface did not increase its thickness in spite of net accumulation. This flow form was also proposed by Demorest (1938), on the basis of relict glacial erosion features in Glacier National Park in North America. Sharp (1948) also mentions ice movements associated with phase changes of water from solid to liquid to gas and back and illustrates the several mechanisms (Fig. 8.7). As mentioned before, ice in a glacier is under pressure, and ice near the base may approach the pressure-melting point.* When geothermal heat encounters such ice it may melt and refreeze (*relegation*) as noted by Flint (1971), and this melting could accentuate basal slippage. Water was encountered at the base of the Antarctic Ice Sheet in one bore hole.

Discussions of glacial movement that resound with words like "plastic" and "flow" tend to convey an image of something like toothpaste and obscure the essential rigidity of most glacial masses which is otherwise indicated by ice character and erosional effects. The flowage involved is analogous to that in rock metamorphism. Fractures, often in the form of thrust planes or expressed by open crevasses, are one expression of ice rigidity, but alpine crevasses are never very deep—the deepest recorded by Strieff-Becker (1938a) was 30 m. Faceted rocks transported by glaciers (Fig. 8.8A) clearly express oriented abrasion of the faces in question while the

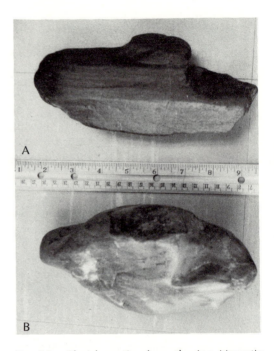

A

B

Fig. 8.8. Glacial erratics from the Lac Megantic region of southeastern Quebec, Canada. A is characterized by slightly grooved, flat facets developed on a piece of quartzite which was presumably held immobile during the process. B is highly polished and evidently shaped from a piece of serpentinized peridotite but has a torpedo shape lacking flat surfaces; scale, inches and centimeters.

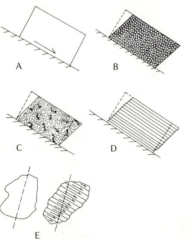

A   B

C   D

E

Fig. 8.7. Sketches of possible mechanisms of glacier motion. (A) Sliding on base; (B) intergranular adjustments; (C) phase change; (D) internal slip planes; and (E) intragranular gliding and recrystallization. (After R. P. Sharp, 1948.)

clast was held essentially immobile—an effect hardly possible for an object supported by a plastic medium. Furthermore, effective abrasion of rock floors beneath glaciers by transported debris could hardly occur where the moving clasts were free to shift readily away from bedrock obstructions. On the other hand, some *glacial erratics* have acquired striking torpedo shapes, with the streamlining and polish extending even into recesses and hollows (Fig. 8.8b). The latter effects could only be brought about if the abrasive medium completely molded itself to the contour of the clasts in question—a strong indication of plasticity.

Perutz (1940, 1948) was among the first to note that neither glacial erosion

---

* According to Porter (personal communication, 1973), in a temperate glacier it is AT the melting point.

nor glacial speed-sensitivity to thickness could be explained by treating a glacier as a Newtonian fluid of constant viscosity. Thickness can, of course, vary climatically.* Vertical velocity data within flowing ice were needed and were first obtained by boring a hole through the Jungfraujoch Glacier with a heating element. In the 400 ft (125 m) Jungfraujoch ice thickness (from August 15, 1948 to October 12, 1949) the top 165 ft (50 m) of ice advanced 125 ft (38 m) with decreasing flow rates recorded farther down and a minimum movement of 46 ft (14 m) in the basal ice layer (see Fig. 8.9). Later studies of flow in the Malaspina Glacier in Alaska (Sharp, 1958) and the Saskatchewan Glacier in Alberta, Canada (Meier, 1960) yielded similar results. As noted by Perutz, the result is consistent with the assumption that the rate of shear strain is proportional to some power of the shear stress, instead of varying linearly with it as in a classic fluid. The results can be expressed as

$$\frac{dx}{dt} = \frac{r^{2.3}}{550}$$

where x equals shear strain, and r (kg/cm^2) equals shear *stress* which approximates Glens Flow Law (cf. Glen, 1955).**

In a flowing glacier, gravity sets up an external stress (force per unit area) which forces the ice to change shape, volume, or both. The change in shape of the mass can be expressed in a ratio termed *strain*

$$S = \frac{a - b}{a}$$

where a is the original shape or volume, and b is the changed shape or volume. A moving glacier is under constant strain and its rate governs velocity (V). This velocity can be expressed in the relation

$$V = \frac{(pg \sin a)^{2.3} h^{3.3}}{C}$$

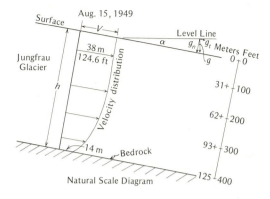

Fig. 8.9. Plot of the velocity profile obtained from the Jungfrau Glacier by boring a hole to bedrock with a heating element. Also indicated are the components of gravity expressed in the recorded velocity V, with respect to glacier thickness h, and slope angle α.

where p (g/cm^3) is the ice density, g (cm/sec/sec) is the acceleration of gravity, a is the angle of slope, h (cm) is the thickness of the ice, and C is a constant which for most glaciers works out to be 0.68×10^{-18}. Surface velocities calculated from the previous equation have been found to agree with observed values for shallow glaciers of large area but are too high for deep valley glaciers, probably because no allowance is made for side friction (cf. Fig. 8.9). From the foregoing it is also evident that a glacier's velocity is extremely susceptible to changes in thickness and slope. It is clear also that a glacier's flow rate constitutes a check on its thickness, a relation alluded to by Flint (1971) as a possible thickness equalizer where he notes the comparable thicknesses of the Greenland and Antarctic Ice Sheets.

PATTERNS OF EROSION AND DEPOSITION

The geomorphologic effects of glaciation relate to several interrelated factors, foremost among which are (1) ability of mov-

---

* In general, thick ice masses move most rapidly, though ice tongues moving into unfavorable climes (warm) may thin and slow.
** The values indicated are not absolute as noted by Lustig (personal communication, 1973) and can be expressed as $dx/dt = ar^b$ in which a and b depend upon several factors, and have been found to be "about" 1/500 and 2.3, respectively, in several cases.

ing ice to erode, (2) direct deposition of glacial detritus where the ice wastes away, (3) inducement of freeze-thaw phenomena and related effects beneath and adjacent to glacial ice masses, and (4) meltwater effects. No one of these factors is totally independent of the others. And their interaction is compounded in regions where glacial conditions reoccur several times over the same general land area. In their hydrographic origins and behaviors, glaciers are in certain respects similar to running water, sharing, as it were, some attributes of water in deserts and some of those in humid lands.

The glacial climatic nucleus resembles the precipitation catchment area of a humid drainage net or an upland precipitation locus in a desert in that all such areas are also foci for erosional events. Thus, whether the glacier is a continental or an alpine type, there is an area under the ice including the area of net accumulation, and at least part of the area of net ablation where erosion dominates (Fig. 8.10). A more peripheral zone is dominated by deposition from the ice due to ablation, and, just as in an ARID FLUVIAL EVENT, there is an erosional interval and a depositional interval. These remarks are applicable not only during the peak of a major glacial expansion but also during growth and recession, simply because glaciation is "erosionally" both a difficult phe-

nomenon to maintain and a "hard act to follow," insofar as ease of sediment acquisition is concerned.

The rationale behind the foregoing remarks is rather simple. It is a matter of supply. Weathered debris is much easier to remove than consolidated bedrock, even for a glacier, and, though the capacity of a glacier is very great, it seems evident that the amount of rock being transported by a glacier newly formed on a weathered surface will diminish as the weathered zone is stripped away to bare bedrock. Shallow bedrock, with open fracture systems, tends to be eroded next. Moreover, repeated glaciation in a region with only intermittent, relatively brief exposure of bedrock to weathering means that later glaciers will carry less rock than earlier ones. This is in part due to prior stripping and ineffective weathering in the glacial nucleus and in part to progressive removal of the more readily eroded bedrock types. Thus, eastern Canada and much of Scandinavia are examples of regions which, in terms of regolith, "have given their all," so to speak. Each area is surrounded by a depositional "halo," where transport by ice has weakened and ultimately failed. There is reason to believe in fact, that once developed in a particular site, glaciers tend to expend a great deal of energy shifting around debris left in the "marginal halo" by previous glaciers.

Meltwater effects of geomorphic significance appear to begin mainly in the ablation zone of a glacier and to extend radially outward—downslope where possible—and apparently at all stages of glacier development and recession. Therefore, *glacio-fluvial* effects are superimposed on and admixed with effects achieved by direct glacier ice erosion and deposition. And some such aqueous effects may extend at a distance down associated watercourses. Thus, unlike the humid situation where the main geomorphic agency (water) often acts beyond the environmental limits, or the arid situation where water rarely reaches the regimen limits, the ice of a glacier loses its geomorphic effectiveness only to melt

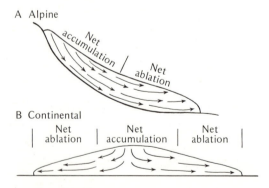

Fig. 8.10. Sections of (A) alpine and (B) continental glaciers showing relationships between accumulation areas, regions of wastage (ablation), and generalized flow; not to scale.

and reappear, phoenix-like, in liquid form, whence the glacial environment's existence can be expressed elsewhere.

Of course, glacial meltwaters must respond to the same constraints of hydraulic geometry that rule running water in other settings (see Chapters 5, 6, and 7). In a sense, meltwater is trans-environmental runoff (cf. Chapter 5). But again a major difference emerges. Whereas humid runoff generally maintains its transportative power, and arid runoff loses this power after each precipitation event, glacial runoff diminishes seasonally and even diurnally with the onset of freezing temperatures. This, of course, means that seasonal sedimentary deposits tend to develop in many pro-glacial lakes (*varves*) and meltwater streams (*outwash gravel*). It follows that pro-glacial valleys generally experience diminished flow continuity in comparison to perennial humid rivers, subject, of course, to possible ice dam ruptures (Fig. 7.60) or accelerated melting during deglaciation. Nor do these comments take into account the character of near-glacial land environments which may alter the nature of meltwater movement just as a desert alters runoff from a humid vegetated region.

Before considering glacial erosion and its topographic effects in more detail, we should briefly mention the fourth factor of glacial environments which enters into the over-all morphogenesis. Ground beneath glaciers tends to be more or less deeply frozen and so does that bordering glaciers in high latitudes and occupying high-latitude regions where glaciers formerly spread.* The present 32° F (0° C) (mean annual) isotherm does not coincide with extant areas of perennially frozen ground (*permafrost*) (cf. Fig. 8.11), and a good bit of such perennially frozen terrain in northern North America and Eurasia can be considered to be relict from former, more extensive subfreezing mean annual temperatures, if not actual glacial conditions as noted by MacKay *et al.* (1972).

The perennially frozen ground adjacent to high-latitude glaciers is compatible with their environmental surroundings in the Antarctic and northern Greenland. But the ability of a glacier to induce a cold peripheral environment probably diminishes as the ice mass invades warmer climes and may terminate completely under some circumstances. Nevertheless, glaciers probably tend to advance across frozen ground over much of the realm where they can develop and spread. This spatial relation must certainly affect the character of glacier erosion as well as the nature of the material being eroded. Also, the already-frozen ground left exposed to the atmosphere by deglaciation is a realm for a wide range of permafrost phenomena in the *cryogenic geomorphic system*, to be discussed in a later section.

Glacier Erosion

To begin with, we will give the broad outlines of erosional mechanisms by glacier ice more or less irrespective of alpine versus continental settings. The acquisition of rock detritus by a glacier can be separated into two simple processes—both rather well understood and agreed upon. First, glacier ice may freeze to rock material, thereafter plucking out the particles in question in a "quarrying" action (cf. Crosby, 1928). This is true erosion with the ice playing an active acquisitive role. Second, rock debris may fall upon the glacier and become a part of the sediment load. The latter process is most common to valley glaciers where much high ground surmounts the ice mass (Fig. 8.12), but it is known from ice sheets like those in Antarctica and Greenland where mountain peaks (*nunataks*) rise above the ice surface. There are details relating to both types of sediment-acquisition processes that require our further attention, however. This section will feature the truly erosional aspects.

---

* Meltwater encountered at glacier bases might indicate otherwise, but "veins" of water are encountered in areas of permafrost, even in winter.

Unlike water, glacier ice does not apparently acquire a sediment load in any immediate volume or velocity relationship. In spite of much relatively recent detailed work by glaciologists such as Demorest (1939) and Carol (1947) and earlier speculation by Hobbs (1911) and Chamberlin and Salisbury (1927), it is difficult to improve on the early appraisal of Russell (1895) who notes that the quantity of rock material a glacier can carry (*capacity*) bears little or no relationship to flow velocity but is related to an extremely high power of the ice volume. As to clast size,

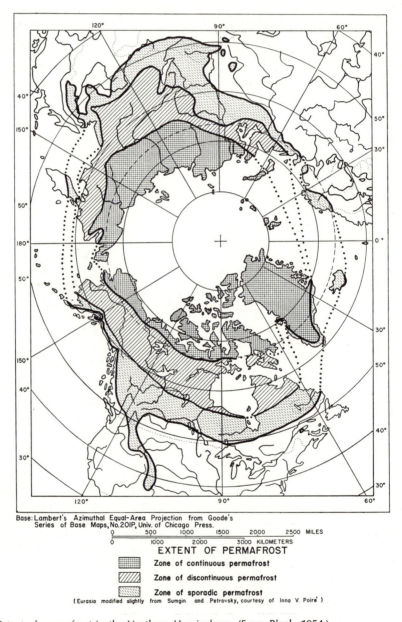

Base: Lambert's Azimuthal Equal-Area Projection from Goode's Series of Base Maps, No. 201P, Univ. of Chicago Press.

0   500   1000   1500   2000   2500 MILES
0   1000   2000   3000 KILOMETERS

EXTENT OF PERMAFROST

Zone of continuous permafrost

Zone of discontinuous permafrost

Zone of sporadic permafrost

(Eurasia modified slightly from Sumgin and Petrovsky, courtesy of Inna V. Poire')

Fig. 8.11.   Extent of permafrost in the Northern Hemisphere. (From Black, 1954.)

Fig. 8.12.   View of the 20,300-ft high crest of Mount McKinley, Alaska, and its southeast face rising above Ruth Glacier. Though potentials for rock falls onto the glacier surface are clearly present, a recent snowfall has covered the morainal debris. (Bradford Washburn.)

as noted by Chamberlin and Chamberlin (1911), the *competency* of glacial ice also has little or no relationship to velocity, and, indeed, upper clast size limits appear to be the largest pieces into which rocks can break and retain their identities during transport. One *erratic* on record in the northeastern United States supplied the only rock for a quarry operation for several decades.

When a glacier erodes it is a case of a solid or a "plastic" (in the crystalline sense of that term) picking up another solid (cf. Finsterwalder, 1950). Carol (1947) observed sub-glacial ice of a cheesy consistency which could have molded itself around a clast. But "solid" ice cannot readily engulf a rock fragment imbedded in other rock fragments or a rigid lithic matrix. Thus, it has been found that glaciers erode somewhat irregular surfaces more effectively than smooth or smoothly undulatory surfaces which often acquire a polish instead (Fig. 8.13). Relatively rigid ice passing over a ground irregularity often cannot conform to the shape of the downflow (*lee*) side of a bedrock knob. The upflow (*stoss*) side becomes rather rounded by abrasion, whereas blocks wedged loose by frost action and plucking on the lee side can become attached to the ice mass and carried away to leave partly rounded, partly angular hillocks called *roches moutonnées* (Fig. 8.14). Carol (1947) claims

Fig. 8.13. Glacial polish just beginning to peel away by exfoliation from the top of basalt columns at The Devil's Postpile National Monument, California. (Hal Roth.)

that the plastic ice phase reflects pressure fluctuations on the stoss sides of rock irregularities.

With regard to glacial plucking action, it seems important to note that rock fragments are not only pulled up into the glacier mass but are commonly, and perhaps even usually, also heaved upward from below or

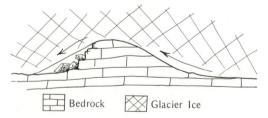

Fig. 8.14. Cross section of ice movement during formation of a roché moutonnée. The upflow (stoss) side becomes rounded and polished, whereas the rigidity of the ice inhibits conformance to the downflow (lee) surface form.

to the side. Meltwater commonly penetrates to the base of a glacier and freezes there. As noted by Stephen (1930) and Riley (1935), wedging and heaving by ice formation in rock fractures can shift blocks many tons in weight. Fragments of rock incorporated into the glacier increase its mass and also serve as tools of corrasion which scratch, scrape, abrade, groove, scour, and otherwise shape the land surface. How the moving clasts are held, and whether they can shift position or not, is sometimes indicated by the kinds of marks made, and this means we must rely to some extent upon the record left by past glaciers.

Data so far gathered on present-day glacial erosion by intrepid glaciologists like Carol (1947) and Demorest (1939) are all too few. These workers used alpinist techniques to go into crevasses and get "under" alpine glaciers, and the search for more information cost Demorest his life.

Apart from plucking and quarrying action, glacier erosion tends to approximate a giant abrasion mill operated by a careless lapidary enthusiast who mixes abrasives of several different hardnesses and grades in order to shape surfaces of several compositions and textures. Glacially, the problem involves two issues—(1) whether the ice is "plastic" enough to conform to the shape of the surface and (2) whether the abrasives carried by the ice are small enough to enter existing (or developing) depressions. Thus, as noted by Flint (1971), there is a hierarchy of linear markings arranged more or less parallel with the directions of ice movement. The finest-grade abrasives (silt, clay) impart microscopic lines to whatever mineral and rock surfaces will take a polish and megascopically appear smooth excepting the "odd" *scratch* or *striation* left by the equally "odd" outsize clast of abrasive character (cf. Fig. 8.13). Glacier loads are not noted for the excellence of their internal sediment sorting. And it shows.

The enlargement of striae or lithically "weak" zones of bedrock into *grooves* or furrows is partly a matter of differential hardness and partly a matter of ice rigidity. Where ice can mold itself to a depression, and where abrasive fragments are small enough to enter a scratch, it may expand into a striation which may, in turn, be enlarged into a groove (Fig. 8.15). Some grooves have overhanging walls that are striated on all surfaces, and they range

Fig. 8.15. Glacial grooves developed in Precambrian granite-gneiss near Neyes Provencial Park on the north shore of Lake Superior. The central groove is up to 3 ft (1 m) wide and locally 5 ft (1.5 m) deep.

in size from a few centimeters wide and deep to the Mackenzie River examples cited by Thornbury (1969) and Flint (1971) that approach 1 mi (1.5 km) in length and 100 ft (30 m) in depth. Also, as would be expected, the moving clasts that serve as an abrasive are themselves abraded, scratched, and grooved as well as faceted (cf. Fig. 8.8).

Depressions scoured out by glacier abrasion also have their relatively raised counterparts on several different scales. Of course the location of an eroded depression may merely signify the passage of a solitary abrading clast. However, it could also signify the only one of several abrading clasts sufficiently hard to mark the substance in question, or the only one of many clasts held with sufficient rigidity, or the lithically weakest zone in a surface being scoured. And it should be kept in mind that the creation of one or more depressions automatically generates raised areas between or on either side for the same general reasons just listed. Thus, ridges may simply occur between two closely spaced scratches. Or a ridge may be all that remains when the remainder of a surface being abraded is lowered. For example, in the Lac Megantic area of Quebec, Canada, glacial abrasives able to rather smoothly reduce siltstone surfaces encountered pyrite cubes embedded in the siltstone measuring up to ¼ inch (5 mm)

in diameter and left 4- to 5-inch (10–12.5 cm) tapered ridges on what are assumed to be the lee sides (Fig. 8.16).

In addition to the features already discussed (e.g. *rochés moutonnées*), raised features of several sizes left in areas of glacial erosion generally assume a variety of rather streamlined "drumlinoid" forms, or so-called whaleback shapes. When these features are composed entirely of bedrock, they can usually be assumed to be erosional in origin, but it is often uncertain whether larger-shaped hillocks are entirely of glacial scour origin or merely glacially shaped pre-glacial features. Many *drumlins* have been found to combine rock cores and glacial drift in a relation similar to the teardrop-shaped *crag-and-tail* features (cf. Figs. 8.17; 8.18), and both deposition and erosion are apparently involved in their genesis. Some streamlined hillocks of erosional origin express differential bedrock resistance as do many scour depressions—resistance or lack of resistance being determined by such factors as fracture spacing in areas of otherwise homogeneous lithology or compositional or textural differences in regions of variable rock types.

Directional Indicators

Many of the erosional features generated by glacier ice also indicate ice-movement directions. In pioneer studies of glaciation, such movement indicators were used to establish the fact that glaciers came from the North in the Northern Hemisphere or from the higher parts of mountain ranges.* At present, in conjunction with glacier deposit characteristics, the same types of markings are primarily used to unravel the intricacies of local glacier ice movements and establish environmental-stratigraphic sequences as well as over-all glacial histories. In this context it is worth noting that scratches and striations can originate in several manners apart from glacier ero-

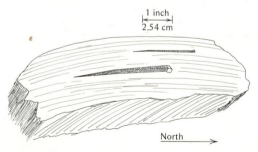

Fig. 8.16. Sketch of mineature craig-and-tail features developed on siltstone near Lac Megantic, Quebec. Pyrite cubes protruding from the silt are responsible for small leeside ridges up to 0.25 inch (0.4 cm) high.

* Even now direction is hardly a trivial matter and is used in paleo-directional problems as in the Ordovician glaciation of the Sahara, the Permian glaciation of Australia and India, and elsewhere (cf. Fig. 11.1).

Fig. 8.17.   Roché moutonnée developed from Precambrian granite gneiss along the north shore of Lake Superior, Neyes Provencial Park. Feature is about 100 ft (30 m) long and ice movement was from right to left (north to south).

sion, including markings left by floating ice, by ice-rafted rock particles, by various types of avalanches, and by debris flows. Even some flutings cut by blown sand and solutional *lapis* are similar. But most of these nonglacial markings lack the consistency of direction of their counterparts, and, in any case, striae should be evaluated for origin in terms of over-all setting and associated deposits.

Of themselves, glacier striations are rather ambiguous indicators of ice movement direction. Even the direction in which they taper is an uncertain clue, though *orientation* can always be determined. Orientations can be rendered dubious where two or more are superimposed on the same rock surface. Multiple striation orientations may signify multiple glacial episodes or merely flow direction variations during a single episode. The latter can re flect local topographic flow responses, shifts in the location of ice accumulation

centers, and similar factors (cf. Flint, 1971). *Sense of movement direction* can be gained by combining striation data with evidence provided by crescentic marks of several kinds (Fig. 8.19) as well as by lee-stoss relations of larger streamlined landforms and studies of till-fabrics (Fig. 8.20). Crescentic marks appear to originate from large rocks being carried in the base of a glacier. In his more detailed dis-

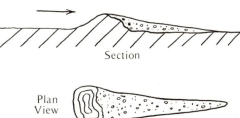

Fig. 8.18.   Crag-and-tail feature depicted (above) in section and (below) in plan view. Bedrock knob induces leeward deposition of drift (arrow indicates ice movement direction).

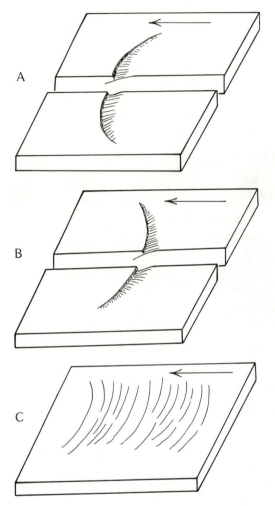

Fig. 8.19. Arcuate markings made on comparatively brittle bedrock, presumably by clasts held in moving glacier ice (arrows). (A) Crescentic gouge; (B) lunate fracture; and (C) Crescentic fractures; (C) is not associated with removal of rock chips. (After Flint, 1971.)

cussion of such features, Flint (1971, pp. 95–97) concludes that the cause was a sliding object but that the marks made can be directionally ambiguous.

Glacier Transport, Mass Wasting, and Deposition

A good bit of the rock debris moved by glaciers is not actually eroded by them.

When a man jumps into a stream we do not say, "He has been eroded." And when a rock falls from a valley side or a nunatak onto a glacier, the fall is often only indirectly a function of the glacier's presence. Load acquisition by glaciers and transportation of this rock is best viewed in terms of the *glacial geomorphic system* which includes not only the moving ice mass proper but also the freeze-thaw environment normally associated with it.

It has already been pointed out that the erosive acquisition of rock debris by basal glacier ice (*subglacial load*) may be abetted by ice heaving. Ice sheets have most of their sediment load at their base as have a good many valley glaciers. Shattering of exposed rock by frost action is, in fact, a normal part of the glacial environment, and its effectiveness is generally expressed by the number of times each year the temperature rises above and falls below freezing. Ice wedging is a major aspect of mass wasting in glacial realms, and, via this mechanism, rock falls and avalanches can feed large amounts of detritus to glacier surfaces (*supraglacial load*). Naturally, the requirement that terrain yielding such debris lie above the glacier surface makes this means of sediment load acquisition more important in piedmont and valley glaciers than elsewhere (Fig. 8.21).

As noted long ago (Chamberlin and Salisbury, 1927), a glacier's load, once acquired, undergoes inevitable mixing due to movement and internal shearing. Through internal thrusting a glacier's subglacial load of rock can be shifted upward and transported within the ice mass (*englacial load*) see Fig. 8.43). No appreciable sorting is generally attributed to the ice proper, but the deposits made by a particular glacier tongue or ice stream will usually give some indication of *provenance* by lithologic indicators. Also material transported on the glacier surface apparently tends to stay there, where it can be identified as *moraines* with various positional terms as prefixes (e.g. *medial, lateral*). Rock material acquired by a glacier naturally adds to the over-all mass and density and thereby must affect such relationships as erosional effectiveness and depth of erosion—both on land and under water. There are a variety of features created by ice thrusting and loading, many of which are not expressed topographically, which the reader may wish to investigate further (see Flint, 1971, pp. 121–126).

Transport and deposition of rock waste by glacier ice, for reasons already outlined,

cannot be viewed as a continuing process of a single nature any more than a semi-arid episode following a humid one can be treated geomorphically the same as a semi-arid period following aridity. Whether glaciers are commonly maintained for long periods in a purely "subsistence" state without either advancing or diminishing is very uncertain. But it is certain that glaciers do grow and shrink, and a discussion of

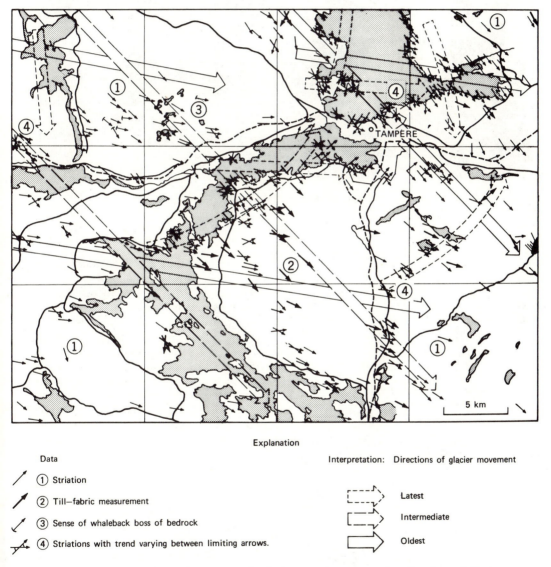

Explanation

Data

⟋ ① Striation

⟋ ② Till—fabric measurement

⟋ ③ Sense of whaleback boss of bedrock

⟋ ④ Striations with trend varying between limiting arrows.

Interpretation: Directions of glacier movement

⟶ Latest

⟶ Intermediate

⟶ Oldest

Fig. 8.20. The Tampere area, southwestern Finland, showing three successive directions of glacier movement (presumably not separated in time by deglaciation). (From Virkkala, 1960.)

Fig. 8.21. A pair of alpine glaciers near the base of the Matterhorn at Gornergrat, Switzerland. Note that the glacier to the left displays two medial moraines (dark ridges) from debris which has fallen on the ice. (W. N. Gilliland.)

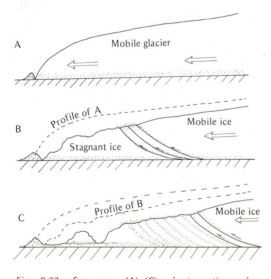

Fig. 8.22. Sequence (A)–(C) of stagnation of a marginal ice mass followed by overriding along imbricate shear zones and related drift accumulation in terminal and ground moraine. If concentration of drift along shears is pronounced, resulting terrain may be hummocky. (After Bishop, 1957.)

sediment transport and deposition by them must proceed in such a context.

The ice and rock an advancing glacier is carrying can move toward and to the glacier edge, where, if the rock is deposited, it may shortly be re-eroded as the ice front advances. Sediment processing during glacial recession is another matter. The event used to be called "glacial retreat" and an imaginative soul could almost "hear the bugles blow." More recently it is appreciated that a combination of reduced flow and accentuated ablation would cause an ice mass to stagnate. Under the latter conditions, transport distances from accumulation centers must gradually diminish, and stagnant ice masses may be overridden periodically by slight re-advances as noted by Bishop (1957) (cf. Fig. 8.22). The bulk of the material carried by glaciers is dumped directly as the ice melts, and, since the latter process is most effective during glacial recessions, much of this deposition is from stagnant ice masses (cf. Hartshorn, 1968). Water begins to have an effect here, also, of course. And the configurations of the deposits and related erosional landforms had best be considered in the two major realms frequented by glaciers.

ALPINE GLACIAL LANDFORMS

Though they share many of the physical attributes of continental ice sheets, *alpine glaciers*, such as those in the Alps and Alaska, exist in a distinctly different hydrographic configuration. A mountain peak penetrates a sufficiently cold atmospheric layer. Water solidifies and, as snow, lodges in pre-existing topographic catchment configurations. Studies in the Northern Hemisphere indicate that depressions on the north and northeast sides of peaks may be favored, but wind-moisture-direction relations may supersede the effects of exposure to the Sun. Any descent below the snow line by flow injects the ice mass into a hostile environment, and, because a valley glacier is linear in shape, the amount of heat it can drain from a marginal or terminal environment is notably less than for a continental ice sheet.

Of course, glaciation may occur on a previously glaciated mountain mass, and developing ice masses in such a setting would necessarily accommodate themselves to the relict glacial terrain. But it is widely conceded that glaciers in mountainous areas commonly develop depressions previously eroded by running water under what qualifies as humid conditions (Fig. 8.29). And the same thing happens on lower portions of continents. As with flooding, the amount of ice required to bury a land surface depends upon the relief of that surface. Valley glaciers, like rivers, are enclosed by the depressions in which they flow. So the erosional and depositional effects of such glaciers tend to be down in these depressions or in subjacent areas contiguous thereto. But some mountains have been covered by ice (see Fig. 8.31).

Cirque Development

Snow catchment areas on mountains above the snow line are sites of initial erosion when a glacier develops and begins to flow. Through the combined effects of mass wasting, quarrying, plucking, ice heaving, and related phenomena, an ovate depression commonly forms which gradually deepens into the semicircular, basin-shaped feature called a *cirque* (Fig. 8.23). Though not restricted to the heads of glaciated valleys, cirques are best developed and most numerous in such settings. Typically they are bounded on three sides by steep or even vertical rock walls (*headwall*). In empty cirques, the headwall descends to an area commonly occupied by an ovate lake (*tarn*) which attests to the second major morphologic part (*basin*). As noted by Derbyshire (1968), some cirque floors slope continuously into subjacent valleys but most are separated from such valley bottoms by the basin lip (*threshold*). In size, cirques range from a few meters across and deep (nivation basins of some authors) to several kilometers across with headwalls thousand of meters in height. Walcott Cirque in Victoria Land, Antarctica is reported to be on the order of 10 mi (16 km) across and 10,000 ft (3 km) deep.

Where occupied by a glacier, the headwall of a cirque is commonly separated from the ice mass by a deep, arcuate crevass (*bergschrund*) which may extend to the cirque floor (Fig. 8.24). The absence of a bergschrund from some cirque glaciers suggests that it is a transient feature related to cirque development. Since the recognition of cirques by A. C. Ramsey in the mid-1800's, several concepts have been developed as to their precise origin. Flint (1971) cites two main processes he believes to reflect stages of cirque development: (1) an initial phase of frost wedging accompanied by mass wasting and (2) glacial erosion and transport. The first of these processes has been termed *nivation* by Matthes (1900), but, though it may reflect glacier development at a valley

Fig. 8.23. Cirque developed on the east face of the Teton Range above Jackson Hole, Wyoming. The basin is occupied by a cirque glacier beneath an essentially vertical headwall and lies behind a distinct threshold.

head, nivation does not account for some other aspects of cirque formation.

Cirque erosion by glacier ice has been proposed in several forms. Johnson (1904) developed the "bergschrund theory" which emphasized a plucking action at the base of the headwall crevass, and he was much impressed by the frost-shattered rock he observed while dangling on the end of a rope in an open bergschrund. Perhaps the most plausible explanation of cirque formation is in the "rotational slip" concept of W. V. Lewis expressed in the joint session of the Royal Astronomical Society and the British Glaciological Society (cf. Nye and Orowan, 1950). In point of fact,

periodic closing of the bergschrund implies a downward and even backward motion of the ice at the headwall, whereas the presence of a lip or threshold at the outer margin of the cirque basin indicates an upward movement component for the ice there similar to a rotational landslip (Fig. 8.25). The two movement increments are compatible with periodic rotation of the ice mass in the cirque with the maximum velocity and most effective erosion near or at the ice base. Rotational slippage coupled with mass wasting could comprise the essence of cirque erosion (Fig. 8.26).

Where glacier ice is confined to a cirque (*cirque glacier*), ice movements can be dominantly rotational or merely limited to firn creep. Where rock fragments freed by mass wasting on a headwall or other free face slide down across ice, a firn mass, or snowbank, a ridge of rock debris called a *protalus rampart,* up to several meters high often accumulates (Fig. 8.27). Both direct rock falls and rotational thrusting may be involved in the genesis of such ridges, thereby lending a morainal character, and the same debris may later comprise part of the cirque threshold where the glacier expands and extends down valley. Where the glacier in a cirque continues downslope as a *valley glacier,* a mechanism exists to regenerate the berg-

Fig. 8.24. A glacier separated along its upper margin from the steep flank of Mount Assiniboine by an open fracture (bergschrund). Peak is in the Canadian Rocky Mountains. (Austin Post, University of Washington.)

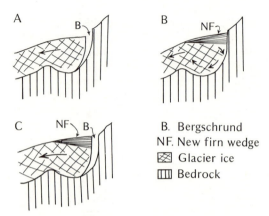

B. Bergschrund
NF. New firn wedge
▨ Glacier ice
Ⅲ Bedrock

Fig. 8.25. Theoretical diagram showing phased development of a bergschrund in cross section. (A) Down-valley slippage of an alpine glacier opening up a crevass at the cirque headwall (bergschrund); (B) rotational slippage of the glacier ice in the cirque during accumulation of new firn, possibly over a period of years; (C) renewed down-valley slippage of the glacier opens up a new bergschrund.

schrund because there is an ice-movement increment normal to the cirque headwall when the glacier slips bodily down valley as most periodically seem to do.* If the movement relations implied here are correct, rotational slippage in cirque ice probably alternates with down-valley movements, and bergschrund development would be sporadic (Fig. 8.25).

Cirque-Shaped Landscapes

Modifications of land surfaces by cirques range from isolated depressions on otherwise smooth uplands to serrated and scalloped mountainous terrain (Fig. 8.28) sometimes called biscuit-board topography. The widespread occurrence, and recognition, of cirques is indicated by the many languages that contain an equivalent term

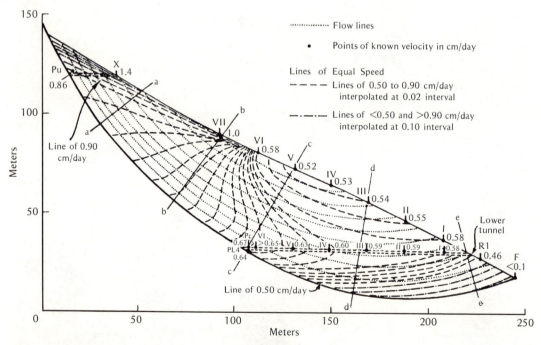

Fig. 8.26. Velocity distribution in a cirque glacier (Vesl-Skautbre, Norway) shown by lines of equal velocity. [After McCall (in Lewis, 1960).]

* Probably glacier surges and the bergschrund are the two main types of evidence of "body" slippage.

to the French word *cirque*. Derbyshire (1968) lists six, Thornbury (1969) five, and Flint (1971) five. Not to be outdone we will note the German *kar*, Spanish *hoyo* and *circo*, Welsh *cwm*, Scotch *corrie*, Norwegian *botn*, Swedish *nisch*, and Scandinavian *kjedel*. And in the regions linguistically implied, cirque development locally reduces mountain peaks to sharp ridges (*aretes*) between adjacent cirques or cirque-shaped peaks (*horns*) called *monuments* by Hobbs (1911) (cf. Figs. 8.29 and 8.30).

Reduction of an upland by cirque erosion can be expressed in Davisian stages of development (cf. Hobbs, 1911), beginning with a "pocked" upland dotted by isolated cirque basins, evolving by cirque

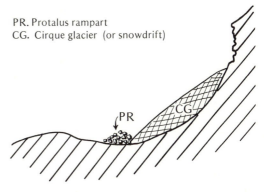

PR. Protalus rampart
CG. Cirque glacier (or snowdrift)

Fig. 8.27. Development of a protalus rampart in schematic cross section. Rock debris breaking away from an escarpment above a cirque glacier (or snowdrift) falls and slides to the distal edge to form a ridge.

Fig. 8.28. Looking much like scraps of dough on a cutting board, the Uinta Mountains in Utah following a light snow display a classic "biscuit-board topography" through the development of many, partly intersecting cirques. (John S. Shelton.)

A Humid (Selva) Mountain Topography

B Glaciation

C Post-Glacial-Humid

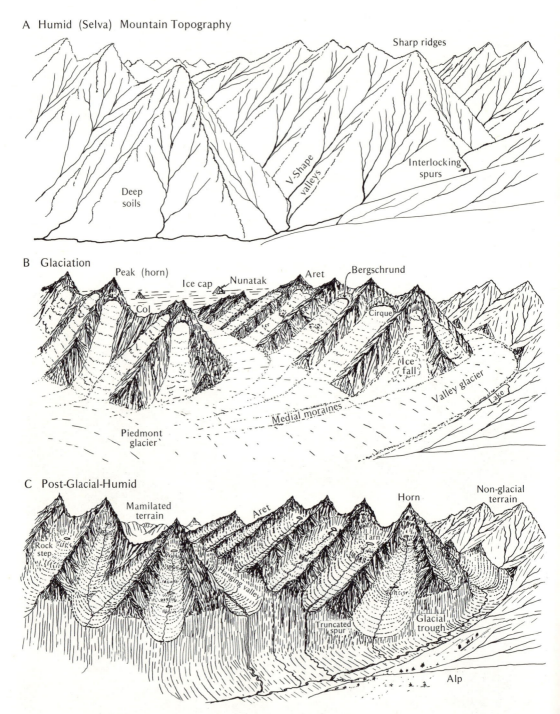

Fig. 8.29. Sequential terrain sketches approximating glaciation following humidity and the opposite. (A) Humid, selva topography with sharp ridges and V-shaped valleys under deep soil with interlocking ridge spurs; (B) widespread glaciation of areas favored by hydrography; (C) post-glacial humidity and re-establishment of fluvial activity. (Modified in part from W. M. Davis.)

Fig. 8.30. Mount Assiniboine, a glacial horn developed by cirque erosion, rises above adjacent peaks in the Canadian Rocky Mountains, Alberta, Canada (cf. Fig. 8.24). (Austin Post, University of Washington.)

enlargement through multiple or compound cirques separated by lowered aretes (*cols*), and terminating in a lowered upland topped by isolated horns. Though the passage of time is certainly implied by such stages, the rates of change involved are very problematic and would certainly require several glacial episodes in the lower latitudes. This in turn implies periodic modifications by nonglacial processes. As

noted by Derbyshire (1968), cirque development is not equal, even in mountainous areas that appear to have been subjected to similarly intensive glaciation, and there appear to be a number of variable factors involved including bedrock homogeneity, fracture spacing, directional exposure, pre-glacial topography, and elevation. Mountains overtopped by ice sheets like many in the northern Appalachian region, near Yosemite Valley, California and the Coastal Mountains and North Cascades of the Pacific Northwest have their peaks ground down and rounded off (Fig. 8.31).

Valley Glaciation

Actually, of course, most cirques are situated at valley heads, and their development in many instances is simply at the point of initiation of valley glaciation. Down-valley movement of glacier ice has long been known to modify pre-existing elongate depressions into glacier troughs (Fig. 8.32). Most such pre-glacial depressions were stream valleys, and these usually have their more or less V-shape cross-profile modified into the classic U (cf. Fig. 2.11). As noted by Thornbury (1969), the usual glacier trough heads at the outer base of a cirque threshold (*trough head-wall*) and the trough floor may be considerably below the level of the cirque basin. The re-shaping of a valley to develop the trough form is generally conceded to occur by a combination of glacier scour, rasping, and plucking. But even in rocks of apparently uniform lithology and structure, the result is not usually a smooth declivity like a stream profile but rather a series of basin-like depressions of vary-

Fig. 8.31. Sketch of the terrain in Yosemite Valley and vicinity. (See also Fig. 8.34.) (After F. Matthes.)

ing distinctness; these often consist of a series of giant treads or steps (*glacial steps*), the whole sometimes being called a *glacial stairway* (Fig. 8.33).

The presence of rock steps in glacier troughs implies an erosional mode by ice that is far from understood. In deglaciated valleys, the *basins* on the treads are commonly occupied by a series of lakes (*chain lakes*) connected by a stream having rapids where a ridge (*riegel*) occurs on the top of the step risers. The generation of this type of terrain is at once puzzling and intriguing. Though theorists have attempted to associate the steps with valley constrictions and resultant ice deepening, varying rock resistance, pre-glacial irregularities (knickpoints in streams), fracture spacings (cf. Fig. 8.33), and sapping on lee sides of irregularities, there are examples that appear to exclude each of these possibilities (cf. Cotton, 1942, 1968). Cotton (1968) states, "Steps and other irregularities in profiles of glacier-valley floors are due for the most part to vertical corrasion by thick and heavy ice streams . . . ," noting elsewhere that, "There is commonly also a step down in the floor of a trough just below the point at which a trunk glacier is joined by a large secondary or tributary glacier, giving it a sudden accession of volume."

The foregoing remarks still leave us to account for steps that do not occur "in all the right places" or that are developed with surprising regularity though they are not apparently related to bedrock character or valley shape. In this connection the rotational slip theory of Lewis (1954, 1960) may prove to be significant, for the basin-riser configuration of glacial troughs is not totally dissimilar from cirques (cf. Figs. 8.25; 8.33). In his earlier work, Lewis (1954) notes the possibility of rotational slip in the central ice stream of a valley glacier, the ice in a given basinal depression resembling a cylinder with a horizontal axis, at right angles to the direction of flow and above the glacier base (cf. Fig. 8.33). In this context, at a riser, movement of the ice would be downward (per-

Fig. 8.32.   Navajo Basin, a U-shaped glacial trough associated with Mount Wilson, Dolores County, Colorado, as seen from the divide extending west to Dolores Peak in 1899. (W. Cross, U. S. Geological Survey.)

haps coinciding with an ice fall) and then outward and upward over the riegel. The condition for basal slip therein would be that the couple exerted by any given portion of the ice about the axis of bounding cylinders exceed the limits set by limiting shear stress of ice (about 10^6 dynes/cm^2) and the radii of the cylinders. Lewis's calculations showed that the process was possible for glaciers with 20° or more of bed slope, inclinations common along risers.

Ice flow patterns corroborative of Lewis's rotational slip concept of trough erosion have been recorded in the Skauthoe Cirque Glacier in Jotunheim, Norway and in the Juvashytt Glacier also in Norway. More to the point, at the foot of the Mount Collon

ice fall of the lower Arolla Glacier in Switzerland, a horizontal tunnel was bored 205 m along the steeply dipping bed of the glacier. In the tunnel, ice bands under the

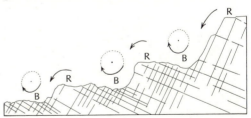

Fig. 8.33.   Longitudinal section through a series of glacial steps consisting of basins (B) and riegels (R), where the ends of theoretical cylinders of rotation are indicated for ice movement of that type (cf. Lewis, 1952).

ice fall visible by contained dirt dipped 30° downstream indicating downward movement at that point, the slope of the bands diminishing downstream and finally reversing to incline 70° upstream near the tunnel mouth. Rock slabs were observed to parallel the attitude of the bands, and there was evidence of recent formation of the features recorded with a rotational slip pattern.

A discussion of valley glaciation would be incomplete without mentioning several features that develop as a result of differential valley deepening and widening. The accordance of junctions between flow lines of streams that caught Playfair's eye so long ago does not consistently apply to glacial valleys where trunk flow lines are commonly but not invariably deepened much more than tributary routes. The latter are expressed as hanging valleys (cf. Figs. 8.34; 8.29), usually with the same U-shaped cross-profile as the one being joined, if both have been glaciated. Postglacial streams flowing in *hanging valleys* commonly join others in falls. As noted by Matthes (1930) and others including Thornbury (1969), hanging valleys also exist due to faulting. They are also often present where perennial rivers have intermittent or ephemeral tributaries (Fig. 2.50). The trough morphology of the glacial variety usually distinguishes them from other types.

Fig. 8.34. A winter view of Yosemite Valley, California, in the Sierra Nevada Mountains. In the background, center, is Half Dome, and to the right is a hanging valley. The glaciated main valley has its floor flattened by an alluvial fill (cf. Fig. 8.31). (U. S. Forest Service.)

Fig. 8.35. Glacially truncated valley spurs in the valley of the Urubamba River, Ollantaytambo, Peru, in the eastern Andes Mountains. The alluvial valley flat is partly glacial outwash, and this modifies the U-shaped, glacial cross-valley profile. (Sketch based on a photo.)

Finally, before we discuss alpine glacier deposition, we should mention two other effects of valley glacier modification of former stream valleys. First, the glacier ice in a valley often displays a fine disregard for many of the relict landform intricacies created by water flow, to the extent that interlocking fluvially eroded spurs of tributary valleys are commonly amputated by trunk ice streams as the latter "straighten out" former drainage net irregularities (Fig. 8.29). The *truncated* or *faceted spurs* are common along glaciated valleys (Fig. 8.35) but can be simulated by faulting (Fig. 6.25). A second aspect of valley glaciation meriting discussion concerns instances where ice tongues enter water bodies to create *fjords* (Fig. 8.36). Technically, fjords are glacier troughs reoccupied by water following glacier recession. The water is usually that of the ocean, and fjords are common to many high-latitude lands including Labrador, Greenland, Norway, Chile, and Alaska. Be-

cause of the rock load valley glaciers carry, fjords may be erosionally deepened far beyond limits dictated by the normal buoyancy factor of ice. Also, of course, times of glaciation generally coincide with lowered sea level. In regard to this array of interacting factors, Flint (1971) cites examples of fjords ranging in depth from about 3,500 ft (780 m) to some 6,400 ft (1,933 m)—with the bedrock not necessarily coinciding with water depths. According to Crary (1966), valley width which influences side wall shearing stresses is more important than ice thickness and flow volume in determining erosional depth in fjords. This relation helps explain why fjords commonly display increased water depth inland.

Valley Glacier Deposition

Rock debris carried by valley glaciers is deposited by glacier ice and by meltwater.

Fig. 8.36. Oulet glaciers from the Greenland Ice Sheet entering the sea, northwest Greenland. The drowned valleys left by melting of the ice tongues are fjords. Note the whale-back erosional form of the foreground peninsula. (The Geodetic Institute, Copenhagen.)

Fig. 8.37.   Glacial till (right) in direct contact with the ice of the Sólheimajökull Outlet Glacier from the Myrdalsjökull Ice Sheet, southern Iceland. (W. N. Gilliland.)

The intimacy of their interaction in this matter is far too great to allow a separate lucid discussion of these phenomena, even though it is possible to point to "this" feature deposited by ice and "that" one laid down by water. Our concern here is with the resulting landforms and with the deposits to the extent that their study lends added understanding of those landforms and the history of their development. Detailed studies of the same subjects comprise entire books.

The term *drift*, or, more formally (*glacial drift*), is widely applied to rock transported and deposited by glacier ice or its meltwater, whether on land or water. Its origin presumably goes back to the early 1800's when *erratics* of evident foreign derivation in northern Europe and Britain were ascribed to rafting by icebergs presumably "drifting" about on Noah's Flood (cf. Lyell, 1830).

Apart from earlier theological connotations, usage of the term drift persists, with the term *till* being usually reserved for the non-sorted, non-stratified heterogeneous variety presumably reflecting direct ice deposition (Fig. 8.37). *Stratified drift*, a widely used expression for glacial deposits in which particle sorting has occurred, is expressed by layering (Fig. 8.38). The stratification and implied sorting is usually taken to connote effects brought about by either standing or running water or the wind,* though some of the intricacies of glacier flow and oscillation may create

---

* Eolian deposits are not usually recognizably glacial, though eolian erosion could conceivably induce stratification.

Fig. 8.38.   Crudely stratified drift comprising a part of the Wisconsin terminal moraine which is Long Island, New York, is exposed here at Montauk Point in a sea cliff at the east end of the island. (Jack Mazer.)

crude layering by ice alone. As noted by Flint (1971), there is essentially a complete gradation between till and stratified drift. And, where possible, analysis of the layered variety should include an attempt at more definitive labeling by expressions such as *glacio-fluvial*, *glacio-lacustrine*, *glacio-eolian*, or *glacio-marine*.

Drift Lithology

In essence, drift composition and texture reflect two distinct types of origins. One, expressive of the below-freezing temperatures of the glacial environment and the mechanical rasping and grinding action of glacier erosion and transport, is essentially mechanical in character (e.g. *rock flour*). The other, in contrast, is environmentally polygenetic in almost every instance. It is expressive of the fact that glaciers acquire their loads by eroding rock material weathered under a variety of conditions. Weathering conditions that normally contribute to glacial drift lithologies include the frost and freeze-thaw activity that characterizes many mountain peaks (Fig. 8.1) and the borders of high-latitude glaciers and generate ice-shattered talus and stony tundra soils (cf. Chapter 5). Also included are the *pedalfer* and even *pedocal* soil residues and lacustrine deposits generated under relatively warm, humid conditions at lower latitudes and elevations; these soils include a relatively high quantity of clay minerals and semi-insoluble minerals reflecting intense chemical activity. Naturally, depending upon glacial setting, geologic circumstance, and environmental disposition, glacial drifts include various amounts of these two types of material (Fig. 8.39).

According to Flint (1971, p. 149) mechanically broken or abraded materials in drift far outbulk those that have been decomposed. This is as it should be, in one sense, for glaciers (particularly valley glaciers) naturally derive their drift from areas where the cold glacial environment predominates. Yet if a glacier were to presently form in northern North America or Eurasia and move south, it would (among other things) erode a considerable variety of soils, all more or less chemically decomposed. There seems little reason to question that the quantity of such materials incorporated in any given glacial drift would depend upon the duration, intensity, and, hence, extent of weathering of the area over which the ice advanced. It is therefore not surprising (cf. Willman *et al.*, 1963; 1966) that some till is composed of dominantly coarse and fragmented detritus, some shows a great range in clast size and character (BOULDER-CLAY), and some is dominated by fine-grained material and contains 40% or more clay (note grain-size variations in stratified drift in Fig. 8.39).

Not only may a considerable amount of drift material be provided by glacial erosion of soil, other fractions may be derived from lacustrine deposits and alluvial gravels in stream coarses traversed by glacier ice. Nor should we omit previously deposited glacial drift in its various guises. The array of source lithologies is strictly limited in most valley glacier deposits, and provenance is rarely in question in such cases. Moreover, the extent and variety of weathering environments that may be encountered by a valley glacier are much more limited than for continental ice

Till	Sample	Montmorillonite	Illite	Kaolinite and Chlorite	Substage
Hipple	P-1317	0	74	26	Buffalo Hart
Funkhouser	P-1321	12	62	26	Jacksonville (?)
Funkhouser	P-1284	17	59	24	Jacksonville (?)
Effingham	P-1274	19	59	22	Jacksonville (?)
Chapin	P-2099	42	38	20	Liman

Fig. 8.39. Clay composition and nature of several unaltered Illinoian glacial tills suggestive of variable weathering histories of the various source materials. (From Willman and Frye, 1970.)

sheets. Possibly reflecting the dominantly cold and waterless, mechanical-weathering environments of high mountains, alpine glacial drift is notably lower in clay fractions than most similar continental deposits. Till composition is also affected by the presence of *lag concentrates*—materials left behind in drift deposits by such processes as sheet wash and eolian deflation which tend to remove fine-grained fractions. The latter process often reflects post-glacial aridity by the presence of abundant ventifacts such as those that mantle part of the Kansan Till in Iowa and the Wisconsin Till described by Clark and Elson (1961) in the St. Lawrence River Valley (Fig. 8.40).

Fig. 8.40. Ventifacts from stratified drift, Charette, Quebec, in the St. Lawrence Lowland, southeast Canada. Arrows indicate inferred wind directions and lines delimit exposed areas. (See Clark and Elson, 1961.)

Detailed discussions of drift and till lithology appear in many places, and excellent discussions are offered by Holmes (1941), Gravenor (1951), Willman *et al.* (1963), Harland *et al.* (1966), German (1968), and Flint (1971). One of the most promising areas of study for purposes of distinguishing till fragments from similar clasts of other origin is that of microclast markings as studied by Krinsley and Donahue (1968). Using magnifications of $\times$ 5,000 or more, they describe eight textural variations between glacial clasts and clasts moved by littoral or eolian agencies. All lithologic aspects of glacial drift cannot possibly be considered in this section, just as all the various ramifications of glacial geomorphology cannot be fully covered in this chapter. Entire volumes (Cotton, 1942; Flint, 1957; 1971; Wright and Frey, 1965) and multi-volume works (Charlesworth, 1957) are devoted to these subjects.

Moraines

Probably the outstanding group of depositional landforms created more or less directly by valley glaciers and ice sheets are the moraines. These are essentially unsorted masses of glacial detritus dumped more or less directly from the ice in configurations ranging from ridge-like to tabular. Glaciers in alpine settings usually generate a greater variety of these features than do continental glaciers simply because of their elongate shape and because of the greater amount of debris that falls on the ice.* Morainal material undergoing transport is usually clearly visible along the upper margins of valley glaciers (cf. Fig. 8.21), and, of course, where two such glaciers merge, their adjacent margins become medial in position. Where deposited, these materials form elongate ridges of drift for which *lateral* and *medial moraines* are the logical designations (Fig. 8.41). Medial moraines are usually the least sharply defined as landforms and

---

* Several different continental glacier lobes have produced a large number of *end* moraines but, of course, not other types of moraines.

Fig. 8.41.   Sharp-crested ridge of the lateral moraine of Sólheimajökull Glacier, southern Iceland. Present outwash valley is to right. (W. N. Gilliland.)

may exist only on the upper glacier surface, in part, because the debris gathering-distance is necessarily shorter, in part, because of meltwater effects, especially where this follows the valley median during and after deglaciation, and, in part, because of the blanketing effect of drift deposited by the main ice mass. The latter detritus usually accumulates in a more or less tabular and continuous layer; this is a *ground moraine* situated between lateral moraines. Ground moraines form gently undulating *till plains* characterized by what is often referred to as "swell-and-swale" topography which generally lacks linear elements. Topographic maps indicate a maximum relief of about 20 ft (6 m).

The line of maximum valley glacier advance is also usually characterized by a ridge of drift termed an *end moraine* or *terminal moraine.** Terminal moraine is perhaps more commonly applied to the ice margin deposits of continental glaciers that often lack distinct sides (Fig. 8.42). Figure 8.42 shows the spatial relations of the main types of valley glacier moraines together with certain types of glacio-fluvial materials, to be discussed shortly. In general, moraines achieve their morphology through direct deposition from the ice, and they are composed mainly of till.** Still-stands of an ice front or ice re-advances short of former maxima may produce one or more moraines (*recessional moraines*) in back of the terminal moraine. As noted by Thornbury (1969, p. 366), working out

* In a sense, lateral moraines are also end moraines.
** Often washed, ice-contact stratified drift or flow till.

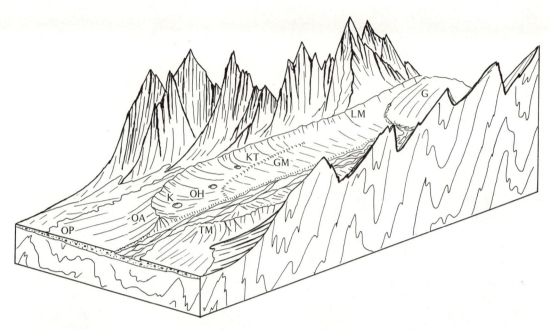

Fig. 8.42.   Block diagram of common deposits made by a valley glacier (G) during its recession. Illustrated are lateral moraine (LM), kame terrace (KT), ground moraine (GM), kettle (K), outwash head (OH), terminal moraine (TM), outwash apron (OA), and outwash plain (OP).

morainal history is one of the more important and difficult aspects of geomorphic analysis. End moraine sequences are only encountered "intact" in areas where there has been essentially continuous glacial recession, since ice readvances tend to destroy or obscure the record.

Moraine Composition and Formation. It has previously been noted that moraines consist mainly of till. Ground moraine is less extensive from valley glaciers than from continental types, since it occurs in a concentrated meltwater path when deposited in a valley. But the basic constitution and formation of ground moraine is similar in each case. Ground moraines often consist of a basal till layer of material presumably deposited from moving ice (*lodgment till*) which, coming as it does from the subglacial load, contains both the fine and coarse fractions that are being

transported. The lodgment till may be built up to a great thickness and is often overlain by a thinner till enriched in coarser clasts and presumably lowered from the glacier's englacial and supraglacial loads by ablation (*ablation till*)—see Gillberg (1955) for a good comparative description (Fig. 8.43). In lodgment till, stones tend to be lodged with their long axes parallel to flow directions.* From being "smeared on" by the overriding ice, such till may be very compact, and, as noted by Flint (1971), fissility may be developed in argillaceous areas. Ablation till is lowered rather gently to the ground by wasting ice and is generally loose. If not overridden by an ice re-advance, it is generally noncompact, less strongly abraded than lodgment till, and shows no fissility (Fig. 8.44).

Formation of lateral and terminal moraines is rather more complex than that of ground moraine, in large part because of

* Till fabric studies can be made based on clast orientations in lodgment till (cf. Holmes, 1941; Glen *et al.*, 1957; Mark, 1971). Used with other directional criteria, till fabric studies show much promise for determining basal ice-flow directions.

the greater potential role of meltwater and possible ice-push phenomena. Meltwater streams often develop between lateral moraines and valley sidewalls, and occasionally water becomes impounded in this *fosse* (see Fig. 8.52). Most lateral moraines are dominantly composed of ablation till, possibly somewhat re-worked by meltwater, the extent of the reworking being generally indicated by the amount of stratified drift present. Terminal moraines in valley glaciers differ from lateral moraines in lying athwart a "most-probable" meltwater path and in representing a dynamic aspect of glacial deposition rather than a static one. The lateral moraine (as a landform) develops as the glacier wastes away and is thus essentially a recessional feature. The terminal moraine reflects a supply of drift being brought to a glacier margin by ice flow, in most cases, and terminal moraines of valley glaciers may be both high and steep-sided. They ordinarily consist of varying amounts of stratified drift and till, but, regardless of proportions, moraine formation must represent an interval during which the ice front wastage (ablation) balanced flow-debris input into the ice margin.

Bulldozing by an advancing ice front may add to or modify an end morainal depositional complex as noted by Chamberlin (1894, p. 525) and recently described in some detail (Bayrock, 1967). In addition, the general "ridge" topography may be modified by drainage lines leaving the ice front, by meltwater deposits developed between the moraine and the glacier (*outwash head*) or on the distal side of the moraine (*outwash apron*), and by debris flows or mudflows which generate scars on the one hand and a deposit called *flowtill* on the other (cf. Hartshorn, 1958). Flowtill usually has a high clay content. Melting of ice blocks within the morainal mass or outwash deposits may create ovate depressions (*kettles*) which may be occupied, in turn, by water bodies for short periods. In addition, *kames* and *kame terraces* (to be discussed shortly) may be

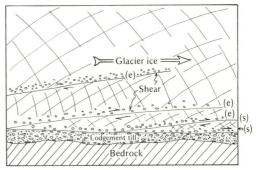

A Subglacial (s) and englacial (e) drift in transit

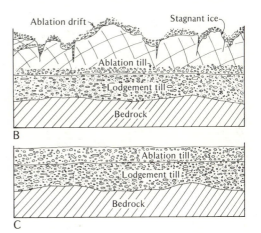

Fig. 8.43. Cross sections showing development of lodgment and ablation till. (A) During ice movement, lodgment till is plastered down on the subjacent surface with many oriented clasts from the subglacial load (s) while englacial loads (e) develop from internal shearing; (B) wastage of stagnant ice lowers the englacial load to form a layer of ablation till with non-oriented clasts—in valley glaciers some of the ablation till could come from supraglacial debris; (C) post-glacial accumulation configurations exposed to the atmosphere.

added to the complex to generate so-called knob-and-basin topography in the over-all glacio-fluvial sediment-landform association termed by Penck and Brückner (1909) the Glacial Series (Fig. 8.45). Of course, as noted earlier, a glacial re-advance may smear away surficial irregularities of an end moraine and plaster on one or more additional increments of lodgment and/or ablation drift, because, a moraine, once formed, is an obstacle to glacier flow and may be overridden.

Fig. 8.44. Lodgment till (lower light-colored interval with some alignment of flattened stones) beneath ablation till in a roadcut near the headwaters of the Arkansas River in the Salida Basin, Colorado.

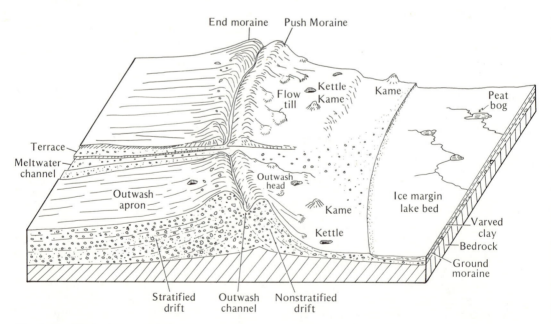

Fig. 8.45. Block diagram of the glacial sedimentary facies and topography associated with a terminal moraine. (After Penck and Brückner, 1909.)

Meltwater Deposits

It is a rare glacial deposit that goes untouched by water during its accumulation. The contact may be limited to meltwater soaking through and into a ground moraine. But appreciable rock material may be shifted hydraulically as a glacier is sustained and ultimately wastes away. Some such effects have already been mentioned in connection with moraines, and those remarks as well as the ones that follow apply to some degree to the effects of continental ice sheets as well as the valley glaciers of immediate concern. Meltwater deposits tend to occur as "ice contact" accumulations of rock detritus, sometimes made in standing water, and as "outwash" deposits made by running water leaving the general area of the glacier. These *glacio-fluvial deposits* naturally retain some of the characteristics of till but have ordinarily undergone a much higher degree of sorting and shaping—depending on distance and mode of water transport. Holmes (1960) has been able to show that softer rock types may undergo considerable shaping through glacial transport (Fig. 8.46).

Meltwater deposits from valley glaciers differ from some continental counterparts in that the former, by definition, are already in a valley. Thus, the development of *valley trains, eskers, kames, outwash aprons (fans)*, or *deltas* is largely associated with runoff or impounded water confined by valley walls. Juxtaposition of meltwater routes and deposits is less clear cut in many continental glacier relations.

Valley Trains. The single most outstanding glacio-fluvial feature of most glaciers and certainly of most valley glaciers is the *valley train.* This is an outwash deposit spread by meltwater over valley floors, often for many miles downslope from glaciers. They appear to have several distinctive characters. (1) Those presently being formed (Fig. 8.47) are apparently being deposited by meltwater flowing in braided form. This is compatible with a high sedi-

Fig. 8.46. Pebbles of limestone [average diameter ~0.5 inch (1.2 cm)] collected from till south of Syracuse, New York, which show progressive abrasion with increased distance from source outcrop. Top, 3 mi (5 km); center, 7.8 mi (13 km); bottom, 11.4 mi (19 km). (C. D. Holmes, 1960.)

ment-injection rate (cf. Chapter 7) which is, in turn, commensurate with the ready availability of essentially all sizes of rock detritus at a glacier terminus. (2) The rock debris in valley trains reflects the mechanical fragmentation process characteristic of the sponsoring regime, especially since it tends to be deposited by through-flowing runoff in an open sedimentary system which tends to flush away fine-grained fractions. (3) Dominant valley train constituents are sand and gravel, often cross-bedded, of a shaping and sorting generally commensurate with the distance from the

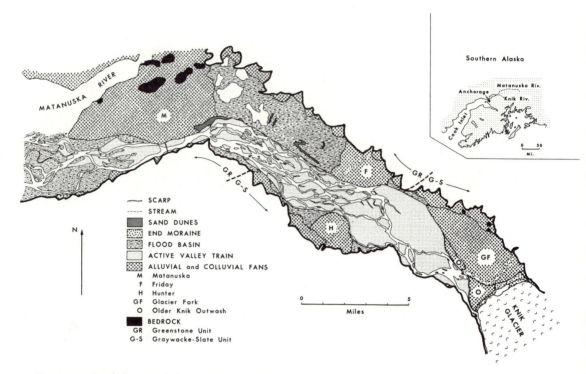

Fig. 8.47. Braided outwash below the Knik glacier, Alaska, in relation to surficial and bedrock geology of the Knik River Valley bottom. Tributary streams are indicated by fans M, F, H, GF, and O. (From Bradley *et al.*, 1972.)

outwash apron or end moraine where fluvial transport generally begins (Fig. 8.48).

Typical of valley train deposits are those in the Mississippi, Volga, and Mackenzie river valleys. The Mackenzie Valley still generally displays its braided flow pattern, and a former similarly widespread pattern for the Mississippi Valley is inferred by Fisk (1944) from stratification characters and channel patterns. Flint (1971) observes that materials in the Mississippi trench attributed to outwash are 25% gravel in a region extending to Louisiana where the Mississippi presently moves silt and fine sand (Fig. 8.49).* Note that for similar travel distances glacial outwash left in valley trains is chemically less mature than gravels left in streams draining humid sources which dry up. Also, of course, meltwaters from most glaciers cease to flow for several months of each

year. When this happens, all the material in transport is deposited—temporarily. Much glacial debris is too coarse for usual runoff to move and is thus bypassed.

In addition, one must consider the environmental conditions attendant with the onset of glaciation and its converse. If the analysis of Ouachita River gravels by Saucier and Fleetwood (1970) is valid in its Pleistocene climatic implications, runoff through valleys like those of the Missouri, Ohio, and Mississippi was erratic in volume. If, as indicated by these authors, increased precipitation marks the onset of glacial conditions, high-runoff–volume, low-sediment–volume, perennial, meandering flow could characterize previously alluviated valleys like those of the Mississippi River basin before the arrival of the actual ice front. Intermittent (seasonal) low-runoff–volume, high-sediment–volume flow

* Paleofluvial relations noted in Chapter 11 suggest that this may be a significant hydraulic observation.

would coincide with glacial self-starvation, braided flow, and valley train formation. Later, aridity in areas near to glaciers would presumably be countered by high-volume runoff of meltwater during deglaciation, probably with a reversion to a meandering flow pattern. Subsequent precipitation accentuation could take up where the meltwater left off and either adjust or maintain the meander pattern; this is presently occurring along the Mississippi River. Meander amplitudes and related parameters recorded on many relict valley fills indicate former runoff volumes both above and below those of the present (cf. Chapter 7).

Outwash Aprons. Otherwise also known as *fans, outwash aprons* are wedges of glacio-fluvial material, spread away from end moraines by meltwater (cf. Fig. 8.45), which commonly merge downslope with valley trains. Their upper surfaces may be rather steeply inclined ($4°-7°$), but on the glacier side of a moraine, removal of the glacier-ice support by melting may permit the deposit to collapse into what Flint (1971) terms an *outwash head* (see Fig. 8.42). Not uncommonly, they contain very large erratic blocks and are occasionally pitted where large pieces of buried ice have melted—this is particularly true on the inner margin of end moraines.

Eskers. Sinuous, often discontinuous ridges of more or less stratified and sorted sand and gravel occur in areas formerly buried by glacier ice of both valley and continental glaciers (Fig. 8.50). These *eskers* are widely believed to represent

Fig. 8.48. Braided outwash meltwater pattern (left), terminal moraine ridge (center), and ice (right) of the Sólheimajökull Outlet Glacier from the Myrdalsjökull Ice Sheet, southern Iceland. (W. N. Gilliland.)

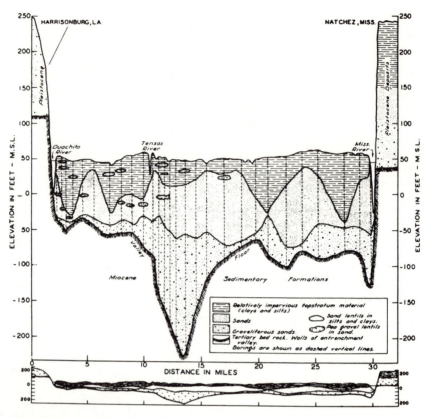

Fig. 8.49. The Mississippi Valley near Natchez, Mississippi, showing the alluvial fill and the configuration of the pre-Pleistocene bedrock in cross section. (After H. N. Fisk.)

glacio-fluvial and/or glacier deposition in contact with the ice—i.e. beneath, within, or on top. Actually, there is some question as to the precise nature of the ice contact. Crests are smooth, hummocky, or pitted by kettles. Eskers are notably steep-sided, even though it is apparent that the gravel in question has slumped somewhat since it accumulated. In fact, side slopes notably approximate repose angles for the material in question. Flint (1971) properly takes note of Hummel's early suggestion (1874) that the esker deposits were confined by walls of ice when accumulated. Flint's summary of early esker theories is excellent.

Because of their form and position there can be little question that eskers are a stagnant ice deposition feature in most instances, where the esker fails to parallel ice flow directions. Those that are distinct and coherent in outward form as well as internal stratification can hardly have been lowered from positions much above the glacier floor. Most commonly, accumulation of esker glacio-fluvial detritus seems to have occurred in subglacial tunnels. Penetrating downwards through a variety of cracks, crevasses, and side-wall openings, meltwater from valley glaciers tends to penetrate to the lowest portion of the valley floor. Flow directions of similar fluids under continental glaciers are presumably somewhat more random. In any event, this water, confined and presumably therefore under a hydraulic head, is forced to deposit gravel in the channelway to some considerable depth. Hydraulic head relations are somewhat confirmed by eskers that climb and cross divides, since only confined flow would move upward. Eskers range from a few meters wide and high to more than 650 ft (200 m) wide. Lengths vary from 325 ft (100 m) or less up to many kilometers.

There is considerable coincidence between ice flow direction and esker trend and between eskers and ice-parallel valley depressions (cf. Prest *et al.*, 1968; Ives, 1960). Esker materials normally range from sand to boulders and are commonly cross-stratified in the apparent flow direction. The constituents are often compositionally close to related till deposits. However, none of these interrelated factors explain how water under a hydraulic head

should be forced to deposit sediment. Some short esker segments do become fine-grained downstream and terminate in deltas. Plugging of the flow conduit is thus implied (cf. De Geer, 1897; Hershey, 1897). In fact, there is reason to believe that free, open flow under hydraulic head would be decisively erosional rather than depositional. Two mechanisms seem to be implied here. One is that esker deposition may occur under diminished flow rates when hydraulic head has been lost, possibly in late fall when melting is about over. Another is that under diminished flow rates, at least a partial plug must form in the conduit, perhaps by downflow

Fig. 8.50. Features at the terminus of Woodworth Glacier (right background), Tasnuna Valley, Alaska, 1938. Sinuous ridge (middle ground) is an esker feeding a fan adjacent to a fluted outwash. Pond (upper left) is in a kettle. (Bradford Washburn.)

deposition in a pond or by an ice block or large erratic against which sediment could back up. Many eskers may have formed in a single season (cf. Fig. 8.51).

Kames and Kame Terraces. Depositional features more commonly associated with stagnant than with moving ice (like eskers) include *kames* and *kame terraces*. Kames are ice-contact phenomena generally considered to form by the accumulation of glacio-fluvial debris in a fissure or hole in glacier ice. When the ice melts, the mass of gravel collapses into a roughly conical heap of rubble. Such "cones" of gravel constitute one of the main types of "knob" in the "knob-and-kettle" topography that typifies many end moraines. A variety of origins are certainly involved in the features that pass for kames. Along valley glaciers, some kames may originate as talus cones. Most consist of water-worn gravel, however. Both Thornbury (1969) and Flint (1971) concede that there are gradational morphologies between kames, kame terraces, and certain types of deltaic features.

Kame terraces are rather more distinctive morphologically than are kames (cf. Black, 1969). They tend to be elongate and essentially parallel with side walls of glaciated valleys. Consisting mainly of stratified drift, kame terraces appear to be mainly glacio-fluvial deposits made between a valley glacier and the adjacent valley wall. Usually, after the ice disappears, the material in question slumps

somewhat, as noted by Rahn (1969). Superficially, kame terraces resemble stream terraces resulting from the dissection of a formerly more extensive valley fill, but the ice-contact feature was never more extensive (Fig. 8.52). Commonly, kame terraces are pitted by kettles. They may occur in pairs on opposite sides of the valley, and some are developed at several levels as the valley ice thins.

Deltas and Lacustrine Features. Ice-contact fans or *deltas* are landforms that form a gradational morphologic series, on the one hand merging in shape with features such as outwash aprons, and on the other blending in form with shapes of actual lacustrine deltas developed in association with ice-margin ponds. As would be expected, most deltas consist of sand and gravel, and most appear to owe their existence to the irregular depositional topography generated by the glacier ice as it wastes away and by the ice mass itself which simultaneously generates meltwater from time to time and otherwise often blocks drainage. The so-called "fans" are mainly braided stream creations, well on their way toward becoming parts of deltas, kame terraces, kames, or outwash aprons. Most display some cross-stratification. And the lacustrine variety commonly grades laterally into lake-clay deposits.

Lakes associated with valley glaciers are comparatively limited in scope if not in number. Most relate to cirques, basins on

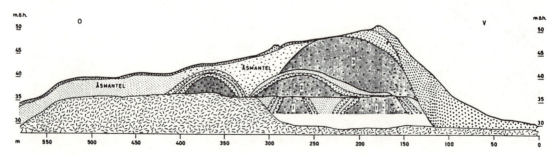

Fig. 8.51. Composition of a multiple esker ridge near Stockholm, Sweden. Three years' accumulation is represented (1, 2, 3). The cores are coarse gravel and boulders (summers) covered by finer debris (winters). The entire feature is blanketed by secondary slump ("Åsmantel") and soil. (After Eriksson, 1960.)

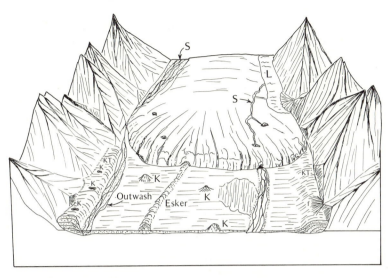

Fig. 8.52. Block diagram of a wasting valley glacier showing development of assorted ice-contact features characterized by stratified drift. Before melting, the ice acts as a retaining wall for sediment being deposited from meltwater, but melting destroys the support and permits the deposits to slump and be deformed. Features illustrated include kames and kettles (K), eskers, streams (S) with associated outwash, fans (f), ice margin lakes (L), and kame terraces (KT).

valley steps, depressions made by melting ice blocks, low areas behind terminal and recessional moraines, and, occasionally, depressions between lateral moraines and valley walls. Most such lakes are quickly sedimented and drained to form lacustrine flats, and a few are extensive enough to be termed plains. The lake sediments are usually some combination of stratified drift and clay, the latter often varved. Since the normal runoff paths of meltwater from valley glaciers often encounters related pro-glacial lakes, they probably fill and drain more quickly than similar depressions formed in association with continental ice sheets. Their planar deposit forms contrast with most montane, glacial topographic settings even though the lake areas involved are normally not great.

CONTINENTAL GLACIERS AND LANDFORMS

It would be surprising indeed if many similarities and parallels did not exist between alpine glacial landforms and land-forms developed by continental ice sheets. Many remarks made in the foregoing sections dealing with alpine glacial phenomena, therefore, can be readily applied to the sections that follow. But there are differences in magnitude, genesis, and consequences that are important. At the peak of Pleistocene glacial advances, it is estimated that some 30% of the earth's land surfaces were under ice. Probably no more than 15 to 20% of that total was in the form of alpine glaciers. Also, though one can argue that any glacier requires a starting point, continental ice sheets apparently reflect more of a geographically localized hydrographic condition than a topographic shelter. The major centers of continental glacier spreading probably reflect regional areas of low temperature and ready access to precipitation. Terrain configurations may have initially influenced the locations, but, once the land was buried, it was the glacier configuration that determined subsequent atmospheric associations. Certainly, buried terrain often exerted a strong influence over flow directions.

The Pleistocene ice masses involved were enormous. Many are estimated at more than two mi (2–3 km) thick; each ice sheet covered hundreds of thousands of square miles, and flow distances involved hundreds of miles. Presumably, the ice movement was radially outward from the accumulation centers and, as in the present ice sheets, more or less equally unless one flank received greater precipitation than another. But one thing must always be kept in mind. "The spreading center of a glacier doesn't 'KNOW' the condition the terminus is in." This teleological remark is also applicable to alpine glaciers, of course. But a continental ice sheet is rather like a dinosaur with an overlong nervous system. An ice sheet has a great deal of momentum (volumetrically speaking), and it may be a very long distance and much time from the central nervous system (spreading center) to a toe that has been stepped on (or melted). Therefore, long after a continental ice sheet has advanced into a realm where ice survival is strictly "marginal," the snow gathering ground may be continuing "business as usual." There is also an analogy that can be drawn between glaciers and streams flowing into tidal estuaries. No matter which way the fresh-salt water margin is moving, the fresh water always flows down. Thus, a marginal environmental change may accelerate ablation at a glacial terminus, or it may reduce ice thicknesses to a depth subcritical for flow purposes over a belt several hundred miles wide. In either case, where a sufficiently thick glacial nucleus remains, flow will continue therein.

Erosional Terrain Morphology

Areas suffer severe glacial abrasion and virtually no deposition where snowfall concentrates to form continental glaciers. Indeed, there is probably some critical area centered under an ice sheet where outward flow and erosion is nil (cf. Thornbury, 1969, p. 377) which is more or less equivalent to a non-erosion area with a radius X_c

for fluvial erosion (cf. Chapter 5). Nevertheless, it is true that several types of antecedent terrain are involved in glacial climatic nuclei. Apart from the Laurentian Upland, where ancient crystalline knobs have been ground down and rounded off, much of glaciated eastern Canada is rather well described by Thornbury's term, *ice-scoured plain* (1969) (cf. Fig. 3.64). Not, as he was quick to point out, that the region is truly smooth and even in contour. But it is generally devoid of pronounced highlands, is dominantly erosional in origin, and consists of beveled, scoured, grooved, and scratched rocks in knobs (rochés moutonnées) and basins with odd patches of glacial drift, all of one general level (Fig. 8.53). Parts of Scandinavia and Labrador are notably mountainous in a degree approaching that of the northern Appalachians. But they too have been ground down and rounded off. If cirques were there to begin with, they have since been destroyed. The latter rounded terrain configuration has been termed a *mammillated surface* (Fig. 8.29). Some of the Scandinavian area is also in the form of ice-scoured plains.

Glacial Drainage Derangement

Regions where glacial erosion has largely deranged preglacial drainage are the areas where that erosion was most severe and most extensive in Scandinavia and northeast Canada. The phenomenon was possibly first cited by Davis and Snyder (1898) and has later been confirmed by many geomorphologists. The writer (1966) called attention to similarities in derangement between shield regions affected by glaciers and those altered by aridity in low latitudes. In general, the glacial effects in much of northeast Canada that relate to drainage derangement are mainly erosional, though there is no question that an occasional patch of drift has had an effect, often in the form of an esker or moraine. Technically, the consequences are the result of alternating glacial and humid or cryergic geomorphic systems. The drainage pattern

Fig. 8.53. The "Barrens," an expanse of glaciated shield, Canada, between Artillery Lake and Thelon River. The sinuous, light-colored feature trending diagonally across the picture is an esker. (National Air Photo Library, Ottawa.)

results (Fig. 8.54) appear to reflect localized scour along former drainage lines to create depressions later occupied by lakes following deglaciation. In addition there seems to be some tendency for elimination of former drainage divides, for, in numerous instances, lakes on divides have multiple outlets that yield water to more than one drainage system. Also, due at least in part to vagaries in slope of the ice-scoured terrain, channels divide around bedrock obstructions; some of these channels reunite farther down slope, whereas others attain base level separately.

In spite of the apparent similarity between glacially deranged drainages and those deranged by aridity (cf. Figs. 7.46; 8.54), there are some differences. For one thing the glacier ice also has the ability to deepen pre-existing drainage lines by erosion as well as to locally aggrade or abrade them away. Thus, many of the deranged drainages in glaciated areas include reaches of pre-glacial drainage. Also, there do not seem to be as many barbed tributaries or clearly defined centripetal drainage patterns as were recorded in shield regions subjected to arid drainage de-

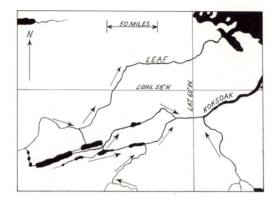

Fig. 8.54. Glacially deranged drainages of the Leaf and Koksoak rivers which flow northeast into Ungava Bay on the Labrador Peninsula. Common to such configurations are distributaries, scour depressions, and multiple débouchérs.

rangement (see Figs. 7.45 and 7.51). In the last analysis, however, as Davis and Snyder (1898) were careful to point out, the multiple channel drainage form is inherently unstable. This applies to glacial-humid as well as arid-humid terrain. Eventually, as the lake depressions are filled with sediment, and channels are selectively deepened, the trunk-stream-tributary configuration usually associated with a "normal" drainage net should develop. These adjustments are presently in progress in the humid parts of North America and Eurasia that were formerly subject to severe glacial erosion and various but generally minor degrees of deposition. They are also occurring in the humid tropics in regions that were previously arid (cf. Chapter 7).

Glacier erosion in regions like eastern Canada north of the St. Lawrence Lowland has stripped the region of most antecedent soil and other regolith. Most former boulder residuals (tors) are now gone (cf. Chapter 3). And a formerly much more extensive cover of Paleozoic sedimentary rocks has also been almost completely removed down to Precambrian crystalline types. In large measure because of glaciation, the current assets of much of the Canadian Shield lie mainly in hydroelectric power, water, timber, mineral wealth,

and tourism related to the wilderness and glacial lakes. Agricultural activities have been greatly limited by the absence of widespread fertile soils, though the industry of sturdy farmers in southeastern Quebec in removing glacial stones from their fields is much to be admired. Technically, however, the latter area falls in the depositional "halo" of North America's continental glaciers. We will discuss this now.

The Depositional Halo

The main catchment and spreading centers of each Pleistocene continental ice sheet, represented now by such features as ice-scoured plains, is ringed by an essentially continuous "fringe" or halo of glacial deposits. In North America, the zone between areas dominated by erosion and areas characterized by deposition trends roughly along the United States-Canadian border, sometimes to the north as in Quebec and sometimes to the south as in northern Wisconsin and Minnesota. The transition between the two types of area is not sharp, but it is unmistakable on a regional basis. The glacial depositional halo begins on the glacial side with polished bedrock swells surrounded by patches of glacial drift and grades (southward) into essentially continuous ground moraine, eventually to end morainal topographic configurations and outwash features.

Mid-continental regions where the continental glacier drift sheets are discontinuous are usually associated with scour pockets in bedrock and other kinds of glacial depressions which later form lakes in regions like southern Finland and Sweden and much of Minnesota. Lakes also occur in areas of dominant glacial erosion for mainly erosional reasons, of course. They are uncommon beyond glacial borders and not particularly abundant over most *till plains*. In North America, a broad belt of land dominated by ground moraine extends from the Dakotas, Kansas, and Nebraska, eastward, into Ohio. A similar depositional landscape extends roughly east-west through central Europe. Characterized by very subtle swell-and-swale topography, these till plains or drift sheets are remarkably level and even (Fig. 8.55). Major departures from this planar configuration take the form of post-glacial stream valleys and morainal topographic complexes. Thick soils developed on these glacial deposits (supplemented in North America and part of Eurasia by wind-blown dust) form the basis of exceptionally rich agricultural regions; North America's so-called "corn belt" is an example. The United States and the U.S.S.R. are unsurpassed in the areas of till plain each possesses. Drift thickness on such plains averages about 100 ft (30 m).

The outer, or distal edge, of deposition by the continental glaciers was determined mainly by the flow rates and patterns of ice movement in combination with ablation and outwash effects. As noted by many researchers, on land the major marginal ice-flow pattern of continental ice sheets took the form of lobes. Some of the ice lobes were influenced by pre-glacial lowlands, whereas others probably reflect divergences around stagnant ice masses. The Driftless Area of Wisconsin may represent one of the latter. It is often reasoned that the ice margin of a glacier is mainly a reflection of flow-rate balance by marginal wastage. In some instances, this is probably true. But our knowledge of the precise conditions that favor glacial growth is incomplete, and our understanding of the relationship between ice-margin environments and accumulation area flow rates in the comparatively low-latitude ice sheets of the Pleistocene is also meager. Many ice lobes may have merely advanced until spreading, a resultant thinning, and an increased basal load and friction created a stagnant ice obstruction out of the lobe, one with an immobility not closely related to marginal environment. Intense wasting may have come much later in some cases.

In any event, most Pleistocene continental glacial margins were roughly lobate as

Fig. 8.55.   Till plain developed on glacial drift of Wisconsin age in South Dakota west of Sioux Falls.

indicated by the generally arcuate disposition of end moraines and related ice-margin deposits (Fig. 8.56). In addition there were minor salients and re-entrants. End moraines of continental glaciers are generally higher in clay and less stony than those of alpine glaciers which, in part, probably accounts for the usually greater height and steepness of the alpine types. Though the end moraines of continental glaciers show up well on aerial photographs and topographic maps, many are not particularly obvious to the untrained eye at ground level. They are also discontinuous because of local fluvial dissection and removal by meltwater, lateral variations in drift amounts and clay contents, and local overriding by ice. Major glacial lobes indicated by morainal topography in North America bear such names as Dakota, Des Moines, Superior, and Michigan (cf. Fig. 8.61).

Topographic deviations from the till plain-moraine depositional configuration of continental ice sheets take several forms, though all reflect aspects of the depositional halo. In several areas or belts of both North America and Europe, there are "fields" of *drumlins* (Fig. 8.57). These

groups of streamlined hills tend to be of a single general size in a given group, and Flint (1971, p. 101) lists their "ideal" dimensions as ". . . length 1 to 2 km, width 400 to 600 m, and height 5 to 50 m." Most fields are situated on areas that would otherwise qualify as till plains, though a few groups occur in hilly areas (Fig. 8.58). Most individual drumlins studied consist of clay-rich till, and there is some consensus that individual fields were formed under a single set of conditions involving a balance between load acquisition and deposition as proposed by Smalley (1966) and Smalley and Unwin (1968).

An additional topographic aspect of the depositional halo formed by continental ice sheets exists where hilly or mountainous areas have been invaded by the ice and more or less buried. The northern Appalachian Mountains and adjacent areas of western New York and Pennsylvania are perhaps typical of this relation. In most of these regions, there were distinct preglacial valleys so that many aspects of glacial advance and wastage resemble those of alpine glaciers. This is particularly true where the topography influenced ice flow directions as well as erosional and

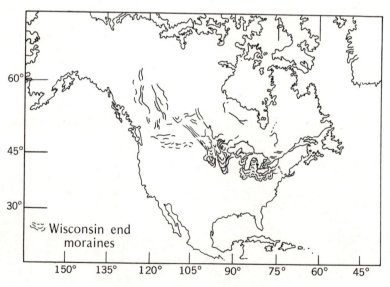

Fig. 8.56. General distribution of major end moraines associated with the Wisconsin glaciers. Note the lobate character of the margins (see also Fig. 8.61).

Fig. 8.57. An archipelago of drowned drumlins comprising part of a drumlin field, Strangford Lough, Northern Ireland, which has become submerged during the post-glacial rise in sea level. Presumed ice movement was from upper left to lower right. Some glacio-isostatic rise of the land may also be indicated by the wave-cut benches and cliffs indicated by shadows to the right which modify the otherwise streamlined hill form developed on boulder-clay. (Aerofilms Ltd.)

depositional patterns. *Mammillated terrain* is common. In general, drift deposits in this region are less continuous, being generally thickest in valleys and thin or absent from many upland crests. The ice moved along pre-existing valleys in many cases, but, in some, flow was transverse and some topographic asymmetry resulted. Naturally, this higher-relief terrain influenced meltwater flow, mainly, just as a valley glacier's meltwater flow is directionally affected.

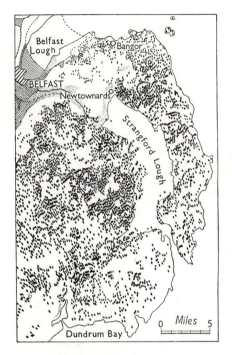

Fig. 8.58.   Part of a drumland tract in County Down, Northern Ireland. (After J. K. Charlesworth.)

Glacio-Depositional Drainage Effects

In the comparatively extensive regions where the landscape is dominated by expanses of glacial ground moraine, a gentle "swell and swale" topography has been the scene of post-glacial drainage development. Prior drainage configurations are buried by glacial drift over vast areas in the Dakotas, Iowa, Illinois, and Indiana, though former regional slopes developed with respect to large drainage basins persist to be expressed on the surface through several tens of feet of glacial debris and thereby impart what amounts to a relict slope to the ground moraine of many areas. Even so, the ground moraine surface was apparently undissected by glacio-fluvial action over wide areas, and subsequent environments have developed their own characteristic drainage configurations on the relict, granular surface of low re-

lief. The setting, at least where it is humid, is almost an ideal natural site for examination of drainage net development in accordance with the principles set forth by Horton (1945) and discussed in Chapter 5.

In theory, drainage nets developed on granular materials of similar inclination and composition should evolve through a rill phase into a dendritic drainage net trending down the regional slope. Moreover, if glacial drifts of more than one age exist in the same immediate area, the oldest drift areas should exhibit the highest drainage densities and the youngest drift areas should have the lowest drainage densities assuming, of course, that subsequent conditions were similar. Let us examine a case history from Iowa outlined in some detail by Ruhe (1969) and illustrated here (Fig. 8.59). In three adjacent Iowa counties, we find a small area of Kansan or Illinoian age till to the west (Cherokee County), locally overlain by Wisconsin-age loess ranging in age from 30,000 years at its base to about 13,000 years at the top. The dates are based on C_{14} analyses. The till in the same area yielded two radiometric dates, both greater than 39,900 years. The Illinoian or Kansan would both be older than this, but so would the Permian and Cambrian. It is not a definitive date. An erosion surface developed on this older drift [termed Iowan Surface by Ruhe (1969)] is here designated Cherokee Surface for the county in which it occurs (cf. Fig. 8.59).* Drainage densities for the area of older drift based on the point-intercept measurement technique (cf. Ruhe, 1969) are 7.7 and 6.1 for the Cherokee Erosion Surface and 10.2 and 7.9 for the area of Kansan(?) drift incised below this surface. It should be emphasized that the two lower drainage-density values are from the erosion surface that truncates the older drift, though the higher of these "low" values clearly differs little from the lower of the two "high" values.

In eastern Cherokee County and western

---

* This may help avoid the confusion that otherwise develops because of the former use of the term Iowan for a glacial deposit that has since been disavowed (cf. Ruhe, 1969).

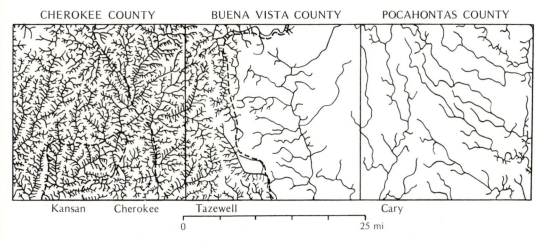

Fig. 8.59. Drainage nets developed on Kansan drift, Cherokee erosion surface, and Tazewell and Cary drifts, northwest Iowa. (After Ruhe, 1969.)

Buena Vista County there is an area of older Wisconsin drift (Tazewell) which has yielded C_{14} dates ranging from 22,000 to 14,000 years and on which there are drainage densities from 5.4 to 4.7. The age of a younger Wisconsin drift (Cary) in eastern Buena Vista County and Pocahontas County ranges from 13,000 to 14,000 years; drainage densities there range from 2.1 to 1.9. Thus, in general, the younger the drift sheet the lower the drainage density. However, the drainage densities on the Cherokee Erosion Surface and on the Tazell drift are very similar considering their supposed age differences. Ruhe (1969) has plotted the stream order—drainage density—time relationships (Fig. 8.60) and observes that the uncontrolled use of morphometric analysis may be misleading, insofar as the curves indicate accelerated drainage in the first 20,000 years and slower development subsequently as interpreted by Leopold, Wolman, and Miller (1964, p. 424), whereas "field knowledge indicates rapid development between 14,000 and 20,000 years ago and relatively slower [development] since 14,000 B.P." It should be pointed out, however, that conditions for drainage develop-

ment have not remained constant.* In addition to the older Wisconsin glacial episode, there is a humid-type soil developed on the Kansan(?) drift (Sangamon ?) overlain by Wisconsin loess which was deposited on the widespread, low-relief Cherokee Erosion Surface previously mentioned (cf. Fig. 8.59). It is on this same erosion surface that many ventifacts are encountered, as noted by Flint (1971, pp. 169–170, 248), which would presumably indicate an arid planation and deflationary period between the time of accumulation of the younger Wisconsin drifts and the older drift (Illinoian—Kansan?).

If analytic techniques and principles set forth in earlier chapters are applied to the above sequence of dated events, one would conclude that we are not merely viewing a record of glaciation versus humidity. If the incision of drainage nets is a generally humid phenomenon over areas of rather low relief, then the oldest humid incisional phase presumably followed the accumulation of the Illinoian (?)-Kansan(?) drift and is equivalent to the related soil. Formation of the Cherokee Erosion Surface strongly implies an episode of arid planation.

* The problem is the assumption that drainage development is solely a time-dependent phenomenon. The morphometric data are presumably valid. Thus, it is an interpretation that is needed.

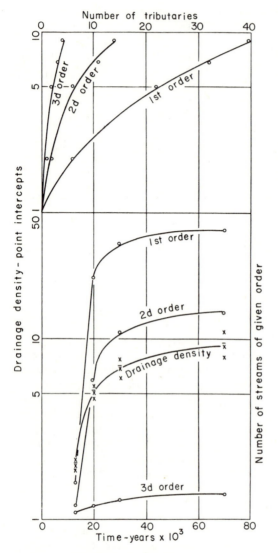

Fig. 8.60.   Relation of number of streams of a given order to drainage density and of drainage density to number of streams of a given order to time based on areas illustrated in Fig. 8.59. All curves, Cary at left to Kansan at right. (After Ruhe, 1969.)

may have persisted during the first (Wisconsin) glaciation in Iowa (Tazwell), or part thereof (Fig. 8.59), and may further have persisted during all or part of the second Wisconsin Glacial advance in that area (Cary). Drainage development since the Cary has been meager, unless one stops to consider the probability that drier conditions may have retarded incisional processes considerably during the Jewell Interval (4,000–7,000 years B.P.). If the proglacial climatic analyses of Saucier and Fleetwood (1970) are applicable to the Iowa situation, and if late-glacial-early-post-glacial times were arid, most Cary drainage would be a product of the last 3,000–4,000 years. In addition, most Tazell and pre-Tazwell drainage density presumably was being accentuated at the same time. The Cary glacial lobe deposition (Fig. 8.59) to which Ruhe (1969) assigns a duration of 1,000 years, may merely represent an adventitous ice advance or "surge" along a fortuitous avenue of egress from a dominantly wasting continental ice sheet and may not actually correspond to a particular adjustment in the North American glacio-climatic pattern.*

Ruhe (1969) makes the observation that, this, the example just cited, establishes that first- and second-order streams are instrumental in the development of an area into a watershed or drainage basin. Also, if one takes into account the complexities of climatogenic drainage relations, the situation offers reasonable solutions for some apparent contradictions. For example, Ruhe explains the slight differences between Cherokee Erosion Surface and Tazwell drift surface drainage densities by concluding that drainage on the Cherokee Surface was somewhat advanced by the time of Tazwell glaciation (20,000 years B.P.) and that the same drainage extended and integrated on the Tazwell surface after loss of ice cover, thereby producing a similar drainage density on the two areas. If these were the

We are left to conclude that the Cherokee planation episode (? = Yarmouth-Illinoian or Sangamon-Wisconsin) terminated stream incision until the onset of a later humid (early glacial?) phase. This later (Wisconsin) drainage net development

* However, Porter (personal communication, 1973) notes that ice advances in the Yukon, Washington State, New Zealand, and the southern Andes correlate with the Cary advance.

developmental relations, it would seem that the Cherokee Erosion Surface drainage densities should be much greater than those of the Tazwell. It appears to be more reasonable to assume that there was little or no integrated drainage net on the Cherokee Surface at the beginning of the Tazwell glacial advance and that pre-Cherokee drainage nets were deteriorating because the whole region was being aggraded and pedimented under arid conditions. If so, this would mean that Tazwell, Cherokee, and pre-Cherokee drainage nets all largely developed in post-Tazwell time. This would explain their very similar drainage densities—the somewhat higher values of part of the pre-Wisconsin drift area being explicable as a re-expansion of an already-established net that had been somewhat alluviated adjacent to a pediplain.

A thesis developed in several earlier chapters and applied to the previous glacio-depositional situation is that drainage net development and extension is mainly a humid phenomenon. Gullying can occur by localized precipitation in desert uplands and, WHERE FAVORED BY WEAKLY CONSOLIDATED FINE-GRAINED DEPOSITS, may generate badlands. Downstream reaches of the same drainage nets would tend to be simultaneously aggraded. A planar surface in a desert approximates an equilibrium landform, and a glacial ground moraine is a nearly planar landform. It follows that regions in which planar glacial deposition is succeeded by aridity will tend to remain fluvially almost unchanged unless or until such time as humid incision develops. However, the glacio-depositional surface would be exposed to desert sheetwash as well as eolian deflation and could thereby acquire a lag gravel concentrate along the surface, just as on the Cherokee Erosion Surface previously discussed. In instances of severe aridity, the surface gravels might be ventifacted (cf. the Cherokee Erosion Surface) and, in any case, if later buried, the lag gravels would appear as *stone lines* in exposed sections (see Ruhe, 1959 and Biga-

rella *et al.,* 1965 for other modes of stone line genesis).

In general support of the foregoing contentions that planar glacial deposits would be little altered in dominantly dry regions we can point out the comparatively slight degree of incision of late Pleistocene drift sheets in the presently semi-arid Dakotas as compared to dominantly more humid regions farther east (cf. quadrangle maps of Todd, 1909). In Todd's summary discussion of deposits made by the Dakota Lobe of the Wisconsin Ice Sheet in the Aberdeen-Redfield area of South Dakota (Fig. 8.61), he observes that the deposits of the several glacial advances tend to be separated mainly by a planar surface marked by striated boulder pavements (i.e. little or no interglacial weathering) in which there is evidence that sheetwash has removed fine detritus from the gravelly surface (Fig. 8.62).

Poorly defined drift contacts are essentially what one would expect if glacial ice advances alternated mainly with arid conditions. Few streams in the same region are presently perennial, and the drainage densities for the Northville, Aberdeen, Redfield, and Byron 30-minute quadrangles are far lower than for the youngest Wisconsin (Cary) deposits in Iowa. Drainage density for the Northville quadrangle in Dakota is approximately 0.04, if only perennial watercourses are used in the calculation; the value increases to 0.09, if those ephemeral watercourses expressed by the 20-ft contour lines are included (Fig. 8.63). This density compares with drainage densities on the Cary drift of 2.1 and 1.9 (cf. Ruhe, 1969).

It seems important to note also that the less dissected Dakota tills are very fine grained (cf. Todd, 1909), probably finer grained than correlative Iowa tills. The lack of incision of the Dakota till is therefore more probably a climatic effect than a lithologic one. The lithic factor would favor a higher drainage density in South Dakota instead of the low one recorded.

Discussions of glacial drainage effects bring out the fact that post-glacial and interglacial drainage development (or lack of development in an arid setting) is usually accompanied by corresponding variations in soil development and genetically compatible geomorphic phenomena which discussions in Chapters 5 and 6 to a degree anticipate. Our legitimate interest in the complex series of glacio-stratigraphic sequences on the various continents is limited to the light these sequences may shed on landform genesis, compositionally, or in the form of materials that indicate age or environment. It must be kept in mind that geomorphic events can only be compared from place to place to the extent that they can be isolated in terms of time. Thus, a soil developed on an early Pleistocene drift such as the Kansan now at the surface, as it is in southern Iowa, is not particularly well time isolated. The soil could have formed at any time pedogenesis was favored since the accumulation of the drift, and, under the present temperate humidity, it is forming again. Such a soil could be compared with other presently developing soils; it is a part of the landscape, but it clearly may reflect a number of relict conditions. In contrast, a soil formed during a particular finite episode of time (e.g. post-Tazwell, pre-Cary) can be related to a more precise time-environmental framework and compared with other materials of similar age, some of which may not be soil. It is the latter type of comparison that glacial stratigraphers may profitably make; various examples are

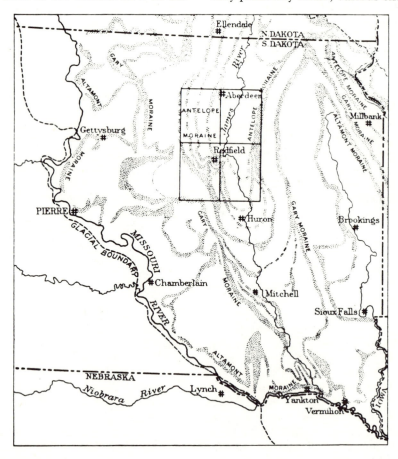

Fig. 8.61. The limits of Wisconsin ice shown by the distribution of moraines of the Dakota Glacial Lobe, South Dakota. (From J. E. Todd.)

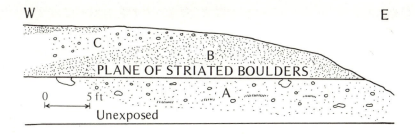

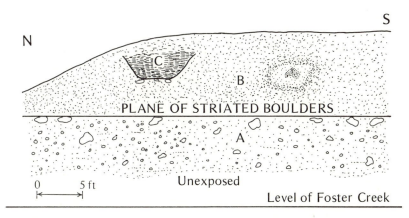

Fig. 8.62.   Sections of two South Dakota, Wisconsin age tills from Byron Quadrangle based on illustrations of Todd (1909). Shown above, near the southwest corner of sec. 9, T. 113 N, R. 60 W, in Beadle county is a section including a plane of striated boulders where A is yellowish till, darker above, with wet sand pockets whose upper surface is a striated boulder pavement and B is dark gumbo-like till under Cm a buff till with few pebbles which darkens upward. Below is a section from the NW corner of sec. 9, T. 113 N, R. 61 W, in which A is a yellowish till with wet sand pockets and traces of stratification topped by a striated boulder pavement, B is dark, clayey loam and sand with vertical joints containing loess-like pockets and gray till, and C is a channel filled with blackish clay.

considered in some detail in several major works (cf. Wright and Frey, 1965; Flint, 1971). The direct geomorphic implications of such correlations may be few indeed and are in general beyond the scope of this volume.

The glacio-depositional sequences of North America and Eurasia were created by repeated ice advances and consist mainly of drift, soils, loess, and varieties of glacio-fluvial materials. The bulk of these materials have accumulated by a process that disturbs prior deposits and levels old landforms even as it "may" create new ones. The study of recently developed terrain configurations of either glacial or subaerial origin is geomorphol-ogy. So too is the study of such ancient deposits as may assist our landform analysis. In general, the geomorphic aspect is most profitably pursued in deposits that are also part of extant landforms. Other deposits must either be considered in the context of stratigraphy, or, if the issue is one of buried terrain, in the context of paleogeomorphology. Paleogeomorphology will be considered in some of its major aspects in Chapter 11.

Ice-Margin Streams

The general lack of topographic confine-ment of continental ice sheets compared to valley glaciers has already been noted. In

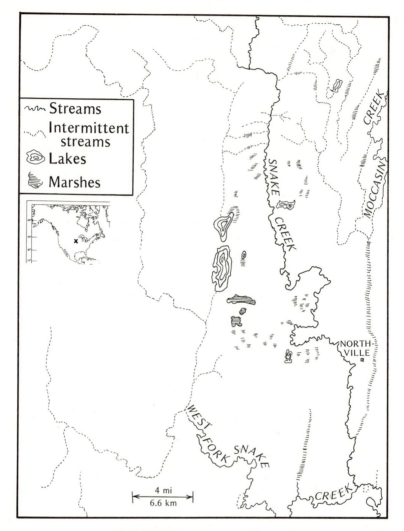

Fig. 8.63. Drainage map of the Northville Quadrangle, South Dakota (map inset), on a till plain of Wisconsin age. Compare the density indicated visually and calculated (text) with that illustrated in Figs. 8.59 and 8.60.

spite of this relationship, more often than not, the land adjacent to the Pleistocene ice sheets sloped toward the ice. Several factors are involved, but the major ones include the tendency of ice lobes to follow lowlands or depressions, plus the long-noted tendency of large ice masses to depress the earth's crust. One major consequence of these relations is that glacial meltwaters often tended to flow along the ice margin and to develop channels and

valleys there. In North America, the Ohio and Missouri rivers are the two most striking examples of drainages in which the courses appear to have been largely determined with relation to ice-sheet borders. The Mississippi and Columbia rivers are important as proglacial drainages affected by meltwater and outwash deposits, but their channels were not localized by ice margins. For more detailed consideration of glacial drainage influences on the Missouri River, the reader is referred to studies by Todd (1914), Flint (1949, 1957, 1971), and Warren (1952). Shifts in the Ohio River and later development are the subjects of study of many researches including Horberg (1945, 1950), Wayne (1952), and Thornbury and Deane (1955). Flint (1971) gives an excellent summary of much of this work (Fig. 8.64) and further notes the development of several of the European ice-margin drainages.

*Ice-Margin Lakes
and Proglacial Floods*

Essentially the same conditions that lead to the development of ice-margin streams also cause the formation of lakes between the ice front and adjacent land slopes. Some of the lakes so formed drain tranquilly away as the ice disappears and some appear to have overflowed rather catastrophically. The possible extent of ice-margin lakes along the borders of continental glaciers is enhanced by the fact that the largest amounts of meltwater are generated during deglaciation while the ice mass is receding into the depression previously created by isostatic subsidence of the crust under ice weight. Such relations apparently account for the development of Glacial Lake Agassiz which left lacustrine deposits over an area in excess of 100,000 sq mi in parts of North Dakota, Manitoba, Saskatchewan, and Ontario and more than forty sets of beach deposits (Fig. 8.65).

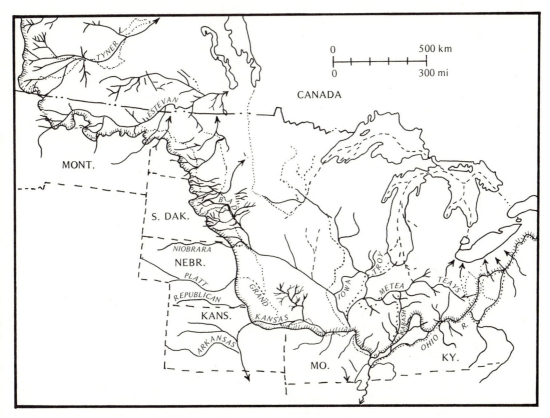

Fig. 8.64   Ice-margin drainages associated with the North American mid-continental glaciers whose outer limits are indicated by the hachured line. Valleys shown antidate at least one major episode of glaciation, and no suggestion of contemporeneity is made. If not occupied by existing rivers, drainage lines shown are mainly buried by drift. Dotted lines are those of existing rivers. Compiled from various sources including Horberg (1945), Wayne (1952), and Flint (1955, 1971). (For greater detail, see Flint, 1971, Fig. 9.4.)

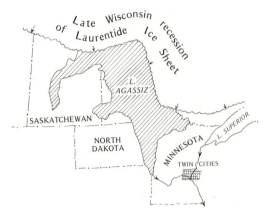

Fig. 8.65. General location of glacial Lake Agassiz during waning phases of the Laurentide Ice Sheet in North America. Southward drainage (shown) was later reversed as the ice barrier disappeared.

Though the Great Lakes have been occasionally discussed in the context of ice-margin relationships (cf. Thornbury, 1969), they certainly owe much of their existence to deep glacial erosion (see Leverett and Taylor, 1915; Hough, 1958; White, 1972). In this regard, the Great Lakes are more closely related to many hundreds of similarly erosional lakes scattered across much of glaciated North America and Europe than to impoundments against ice barriers. Needless to say, there are records of countless ice-margin lakes of the "Agassiz type" associated with both valley and continental glaciers.

There is an occasional indication of comparatively high-volume runoff associated with glacial meltwater, and at least some of these high discharges seem to relate to the removal of ice dams on ice-margin lakes. Probably the outstanding example of this kind of situation, and one that generated a truly spectacular suite of erosional and depositional landforms, is the Channeled Scabland area of Washington (see especially Bretz, 1923; Bretz *et al.*, 1956). This example was earlier mentioned (Chapter 7, Fig. 7.60) in a separate context, but it merits an additional few words. One of the best pictoral summations and recent synopses is by Shelton (1966).

From Bretz's original account (1923), and many subsequent studies (Baker, 1971), it has become apparent that a great volume of runoff was released across the broad area southwest of Spokane, Washington not far from the southern margins of the Cordilleran Ice Sheet (Fig. 7.60). An area of more than 10,000 sq mi was more or less simultaneously inundated by runoff, thereby creating enormous channels, giant waterfall scars, *mega-ripples*, and anastomotic erosion patterns (Fig. 8.66). Recent estimates (calculated on the basis of a gradient of 20–25 ft/mi and some of the features developed) indicate flow velocities of 17.5–20.6 ft/sec (20 ft/sec = 13.6 mph), discharge volumes of as much as 10 cu mi/hour, and water depths of several hundred feet. Some of the cataract scars, up to 1.5 mi (2 km) wide and 200–400 ft (60–120 m) high, are larger than similar features associated with the largest extant falls (Fig. 8.66), and mega-ripples are present with chords that measure 60–425 ft (20–130 m) and that are 15–50 ft (5–15 m) high (Fig. 8.67). Baker (1971) lists mean heights of 1.5–22 ft. Other explanations of the scablands have been offered (cf. Flint, 1938), but none appears to conform so well with the evidence as that of Bretz (1923).

In earlier explanations of the scabland phenomenon, Bretz (1923) insisted upon only a single flood, though later research resulted in a qualification that seven or possibly eight separate floods probably occurred (cf. Bretz *et al.*, 1956; Bretz, 1969). In Chapter 7 it was pointed out that the scabland region is presently a desert and shows evidence of having been alluviated prior to the flooding. In 1951, the writer recognized the similarity between the scabland flow pattern and flooding patterns in South America on what are now believed to be formerly desert plains become humid (cf. Garner, 1958; 1966; 1967). Arid alluviation has been shown to reduce relief and thus make a region more susceptible to flooding. There is little question that the water origins for the Channeled Scablands and those for most other regions with anas-

Fig. 8.66.   Dry Falls, a relict cataract in the southern part of Grand Coulee in the Channeled Scabland of Washington shown in a sketch based on an aerial photo. Though only partly visible as shown, the falls escarpment is some 400 ft high and almost 3 mi wide, dimensions, respectively, 2½ and 5 times as large as Niagara Falls today.

tomosing channel systems are generally different. The low-latitude examples seem to require an arid-humid climate change. Following the suggestion of Pardee (1942), it is now widely acknowledged that episodes of scabland flooding probably relate to eight glacial advances and intervening ice dam ruptures that emptied Glacial Lake Missoula situated in western Montana (cf. Bretz, 1969). Baker (1971) places the early Pinedale flood at 22,000 B.P. with a duration of a week or two during which one reach (Rathdrum Prairie) transmitted 752×10^6 cu ft/sec of discharge. Rather curiously, an anastomosing channel system has been discovered in Antarctica (Fig. 8.68), with Smith (1965) maintaining that pro-glacial flooding is the best probable explanation, and Cotton (1966) questioning

his stand and perhaps leaning somewhat toward a subglacial origin. Selby and Wilson (1971) postulate salt weathering as an alternative genesis.

Glacial Fringe Benefits

A geomorphic account of continental glaciation can hardly be considered to be complete if it fails to note glacio-isostatic effects, particularly crustal rebound, and glacio-environmental effects, for example ice-margin environments and biologic refuges. Glacio-eustatic relations have already been mentioned (i.e. Chapter 7), and, insofar as they relate to coastal geomorphology, they will be discussed further in Chapter 9.

Glacio-isostatic Rebound

Our interest in crustal unloading and uplift following glacial recession is mainly concerned with consequent landforms, and particularly raised beaches. Of course, with glaciers melting, sea level is also rising. But in the actual areas occupied by continental ice sheets, the crustal rebound exceeds the sea-level rises. Thus, land water dispositions are changed, and, with this altered geography, some sea-floor areas are raised above sea level. The outstanding example of this relation in North America is the Hudson Bay region where a former ice-covered area now exhibits a series of raised beaches, each presumably representing a stillstand of land relative to sea level (Fig. 8.69). In view of the setting and circumstances, it would appear that each beach would relate to an interval when land and sea-level movements coincided. Similar relations are also recorded from the Scandanavian Peninsula (for

Fig. 8.67. Mega-ripples on the inside of a bend of the Columbia River near Trinidad, Washington. The ripples relate to the "Spokane Flood" described by Bretz and others and are found in several Scabland areas. (John S. Shelton.)

Fig. 8.68. Oblique, eastward looking aerial photo of the scabland terrain of Wright Dry Valley, Antarctica. Similar labyrinthine features occur in Washington State and are related to flood runoff (cf. Figs. 8.66 and 8.67). (U. S. Geological Survey.)

rates see Chapter 2, Fig. 2.33). One result, of course, is a relative increase in land elevation.

Glacio-environmental Margins

The edge of continental ice sheets during the Pleistocene glacial maxima must have been fascinating places. This is the *periglacial realm* of some authors. A vast array of boreal animals were forced to migrate southward, often entering into direct competition with more temperate species which had no place to go. Little by little, plants were also displaced into various refuges, some of which were in the same areas occu-

Fig. 8.69. A series of elevated shoreline deposits (storm beaches?) near Ft. Severn, Canada, related to glacio-eustatic sea level adjustments and associated glacio-isostatic rebound along Hudson Bay near lat. 55° 50′, long. 87° 00′, Canada. (National Air Photo Library, Ottawa.)

pied by the animals. The present edge of an ice sheet in Greenland or the Antarctic is a cold, barren, generally inhospitable place such as is generally visualized by many who use the term periglacial. But the distal edge of an ice sheet in southern Illinois or Missouri in the summer during the Pleistocene was probably a much different place (cf. Black, 1964; Grüger, 1972). There is evidence that plants grew right up to the edge of the ice (Braun, 1951). In other words, sometimes the ice margin environments were humid. They may have been arid at times also, to judge from ventifacts, loess deposits, and related features (cf. Smith, 1962). Thus, outwash deposits were probably extended into different kinds of environments, some tending to perpetuate their transport and some tending not to. The winters may have been cold, possibly augmented by the chill katabatic winds presently known to pour down from ice sheets. And some freeze-thaw activity can safely be assumed at those times.

The point of much of the foregoing description is the great abruptness of the contact between the margin of a low-latitude continental glacier and other essentially distinct types of geomorphic system.

The gradual environmental transitions, found elsewhere, between conditions such as aridity and humidity are missing. A glacier can produce an environmental and geomorphic change merely by advancing across an expanse of ground. A shift in the other direction might just require melting, but it might also require re-vegetation. At higher latitudes, the same glaciers—more receded—alternate with cryergic conditions and the geomorphic system related thereto. It is this last relationship to which we now turn our attention.

CRYERGIC GEOMORPHIC EFFECTS

The *cryergic geomorphic system* is dependent upon the presence of varying amounts of ice in the ground (cf. Guillier, 1949). Conditions range from a few inches of winter frost penetration (here the effects are mainly subtle pedologic changes and soil creep in places like central Europe or the United States) to perennially frozen ground. Probably half of the earth's land surfaces are affected by cryergic conditions. The ice, in turn, is evidently dependent upon certain atmospheric conditions generally related to sub-freezing temperatures [*Kryos*(Gr.) = cold]. The same, or only slightly intensified, conditions above ground level apparently favor development of glaciers, so the two conditions are commonly associated with what Dobrolowski calls the *Cryosphere* (1923). The expression *cryergic activity*, advocated by Baulig (1957), is preferred here to "periglacial activity," since conditions adjacent to ice sheets appear to vary so markedly, and not all seem to involve appreciable ice in the ground.

The cryergic geomorphic system typifies deglaciated high-latitude regions and the margins of high-latitude and certain mountain glaciers; it is characterized by low mean annual temperatures, surficial thawing in the summer, and, over barren ground, is often accompanied by considerable strong wind erosion and deposition. Where permafrost has melted out, the resulting pitted, hummocky terrain is termed *thermokarst*. The surface may otherwise be either wet or dry. Groundcover thereon, in high latitudes, if there is any, is typically tundra vegetation consisting of dwarfed trees and shrubs plus mosses and lichens, but there are also local developments of northern spruce forest (taiga). As noted by Thornbury (1969), there is little evidence of extensive tundra in the United States during the Pleistocene, though, as we shall see there is evidence of cryergic activity. Since the southernmost glacial margins in North America were not everywhere characterized by intensive development of ice in the ground, most of the cryergic features on record are in more northern sites (cf. Frye and Willman, 1958; Black, 1965), relate to colder, mountainous areas, or can be assumed to have been frozen while buried beneath glacier ice.

The last point in the previous paragraph has to do with the problem of GENESIS of ice in the ground versus PERPETUATION of permafrost (perennially frozen ground) once it forms. To some extent, cryergic activity is dependent upon the development and presence of perennially frozen ground. But, to freeze ground deeply, the surface temperature must remain very low for a long period, and it is much easier to maintain requisite low temperatures under a polar glacial ice mass than under a temperate glacier or in contact with the atmosphere at all but very high latitudes. Summer thaw depths at present over permafrost range on the average from 3.5–10 ft (1–3 m) under the atmosphere to zero under permanent ice or snow cover. Winter freeze depths would thus have to exceed this (*active-layer*) depth for existing permafrost to spread or for some to develop where there was none (under present thaw conditions). In effect, the active-layer depth on permafrost defines the minimum depth at which temperatures are always below freezing. Once formed to a depth in excess of the summer thaw depth, permafrost is insulated from heat penetration by the surface layer of rock, soil, vegetation, and water. It may remain stable (relict) below the summer thaw depth long after the actual formative conditions have disappeared as long as thawing depths do not increase. In fact, it is the summer thaw depth rather than the 32° F (0° C) isotherm that directly determines the present near-surface extent of perennially frozen ground. If one assumes a top ice temperature of −10° C and a downward tem-

perature increase (*geothermal gradient*) of 1° C/100 ft (30 m), the permafrost layer should be 1,000 ft (300 m) thick (Fig. 8.70). It should be emphasized that not all cryergic geomorphic effects depend upon the existence of permafrost. Frost action including ice wedging and heaving can relate to shallow-depth, freeze and thaw phenomena and various types of weathering and mass wasting. But these effects are probably accentuated where there is a perennial ice substrate. We are concerned with all of these processes and phenomena because a variety of landforms and deposits result from them and because the cryergic geomorphic system reflects a common environmental condition that alternates with glaciation and, hence, through relicts predisposes many of the effects achieved by glaciers.

Permafrost

Perhaps because they have more of it than anyone else, Russian scientists have led in the study of perennially frozen ground (cf. Muller, 1947). *Permafrost* is not an agency or a process. It is a condition. And that condition may activate certain agencies and processes in cryergic realms. Not counting discontinuous patches, permafrost underlies about 15% of the Earth's land surfaces (cf. Black, 1954), in places to depths of more than 3,000 ft (1,000 m) and with thicknesses over wide areas averaging a few hundred feet (Fig. 8.11). There is reason to believe that a permafrost layer acts as a seal over some oil reservoirs recently found on the Alaskan north continental slope. Flint (1971) lists ice temperatures within permafrost as ". . . between −5° and −12° centigrade." According to Purington (1912), freezing and thawing have been observed to be effective to depths of over 100 ft (30 m). Muller (1947) notes that freezing and thawing in the active layer may develop stresses in excess of 28,500 lb/sq in (2,000 kg/sq cm).

Internal stresses in permafrost can

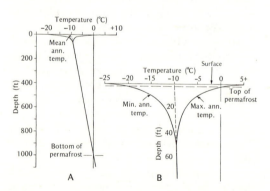

Fig. 8.70. Relationships between temperature and depth showing how upper and lower limits of permafrost are established, assuming one-dimensional steady heat flow in a homogeneous medium. (After Lachenbruch, 1968.)

cause *"swelling"* or *"frost-heaving,"* the former term being preferred by the Russians. This occurs in winter and may be distructive of man-made structures as may the equally severe settling, caving, and sloughing of ground during summer thaw. One result is *pingos* (Fig. 8.71). These dome-shaped hills, named by Eskimos, are apparently uplifted by groundwater freezing in large ice lenses. They may be 150 ft (50 m) or more high and several hundred meters in diameter. In some cases the lifting force may be partly artesian. According to Pissart (1968), relict pingos consist of ovate depressions fringed by a rampart with ice-deformed sediments (Fig. 8.72).

According to Muller (1947), permafrost is defined on the basis of temperature, irrespective of texture, lithology, induration, or water content. Ground containing no water, if frozen, is termed *"dry frozen ground"* and if perennially thus, *"dry permafrost."* Dry permafrost is immobile. According to Sumgin and Demchinski (1940), the amount of deformation in ice in the ground is directly proportional to the load and inversely proportional to the coefficient of viscosity—the latter factor being found to vary markedly with temperature (load in this case being any unevenly distributed surface load). Muller (1947) states that frozen ground (under load) with pores completely filled with ice becomes increasingly deformed as temperatures rise, ice plasticity increases, and flowage becomes more pronounced. Actual thawing further intensifies deformation even without additional stress. Lower temperatures render permafrost more stable.

According to Muller (1947), frost-heaving is only one component of movement implied by swelling (upward) with swelling alluding to volume increases and with lateral and even downward directional components. It appears to be caused by (1) hydrostatic pressure of water, (2) increased volume when water is changed to ice, (3) force of crystallization of ice, and (4) any combination of these. Note that fissures may open up due to stresses in frozen ground, and some of these may

contain water, even in the winter. In Siberia, and along the north slope of North America, water in ice-covered, northward-flowing rivers under hydrostatic pressure and hydraulic head may be forced to penetrate fissures in permafrost due to the *icing* that constricts flow in main channels. Breakthrough may occur as a river bank seep, as open flow through a floodplain opening, or, with explosive force, hurling blocks of ice and rock many meters through the air. A likely place for such an event would be a spot warmed and thawed by a pot-bellied stove in a trapper's cabin, or a PIPELINE CARRYING HEATED PETROLEUM. Reimnitz and Bruder (1972) also note that rivers flowing north into the Beaufort Sea off Alaska in spring may inundate sea ice with fresh water up to 14 mi (10 km) offshore. Drain holes through the ice (*strudel*) overlie sea bottom scour depressions (Fig. 8.73).

In general the bottom topography of a permafrost body mirrors the surface land topography, rising under hills and descending under low land (Fig. 8.74). From the figure it is also apparent that surface insulation (e.g. a deep lake that never freezes) inhibits subjacent freezing. The surface regolith structures and landforms are created through phase changes in ice in the ground relative to the *active layer* in permafrost and through comparatively shallow, more isolated effects of wetting, drying, ice, and frost elsewhere. The results appear to vary with several factors including (1) degree of water saturation, (2) relief and slope, (3) regolith texture, (4) freeze-thaw frequency, (5) freeze-thaw depth, (6) temperature maxima and minima, (7) type of groundcover, if any, and (8) depth to and nature of bedrock lithology and topography.

Cryergic Process Associations

In the cryergic setting, a variety of geomorphic processes have been identified that function together in associations in some circumstances and more or less inde-

Fig. 8.71.   A pingo on the rather featureless surface near Tuktoyaktuk, Northwest Territories from the top of a similar pingo. These are similar in size and shape to many others found on the floor of the Beaufort Sea (background) which average about 1,300 ft (400 m) in diameter and some 100 ft (30 m) in height. [Scientific Photo Unit, Atlantic Oceanographic Laboratory, Bedford Institute of Oceanography. (Roger Bélanger.)]

pendently in others. These include WEATH-ERING in several forms; *frost action* resulting in MASS WASTING of several kinds, plus *patterned ground* via such effects as *frost cracking, frost heaving, frost thrusting,* and *frost sorting;* FLUVIAL EROSION in the form of intermittent streams carrying permafrost- or snow- or glacial-meltwater during warm months; EOLIAN ACTIVITY in areas of barren ground; and, in very high altitudes and latitudes, GLACIATION. At a glance, we would appear to be dealing with a potentially complex set of geomorphic consequences.

Space limitations here forbid that we more than briefly summarize the major aspects of those geomorphic phenomena not already considered previously. Also, it seems important to emphasize the apparently ambivalent character of cryergic environments. This ambivalence is tentatively expressed by the summer-thaw–winter-freeze situation, but can be more decisively contrasted in terms of the long-term, extremely cold, dry boreal realms such as the Antarctic and extreme northern Greenland which are technically deserts and in which water in liquid form plays a meager role, and the long-term, dominantly cold regions that nevertheless have pronounced summer melting and some vegetation.

Cryergic Weathering

In the relatively moist cryergic environment studied by Washburn (1967, 1969) in northeast Greenland, there is some evidence of periodic recent climatic fluctuation apart from that expressed glacially. Washburn lists *oxidation,* expressed as weathering rinds on arkosic sandstone; minor carbonate solutioning and deposition with *leaching* being limited to very shallow depths in immature soils; possible *salt (crystal-growth) wedging* and secondary aragonite growth causing *spalling* in a granite; *desert varnish* on basalt and quartzite clasts, apparently due to local moisture concentrations (cf. Blackwelder,

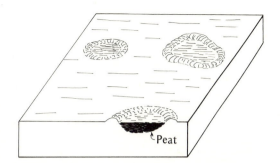

Fig. 8.72. Scars produced by pingos (relict) indicating swampy interiors and peat development. (After Pissart, 1965.)

1954) and approaching ferric chamosite in composition according to X-ray deffractometer studies; *case hardening,* particularly on basalt knobs; *granular disintegration,* apparently more effective on coarse-grained trap rocks (basalt, diabase, or dolerite) than on granite in most cases, but notably effective on arkosic sandstone and conglomerate, possibly due to combined thermal-expansion and chemical effects; *exfoliation,* on both trap rock and granite, apparently in association with case hardening and *grus* formation—the exfoliation was clearly destroying glacial polish; *cavernous weathering,* involving pitting and honeycomb surfaces of particularly exposed faces, possibly as combined salt weathering, hydration, and eolian deflation; *frost cracking* (defined by Washburn as fracturing due to low-temperature contraction), particularly resulting in split stones; and *frost wedging*—previously discussed. Washburn's studies rank as classics in the matters of mass wasting, weathering, and the formation of patterned ground, both qualitatively and quantitatively.

In an intensely cold and dry cryergic realm, such as that bordering the Antarctic Ice Sheet, moisture effects in weathering have been found to diminish, whereas evaporite effects become more important in several ways. Included are mirabilite ($Na_2SO_4 \cdot 10\ H_2O$) deposition and *salt wedging* which cause granular disintegration through crystal-growth pressures (see

Black and Bowser, 1967). As noted by Wellman and Wilson (1968), rocks such as granite encountered in Wright Dry Valley, South Victoria Land, Antarctica have also been *salt fretted*, a process that produces a thin layer of rock grains held together by salts. Frost cracking may also be of increased importance where tempera-

tures are extremely low. An account of R. H. Kilgour (in Washburn, 1969, p. 32) seems worth repeating in this context. "At Letty Harbour in the Canadian Western Arctic in early February 1933, at an air temperature of about −43° (−45°F) while the sun was shining on a trap block some 60 cm (2 ft) in diameter, located 4–5 m

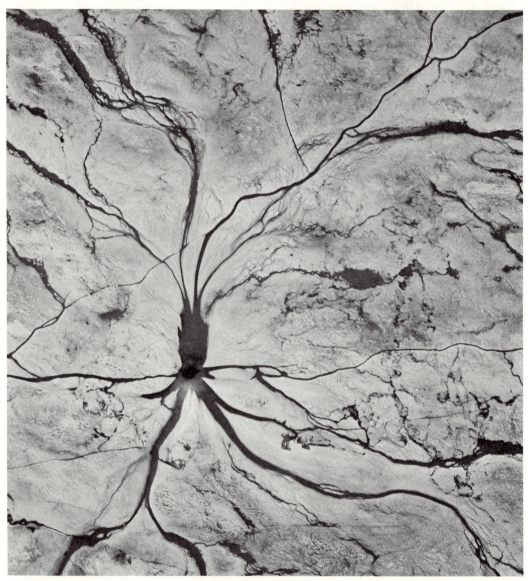

Fig. 8.73. Strudel (whirlpool) developed by flow of land runoff of Ikpikpuk River onto Arctic sea ice of Smith Bay. Major holes in this "fast" ice coincide with scour pockets on floor of Arctic Ocean. Pictured area is about 250 m wide. (Erk Reimnitz.)

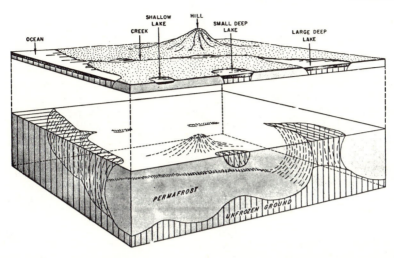

Fig. 8.74. Schematic representation of effects of surface features on permafrost distribution in a zone of continuous permafrost. (After Lachenbruch, 1968.)

(15 ft) from Mr. Kilgour, he heard a noise like a 22-caliber rifle shot from the direction of the block. He found rock dust and sharp-edged fragments scattered over the snow for a radius of about 1.5 m (5 ft) around the block and a fresh-appearing crack about 2.5 cm (1 in) wide at the top." In realms where temperatures are almost always below freezing, many permafrost phenomena and weathering effects dependent upon periodic thawing are presumably much less important geomorphically.

Cryergic Mass Wasting

Mass wasting in a cryergic environment differs from that in deserts or relatively warm humid regions mainly in the periodic freezing of water and subsequent thawing, both as a means of rock rupture and as an accessory of gravitational force in downslope movements of rock waste. It has already been noted that stresses generated by freezing water may be expressed in virtually all directions. On low-relief, nearly horizontal surfaces, the resulting shifts of earth materials generate several essentially equi-dimensional forms of what Washburn (1969) terms *patterned ground*,

mostly characterized by comparatively small horizontal movement increments, and sometimes accompanied by sorting effects. On inclined surfaces, however, horizontal stress releases produce major lateral shifts in detritus, the ground often shows linear patterns, and mass wasting as an erosive agency becomes truly effective (cf. Russell, 1965).

In his instrumental observations of mass wasting in northeast Greenland, Washburn (1967) cites several modes of movement, at least some of which we have already discussed (see Chapters 5, 6, and 7). He cites *creep*, noting the classic description of the author of that term (Davidson, 1889) as dealing with the role of frost action in which there is a heaving of particles at right angles to a slope and near-vertical collapse following thawing. Such effects appear to exist essentially throughout the cryergic realm in varying degrees. Other forms of creep (*rock creep, rock glacier creep, talus creep,* and *soil creep*) are cited by Sharp (1938), and some forms of creep, in other environments, are largely motivated by other factors (i.e. wetting and drying).

Washburn also cites cryergic *solifluction*, noting that the water-saturated mass of waste slowly flowing from higher to lower ground may derive its moisture content from rain, snow, or the thawing of frozen ground. Any temptation to restrict solifluction to areas of thawing ice in the ground should be resisted, as Washburn observes. In the southcentral United States where frost action in the ground is nil most years, the Spring-rain–Fall-drouth relationship finds relict desert waste sheets on hillslopes alternately immobilized by desiccation and flowing from water saturation. The term *gelifluction* may be applied to instances of earth flow where it can be ascertained that associated frozen ground influenced flow texture and water content.

A variety of mass-wasting phenomena are intensified by the cryergic environment, particularly in areas of high relief.

Ice wedging figures prominently in such activity, as has already been noted with respect to valley glaciers. *Protalus ramparts* are one example of a resulting landform (Fig. 8.27). *Block fields* (Fig. 8.75) appear to be the result of rock creep during times of intensified freeze-thaw activity (cf. Potter and Moss, 1968) though some are on very low-inclination surfaces. Active examples occur in some permafrost areas, but most fields in warm or dry temperate areas are relict as noted by Flint (1971). Another mass-wasting landform, the existence, or at least development of which appears to depend upon the cryergic environment, are *rock glaciers* (Fig. 8.76). They are lobate accumulations of extremely angular boulder rubble originating

Fig. 8.75. Blue Rocks block field, Berks County, Pennsylvania. Though often developed over very low slopes, some lateral transport is indicated as in the case of rock glaciers. (Noel Potter, Jr.)

Fig. 8.76. Galena Creek rock glacier, an ice-cored example from the northern Absaroka Mountains, Wyoming. (Noel Potter, Jr.)

against the free faces of some escarpments and cliffs. Active examples are never far from the snow line, and there is more than a suspicion that ice in the ground is involved in movement. Most of the fragments are joint-bounded clasts related to bedrock exposures directly upslope, with maximum dimensions in excess of 35 ft (10 m). It has been argued that they evolve from cirque glaciers (cf. Kesseli (1941), but Sharp (1938) insists that they reflect rock creep. Large examples are more than 1 mi (2–3 km) wide, some 0.75 mi (1,000 m) long, and up to 160 ft (50 m) thick. Pillewitzer (1957) cites a Tyrolian Alp example which was studied photogrammetrically for 17 years (1938–1955). It showed yearly motions on the order of 5 ft (1.5 m) in medial areas and 10–12 ft (3–4 m) distally with notable marginal retardation.

Patterned Ground

We will devote some words to the subject of patterned ground under the over-all heading of mass wasting, even though some examples on level ground involve little actual net transport. In his definitive work on the subject (1969), Washburn notes that patterned ground includes many dissimilar forms obviously of polygenetic origin. As basic processes he lists (1) desiccation cracking, (2) frost cracking, (3) dilation cracking, (4) sedimentation, (5) mass displacement, sometimes in combination with gelifluction, (6) rillwork, and (7) differential mass wasting including frost wedging and frost heaving.

Forms on level ground are comparatively regular, circular, or polygonal patterns on the regolith surface. According to Jahn (1968), "Fully developed polygons are characterized by well-sorted stone borders and a central area which is free from coarser material." These are termed *stone nets*, or, if isolated, *stone circles*. In regolith of a homogeneous nature, frost-crack polygons develop which are, of course, non-sorted (Fig. 8.77). This seems particularly true of fine-grained regolith. There

are overtones here of a developmental stage. Jahn uses the expressions "developmental" and "declining" with respect to sorted polygons (Fig. 8.78). The initial developmental unit seems to be the crack or crack pattern. A crack is a potential water receptacle. When the water freezes and expands so does the crack (Fig. 8.78). If the crack is part of a polygonal system, its downward thickening puts material within the polygon under effectively radial compression which necessarily forces the crack-enclosed material upward—larger clasts being moved most effectively. Circulation of material within the polygon (three dimensionally it is a "polygonal cell") is essentially analogous to convectional circulation of water in a heated cauldron (Fig. 8.79). The precise relationships between this probable set of formative relationships, the morphologies illustrated by Jahn (cf. Figs. 8.78 and 8.79), and the actual cryergic environmental fluctuations remains uncertain at this writing. Jahn points out that frost-crack filling may be sand (common in the Antarctic which is dry) or ice (common to Alaska and Siberia).

Variations on the polygonally patterned ground include *turf hummocks* (Russian *bugor;* Icelandic *thufur*) which are perhaps developed best in fine-grained materials (clay; loam) especially where water-saturated and boggy ground exists; in Alaska they are called *tussocks* or *tussock rings* and consist of small mounds of turf surmounted by a cap of mosses and possibly an admixture of vascular plants typical of the tundra. Large tussocks 45 ft (15 m) high may have ice cores. According to Jahn (1968), if the tussocks are pierced in summer, a mixture of clayey liquid erupts; it is thus apparently under pressure.

On slopes in excess of about 2°, the circular or polygonal patterns are modified into elongate patterns consisting of *stripes*. Some of these are merely fractures and others are sorted concentrates of coarse rubble separated by linear areas of finer material (Fig. 8.80). Variously termed *soil*

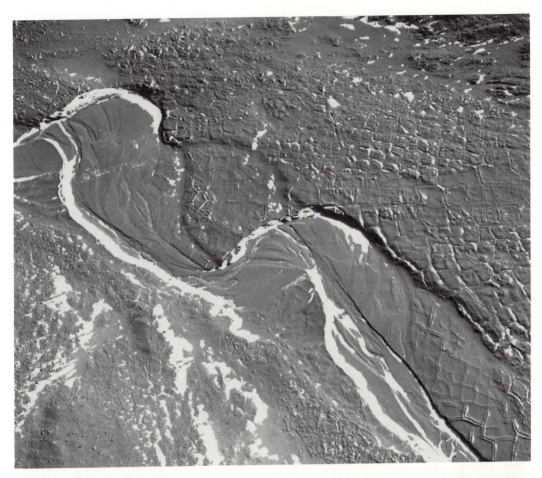

Fig. 8.77. Nonsorted frost polygons in the eastern Wright Valley and along and in the Onyx River. Surficial expression ranges from newly formed narrow troughs in the active floodplain, through double raised rims adjacent to troughs in the older floodplain, to older, wide troughs surrounding high centers in more elevated areas. Approximate scale (center) is 16 m/cm. (R. F. Black.)

stripes, stone stripes, or *striped soil,* a combination of processes is probably involved including frost sorting and solifluction. Vegetation commonly takes root in the finer-grained material, and downslope movements of the mass commonly produce an undulatory surface.

Eolian and Fluvial Activity

The effects of wind and water in cryergic realms tends to reflect the two major environmental variations previously mentioned. In vegetated areas, eolian effects are nil, whereas in cold, dry polar areas, wind erosion can be severe, and sandblasted surfaces and ventifacts are common in northern Greenland and the Antarctic. In general, the same type of relation holds for running water. Glacier meltwater-outwash has already been discussed. In tundra areas, and other cryergic settings where runoff can come from ground ice thaw, snow meltwater, or rain, drainage systems approaching the configura-

Fig. 8.78. Ice wedges developed in lake silts truncating glacially deformed Pleistocene sediments more than 40,000 years old on Garry Island, Northwest Territories, Canada. At least some of the permafrost is relict. (J. M. Mackay.)

tions of temperate humid rivers are common. Solifluction, as well as creep, is a common mechanism to feed frost-shattered detritus into watercourses.

Cotton has discussed the effects of solifluction and fluvial activity upon landscape (Cotton 1942; Cotton and Te Punga, 1955), with the general conclusion that a subdued dissected terrain results (Fig.

8.81). A number of writers including Bowman (1916), McLaughlin (1924), and Dresch (1958) have argued that accelerated movement of waste on slopes under cryergic conditions is sufficient to account for alluviation of valleys (see Garner, 1959b). Studies of solifluction (Washburn, 1967) demonstrate that frost creep and creep due to wetting and drying have an

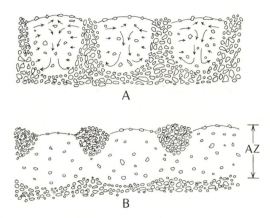

Fig. 8.79. Sequential cross sections of sorted stone polygons (A) showing flow and material transfer directions caused by ice wedging in which stones within the active zone of the permafrost (AZ) have not yet been brought to the surface and (B) illustrating "floating" stone borders where most rocks within the active zone have surfaced (cf. Fig. 8.80).

Fig. 8.81. Trough-shaped upper valley of the River Lech, Edge Island southeast of Spitsbergen taken by Cotton to exemplify cryergic-fluvial valley development. The valley is incised into a plateau of rather resistant rocks. Assumed developmental stages are indicated (inset). (After Cotton, 1964.)

associated retrograde effect that reduces net movement. Also, in solifluction, moisture can be more of an influence than slope or vegetation, and snow melting is a main source of moisture. Dry deposits move slowly or not at all.

Washburn (1967) concludes that the observed axial rates of gelifluction lobes ranged from at least 1.5 inches (3.1 cm)/ year on a gradient of 3°, to at least 6 inches (12.4 cm)/year on a gradient of 12°. Absolute values of mass wasting due to frost creep and gelifluction on a gradient of 10°–14° ranged from a mean of

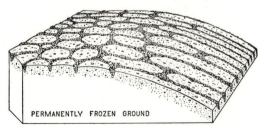

Fig. 8.80. Diagram illustrating development of stone-stripes on a hillslope in relation to stone-rings on an upland flat surface (C. F. Stewart Sharp, 1938).

about 0.5 inch (0.9 cm)/year in sectors subject to summer desiccation, to a mean of about 1.5 inches (3.7 cm)/year where saturation persists. These rates are of the same order of magnitude as those indicated for the southern Ozarks (1.3–1.8 cm/year) where wetting and drying probably supersedes frost action (cf. Chapter 1). Washburn adds that if these rates are maintained for 1,000 years, they would result in movements of about 20–120 ft (9–37 m). To these figures we should add the comment that, on an annual basis, the mass-wasting processes tend to cease with the winter "freeze-up" even though streams continue to flow throughout the winter in many areas. Conversely, the peak downslope waste movement interval in summer tends to coincide with times of peak stream discharge. None of the foregoing relationships lend credance to the notion that periglacial (i.e. cryergic) weathering and valley-side mass wasting could cause overloading of a perennial stream and thus cause valley alluviation. Valley fill in cryergic realms is therefore more probably related to glacial outwash situations or is relict from some other type of prior geomorphic system.

SUMMARY

Glacial and cryergic geomorphic systems compliment and supplement one another in an intimate association of environments, commonly juxtaposed in space, but often widely separated in time. Strictly defined in terms of present glacier ice locations, the glacial realm is readily delineated. But, defined according to peripheral glacial influences and areas where glaciers have recently been and could presumably be again, the glacial environment is a vastly more extensive and vaguely circumscribed regime. So it is with the cryergic system. Areas related to perennially frozen ground can be denoted fairly precisely, but areas subject to geomorphically significant frost action are, again, more vast and vaguely delimited.

As in the case of arid and humid environments, the distal portions of glacial and cryergic geomorphic systems interact most significantly with relicts from other systems. And, of course, a glacier maintains a constant environmental interface with the cryergic realm at its base, just as the sea shares a similar interface with coastal land. Thus, the cryergic realm is always more extensive areally than the glacial. This means that there are areas with cryergic landforms and deposits that have never been glaciated, but it is doubtful if there have ever been extensive regions glaciated that were not first affected by the cryergic geomorphic system.

Glacial and cryergic landforms and deposits tend to reflect the just-mentioned environmental arrangement, and so do the over-all related landscapes. In a glacial advance across cryergic terrain, erosion is undoubtedly modified by the tendency for the ground to be somewhat (if not completely) frozen, and, certainly, glacial sediment loads must be materially affected by the cryergic weathering and mass-wasting effects that to some degree antedate an ice advance. This set of relationships is no doubt instrumental in the preservation of the numerous unconsolidated soil profiles and similarly fragile structures that have been overridden by continental ice sheets, and their existence makes possible the detailed studies of Quaternary glacial-interglacial stratigraphy now in progress.

With glacial and cryergic geomorphic systems, we have completed our descriptions of widespread surficial mechanisms capable of generating landforms and landscapes—with one major exception. Coasts. Like landforms generated by ice, coastal landforms are in many ways distinctive and even unique. Even so, a good number of the resulting landforms in both instances rather closely resemble those created under other conditions. Careful study is often required to identify environments and agencies of origin precisely. This type of problem becomes more complex where glacial and cryergic conditions have alternated with others. Similar complications exist where coastal conditions have been laterally displaced into or out of other environments.

We will consider coastal geomorphology in Chapter 9 as the first of two main groups of polygenetic landform assemblages, one where the interface is everything. Then we will consider orogenic geomorphology, the polygenetic landscapes of probable greatest complexity, mountains, all in Chapter 10. Chapters 11 and 12 will, respectively, treat geomorphology's major areas of application, paleogeomorphology and environmental geomorphology. Therein we will be using the present to understand the past and our knowledge of the past to comprehend the present.

REFERENCES

Bader, H. and Haefeli, R. (1939) "Der Schnee und Seine Metamorphose, *Beitr. Geol. Schweiz, Geotech. Serie Hydrologie*, pt. 3, Bern, 340 pp. (Translated into English by U. S. Army Engr. Corps Snow, Ice and Permafrost Res. Estab., Res. Dept., Rept. Trans. 14, 1954.)

Baker, V. R. (1971) "Paleohydrology of Catastrophic Pleistocene Flooding in Eastern Washington," *Geol. Soc. Amer.* (Abstract) pp. 479.

Baulig, Henri (1957) "Pediplains or Peniplains, *Geol. Soc. Amer. Bull.*, v. 68, pp. 913–930.

Bayrock, L. A. (1967) "Catastrophic Advance of the Steele Glacier, Yukon, Canada," *Univ. Alberta Boreal Inst., Occasional Pub.* 3, 35 pp.

Bishop, B. C. (1957) "Shear Moraines in the Thule Area, Northwest Greenland," *U. S. Army Engr. Corps Snow, Ice and Permafrost Res. Estab., Res. Dept. Rept.* 17, 46 pp.

Black, R. F. (1954) "Permafrost—A Review," *Geol. Soc. Amer. Bull.*, v. 65, pp. 839–856.

Black, R. F. (1964) "Periglacial Studies in the United States 1959–1963," *Biul. Peryglacjalny*, no. 14, pp. 5–29.

Black, R. F. (1965) "Ice-Wedge Casts of Wisconsin," *Wis. Acad. Sci., Arts and Letters*, v. 54, pp. 187–222.

Black, R. F. (1969) "Glacial Geology of Northern Kettle Moraine State Forest, Wisconsin," *Wis. Acad. Sci., Arts and Letters*, v. 57, pp. 99–119.

Black, R. F. and Bowser, C. J. (1967) *Salts and Associated Phenomena of the Termini of the Hobbs and Taylor Glaciers, Victoria Land, Antarctica*, Snow and Ice Comm., Bern, pp. 226–238.

Blackwelder, Eliot (1954) "Geomorphic Processes in the Desert, *Calif. Div. Mines Bull.*, v. 170, pp. 11–20.

Bowman, Isaiah (1916) *The Andes of Southern Peru*, Henry Holt, New York, 336 pp.

Braun, L. E. (1951) "Plant Distribution in Relation to the Glacial Boundary," *Ohio Jour. Sci.*, v. 51, pp. 139–146.

Bretz, J. H. (1923) "The Channeled Scablands of the Columbia Plateau," *Jour. Geol.*, v. 31, pp. 617–649.

Bretz, J. H. (1969) "The Lake Missoula Floods and the Channeled Scabland," *Jour. Geol.*, v. 77, pp. 505–543.

Bretz, J. H., Smith, H. T. U. and Neff, G. E. (1956) "Channeled Scabland of Washington: New Data and Interpretations," *Geol. Soc. Amer. Bull.*, v. 67, pp. 957–1050.

Carol, H. (1947) "The Formation of Rochés Moutonnées," *Glac. Jour.*, v. 1, pp. 58–59.

Chamberlin, T. C. (1894) "Proposed Genetic Classification of Pleistocene Glacial Formations," *Jour. Geol.*, v. 2, pp. 517–538.

Chamberlin, T. C. and Chamberlin, R. T. (1911) "Certain Phases of Glacial Erosion," *Jour. Geol.*, v. 19, p. 193.

Chamberlin, T. C. and Salisbury, R. D. (1927) *The Work of Glaciers*, v. 1, Univ. of Chicago Press, Chicago, 281 pp.

Charlesworth, J. K. (1957) *The Quaternary Era, With Special Reference to its Glaciation*, Edward Arnold, London, 2 vols., 1700 pp.

Clark, T. H. and Elson, J. A. (1961) "Ventifacts and Eolian Sand at Charette, P.Q.," *Roy. Soc. Canada Trans.*, v. 55, pp. 1–11.

Cotton, C. A. (1942) *Climatic Accidents in Landscape Making*, Whitcomb and Tombs, Christ Church, N. Z., 354 pp. (Reprint, John Wiley and Sons, New York).

Cotton, C. A. (1966) "Antarctic Scablands," *New Zealand Jour. Geol. and Geophys.*, v. 9, pp. 130–132.

Cotton, C. A. (1968) "Mountain Glacier Landscapes" [in] *Encyclopedia of Geomorphology*, R. W. Fairbridge, ed., Reinhold Book Corp., New York, esp. pp. 739–745.

Cotton, C. A. and Te Punga, M. T. (1955) "Solifluction and Periglacially Modified Landforms at Wellington, New Zealand," *Roy. Soc. N. Z. Trans.*, v. 82, pp. 1001–1031.

Crary, A. B. (1966) "Mechanism for Fiord Formation . . .," *Geol. Soc. Amer. Bull.*, v. 77, pp. 911–930.

Crosby, W. (1928) "Certain Aspects of Glacial Erosion," *Geol. Soc. Amer. Bull.*, v. 39, pp. 1171–1181.

Damuth, J. E. and Fairbridge, R. W. (1970) "Equatorial Atlantic Deep-Sea Arkosic Sands and Ice Age Aridity in Tropical South America," *Geol. Soc. Amer. Bull.*, v. 81, pp. 189–206.

Davidson, Charles (1889) "On the Creeping of the Soilcap through the Action of Frost," *Geol. Mag.*, v. 6, pp. 255–261.

Davis, W. M. and Snyder, W.H. (1898) *Physi-

cal Geography, Ginn and Co., Boston, 428 p.

De Geer, Gerhard (1897) "Om rullstensåsarnas bildningssätt," *Geolog Fören. i Stockholm Förh.,* v. 19, pp. 366–388.

Demorest, Max (1938) "Ice Flow as Revealed by Glacier Striations," *Jour. Geol.,* v. 46, pp. 700–725.

Demorest, Max (1939) "Glacial Movement and Erosion," *Amer. Jour. Sci.,* v. 237, pp. 594–605.

Derbyshire, E. (1968) "Cirque Glaciers" [in] *Encyclopedia of Geomorphology,* R. W. Fairbridge, ed., Reinhold Book Corp., New York, esp. pp. 123–126.

Dombrowolski, A. B. (1923) *Historja Naturaina Loda,* Warsaw, 940 pp.

Dresch, Jean (1958) "Problems Morphologiques des Andes Centrales," *Geogr. Ann.,* v. 67, 130–151.

Ewing, Maurice and Donn, W. L. (1956) "A Theory of Ice Ages," *Science,* v. 123, pp. 1061–1066.

Finsterwalder, R. (1950) "Some Comments on Glacier Flow," *Glac. Jour.,* v. 1, pp. 383–388.

Fisk, H. N. (1944) *Geological Investigation of the Alluvial Valley of the Lower Mississippi River,* U. S. Army Engr. Corps (Miss. R. Comm.), 78 pp.

Flint, R. F. (1938) "Origin of the Cheney-Palouse Scabland Tract, Washington," *Geol. Soc. Amer. Bull.,* v. 49, pp. 461–524.

Flint, R. F. (1949) "Pleistocene Drainage Diversions in South Dakota," *Geograf. Annaler,* v. 31, pp. 56–74.

Flint, R. F. (1957) *Glacial and Pleistocene Geology,* John Wiley and Sons, New York, 553 pp.

Flint, R. F. (1971) *Glacial and Quaternary Geology,* John Wiley and Sons, New York, esp. pp. 1–243.

Friedman, G. M. and Saunders, J. E. "Coincidence of High Sea Level with Cold Climate and Low Sea Level with Warm Climate: Evidence from Carbonate Rocks," *Geol. Soc. Amer. Bull.,* v. 81, pp. 2457–2458.

Frye, J. C. and Willman, H. B. (1958) "Permafrost Features near The Wisconsin Glacial Margin in Illinois," *Amer. Jour. Sci.,* v. 256, pp. 518–524.

Garner, H. F. (1958) "Climatic Significance of Anastomosing Channel Patterns Typified by Rio Caroni, Venezuela, *Geol. Soc.*

Amer. Bull. (Abstract) v. 69, pp. 1568–1569.

Garner, H. F. (1959a) "Recent Climate Change Patterns—A Mode of Control," *Geol. Soc. Amer. Bull.* (Abstract) v. 70, p. 1607.

Garner, H. F. (1959b) "Stratigraphic Sedimentary Significance of Contemporary Climate and Relief in Four Regions of the Andes Mountains," *Geol. Soc. Amer. Bull.,* v. 70, pp. 1327–1368.

Garner, H. F. (1966) "Dérangement of the Rio Caroni Venequela," *Révue de Géomorphologie Dynamique,* no. 2, pp. 50–83.

Garner, H. F. (1967) "Rivers in the Making," *Scientific American,* v. 216, pp. 84–94.

German, R. (1968) "Moraines" [in] *Encyclopedia of Geomorphology,* R. W. Fairbridge, ed., Reinhold Book Corp., New York, esp. pp. 710–717.

Gillberg, Gunnar (1955) "Den glaciala utvecklingen inom Sydsvenska Höglandets västra randzon. I. Glacialerosion och moränackumulation," *Geol. Fören. i. Stockholm Förh.,* v. 77, pp. 481–524.

Glen, J. W. (1955) "The Creep of Polycrystalline Ice," *Roy. Soc. London Proc.,* v. 288, pp. 519–538.

Glen, J. W. and others (1957) "On the Mechanism by which Stones in Till Become Oriented," *Amer. Jour. Sci.,* v. 255, pp. 194–205.

Gravenor, C. P. (1951) "Bedrock Source of Tills in Southwestern Ontario," *Amer. Jour Sci.,* v. 249, pp. 66–71.

Grove, A. T. and Goudie, A. S. (1971) "Quaternary Lake Levels in the Rift Valley of Southern Ethiopia and Elsewhere in Tropical Africa," *Nature,* v. 234, pp. 403–405.

Grüger, Eberhard (1972) "Pollen and Seed Studies of Wisconsinan Vegetation in Illinois, U.S.A.," *Geol. Soc. Amer. Bull.,* v. 83, pp. 2715–2734.

Guillier, Y. (1949) "Gel et dégel du sol: les Mechanisms Morphologique," *L'Information Geogr.,* Paris, v. 13, pp. 104–116.

Harland, W. B. and others (1966) "The Definition and Identification of Tills and Tillites," *Earth Sci. Rev.,* v. 2, pp. 225–256.

Hartshorn, J. H. (1958) "Flowtill in Southeastern Massachusetts," *Geol. Soc. Amer. Bull.,* v. 69, pp. 477–482.

Hershey, O. H. (1897) "Eskers Indicating Stages of Glacial Recession in the Kansan Epoch in Northern Illinois," *Amer. Geologist,* v. 19, pp. 197–209, 237–253.

Hobbs, H. G. (1911) *Characteristics of Existing Glaciers*, Mich. Univ. Pub., Ann Arbor, 301 pp.

Holmes, C. D. (1941) "Till Fabric," *Geol. Soc. Amer. Bull.*, v. 51, pp. 1299–1354.

Holmes, C. D. (1960) "Evolution of Till Stone Shapes in Central New York," *Geol. Soc. Amer. Bull.*, v. 71, pp. 1645–1660.

Horberg, Leland (1945) "A Major Buried Valley in East Central Illinois and its Regional Significance," *Jour. Geol.*, v. 53, pp. 349–359.

Horberg, Leland (1950) "Bedrock Topography of Illinois," *Ill. Geol. Surv. Bull.*, v. 73, pp. 67–72.

Hough, J. L. (1958) *Geology of the Great Lakes*, Univ. of Illinois Press, Urbana, 313 pp.

Hughes, R. and Seligman, G. (1938) "The Temperature, Melt Water Movement and Density Increase in the Névé of an Alpine Glacier," Pub. 2, Jungfraujoch Research Party, *R. A. S. Geophys.* Supp. v. 4.

Hummel, D. (1874) "Om rullstenbildningar," *K. Svenska Vetenskapsakad., Bihang till Handl.*, v. 2, 35 pp.

Ives, J. D. (1960) "Former Ice-Damned Lakes and the Deglaciation of the Middle Reaches of the George River, Labrador-Ungava," *Geogr. Bull.*, no. 14, pp. 44–70.

Jahn, A. (1968) "Patterned Ground" [in] *Encyclopedia of Geomorphology*, R. W. Fairbridge, ed., Reinhold Book Corp., New York, esp. pp. 814–817.

Johnson, D. W. (1904) "The Profile of Maturity in Alpine Glacial Erosion," *Jour. Geol.*, v. 2, pp. 569–578.

Kesseli, J. E. (1941) "Rock Streams in the Sierra Nevada," *Geogr. Rev.*, v. 31, pp. 203–227.

Krinsley, D. H. and Donahue, Jack (1968) "Environmental Interpretation of Sand Grain Textures by Electron Microscopy," *Geol. Soc. Amer. Bull.*, v. 79, pp. 743–748.

Leverett, Frank and Taylor, F. B. (1915) "The Pleistocene of Indiana and Michigan and the History of the Great Lakes," *U. S. Geol. Surv. Mono.* 53, pp. 316–518.

Lewis, W. V. (1954) "Pressure Release and Glacial Erosion," *Glac. Jour.*, v. 2, pp. 417–422.

Lewis, W. V. (1960) "Norwegian Cirque Glaciers," *Roy. Geogr. Soc. Res.*, Ser. 4, 104 pp.

Lyell, Charles (1830–1833) *Principles of Geology*, J. Murray, London, v. 1, 1830, 511 pp.; v. 2, 1832, 330 pp., v. 3, 1833, 398 + 109 pp.

Mackay, J. R., Rampton, V. N. and Fyles, J. G. (1972) "Relic Pleistocene Permafrost, Western Arctic, Canada," *Science*, v. 176, pp. 1321–1323.

Mark, D. M. (1971) "Rotational Vector Procedure for the Analysis of Till Fabrics," *Geol. Soc. Amer. Bull.*, v. 82, pp. 2661–2666.

Matthes, F. (1900) "Cause of Glacial Motion," *U. S. Geol. Surv. Ann. Rept.* 21, pt. II, pp. 167–190.

McLaughlin, D. H. (1924) "Geology and Physiography of the Peruvian Cordillera, Departments of Junin and Lima," *Geol. Soc. Amer. Bull.*, v. 35, pp. 591–632.

Meier, M. F. (1960) "Distribution and Variations of Glaciers in the United States Exclusive of Alaska," *Internat. Assoc. Scientif. Hydrol. Pub.* 54, pp. 420–429.

Mousinho, M. R. De Meis (1971) "Upper Quaternary Process Changes of the Middle Amazon Area," *Geol. Soc. Amer. Bull.*, v. 82, pp. 1073–1078.

Muller, S. W. (1947) *Permafrost or Permanently Frozen Ground and Related Engineering Problems*, Edwards Bros., Ann Arbor, Mich., 231 pp.

Nye, F. and Orowan, E. (1950) "Accentuation of Glacial Erosion and Glacial Movement," *The Observatory*, v. 70, pp. 63–69.

Pardee, J. T. (1942) "Unusual Currents in Glacial Lake Missoula," *Geol. Soc. Amer. Bull.*, v. 53, pp. 1569–1600.

Paterson, W. S. B. (1969) *The Physics of Glaciers*, Pergamon Press, New York.

Penck, A. and Brückner, E. (1909) *Die Alpen im Eiszeitalter*, Leipzig, Touchnitz, 3 vols., 1199 pp.

Perutz, M. (1940) "Mechanisms of Glacier Flow," *Phys. Soc. Proc.*, v. 52, p. 132.

Perutz, M. (1948) *Glaciers—A Development of Glacier Physics*, Science News (Penguin Books), London, no. 6, pp. 105–127.

Perutz, M. (1950) "Direct Measurements of the Velocity Distribution in the Vertical Profile through a Glacier," *Glac. Jour.*, v. 1, pp. 382–383.

Perutz, M. and Seligman, G. (1939) "A Crystallographic Investigation of Glacier Structure and the Mechanism of Glacier Flow," *Roy. Soc. London Proc.*, v. 177, pp. 335–336.

Pillewitzer, W. (1957) *Untersuchnugen and*

Blockströmer der Ötztaler Alpen, Abhandl. Geogr. Inst. Fr. Univ. Berlin, pp. 37–50.

Pissart, A. (1968) "Pingos, Pleistocene" [in] *Encyclopedia of Geomorphology,* R. W. Fairbridge, ed., Reinhold Book Corp., New York, esp. pp. 847–849.

Potter, Noel and Moss, J. H. (1968) "Origin of the Blue Rocks Block Field and Adjacent Deposits, Berks County, Pennsylvania," *Geol. Soc. Amer. Bull.,* v. 79, pp. 255–262.

Prest, V. K. and others (1968) *Glacial Map of Canada,* Scale 1:5,000,000, Geol. Surv. Canada, Map 1253A.

Purington, W. (1912) "Depth of Frost," *Mining Mag.,* v. 6, p. 50.

Rahn, P. H. (1969) "The Relationship between Natural Forested Slopes and Angles of Repose for Sand and Gravel," *Geol. Soc. Amer. Bull.,* v. 80, pp. 2123–2128.

Reimnitz, Erk and Bruder, K. F. (1972) "River Discharge into an Ice-Covered Ocean and Related Sediment Dispersal, Beaufort Sea, Coast of Alaska," *Geol. Soc. Amer. Bull.,* v. 83, pp. 861–867.

Riley, C. (1935) "The Granite Porphyries of Great Bear Lake," *Jour. Geol.,* v. 43, esp. pp. 499.

Russell, I. C. (1895) "The Influence of Debris on the Flow of Glaciers," *Jour. Geol.,* v. 3, pp. 823–833.

Russell, R. J. (1965) "Mass-Movement in Contrasting Latitudes," *VIth Int. Quaternary Cong., Warsaw (1961),* v. 4, pp. 143–153.

Saucier, R. T. and Fleetwood, A. R. (1970) "Origin and Chronologic Significance of Late Quaternary Terraces, Ouachita River, Arkansas and Louisiana," *Geol. Soc. Amer. Bull.,* v. 81, pp. 869–890.

Selby, M. J. and Wilson, A. T. (1971) "Origin of the Labyrinth, Wright Valley, Antarctica," *Geol. Soc. Amer. Bull.,* v. 82, pp. 471–476.

Seligman, G. (1949) "The Growth of the Glacier Crystal," *Glac. Jour.,* v. 1, pp. 254–266.

Sharp, C. F. S. (1938) *Landslides and Related Phenomena,* Columbia Univ. Press, New York, 137 pp.

Sharp, R. P. (1948) "The Constitution of Valley Glaciers," *Glac. Jour.,* v. 1, pp. 182–189.

Sharp, R. P. (1958) "Malaspina Glacier, Alaska," *Geol. Soc. Amer. Bull.,* v. 69, pp. 617–646.

Shelton, J. S. (1966) *Geology Illustrated,* W. H. Freeman, San Francisco, esp. pp. 338–351.

Smalley, I. J. (1966) "Drumlin Formation: A Reological Model," *Science,* v. 151, p. 1379.

Smalley, I. J. and Unwin, D. J. (1968) "The Formation and Shape of Drumlins and Their Distribution and Orientation in Drumlin Fields," *Glac. Jour.,* v. 7, pp. 377–390.

Smith, H. T. U. (1962) "Periglacial Frost Features and Related Phenomena," *Biul. Peryglacjalny,* no. 11, pp. 325–342.

Smith, H. T. U. (1965) "Anomalous Erosional Topography in Victoria Land, Antarctica," *Science,* v. 148, pp. 941–942.

Stephen, T. (1930) "The Mechanics of Frost Heaving," *Jour. Geol.,* v. 38, pp. 313–317.

Streiff-Becker, R. (1938a) "Eisbewegung im Firn und Gletscher und ihre Wirkung," *Schwizer, Naturf. Ges. Vern.* 119, pp. 128–130.

Streiff-Becker, R. (1938b) "Zur Dynamik des Firneises," *Zeits. für Gletscher,* v. 26, pp. 1–21.

Sumgin, M. and Demchinski, B. (1940) *Oblast Vechnoi Merzloty, (The Permafrost Province),* Glavesvmorput 'Izd., Moscow, 238 pp.

Thornbury, W. D. (1969) *Principles of Geomorphology,* John Wiley and Sons, New York, esp. pp. 345–419.

Thornbury, W. D. and Deane, H. L. (1955) "The Geology of Miami County, Indiana," *Ind. Geol. Surv. Bull.,* 8, 49 pp.

Todd, J. E. (1914) "The Pleistocene History of the Missouri River," *Science,* v. 39, pp. 263–274.

Warren, C. R. (1952) "Probable Illinoian Age of Part of the Missouri River," *Geol. Soc. Amer. Bull.,* v. 63, pp. 1143–1156.

Washburn, A. L. (1967) "Instrumental Observations of Mass-Wasting in the Mesters Vig District, Northeast Greenland," *Meddelelser om Grønland,* v. 166, 296 pp.

Washburn, A. L. (1969) "Weathering, Frost Action and Patterned Ground in the Mesters Vig District, Northeast Greenland," *Meddelelser om Grønland,* v. 176, 303 pp.

Wayne, W. J. (1952) "Pleistocene Evolution of the Ohio and Wabash Valleys," *Jour. Geol.,* v. 6, pp. 575–585.

Wellman, H. W. and Wilson, A. T. (1968) "Salt Weathering or Fretting" [in] *Encyclopedia of Geomorphology,* R. W. Fairbridge, ed., Reinhold Book Corp., New York, esp. pp. 968–970.

White, W. A. (1972) "Deep Erosion by Continental Ice Sheets," *Geol. Soc. Amer. Bull.*, v. 83, pp. 1037–1056.

Willman, H. B., Glass, H. D. and Frye, J. C. (1963) "Mineralogy of Glacial Tills and Their Weathering Profiles in Illinois, Pt. I, Glacial Tills," *Ill. Geol. Surv. Circ.* 347, 55 pp.

Willman, H. B., Glass, H. D. and Frye, J. C. (1966) "Mineralogy of Glacial Tills and Their Weathering Profiles in Illinois, Pt. II, Weathering Profiles," *Ill. Geol. Surv. Circ.* 347, 76 pp.

Wright, H. F. and Frey, D. G. (editors) (1965) *The Quaternary of the United States,* Princeton Univ. Press, Princeton, N.J., 922 pp.

9 Polygenetic landscapes/coasts

It is by now perhaps obvious that the earth generates its landforms in environmentally very complex places, even when one makes no mention of former changes in environmental dispositions or things like bulldozers and dynamite. In the foregoing chapters, we have attempted to delineate the realms in which the major landform-making environments hold sway. The boundaries we have indicated between environment types suffer the inevitable penalties imposed on arrangements that are man-devised and hence somewhat arbitrary. Yet as long as we cling to the same essential framework, the arbitrariness should be acceptable if we can use our categories to communicate. Thus far, we have recognized (1) an ENDOGENIC REALM of tectonism mainly expressed geomorphically by bedrock lithology, structure, and crustal movements, (2) a MESOGENIC REALM characterized geomorphically by effects of gravity and volcanism, and (3) an EXOGENIC REALM where surficial landscape changes induced atmospherically relate to (a) HUMID GEOMORPHIC SYSTEMS, (b) ARID GEOMORPHIC SYSTEMS, (c) GLACIAL GEOMORPHIC SYSTEMS, and (d) CRYERGIC GEOMORPHIC SYSTEMS.

All of the foregoing geomorphic systems share interfaces to some degree, though some must undergo considerable displacement through space in order to interact appreciably over a period of time and so generate a landscape. In contrast, there is one environmental interface that is far more linear than areal in dimension and along which notable landforms and deposits of striking character develop. These are generated by the *coastal geomorphic systems* which exist where the sea embraces the land, lingeringly, and often with what might well pass for passion. We will begin by considering the geomorphic effects of this land-sea-air contact, in what must qualify, environmentally, as a polygenetic relationship. Thereafter, we will turn to another type of polygenetic landscape where several distinct geomorphic systems interact to generate landforms and landscapes along the world's mountain belts. There, endogenic, mesogenic, and exogenic systems all vie for supremacy, each with some measure of success.

COASTAL GEOMORPHIC SYSTEMS—ENVIRONMENTS IN CONTACT

Land-water interfaces are shared by oceans, lakes, and frog ponds. So are many of the same interface phenomena. But the great majority of the coastal geomorphic effects of major import relate to water bodies endowed, because of their size, with impressive powers of erosion through waves and currents. Most of our remarks hereafter will therefore concern the coasts of the oceans, seas, and larger lakes that have appreciable depth and bottom inclinations. For there is considerable reason to believe, as noted by Shaw (1964), that A VERY SHALLOW SEA—no matter how broad—may have a great deal in common with a frog pond.* And it should be emphasized here, that our interest in standing water bodies concerns the land-water-air interface and not their potential simply as a receptable for sediment. Marine plains of deposition reflecting mainly depositional effects were considered in Chapter 3. Deltas came under some scrutiny in Chapter 7, and glaciated coasts and emergent beaches were mentioned in Chapter 8. Here, shorelines away from immediate débouchér areas of major rivers are the main interest.

COASTAL ENVIRONMENTS

Our discussion of coastal geomorphic systems cannot take place in an environmental vacuum, with respect to either time or space. Apart from the manner in which they interact, there are several ingredients in a given coastal environment, and each ingredient can assume one of several possible guises. The geomorphic result will tend to vary as the components of the coastal environment differ and change through time. Let us begin by considering some of the major variants of a coastal geomorphic system.

Water Conditions

Water conditions that contribute to coastal geomorphology mainly concern salinity (fresh-brackish-saline), turbulence, and temperature. Turbulent conditions range from the tranquility that develops in sheltered or lee inlets and shores like Florida's northwest coast (cf. Tanner and Bates, 1965), to the powerful surfs that characterize certain exposed shores on the North American west coast, northern Hawaii, parts of the east Australian coast, and the British North Sea coast, particularly during times of storm. On islands, great contrasts are often present in short distances as, for example, the towering Atlantic Ocean surf of the east coast of Barbados and the tranquil swell on the Caribbean strand four miles to the southwest. Similar contrasts are common along coasts of great irregularity.

Water temperature is a second factor of great geomorphic importance along coasts. It affects both solute saturation levels and degrees of biotic activity related to this saturation. Conditions vary from cold, polar seas and currents originating therefrom ($\sim$32°F, 0°C) with related high O_2 contents, to tepid, tropical seas and related currents (75–80°F, 22–28°C), often nearly saturated or even hypersaturated with solutes, particularly $CaCO_3$. Such water conditions can clearly affect the erosive ability of the sea, particularly as a solutioning agent, since a saturated solution cannot dissolve more (cf. Fig. 9.28). It would further influence biologically related sedimentation, since lime-secreting organisms do best in abundant solutes. Another major aspect of water conditions affecting both turbulence and temperature is water level as compared to a historic mean. Is it relatively high or low? Russell (1967) notes that *most* coasts are now

* For a more complete discussion of this analogy see the Chapter 11 section, "The Tideless Sea and the Strand."

newly submerged because of glacier melting and that coastal geomorphology must respond to this situation. Some deglaciated coasts, however, are clearly uplifted. Local crustal movements to the contrary, many coastal topographic relations are the hardly modified expressions of submerged subaerial terrain of various kinds.

Organisms

The role of organisms along coasts is surprisingly similar to their role in subaerial environments. Coastal conditions may favor organisms to the extent that they are able to occupy the land-water interface and thus influence morphogenesis. The main associations involve organic reefs and mangroves, and the latter occur within the areal distribution of the former. However, along about half the world's coasts, these phenomena are not important.

Generally speaking, the cut-off line for reef-building organisms is the northern and southern 68°F (20°C) surface water isotherm. Growth extends farther poleward where warm currents such as the Gulf Stream invade cooler waters (northern Florida is the approximate limit on the North American east coast).* The converse is also true, and there are few reef-building organisms along cold current coasts (e.g. southwest Africa; southwest South America), even approaching the equator. At higher latitudes, organic influences along coasts are limited to the few plants with adequate salt-sand-eolian tolerances on beach backshores (salt cedar; poison ivy; holly, on the New Jersey coast),** plus the occasional shell of a marine creature that becomes incorporated into the beach. Sea-floor kelp beds are not confined to coasts, and their influence is limited in any case.

In tropical climes, mangroves and reef organisms will flourish where their ecologic requirements are met, which, for modern reefs, include warm, clear, shallow, well-aerated (agitated) water. Coasts are common reef sites. Organic reefs are geomorphically important in terms of their own basic morphology and through their occasional presence along the land-water interface. In essence they comprise a wave-resistant, sediment-binding coralline framework, presently dominated by Madreporian or Scleractinian colonial corals, with an association of calcareous algae and bryozoa, plus a variety of reef-dwelling phyla including molluscs, echinoderms, protozoans, arthropods, and vertebrates. It is the reef framework that determines the interface function of reef organisms, for just as a continuous cover of land plants modify effects of wind and running water, so a continuous organic presence on or along a coast modifies the effects of ocean wind, waves, and currents.

Subaerial Land Conditions

Environmental conditions on a coastal land are easily as variable as those of the adjacent water and probably more so. To begin with, any of the subaerial geomorphic systems in any stage of development may be sharing a coast with an ocean. Thus, the subaerial land condition may reflect a recently developed climate of some sort or one long established, with all the erosional and sedimentologic implications pertaining to such circumstances (cf. Chapters 5 to 8). In addition, of course, the land environment immediately adjacent to a sea is commonly strongly modified by adjacent oceanic effects. The latter effects may range from sea breezes or wind-blown salt or sand to over-all land-moisture distribution and amounts (cf. Chapter 4). In addition, latent heat retention by water in some climes (the North American Mid-Atlantic Coast) keeps coastal air temperatures higher than those in-

---

* There are ahermatypic (cold-water) corals, but they are not reef-building, coastal types.
** Vegetated coastal dunes are often somewhat immobilized by Marram grass (*Ammophilia brevilugulata*), particularly on the east and Gulf coasts of the United States.

land in early winter and lower than those inland in early summer.

Sedimentary Effects

Quite apart from an ocean's ability to acquire sediment by coastal erosion, coasts often reflect sediment influxes in relatively distant areas or as the result of some previous environmental episode. Remarks already made here about the relict nature of alluvium that many streams are reworking in their valleys also apply to detritus being moved by ocean waves. We will elaborate on this later. Sufficient to the moment is our awareness that *some* coasts are dominated by an abundance of sediment on both the landward and seaward sides of the strand. The Texas Gulf Coast is an example. Other coasts are characterized by extensive bare bedrock exposures with only scattered patches of beach sand or gravel. The Big Sur Coast of California is one example, and the seaward coast of Barbados is another (Fig. 9.1). Expanses of the glaciated Maine coast are similar. Where sediment is abundant, waves and currents complimented by sea level changes may generate a wide variety of landforms. As noted by Bajorunas and Duane in their study of Lake Superior inlet sedimentation (1967), large lakes have many depositional attributes of oceans.

Actually, the examples cited in the previous paragraph represent an interplay between two types of sediment-generating systems, the sporadic and the continuous. The subaerial geomorphic systems produce and yield sediment sporadically (cf. Chapters 7 and 8). Marine erosion processes, on the other hand, act more continuously to generate a certain, relatively constant, though often small increment of detritus

Fig. 9.1. Block caving along Barbados sea cliff due to undercutting of a Tertiary marl from beneath the Coral Rock Formation. Small, light-colored specks on cliff top (center right) are goats.

along coasts. Where the aqueous side of the strand is dominantly detrital, there is usually evidence of prior or continuing sediment influx, probably via rivers. Where the terrestrial side of the strand is detrital, there is usually evidence of aridity, glaciation, sea-floor emergence, or pyroclastic volcanism. The latter, of course, relates mainly to the subaerial environmental situation (cf. Chapters 5–8).

Tectonic and Volcanic Effects

Coasts originating through volcanism and tectonism are usually rather distinctive, and there are probably far fewer tectonic coasts than was supposed before the ad-

vent of plate tectonics concepts (cf. Chapter 3). Volcanic coasts on continents can often be related to some specific vent eruption or fissure flow terminating in the sea (Fig. 9.2), and of course there are many volcanic islands. However, it is no longer necessary or realistic to relate every high sea cliff to a "probable" fault. Red Sea coasts and the coast of the Afar Triangle (cf. Fig. 9.3), which are clearly associated with spreading oceanic ridges, may reasonably be designated tectonic (faulted), as may a part of Californian coast where transform faulting is apparently influencing shape. Post-glacial isostatic rebound usually induces more subtle changes in the coastal regions where it occurs, and lands bordering subduction zones like western

Fig. 9.2. A "lava delta" prograding and establishing a shoreline near Barcena on the island of San Benedicto, Mexico. Flow is from a side fissure which breached a sea cliff eroded into pyroclastic debris. Note the fine, nearly parallel rill structure eroded by running water in the cone slope. (Paul Popper Ltd.)

Fig. 9.3.   Block faulting along the eastern coast of Lake Guilietti, in the Afar Triangle at the head of the Gulf of Aden. The stair-step landscape is characteristic of topography developed in the vicinity of mid-ocean ridge spreading and fracturing. (Haroun Tazieff.)

South America are probably subject to sporadic vertical movements apart from initial orogenic uplift (see Fig. 9.8).* Coastal folding is rare, though warping is not uncommon. Key Minor Island in the Moluccas on the western Pacific rim (Fig. 9.4) is apparently due to recent folding (see Cotton, 1961). Salt domes may also affect coastal form as in the Persian Gulf (Twidale, 1971, plate 50), and of course coasts that have been uplifted like the Oregon coast often exhibit raised marine terraces (Fig. 9.5).

Early classifications (cf. von Richthofen, 1886; Gulliver 1899; Johnson, 1919) tended to reflect supposed morphologic responses to crustal movements or water level changes, and coasts were designated

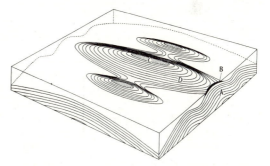

COASTAL DEVELOPMENT AND CLASSIFICATIONS

In theory, a classification of coastal landforms should be essentially compatible with modes of coastal morphogenesis.

Fig. 9.4.   An island of anticlinal upheaval exemplified by Key Minor in the Moluccas. Coastal configurations of a similar nature and origin occur along the Dalmatian Coast of Yugoslavia. (A)–(D) Folded sediments and capping reef deposits. (After Cotton, 1961.)

* See the Chapter 10 discussion of cordilleran-type orogeny.

Fig. 9.5. View to southeast of a set of elevated marine terraces in the vicinity of Slate Springs, California, some 55 mi (88 km) south of Monterey. These terraces are mainly due to relative land uplift of what must rank as an erosional coast of disequilibrium. (B. Willis, U. S. Geological Survey.)

by Johnson and several others as EMERG-ENT, SUBMERGENT, or STABLE in this context. More recently (1968), Shepard has suggested a classification based on his two earlier efforts (1937; 1963) to respond to some realities of the post-Wisconsin rise in sea level as well as to previous types of land genesis. He recognizes PRIMARY COASTS where the configuration reflects submergent terrestrial topography and SECONDARY COASTS where the form relates to marine agencies (Fig. 9.6). There is a great deal to be said for this approach, but there are indeed few coasts that fail to qualify for both categories in SOME degree, and Shepard's emphasis on primary sub-aerial topography may unduly subordinate the fact that we are dealing with coastal phenomena evolving in that environment (cf. Thornbury, 1969, p. 438). This also applies to McGill's (1958) effort. In some respects, Valentin's (1952) grouping of coasts into categories that have advanced (seaward) versus those that have retreated (landward) is preferable. Bloom (1965) prefers to include some parameter expressive of each potential variable in his explanatory description system for coasts. The result lacks the genetic thrust we seek here, particularly with respect to "interface" evolution, for there are other considerations in devising coastal classifications besides those previously employed.

As noted by King (1959, 1972), Kidson (1968), and others, coasts have a well-known tendency to evolve to a rather

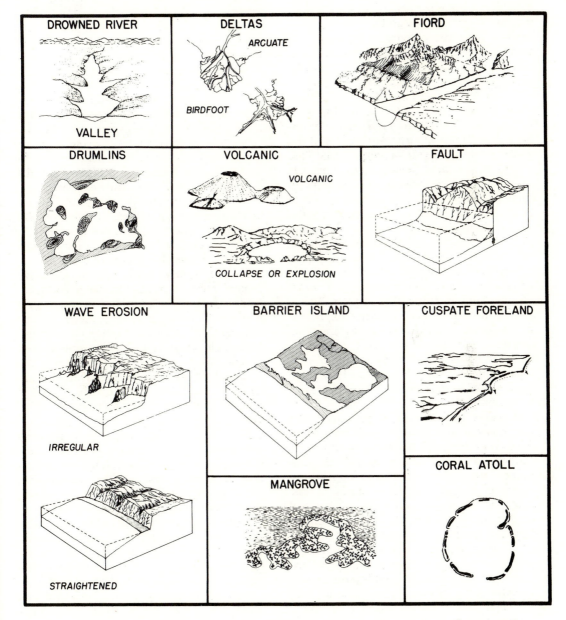

Fig. 9.6. Thirteen types of coast illustrating the coastal classification of Shepard. (From Shepard, 1968.)

smooth, uncomplicated regional outline in plan view, presumably through the erosional recession of terrigeneous headlands. Dolan (1971) and others see a hierarchy of arcuate beach patterns being superimposed on this regional scheme through deposition and possibly erosion. In addition, there is an evident and much-discussed tendency for oceans to sweep their margins clear of loose terrigenous sediment, a progressive change with respect to times of maximum sediment influx, which can be viewed in much the same light as the previously noted tendency for rivers to periodically eliminate alluvium from their channels and valleys (cf. Chapter 7).

Both the changes in areal coastal configuration (STRAND MORPHOLOGY) and the changes in STRAND COMPOSITION are clearly interrelated developmental aspects of present coasts which vary greatly in their degree of attainment from place to place. But there is a third factor. ORGANISMS along the land-water interface may nullify primary morphologic and compositional considerations in instances where coastal evolution would otherwise progress by erosion of terrigenous material. It must also be kept in mind that marine hydrologic equilibrium along coasts, expressed in terrigenous strand morphology and composition (and possibly bottom configurations), and coastal organic equilibrium, once achieved, should be essentially independent of associated subaerial coastal topography.

It seems best to admit at the outset that our coastal geomorphic problems often have very little to do with equilibrium. Rather, as in the case of subaerial landscapes, we are beset with a superabundance of coasts in various states of disequilibrium, in form, in composition, and in organic development. High contemporary sea level has inundated a vast array of subaerial topographies. If sea level were even 100 ft (30 m) lower than it is, we would no doubt gasp at the "comparative" morphologic monotony of oceanic strands developed on marine deposits of RELA-TIVELY subdued relief on continental shelves (cf. Cotton, 1966; 1968). As it is, what we see on many coasts is a group of gross coastal configurations expressing a myriad of partially submerged terrestrial landscapes whose modification by marine erosion and deposition tends to be superficial. Included among the drowned landscapes are fluvially incised, dominantly humid selvas, desert plains, glacial terrains of both erosional and depositional character, cryergic areas, combinations of some of these, and, of course, tectonically and volcanically active crustal segments that have moved both up and down. Some have much associated sediment. Some have little. Most have some. In addition, recent sea level rises have opened up extensive ecologic vistas to marine organisms under shallow neritic conditions which the organisms have not yet fully exploited.

Difficulties in classifying coasts are indeed formidable, and certainly no effort will go entirely unchallenged. Nevertheless, in accordance with the foregoing paragraphs and the geomorphic theme of earlier chapters, there seems to be a need to categorize coastal types in terms of the main interacting factors: (1) relict terrain morphology, (2) marine modifications, the latter tending toward some equilibrium expression, relating to (A) the presence or absence of abundant organisms along the land-water interface, (B) the relative abundance of sediment, and (C) the interplay between these elements of coastal geomorphic systems.

Where the sea has invaded the land, as has so widely and recently been the case, the submerged configurations are relicts now expressed as bottom topography, shoreline morphology, and strand composition. Strand composition may also reflect contemporary factors. In other instances, the coastal terrain relicts relate to tectonic, volcanic, or subaqueous effects. Our classification, as well as the geomorphic discussion that follows, should therefore reflect the modification of relict terrain by mainly aqueous processes as regionally modified by organisms, sediment, and, occasionally by eolian activity. In those comparatively isolated instances where sea or lake floors and strands have become emergent, the focus of our coastal geomorphology will remain on the strand and immediately contiguous zones. The emergent features may or may not be coastal. And an elevated strand situated well inland, like any other relict, is in a new environment and will probably change accordingly.

In conformance with the foregoing remarks, we will now try to learn from classification prototypes by Johnson (1919), Valentin (1952), Shepard (1968), and others and make our own effort. We recognize (1) TERRIGENOUS COASTS OF PRIMARY MORPHOLOGIC DISEQUILIBRIUM, (2) TERRIGENOUS COASTS OF PRIMARY MORPHOLOGIC EQUILIBRIUM, (3) ORGANIC COASTS, and (4) EQUILIBRIUM COASTAL SEGMENTS. A detailed outline is as follows.

1. Terrigenous Coasts of Primary Morphologic Disequilibrium
   Submergent Coasts of:
   Fluvial Incision (Carolina Coast; Virginia Coast; Venezuelan Coast—part)
   Localized Glacial Scour (Maine Coast—part; Labrador; Norway; Finland; Chile)
   Eolian Deflation (Carolina Bays, cf. Chapter 4, Figs. 7.56; 6.53; 6.54)
   Localized Deposition
   Fluviatile Irregularities (Mississippi)
   Glacial Irregularities (Long Island Moraine; Strangford Lough, Ireland—drowned drumlins, Fig. 8.57)
   Eolian Irregularities (Mauritania Dunes)
   Structural Deformation (Persian Gulf Coast—salt dome topography)
   Emergent Coasts of:
   Fault-Block Topography (Afar Triangle; Red Sea Coast)

   Folded Topography (Dalmatian Coast, Yugoslavia; Key Minor Island, West Pacific)
   Volcanic Topography (Iceland—part; Hawaii—part; Mexico—part)
   Subaqueous Channeling (Rare at present but probably the most common type with emergent continental shelves and upper slopes)
2. Terrigeneous Coasts of Primary Morphologic Equilibrium
   Submergent Coasts of:
   Desert Plains (Southwest Gulf Coast U.S.; Persian Gulf Coast, Arabia)
   Glacial Ground Moraine (South Baltic Coast)
   Ice-scoured Plains (Canadian Arctic Coast)
   Emergent Coasts of:
   Subaqueous Deposition (Motouanu, New Zealand; Salton Sea Coast; West Caspian Coast)
3. Organic Coasts
   Reef-buttressed Coasts (North Cuba; East Florida)
   Mangrove Coasts (Southwest Florida)
4. Equilibrium Coastal Segments
   Cliffed Bedrock Coasts or Headlands
   (Sussex Coast—part)
   (Peru Coast—part)
   Reef-straightened Coasts and Atolls (Northeast Australia)
   Zero-Energy Coasts (Northwest Florida)

At this stage in our discussion, there seems little point in singling out for mention those coasts where wave action, current deposition, organic activity, eolian effects, or similar factors have dominated the morphogenesis. We will discuss these in due course. It remains to emphasize only that the TERRIGENOUS COASTS OF PRIMARY MORPHOLOGIC DISEQUILIBRIUM were initially irregular morphologically, regardless of the origin of that irregularity. Erosionally stabilized portions would be classified under coastal equilibrium segments. TERRIGENOUS COASTS OF PRIMARY MORPHOLOGIC EQUILIBRIUM originally exhibited relatively smooth areal configurations. They appear to evolve little morphologically beyond the acquisition of an array of marine depositional characters. ORGANIC COASTS are those with a fringe of organisms along the land-water interface that governs the type of subsequent morphologic and sedimentary adjustments in stable subaqueous and subaerial environments. Stable forms, which are more nearly straight and especially if devoid of unconsolidated debris, would be classed as COASTAL EQUILIBRIUM SEGMENTS.

As noted by Wooldridge and Morgan (1937), it is difficult to meaningfully distinguish between coast and shoreline (cf. Lucke, 1938). However, in the discussions that follow, it is often possible to distinguish between relict coastal morphology and developing coastal attributes. The coastal morphologic equilibrium earlier noted by Kidson (1968), applicable to terrigeneous coasts, is often a matter of interaction between relicts and developing erosional and depositional features. Organic equilibrium also has its own particular morphologic expression. Such effects are restricted to comparatively short reaches of most coasts and are here classified under coastal equilibrium segments. Entire coasts in a state of equilibrium probably do not exist above an insular level. But then, that is what one would expect if most of the world's coasts reflect the recent inundation by both sea water and sediment from the land. That said, it now seems appropriate to examine the character of coastal geomorphic systems beginning with their most active components, water and wind. Specific effects of more passive ingredients, including sediment, organisms, and relict morphologies, will be taken up as appropriate.

COASTAL WAVES AND CURRENTS

The major bodies of "standing water" are, of course, "in motion." They are "standing" only in the sense that the mass of water as a whole is not moving in a particular direction, as it is in a stream or river. However, in all of the oceans and larger lakes, a number of factors combine to move water about, the most important of which are wind stress, tidal attraction, earth rotation, and density variations. Our immediate concern is with water movements capable of modifying the shape of a coast. Mainly, this would appear to involve the zone between high and low tides (littoral zone of some authors) and beyond into the *shallow neritic zone* (0–100 fathoms) to the depth where ocean waves normally feel bottom (*wave base*).* Zenkovitch equates the *littoral zone* with the interval from the storm tide level down to a depth of one-third storm wave length (1968) and that will be essentially our definition of the expression. Deitz and Fairbridge (1968) illustrate several alternative theoretical morphologic expressions of wave base on coasts (Fig. 9.7).

Actually, of course, there is a "zone" within which waves make contact with the bottom, and whether they are erosionally effective or not is largely dependent upon what type of bottom material is encountered. Though soft sediment may be stirred up at great depth by storms, such activity can hardly be regarded as vigorous abrasion. The depth at which such activity occurs is generally determined by the wavelength L, since the orbital motion of the related water particles diminishes ALMOST to zero at a depth equal to $L/2$ as illustrated (Fig. 9.9). In this sense, as widely used by stratigraphers, wave base is the shallowest water in which fine-grained sediment can accumulate without displaying littoral wave features. Of more geomorphic significance is the base of severe wave abrasion established by waves of

* Goldsmith (personal communication, 1973) observes that a 14-second wave off the Virginia coast would touch bottom at about 300 ft (60 mi off shore) whereas a 6-second wave would touch about 1 mi off shore. Both types are common.

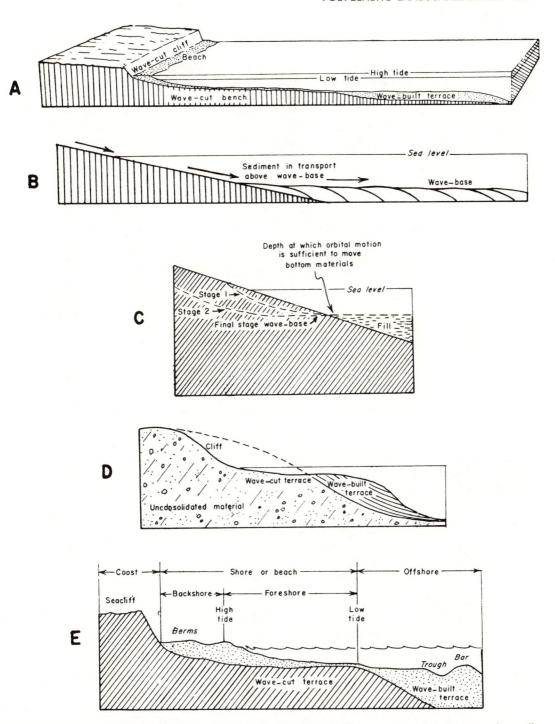

Fig. 9.7. Several views of the origin of the wave-cut and wave-built terraces in relation to wave base, all of which imply a controlling effect by the wave base. (A) After Longwell, Knopf, and Flint (1948); (B) after Clark and Stearn (1960); (C) after Garrels (1951); (D) after Von Engeln (1942); and (E) after Leet and Judson (1958). (From Deitz and Fairbridge, 1968.)

translation within the intertidal belt (cf. Dietz and Fairbridge, 1968). Strong abrasion extends to the depth at which waves begin to peak, and, under storm conditions, break, according to these writers (see Fig. 9.33). They maintain that more than 95% of wave energy is dissipated between a depth of about 10 m and the shore.

Ordinarily, the major ocean current gyres (discussed in Chapter 4) are not direct agencies of coastal morphogenesis, though their sediment transport capacities and flow velocities approach those of terrestrial rivers (cf. Schmitz *et al.*, 1970). However, along coasts that lack appreciable continental shelf, with deep water close inshore, as on the west coast of South America (Fig. 9.8), the ocean current gyres may NEARLY impinge on the shore and could augment the net water and sediment movement along the coast normally accomplished by wave-generated, near-shore currents. Ultimately, of course, these ocean current gyres are intrinsically involved in long-term sediment movements, both along

coasts and out into the deep ocean basins. The latter relation appears to result from an apparent tendency for *longshore currents* to shift sediment into submergent transversely oriented relict drainage depressions crossing the shelf or into heads of submarine canyons.

Many of the water movements in a large lake or sea do not directly affect the shore. Water movements due to density variations and deep-flowing geostropic currents are examples of currents that hardly affect coastal landforms. Open ocean waves (*waves of oscillation*) are similarly nonerosional with respect to the strand, though, as noted by Johnson (1966), even though the water mass as a whole does not advance with the wave form, about 1/20 to 1/30 of the wind stress is expressed in a *net drift*, to account for the buildup of water on a windward coast toward which the waves of oscillation are moving. This is explicable if the orbital paths followed by the water particles are open rather than closed. Ultimately, of course, where these

Fig. 9.8. Sea cliffs interrupted by a dry valley along the north coast of Peru near Talara. The cliffs are 150–200 ft (45–60 m) high and locally strewn with dune sand blown up from the beaches.

oscillatory wave forms contact the sea floor, they take on a lateral movement increment (*waves of translation*) which is geomorphically very important. For a quantitative treatment of wave theory see Wiegel (1964, 1966). Since waves of translation are directly involved in coastal erosion and deposition and also indirectly generate a variety of near-shore currents, we shall consider them first.

Waves of Translation

Waves displaying notable translatory motion have already made contact with the sea floor. The depth at which this occurs, and the height and force of the resulting wave, is a function of the orbital path dimensions of the moving water particles. These water-particle orbits, in turn, reflect the oscillation wavelengths and amplitudes developed mainly by wind in the open ocean (Fig. 9.9). As previously noted (Deitz and Fairbridge, 1968), several workers place effective wave base (FROM THE STANDPOINT OF SHORELINE EROSION) at a depth of about 35 ft (10 m) on the average. Storm swell (because of greater wavelength) may stir up sediment on occasion at ten times this depth (cf. Fairbridge, 1952; Bradley, 1958). Open-ocean wave dimensions are determined by the distance of unobstructed open water across which the wind can blow (*fetch*), by wind velocity, and by wind duration. Regions of

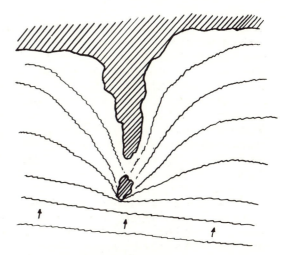

Fig. 9.10.   Wave train approach and refraction on an irregular headland and coast. The tendency is for parallelism with the shore, but such ideal relations are not always achieved.

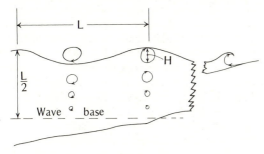

Fig. 9.9.   Cross section of wave forms showing the orbital motion of water particles to a depth of *L*/2, the wavelength *L*, and the wave height *H*. In theory, the orbital motion continues below a depth of *L*/2 but is not erosionally effective at greater depths which thereby establishes the base of wave erosion.

ideally long fetch exist in the Pacific Ocean where Trade Winds can blow without obstruction for thousands of miles. As any avid surfer knows, Pacific waves coming in on the California coast commonly form breakers which run 6–7 ft (2 m), whereas those on the Atlantic Coast of the United States rarely exceed 3 ft (1 m). Of course, storm waves, particularly those augmented by high tides on a windward shore, are another matter (cf. Bretschneider, 1966). Thornbury (1969) states that the greatest accurately measured wave height is on the order of 50 ft (16 m). Other estimates are greater. Most coastal morphology developed on a day-to-day basis is of course due to much smaller waves. As noted in Chapter 1, however, a hurricane's waves may in a day completely alter coastal forms developed in unconsolidated sediment over a period of several decades (cf. Fig. 2.1).

Translatory waves achieve their geomorphic effects in several manners closely related to the nature of water movements in the wave form (cf. Figs. 9.9; 9.10). As noted previously, water particles in an oscillatory wave form follow orbital paths. Where these orbital paths are intersected by the sea floor, as the wave form enters shallow water, the base of the form is retarded.* In plan view, this contact is expressed where the wave trains bend to more nearly parallel the shore (*wave refraction*)—i.e., when they are normal to bottom slope (Fig. 9.11) (cf. Silvester, 1959, 1963, 1966). Some wave crowding also occurs, and this is intensified in the shallowest areas. As the orbiting water masses advance into the shallows, the water particle orbits flatten (cf. Wiegel, 1966, Fig. 5), the axis of rotation is forced to rise as the land shelves and wavelength shortens, and the wave form is thus also seen to become slightly elevated as it moves to the strand. Basal friction may also prevent transfer of water to the face of the orbiting mass which in any case becomes higher than water depth; the mass therefore thins, fails to complete the orbit, and causes the wave to appear to curl forward and collapse, or, as we say, "break." The resultant "breaker" or "comber" then sweeps on up the surface of the beach or cliff as "surf." Frequently the surge of landward-moving water curls at its margin of maximum advance to deposit larger sediment grains and flotsam (commonly seaweed) and to form "*swash marks*" on any existing beach; it then flows back into the base of the next-breaking wave as "*backwash*."

The geomorphic effects of translatory waves depend to a great extent upon the type of lithic material comprising the coast, tidal range and direction, bottom depths and configurations, and nature of coastal exposure. In fine-grained, unconsolidated materials, erosion usually begins as soon as the wave form "feels" bottom and continues in various, usually intensified aspects to the point of maximum surge and backwash. Thereafter, currents that are a by-product of the wave activity become erosionally important. The sediment erosion-deposition configurations and organic associations that accompany surf activity on a rising tide differ markedly from those on a falling tide. On a barren rocky coast, the main erosive effects may come as a breaking wave delivers its impact to a beach or sea cliff. In general, a gently shelving bottom tends to dissipate wave energy frictionally before the wave arrives at the shoreline. On the other hand, an oscillatory wave that fails to touch bottom acquires no translatory motion and cannot deliver a laterally directed impact of any great force—as against a steep cliff rising from deep water. With respect to average daily translatory wave erosion and deposition and associated geomorphic effects, we will undertake a detailed consideration of each of the main types of littoral setting.

Short-Lived Wave Phenomena

Several wave types of great geomorphic and sociologic importance are of very brief duration. Included here are *tsunami, storm surges*, and *landslide surges*. *Tsunami* (Jap., *tsu* = harbor; *nami* = wave) are mainly creatures of submarine earthquakes, landslides, or volcanic explosions and have little to do with the tide (cf. Van Dorn, 1965). They are common in oceans like the Pacific where the floors are tectonically active. The originating impulse develops when a tectonically shifted segment of crust attempts to displace an equivalent volume of sea water more or less instantaneously. Since water is essentially incompressible, a wave system is propagated outward at the limiting velocity $c = \sqrt{gh}$ for free waves in water of depth h, where g is gravitational acceleration. This is the velocity formula for shallow water waves defined as those waves in water less than half the wavelength in depth. Since tsunami have wavelengths in excess of 100 mi, they are always in shallow water, even out over the deep ocean basins. Waves traveling 400 mph or more (650–950 kph), with lengths approaching 100 mi (140 km) and amplitudes of as much as 150 ft (50 m) are the usual result.** On low coasts,

---

* According to Yasso (personal communication, 1973), there is no consensus that friction is involved, though it is difficult to comprehend why it would not be.
** The vertical motion increment is, of course, not visible as wave height in deep water but rather only where the wave form enters shoal coastal areas.

Fig. 9.11. Wave refraction at the entrance of Tomales Bay, 40 mi (66 km) northwest of San Francisco, California. Note also the cuspate form of wave-cut cliffs on the peninsula and the bay-mouth spit and associated sand dunes. (Pacific Resources Inc.)

tsunami can constitute a nasty surprise for the itinerent sun bather. Though not capable of sustained sculpture of bedrock, considerable rearrangement of unconsolidated rock waste and shore structures by a tsunami is common; see Coleman (1968). Comparatively fragile, near-shore depositional landforms may be temporarily wiped out by the normal series of three to five major wave oscillations comprising a tsunami wave train, and the ocean may appear disturbed for several days thereafter.

As a matter of fact, all of the wave types listed in the previous paragraph are forms of *gravity waves* which later acquire trans-latory motions in shallow water, and their basic water-movement patterns do not appear to be particularly unusual. What is unusual are the wave dimensions. And the catastrophic changes in coastal configurations that accompany tsunami and the other large-wave types appear largely in a dimensional sense. The "large" waves simply do not fit the erosional and depositional configurations developed along a coast on a day-to-day basis. And as one would expect, the disparity becomes most obvious in depositional landforms such as bars, spits, cusps, and beaches along sedimented coasts.

Storm surges involving rise of ocean level by onshore wind (possibly augmented by tide) are most commonly associated with hurricanes and typhoons (cf. Bretschneider, 1966). The accompanying waves are not as large as tsunami, but they keep coming, sometimes for days. Thornbury (1969, p. 425) echoes others in the idea that storm waves may supersede other wave types in coastal erosion. In their ability to modify coasts drastically with respect to configurations developed gradually, there seems to be much merit in this notion. But in terms of over-all sediment transport along the shore and the long-term removal of such detritus from continental shelf areas, the more average waves and currents may be more important (see next section). For a quantitative consideration of storm surge, see Bretschneider (1958, 1966) and Groen and Groves (1962).

Landslide surges develop where rock masses fall into rather constricted water bodies—estuaries, bays, and lakes. Like tsunami, landslide surges are essentially one-time affairs, and have a geomorphic potential relating to this, but they are the biggest waves on record. Miller (1961) records a wave generated when a 40,000,000 cu yd (30,640,000 cu m) mass of rock broke loose from an Alaskan mountainside and plunged 3,000 ft (1,000 m) into Lituya

Bay on July 9, 1958. The wave surged across the bay and sheared away soil and vegetation to a height of 1,720 ft (510 m) and then, under the pull of gravity, exited out of the bay at an estimated velocity of 100 mph (140 kph). A similar wave was generated in an artificial reservoir by a 312 million cu yd (240 million cu m) avalanche in Vaiont Canyon, Italy. The subsequent gravity wave over-topped the associated dam 300 ft (100 m) and moved down valley at a speed of 2 mi/min (3 km/min) with disasterous effects on at least 2,500 valley-dwellers.

Near-Shore Currents

An oriented, relatively localized movement of water or gas in a particular direction is a *current*. Water currents of geomorphic significance along coasts mainly relate to wind, waves, and gravity. Wind determines the direction from which wave trains approach a given shore, and regional winds also determine the main oceanic current gyres and general circulation direction along shore. The wave-induced currents that result may compliment and strengthen the general circulation or may oppose and weaken it. Gravity has several effects. It is the ultimate cause of wave breaking, of course. Furthermore, it directs the flow of backwash, and in conjunction with bottom topography, influences the seaward return of water piled up on a shore by the wind. Let us consider each of these variables separately.

In the geometric relationship between wind, wave, and coastal alignment, only the coast is essentially fixed. Waves approach coasts at various angles, commonly obliquely, so that a breaker's surge up the beach is in the direction of wave propagation but the water returns parallel to the greatest beach slope (Fig. 9.12). A vectoral resultant of the zig-zag movement increments involved has a net component along the shore in all instances where waves are approaching the shore obliquely from one main direction. This movement is perhaps

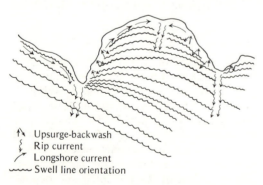

↟ Upsurge-backwash
↧ Rip current
⟋ Longshore current
⌇ Swell line orientation

Fig. 9.12. Plan diagram showing usual relation between swell orientation (perpendicular to prevailing wind) and related long-shore current and rip current development. Upsurge direction usually includes some increment of wind vector, but backwash is always downslope.

best termed *longshore current* (*littoral drift*). Longshore current on a given coast tends to be a product of the most persistent swell approach direction. Where it is in the same direction as the tidal currents, sediment transport is very effective.

Sylvester (1959; 1966) observes that storm waves tend to come from multiple directions, erode the beach, and, in absence of a net littoral drift, they cause an offshore submarine bar to be deposited. Ordinary day-to-day small waves (*swell*) tend to return the offshore bar to the beach over a period of weeks.* Because of littoral drift of sediment occurring during the latter period, the beach may suffer a net sediment loss unless a continuous supply of material is moving into the same coastal area from farther up the coast.

Before we leave the subject of wave-induced currents in the littoral zone we should mention *backwash currents* and the development of so-called *rhomboidal ripple marks* (cf. Guilcher, 1958). The marks in question are more probably a form of *incipient rilling*, a notion expressed speculatively by Holmes (1965, p. 813) as a possible explanation for *beach cusp* development, but not related by him to any particular rill feature. In several earlier chapters (especially Chapter 5), we noted the apparent tendency of, and theoretical justification for, a granular surface under a sheet of water to develop a rill pattern. As it happens, a breaker's surge up a beach spreads a thin sheet of water across an inclined sand surface of unusual smoothness. On very fine-grained sand beaches (i.e., Wildwood Beach and Cape May Beach in southern New Jersey) the film of backwash can be observed to break up into a mosaic of incipient rills which very slightly groove the beach before the water dissipates (Fig. 9.13). The incipient rills are often of such low relief that they are only visible in very low-angle lighting directed along the beach. The material in question is of very uniform texture, and, of course,

the next breaker surge often wipes out the previous rill texture (cf. Fig. 5.32).**

Our final remarks in this section should be directed to gravity-propelled water movements, particularly those related to tides. In the gravity situation, one of the most pronounced effects relates to water accumulating to greater than normal depth (height) against a downwind shore due to seabreeze. The *net drift* of water associated with oscillatory waves against such coasts must be accommodated by seaward water movements. The predominant form of movement appears to be expressed in linear seaward currents usually developed where littoral drift and longshore currents are deflected seaward, particularly where incoming wave trains are broken up by shore configurations or bottom topography. These *rip currents* are often not related to any visible coastal deflection but they are commonly situated on piers, jettys, points of land and off major peninsulas. Strong currents with "rip" and "tidal" aspects occur on the south New Jersey extension between the Atlantic Ocean and Delaware Bay (Fig. 9.14). Rip currents are commonly denoted by series of visible standing waves where shoals occur, and considerable seaward sediment transport is associated with them as illustrated by Thornbury (1969, Fig. 17.1).

Localized seaward movement of orbiting water particles on the bottom of a breaking wave has acquired the colloquial label *undertow*, and with full cooperation of the landward surging water at the wave crest, such water movement may exchange positions of an unwary sea bather's head and feet. However, the orbital water movements probably inhibit any consistant seaward transfer of water and sediment. Rip currents surging seaward are often casually termed undertow, but certainly are another matter. A bather caught in a rip current may notice the land disappearing over the horizon (velocities approach 3 ft/sec). If so, he should resist the temptation to swim

---

* Yasso (personal communication, 1973) observes that these waves are "small" in amplitude but long in period which is why they move the bar landward.
** For further discussion, see the section on cuspate coastal configurations and beaches in this chapter.

directly shoreward, turning aside instead until out of the strong flow before attempting to make land. Rip currents often coincide with shallow channels in the bottom trending seaward, breaks in wave trains, and strips of foamy or white water.

Tidal currents constitute the other major type of near-short current of potential geomorphic import. On broad, open, shelving coasts, effectively eroding currents rarely develop with tidal rises and falls since the over-all twice-daily shift in water mass is dispersed rather than localized. However, in bays, estuaries, inlets, and some river

mouths, tidal water movements may become sufficiently localized and intensified to erode and become geomorphically important (see Tricker, 1964). And the morphologic consequences are largely due to water volume accommodations where embayments opening on an ocean or sea narrow very gradually in a landward direction (Fig. 9.15). As shown in the figure, a given water volume is admitted to the embayment as determined by the tidal rise and cross-sectional area. Where the embayment is narrowed to one-half the original aperture width, the same volume must

Fig. 9.13. Oblique view along Cape May Point beach, New Jersey, at 8 A. M. showing rhombic rill pattern developed under wave backwash on a falling tide (upslope, left). The rhombic rills merge downslope to form deeper dendritic and anastomotic flow routes in the very fine sand, aided by surface return of water which infiltrated during wave uprush. On beaches with cusps, such rhombic rills develop on both horns and in bays. The coin is a 25-cent piece. (See also Fig. 5.32.)

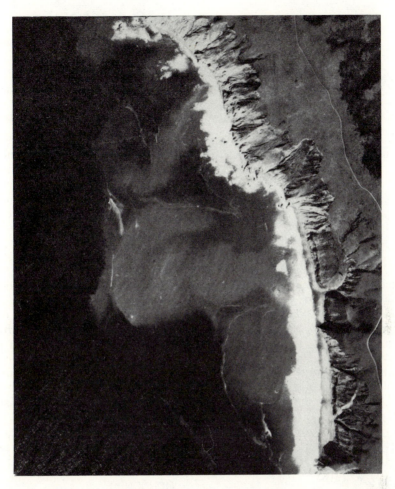

Fig. 9.14. Three rip currents along McClures Beach near Tomales Point, Marin County, California. Largest surge (cloudy water area) extends just over 1,600 ft (500 m) from the beach to a water depth of about 75 ft (20 m). North is toward the top. (Harold Wanless.)

pass, so the depth of water must double (if velocity does not increase) in order to accommodate it. This continues on to the recesses of the embayment. The increased depth may be disclosed by a detectable face of advancing water termed a *tidal bore*.

The cumulative heights of tidal bores and the intensity of related erosive effects are related to tidal change intensities. Those of the Bay of Fundy between New Brunswick and Nova Scotia average about 40 ft (12 m) and range from 30–50 ft (9–15 m). These bore heights relate to several individual waves and to tidal rises at the

bay mouth a third or less as high. Strong tidal shifts are known from many regions including the River Severn in England (cf. Holmes, 1965, Fig. 582; Dury *et al.*, 1972), the Rio Guayas in Ecuador, and the Amazon River where tidal bores up to 15 ft (5 m) are on record. Though impressive in appearance as walls of water several feet high, and a true surprise to the preoccupied clam digger, it has been questioned that tidal bores are erosionally very effective (Johnson, 1925). It is nonetheless true that many tidal estuaries and rivers that lack bores, but have strong outgoing tidal flows, typically exhibit a funnel or

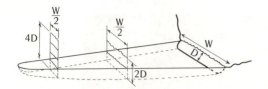

Fig. 9.15. Diagram of the manner in which a volume of tidal water entering a constricted estuary may rise to form a tidal bore. In the instance illustrated, initial width *w* is reduced, while depth *D* remains constant along with velocity. If velocity increases, the rise would be less than that shown.

trumpet-shaped plan form. These have been studied by many researchers including Russell, 1967; Geyl, 1968; Dury, 1969; Price, 1963, 1968. Dury has raised several serious questions about the views of Geyl on tidal effects on meanders. It seems sufficient for the moment to point out that the same water volume that enters a narrowing estuary, often very gradually, must also exit under the pull of gravity. This downslope hydraulic surge may help explain the rapid broadening of many tidal rivers (Fig. 9.16). For a detailed hydraulic

discussion, see Dury (1969). Zenkovitch (1967) also treats related littoral processes quantitatively.

MARINE GEOMORPHIC EVOLUTION

Having taken a rather general look at marine agencies of erosion and deposition, we must now consider their more specific effects in certain coastal settings. As previously noted, the apparent long-term tendencies involve elimination of local irregularities and reduction of strand sediment. Just how these effects are accomplished is our immediate concern, but we must also be aware of dominant themes. The vast majority of present marine shorelines are recently submerged because of the last glacio-eustatic sea level rise on the order of 330+ ft (100+ m). Most of the continents were affected mainly in shelf areas where modification by subaerial agencies prior to the last submergence is of unknown but probably considerable extent, particularly by stream incision (see McMaster and Ash-

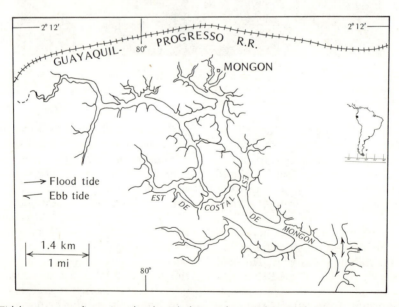

Fig. 9.16. Tidal estuary configuration developed about 5 km southwest of Guayaquil, Ecuador. Flood tides submerge extensive surrounding mangrove forests. Note the rapidly flaring character of the inlets. Only the northwestern-most channel receives appreciable runoff from the mainland.

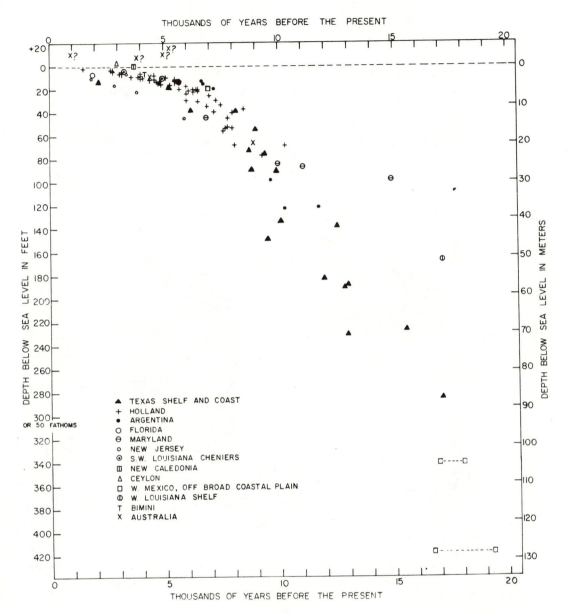

Fig. 9.17. Evidence from C_{14}-dated organisms indicate the rise in sea level in the past 20,000 years. All data are from relatively stable coasts. (From F. P. Shepard, 1968.)

raf, 1973). The rising seas infringed on a variety of terrestrial landscapes in a period of only about 20,000 years (Fig. 9.17), and the transgressive aspect has impressed some special erosional and depositional effects on many coasts. Indeed, a virtual revolution in the subject area of coastal geomorphology in the past quarter century has involved a reassessment of coastal landforms in light of their almost ubiquitous modification by marine submergence. We will refer to these aspects from time to time as we endeavor to outline the main modes of coastal morphogenesis.

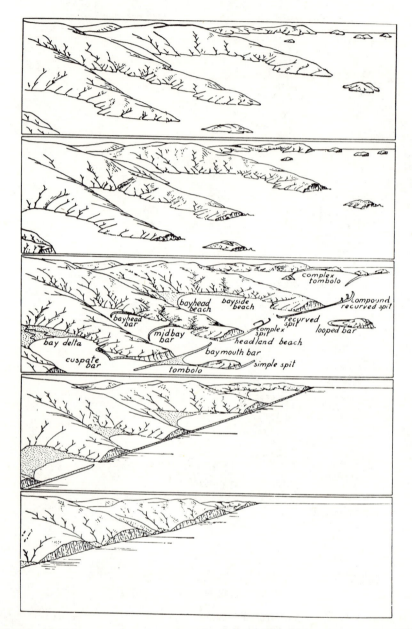

Fig. 9.18. Developmental stages of erosion of a coast of primary morphologic disequilibrium produced by submergence of a fluvially dissected terrain. (After D. W. Johnson.)

Terrigenous Coasts of Primary Morphologic Disequilibrium

Irregular coasts apparently tend to undergo considerable modification by marine agencies, regardless of the origin of the irregu-

larities. Basically this is because the headlands bear the brunt of strong wave and current activity whereas re-entrants, bays, estuaries, and coves are more protected. As a result, headlands tend to be truncated and sediment eroded therefrom plus that introduced by any coastal streams may accumulate in intervening bays. The basic scheme depicted by Johnson (1919), and re-illustrated by Lobeck (1939), appears to be still essentially valid (Fig. 9.18). Details of such changes must, of course, vary with the original morphology of the coast, its composition, and its orientation with respect to prevailing wave and current directions. The coast illustrated (Fig. 9.18) is essentially a coast of fluvial incision which has been submerged like that shown by Fig. 9.19.

Morphologically unstable coasts may be composed of considerable unconsolidated rock detritus or there may be extensive areas of bedrock. The rapidity of strand morphology changes will depend to a great degree upon a coast's lithology. Holmes (1965, p. 806) describes a 2.5- to 3-mi

(3.5–4 km) coastal recession in glacial sediment since Roman times in Great Britain. Erosion can convert unindurated materials into beach detritus very quickly, but related eroded faces tend to subside to repose angles at the same time, and *sea cliffs* related to this are rarely imposing (Fig. 8.38). In his synopsis of littoral processes (1968), Fairbridge categorizes soft rock shores and hard rock shores. Along irregular coasts, bedrock exposures (if there are any) tend to be on headlands. Because of the possible variety of relict lithologies and relict topographic configurations, and the nature of probable subsequent marine modifications, it seems best to divide our discussion of terrigenous coasts of primary morphologic disequilibrium into *erosional coastal segments* and *depositional coastal segments*. Organic reef growth (fringing

Fig. 9.19. View to the east of the Ria Coast, northeastern Venezuela, where deep V-shaped valleys incised by rivers extend into the sea separated by peninsular ridges that form long headlands from what is essentially a drowned selva topography. Organic reefs occur along this coast, especially on the points of headlands exposed to greatest wave action. (W. N. Gilliland.)

reefs) on tropical headlands (Fig. 9.19) may stablize an ordinarily unstable coastal morphology by inhibiting headland erosion.

Erosional Coastal Segments

As one might expect from the foregoing remarks, coastal segments that are dominantly erosional tend to be associated with terrigenous headlands or windward terrigenous shores of offshore islands. Since erosive intensity tends to be severe in such sites, fine-grained detritus is swept away entirely, particularly by winter storms (cf. Figs. 9.20 and 9.21). In coarse-grained materials, however, such as glacial till, winnowing, rounding, and flattening of clasts may leave a coarse beach gravel (*shingle*) as it has in places on the eastern end of Long Island, a part of the Wisconsin terminal moraine, and along certain shores of the Great Lakes (Fig. 9.22). Thornbury

(1969) lists five processes contributing to marine erosion that would presumably be intensified on terrigenous headlands. They include corrosion, corrasion or abrasion, attrition, hydraulic action, and shock pressure of breaking waves. Holmes (1965) includes wave shock pressure as an aspect of hydraulic action.

By *corrasion,* just as in fluviatile settings, we mean erosion by moving rock fragments, in this case, shifted by waves and related currents. Frankel (1958) reports that water-faceted clasts on the South African coast may be a result of corrasional wet-blasting. *Corrosion,* as in other aqueous situations, refers to the solvent and chemical action of water. *Attrition* has to do with the wear of moving rock clasts, and *hydraulic action* with the surge, weight, pressure, and force exerted by the water itself, particularly in the form of breaking waves on a coast. Johnson (1919)

Fig. 9.20.   Summer view of Boomer Beach and a portion of the southern California coast, La Jolla. Note the sandy character in comparison with that in Winter (Fig. 9.21). (John S. Shelton.)

Fig. 9.21. The Boomer Beach at La Jolla, California in Winter; storm waves have swept Summer sand beaches away, in contrast to the situation in Fig. 9.20. (John S. Shelton.)

used a dynamometer on the Atlantic coast of Scotland to determine that average winter waves there developed pressures of over 2,000 lb/sq ft and that storm wave pressures often exceeded 6,000 lb/sq ft. Of course, corrasion, corrosion, and attrition are most important where waves and currents are acting on unconsolidated materials. It is on headlands of consolidated bedrock that hydraulic effects are most pronounced, particularly in rocks with notable partings.

In part, hydraulic action by waves on sea cliffs is, actually, pneumatic in character. The advancing wave pushes a volumn of air before it into existing cavities, cracks, and pores followed closely by wedge-like jets of water backed up in turn by water masses commonly weighing tens of thousands of pounds. Partings are forced open, and air pressures developed in the smaller voids may reach several thousand pounds per square inch. As the wave abruptly recedes, confining water pressures are reduced sharply, and internal air pressures, then unbalanced, are exerted on blocks of rock from within in a quarrying action not unlike frost heaving. High-pressure air in more open cracks often explodes outward and upward to blast lingering water masses to spume and spray. Combined repeated wave bombardment, wedging, and "air-quarrying" may quickly disaggregate already jointed and fractured rock masses on headlands. As one would expect, massive rock types are more resistant to hydraulic wave action than are those with many partings.

As headlands composed of bedrock erode backward, a pronounced *sea cliff* may develop (Fig. 9.23), usually more or less normal to the prevailing wave direction.

Fig. 9.22. Shingle beach along the western shore of Lake Superior, near Thunder Bay, Ontario. Clasts shown are flattened (disks) of assorted igneous and metamorphic lithologies.

Fig. 9.23. Sketch of the southwest facing cliff of Cabo Girão, Madeira Island, based on a photo; the cliff is locally believed to be the world's highest sea cliff, 1,750 ft (545 m). Note town, right foreground. (From Cotton, 1969.)

Darwin (1844, 1846, 1891) was an early student of marine cliffing mechanisms, and he believed that sea level had to be rising for extensive cliff recession to occur through wave action (cf. Cotton, 1969). Where the materials in a cliff undergoing erosion vary in their resistance to the erosional mechanisms, a part of a headland may become isolated as a *pedestal rock* or *stack* (Fig. 9.24). Along the coast of Barbados, undercutting of a Tertiary marl by waves allows huge blocks of Pleistocene reef limestone to collapse into the surf zone (cf. Fig. 9.1). Exploitation of "weak" rock zones by waves and currents also causes *sea caves* to develop (Fig. 9.25) as

Fig. 9.24. Sea stacks eroded from metamorphic rocks along the north coast of Venezuela near the town of La Guaira. (Sketch based on a photo.)

Fig. 9.25. A sea cave and a sea stack eroded from columner basalt, the southern coast of Iceland. (W. N. Gilliland.)

Fig. 9.26. A sea arch developed on a headland on the south Mediterranean coast, Algeria. Scale shown by buildings on the cliff crest.

well as various types of *arches* (Fig. 9.26), where caves pierce headlands. Also, as the headland recedes, wave and current action commonly bevel the coastal bedrock at the level of most intense erosion* to produce a *wave-cut platform* or *terrace* (cf. Figs. 9.7; 9.27). The development of wave-cut terraces focuses on the most intense zone of wave erosion. Most such terraces are slightly below sea level at high tide; most sea caves are so situated and extend somewhat above high tide. Thus, as noted by Holmes (1965, pp. 798–806), wave erosion tends to undermine wave-cut cliffs (Fig. 9.28), sometimes with the development of upward vents for air escape (blow holes) with associated collapse or subsidence.

With headlands composed of bedrock, the development of a wave-cut platform forces waves approaching a sea cliff to cross an increasingly broader expanse of shallow water (cf. Cotton, 1969). Reduced rates of sea-cliff recession result (in terms of direct wave erosion of bedrock), and the nature of subaerial weathering becomes increasingly important as an aspect of scarp retreat. In humid coasts, subaerial solutioning effects may be added to marine corrosion (Fig. 9.29). Frost action may also play a part. Where rocks composing such cliffs are decomposing granularly, a relatively weak surf may keep the platform swept clear of the detritus that would otherwise become a protective scree apron. Holmes (1965) cites platforms west

* Such beveled rock terraces may constitute the most tangible evidence we have of an "erosional" *wave base*.

and northwest of Norway up to 37 mi (50 km) wide, relating them to the foregoing process, in association with frost and thaw weathering. In south coastal Ecuador, the Paleocene Azucar Sandstone in sea cliffs tends to decompose granularly in an arid environment above wave-cut platforms several hundred feet wide. Curiously, a local cobble beach at some cliff bases is composed of tough, ferrugenous sandstone nodules freed by the granular disintegration (Fig. 9.30).

Along terrigenous coasts of morphologic disequilibrium, the long-term tendency of coastal evolution seems to be to reduce the prominance of headlands and, as they recede, to increase the linear extent of erosional coastal segments at the expense of intervening depositional areas (cf. Figs. 9.12 and 9.24). In those in-

stances where the process goes to completion, a coast of equilibrium (or segment thereof) is the result (Fig. 9.31). Often, in such cases, ancillary marine processes have disposed of the sediment in the ocean depths. Unless the coastal rocks in question are readily eroded, or the littoral erosion is very intense, or both, such a process may require a very long time. Furthermore, it is worth noting that we are, in fact, discussing a dynamic equilibrium (climax) landform expression of one type of coastal environment. The result, like the equilibrium expressions of other environments cited in earlier chapters, appears to involve a least-work, equal-area energy expenditure relationship. Not only is the land-sea interface reduced to an areal minimum, sediment is limited to that brought along the coast or that yielded through local

Fig. 9.27 A wave-cut bench truncating inclined strata of the Cretaceous Barranquin Group near Playa Colorado, Venezuela, along the north coast. The platform has a veneer of living coralline reef growth. (W. N. Gilliland.)

Fig. 9.28.   Marine erosion of El Cantil Limestone, on the north coast of Venezuela. The notch just above water level extends 2 to 3 ft back from the cliff face; note the small sea cave. (W. N. Gilliland.)

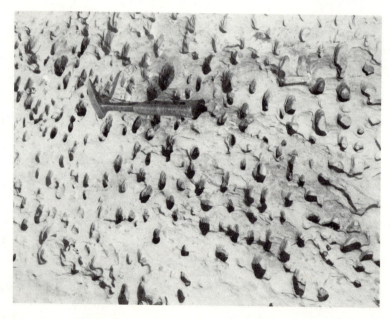

Fig. 9.29.   Wave-splash fretwork developed in medium- to fine-grained schistose sandstone, just east of Newport, Rhode Island. No mineral or structural lineations were observed. Sketch shows shore relations. (S. L. Agron.)

Fig. 9.30. Wave-cut terrace developed on steeply dipping massive conglomerates and sandstones of the Azucar, Paleocene series, the south coast of Ecuador.

weathering, and coastal erosion is inhibited by wave-cut platforms. We should now consider the depositional geomorphic expression that complements erosion along terrigenous coasts of primary morphologic disequilibrium (cf. Fig. 9.18).

Depositional Coastal Segments

Enough has heretofore been said about terrigenous coasts of primary morphologic disequilibrium to prepare the reader for the idea that depositional phenomena on such coasts often is related to various types of embayments. Sediment eroded from headlands as well as any detritus being shifted down a coast by longshore currents may come to rest, via a complex of cur-

rents, waves, and relatively quiet water, in bays and estuaries. Apart from the occasional situation that lends itself to tidal bores, most waves and currents in bays are weaker than their counterparts in more exposed, seaward portions of irregular coasts. And, in a sense, the resulting landforms and deposits are less spectacular than those of headlands.

In general, sediment that is most readily moved away from eroding headlands consists of the finer-grained materials present, and sediment that is shifted back into bays tends to become increasingly fine as waves and currents there weaken. Beaches of detritus on headlands (if any) tend to consist of coarse shingle (cf. Fig. 9.22). *Spits* extending down-current from head-

Fig. 9.31. White chalk cliffs along the Sussex coast facing the Channel and France. Erosion here has reduced headlands and straightened the coastal profile by truncating a group of hills known as the Seven Sisters (see also Fig. 2.17). (British Tourist Authority, New York.)

lands may reflect longshore currents as suggested by Johnson (1919), littoral-drift outbuilding (Steers, 1953; Hoyt, 1967), or spasmodic propagation by obliquely impinging storm waves as described by Lewis (1931, 1932). Spits are more likely to be dominated by sand than gravel (Fig. 9.32), but there are exceptions. And beaches developed along the margins of coves com-

monly consist of either sand or mud. Coarse sediment provided by a stream at the head of a bay or through weathering of bay-side rocks may, of course, alter the patterns envisioned. Figure 9.18 illustrates a number of the possible depositional landforms including *bayside beaches, bay deltas, baymouth, midbay,* and *bayhead bars, simple* and *recurved spits (hooks),*

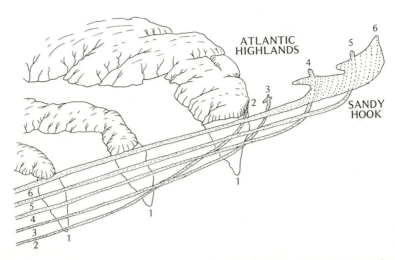

Fig. 9.32. Stages of development of the Atlantic Highlands shoreline of New Jersey and Sandy Hook, a recurved spit, from oldest to youngest (1–6). Longshore current moves left to right (south to north). (After Johnson, 1919.)

and various islands tied to land by spits (*tombolos*).

Not uncommonly, a spit will be extended by deposition of a longshore current between two headlands, thereby forming a *baymouth bar* (*barrier-bar*) and converting the intervening bay into a *lagoon*. The resulting quiet water body may become filled with sediment brought in by bayhead streams, and, through establishment of vegetation, be converted into a *tidal marsh*. Numbers of such marshes occur along the New England coast where they are an ob-

ject of continuing study by several geomorphologists including Bloom and Stuvier (1963) and Bloom and Ellis (1966). Tidal marsh development can indeed document (in the form of deposits) the correlative recession of adjacent headlands, though there seems to be no particular reason for the processes to match precisely. Sediment cores from such marshes frequently record more than one fluctuation in sea level, and they have commonly been used to gauge glacio-eustatic sea level relations where augmented by C_{14} dating (Fig. 9.33).

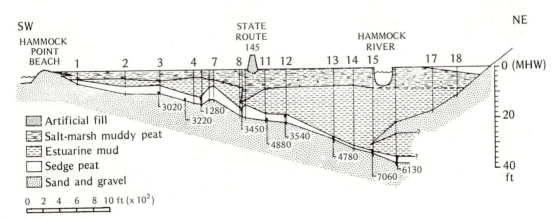

Fig. 9.33. Cross section of the Hammock River tidal marsh near Clinton, Connecticut. (From Bloom and Ellis, 1966.)

In many tidal bays and estuaries of the tropics, the marshes are mangrove swamps. In the final analysis, deposits made along an irregular coast are seen as a waning phenomenon as long as erosion of headlands progresses. Many irregular coasts presently display a mixture of erosional and depositional configurations (Fig. 9.34). This is particularly understandable when one considers that the marine modification of most subaerial landforms virtually ceases with each glacio-eustatic drop in sea level. In other words, relative to Pleistocene Glacial fluctuations, coastal erosion and deposition along the present coastline probably has only occurred about half or less of the time. There is, in fact, no particular reason to believe that the long-term Pleistocene mean sea level is even close to the inner margin of the continental shelf.

A rather different type of coastal development than that just described may occur along irregular coasts characterized by abundant unconsolidated sediment. The Carolina Capes, for example, are deposits of distinctly regular arcuate form associated with a relict coast of notable irregularity (Fig. 9.35). Previous explanations for the depositional characters of the capes in question include a genesis by eddy currents related to the Gulf Stream (cf. Tuomey, 1848; Abbe, 1895), hardly confirmed by recent current studies (Bumpus, 1955; Gray and Cerame-Vivas, 1963), and a genesis by waves moving sediment from embayments to capes to shoals (cf. Cooke, 1936; Zenkovitch, 1964), a notion generally counter to the concept of headland erosion (Hoyt and Henry, 1971). In addition, Dolan and Ferm (1968) suggest that

Fig. 9.34. A cove and adjacent headlands developed on the Atlantic coast of Morocco south of Tangier. Sediment eroded from the headland or shifted along the coast tends to be deposited in the bay with consequent straightening of the coast. (W. N. Gilliland.)

long-term variations in ocean state may involve a mechanism similar to that which produces beach cusps (cf. Komar, 1971). And White (1966) suggests that relict capes have perhaps localized more recent examples.

More recently, Hoyt and Henry (1971) combine the requirement for headland erosion on irregular coasts with the concurrence of Carolina Capes and rivers, noting that the capes probably developed initially as shelf deltas during low sea level stands (Fig. 9.36). In this intriguing view, the sediment disposition is largely as a relict deltaic phenomenon, and the present cape configuration represents a progressing erosion of an irregular soft sediment coast in a relatively traditional fashion; present beach-dune ridges and spits now being either, in their words, ". . . destroyed or transitional deposits which accumulate during periods of erosion." The same coast shows irregularities which may be due to eolian erosion and deposition (cf. Price, 1968b).

Terrigenous Coasts of Primary Morphologic Equilibrium

In some cases, where land and sea come together they are, as the saying goes, "well met." As indicated in the coastal classification outline (pp. 545), there are several possible origins for the "equilibrium" lands. But the soft sediment examples all share the general attribute of being nearly planar, possibly in a broadly undulatory fashion, so that the juncture between land and water is comparatively smooth, sometimes gently arcuate, or curvilinear.* Initial shoreline configurations developed where a planar landform intersects the sea offer few outstanding targets for coastal erosion (i.e., headlands). Thus, few pronounced irregularities exist to be smoothed out by deposits. And since many planar landforms are of depositional origin, sedi-

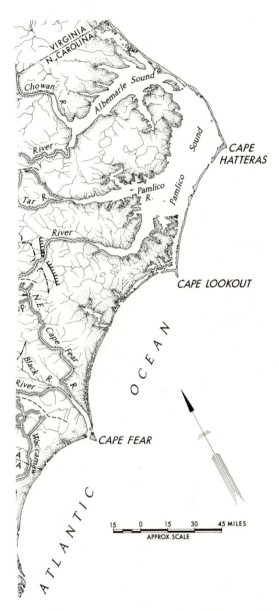

Fig. 9.35. Broadly arcuate coastal configuration of the Carolina coast near Cape Hatteras compared with the enclosed, irregular coastal configuration situated landward. (After White, 1966.)

ment, apart from that supplied by rivers, is often readily available.

Coasts that are comparatively regular

---

* One or another of the hierarchy of cuspate forms noted by Dolan and Ferm (1968) may secondarily modify this primary form.

but that lack appreciable unconsolidated sediment are rather rare. The southern coast of Sweden that presently rather precariously balances a glacial plain in a state of post-glacial isostatic rebound against a glacio-eustatic rise in sea level comes rather close. Sediment is not particularly abundant, but scour has generated sufficient glacial relief, now expressed as islands, to place the coast in the disequilibrium category. Coastal regions in Arctic Canada display similar traits. It is in depositional surfaces such as the emergent sea floors of South Island, New Zealand (Fig. 9.37), and the submerged desert coasts of Iran or Texas (Fig. 9.38) that we begin to see true primary strand-line morphologic simplicity. But with it comes sediment. In the Persian Gulf the coastal materials include much unconsolidated detritus. And in the northern Gulf of Mexico, rivers such as the Rio Grande and Mississippi continue to make deposits of

mud and sand as if there was no limit to the supply of these substances. There is. But the beaches are now composed of the same materials.

Barrier Beaches and Islands

The extent to which the morphology of soft sediment coasts is a relict phenomenon as opposed to a marine adjustment is a subject for open debate. A glance at a map of coastal Texas and many other similarly arid or semiarid coasts is sufficient to establish that several formerly sizable river débouchérs are represented by silted-up bays behind *barrier beaches* in places such as Corpus Cristi and Galveston. *Barrier beaches* are elongate sand ridges bordered on the landward side by a lagoon. They commonly occur in linear groups (*barrier chains*) with segments separated by *tidal inlets* (Fig. 9.39). If Hoyt and Henry (1971), are correct and the Carolina

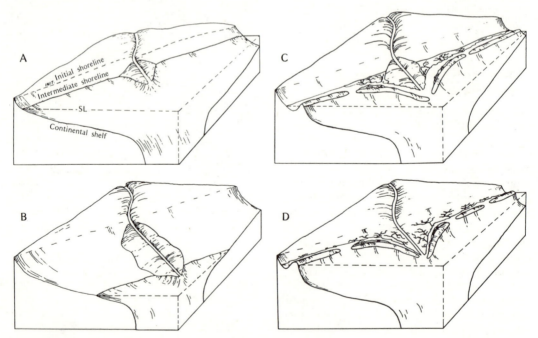

Fig. 9.36. Stages of development of the Carolina Capes involving (A) deposition of sediment at river mouth during lowered sea level; (B) development of deltaic ridge on continental shelf from abundant river sediment; (C) development of barrier islands along deltaic ridges by submergence of dunes; and (D) continued submergence, headland erosion, and barrier island retreat leading to present configuration. (From Hoyt and Henry, 1971.)

Fig. 9.37.   Newly emergent Motunau Coast of North Canterbury, New Zealand, displaying virtually un-modified depositional topography (undissected) with a veneer of recent shelly deposits. Notched, former headlands, backed by more irregular topography, are visible (right foreground). (V. C. Browne.)

Capes represent reworked shelf deltas, then *barrier beaches* and *barrier islands* that smoothly approximate the outer contour of the mainland coast must reflect some other type of genesis (cf. Hoyt, 1967). One could readily argue that former Texas rivers generated no deltas or that they did but the "cape phase" is completed. Either notion would be extremely difficult to prove or disprove. However, the subaerial climatic histories of the Carolinas and of southwest Texas differ greatly, and this probably has influenced both river flow and delta building. But let us backtrack and consider the theories and evidence pertaining to what used to be called *offshore bars* (cf. Price, 1951; Shepard, 1952).

In his study of raised shorelines of Lake Bonneville in the western United States, Gilbert (1890) noted that the relict offshore bars he termed *barrier bars* might have been built by shore drift. Johnson (1919), viewing offshore bars as features of an emergent shoreline, borrowed an earlier idea from De Beaumont (cf. Thornbury, 1969, p. 435) and said that such bars were wave built from material eroded from the sea floor. Holmes (1965) points out the difficulty of such a bar ever being raised above sea level, an idea elsewhere echoed by Price (1947, 1954, 1968) and Shepard (1960, 1963). In his recent synopsis of barrier features, Price (1968c) states that barriers begin as a single water-laid beach ridge, sometimes vegetated and raised by eolian sand dunes. The addition of *beach ridges* on the seaward side may prograde the barrier, presumably where there is abundant sediment supply. Price (1968) does not indicate the precise origin of the initial "water-laid beach ridge" other than to note that it is backed by a lagoon

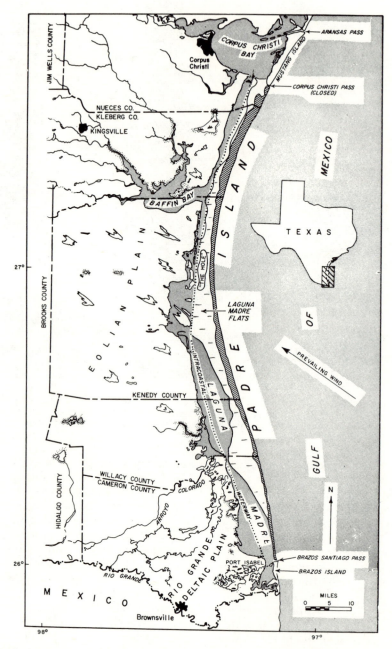

Fig. 9.38. Padre Island, an offshore barrier bar along the south Texas coast. (After H. N. Fisk.)

and may be formed as a longshore bar created at storm surge level (cf. Price, 1962). He also maintains that work by Marmer (1951) and McKee and Sterrett (1961) fail to prove that longshore bars could form barriers through tidal action. Price takes the view that barriers are equilibrium structures, built up and prograded during a marine transgression of Flandrian Time (cf. Chapter 2), though not recently expanded and not destined for eventual destruction.

It is difficult to reconcile the final statement in the last paragraph with the realities of coastal sedimentation in a world of changing climates. The quantity of coarse sediment (sand) required for barriers and beaches elsewhere and introduced into the seas in late- and post-Wisconsin time was finite. Present rates of influx along many coasts with barriers are much reduced for a variety of reasons. In the Texas region the main cause of slower sediment influx is present aridity; in the Carolinas, there

is reason to believe that prolonged upland humidity is involved, coupled with an elimination of relict arid coastal valley alluvium; and, along the New Jersey shore, many drainages are characterized by coarse glacial alluvium now stabilized by fluvial bypassing. There is also the problem of relicts. Barriers are known from many coasts including those of New Jersey, Florida, northwestern Gulf of Mexico, Iceland, Australia, and Zululand, in south Africa. In South Africa, Antony Orme (in Coates, 1973) describes a barrier system surmounted by dunes up to 550 ft (180 m)

Fig. 9.39. Moriches (tidal) Inlet cutting through the offshore barrier bar, Long Island, New York. Note the large amounts of sand carried into the lagoon by tidal currents.

high, backed by lagoons, swamps, and relict dune complexes. Contemporary sedimentation there is man-accelerated and probably can maintain barriers at present, but in many of the other regions, gradual removal of sand by the sea should ultimately reduce the barriers. Furthermore, insofar as can be determined, most barrier systems are on coasts of submergence; at least some could be built on depositional nuclei created at lower stands of the sea and since built up as they were progressively drowned (cf. Hoyt, 1967; Otvos, 1970). Other barriers may relate to down-current deposition on spits tied to headlands or islands.

Cuspate Coastal Configurations and Beaches

There are clearly many kinds of beaches. Some types have already been discussed herein, and, apart from gross areal and cross-profile morphology, they vary mainly in composition—shingle, sand, etc. The dominantly sandy ones concern us here, mainly because their morphology relates most closely to the waves and currents that move along a shore and to the wind and tides that variously work around, through, and over the deposits in question. Thornbury (1969) alludes to the temporary character of the sedimentary veneer that comprises a beach. Most extensive beaches share a basic morphology which can be expressed in cross-profile or block diagram (Fig. 9.40). As noted by Shepard and LaFond (1940), however, individual attributes of beach morphology vary with the seasons (cf. Figs. 9.20; 9.21; 9.46). Seasonal changes are mainly a matter of rather long-term variations in wave and current directions and intensities. Also, storms can induce similar changes on a short-term basis. In addition, tidal level affects beach morphology, and beach shape is observed to change on a small scale as the tide rises and falls.

Beach cusps are one of the most notable transient morphologic aspects of beaches.

Genetically, cusps are also of considerable interest (cf. Figs. 9.40; 9.41). They develop on the seaward slope of the *berm* (Fig. 9.39) as alternating arcuate shallow swales and low, seaward-pointing ridges (*horns*) under certain sea and wave conditions. Ridge heights range up to 20 ft (7 m) and spacings vary from 15–220 ft (5–77 m). Within limits, the height and width of the cusps relates to wave height and extent of uprush (see Longuet-Higgins and Parkin, 1962). Storm waves coming from many angles tend to obliterate cusps as do constantly oblique waves according to Holmes (1965) who states, "Consequently, cusps are formed and maintained only by waves advancing with their crestlines parallel to the shore."

Opinions differ as to how beach cusps are formed (cf. Jefferson, 1899; Holmes, 1965; Russell and McIntire, 1965; Dolan and Ferm, 1968; Dolan, 1971), and, if destroyed by unfavorable water conditions, how reformed. There is some suggestion that *spilling breakers* as defined by Lewis (1938) provide a more favorable climate for cusps than *plunging breakers* (Fig. 9.42). Studies of dyed pebbles show that once the cusp form develops, it tends to concentrate backwash and cause erosion in the embayments with deposition on the ridges. Holmes (1965) suggests that a possible explanation for initial cusp development is first that wave orientation is parallel to shore, i.e., their approach is at a 90° angle, and that, starting with a smooth slope, the backwash acts like a sheet-flood and produces a series of rills or gullies. The writer previously mentioned such a rill pattern (Fig. 9.13) associated with beach cusps on the New Jersey coast. The same features have been termed rhomboidal ripple marks (Guilcher, 1958), apparently under the assumption that they reflect rhombic interference ripple patterns like those commonly developed between a longshore bar and the low-tide terrace of the foreshore at times of low water and slack tide (see Fig. 9.40). Holmes goes on to note that enlargement of the rill systems

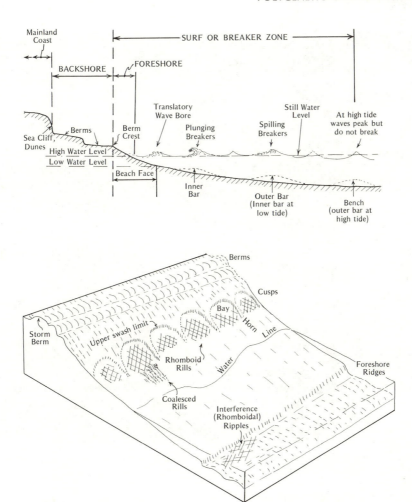

Fig. 9.40. Cross section and block diagram showing terminology of beach areas. The block diagram focuses on the shore and indicates development areas of cusps, rhomboidal rills, and interference (rhomboidal) ripples. (Modified from several sources including Guilcher, 1958, and Le Méhauté, 1968.)

(presumably in response to the water volume to be handled) could produce cusp depressions. Presumably the spacing would be governed by backwash volume which would, in turn, relate to beach slope, wave height, and related factors. Price and Zenkovitch (1964) relate cuspate sand spits to variations in shoreline curvature and wave fetch which affect sediment transport efficiency.

More recently, Dolan and Ferm (1968) and Dolan (1971) have argued that there is a hierarchy of cuspate migrating *sand*

waves along beaches which because they are larger should be distinguished from beach cusps (Fig. 9.43). This notion has been echoed by Komar (1971) who attributes the features to "cell circulation," rip current activity, and associated longshore currents. He also notes that Shepard (1963, p. 195) previously recognized these features as *giant cusps*. Dolan's measurements showed cusp spacings of 500 to 3,500 ft (150 to 1,000 m) with most between 1,500 and 1,800 ft (500 and 600 m) and horns projecting 50 to 85 ft (15 to 25

Fig. 9.41. Beach cusps developed along the northern Lake Superior shore in coarse sand. Distance from horn to horn averages about 7 ft (2 m).

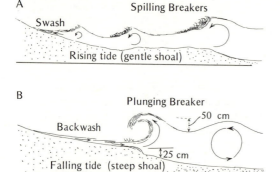

Fig. 9.42. Diagrammatic sections through spilling breakers (A) and plunging breakers (B). The type that develops depends upon several factors including wind and tide direction, bottom slope, and water depth. In general, the spilling type is favored by a gradual shoal, whereas the plunging type often occurs along a steep beach. Dimensions given are those from examples along the New Jersey shore.

m) seaward from embayments. Komar (1971) concludes from wave tank experiments and actual observations that cusps develop in lee of rip currents in some cases (Fig. 9.44), either in isolation or in a rhythmic series. The available evidence suggests that we may be dealing with at least two types of features, since observed flow relations in beach cusps do not relate to rip currents off cusp horns (cf. Figs. 9.44 and 9.45). It seems clear that more research is required before our understanding of cuspate coastal features is complete.

Beach morphology, apart from cuspate phenomena, appears to relate to several types of fluctuations in water conditions previously noted. For a quantitative treatment of this subject the reader is referred to works by Bagnold (1946), Longuett-

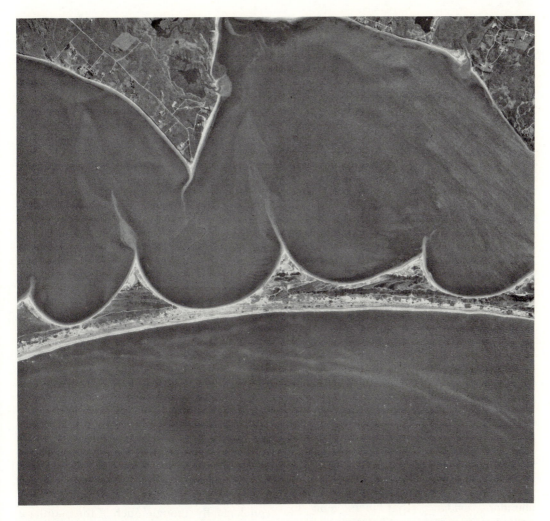

Fig. 9.43. Giant cusps on the lagoonal side of a barrier island off the tip of Nantucket Island near the Massachusetts coast. The two larger cusps measure approximately three-quarters of a mile (1.2 km) from horn to horn. A hierarchy of such cyclic beach forms may exist along coasts. (U. S. Geological Survey.)

Higgins (1953), Bruun (1962), and Le Méhauté (1968). For the most part, shores suffer retrogradation and steepening apparently due mainly to winter storms. Under quieter summer conditions the beaches may be built outward (progradation) and the profile lowered and flattened (cf. Schofield, 1968a; 1968b) (Fig. 9.46). Dune building in the backshore areas is largely a matter of wind intensity and sand supply coupled with variations in vegetal conditions and overwash fan development, as noted by Price (1968).

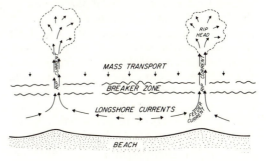

Fig. 9.44. Near-shore cell circulation system of rip currents and associated longshore currents. (From Komar, 1971.)

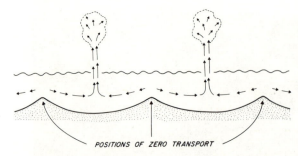

POSITIONS OF ZERO TRANSPORT

Fig. 9.45. Cuspate shoreline produced by sand transport under currents of the cell circulation (Fig. 9.44). Cusps would occupy positions of zero sand transport with rip currents developed in embayments. (From Komar, 1971.)

Organic Coasts

Organic coasts are more intrinsically dependent upon particular marine water environments than they are upon the relict terrain morphologies from which a coast is being developed. Where conditions for reef or mangrove development are met along a coast, subsequent morphologic and sedimentary developments may be essentially independent of the coastal changes one might expect to occur along a terrigenous coast. This is particularly true of reef-buttressed coasts where open-ocean wave and current effects are generally prevented from reaching the mainland shore. In accordance with the ecologic conditions set forth in an earlier section, there is abundant evidence that reefs tend to assume a protective configuration, either enclosed or linear, as an equilibrium morphology. This tendency may be accompanied by a progradation of the strand in the form of reef rock, often in headland areas where terrigenous coasts would be most expected to recede by erosion. Under ideal conditions, breaks in reef fronts tend to be limited to those openings or channels required for tidal flood and ebb. Unlike plants on land, reefs tend to leave lithic relics. Most reefs were presumably killed to levels some 300 ft (100 m) lower than those of the present some 20,000 years ago

(Fig. 9.17). Uninhabited relict reefs of this origin are still common on many coasts (cf. Barbados), and erosional breaches through many reefs are not yet healed by new growth.*

Reef Buttressed Coasts

Many kinds of coastal configurations are either composed of or garnished by organic reef complexes. As noted in the section on coastal environments, and later in our discussion of erosional coastal segments, some morphologically unstable coasts may be reinforced by reefs and so stabilized, though, in effect, an equilibrium adjustment has been made between the organic assemblage of the reef community and the adjacent sea. Most of the reefs that function this way in an early stage of development are fringing reefs (Fig. 9.47) which lie near to or even on the shore and areally reflect the mainland morphology closely (cf. Fig. 9.19). In other instances, as along the northeast Australian coast and a number of other tropical strands, reef developments have extended seaward, straightened, and coalesced to form an essentially complete organic layer along the land-water interface. The Australian Great Barrier Reef (Fig. 9.48) is a striking example of a regional coastal equilibrium established organically and reflected morphologically, though much of the mainland coast behind the reef is highly irregular.

As noted by McNeil (1954, 1968), an

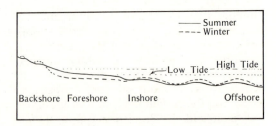

Fig. 9.46. Cross section of seasonal beach sediment redistribution resulting in profile changes.

* At least some of the "beachrock" (coral framework debris) that litters shores in the West Indies may relate to relict fringing reef deposits.

Fig. 9.47. Waves breaking on a fringing reef, this one about 300 ft (100 m) from the shore near San Juan, Puerto Rico.

Fig. 9.48. Aerial photo of a reach of Australia's Great Barrier Reef off the northeast coast. (R. W. Fairbridge.)

atoll ". . . is an annular organic reef enclosing a lagoon. The shape ranges from circular to very irregular. The reef may or may not have gaps or passages." Theories as to atoll development include work by Darwin (1838), who recognized the role of submergence, Daly (1915), who noted prior glacio-eustatic sea level lowering and erosion, and McNeil (1954), who cited evidence for notable subaerial karst modifications prior to recent development (Fig. 9.49). The tendency for the atoll to assume an enclosed circular form is here taken to express a general protective morphology reflecting an equilibrium interaction between organisms and waves in the open ocean. Plant clumps in deserts exhibit an analogous morphology for analogous reasons (cf. Chapter 6). Some atolls relate to submerged volcanic peaks, and some have central, inorganic islands. Many have grown up from submergent platforms, and large numbers of the central lagoons are open. An emergent atoll is often expressed as an island with a raised, parapet-like wall around it, termed *makatea* (Fig. 9.50). The former lagoon is often a swampy area, and, as in the case of atolls, there may or may not be a volcanic core area.

Mangrove Coasts

In many protected tropical coves, bays and estuaries and along coasts characterized by waves of low energy like that of southwest Florida, the strands may exhibit

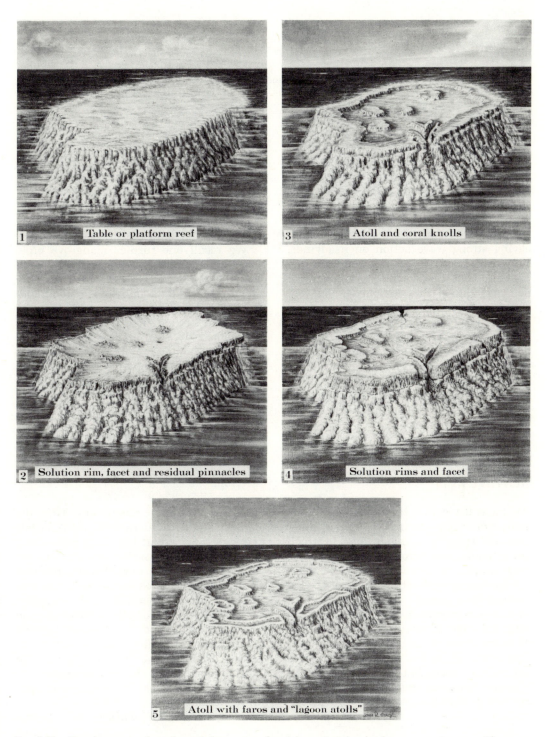

Fig. 9.49. Development of atolls and faros on subaerial erosion forms shown as cutaways with ocean water removed. (From MacNeil, 1954.)

Fig. 9.50. Cross section of a segment of Atiu Island in the Cook Group (20°2′ S, 158°7′ W) showing emergent atoll topography (Makatea) around a volcanic core. Length of section is about 1 mi. (After Marshall, 1930.)

tidally submerged mangrove swamps. Scholl (1963, 1964–65, 1968) maintains that they are the modern counterpart of the marine and brackish-water coal swamps of the late Paleozoic.* Mangrove swamps are restricted to protected inlets where they are associated with disequilibrium coasts having strong wave action on headlands. Mangrove roots, of course, tend to trap mud and hold it, and some shore progradation may thus occur, but only under very quiet water conditions. Mangroves are salt tolerant (halophytic), naturally, but cannot root in water deeper than 2 ft at low tide (cf. Guppy, 1906; Ridley, 1930; Scholl, 1963). Therefore, mangroves are exposed fixtures of equilibrium coasts only where the equilibrium involves a very low energy level, very fine-grained sediments,

and shallow water. Scholl (1968) observes that our knowledge of mangrove coasts is inhibited by their tropical remoteness, the complexities of estuarine labyrinths, high humidities and air temperatures, biting insects, tropical diseases, impenetrable vegetation, and malodorous mud.

We recall one further mangrove problem in the Estero Salado of south Coastal Ecuador. A man with a surveying rod stepped out of a dugout canoe onto a mudbank, and, had he not been holding the graduated 4 m length of wood horizontally, he might only have required a tombstone —which would no doubt have sunk also. As it was, he was only buried to his armpits in muck having the general appearance and consistency of old, warm axle grease.

SUMMARY

We have devoted a brief chapter to the subject of coasts which, like so many other geomorphic topics, includes items to which entire books are devoted. For the beginning student of geomorphology, we have tried to convey some of the scope of coastal geomorphic problems and their complexities. A retrospective look at what has been said gives the impression that the major problem areas have been touched upon. The treatment is less quantitative than it might have been, but some study of the literature suggests that this is even more an area for specialists than is fluvial hydrology.

In our synoptic examination, as in the detailed studies of many researchers, coasts "emerge" as one part land, one part water, and a continual series of adjustments along an interface toward some form of equilibrium. The equilibrium apparently develops most sequentially with respect to relict configurations, both morphologic and sedimentologic, and in evolving organic relationships. Equilibria are evidenced in several kinds of areal morphology, in strand composition changes, in organic associations, and in levels of energy expenditure and balance.

In general, our earlier classification

* Data presented in Chapter 11 suggest that this analogy is somewhat imperfect.

seemed to function reasonably well as a basis for discussion categories. In retrospect, however, we found that it was necessary to add a couple of categories presently not notable but undoubtedly important in times past, i.e., subaqueously channeled coasts such as would exist if the continental shelves and upper slopes were emergent.

Coastal studies are a particularly intriguing aspect of geomorphology from several standpoints. The agencies are more or less continuously active, and environmental displacements are continuous rather than a theory which needs to be proven as in many continental situations. Also, the environmental interface is notably sharp, and the association of interacting processes is readily available for observation. Complexities seem to stem mainly from sometimes unrecognized relict relations, both in terrigenous morphologies and sea level stands. In addition, the geomorphology of a coast is clearly related to a global aqueous geomorphic system where the modification of any segment often produces recognizable consequences elsewhere in landforms, deposits, or agency behaviors. The notion of "system" in a geomorphic sense is perhaps better learned along a coast than anywhere else.

With the strand behind us, it seems rather appropriate to turn our attention to mountains, to exchange, as it were, the cry of the gull for the scream of the eagle. There, in addition to the several interchangable SURFICIAL geomorphic systems, we encounter the culmination of ENDOGENIC forces expressed as relief and elevation, often supplemented by MESOGENIC events. There are many mountain ranges on this earth, too many to discuss each one. So Chapter 10 is devoted to the development of landscapes in the two main kinds of mountains discussed in Chapter 3. The kinds that relate to crustal subduction. One of these, the cordilleran-type mountain system, defies all prior erosional theories by resisting erosional reduction. The other, collision mountains, will also be considered in some detail. Erosional (residual), faultblock, and volcanic types have already been discussed.

REFERENCES

Abbe, C. J. (1895) "Remarks on the Cuspate Capes of the Carolina Coast," *Bost. Soc. Nat. Hist. Proc.*, v. 26, pp. 489–497.

Bagnold, R. A. (1946) "Motion of Waves in Shallow Water," *Roy. Soc. London Proc.*, v. 187, Ser. A.

Bajorunas, L. and Duane, D. B. (1967) "Shifting Offshore Bars and Harbor Shoaling," *Jour. Geophys. Res.*, v. 72 pp. 6195–6205.

Bloom, A. L. (1965) "The Explanatory Description of Coasts," *Zeits. für Geomorph.*, Bd. 9, Hft. 4, pp. 422–436.

Bloom, A. L. and Ellis, C. W. (1966) *Post Glacial Stratigraphy and Morphology of Central Connecticut*, Conn. Geol. and Nat. Hist. Surv. Guidebook I.

Bloom, A. L. and Stuvier, Minze (1963) "Submergence of the Connecticut Coast," *Science*, v. 139, pp. 332–334.

Bradley, W. (1958) "Submarine Abrasion and Wave-Cut Platforms," *Geol. Soc. Amer. Bull.*, v. 69, pp. 967–974.

Bretschneider, C. L. (1958) "Engineering Aspects of Hurricane Surge," *Amer. Meteoro. Soc. Tech. Conf. on Hurricanes Proc.*, Miami.

Bretschneider, C. L. (1966) "Storm Surge" [in] *Encyclopedia of Oceanography*, R. W. Fairbridge, ed., Reinhold Book Corp., New York, esp. pp. 586–860.

Brunn, Per (1968) "Sediment Transport—Fluvial and Marine" [in] *Encyclopedia of Geomorphology*, R. W. Fairbridge, ed., Reinhold Book Corp., New York, esp. pp. 982–984.

Bumpus, D. F. (1955) "The Circulation over the Continental Shelf South of Cape Hatteras," *Amer. Geophys. Union Trans.*, v. 36, pp. 601–611.

Coates, D. R. (1973) "Coastal Geomorphology," *Geotimes*, v. 18, pp. 22–23.

Coleman, P. J. (1968) "Tsunamis as Geological Agents," *Aust. Geol. Soc. Jour.*, v. 15, pp. 268–273.

Cooke, C. W. (1936) "Geology of the Coastal Plain of South Carolina," *U. S. Geol. Surv. Bull.*, v. 867, 196 p.

Cotton, C. A. (1961) "Growing Mountains and Infantile Islands on the Western Pacific Rim," *Geogr. Jour.*, v. 127, pp. 209–211.

Cotton, C. A. (1966) "The Continental Shelf," *N. Z. Jour. Geol. Geophys.*, v. 9, pp. 105–110.

Cotton, C. A. (1968) "Relation of the Continental Shelf to Rising Costs," *Geogr. Jour.*, v. 134, pp. 382–389.

Cotton, C. A. (1969) "Marine Cliffing according to Darwin's Theory," *Roy. Soc. New Zealand, Trans.*, v. 6, pp. 187–208.

Daly, Reginald (1915) "The Glacial Control Theory of Coral Reefs," *Amer. Acad. Arts & Sci., Proc.*, v. 51, pp. 155–251.

Darwin, Charles (1838) "On Certain Areas of Elevation and Subsidence in the Pacific and Indian Oceans, as Deduced from the Study of Coral Formations," *Geol. Soc. London Proc.*, v. 2, pp. 552–554.

Darwin, Charles (1844) *Geological Observations on Volcanic Islands*, London.

Darwin, Charles (1846) *Geological Observations on South America*, London.

Darwin, Charles (1891) *Geological Observations*, Smith Elder, London, 664 pp.

Deitz, R. F. and Fairbridge, R. W. (1968) "Wave Base" [in] *Encyclopedia of Geomorphology*, R. W. Fairbridge, ed., Reinhold Book Corp., New York, esp. pp. 1224–1228.

Dolan, Robert (1971) "Coastal Landforms: Crescentic and Rhythmic," *Geol. Soc. Amer. Bull.*, v. 82, pp. 177–180.

Dolan, Robert and Ferm, J. C. (1968) "Crescentic Landforms along the Mid-Atlantic Coast," *Science*, v. 159, pp. 627–629.

Dury, G. H. (1969) "Tidal Stream Action and Valley Meanders," *Austral. Geogr. Studies*, v. 7, pp. 49–56.

Dury, G. H., Sinker, C. A. and Pannett, D. J. (1972) "Climatic Change and Arrested Meander Development on the River Severn," *Inst. Brit. Geographers, Area*, v. 4, pp. 81–85.

Fairbridge, R. W. (1952) "Marine Erosion," *7th Pacific Sci. Cong. Proc.*, v. 3, pp. 347–358.

Fairbridge, R. W., ed. (1968) "Littoral Processes—An Introduction" [in] *Encyclopedia of Geomorphology*, Reinhold Book Corp., New York, esp. pp. 658–662.

Frankel, J. J. (1958) "Water-Faceted Pebbles from the Natal Coast," *So. Afr. Jour. Sci.*, v. 54, pp. 35–36.

Geyl, W. F. (1968) "Tidal Stream Action and Sea Level Change as One Cause of Valley Meanders and Underfit Streams," *Aust. Geogr. Studies*, v. 6, pp. 24–42.

Gilbert, G. K. (1890) "Lake Bonneville," *U. S. Geol. Surv. Mono.* 1, esp. p. 40.

Gray, I. E. and Cerame-Vivas, M. J. (1963) "The Circulation of Surface Waters in Raleigh Bay, North Carolina," *Oceano. Limnol.*, v. 8, pp. 300–337.

Groen, P. and Groves, G. W. (1962) "Storm Surges" [in] *The Sea*, M. N. Hill, ed., v. 1, John Wiley and Sons (Interscience), New York, esp. Chapt. 17.

Guilcher, A. (1958) *Coastal and Submarine Morphology*, Methuen and Co., London, 274 pp.

Gulliver, F. P. (1899) "Shoreline Topography," *Amer. Acad. Arts & Sci., Proc.*, v. 34, pp. 149–258.

Guppy, H. B. (1906) *Observations of a Naturalist in the Pacific Between 1896 and 1899; v. II, Plant Dispersal*, Macmillan and Co., London 627 pp.

Holmes, Arthur (1965) *Principles of Physical Geology*, Ronald Press Co., New York, esp. pp. 782–840.

Hoyt, J. H. (1967) "Barrier Island Formation," *Geol. Soc. Amer. Bull.*, v. 78, pp. 1125–1136.

Hoyt, J. H. and Henry, V. J., Jr., (1971) "Origin of Capes and Shoals along the Southeastern Coast of the United States," *Geol. Soc. Amer. Bull.*, v. 82, pp. 59–66.

Jefferson, M. S. W. (1899) "Beach Cusps," *Jour. Geol.*, v. 7, pp. 237–246.

Johnson, D. W. (1919) *Shore Processes and Shoreline Development*, John Wiley and Sons, New York, 584 pp.

Johnson, J. W. (1966) "Ocean Currents, Introduction" [in] *Encyclopedia of Geomorphology*, R. W. Fairbridge, ed., Reinhold Book Corp., New York, esp. pp. 587–590.

Kidson, C. (1968) "Coastal Geomorphology" [in] *Encyclopedia of Geomorphology*, R. W. Fairbridge, ed., Reinhold Book Corp., New York, esp. pp. 134–139.

King, C. A. M. (1959) *Beaches and Coasts*, Edward Arnold, London, 403 pp.

King, C. A. M. (1972) *Beaches and Coasts*, St. Martins Press, London.

Komar, P. D. (1971) "Nearshore Cell Circulation and the Formation of Giant Cusps," *Geol. Soc. Amer. Bull.*, v. 82, pp. 2643–2650.

Le Mehaute, Bernard (1968) "Littoral Processes—Quantitative Treatment" [in] *Encyclopedia of Geomorphology*, R. W. Fairbridge, ed., Reinhold Book Corp., New York, esp. pp. 667–672.

Lewis, W. V. (1931) "The Effect of Wave Incidence on the Configuration of a Shingle Beach," *Geogr. Jour.*, v. 78, pp. 129–143.

Lewis, W. V. (1932) "The Formation of Dungeness Foreland," *Geogr. Jour.*, v. 80, pp. 309–324.

Lewis, W. V. (1938) "The Evolution of Shoreline Curves," *Geol. Assoc. Proc.*, v. 49, pp. 107–127.

Lobeck, A. K. (1939) *Geomorphology*, McGraw-Hill Book Co., New York, esp. pp. 329–366.

Longuet-Higgins, M. S. (1953) "Mass Transport in Water Waves," *Roy. Soc. Phil. Trans.*, A, pp. 245, 535–581.

Longuet-Higgins, M. S. and Parkin, D. W. (1962) "Sea Waves and Beach Cusps," *Geogr. Jour.*, v. 128, pp. 194–201.

Lucke, J. B. (1938) "Marine Shorelines Reviewed," *Jour. Geol.*, v. 46, pp. 309–324.

Marmer, H. A. (1951) "Tidal Datum Planes," *U. S. Coast & Geodetic Surv., Spec. Pub.* 135, 142 pp.

McGill, J. T. (1958) "Map of Coastal Landforms of the World," *Geogr. Rev.*, v. 68, pp. 402–405.

McKee, E. D. and Sterrett, T. S. (1961) "Laboratory Experiment on Form and Structure of Longshore Bars and Beaches" [in] *A.A.P.G. Geometry of Sandstone Bodies*, Peterson and Osmund, eds., Tulsa, pp. 13–28.

McMaster, R. L. and Ashraf, A. (1973) "Sub-Bottom Basement Drainage System of Inner Continental Shelf off Southern New England," *Geol. Soc. Amer. Bull.*, v. 84, pp. 187–190.

McNeil, F. S. (1954) "The Shape of Atolls: An Inheritance from Subaerial Erosion Forms," *Amer. Jour. Sci.*, v. 252, pp. 407–427.

Miller, D. J. (1961) "Giant Waves in Lituya Bay, Alaska," *U. S. Geol. Surv. Prof. Pap.* 354-C, pp. 51–83.

Otvos, E. G. (1970) "Development and Migration of Barrier Islands, Northern Gulf of Mexico," *Geol. Soc. Amer. Bull.*, v. 81, pp. 241–246.

Price, W. A. (1947) "Equilibrium of Form and Forces in Tidal Basins of the Coast of Texas and Louisiana," *Amer. Assoc. Pet. Geologists Bull.*, v. 31, pp. 1619–1663.

Price, W. A. (1951) "Barrier Island, not 'Offshore Bar'," *Science*, v. 113, pp. 487–488.

Price, W. A. (1954) "Dynamic Environments—Reconnaissance Mapping, Geologic and Geomorphic, of Continental Shelf of Gulf of Mexico," *Gulf Coast Assoc. Geol. Soc. Trans.*, v. 4, pp. 75–107.

Price, W. A. (1962) "Origin of Barrier Chain and Beach Ridge," *Geol. Soc. Amer. Abs. for 1962*, p. 219.

Price, W. A. (1963) "Patterns of Flow and Channeling in Tidal Inlets," *Jour. of Sed. Pet.*, v. 33, pp. 279–290.

Price, W. A. (1968a) "Tidal Inlet" [in] *Encyclopedia of Geomorphology*, R. W. Fairbridge, ed., Reinhold Book Corp., New York, esp. pp. 1152–1155.

Price, W. A. (1968b) "Carolina Bays" [in] *Encyclopedia of Geomorphology*, R. W. Fairbridge, ed., Reinhold Book Corp., New York, esp. pp. 102–109.

Price, W. A. (1968c) "Barriers—Beaches and Islands" [in] *Encyclopedia of Geomorphology*, R. W. Fairbridge, ed., Reinhold Book Corp., New York, esp. pp. 51–54.

Price, W. A. and Zenkovitch, V. P. (1964) "Cyclic Cuspate Sand Spits and Sediment Transport Efficiency," *Jour. Geol.*, v. 72, pp. 876–880.

Richthofen, F. F. von (1886) *Führer für Forschungsreisende*, Robert Oppenheim, Berlin, pp. 294–315.

Ridley, H. N. (1930) *The Dispersal of Plants throughout the World*, L. Reeve and Co., London, 744 pp.

Russell, R. J. (1967) "Aspects of Coastal Morphology," *Geografiska Ann.*, v. 49, pp. 299–309.

Russell, R. J. and McIntire, W. G. (1965) "Beach Cusps," *Geol. Soc. Amer. Bull.*, v. 76, pp. 307–320.

Schmitz, W. J., Jr., Robinson, A. R. and Fugilister, F. C. (1970) "Bottom Velocity Observations Directly under the Gulf Stream," *Science*, v. 170, pp. 1192–1194.

Schofield, J. C. (1968a) "Prograding Shoreline" [in] *Encyclopedia of Geomorphology*, R. W. Fairbridge, ed., Reinhold Book Corp., New York, esp. pp. 894–896.

Schofield, J. C. (1968b) "Retrograding Shoreline" [in] *Encyclopedia of Geomorphology*, R. W. Fairbridge, ed., Reinhold Book Corp., New York, esp. pp. 940–942.

Scholl, D. W. (1963) "Sedimentation in Modern Coastal Swamps, Southwestern Florida," *Amer. Assoc. Pet. Geologists Bull.*, v. 47, pp. 1581–1603.

Scholl, D. W. (1964–65) "Recent Sedimentary Record in Mangrove Swamps and Rise in Sea Level over the Southwestern Coast of Florida," *Marine Geology*, v. 1, pp. 344–366; v. 2, pp. 343–364.

Scholl, D. W. (1968) "Mangrove Swamps: Geology and Sedimentology" [in] *Encyclopedia of Geomorphology*, R. W. Fairbridge, ed., Reinhold Book Corp., New York, esp. pp. 683–688.

Shaw, A. B. (1964) *Time in Stratigraphy*, McGraw-Hill Book Co., New York, esp. pp. 1–81.

Shepard, F. P. (1937) "Revised Classification of Marine Shorelines," *Jour. Geol.*, v. 45, pp. 602–624.

Shepard, F. P. (1952) "Revised Nomenclature for Depositional Coastal Features," *Jour. Geol.*, v. 45, pp. 602–624.

Shepard, F. P. (1960) "Gulf Coast Barriers" [in] *A.A.P.G. Recent Sediments NW Gulf of Mexico*, F. P. Shepard *et al.*, eds., Tulsa, pp. 197–220.

Shepard, F. P. (1963) *Submarine Geology*, Harper and Row, New York, 557 p.

Shepard, F. P. (1968) "Coastal Classification" [in] *Encyclopedia of Geomorphology*, R. W. Fairbridge, ed., Reinhold Book Corp., New York, esp. pp. 131–133.

Shepard, F. P. and La Fond, E. C. (1940) "Sand Movements along the Scripps Institution Pier," *Amer. Jour. Sci.*, v. 238, pp. 272–285.

Silvester, Richard (1959) "Engineering Aspects of Coastal Sediment Movement," *Waterways Harbors Div., Jour.*, v. 85, pp. 11–39.

Silvester, Richard (1963) "Design Waves for Littoral Drift Models," *Waterways Harbors Div., Jour.*, v. 89, pp. 37–47.

Silvester, Richard (1966) "Wave Refraction" [in] *Encyclopedia of Oceanography*, R. W. Fairbridge, ed., Reinhold Book Corp., New York, esp. pp. 975–976.

Steers, J. A. (1953) *The Sea Coast*, Collins, London, 276 pp.

Tanner, W. F. and Bates, J. D. (1965) "Submerged Beach on a Zero-Energy Coast," *Southeastern Geol.*, v. 7, pp. 19–24.

Thornbury, W. D. (1969) *Principles of Geomorphology*, John Wiley and Sons, New York, esp. pp. 420–444.

Tricker, R. A. R. (1964) *Bores, Breakers, Waves and Wakes*, Amer. Elsevier. N. Y., 250 pp.

Tuomey, M. (1848) *Report on the Geology of South Carolina*, A. S. Johnson, Columbia, S. C., 293 pp.

Valentin, H. (1953) "Die Küste der Erd, *Petermanns Geogr. M. H.*, v. 246.

Van Dorn, W. G. (1965) "Tsunamis," *Hydroscience Advan.*, v. 2, Academic Press, New York.

White, W. A. (1966) "Drainage Asymmetry and the Carolina Capes," *Geol. Soc. Amer. Bull.*, v. 77, pp. 223–240.

Wiegel, R. L. (1964) *Oceanographical Engineering*, Prentice-Hall, Englewood Cliffs, N. J., 352 p.

Wiegel, R. L. (1966) "Wave Theory" [in] *Encyclopedia of Oceanography*, R. W. Fairbridge, ed., Reinhold Book Corp., New York, esp. pp. 977–986.

Wooldridge, S. W. and Morgan, R. S. (1937) *The Physical Basis of Geography*, Longmans, Green, London, 361 pp.

Zenkovitch, V. P. (1964) "Cyclic Cuspate Sand Spits and Sediment Transport Efficiency: Discussion," *Jour. Geol.*, v. 72, pp. 879–880.

Zenkovitch, V. P. (1967) *Processes of Coastal Development*, Oliver and Boyd, London, 738 pp.

10 Polygenetic landscapes/ mountains

Mountains are the most complex land-forms on earth, and this is particularly true of the cordilleran- and the collision-type mountain systems to be discussed in this chapter (cf. Chapter 3). Erosion of fault-block type mountains is discussed in Chapter 6. Montane geomorphic complexity has several origins. The complexity is partly tectonic, in the litho-structural sense that rocks encountered in collision mountain systems are often deformed and possibly intruded by igneous rock bodies and/or metamorphosed. In addition, there is the matter of dynamic tectonic relief generated in those orogens now under lateral stress and still being deformed. Also, of course, there is isostatic relief relating to accentuated crustal thickness in orogens and the interplay between these forces of uplift and those of denudation in the genesis of montane terrain. There are, moreover, the periodic influences of volcanism plus the mosaic of subaerial environments expressed in the atmosphere's gross fabric and related arid, humid, glacial, and cryergic geomorphic systems. All these factors can contribute to the complexities of montane geomorphology, and, in the world's great mountain systems, it is usually their combined interaction that produces the over-all landscape.

Because mountains incorporate the world's greatest extremes of subaerial relief and the greatest environmental contrasts in the smallest area, they provide what many regard as our most scenic landscape vistas (Fig. 10.1). No one who has hauled his body through thin air to a mountain's top and gazed across snow-encrusted slopes, icy peaks, sheer rock walls, evergreen forests, and steeply cascading streams is likely to forget what he has seen or fail to become somewhat intrigued as to how it all came about. Through a proper blend of fact and theory, we shall attempt to define the major aspects of mountain geomorphology.

Chapter 3 outlines the broad tectonic and lithologic aspects of collision mountain systems, so we will here get down to cases in terms of actual terrain, beginning with ideas about the origin and persistence of mountainous elevations and relief. Then we will consider some theories of erosion in developing orogens, mainly with respect to subaerial conditions but also with due regard for tectonic denudation. Subaerial conditions involve the hydrographic settings of orogeny and the geomorphic effects where a mountain mass rises higher and higher and higher . . . , in terms of the kind of conditions it encounters and

Fig. 10.1. Mount Gardner, Ellsworth Mountains, Sentinel Range at lat. 88° W, Antarctica. (Thomas Bastien.)

induces. Some attention will then be given to those very high mountain systems whose extreme uplift generates geomorphic environments that otherwise would be encountered mainly near the poles. Of course, we must also consider the sequential development of montane environments and the reflections of this sequence in drainage forms, glaciation, and the asymmetric hydrographic character of many mountain systems. Finally, we will consider the overall consequences of these tendencies in terms of late stages of mountain denudation and relief evolution in various environmental settings.

OROGENIC ELEVATION AND RELIEF

In general, the highest places on the earth's surface are underlain by comparatively thickened segments of crust and most are in isostatic balance in terms of the base of the continental crust defined seismically (cf. Chapter 3). Some mountain systems, including portions of the Andes and the Himalayas, appear to occupy collision positions between two crustal plates. Since they are situated in a lateral stress setting, a truly bouyant isostatic adjustment may not be possible, though whether these mountains would be higher or lower than at present in the absence of continuing collision is difficult to say. The Andean and Himalayan situations are hardly analogous in any case, since the former is an island-arc-subduction zone adjacent to a continent (cordilleran type) and the latter is a continent-to-continent (collision) relation.

The general position taken in Chapter 3 was that orogens probably begin to acquire mountainous elevations as crustal thickening occurs, that is, during the collision phase when sediment wedges accumulated next to continents or island arcs are compressed, deformed, and possibly intruded. Our concern is mainly with the isostatic-dynamic rise of the thickened crustal segment. The isostatic balance in question is one of rock masses, not of surficial relief. Thus, the same amount of thickened crust may isostatically be expressed as a plateau averaging 15,000 ft (4,500 m) in elevation, or a mountain range a third again as high (cf. Fig. 10.2). Furthermore, as noted by Holmes (1965, p. 576), localized erosion, for a time, will induce uplift to levels higher than those initially caused by crustal thickening (Fig. 10.3).

The amount of crustal thickening and resultant mountainous elevation that develops during a given orogeny is, of course, a volume-density problem. In general a thicker cork will float relatively higher than a thinner cork. Where continents collide, the volume of new continental crust is largely a matter of the size of the sediment wedge that accumulated before the collision. In island arc-trench settings, the conveyor belt is presumably adding seafloor sedimentary material more or less constantly along with an occasional continental sliver or "other island arc" or "guyot." Limiting factors on relief in the latter situation would appear to be how long the system continues to function and what is on the subduction "menu." It has

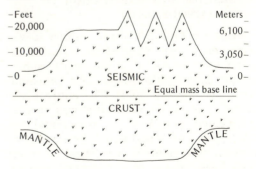

Fig. 10.2. Relations between relief forms on the seismically defined crust. The plateau to the left would possess the same mass above the base line as the mountains to the right but with less extremes of elevation.

A

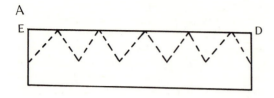

B

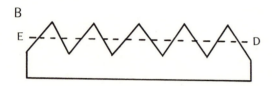

Fig. 10.3. Diagram illustrating (A) uplift of a segment of the earth's crust to an elevation datum (ED), followed by incision of a valley system (dashed line). Loss of mass (B) causes renewed uplift of peaks to altitudes even greater than the original datum. (After Holmes, 1965.)

been suggested that the Tibetian-Himalayan region is so high because there is a double thickness of continental crust (part India, part Asia). But parts of the Andean —Bolivian mountain and plateau complex achieve comparable heights without another continent in sight. If any continents there have been subducted and assimilated in times past, one would expect them to reappear later as batholiths (cf. Sierra Nevada?).

CHARACTER AND MAINTENANCE OF MOUNTAINOUS UPLIFT

During the first half of the century it was common to read discussions of mountain erosion in terms of initial uplift, subsequent downwearing, and later rejuvenation, i.e. renewed uplift. The entire hypothetical sequence of events was largely predicated upon W. M. Davis's normal erosion cycle which postulated these kinds of crustal movements on the basis of certain presumed phases of the normal erosion cycle (Davis, 1899) and assumed landform consequences. In a brief summation of ideas treated in Chapters 1 and 5, streams were assumed to incise their

courses during uplift and thereafter until erosional base level was reached, whereupon lateral erosion generated valley "flats" and ultimately "peneplains." Re-incision of the planar results of this hypothetical sequence was presumed to occur by renewed uplift. Proof of the validity of the assumed sequence was usually sought—and found— in the assemblage of escarpments and planar areas in various mountain systems including the Appalachian and Rocky mountains. The over-all notion came to be known as the "polycyclic erosion theory" of mountains (cf. Cotton, 1960).

Objections to multiple uplifts of a mountain system following an orogenic episode have come from two general areas of geologic research. Geophysical studies have gradually documented a state of isostatic balance for those mountain systems not presently in a "collision" relationship (Woollard, 1969). In effect, this condition of isostatic adjustment in most mountain systems proves that these systems are free to adjust vertically by uplift when they lose rock mass through denudation. One might say, "There's no finger on the floating cork." This last remark refers back to the Davisian notion that a land could be eroded down to a base level of erosion near sea level without an accompanying isostatic adjustment. Removal of mass from the surface of a crustal segment without allowing compensatory uplift should create an area of isostatic imbalance (i.e. an anomaly). It would be analogous to holding a floating cork below its normal buoyant elevation with a finger. However, apart from plate collision zones, geophysical surveys make no record of crustal areas that are both low and very thick. "There's no finger on the cork." Crustal areas thus tend to rise to a balanced level related to their thickness and the densities of the material involved. Most geophysical documentation of the foregoing ideas has been made since the 1950's, with converging evidence being assembled from other sources.

Research discussed in Chapters 5, 6, 7, and 8 makes it clear that linear incision by running water is a matter of runoff flow continuity with the sediment-water volume relationships normally encountered under humid conditions, sometimes related to meltwater, and in a few desert uplands. Planar landforms generally relate to processes of deposition and planation, mainly as developed in deserts and elsewhere by continental glaciers and subaqueous processes. As a consequence, the landform configurations used by Davis and his followers to document polycyclic orogenic uplife are seen to have little or no relationship to tectonic movements—in any direction. The writer (1959) called attention to these facts in a study of the Ecuadorian and Peruvian Andes Mountains where there is evidence for deformation and uplift beginning at least as early as Paleocene time. In that report it was asserted that the present depth of valley incision was no particular measure of recent uplift but only a reflection of a proper fluvial environment and a terrain situated well above regional base level.

In the same Andean situation just mentioned, Dresch (1957, 1958) had only just applied the Davision polycyclic erosion-uplift rationale to orogenesis, a fact not overlooked by Cotton who compared and contrasted the two opposing views (1960). The writer responded in kind (1963a), and, from these published exchanges and ensuing correspondence, it became evident that in the Andes the geomorphic evidence in favor of multiple, high-intensity uplifts following erosional base leveling left a great deal to be desired. Principally at issue here, and advocated by Dresch, Cotton, and several previous workers in the area, was a matter of 12,000 ft (3,500 m) of postulated vertical uplift in the Pleistocene REQUIRED BY THE INTERPRETATION OF THE BOLIVIAN-PERUVIAN PLATEAU (ALTIPLANO) AS A PENEPLAIN DEVELOPED NEAR SEA LEVEL AT THE END OF THE TERTIARY PERIOD. Evidence for the aforementioned uplift was mainly an unverified pre-Pleistocene beach, some unstudied fossils, and a sea horse (*Hippocampus* sp.) supposed to inhabit Lake Titicaca since the latter was an arm of the Pacific Ocean in the "Late Tertiary?" (cf. Welter, 1947). It should be mentioned that the geophysical arguments against such an uplift were never marshalled at the time of the debate in question, though they seem rather overpowering in retrospect. It should also be noted for the record that a team of divers led by Jacques Costeau thoroughly explored Lake Titicaca in the mid-1960's for a variety of purposes including some archaeologic pursuits. In the course of their explorations the divers documented for a large TV audience the only apparent animal life—a species of frog.

In sum, it would appear that mountain systems tend to achieve an initial elevation commensurate with the related crustal thickness induced by orogenesis. In general, and with respect to various orogenic types discussed in Chapter 3, two main types of geomorphic relief situation appear to exist. (1) In island-arc-trench-continent settings like the Andes Mountains, where there is more or less continual crustal input via subduction, initial relief is periodically or continuously augmented by added crustal mass and thickness as subduction persists. After initial uplift, tendencies for denudational lowering of the range may be countered by subductive additions to the mountain root and related essentially continuous uplift—high montane elevations being maintained throughout the subduction episode (Fig. 3.17). (2) In continent-to-continent collisions (p. 179), the intervening trench disappears, and no further additions of crust are made to the mountain root following orogenesis. Denudational losses of mass from the mountain range must then be compensated for by isostatic uplift with resultant absorption of the mountain root as the range is erosionally lowered (Fig. 10.4).

In both the foregoing situations, only the initial orogenic uplift topography (Fig. 9.4) would tend to be carried to higher montane elevations, provided, of course, that it was not otherwise destroyed. Once the maximum montane uplift is established for a given orogenic system, landforms found thereon would tend to be

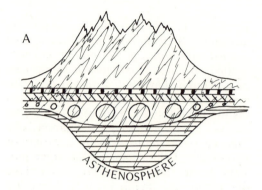

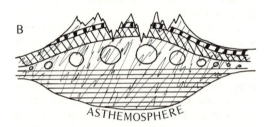

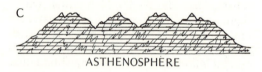

Fig. 10.4. Cross section illustrating transfer of root material during erosional lowering of a collision-type mountain system, depicted in three stages. (A) Initial compression and uplift; (B) upward displacement of deep-seated rocks as crestal mass is lost; (C) exposure and lowering of the root zone typical of many shield regions.

situated near the elevations at which they were formed. Not long after the writer argued for high-level development of high-level planar landforms in the Andes (1959), Mackin (1959) made a similar appeal for the origin of such features in the Rocky Mountains. There seems to be little reason to doubt that a similar mode of genesis applies to many of the planar landforms of other mountain systems. This means, of course, that a new rationale must be sought to explain various aspects of montane terrain. For example, low-inclination surfaces in mountain systems can hardly be graded to regional base level as King (in Cotton, 1960) insisted and Davis (1905)

earlier acknowledged. Another erosional datum must be active (cf. Garner, 1965). We will shortly examine these and many of the related geomorphic details of mountain systems and their principal variants. First, however, we must survey some of the ideas that have been advanced for early orogenic denudation apart from those outlined in some detail in Chapter 3.

EARLY OROGENIC DENUDATION

It seems logical to begin our consideration of the various modes of montane sculpture with the inception of uplift. Denudation in one form or another probably begins with the creation of one or more elongate positive crustal areas which, depending upon setting, may be either subaerial or subaqueous. The tectonic forms of dunudation that undoubtedly contribute greatly to initial aspects of mountainous terrain are discussed in some detail in Chapter 3. Here, our immediate preoccupation will be with the mainly subaerial forms of erosion and deposition that begin to shape mountains even as they are first being uplifted. Several efforts have been made to relate denudation rates to orogeny (Cotton, 1963; Schumm, 1963; Ritter, 1967), for the most part without attempting to consider variations in orogenic environments or environmental interactions (cf. Schumm, 1968; Janda, 1971). These early erosional events must set the pattern for subsequent terrain changes and are presumably reflected in the sedimentary record of progressing orogenesis, both at present and in times past.

A number of early geomorphologists, including Davis (1889) and Penck (1924), proposed developmental erosion schemes for uplifted crustal areas, mainly according to their own notions on landscape development. Penck visualized sporadic uplift reflected by structural scarps and erosional flats (cf. Fig. 1.11), whereas Davis conceived an elaborate staged form of fluvial incision to explain structuro-topographic *relief reversal* in folded mountains such as part of the Appalachians. In the latter ranges, many present mountains are formed on synclines, and adjacent valleys are anticlines (Fig. 10.5); this same phenomenon is common elsewhere. The applicability of Davis's scheme depends upon a number of factors, not the least of which are induration of materials, effects of tectonic denudation as a probable early erosion form, and the character of later subaerial environments.

Diagenesis has to do with changes in sediments after they have accumulated, and it is unfortunate that we cannot go deeply into the subject here, for much of the material that was initially removed from rising mountain chains was sedimentary, and it would help to know what condition the material was in, and particularly the degree of lithification. Perhaps it will be sufficient for us to here rely on pertinent remarks in Chapter 3 dealing with the subject of orogenic lithology, and to additionally observe that the "first-uplifted" materials in an orogen were axiomatically the last to accumulate prior to uplift, the least deeply buried, and therefore, often, little consolidated. Studies of many deposits made during orogenic uplift show that they frequently include such materials as mudstones, estuarine and lacustrine clays and marls, continental sands and gravels, pyroclastic debris, and similar material. These rock types in stratified sequences commonly include evidence for penecontemporaneous slumping, sliding, and contortion. It is such lithic materials that often form large fractions of allochthonous slide masses involved in tectonic denudation. That such allochthons also often include large bodies of igneous and metamorphic rocks is an indication of the extent to which tectonic denudation may expose older and better consolidated rocks in an orogen. In general, it may safely be assumed that subaerial agencies denuding a

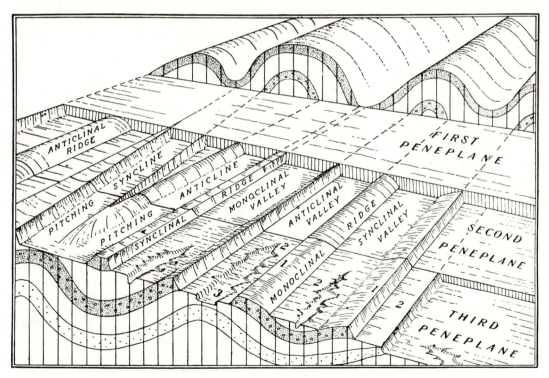

Fig. 10.5. Illustration of erosional modification of a folded mountain belt following the erosional schemes of W. M. Davis. (After Lobeck, 1939.)

rising orogen have erosional access to a wide range of lithologic types, including both metastable and semi-insoluble rocks and minerals, in suites that include appreciable, well-indurated bedrock.

The classic ideas about denudational products eroded from a rising crustal area, as in so many other facets of geomorphology, are found to go back to Davisian notions of landform development, and particularly to Davis's remark (1899) earlier quoted to the effect that eroded sediments would become coarser as relief increased. With modifications, this idea came to dominate geologic thought, and, as noted by the writer (1959), a standard element of earlier tectonic and stratigraphic reports was the equation of coarse-grained sediments with high relief, high elevation, nearness, and tectonic instability (uplift) of source area. Conversely, fine-grained materials, particularly argillites and solutes, were attributed to sources that were presumably low, distant, and stable. There is, of course, the observed relation of coastal sediments that beaches are coarsest and offshore deposits most fine as an expression of energy-sorting relations (cf. Chapter 8). Once again we are involved with geomorphic sedimentology. The reader should keep in mind that sediment studies provide a major basis of orogenic theory, not so much in the actual studies of existing mountains (cf. Garner, 1959; Oberlander, 1965), but rather in the assumption that conglomerates deposited near known orogens reflect periods of instability (cf. Wegman, 1957).

As pointed out in Chapter 5, the logic behind the relationship of coarse sediment and high relief and the converse appears to be faulty. So does the logic that invariably relates coarse debris to tectonic instability and vice versa. Certainly it would seem unwise to argue that a joint block dropped from a 1,000-ft (300-m) cliff would shatter at the bottom into larger pieces than one of similar size that is gradually wedged loose by weathering at the base of the same cliff. And, granting that crustal movements may shatter bed-

rock, the more intense stresses and closer fracture spacings in orogens should produce smaller clasts there than the widely spaced fractures in low-elevation cratons. Certainly, sedimentary transport tends to reduce clast size. Though, as noted in Chapters 5, 6, and 7, and in the forthcoming section on paleogeomorphology, some transport media appear to exact a lower clast-size-reduction per unit distance traveled than others. The occasional instance of tectonic denudation is usually identifiable in alloththon structures. Depositional fault breccias are comparatively rare. For these reasons—set forth in some detail by Garner (1959, 1967), Van Andel and Curray (1960), Trask (1961), Moberley (1963), Crook (1965), and Krook (1968)—the sedimentary record of orogenesis will be presumed to mainly reflect subaerial conditions of weathering and erosion, interpersed in some instances by products of tectonic denudation, volcanism, turbidity currents, and similarly transient phenomena. We will therefore turn our attention to environments of orogenic uplift.

ENVIRONMENTS OF OROGENIC UPLIFT

In our discussion of uplift environments in orogens we must make several basic asumptions. First, where a deposit made adjacent to a rising mountain system seems genetically compatible with a particular subaerial environment and geomorphic system, it seems reasonable that landforms compatible with the system are developing correlatively. Second, where a particular environment appears to dominate during an entire orogeny, the related mountains will reflect this domination in landforms and deposits of related type. Third, an environmental sequence induced on a rising mountain mass by uplift will tectonically place climatic relicts in new geomorphic systems much as an atmospheric climate change achieves similar effects in non-orogenic settings. Fourth, sedimentary records of orogenesis may reflect the order in which rocks are exposed in an orogen, the order in which tectonically induced climates develop, and shorter-term environmental fluctuations, but only rarely will lithology be a direct expression of crustal movement—and that last not ordinarily, regularly, or cyclically repeated.

Much of Chapter 4 is devoted to land-environment distributions, particularly as reflected in amounts of moisture (hydrography) and correlative geomorphic systems. The detailed characteristics of the major types of geomorphic systems are set forth in Chapters 3 through 9. Here we are concerned with geomorphic effects where a deforming mountain belt encounters one or more of the earth's subaerial environments. And it seems best to consider the comparatively simple situation where a rising mountain chain encounters only one subaerial geomorphic system throughout the main initial episode of uplift. This would be equivalent to orogenesis in a climatic nucleus. Actually there are few mountain systems that fall entirely within a single geomorphic environment. But there are many mountain systems with segments within one general geomorphic system from base to top—at least on one side. The general atmospheric zonation that would be encountered by a rising mountain mass is illustrated herein (Fig. 4.7).

A thoughtful review of remarks made in earlier chapters about the relationships between tectonism and climate leaves the impression that climates change much more frequently than portions of the earth's crust move. In general, this seems to be true. But it must be kept in mind that hydrography is in large part a creature of geography. Thus, the formation of a land area (even a long, low, narrow one) in the eventual geographic position of a mountain system will tend to modify basic patterns of moisture distribution through its effects on ocean currents and air movements. Fluctuations in air and water patterns and heat may induce climate changes. But they will be changes in the long-term

pattern induced by the land-water disposition. And if the long-term pattern favors moisture or dryness in a certain area, then shorter-term effects will be geomorphically subordinate. Moreover, once a mountain system is elevated and, in any appreciable measure, transverse to a global wind system, conditions on opposite sides will tend to differ (cf. Garner, 1959) and so will landscapes on those opposite sides (cf. Figs. 2.49 and 2.50).

In conformance with previous remarks, forthcoming statements (to the effect that a particular mountain system entered a particular environment during uplift) may only apply to one side of a mountain system, or even to only a part of one side. Mountains situated in the Doldrum Belt (New Guinea), or essentially parallel to a prevailing wind system (Brooks Range, Alaska), tend to be bilaterally symmetric in a climatic sense. Those more or less transverse to prevailing wind systems are climatically asymmetric as indicated above. With due respect to these variants, we will first consider a case of humid orogenesis (Fig. 10.6).

Humid Orogenesis

Orogenic belts characterized by selva-type topography over more or less extensive areas are not particularly uncommon. Several islands of the East Indies including parts of Sumatra, Borneo, Celebes, and New Guinea exhibit this type of terrain, essentially from sea level to the peaks of the highest mountains (cf. Sapper, 1935; Ruxton, 1967, 1968). The same relationship is found on the north coast of Ecuador adjacent to the Andes Mountains (Fig. 10.7) and along much of the Sierra Del Norte of Venezuela facing the Caribbean Sea (Fig. 10.8). In the latter instance there is a narrow coastal desert with arid influences that now extend up-slope only a few tens of feet (cf. Fig. 2.30). Even though dry conditions may have affected the north face of the humid Venezuelan range in times past, their effects are sub-

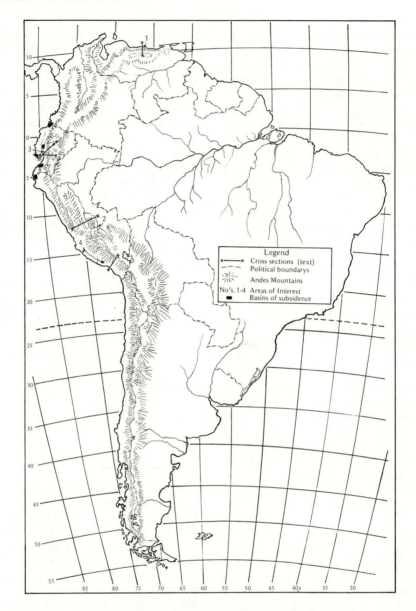

Fig. 10.6. Location map showing Andes Mountains areas of interest, including cross sections and block diagrams (Figs. 10.8, 10.12, 10.14, and 10.41) and basins of subsidence during the Tertiary which from south to north are, respectively, at Talara, Tumbez, Guayas, Daule, and Esmeraldas (cf. Figs. 10.9, 10.10, 10.15, and 10.16). Areas of interest include (1) Venezuelan Sierra del Norte, (2) Peruvian Sierra Oriental, (3) Ecuadorian Sierra Occidental, and (4) Peruvian Sierra Occidental.

ordinate to the effects of chemical weathering and fluvial incision which now prevail (cf. Garner, 1959). Outcrops are sparse, zones of decomposition are locally thick (5–6 m), and all slopes are contiguous with extant drainage lines so that there is

no undrained land. Similarly, the valleys contain no alluvial fills (Fig. 5.45). The north coastal range of Venezuela has received a recent geomorphic appraisal from Maloney (1965) who maintains that what little coastal plain there is relates to wave erosion. However, Royo y Gomez (1956) has observed that the beach terraces there merge with alluvial fans at river mouths which Maloney relates to glacial low sea levels. The same episodes have been equated with times of low elevation aridity in this part of the tropics (cf. Garner, 1966; Krook, 1968).

The reader may rightfully now say, "Well and good, the Venezuelan coastal mountains are now humid, but what assurance is there that it was thus during uplift of the range." The immediate answer would have to be, "None." But that is not the point. The mere fact that a mountain front more than 10,000 ft (3,500 m) high can be dominantly humid means that an orogenic uplift in such a setting could also occur under dominantly humid conditions. The same type of humid, vegetated selva topography exists from sea level to elevations approaching 15,000 ft (5,000 m) on the west flank of the Cordillera Occidental in northern Ecuador and southern Colom-

bia (cf. Fig. 10.7). In the latter area there is presently no coastal desert, so the possibilities for uninterrupted humid uplift to high elevations are even greater. What is more, the Tertiary stratigraphic record in northern Ecuador suggests that this is precisely what happened.

At the mouth of the Esmeraldas River, in the Ecuadorian province of the same name along the Colombian-Ecuadorian border, there is a sedimentary basin of deposition that filled during and after uplift of the Andean orogen, beginning in Paleocene time (Fig. 10.9). It will be recalled from an earlier discussion (Chapter 5) that a region of prolonged humid morphogenesis should yield a sedimentary suite dominated by sand, argillites, and solutes. Also, in a tropical setting where silica could readily dissolve, the sand fraction might be absent (cf. Garner, 1959). The basal Tertiary deposits in the Esmeraldas basin are a siliceous, Eocene limestone overlain in shelf areas by alternating

Fig. 10.7.   Sketch based on a photo of selva topography under rain forest on the west flank of the Andes Mountains, Ecuador, elevation about 6,500 ft (2,000 m), along the road from Duran to Tambo. Similar topography slopes directly to the sea along many humid mountainous coasts.

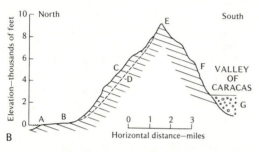

Fig. 10.8. Block diagram (A) and cross section (B) of the Sierra del Norte, Venezuela, showing selva topography and regional slope relations (cf. Fig. 10.6 for location). Presently, the north side of the range is humid except for a narrow, arid coastal strip, whereas the south side is semiarid to arid. Vertical exaggeration of (A) ×3.

carbonate deposits and shale but elsewhere entirely composed of shale aggregating some 10,000–12,000 ft (3,500 m) of thickness (see Fig. 10.10). Only in the younger Tertiary deposits (Mio-Pliocene) do gravel and sand occur in appreciable amounts. The major Andean uplift in this area was certainly accomplished by Oligocene time, and its sedimentary counterparts are what one would expect from a dominantly humid tropical region, whether mountain segment or not.

Before we judge the geomorphic significance of the Esmeraldas Tertiary sediments and associated selva mountain front, it would be well to examine the terrain-deposit relations in other areas of the same mountain system where similar tectonic age, source elevation, composition, and proximity are involved so that any deposit composition variation must be related to environment. The reader may perhaps feel that undue attention is being paid to the Andes Mountains. But one must evaluate geomorphic relations where one has the data and the control and can minimize the variables. The Andean relations ascer-

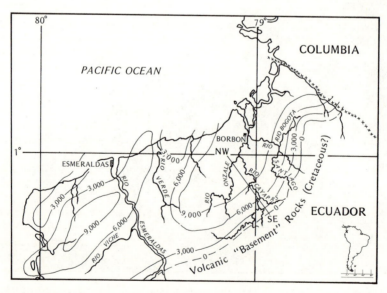

Fig. 10.9. Isopach map of the Esmeraldas Sedimentary Basin showing thickness distribution of Tertiary sediments (in feet) in this north Ecuadorian setting (map inset). Line NW-SE is cross section on Fig. 10.10.

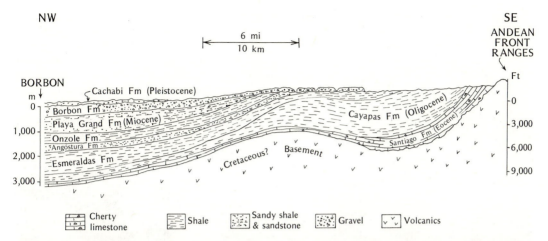

Fig. 10.10. Northwest-southeast stratigraphic cross section of the Esmeraldas Basin (cf. Fig. 10.9) depicting the dominantly argillaceous rocks. Even the thin sandy zones give way to shale to the southwest along the Esmeraldas River in the other illustration cited. (Based in part on data collected by T. Scott and K. Harris.)

tained as a model on the only major extant, cordilleran-type mountain system with good control of variables should serve as analogs in assessing the geomorphic histories of other similar orogenic belts.

Arid Orogenesis

There are a number of mountain ranges that are presently arid from base to crest, but most show some signs of climatic variations in times past, and many have been elevated to the point where their crests now incorporate glacial or cryergic conditions. Examples of arid mountain ranges include the Mccdonnell Ranges of central Australia (cf. Mabbutt, 1966a), the southeast Zagros Mountains in Iran and extensions in the arid ranges of Afghanistan (cf. Oberlander, 1965), the High Atlas Mountains of Morocco and Algeria (cf. Joly, 1952 and Dresch, 1941; 1952), and the western Andean Front Ranges in southern Peru and Northern Chile (cf. Jenks, 1948; Jenks *et al.*, 1956; Garner, 1959). In addition, of course, there are the arid fault-block mountains of the southwestern United States discussed in Chapter 6, and a large number of other desert mountain ranges in central Asia and Australia plus

the southeastern Andes Mountain flank in Argentina.

We obviously cannot consider each of these arid montane situations, and, in many, the record of orogenesis or the correlative deposits are too poorly known for our purposes. But there are other problems related to different loci of deposition under different environments. As pointed out long ago by Johnson (1901), an arid mountain range tends to bury its own base (Fig. 10.11). Montane arid residuals thus tend to accumulate as a continental facies, as recorded by Gignoux (1955), whereas the products of humid fluvial erosion tend to move to regional base level and accumulate as marine (or lacustrine) facies, as in the Esmeraldas example just considered. It is therefore not surprising that the intermontane depressions in the southwestern United States are shown by borings to contain fills of *fanglomerate* deposits with intercalated lacustrine (playa) materials. However, the relation there between the faulting, uplift, subsidence, and deposition is only correlated in a general way. We can postulate that the coarse, metastable fanglomerates reflect rigorous desert weathering, whereas finer-grained material and clay-rich argillites are favored by humidity as is lacustrine deposition. However, the

Fig. 10.11. Part of the Yucca Mountains northwest of Las Vegas, Nevada, one of several small mountain ranges gradually being buried in their own alluvial waste. View is to east across a reach of the Amargosa Desert. (See also Fig. 6.4.)

tectonic-geomorphic picture remains unclear without more precise dates of crustal movement and sedimentation.

On a dominantly arid mountain system one would anticipate evidence of extensive alluviation and associated planation terrain features as well. These exist on the western Andes Mountain flank in southern Peru (cf. Figs. 10.6; 10.12), though the geomorphic record is complicated by extensive volcanism during the same Tertiary period. At lower elevations, the record of non-marine aggradation is impressive all along the west flank of the Peruvian Cordillera Occidental (cf. Bowman, 1916;

Bosworth, 1922; Fenner, 1948; Jenks, 1948; Garner, 1959). A well was drilled in a supposed marine basin in the northern Tumbes Desert in the mid-1950's by people who wanted oil. Instead of oil, they got experience. The drill penetrated only a thick sequence of continental sand and gravel. In more southerly areas of Peru, and along the few major valleys, aggradation by coarse sand and gravel began during the Teritary Period and continues at present. Few streams reach the sea (cf. Bowman, 1916). Newell (1949) and Garner (1959) are among those who have emphasized the present retention of alluvium by the Atacama and Peruvian deserts and the slow correlative deposition in the Peru-Chile Trench of the adjacent Pacific Ocean Basin (Fig. 10.13). As would be expected with a climatic control of sediment character, in spite of the orogenic setting, the trench deposits are fine grained (cf. Trask, 1961). Farther south, off the presently humid and glaciated coast of Chile, as one would expect, the depositional focus is in the ocean, and, as noted by Lister (1971), the Chile trench has a substantial sediment fill. The montane desert region to the north has a record of climatic fluctuations (cf. Garner, 1959), and the occasional episode of fluvial incision is recorded by marine deposits in coastal sedimentary basins, such as that near Talara, or presumably in the ocean trench.

As in the instances of humid orogenic uplift, the fact that a mountain range is arid from sea level to crestal areas 15,000–20,000 ft (5,000–6,000 m) high is taken to be a most significant geomorphic relation. Evidence for climate changes involv-

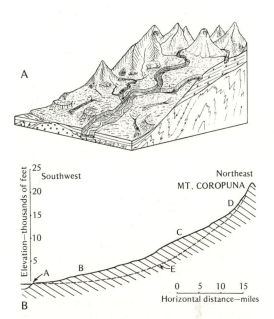

Fig. 10.12.   Block diagram (A) and cross section (B) of the Andes Cordillera Occidental, Peru (cf. Fig. 10.6 for location); note especially volcanic upland, arid pediments and plains, and regional slope relations. Vertical exaggeration of (A) ×10.

Fig. 10.13.   Sketch based on a photo of the Amatope Mountains, a foothill range of the Andes, northern Peru, in the Tumbez Desert. Foreground rocks are Cretaceous limestone which is fragmenting mechanically, whereas background uplands are Paleozoic metamorphics which have alluviated slopes and valleys.

ing humidity at high elevations increases northward into Peru, whereas glacio-humid effects dominate Chile to the south. Again, in essentially all the arid orogens cited, one cannot prove that initial orogenesis, uplift, and earlier montane sculpture occurred under desert conditions. But such a sequence would clearly be possible in an arid climatic nucleus such as that of the Atacama Desert of Peru and Chile. Before passing further geomorphic judgments on the significance of the desert terrain and coarse clastic continental deposits of the Atacama arid climatic nucleus and related orogen, we should search for a middle ground in the same type of tectonic setting. The climatic transition zone between the deserts of Peru and the humid selvas of Ecuador appears to be a place where the entire arid-humid climatic-tectonic relationship can be clarified. It is a region where alternating aridity and humidity are a part of the subaerial tectonic record (cf. Garner, 1959, 1965, 1967, 1968).

Arid-Humid Orogenesis

The Andean Sierra Occidental of northern Peru is replete with valleys which are extensively alluviated, and these conditions persist into adjacent Andean ranges of southern Ecuador where the desert alluvium is presently found under rain forest at elevations above 1,000 ft (300 m) (cf. Figs. 10.6 and 10.14). As illustrated, alluvial deposits clearly derived from the mountains are as yet incompletely removed from montane valleys and occur in remnant patches on the adjacent coastal lowland. Regolith presently developing under humid, tropical conditions includes saprolite and laterite representing thorough chemical decay of bedrock. Depths of soil weathering increase from south to north and away from present desert areas. It should be emphasized at this point that the Andean orogeny is not believed to differ greatly in age from any of the western South American areas discussed here.

Along the coastal lowland of Ecuador and northern Peru, there are a number of structural depressions, some of graben-like character, which contain sediments correlative with the associated Andean orogeny. The Esmeraldas Basin already discussed is an example, and it will be recalled that its argillite and exsolutional rock content matches an adjacent selva mountain landscape. In coastal Ecuador 250 mi (350 km) to the south is the Guayas Basin, a major area of sedimentation throughout the Tertiary (Fig. 10.15). Situated near the present mouth of the Guayas River, this basin area first received an allochthonous mass of Paleocene sandstone and conglomerate due to tectonic denudation of the rising Andean orogen more than 100 mi (150 km) to the east (cf. Chapter 3, mesogenic landforms). Blocks of the allochthon now in the form of horsts outline the main areas of Tertiary sediment accumulation (cf. Fig. 3.85).

As shown very generally in cross section (Fig. 10.16), the oldest non-tectonic Tertiary deposits are middle Eocene Seca Shale, Socorro Sandstone, and San Eduardo Limestone, the latter, as far as is known, constituting one of the only two organic limestone reef deposits of the entire upper Cretaceous-Tertiary sequence of sediments south of the equator in west-coastal South America. Non-marine Eocene sand and coarse gravel is overlain by the Eocene-Oligocene Las Canas Formation boulder-clay, deltaic deposits interfingered with foraminiferal marine sand, and clay which gives way upward to the Dos Bocas Shale of Oligocene age. Some depositional fault breccias are known from the Eocene-Oligocene deposits. Later, Oligocene and Miocene cobble-to-boulder fanglomerates alternate with deltaic marine boulder clays, Mio-Pliocene diatomites, clays, and Pliocene Progresso Formation estuarine-marine calcareous sandstone and siltstone, followed, of course, by the aforementioned continental Pleistocene gravels and the "other" marine limestone (Tabalzo Formation) which was apparently in part deposited on a pediment. Tertiary sediment thicknesses aggregate 12,000–15,000 ft (3,700–4,600 m).

The foregoing description of Guayas Basin deposits stands in stark contrast to the one previously cited for the Esmeraldas Basin in northern Ecuador some 250 mi (325 km) away. Both differ markedly from the continental gravels of western Peru and Chile farther to the south. Each Ecuadorian site is about 1.5° from the equator. Curiously, the northern basin with the fine-gained and exsolutional rocks lies immediately adjacent to the mountain front, whereas the southern basin with its sporadically coarse conglomerate filling lies 50–75 mi (70–100⁺ km) from the mountain front that shed much of the rock waste. All segments of the range are somewhat volcanic (cf. Chapter 3) and so are parts of the adjacent basin deposits of essentially all ages. The ranges are also of essentially equal elevation and lithologic

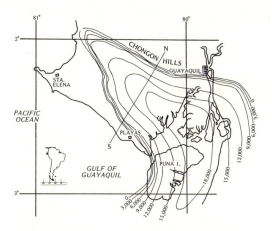

Fig. 10.15. Isopach map showing thickness of Tertiary sedimentary rocks in the Guayas Basin, southern Ecuador (map inset), by contour (3,000 ft interval). Cross section of lithology (NS) is Fig. 10.16.

diversity. The sole major variable that remains to account for the differences in landscape and derivative deposits along the western Andes Mountain front would appear to be subaerial "geomorphic" climate. And in the instances discussed in some detail, the tectono-climatic regime is expressed mainly in terms of climatic stability or variability.

In brief summation of the three main types of tectono-climatic setting so far discussed, we can recognize:

(1) A humid orogenic setting typified by selva topography with narrow V-shaped valleys and knife-edge ridges usually mantled by chemically decomposed regolith and drained by rivers carrying mainly fine-grained sediments and solutes (Fig. 10.7).

(2) An arid orogenic setting typified by pediment, pediplain, bench-and-bluff terrain, and occasional badland topography augmented by eolian deflation and deposition effects and traversed by occasional rivers flowing intermittently from sporadically more moist uplands. Few drainage lines are deeply incised. Weathering modes are dominantly mechanical (in effect), and sediment derivatives, where not eolian, are mainly dependent upon mass wasting and an occasional flash flood extending to regional base level (Fig. 10.12).

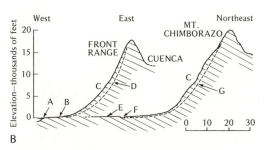

Fig. 10.14. Schematic block diagram (A) and cross sections (B) of the Andes Mountains, western Ecuador, show relict arid alluvial valley fills in ranges rising above the Gulf of Guayaquil, plus mountain front-coastal lowland slope relations. Adjacent sedimentary basins filled during uplift of the range contain mixed coarse and fine-grained sediments (cf. Figs. 10.15 and 10.16). Vertical exaggeration of (A) ×8.

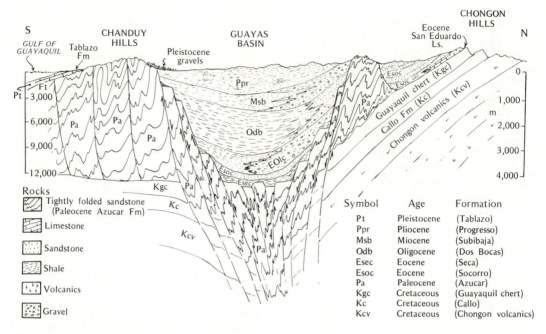

Fig. 10.16. Structural-stratigraphic cross section of the Guayas Basin along line NS on Fig. 10.15. Note especially the tightly folded Azucar allochthon (Pa)—relative motions along shears in this unit are only indicated in relation to block faulting which followed sliding.

(3) An arid-humid orogenic setting typified by ridge-ravine topography modified by alluvial fills in valleys. Drainage lines in the dominantly arid (low-elevation) portions are gravel strewn, and valleys are broader and more open than valleys in selva (cf. Fig. 7.17) and tend to narrow and deepen upslope where humidity is more nearly dominant (Fig. 10.14).

The north coastal range of Venezuela mentioned previously is more than 10,000 ft (3,000 m) high and is still humid on ridge crests thereby sharing a nearly total humid orogenic topography with terrains described from Ecuador (cf. Garner, 1959), from the Bewani area of New Guinea by Simonett (1967), and from Papua by Ruxton (1967); and these terrains occur elsewhere in Burma, Thailand and Indochina, and the Philippines. The higher ranges of the main Andean chain and those of other mountains elsewhere, including the Rockies and Himalayas, rise higher than those elevations that are normally humid or subject to humid-arid cli-

mate changes (cf. Fig. 4.7). Before turning to a discussion of montane topography related to extreme conditions of uplift, we should take time to compare and contrast the landscapes of our previous Andean examples. Several aspects should be emphasized.

Climatic Orogenesis at Moderate Elevations—A Synopsis

A major point of probable geomorphic significance in comparing tectono-climatic landscapes formed at low and intermediate elevations is the fact of humid orogenic zones lacking coastal low lands for the most part. The majority have mountain flanks which slope almost without interruption to the strand. A sole exception of note is part of the Sierra Del Norte of Venezuela, and it is here taken to be significant that the narrow Venezuelan coastal plain is occupied by an equally narrow coastal desert. In most of the other cases, including our northern Ecuadorian example, many of the humid selva mountains slope into the sea.

The arid Peruvian and Chilean Andes mountains are separated from the sea by a coastal pediplain or a series of pediment risers separated by low escarpments. Either the pediplain or pediment complex or both may be many tens of miles wide, being broadest opposite re-entrants in the western mountain front and narrowest next to salients extending seaward (Fig. 10.17). Where present, the coastal pediplain might be mistaken for a plain of marine erosion (cf. Chapter 9), but it is generally surmounted by gibber gravels, the gravels are often ventifacted and are devoid of marine deposits or fossils, except where they are capped by the locally deposited Tablazo Limestone Formation of Pleistocene age in northern Peru and southern Ecuador. The local cap of Pleistocene limestone indicates recent uplift of the pediplain averaging about 200 ft (60 m) in the Talara area

(Fig. 9.8). Elsewhere, according to Jenks (in Garner, 1959, p. 1331), similar surfaces have been uplifted by faulting 500–600 ft (150–175 m), others have tilted down, and some have been left apparently unmoved. Original erosional grading to a level near that of the strand is implied for some of these planation surfaces, but contemporary retreat of escarpments at the pediment heads is no longer dependent upon a distal base level—if it ever was. It is important to note that the lithologies beveled by the coastal pediplains of Peru and Chile are of the some general range and type as those in adjacent mountains.

The broadest coastal lowland next to the Western Andes Mountains is in southern Ecuador where we have a record of orogenesis during alternating aridity and humidity, not to mention high-altitude environmental effects and volcanism. At the latitude of the city of Guayaquil, about 1° south of the equator, it is about 150 mi (220 km) from the base of the mountain

Fig. 10.17.   The coastal pediplain of the Tumbez Desert, northern Peru, where foothills of the Amatope Mountains extend coastward. Similar planation surfaces typify much of arid, coastal Ecuador, Peru, and Chile.

front to the sea (Fig. 10.6). The Ecuadorian lowland is not a pediplain, however. Nor, with the climatic history indicated thereon, would we expect it to be if the region has been sporadically both humid and arid. Moreover, the area is structurally complex. Upper Cretaceous strata, including siliceous shale, graywacke sandstone, pillow basalt flows, volcanic ash and agglomerate sequences, massive chert formations, and local andesitic volcanic necks, have been involved in broad folding and are imperfectly truncated. More resistant Cretaceous lithologies and massive early Tertiary sandstone and conglomerate form localized, gullied uplands, averaging about 1,200 ft (375 m) high, surrounded by pedimented lowlands and patches of alluvium. An inland area of subsidence (Daule Basin) northeast of the Guayas Basin and not far above sea level was found by drilling to contain more than 7,000 ft of lacustrine clay and sand. The Guayas River "delta" and adjacent piedmont areas are extensively alluviated, and many areas are swampy. The coastal lowland narrows to the north, a direction in which humidity now intensifies, and our previous discussions of northern Ecuador indicate a history of similar conditions.

The three main types of tectono-climatic terrain found in the western Andes are sufficiently distinct so that we should be able to recognize their counterparts elsewhere (cf. Figs. 10.7, 10.12, and 10.14). It is somewhat more difficult to assess the significance of the various coastal lowland developments (or non-developments). The almost world-wide absence of coastal plains adjacent to humid orogens may signify an absence of rapid mountain flank recession under humid conditions. This notion is compatible with landslopes contiguous with drainage lines that are still being deepened (cf. Chapter 6)—a condition that holds for elevated drainage lines in the main ranges. One might also conclude that the broad lowland in southern Ecuador relates to rapid backwearing under alternating climates with periodic planation assists from pedimentation. The pediments in the

deserts to the south may be self-explanatory.

Many of these ideas are appealing. But the relationships involved are probably not that simple. Since the Andes Mountains occupy a continent-ocean trench margin, they are theoretically more or less continuously being replenished by mass acquired through subduction of their own denudation products. The Andean ranges are literally "feeding on themselves." This added mass is presumably being expressed by rather recent, local uplift along the Peru Coast, but, along the Ecuadorian coast, uplift seems to focus on the mountain mass proper. The lowland locally subsides. Also, during the Tertiary while the Ecuadorian Andes were being uplifted and were acquiring a snowy crown of 20,000-ft (6,000-m) volcanoes, a number of adjacent coastal graben subsided 10,000–15,000 ft (3,050–4600 m) (Fig. 10.6). The intervening horsts did not apparently rise commensurately, so the rock volumes that subsided displaced Cretaceous and older rocks on a fantastic scale and must be expressed in adjacent Andean uplift and volcanism, since they do not appear to have affected the adjacent trench. The Guayas Basin subsidence alone displaced some 6,500 cu mi (12,300 cu km) of rock, the continental Daule Basin to the northeast accounts for some 7,500 cu mi (14,500 cu km), the Esmeraldas Basin involves an estimated 5,000 cu mi (9,800 cu km), and, if the Gulf of Guayaquil is a submergent graben (as appears probable), at least another 8,750 cu mi (17,000 cu km) of rock is involved there. In all, more than 30,000 cu mi (50,000 cu km) of rock was "subducted" from the bases of the grabens in question apart from what may have been added by sedimentation in the adjacent trench. The coastal lowland of Ecuador is therefore crudely planar mainly because of deposition of sediment in areas of subsidence marginal to the main Andean uplift zone. The lowland deposition was able to fill most coastal depressions while adjacent positive areas were being alternately incised and pedimented.

It would therefore appear that the Ecuadorian coastal lowland is not mainly a product of backwearing of the Andean frontal ranges. True, there is evidence of some recession. However, the Mesozoic rocks of the coastal lowland are neither intensely deformed nor widely intruded and they are not metamorphosed. It is doubtful if they were ever greatly uplifted. As soon as one crosses from the lowland into the mountains there is an abrupt transition to intensely altered, intruded, and metamorphosed materials which have clearly been brought up from considerable depth. In contrast, as noted before, the bulk of the Peruvian coastal lowland bevels igneous and metamorphic rocks of the same types found in the sierras. There is little doubt but that the western Andean front in Peru and Chile has receded eastward through arid morphogenesis accompanied by pediplanation and aggradation.

UPLIFT MORPHOGENESIS: A THEORETICAL SYNTHESIS

It is not our intention here to duplicate descriptions and analyses set forth in Chapter 8 on the geomorphic results of alpine glaciation and related effects. Nor do we intend to recapitulate in any detail effects of desert or humid geomorphic systems outlined in Chapters 6 and 7. Instead we want to relate the specific topographic consequences of these kinds of geomorphic systems to the over-all character of montane sculpture during orogenesis. Since it is orogenic uplift that is the "prime mover," we can start at the bottom and work our way up. We will first embark on a largely theoretical discussion of the probable sequence of environments encountered by an uplifting orogenic belt. Later we will take up a series of actual examples of climatic orogenesis. It should be kept in mind at this time that none of the subaerial environmental conditions in question is entirely immobile. Any could be atmospherically displaced up or down onto terrains developed under other conditions, thereby causing rapid climate changes. Also, depending upon hydrographic setting and amount of uplift of a given mountain range, any environmental stage could be omitted.

Low-Level Planation and Aggradation

We should begin this discussion by noting that we are not for the moment concerned with collision mountain systems involving continents (see Chapter 3). When two continents collide they generally seem to destroy intervening oceans. What was coastal becomes inland. The Appalachians, for example, at the end of the Paleozoic just after the final orogeny may well have been 3,000 mi (4,500 km) from the nearest ocean throughout most of their length. And mainly because of distance to evaporative sources, they probably remained arid until the Atlantic Ocean reopened during the Jurassic and Cretaceous or until uplift raised peaks high enough to catch remnant moisture. It is unclear which occurred first. We will first consider coastal mountains, such as the Andes, the Atlas, or the Sierra Nevada that emerged or were initially uplifted near an ocean and that can be related to the kind of layered atmospheric moisture configurations characteristic of such settings (Fig. 4.7). Continent collision-type mountains probably develop the same environmental stratification when later they are hydrographically associated with an ocean (i.e. the Alps and Zagros mountains), but they would probably not be exposed to such detailed moisture stratification during uplift in an inland site.

As noted in Chapter 4, those lands emerging over a wide belt within some 35–40° of the equator, but not on the Doldrum Belt, would probably experience initial aridity. For an incipient mountain range, this may well mean an initial phase of tectonic denudation followed (or accompanied) by desert planation and aggradation. Actual terrain modifications are highly conjectural, since the duration of exposure to the arid conditions could vary widely. It could include development of an extensive pediplain in extreme cases. For the sake of further discussion, let us assume a measure of desert terrain development and alluviation in many instances. It is interesting to note that correlative marine deposits to such an arid emergence would probably be eolian dust and possibly some sand with only minor fluvatile additions. Of course, in the Doldrum Belt, and around 40° north or south of the equator, initial emergence of a land would probably be into a humid environment, and there would be no initial planation episode. Even so, a land raised above sea level in conditions presently prevailing in south Ecuador 1.5° from the equator would remain arid until it was about 200 ft (60 m) above sea level. In northern Peru (5° south latitude) some 500–1,000 ft (150–300 m) of uplift would be under arid conditions, and in southern Peru (15°–20° south latitude; Arequipa) some 8,000+ ft (2,450 m) of uplift would be arid. These vertical intervals can be expressed in terms of probable times of exposure to the arid geomorphic system in the following manner.

From the last marine submergence of the Zagros Mountains of Iran in the late Miocene until the accumulation of some 15,000 ft (4,600 m) of intermontane gravels in mountains of comparable relief and elevation in late Pliocene time was about 10 million years. This calculates to about 660 years/ft of uplift (2,165 years/m). Applying this uplift rate to vertical arid intervals previously discussed in the western Andes, works out to 132,000 years of arid exposure in south Ecuador, 330,000–660,000 years in northern Peru, and 5,280,000 years in southern Peru, assuming Andean uplift under present hydrographic configurations. The intervals are all of sufficient duration to induce notable arid planation, judging from erosional effects in much shorter time spans during the Pleistocene (cf. Chapter 8). Then, in our hypothetical instances, the rising mountains would enter more humid realms (cf. Fig. 4.7).

Intermediate-Level Fluvial Incision

A rising mountain chain passing through an initial arid phase would eventually be uplifted into a humid realm in most settings. The probable effects on landscape and derivative deposits are interesting. Inception of humid conditions on more elevated areas should cause drainage net development on the relict arid and tectonic terrain. This could well induce the first major influx of terrigenous sediment into an adjacent marine basin, provided, of course, that atmospheric adjustments had not caused such effects one or more times already. There is every reason to believe that the subsequent denudation and montane sculpture would follow one of the possible arid-to-humid geomorphic patterns outlined in Chapter 7, characterized by removal of relict alluvium and increasing incision within the humid atmospheric interval being encountered by the rising mountain mass. It seems clear that, in some instances, a very slightly developed, low-level desert landscape configuration could be thoroughly destroyed by a more elevated humid geomorphic system long before extreme elevations were achieved by uplift. ALTERNATIVELY, a very extensively developed low-level arid terrain might undergo little or no alteration in a weakly developed humid zone at intermediate elevations before being raised to a very high level. There are clearly a number of landscape possibilities (Fig. 10.18).

High Montane Environments

Having discussed some of the tectonic-climatic interactions that occur (and could occur) at relatively low elevations along orogenic belts [i.e. 10,000–12,000 ft (3,000–3,500 m)], we should now consider the topographic consequences of extreme montane uplift. Several of the world's mountain systems include appreciable land areas situated at elevations between 10,000 and 20,000 ft (3,000–6,000 m). Notable among these are the Rocky Moun-

Fig. 10.18. Semi-block diagrams of climatic orogenesis under (I) Humid, (II) Arid, and (III) sequential environments. Sequence I (A)–(E) shows the theoretical effects of mountainous uplift and degradation in an enduring humid condition, proximate to standing water, probably in an insular or peninsular setting or between seaways at relatively low latitudes. Sequence II (A)–(E) shows the postulated geomorphic sequence of a mountainous uplift and degradation in an enduring arid setting, possibly mid-continental, following a plate collision. Sequence III (A)–(E) shows the probable sequence of geomorphic events where an uplifted orogen encounters low-level aridity, intermediate-level humidity, high-level aridity, alpine glaciation during uplift, and the converse during downwearing. In all instances, glaciation may accompany extremes of uplift, and in many instances orogenesis under only a single subaerial environment would be restricted to segments of major mountain systems or only one flank of a range.

tains in association with the Colorado Plateau; the Himalayas, Parmirs, Hindu Kush, Kunlun, Karakorum, and related ranges of central Asia, plus the Afghanistan, Tibetan, Mongolian, and other associated plateaus; and the Andes Mountains and contiguous Bolivian Plateau and Peruvian Altiplano. Two main kinds of subaerial environments are related to such high montane settings in general conformance with the atmospheric fabric illustrated earlier (Fig. 4.7).

High-elevation deserts are associated with the ZONE OF MOISTURE-DEPLETED AIR that occurs mainly above 10,000–12,000 ft (3,000–3,500 m). The atmospheric zone (and the related deserts span a vertical interval ranging in thickness from zero (within 2°–3° of the equator) to several thousand feet at latitudes (20°–25° north or south). Above the atmospheric zone of moisture-depleted air is the ZONE OF FROST. The zone of frost is delimited by mean air temperatures below 32° F (0° C), and its lower limits generally coincide with the snow line which on the equator occurs at elevations of 14,500 to 15,000 ft (4,575–4,610 m) depending upon seasonal exposure (cf. Fig. 8.2). High-altitude cryergic and glacial environments coincide with the zone of frost, and the zone descends toward sea level at higher latitudes (cf. Chapter 8).

Montane Plateaus

Many high mountain areas have tabular regions, often termed *plateaus,* of uncertain origin. In general, genesis theories have been of two types (1) formation at a low level followed by uplift and (2) formation at high elevations. With the exception of those segments of mountain systems being elevated in arid climatic nuclei, most ranges being raised into the high-elevation zone of moisture-depleted air have previously been exposed to a humid geomorphic system at lower elevations—perhaps long enough to develop a selva topography. Disregarding for the moment any previous morphologic effects of tec-

tonic denudation or low-level arid morphogenesis, some interesting events should occur. For example, a group of rising, humid, vegetated ridge crests entering a high arid zone would no longer catch moisture. This is presently true of mountain peaks that penetrate the high arid zone in question (Fig. 10.19). The result would be to depress the drainage basin precipitation catchment areas farther and farther below the drainage divides and thus reduce runoff from the rising mountain range. In effect, the critical runoff flow distance X_c, required before fluvial erosion can occur (discussed in Chapter 5), would thereafter have to be measured from the lower limit of the arid zone rather than from the "relict" drainage divides. Fluvial erosion related to elevated desert areas often deposits rock waste within such deserts because ephemeral flow there dries up before it reaches perennial drainage nets situated downslope (cf. Chapter 6). Loss of vegetation and soil due to an orogenically induced climate change from humid to arid would presumably expose superjacent slopes to backwearing above elevated depositional areas in a manner similar to that previously discussed in low-elevation deserts (Fig. 10.20). The development of a high-elevation plateau in a mountain system could well begin in this fashion.

We would not wish to create the impression that montane plateaus are entirely planation surfaces, although known examples include pediments and related planar landforms (Fig. 10.21). However, as noted by Newell in his studies of the Lake Titicaca area and Peruvian-Bolivian Plateau (1949; 1956), and similar reports relating to the Mexican and Tibetan plateaus, many planar montane uplands are composite features involving a combination of "basin" filling and escarpment retreat, not uncommonly with some structural modification. Superjacent mountains often exhibit erosionally truncated spurs and fringing pediments (cf. Mackin, 1959). Dana (1895) recognized MARGINAL and INTERMONTANE plateau categories which are positional rather than genetic designations. The intermontane plateaus that appear to have formed during or after uplift usually appear to involve the entrapment of lacustrine and mass-wasting materials in undrained or poorly drained depressions—some are structural depressions, but some are valleys of fluvatile origin presumably incised in moist environments developed at moderate elevations (cf. Oberlander, 1965).

In terms of arid, high-elevation sculpture of a rising orogen, it is well to keep in mind that peripheral humid environments often do not penetrate to valleys in the heart of a mountain range. Valley incision in rising orogens creates "internal rain shadows," and many intermontane depressions are notably arid in their deeper portions, even in mountain systems that are externally very humid. Fenner (1948) was among the first to note the geomorphic effects of prolonged internal montane aridity. For a more detailed discussion, see Garner (1959, p. 1364). Most dry intermontane valleys carry runoff from more moist or glaciated uplands, but it is apparent that neither moisture source would exist for a range being elevated into a high-elevation dry zone. An arid intermontane valley deprived of upland runoff would begin to fill with sediment by arid fluviation and mass wasting. The aggradation might continue until such a fill level was attained

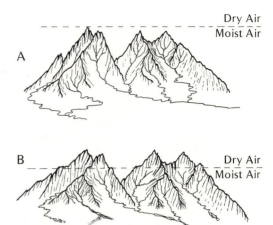

Fig. 10.19. Diagram illustrating the effects of montane uplift through a humid low-level atmospheric zone into a higher and drier situation. (A) Selva topography and integrated drainage nets extending to divides develop following initial uplift; (B) divides enter dry atmospheric zone with resulting restriction of drainage nets to lower elevations and reduction in over-all runoff to related rivers.

that an aggradational plateau could be said to exist (Fig. 10.22). Lacustrine deposits as well as alluvium are often present in such basin fills. Related planar surfaces, whether aggradational or planational, would not slope to regional base level.

It would seem that the development of montane plateaus represents a virtually invariable geomorphic event in a major oro-

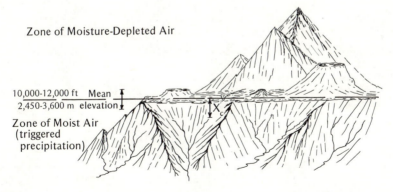

Fig. 10.20. Diagram of the establishment of a planation datum in a high mountain range intersecting the elevated zone of moisture-depleted air. As long as the mean position of the zone remains constant, it may fluctuate in vertical position and still localize deposition of upland runoff by desiccation. The X_c critical erosion distance of drainages in downslope moist regions would be measured from the mean base position (arrows) of the dry zone.

gen uplifted to great heights. If it is not formed at a low level by planation, it could well develop later through basin filling. And, of course, variations on the foregoing scheme are several. Vast expanses of the several central Asian plateaus appear to be planational. The Bolivian Titicaca area of the Andean Altiplano involves considerable alluvial fill, and there has been modification by faulting, but, in Ecuador where the Altiplano is presently humid, the high level surface is closely related to vertical accretion of pyroclastic debris. Over wide expanses of the Colorado Plateau there are very low-inclination stratified rocks which nevertheless appear to have been planated along with more local deformed areas. Not

all of the main examples of montane plateaus are climatic in origin. But even the volcanic example, like the majority of the others, probably owes its extensive planar expanses to the absence of effective fluvial incision. Of course, further uplift of such an orogen into the frost zone would create an upland runoff source previously missing, and we will discuss this shortly.

In spite of anything that may have previously given an impression herein to the contrary, an arid montane plateau AT high elevation may not be in a planation environment in the same sense that the Sahara or the Mojave or the Simpson deserts are planation environments. For one thing, rainfall on montane plateaus is

Fig. 10.21. Pediments on the west side of the San Luis Basin facing the Sangre de Cristo Mountains on the Colorado Plateau at an elevation of about 11,000 ft. The center of the basin has a thick alluvial fill. Note that sage growth in the distance is as discontinuous as that in foreground though it APPEARS to thicken.

much less frequent than it is in lower deserts. The higher the plateau the less the geomorphic role played by direct precipitation in the form of rain. It does rain occasionally on the Colorado Plateau, a not particularly high example. It rarely rains on the Tibetan Plateau. There is snow in winter there. But the snow tends to sublimate rather than melt. Lake Titicaca on the Bolivian Plateau is fed mainly by snow meltwater, but it overflows into a lower, highly saline lake lacking an outlet. Eolian activity at high elevations can be intense. But conditions for thunderstorms or flash floods or mudflows are distinctly marginal in most instances. None of the foregoing observations, by themselves, prove that arid planation cannot occur at high elevations. But they do suggest that desert planation there might be very slow. Development of an intermontane basin fill would create a pedimentation erosional datum. But an extensive planation plateau at a very high level is perhaps more likely to have been generated before uplift then after. Rather interestingly, not only does glacial meltwater often bypass or only locally and linearly incise arid montane plateaus, drainage nets at lower elevations lack the ability to erode into such elevated surfaces by headward extension of tributaries.

Alpine Glacial Effects

Geomorphic environments encountered by a rising mountain range can clearly vary, but the last ones to develop in most cases of extreme uplift are those related to glacial and cryergic effects. Direct landscape effects of both types of systems are discussed in some detail in Chapter 8. We are more concerned with the over-all landscape effects on a mountain system in those instances where uplift causes a more or less extensive cap of ice and snow to develop. Here, the meltwaters from developing alpine glaciers would encounter relict arid alluvium at slightly lower elevations in places like Peru or the east flank of the Sierra Nevada, but would alternatively

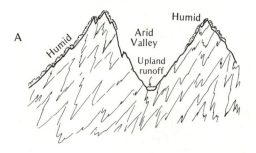

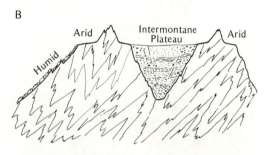

Fig. 10.22. Montane plateau development related to internal and high-level aridity. (A) Uplift of a range in a humid environment where the moisture has no direct access (except via upland runoff) to interior valleys which are arid; (B) a further stage of uplift where the peaks enter the elevated zone of moisture-depleted air and cease to yield runoff to interior valleys which thereupon become aggraded. A high-elevation plateau related to both aggradation and planation may result, while sculpture under more moist conditions persists elsewhere.

flow down into a humid tundra or forest in Ecuador or New Zealand. The potentially variable glacial effects of such circumstances are rather obvious. But other geomorphic concerns must occupy our attention here.

Interior dry valley systems in mountains developing glaciers, acquire major upland sources of runoff which (coupled with abundant abrasive "tools" in the form of bed load) should accentuate and perpetuate valley incision. The potential for very high montane relief is thus accentuated far beyond any actual effects of glacial erosion. Rather curiously, alpine glacial catchments tend to develop at the heads of relict fluvial drainage basins so that erosion and meltwaters related thereto often bypass desert plateaus (Fig. 10.23). And,

Fig. 10.23. Gorge of the Urubamba River Valley at an elevation of about 8,000 ft (2,400 m) on the east flank of the Andes Mountains, Peru. The river has cut more than a mile below the level of the arid Altiplano.

of course, meltwaters from alpine glaciers and snowfields often maintain perennial flow of trans-environmental rivers far across adjacent desert lowlands. Deep valleys such as the Grand Canyon of the Colorado River owe much of their existence to such relationships.

OROGENIC LANDSCAPE CASE STUDIES

We will now consider at some length a few of the major changes in mountain morphology that appear to have occurred in several different types of mountain systems as they were being uplifted and apparently either maintained at some general level (as in the Andes) or worn down (as in the Appalachians or the Sierra Nevada). Several geomorphic tendencies which relate to the tectonically induced sequence of environments appear to have developed in these various orogenic settings. One tendency is toward alluvial burial at some stage. Another tendency is for drainage lines to cross structural trends. A third tendency is for different types of sculpture to occur on opposite sides of a mountain system, and a fourth tendency is for montane mass loss by denudation to be local and linear. Any of these may apparently develop independently or in combination with one or more of the others.

COVERMASSES AND
TRANSVERSE DRAINAGES

A geomorphic characteristic of many folded mountain belts are drainage lines transverse to rock structural trends. If one takes the view that drainage nets normally develop by headward tributary extensions, then *transverse drainages* are difficult to explain since the flow lines often cut across very resistant rock types, and erosion has thus failed to exploit the weakest materials (Fig. 10.24), unless hidden weaknesses exist. Data already presented suggest that many drainages are developed CONSEQUENT

flow routes related to original land slopes and to runoff sources in uplands. Flow routes selected thus by water "seeking" the most direct route downslope could well follow depressions not directly reflective of rock structure and could cut across some features. Also, of course, transverse drainages have been explained by *superimposition* where a drainage line cuts through an alluvial or other sedimentary covermass into buried structures (Fig. 10.25), or by *antecedence* where watercourses are able to maintain their courses across rock structures being developed athwart the flow route (Fig. 10.26), both as suggested by Powell (1875). Most drained lands, whether mountainous or not, include some *subsequent rivers*, i.e. those with courses apparently controlled by rock structure (cf. Jukes, 1862). Perhaps the earliest explanation of transverse drainage was that it followed zones of lithic or structural (fault) weakness (cf. Rodgers, 1858; Thompson, 1939).

Two folded (collision) mountain systems particularly well known for their examples of transverse drainage are the Appalachians and the Zagros, though other cases have been cited—the Alps and the Himalayas and the Cascades and Pacific Coast Ranges of western North America (Fig. 10.27). The eastern North American situation has been intensively studied over the years (cf. Davis, 1909; Johnson, 1931; Meyerhoff and Olmstead, 1936; Strahler, 1945; Meyerhoff, 1972). The Zagros Mountains of Iran were the object of an excellent recent geomorphologic "tour de force" by Oberlander (1965). It should perhaps first be observed that the Appalachian and Zagros mountains both appear to be the product of continent-to-continent collision-type orogenesis (cf. Chapter 3). As previously noted, the hydrography of this type of montane setting favors aridity during orogenesis and thus a simultaneous initial planation and development of an alluvial covermass. This is not to suggest that all drainage configurations of these two ranges can be accounted for by some form of superimposition. However, either direct or

Fig. 10.24. Transverse drainage, here probably related to superposition of the Ab-i-Zal River on the core of the Kialan anticline in the frontal fold belt of the Zagros Mountains, Iran. The covermass on which incision was initiated may have been the Miocene, Lower Fars Formation according to Oberlander (1965). (Aerofilms Ltd.)

indirect evidence of early planation or covermasses would do much to explain many of the transverse drainage effects.

The Appalachians and Zagros are clearly of quite different ages, final deformation of the former occurring in the Late Paleozoic (Permian) and of the latter in the Mio-Pliocene, with some movements continuing at present (cf. Lees, 1955). In the Appalachians, postorogenic denudation has certainly destroyed many features probably analogous to those still preserved in the Zagros. Hydrography for the two ranges also differs greatly at present. The Appalachians are everywhere humid on uplands [average mean annual precipitation 40–50 inches (1,000–1,200 mm)], whereas

the Zagros decreases from a high of over 90 inches (2,400 mm) on northwest peaks to less than 10 inches (225 mm) in the southeast. Adjacent Zagros lowlands are arid, whereas Appalachian lowlands are now humid. As will be shown shortly, the earlier records of both areas probably differed.

Zagros Covermass and Drainage

Evidence for a covermass in the Zagros occurs within the mountain mass as well as in various foreland areas in the form of the Pliocene Bakhtiari Conglomerate (Fig. 10.28). Oberlander (1965) favorably compares this unit with the NAGELFLUH

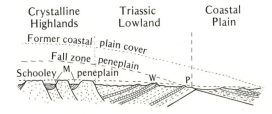

Fig. 10.25. Cross section of superimposition of drainage in the Appalachian Mountains from a theoretical marine Cretaceous covermass according to the theory of Johnson. M, modern valley flats; W, Wachung Mountain Basalts; P, Palisades Diabase. (After Johnson, 1932.)

molasse of Switzerland and the Siwalik deposits of India (see also Fig. 3.77). He notes rapid thickness variations (locally) from inches to 15,000 ft (4,600 m), grain-size variations from siltstone to boulder conglomerate of 3 ft (1 m) maximum diameter which include much angular-to-rounded limestone and other metastable clasts. He further observes that the Bakhtiari conglomerates filled basins, flooded through passes, and blanketed plains in and adjacent to the orogen but that they show not the slightest semblance of stratification. This material may conform closely to certain sedimentary-tectonic conceptions (i.e. molasse), but there can be little doubt that it is largely a product of a rigorous weathering environment, probably arid aggradation, mainly as intermontane basin fills and piedmont deposits similar to the High Plains gravels of the United States. Deformation of these materials indicates that they were deposited during orogenesis. The linear incision of these and subjacent bedrock materials signifies uplift of the range into the present humid upland environment, and/or possibly creation of adjacent moisture sources, mainly the Mediterranean Sea. We suggest this latter idea, for more Zagros moisture comes from the Mediterranean than from the Persian Gulf, and recent study suggests that the Mediterranean may have evaporated to dryness several times during the Pliocene (Hsü, 1972) (see Chapter 11).

Oberlander (1965) incorporates the

covermass and superimposition into his explanation of Zagros transform drainage development and also recognizes consequent, subsequent, and antecedent drainage examples. Remnants of the Pliocene alluvial

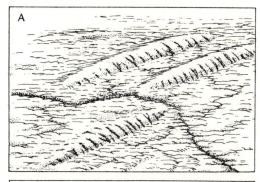

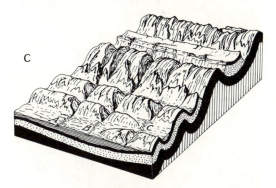

Fig. 10.26. Development of antecedent drainage (A) during initial deformation of folds not fully exposed and (B) following full exposure of the fold structure. (C) Antecedent drainage in the Zagros for streams (A) and (B) originally consequent on a large arch of Cretaceous limestone and antecedent to two younger adjacent anticlines. Stream (C) is neither consequent nor antecedent for the line of flow has no erosional counterpart on the large arch. (From Oberlander, 1965.)

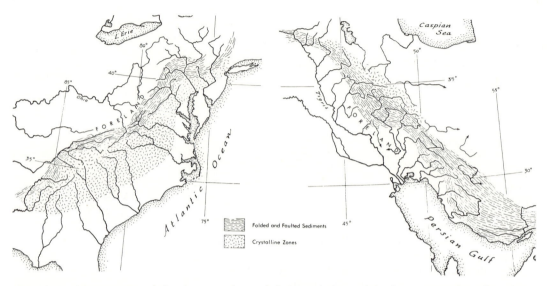

Fig. 10.27.   Location map of the Zagros (right) and the Appalachians (left) for accompanying discussion. (From Oberlander, 1965.)

Fig. 10.28.   The Pliocene Bakhtiari Conglomerate as it appears in Dez River Canyon in the mountain front detrital apron of the Zagros Mountains, Iran. The conglomerate is probably part of an arid covermass. (Theodore Oberlander.)

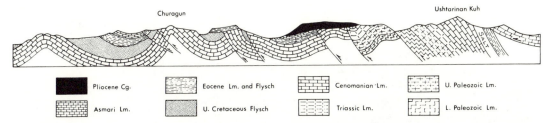

Fig. 10.29.   Structural and stratigraphic cross section in the vicinity of the central Zagros Mountains showing the Tang-i-Bahrein Pliocene covermass (black). (From Oberlander, 1965.)

covermass are still preserved on some beveled uplands (Fig. 10.29), and the resultant montane terrains are truly spectacular (Fig. 10.30; 10.31). Space limitations do not permit a detailed recapitulation of the Zagros montane sculpture. Here we will merely point out that the Zagros covermass was apparently incomplete and that drainage development occured in relation to continuing deformation, uplift, and

probably several alternations between severely arid and equally humid conditions, particularly in the more lofty parts of the range.

Appalachian Covermass and Drainage

Essentially every type of drainage development previously mentioned for the Zagros

Fig. 10.30.   Aerial view of Tang-i-Lailum, a transverse gorge through the Kuh-i-Kialan anticline, Iran. The fold consists of Oligocene-Miocene Asmari Limestone and is flanked by a lowland developed on the Lower Fars Formation (cf. Fig. 2.43). (Aerofilms Ltd.)

Fig. 10.31. Aerial view of a portion of the fold zone of the Zagros Mountains, Iran. Foreground, two unroofed anticlines (Kuh-i-Pabda, right; Kuh-i-Mirali, left) expose upper Cretaceous limestone cores, adjacent annular valleys developed in Cretaceous-Eocene flysch, and flanking ridges and hogbacks of Asmari Limestone. (Aerofilms Ltd.)

region was earlier suggested for one area or another of the Appalachian Mountains. The latter range, as previously noted, is a much older mountain system, so the evidence pertaining to early morphogenesis is meager. Mainly, it takes the form of several levels of rather accordant upland summits (Fig. 2.13) and deposits made in marginal basins and on piedmonts that may be environmentally indicative of former covermasses. There do not appear to be any upland remnants of covermasses,

so suggestions that there may once have been one must rest on some theory or another. An early speculation along this line came from Johnson (1931), who proposed drainage superimposition from a Cretaceous marine covermass following erosion to base level and submergence (cf. Fig. 10.25). There are obvious theoretical difficulties with postulating a sea level rise the required 1,800+ ft (600 m) above that of the present to make the deposit, which may explain why no one has ever suggested this mechanism for covermass formation. But such theoretical obstacles are no less formidable than the geophysical difficulties, already discussed, which face those who wish to "hold down" a mountain range while it is being eroded—and until it can be submerged and buried. In short, it seems doubtful that the Appalachians have been submerged since Late Paleozoic time or that there ever was a post-orogenic marine covermass during the interval in question.

Other types of covermass for the Appalachians may be another matter. There is the lower Permian? Dunkard Group of continental deposits in Ohio and West Virginia and the far more extensively preserved Triassic red beds of the Newark Group along the Atlantic coastal lowland (Fig. 10.32). The latter deposits have been studied extensively by a number of researchers and principally by Krynine (1942, 1949, 1950). To put the matter in better chronologic-tectonic perspective, it should be noted that the Newark continental sediments are preserved in a series of fault-bounded depressions and that they incorporate ferrugenous, lacustrine, silt-

stone, and fluvatile, arkosic sand and gravel plus several diabasic and basaltic igneous rock masses. The intrusive and extrusive igneous suite is clearly referable to an oceanic origin, and the structural style is that of a spreading ridge (cf. Chapter 3, Fig. 3.5). These relations coupled with the discovery of Jurassic salt in the South Atlantic Ocean basin and Cretaceous sediment in the western North Atlantic mark the Triassic volcanism as symptomatic of the impending opening of the ocean basin. In other words, during Permo-Triassic times, there was no Atlantic Ocean, and the Appalachian Mountains lacked a nearby evaporation source.

Krynine's interpretation of the Triassic deposits in the Appalachian area is tied to his earlier conclusion (1935, 1936) that *arkose* and other metastable conglomerates are products of accelerated erosion due to high relief in humid tropical environments. On the strength of Krynine's early findings, a number of workers have interpreted continental deposits similarly (Hubert, 1960a), and this has affected orogenic analyses directly since one is left with the impression that coarse, metastable

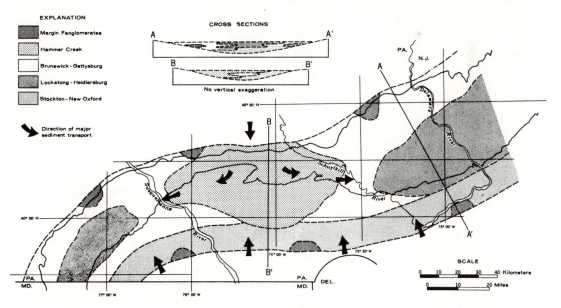

Fig. 10.32.   Paleofacies map of the Newark-Gettysburg basin, Pennsylvania and western New Jersey, shows fanglomerate occurrence and related facies. (From Faill, 1973.)

conglomerates appear in the mountain-building sedimentary record because an unstable rising orogen has attained high relief and humid conditions (see Ryder and Scholten, 1973). Actually, in orogens such as the Appalachians and Rockies which underwent tectonic denudation, the appearance of arkose and similar metastable coarse clastics probably marks the appearance of requisite source lithologies and

Fig. 10.33. Sawed slab of Triassic "border breccia" from the Newark Group of sediments associated with the Rampo fault system near Stony Point, New York. The material is feldspathic though most of the large clasts in the illustrated specimen are siliceous. In other examples, limestone clasts have been found and a rigorous weathering environment is at least sporadically indicated during deposition; scale, inches and centimeters.

a mechanical weathering environment, possibly before the time when excessive amounts of tectonic relief were developed by uplift.

In addition to feldspathic debris, the Triassic conglomerates in New Jersey also include limestone detritus, and once again the probable rigorous environment is that of aridity (Fig. 10.33) as noted by Faill (1973). Otherwise, apart from issues of weathering maturity and red coloration origin (cf. Hubert, 1960b), ONE IS HARD PUT TO EXPLAIN ACCUMULATION OF THE ARKOSIC DETRITUS ON THE LAND AT ALL as noted by Garner (1963b). Faill is careful to note that the Newark deposits in the Gettysburg Basin need not have been made in down-faulted areas and that some in New Jersey may well have accumulated on inclined planar surfaces.

Meyerhoff (1972) has concurred with the earlier assessments of Thornbury (1965) and Mackin (1938) of the impossibility of reconstructing the initial Triassic (postorogenic) surface in the Appalachians. But though these writers are probably correct in terms of detail, the probability of a Permo-Triassic intermontane and piedmont desert with basin deposits and pediplains is extremely great. As a collision-type mountain system, the Appalachians were originally in an arid hydrographic position analogous to several of the central Asian mountain systems (i.e. Tien Shan, Altai, Sayan), though, in tectonic style, the Appalachians are perhaps closer to the Himalayas. In the latter instance, however, the Indian subcontinent to the south is too small and the ocean areas too near to deprive the south Himalayan flank of moisture as the vast land expanses of Africa and North America probably did for the Appalachians. Continental drift reconstructions for North America indicate an early Mesozoic location as much as 15° closer to the Equator than at present, a location favorable for a Horse-latitude Desert. It is also probably significant that Jurassic time ended Mesozoic continental deposition in the Appalachian region, the Jurassic and Cretaceous products of subaerial weathering and erosion moving generally to base level, presumably under dominantly humid conditions.

The last sentence points up a major problem of the sedimentologic basis for a geomorphic evaluation of orogenesis, and it raises the question of why Krynine (1935; 1936) believed in the humid, continental accumulation of arkose and related redbeds. Examination of his data from the Gulf of Campeche region of Mexico (Fig. 10.34) shows clearly that two types of potential sedimentary material are present in the area—weathered regolith developed on granite in the humid, tropical Sierra Madre Oriental, and relict, feldspathic, Pleistocene and Tertiary gravels on the coastal lowland. Present perennial streams draining the mountains and flowing across the lowlands to the sea derive their loads from both sources, depositing some in lacustrine sites and the remainder in the Gulf of Mexico. The arkose stream gravels therefore represent a delayed movement (*sedimentary lag*) of detritus deposited under pre-Recent and probably arid conditions on the land (cf. Garner, 1959, pp. 1365–1366). But the fact also is that such an arkosic deposit, found in a lake or sea, is probably accumulated there mainly under humid conditions and does not (on the whole) reflect a correlative weathering mode.

Remnants of the Appalachian covermass upon which many of the drainages were probably initiated consists of nonmarine fanglomerate material and is only sporadically lacustrine. Unfortunately, the covermass is now almost entirely gone from the uplands, and it can only be studied in marginal areas where depositional settings were probably piedmonts or structural depressions rather than intermontane basins.

The significance of the sedimentary-lag concept in the analysis of developing mountain systems lies in the fact that an initially arid orogen will tend to retain residual arid waste until it becomes humid. In the case of the Appalachians, the waste-retention period spanned the Permo-Triassic interval and may have aggregated

for twenty or so million years. The appearance of this rock waste in a Jurassic marine deposit would not date an orogeny but, rather, would mark re-erosion of the detritus following a climate change to humidity. Conversely, an initially humid orogen should not yield appreciable coarse detritus due to weathering until there is a climate change, either CRYERGIC-GLACIAL, at very high elevations, or ARIDITY, at some intervening elevation or condition during uplift. The geomorphic analyst would therefore do well to note whether sediments related to an orogeny accumulated on land or at sea. To date, the best indicators we have for crustal movements are deformed rocks, and the best tectonic dates are probably radiometric ones based on syntectonic igneous activity.

Ancestral Rockies
Covermass and Drainage

As an example of tectonism resulting in mountains that were probably humid during initial uplift and that later experienced other conditions, the Colorado region is worth considering in some detail (Fig. 10.35). As noted in Chapter 3, the tectonic development of upthrust mountains of the Rocky Mountain type is not fully understood nor are the potentials of such mountains in terms of renewed uplift. The Ancestral Rocky Mountains were deformed near the end of the Mississippian Period, an event marked by an angular unconformity between Mississippian and older Paleozoic carbonates and ortho-quartzitic sandstones (beneath) and the Pennsylvanian Fountain Formation (above) (Figs. 10.36 and 10.37). Modern Southern Rocky Mountain Front Ranges in Colorado and New Mexico partly coincide with the area of ancestral mountains, and structural features generally coincide. The degree of erosional lowering of the older mountains by the end of the Mesozoic is uncertain, but there is overpowering evidence for renewed deformation and uplift at the latter time, during the Laramide Orogeny (Fig. 3.73).

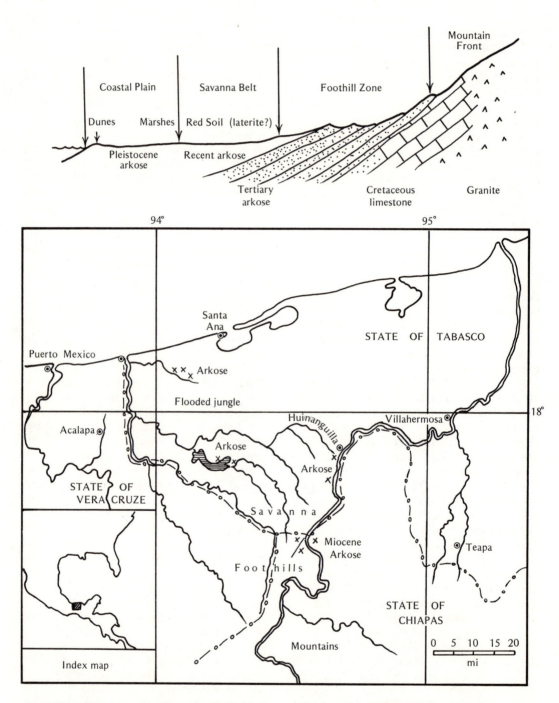

Fig. 10.34. Drawings taken from the P. D. Krynine's original study of arkose in Mexico in which it was concluded that such deposits were humid and tropical. Note the extensive Tertiary and Pleistocene arkose which presently lacks access to mountain drainage but can be re-worked. (After Krynine, 1935.)

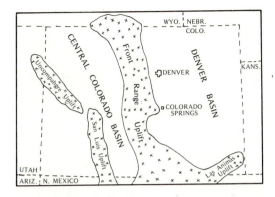

Fig. 10.35. Major areas of Late Paleozoic uplift during formation of the Ancestral Rocky Mountains, Colorado. Laramide (Paleocene) uplifts were approximately coincident.

The orogenic episode of immediate concern here is that which formed the Ancestral Rockies, and it is of special interest because of the development of an extensive covermass some time after initial deformation. Structural evidence of the deformation exists beneath the unconformity, apparently generated through regional warping and minor faulting. Evidence of humid morphogenesis at that time is impressive. Mississippian and older rocks including the Precambrian Pikes Peak Granite were deeply weathered with the development of mature, lateritic-type soils (cf. Wahlstrom, 1948). Also, Mississippian and older carbonate rocks were thoroughly riddled by solution cavities and caverns, in some of which Fountain deposits accumulated (Fig. 10.38). The duration of the humid episode in question is uncertain, but in part it coincides with the accumulation of the marine, Lower Pennsylvanian (Morrowan) Glen Eyrie Shale Member of the Fountain Formation in an adjacent seaway. The Glen Eyrie consists of orthoquartzitic sandstone, black shale, and some crinoidal limestone and minor coal beds together aggregating as much as 360 ft (100 m) thick (see Hubert, 1960b).

Though the precise duration of the humid "Glen Eyrie" interval is uncertain, it was probably long enough to produce a deeply incised selva topography with integrated drainage nets on a montane area that had already undergone considerable uplift. By the end of Glen Eyrie time, a quartzo-feldspathic terrain composed of Pikes Peak Granite was exposed and deeply weathered (cf. Fig. 5.29) and surmounted in some areas by weathered and incised remnants of early Paleozoic sandstone and carbonate deposits. Geometry of subsequent deposits indicates local relief well in excess of 5,000 ft (1,500 m). It was in this setting that the Ancestral Rockies developed a new aspect. A covermass formed in the guise of a series of clastic redbeds (red = hues from 5P through 5R; cf. National Research Council Color Chart —Goddard et al. (1948). Initial deposits comprise the remainder of the Fountain Formation arkose and arkosic sandstone, conglomerate, argillaceous sandstone, and siltstone. The materials were overlain by similarly colored detrital constituents of

PPL	Permo-Pennsylvanian Lyons Sandstone
Pf	Pennsylvanian Fountain Arkose (Glen Eyrie Shale Member)
Mm	Mississippian Madison Limestone
Oh	Ordovician Harding Sandstone
Om	Ordovician Manitou Dolomite
Cs	Cambrian Sawatch Sandstone
Ppg	Precambrian Pikes Peak Granite

Fig. 10.36. Generalized Paleozoic stratigraphic sequence relating to Fountain deposits, central eastern Colorado; AU, the only notable angular unconformity.

Fig. 10.37. Cross-bedded, conglomeratic arkose under arkose conglomerate of the lower Fountain Formation of Pennsylvanian Age, southwest of Colorado Springs, Colorado.

the Lyons and Lykins formations of Permo-Pennsylvanian age (Fig. 10.39).

The Pennsylvanian Fountain Formation accumulated on a mountainous terrain and in adjacent seaways. Most of the deposits on outcrop are continental deposits, with a maximum thickness near Colorado Springs, Colorado of about 4,500 ft (1370 m). Thicknesses are variable just as in the Zagros Bakhtiari Conglomerate, previously discussed, and general thinning toward seaways is indicated by thicknesses of 1,000–1,200 ft (325–375 m) or less where the first marine facies are encountered (Fig. 10.40). The Fountain is known on outcrop in Colorado and Wyoming over a distance of some 215 mi (300 km) north-south with marine extensions in the subsurface. Similar formations in central Colorado (Maroon Formation) and northern New Mexico (Sangre de Cristo Formation) achieve thicknesses of 7,000 to 13,000 ft (2,100–4,000 m) in continental facies (cf. Hubert, 1960b).

In the Fountain Formation we encounter an orogenic-climatic enigma. A highland is documented by rock structure. So, too, some deposits verify the existence of an adjacent ocean. Hubert speaks of marine regression during Fountain time and his diagrams (1960b, Fig. 3) suggest oscillations as do field relationships. A vast bulk of coarse, metastable sediment is nonmarine. It is mainly feldspathic sand and gravel but locally contains fragmental limestone and is certainly indicative of an arid montane environment, probably affecting intermontane basins as well as piedmont areas. Subsurface data and cross sections presented by Maher and Collins (1952), Maher (1953), and Mitchell (1954) suggest that the Fountain arkosic facies makes a subsurface transition some 40 mi (56 km) east of Colorado Springs into a marine sequence consisting of oolitic limestones, green and black shales, minor reddish and greenish siltstones, and silty shales. The lower part of the section is more arkosic in a seaward direction than the upper. We must then ask, if continental arkoses are mainly arid deposits and marine ones humid, how was the Fountain Formation developed?

In general there would appear to be two major possible relationships between the continental Fountain arkosic facies and marine equivalents. In what we may call situation number one, a montane cover-mass develops under arid conditions and is later removed seaward by humid rivers. In this situation the marine deposits should be distinctly younger than the continental equivalents. In situation number two, a montane covermass begins to develop under sporadically arid conditions, occasionally developing soil profiles and losing sediment mass to base level under humid conditions, but there is always a net positive residual arid increment on the land bearing an "overprint" of humid effects. Ultimately, in late Fountain time, humidity predominates over aridity, and, by Permian time, the deposits are all marine. In this instance, marine and non-marine Fountain deposits would be broadly time correlative.

Our chronologic control between the marine and continental facies of the Fountain Formation leaves much to be desired, but what little there is plus general field and subsurface relationships appear to favor situation two. In other words, it would seem that the rising, Ancestral Rocky Mountain orogen, after an extended episode of humidity, became alternately humid and arid, and ultimately more arid than humid. As noted by Hubert (1960b), the Fountain Formation as a whole consists of alternating coarse-grained arkosic sediments (80%) and argillaceous "redstone," hemititic silty sandstones, and pebbly mudstones (17%). The latter lithology, in particular, is explicable as an arid mudflow deposit, and either type exhibits some traits of closed-system sedimentary deposits. The remainder of the formation is

some 3% feldspathic sandstone containing more than 35% quartz deposited in open stream channels, probably by streams variously braiding, meandering, or drying up of the end of various humid episodes (cf. Hubert, 1960a; Garner, 1963; Hubert, 1963).

The ultimate cause for the fluctuating climatic regimen of the Ancestral Colorado Rockies must be sought beyond the confines of the orogen. A hundred or more climatic fluctuations could be indicated by lithic variations in Fountain deposits during a part of the Pennsylvanian Period, characterized in the mid-continental United States by *cyclothem* deposits, repeated broad oscillations of sea level and strand lines, and similarly repeated glaciation elsewhere in a Southern Hemispheric landmass including part of what is now Australia and South Africa (Gondwanaland). Aridity would be favored on a land if the adjacent sea withdrew or if the sea became cold (cf. Chapter 4). Glacial episodes would

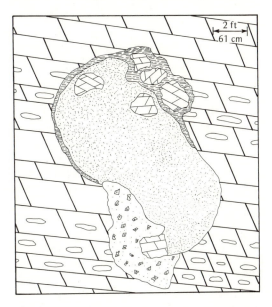

Fig. 10.38. A cavern filling in the Ordovician Manitou Dolomite documents the existence of solutional topography in pre-Fountain (Pennsylvanian) time. The filling is sandstone, probably from the Fountain Formation, which includes boulders of siliceous dolomite and a capping of clay.

Fig. 10.39.   Festoon cross-stratification in the Permo-Pennsylvanian Lyons Formation southwest of Colorado Springs, Colorado. Unit is a typical "red bed deposit," with an iron-oxide coating on well-sorted, well-rounded, frosted, eolian? sand grains. Basal formation contact with Fountain Formation is gradational. The Lyons sandstone is lithologically what one would anticipate from re-working of an erg, possibly by a transgressing sea. See geologic pick (center) for scale.

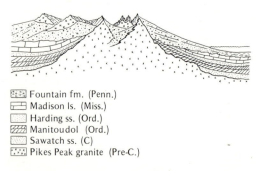

☐ Fountain fm. (Penn.)
☐ Madison ls. (Miss.)
☐ Harding ss. (Ord.)
☐ Manitoudol (Ord.)
☐ Sawatch ss. (C)
☐ Pikes Peak granite (Pre-C.)

Fig. 10.40. Sectional diagram with perspective showing stratigraphic-topographic relations of the Fountain and related older Paleozoic formations, Colorado. The Fountain accumulated as inter-montane and piedmont deposits with minor, lateral marine equivalents in areas subjacent to elevated areas of Precambrian granite much like those today.

tend to induce both arid-marine situations and interglacial times humid-marine. The character of montane sculpture in the Ancestral Rocky Mountains during and after orogeny was apparently influenced very strongly by global climatic systems.

The extent of planation related to cover-mass development is uncertain in the Ancestral Rocky Mountain situation, but the evidence of pre-Pennsylvanian fluvial relief and later sporadic humid incision suggests that original consequent drainage configurations were probably not destroyed. The Southern Rockies are not noted for transverse drainage configurations. And even though there is abundant evidence for post-Laramide intermontane and piedmont (High Plains) aggradation and planation, there is little evidence of drainage derangement thereafter. It would seem, however, that the many pedimented mountain spurs and planated upland remnants developed during and after the more recent orogenic episode. Moreover, they must have formed in hydrographic circumstances involving progressively increased continental emergence, progressively higher montane elevations, and more and more distant oceans in a sequence of relationships extending to the present.

Montane Geomorphic Asymmetry

Climatologists, geographers, and geologists have all long been aware of differences in environment on opposite sides of some mountain systems. One can readily anticipate different kinds of geomorphic systems being associated with these major types of environment as well as distinctive related landscapes. And, in some instances, the contrasts are pronounced. The Appalachian Mountains are a notable exception to the rule in their present orientation and elevation; they display a remarkably uniform hydrography and topography on either side throughout their length. They are presently humid throughout. Several ranges in central Australia (Macdonnell) are presently arid throughout. At the opposite extreme we have cases of single mountains with entirely different geomorphic histories on opposite sides.

One of the striking examples of environmental asymmetry on individual mountains is described by Fenner (1948, pp. 899–903) from the Western Peruvian Sierra. There, two volcanoes, El Misti and Pichu Pichu, show a long series of climatically induced erosional and depositional features (Fig. 2.49) on their southwest sides (toward the Pacific Ocean), and an essentially unmodified volcanic declivity on their north and northeast sides in an arid intermontane pocket receiving essentially no moisture [cf. Rondon (1937, p. 53; in

Fenner, 1948, p. 915]). The writer (1959, p. 1360) observed that similar environmental extremes characterize opposite sides of the Andes Mountains throughout most of Peru and northern Chile, and the montane terrains reflect this difference (cf. Figs. 2.49 and 2.50). So do the drainages. The presently humid, eastern Andean flank appears to have a geomorphic history dominated by that environment and is by far the most deeply and thoroughly dissected side of the orogen. Moreover, the implication of the drainage configurations is that they developed as the range was uplifted. Rivers such as the Ucayali, Marañon, and Urubamba flow hundreds of miles northward parallel to the trend of the Andean orogen before turning east through deep gorges to join the main Amazon trunk drainage (see Figs. 10.6 and 10.41). On the west side, the few major quebradas originating in the mountains flow essen-

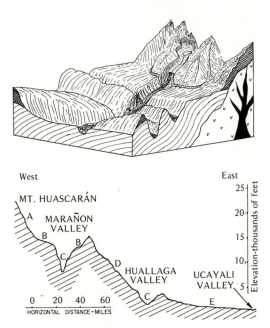

Fig. 10.41. (A) Block diagram and (B) cross section depicting typical topographic and regional slope relations in the Peruvian Andes Cordillera Oriental (cf. Fig. 10.6 for location). Note particularly the arid high-elevation plateaus (ca. 13,000–14,000 ft). Vertical exaggeration of (A) ×7.5. (After Bowman, 1916.)

tially straight to the sea—when they carry water.

In varying degrees a number of other mountain systems show environmental and geomorphic asymmetry. The coastal ranges of Venezuela previously discussed have a humid, tropical selva terrain on their north face toward the Caribbean Sea and a series of semiarid-to-arid, deeply alluviated intermontane basins to the south (Fig. 10.8). The Great Dividing Range of Australia is generally humid on the east side adjacent to the Pacific Ocean and arid on the west, and the Atlas Mountains of Algeria and Morocco are arid on their south sides and more or less semiarid to the north. The Himalayas present a moist flank to the monsoons coming from the south and a dry side to the interior of Asia. And, in each of the examples just cited, the general montane terrain on the moister flank is that of greatest drainage densities and deepest fluvial dissection, whereas the arid side is more or less extensively alluviated.

Major departures from the geomorphic asymmetry of mountains are due to precipitation that is caught in major divide areas; these areas often divert runoff to the dry side of the range and thus provide watercourses there with more moisture than the related environments could generate. One common consequence of this relationship develops where runoff from moist peaks began before the physiographic barrier of the range was completed. A form of drainage antecedence commonly results where runoff from the "dry" side cuts back through the montane barrier to reach the sea and continues to do so during uplift. This happens in the case of the Brahmaputra River which cuts through the Himalayas to reach the Bay of Bengal. A minor river near Caracas does the same thing to reach the Caribbean Sea through the Cordillera de la Costa.

The Sierra Nevada Mountains of the western United States are apparently a block-faulted Jurassic batholithic mass uplifted on the east side and tilted westward.

Opinions differ as to the age of the movements involved. Hudson (1960) suggests 2,000–4,000 ft (600–1,200 m) of uplift in post-Pliocene time on the basis of altered stream gradients, and Axelrod (1962) calculates 6,500–7,000 ft (1,850–2,150 m) of uplift during the same interval on the basis of paleobotanic and paleoclimatic data. The range shows great environmental asymmetry, the east side being decidedly more arid (Fig. 10.42). The range crest was extensively glaciated (cf. Chapter 8, Figs. 8.31 and 8.34), but the west slope with its oceanic exposure, like the Peruvian volcanoes previously mentioned, has a more complex environmental and geomorphic history than the east. Wahrhaftig (1965) calls attention to the stepped erosional topography developed on granites of the western flank of the southern Sierra Nevada (Fig. 10.43). Glaciation has obscured any record of stepped terrain above the elevation of 9,000 ft (2,750 m).

To the north of the Yosemite area, the west slopes of the Sierra Nevada are developed on a more predominantly metamorphic terrain. There, broad, flat-topped interfluves capped by Tertiary gravel remnants ascend gradually and regularly from the floor of the Great Valley to the summit crests of the Sierras (cf. Lindgren and Turner, 1894; Lindgren, 1911). Wahrhaftig (1965) maintains that the stepped surfaces in the southwest Sierras are due to accelerated sub-mantle weathering on granite (cf. Mabbutt, 1966b; see also Chapter 6). The features he compares them with in the Colorado Front Ranges and elsewhere (1965, p. 1187) are mainly pediments and pediplain remnants.

The precise origins of the stepped erosional features and the planar interfluves in the northwestern Sierras depend considerably upon the age of the Sierra Nevada uplift and its westward tilt, but the landforms are all essential planation features, and at least some also show evidence of fluvial incision. A sequence of subaerial environmental alternations is indicated, probably involving both aridity and humid-

Fig. 10.42. A boulder field near Manzanar, California, at the base of the arid east flank of the Sierra Nevada Mountains, was probably produced by deflation of a mudflow on the alluvial fan surface inclined toward the viewer. (Ansel Adams.)

ity; the exact details of this sequence remain to be worked out. In addition to the flank terrain that relates to comparatively low-elevation environments, there are, of course, several of the deepest montane canyons in North America on the "humid" west slope of the Sierra Nevada. These relate in large measure to snow meltwaters coming from the crestal peaks. Notable among several are Yosemite Valley (Fig. 8.34) and the canyons of the Kings, Tuolumne, and Kern rivers, all of which are 4,000–5,000 ft (1,200–1,500 m) deep—the King's River Canyon locally attains a depth of about 8,000 ft (2,450 m).

TERMINAL ASPECTS OF OROGENIC TOPOGRAPHY

The previous paragraph brings out a montane erosional relationship that should be emphasized. The low-density rock mass (mountain root) upon which a high range depends for its initial uplift is apparently

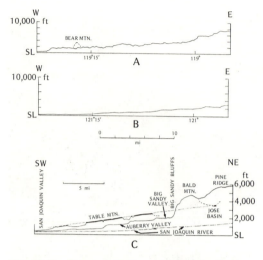

Fig. 10.43. Stepped erosion surfaces on the west flank of the Sierra, Nevada, California. (A) The drainage divide on the south side of Kings River; (B) the drainage divide on the north side of American River; both project to east-west planes in the foothill region. (C) the longitudinal profiles of the basalt-capped table mountains of the lower San Joaquin River, Auberry Valley, and Big Sandy Valley; all project into a plane trending N 52° E. (From Wahrhaftig, 1965.)

essentially continuous along the trend of an orogen, though it may vary in bulk. In mountain ranges like the Andes, where the mass of the root is now apparently being added to rather continuously, the potential duration of the system as a pronounced topographic high is exceedingly great. In such instances Powell's dictum, that "mountains cannot long remain mountains," would hardly be applicable (cf. Bloom, 1971). Even where a root zone is of finite volume following a continent-to-continent plate collision, mountains exhibit considerable persistence, in large measure because the removal of exposed mass by denudation tends to be linear in form (cf. Fig. 10.3). Moreover, montane denudation is limited in subaerial aspects to those episodes of glaciation and humidity when rock waste is actually carried out of the orogen. The notching of a mountain range by running water does little to reduce its over-all mass, even where it happens in several places. And if the range is very

high the problem of its erosional destruction is compounded by the fact that arid uplands do not spawn streams or glaciers. In short, the high-relief phase of a mountain system may last a very long time.

It is also apparent, that collision mountains become lower as their roots are consumed through isostatic uplift in response to reduced load. An accordance of summits (Fig. 1.10) could clearly mean a number of things, dependent in large measure on when and where they are encountered. They could reflect similar uplift of a dissected planar surface. They could reflect fortuitous differential weathering and erosional lowering of ridge crests. They could reflect upland planation and basin filling followed by dissection. And they could reflect gradual lowering under uniform erosion from former accordant upland summits or planar surfaces or covermasses. This does not exhaust the possibilities. It would be foolish to here attempt any thorough assessment of accordant summits in the many mountain chains of the earth where they occur, though each example should be reassessed in light of geomorphic developments of the past quarter-century and particularly recent discoveries in geotectonics, structural geology and geophysics, geohydrology, and climatic geomorphology. Much remains to be done.

It also seems important to call attention to the fact that a mountain system being lowered by erosion and a loss of root mass will gradually withdraw from those very environmental realms that were invaded during orogenic uplift. In a range that is humid or arid from top to bottom, lowering might change little, geomorphically, though we have already pointed out that uplands in deserts tend to catch the most moisture, and their destruction would change this. However, in a mountain with pronounced hydrographic zonation, with isohyats roughly paralleling the contours (cf. Oberlander, 1965), the disappearance of rigorous, high-altitude weathering environments would certainly affect stream loads at lower elevations. In an otherwise humid mountain range, this could mean a shift from corrasional incision to essentially corrosional erosion as in the southern Appalachians and greatly slowed denudation. Finally, of course, in many mountain regions, the final environmental change would be the disappearance of intermediate elevation moist conditions and subsequent arid planation at a low level. Geologists have often been amazed by the planar character of terrain developed on the roots of ancient mountain systems. Low-level aridity is certainly a probable explanation, but a possible alternative at compatible latitudes is erosion under an ice sheet, particularly since the physiographic obstacle to continental glaciation represented by high mountains has already been eliminated.

SUMMARY

With this chapter we end the portion of this work devoted to what might well be termed systematic geomorphology. In the two remaining chapters we will first attempt to apply the concepts so far considered to the understanding of ancient (buried) landscapes, and then we will try to apply such knowledge to some problems of our own evironment.

At this writing, the earth's outward form emerges as complexes of landforms assembled into landscapes by a vast three-way interplay between internal earth processes, the atmosphere, and the oceans. The neat physiographic provences of a half-century ago, in their compartmentalized environments, are gone as a viable avenue to the comprehension of landscapes. So too, perhaps, is the comfortable feeling one sometimes senses from the earlier literature that the major geomorphologic problems were all worked out and all that remained were details. Now, all that remains are major problems. Detailed studies are the weapons. But the studies must be anchored to new developments in structural geology, geophysics, hydrography, hydrology, and climatic geomorphology. Furthermore, these studies must relate to theories that encompass the interaction of the various geomorphic systems in both time and space. We pursue our specialties so that we may ultimately be able to generalize and, having generalized, we often find that our specialties have taken on new meaning and direction.

There is a revolution brewing in geomorphology as profound as the one presently affecting global tectonics. It involves the concept of the environmental interface in a space-time continuum, and it may provide us with the vital conceptual tools we require to deal with our contemporary environment. One such interface is that between the atmosphere and the earth, where events are often so slow as to seem not to be happening at all. Along another interface between land and sea the changes are so rapid and continuous we are hard-pressed to monitor them, even in part. In orogenic belts, we find the ultimate of interaction, the maximum in variables, the culmination of forces. If a coast is environments in contact, a mountain system is environments in conflict; the up-building vies with the down-tearing. And whether it is a mountain range or the human form, everyone knows what happens when support is withdrawn.

There is little denying that gravity will probably have the last word on earth, but present internal activities place such a notion among those concepts which can be contemplated as a remote eventuality. Even so, for mountains, gravity is an actuality magnified by slope and elevation; as real as a boulder wildly bouncing down through space toward a fragile creature of flesh and blood. In the strand we note a change in medium that hardly alters gravity's pull upon the land, a pull effectively begun at the mountain's peak. At the start there is only frost and rock fall—from the "bare bedrock bones" to the rubble heap, and the land form is stark, angular, and conic. The debris slopes repose between truncated spurs to flank the high desert's near-static tablelands, temporarily immune to the erosional ravages wrought by climate change just below. Steep, bounding slopes of the mountains and high plateaus "melt back" under an atmospheric "surf" where climate follows climate, erosion and weathering alter and accelerate with each shift, and then wane, and the energy expended against the crust rivals that of the distant ocean shore.

If the foregoing involves a measure of non-scientific imagery, it is at least imagery anchored in the theories and physical relations worked out by many researchers and outlined in earlier pages. It has been fun. And it has been easy—compared with interpreting buried landscapes which in rock outcrops tend to exhibit only their edges to our views, or their surfaces for reconstruction where pierced by bore holes, or what has been eroded from them, or what is left when they are uncovered. In essence, these things are what one works with in the study of paleogeomorphology, the subject of our next chapter. The alert reader will be quick to observe that a good bit of this kind of maneuvering was involved in our analyses of orogenic geomorphology where we had to work out the form and character of deposits in order to begin understanding mountainous topography. Glacial geomorphology (Chapter 8) is also not without its counterparts. But it is mainly a matter of degree, and in the sections to come we will be forced to largely dispense with the scenic vista and the unobstructed three-dimensional perspective. The geomorphic relationships we have so carefully worked out on the earth's surface must become analogs for the interpretation of ancient landscapes. As in so many ventures, "timing is of the essence."

REFERENCES

Axelrod, D. I. (1962) "Post-Pliocene Uplift of the Sierra Nevada, California," *Geol. Soc. Amer. Bull.*, v. 73, pp. 183–198.

Bloom, A. L. (1971) "The Papuan Peneplain Problem: A Mathematical Exercise," *Geol. Soc. Amer. Abs. With Programs*, pp. 507–508.

Bosworth, T. O. (1922) *Geology of the Tertiary and Quaternary Periods in the North-West Part of Peru*, Macmillan and Co., London, 434 pp.

Bowman, Isaiah (1916) *The Andes of Southern Peru*, Henry Holt, New York, 336 pp.

Cotton, C. A. (1960) "The Origin and History of Central Andean Relief: Divergent Views," *Geogr. Jour.*, v. 125, pp. 476–478.

Cotton, C. A. (1963) "The Rate of Down-Wasting of Land Surfaces," *Tuatara*, v. 11, pp. 26–27.

Crook, K. A. W. (1965) "Tectonics, Climate and Sedimentation," *7th Int. Sedimentological Cong., Aust. Nat. Univ. Press, Pub.* 109, 4 pp.

Dana, J. D. (1895) *Manual of Geology*, 4th ed., American Book Co., New York.

Davis, W. M. (1889) "The Rivers and Valleys of Pennsylvania," *Nat. Geogr. Mag.*, v. 1, 183–253.

Davis, W. M. (1899) "The Geographical Cycle," *Geogr. Jour.*, v. 14, pp. 481–504.

Davis, W. M. (1905) "The Geographical Cycle in an Arid Climate," *Jour. Geol.* v. 13, pp. 381–407.

Davis, W. M. (1909) *Geographical Essays*, Ginn and Co., New York, (Repub. Dover, 1954, pp. 413–484.)

Dresch, Jean (1941) *Recherches sur L'evolution du relief dans le Massif Central du Grand Atlas, Le Haouz et le Sous*, Arrault, Tours, 712 pp.

Dresch, Jean (1952) "Le Haut Atlas Occidentale, *XIX Int. Geol. Cong. Regional Mono.* no. 3," pp. 107–121.

Dresch, Jean (1957) "Les Types de Relief Morphoclimatiques et leur limites dans les Andes Centrales," *Assoc. Geogr. franc. Bull.*, v. 263–4, pp. 2–19.

Dresch, Jean (1958) "Problems Morphogenetiques des Andes Centrales," *Geogr. Ann.*, v. 67, pp. 130–151.

Faill, R. T. (1973) "Tectonic Development of the Triassic Newark-Gettysburg Basin in Pennsylvania," *Geol. Soc. Amer. Bull.*, v. 84, pp. 725–740.

Fenner, C. N. (1948) "Pleistocene Climate and Topography of the Arequipa Region, Peru," *Geol. Soc. Amer. Bull.*, v. 59, pp. 879–894.

Garner, H. F. (1959) "Stratigraphic-Sedimentary Significance of Contemporary Climate and Relief in Four Regions of the Andes Mountains," *Geol. Soc. Amer. Bull.*, v. 70, pp. 1327–1368.

Garner, H. F. (1963a) "Mountains from Mole Hills," *Geol. Soc. Amer. Bull.*, v. 74, pp. 195–196.

Garner, H. F. (1963b) "Fountain Formation, Colorado; A Discussion," *Geol. Soc. Amer. Bull.*, v. 74, pp. 1299–1302.

Garner, H. F. (1965) "Base-Level Control of Erosion Surfaces," *Ark. Acad. Sci. Proc.*, v. 19, pp. 98–104.

Garner, H. F. (1967) "Moorefield-Batesville Stratigraphy and Sedimentation in Arkansas," *Geol. Soc. Amer. Bull.*, v. 78, pp. 1233–1246.

Garner, H. F. (1968) "Tropical Weathering and Relief" [in] *Encyclopedia of Geomorphology*, R. W. Fairbridge, ed., Reinhold Book Corp., New York, esp. pp. 1161–1172.

Gignoux, Maurice (1955) *Stratigraphic Geology*, W. H. Freeman, San Francisco, esp. pp. 1–5.

Goddard, E. N., Trask, P. D., DeFord, R. K., Rove, O. N., Singewald, J. T. Jr., and Overbeck, R. M., (1948) *Rock Color Chart*, Geol. Soc. Amer. spec. pub., New York.

Holmes, Arthur (1965) *Principles of Physical Geology*, Ronald Press, New York, esp. pp. 574–579.

Hsü, K. J. (1972) "When the Mediterranean Dried Up," *Scientific American*, v. 227, pp. 27–36.

Hubert, J. F. (1960a) "Petrology of the Fountain and Lyons Formations, Front Range, Colorado," *Colo. Sch. Mines Quart. Jour.*, v. 55, 242 p.

Hubert, J. F. (1960b) "Syngenetic Bleached Borders on Detrital Red Beds of the Fountain Formation, Front Range, Colorado," *Geol. Soc. Amer. Bull.*, v. 71, pp. 95–98.

Hudson, F. S. (1960) "Post-Pliocene Uplift of the Sierra Nevada, California," *Geol. Soc. Amer. Bull.*, v. 71, pp. 1547–1574.

Janda, R. J. (1971) "An Evaluation of Procedures Used in Computing Denudation Rates," *Geol. Soc. Amer. Bull.*, v. 82, pp. 67–80.

Jenks, W. F. (1948) "Geology of the Arequipa Quadrangle of the Carta Nacionale del Peru," *Geol. Inst. Peru Bol.* 9, 204 pp.

Jenks, W. F. (1956) "Handbook of South American Geology," *Geol. Soc. Amer. Mem.*, 65, 378 pp.

Johnson, D. W. (1931) *Stream Sculpture on the Atlantic Slope*, Columbia Univ. Press, New York, 142 pp.

Johnson, W. D. (1901) "The High Plains and Their Utilization," *U. S. Geol. Surv. Ann. Rept.*, v. 4, pp. 601–741.

Joly, F. (1952) "Le Haut Atlas Oriental," *XIX Int. Geol. Cong. Regional Mono.*, no. 3, pp. 67–80.

Jukes, J. B. (1962) "Formation of River Valleys in the South of Ireland," *Geol. Soc. London Quart. Jour.*, v. 18, pp. 378–403.

Krook, L. (1968) "Origin of Bedrock Values of Placer Deposits," *Econ. Geol.* v. 63, pp. 845–846.

Krynine, P. D. (1935) "Arkose Deposits in the Humid Tropics: A Study of Sedimentation in Southern Mexico," *Amer. Jour. Sci.*, v. 29, pp. 353–363.

Krynine, P. D. (1936) "Geomorphology and Sedimentation in the Humid Tropics," *Amer. Jour. Sci.*, v. 32, pp. 297–306.

Krynine, P. D. (1942) "Differential Sedimentation and its Products during a Complete Geosynclinal Cycle," *1st Pan Amer. Cong. Min., Eng., and Geol. Ann.*, v. 2, pp. 537–561.

Krynine, P. D. (1949) "The Origin of Red Beds," *New York Acad. Sci. Trans.*, v. 11, pp. 60–68.

Krynine, P. D. (1950) "Petrology, Stratigraphy and Origin of the Triassic Sedimentary Rocks of Connecticut," *Conn. Geol. Surv. Bull.*, v. 73, 239 p.

Lees, G. M. (1955) "Recent Earth Movements in the Middle East," *Geol. Rundschau*, v. 43, pp. 221–226.

Lister, C. R. B. (1971) "Tectonic Movement in the Chile Trench," *Science*, v. 163, pp. 719–722.

Lindgren, Waldemar (1911) "The Tertiary Gravels of the Sierra Nevada of California," *U. S. Geol. Surv. Prof. Pap.* 73, 226 pp.

Lindgren, W. and Turner, H. W. (1894) *Description of the Gold Belt; Description of the Placerville Sheet*, U. S. Geol. Surv. Folio no. 3.

Mabbutt, J. A. (1966a) "Landforms of the Western Macdonnell Ranges" [in] *Essays in Geomorphology,* G. H. Dury, ed., American Elsevier Co., New York, esp. pp. 83–119.

Mabbutt, J. A. (1966b) "Mantle-Controlled Planation of Pediments," *Amer. Jour. Sci.,* v. 264, pp. 78–91.

Mackin, J. H. (1938) "The Origin of Appalachian Drainage—A Reply," *Amer. Jour. Sci.,* v. 236, pp. 27–53.

Mackin, J. H. (1959) "Timing of Post-Orogenic Uplift in the Rocky Mountains and the Colorado Plateau," *Geol. Soc. Amer. Bull.* (Abstract) v. 70, pp. 1733–1734.

Maher, J. C. (1953) "Permian and Pennsylvanian Rocks of Southeastern Colorado," *Amer. Assoc. Pet. Geologists Bull.,* v. 37, pp. 913–939.

Maher, J. C. and Collins, J. B. (1952) "Correlation of Permian and Pennsylvanian Rocks from Western Kansas to the Front Range of Colorado," *U. S. Geol. Surv.* Chart OC-46.

Maloney, N. J. (1965) "Geomorphology of the Central Coast of Venezuela, *Inst. Oceanogr. Bol.,* v. 4, pp. 246–265.

Meyerhoff, H. A. (1972) "Postorogenic Development of the Appalachians," *Geol. Soc. Amer. Bull.,* v. 83, pp. 1704–1727.

Meyerhoff, H. A. and Olmstead, E. W. (1936) "The Origins of Appalachian Drainage," *Amer. Jour. Sci.,* v. 232, pp. 21–41.

Mitchell, J. G. (1954) "Lyons Sandstone, a New Objective in Eastern Colorado," *Oil and Gas Jour.,* pp. 106–108, 130.

Moberley, R. (1963) "Rate of Denudation in Hawaii," *Jour. Geol.,* v. 71, pp. 371–375.

Newell, N. D. (1949) "Geology of the Lake Titicaca Region, Peru and Bolivia," *Geol. Soc. Amer. Mem.* 36, 111 p.

Newell, N. D. (1956) "Reconocimiento Geologico de la Region Pisco-Nazca," *Geol. Soc. Peru Bol.,* v. 30, Cong. Nac. Geol. Ann., Pt. I, pp. 261–295.

Oberlander, Theodore (1965) *The Zagros Streams,* Syracuse Geogr. Series, no. 1, 168 pp.

Penck, Walther (1924) *Morphological Analysis of Land Forms,* (Translated in 1953 by H. Czech and K. C. Boswell), Macmillan, London, 430 pp.

Powell, J. W. (1875) "Exploration of the Colorado River of the West," *Smithsonian Inst. Rept.,* v. 153, esp. pp. 163–166.

Ritter, D. F. (1967) "Rates of Denudation

(Short Review Pap. 6), *Jour. Geol. Ed.,* v. 15, pp. 154–159.

Rodgers, N. D. (1858) "The Geology of Pennsylvania," *First Penn. Geol. Surv. Rept.* 2, pp. 895–896.

Royo y Gómez, J. (1956) "Quaternary in Venezuela" [in] *Stratigraphical Lexicon of Venezuela* (English ed.) Geol. Bol., Spec. Pap. I, pp. 468–478.

Ruxton, B. P. (1967) "Slopewash under Mature Primary Rainforest in Northern Papua" [in] *Landform Studies from Australia and New Guinea,* J. N. Jennings and J. A. Mabbutt, eds., Aust. Nat. Univ. Press, pp. 85–94.

Ryder, R. T. and Scholtem, Robert (1973) "Syntectonic Conglomerates in Southwestern Montana, Their Nature, Origin and Tectonic Significance," *Geol. Soc. Amer. Bull.,* v. 84, pp. 773–796.

Sapper, K. (1935) *Geomorphologie der Feuchten Tropen,* B. G. Teubner, Leipzig-Berlin.

Schumm, S. A. (1963) "The Disparity between Present Rates of Denudation and Orogeny," *U. S. Geol. Surv. Pap.* 454-H, pp. 1–13.

Schumm, S. A. (1968) "Speculations Concerning Paleohydrologic Controls of Terrestrial Sedimentation," *Geol. Soc. Amer. Bull.,* v. 79, pp. 1573–1588.

Simonett, D. S. (1967) "Landslide Distribution and Earthquakes in the Bewani and Torricelli Mountains, New Guinea" [in] *Landform Studies from Australia and New Guinea,* J. N. Jennings and J. A. Mabbutt, eds., Aust. Nat. Univ. Press, pp. 64–84.

Simonett, D. S. (1968) "Selva Landscapes" [in] *Encyclopedia of Geomorphology,* R. W. Fairbridge, ed., Reinhold Book Corp., New York, esp. pp. 989–991.

Strahler, A. N. (1945) "Hypothesis of Stream Development in the Folded Appalachians of Pennsylvania," *Geol. Soc. Amer. Bull.,* v. 56, pp. 45–88.

Thompson, H. D. (1939) "Drainage Evolution in the Southern Appalachians," *Geol. Soc. Amer. Bull.,* v. 50, pp. 1323–1358.

Thornbury, W. D. (1965) *Regional Geomorphology of the United States,* John Wiley and Sons, N. Y., 609 pp.

Trask, P. D. (1961) "Sedimentation in a Modern Geosyncline off the Arid Coast of Peru and Northern Chile," *XXI Int. Geol. Cong. Rept.,* v. 23, pp. 103–118.

Van Andel, T. H. and Curray, R. (1960) *Recent Sediments, Northwest Gulf of Mexico*, Amer. Assoc. Pet. Geologists, Tulsa.

Wahlstrom, E. E. (1948) "Pre-Fountain and Recent Weathering on Flagstaff Mountain near Boulder, Colorado," *Geol. Soc. Amer. Bull.*, v. 59, pp. 1173–1190.

Wahrhaftig, Clyde (1965) "Stepped Topography of the Southern Sierra Nevada," *Geol. Soc. Amer. Bull.*, v. 76, pp. 1165–1190.

Wegman, E. (1957) "Tectonique vivante, Dénudation et phenomènes convexes," *Géog. Physique et geol. Dynamique, Rev.*, v. 1, pp. 3–15.

Welter, Otto (1947) "Sobre el Levantamiento Pliocénico-Cuaternario de los Andes Peruanos," *Geol. Soc. Peru Bull.*, toma 20, pp. 5–19.

Woolard, G. P. (1969) "Regional Variations in Gravity" [in] *The Earth's Crust and Upper Mantle*, P. J. Hart, ed., esp. pp. 320–341.

11 Ancient landforms and landscapes

The study of the manner in which ancient topography came into existence goes under the rather euphoniously ponderous name, paleogeomorphology. The main difference between an ancient landscape and a modern one is not a matter of age, though of course there is an age difference. Rather, ancient landscapes have been buried. And the vast majority of them still are. A few have been uncovered, mainly by natural processes, and are said to be *exhumed landscapes* (Fig. 11.1). The remainder lie interred under one or more layers of rock or regolith which really brings us down to the crux of the matter. Because there is virtually an infinite number of buried landscapes in a major sequence of stratified rocks, some subaerial, some subaqueous, one must decide what to study (cf. Martin, 1966).

Ancient landscapes must, in effect, be fossilized so that we can detect them. The mere fact that, in a bed several meters thick, we know that each and every grain of sand and gravel was formerly at the surface, does not help us to reconstruct that terrain as the bed developed. At least it does not unless some larger textural manifestation of the surface configuration also exists. In short, an internally homogeneous rock mass tells us nothing of the terrain conditions that prevailed during its formation. The external form of the same rock mass may clearly indicate some depositional or erosional morphology. But unless the inside discloses something, a series of ripple-marks, some cross-stratification, a lense, something we can accept as a fossilized expression of the earth's surface, the developmental stages of the larger deposit are lost (Fig. 11.2).

UNCONFORMITY AND DEPOSIT ANALYSIS

The paleogeomorphologist, like his counterpart with more contemporary interests, is mainly involved in two types of fossil landforms, erosional and depositional. A buried erosion surface is, by definition, an *unconformity*. Thus, to a great extent, paleogeomorphology is an analysis of unconformities. Some depositional microtopography

Fig. 11.1. An exhumed landscape in the central Sahara at Wadi Taffasassett, near the borders of Algeria, Libya, and Nigeria. Pedimentation has exposed the grooves and striations left by continental glaciers, which moved from south to north across West Africa in the Upper Ordovician. The glaciated surface was later buried by Paleozoic and Mesozoic formations now removed by Cenozoic agencies. The exhumed surface is partly masked by drifting sand and patches of gravel. See footprint and hammer for scale. (R. W. Fairbridge.)

comes under scrutiny also, but our interest here is largely by way of an approach to comprehending larger depositional landforms such as deltas. Of course, structural and volcanic landforms of the past are also of significance.

In one definition of the term unconformity, it is described as a surface of erosion or non-deposition. From earlier discussions, it will be recalled that deposition tends to generate planar surfaces. Thus, a break in deposition not accompanied by erosion, if it is expressed topographically at all, would be an essentially planar feature. A non-depositional break might appear as a bedding plane, in which case it is less likely to have been a "break" than a brief switch to clay accumulation in stratified sequences consisting of sandstone,

limestone, or other similarly non-argillaceous material. Alternatively, a non-depositional break may only be expressed within an otherwise homogeneous rock mass by the detection of an interruption in the fossil record—*paleontologic unconformity*. True paleontologic unconformities are of little interest to paleogeomorphologists since no fossil landform is directly involved. Also, it is difficult to be enthusiastic over other kinds of non-depositional breaks, since there is a limit to what one can say about a plain. That leaves unconformities.

Erosional unconformities have several advantages over the paleontologic variety. (1) Generally, they can be seen on outcrop or otherwise detected because they cut across pre-existing features (Fig. 11.3).

(2) A series of deposits is formed during the creation of an unconformity which (a) are equivalent in time to the development of the surface, (b) may include datable materials and organisms and thereby bridge the depositional gap and lost organic and environmental record, and (c) may express the conditions under which the unconformity developed, more or less in chronologic order (Fig. 11.4).

It should be kept in mind, in any case, that the principles of landscape analysis set forth in earlier chapters also apply to fossil landscapes. Therefore, the surface of an unconformity may reflect a single environment or several geomorphic systems imposed in sequence. Where an unconformity represents a long time interval, only the more recent events may be reflected in its ultimate topography. Earlier episodes may be reflected elsewhere by deposits. But let us stop to consider the im-

plications of the several types of geomorphic environment in terms of the probable nature of their inclusion in the geologic record.

I. In humid geomorphic systems, the land is eroded fluvially and develops an increased relief and a regolith consisting mainly of soil and colluvium under ideal circumstances. The soils are theoretically youngest next to the bedrock from which they are forming and oldest on top if they are not accreted vertically. Deposits are mainly elsewhere, subaqueous in many cases, and made in accordance with the Law of Superposition and the principle of sedimentary lag (Fig. 11.5).

II. In arid geomorphic systems, the land is eroded by both running water and wind, tends to become increasingly planar, and develops an alluvial regolith which is youngest on top and oldest next to the bedrock being buried—derivative deposits

Fig. 11.2. Buried subaqueous topography expressed by a lense of sandstone in the lower Cretaceous, Lakota Formation southwest of Colorado Springs, Colorado.

Fig. 11.3. An impressively angular unconformity between the near-vertical beds of the Jurassic, Twist Gulch Formation and the superjacent Paleocene, Flagstaff Formation exposed along Salina Canyon, central Utah. (W. N. Gilliland.)

elsewhere tend to be minor, mainly eolian, and are usually entrapped by downwind vegetation or standing water (Fig. 11.6).

III. In glacial geomorphic systems, the ice produces ice-scoured surfaces and drift deposits, the latter being oldest at their bases—derivative glacio-fluvial debris deposited elsewhere subaqueously shows similar age relations and eolian effects may somewhat resemble those of deserts (Fig. 11.7).

IV. In coastal and subaqueous geomorphic systems, there are major, essentially continuous sediment input conditions adjacent to humid and glacial environments, particularly under fluctuating environmental conditions—minimal, discontinuous sediment input conditions prevail adjacent to deserts. Coastal erosion conditions range

from high energy surfs of open oceans to near-zero energy strands of sheltered gulfs, coves, bays, and probably a good many epicontinental sea coasts of the past—marine

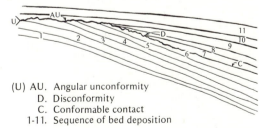

(U) AU. Angular unconformity
    D. Disconformity
    C. Conformable contact
1-11. Sequence of bed deposition

Fig. 11.4. Cross section of an unconformity (U) which passes from an angular unconformity (AU) laterally into a disconformity (D) and finally into a conformable sequence of strata (C). Numbers (1–11) are in order of decreasing age in this idealized sequence.

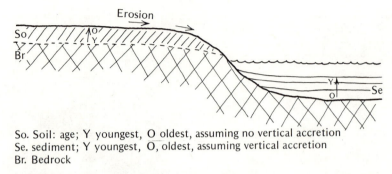

So. Soil: age; Y youngest, O oldest, assuming no vertical accretion
Se. sediment; Y youngest, O, oldest, assuming vertical accretion
Br. Bedrock

Fig. 11.5.   Cross section of age relations within a soil developing in a humid region (So) and derivative fluvial sedimentary deposits (Se).

conditions replace subaerial environments and vice versa through epeirogenic, glacio-eustatic, and evaporative mechanisms (cf. Chapters 3 and 4).

V. Tectonic and volcanic geomorphic systems relate to specific types of geotectonic setting, and from time to time they modify subaerial and subaqueous systems with appropriate styles of deformation and deposition plus distinctive relief reflective of the deformative style.

The first point to be made about these geomorphic associations has to do with the probable location of the resulting unconformity in the event that one or another of these terrain types becomes buried. If a humid area is buried, and if the regolith is reasonably intact, the unconformity will often appear to be at the base of the weathered zone on bedrock. An arid region is likely to be buried by its own detritus, and the unconformity of note appears to

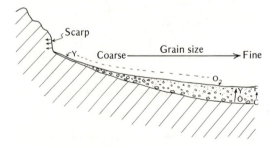

Fig. 11.6.   Cross section of particle size relations during pediment development and scarp retreat (C, coarse; F, fine) in relation to younger (Y) and older (O) portions of resulting deposits.

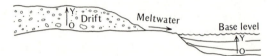

Fig. 11.7.   Cross section of usual deposit time relations between oldest materials (O) and youngest accumulations (Y) in glacial drift and derivative meltwater deposits made in an adjacent lake or sea.

be at the base of the debris if the subjacent bedrock is stratified. Continental deposits frequently incorporate many erosional surfaces within dominantly depositional intervals, but it should be kept in mind that if erosion had been dominant the deposit would not be preserved. In glaciated areas, the pavements are often overlain by either continental or marine deposits, and the unconformity may be very obvious; alternatively, the depositional areas tend to consist of a maze of pedologic, glacio-fluvial, and drift accumulations in which the most abrupt and striking contacts are often between soil horizons, cross-stratification layers, or lacustrine beds.

The marine and coastal terrain situation "paleogeomorphologically" calls for some special remarks in light of our previous discussions. For example, if a desert is dominantly a depositional realm, and adjacent seaways tend to be starved of sediment and thus dominantly erosional, A MARINE TRANSGRESSION OF THE DESERT BY THE SEA REPLACES A DEPOSITIONAL ENVIRONMENT WITH AN EROSIONAL ONE, particularly if the coastal condition is one of high energy. Conversely, if the sea regressed from the same area, and the emergent area remained (or became) a desert, THE EMERGENCE OF AN ARID LAND WOULD INDUCE DEPOSITION RATHER THAN EROSION. These relations are precisely the converse of those usually visualized for the effects of marine submergence or emergence with respect to erosion and deposition. The usually anticipated emergence-erosion versus submergence-deposition relationships are evidently those of an ocean adjacent to a humid region that normally exports sediment to adjacent sub-base–level depositional sites. We will further consider these matters shortly.

A second point relating to the previously cited geomorphic settings has to do with unconformity relief. In theory, those unconformities with notable local erosional relief are most likely to be of humid fluvial genesis. Ideally, aridity should generate low-relief buried surfaces. Unconformities with tectonic relief may display greater roughness but are inclined to be difficult to trace, especially if faulting should occur prior to or during burial. Also, if the maturity of sediment related to a developing unconformity (erosion surface) is largely governed by subaerial environment, then unconformities with high erosional relief should correlate with mature derivative sediments normally yielded by a humid region—it being clearly understood that we are speaking of probabilities. Unconformities with low erosional relief should correlate with immature derivative sediments. Furthermore, crustal movements that do not directly generate breccias or pyroclastic debris should not normally otherwise influence related sedimentary suites apart from the usual considerations of provenance. Just for the record, it should be noted that breccias developed during faulting are not all equally accessible to associated deposits. A thrust fault, for example, tends to bury its own breccia or mylonite zone with the allochthon. A normal fault tends to uncover its breccia zone gradually, usually in increments of small extent and volume. A gravity slide mass (cf. Chapter 3) related to tectonic

denudation may rapidly expose an extensive brecciated and slicken-sided glide zone or surface, particularly if detachment occurs and the allochthon divides into separate fragments (Fig. 11.8).

There are hundreds of major unconformities across the earth's surface within various stratigraphic sequences and of widely differing ages. We have already considered an example from an orogenic setting in our discussion of the Ancestral Rocky Mountains (Chapter 10). Since we cannot deal with each known case, and, to balance our treatment, we will discuss several examples developed in sequence in a cratonic setting. This should give some perspective in terms of variations in tectonic activity, lithology, and environments

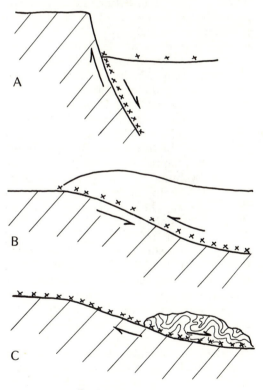

x x x x Fault breccia or gouge

Fig. 11.8. Cross sections showing the relative ability of faults to generate fault breccia which could become incorporated into the sedimentary record. (A) Normal; (B) reverse or thrust; (C) gravity allochthon.

where the principal seas were of the shelf or the epicontinental variety.

BURIED PALEOZOIC LANDSCAPES OF THE OZARK REGION

In the southcentral portion of the North American craton, there are a series of deposits made mainly in epicontinental Paleozoic seaways and on the comparatively low lands that periodically developed adjacent to those seas (cf. Frezon and Glick, 1959). Within the sequence of resulting stratified rocks, which are dominantly marine deposits, there are a number of major unconformities with morphologies that may help us to better understand ancient terrain development in tectonically rather quiescent realms.

In general, there would appear to be two main kinds of continental submergence by sea water. One, which we can term epeirogenic, has to do with broad, essentially vertical continental shifts and warpings or alternate relative adjustments in sea level, and we can relate these to long-term changes in the thickness of continental crust (cf. Chapter 3) versus increases in oceanic volume by lithic dewatering. Crustal thickening or thinning determines the height at which a continent will be isostatically stable, whereas the rate of water production determines whether the oceanic depressions can contain the fluid. If water is produced faster than continents thicken, an epeirogenic submergence is a probable result. If continents thicken faster than water is freed from lithic material, an epeirogenic emergence occurs. As one would expect from such ponderous processes, considerable time is required for a sea level change of this kind—something on the order of tens or scores of millions of years.*

The other type of land submergence is that related to the formation and wastage of continental ice sheets. Glacio-eustatic changes in sea level on the order of 300–400 ft (100–130 m) were common through the Pleistocene every few tens of thousands of years, and it is becoming increasingly evident that similar adjustments characterized much of earlier geologic time when landmasses occupied near-polar positions. Though a sea level rise of the usual glacio-eustatic magnitude would not submerge a high-standing continent, it might well inundate much of a low-standing landmass. Furthermore, the actual development of epicontinental seas in response to tectonic phenomena almost certainly occurred in conjunction with GLACIO-EUSTATIC SEA LEVEL CHANGES in many cases. Of course, it may not always be possible to determine which type of strand-shift mechanism was instrumental in causing a given land-sea relationship. Even so, it is probably reasonable to assume that a very prolonged emergence or submergence is of epeirogenic nature, whereas a series of more brief marine oscillations may be glacio-eustatic.

Basal St. Peter Unconformity

One of the major regional unconformities in the mid-continental United States occurs within the Ordovician System, at the base of the St. Peter Sandstone. As noted by Moore (1933), the sea in which the St. Peter Sandstone was deposited was in the southern Ozark region in lower Ordovician time but gradually transgressed northward and, by middle Ordovician time, had reached the area of Minnesota where the type section occurs along the St. Peter River. On the basis of duration, it would appear to be an epeirogenic-type marine submergence, and a number of studies have been made to account for the origin of the well-rounded and often frosted quartz sand that comprises the unit to the exclusion of almost all other materials (cf. Berkey, 1906; Trowbridge, 1917; Dake, 1921; Born, 1940; Stauffer and Theil,

---

* Since orogeny may thicken continental crust, some related emergence can be anticipated.

1941). This last relationship in association with the basal unconformity is of prime interest.

In the southern Ozark region the St. Peter Sandstone averages about 100 ft (30 m) thick and rests disconformably on the Everton Dolomite which is also of Ordovician age (Fig. 11.9). The disconformity locally shows as much as 10 ft (3 m) of relief, and yet no materials from the subjacent rocks are incorporated in a basal conglomerate, either from the dolomite or from the locally present chert layers from the same formation. In short, the marine transgression of the sea was not accom-

panied by sufficient erosional energy to truncate erosional relief previously developed or to produce coarse detritus. Furthermore, though there is evidence of prolonged emergence of the land before the St. Peter transgression, detritus generated during this interval was apparently not of a coarse fragmental nature. The materials deposited on the unconformity are entirely of sand size or smaller, and are chemically and mineralogically mature material. We will defer a final assessment of the St. Peter unconformity relations until other buried erosion surfaces from the same region have been described. Nevertheless, it can be

Fig. 11.9.   The figure (left foreground) stands just below the unconformable contact between the St. Peter Sandstone and subjacent Everton Formation, both of Ordovician age as exposed northeast of Marshall, Arkansas. Surface irregularity of the disconformity is evident, but there is essentially no associated conglomerate from the subjacent formation.

noted that the marine transgression of an erg would generate the deposit in question, except that there would be a basal conglomerate, at least in the Sahara or Arabia.

Post-Boone Unconformity

In the same portion of the southern Ozarks as the previous case there is a notable unconformity separating the Mississippian age Boone Formation from younger sedimentary rocks (cf. Giles, 1935). Foremost among the latter units are the Moorefield Shale and Batesville Sandstone of upper Mississippian age. The situation has been described in some detail (Garner, 1967) and appears to approximate the ideal of cratonic tectonism, uplift, erosion, and later submergence. The Boone Formation consists of limestone and gray-to-white vitreous chert mentioned here in Chapter 6 in a separate context (see Fig. 3.55). The Boone is an average of 325 ft (100 m) thick and covers most of southern Missouri. In the limestone and chert we have, respectively, metastable and semi-insoluble substances. Before the deposition of younger marine sediments, the Boone Formation was warped, faulted, and locally uplifted and exposed to erosion. There is up to 50 ft (15 m) of local erosional relief on the unconformity (Fig. 11.10).

Theories espousing the tectonic control of erosion outlined in Chapter 10 suggest that sediments derived from the Boone Formation should be immature. Instead, the basal conglomerates of the Moorefield-Batesville interval are mature chert clasts, largely derived from chert beds and nodules in the Boone. Available evidence

(Garner, 1967) indicates that no limestone clasts survived subaerial weathering processes and that none were generated during transgression of the warm, reef-strewn sea that followed. There is also abundant evidence that the disconformity on top of the Boone Formation was generated mainly in a humid, solutional environment. There is no question that the late Boone crustal movements rearranged the local geography, possibly about the time of formation of the Ancestral Rockies to the west. But these tectonic effects did not control the associated sedimentation in the southern Ozarks.

Mississippian-Pennsylvanian Unconformity

The next youngest major stratigraphic break in the southern Ozark region above the Mississippian hiatus just discussed is between the Mississippian and Pennsylvanian systems. The formations in question are the Pitkin Limestone of late Mississippian age as it occurs in the type locality in northwest Arkansas and the overlying Cane Hill Member of the Hale Formation of early Pennsylvanian (Morrowan) age (cf. Frezon and Glick, 1959). In many places the post-Pitkin unconformity would almost pass for a bedding plane, so even is its topography, were it not for the lithologic contrasts that occur above and below and regional field relations and the paleontologic data that demonstrate a notable discontinuity (Fig. 11.11).

The Pitkin Limestone of northwestern Arkansas is a reef and inter-reef shelf carbonate deposit averaging about 30 ft (10

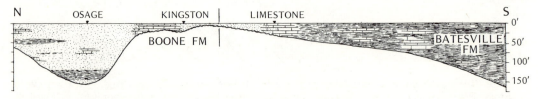

Fig. 11.10. North-south cross section showing unconformity relations between the Boone and Batesville formations, northcentral Arkansas, athwart the Boston Mountains and a related structural rise (cf. Fig. 2.36).

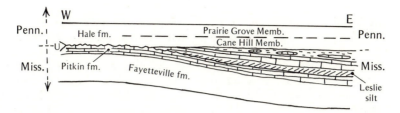

Fig. 11.11. Cross section of the Mississippian-Pennsylvanian unconformity (U), northern Arkansas, and associated rock units. Note that several late Mississippian formations in the east are missing in the west.

m) thick (Figs. 3.53 and 3.54). It includes various massive patch reef core areas and many micro-atolls plus inter-reef detrital, bioclastic limestone which is more distinctly bedded. The overlying Cane Hill unit consists of thin-bedded [2–2.5 inch (5–6 cm)], siliceous, ferrugenous siltstone and fine-grained sandstone some 60 ft (18 m) thick. Apart from impressive paleontologic evidence based on zonations with goniatite cephalopods (cf. Gordon, 1964), it is the field-rock relations that demonstrate the extent of the depositional break between the Pitkin and Cane Hill. In northcentral and northeastern Arkansas, a series of younger Pitkin lithic units are developed which probably never were deposited to the northwest (cf. Easton, 1942). They include a siliceous siltstone unit averaging some 30 ft (10 m) thick (Fig. 3.50), an overlying limestone unit up to 90 ft (25 m) thick, and a superjacent shale, limestone, and sandstone unit about 120 ft (35 m) thick. In all, therefore, the Post-Pitkin unconformity in the northwest developed during the accumulation of some 200+ ft (70 m) of sedimentary rock, mainly argillites and limestone (cf. Fig. 11.11).

Available evidence suggests that the Post-Pitkin unconformity developed not far above sea level, while the foregoing rocks accumulated in a downwarped area to the east and south. The unconformity proper is a planar surface strewn here and there with gravel and includes notable limestone rubble plus odd fragments of shale, ferrugenous nodules, and chert from older formations of the region. The gravel is patchy, in layers up to 6 inches (13 cm) thick. The limestone surface on which the rudites are deposited is generally non-channeled. In eleven field seasons I spent in the region studying the rocks and their relationships, only a single channel was found cut into the Pitkin in what appeared to be a subaerial erosional setting. That channel was about 3 ft (1 m) wide and some 2 ft (60 cm) deep and was filled with 1–2 inch (2.5–5 cm) diameter limestone gravel.

In the Post-Pitkin unconformity situation just discussed, there is little or no evidence of crustal movements apart from local downwarping. The few strand line oscillations are perhaps too prolonged to be explained glacio-eustatically, but the degree of tectonism is certainly far less than that for the Boone situation discussed previously. Even so, the sediments associated with the unconformity are coarse and are dominated by metastable constituents. Also, of course, there is an almost complete lack of erosional relief. This fact, plus the tropical setting indicated by the sporadic reef development and the rigorous weathering required to generate the limestone gravels, suggests a low-level desert environment during development of the Mississippian-Pennsylvanian unconformity. One more Ozark unconformity will be discussed before we turn to other aspects of fossil landscape analysis.

Bloyd-Winslow Unconformity

One of the most interesting unconformities in the late Paleozoic sedimentary rock sequence of the southern Ozarks occurs between the basal Winslow Formation sandstone unit (Atoka of some authors) and various members of the earlier Pennsylvanian and later Mississippian stratal sequence. The sea had withdrawn southward into northern Arkansas, and basal Winslow rocks are conformable with youngest Morrowan strata there (Fig. 11.12). From south to north and northeast, the basal sandstone layer of the Winslow rests on various shale, limestone, and calcareous sandstone members of the Bloyd Formation (Fig. 11.27) and then overlaps onto the sandstone, limestone, siltstone, and shale units of late Mississippian age. In a regional sense, we are therefore dealing with an angular unconformity produced by a transgressive sea. Local erosional relief developed on the surface amounts to several tens of feet.

The basal Winslow sandstone unit previously mentioned is worthy of further comment. Occasionally referred to as the "millstone grit," it often takes the form of a single, cross-stratified sandstone layer 60–90 ft (20–30 m) thick (Fig. 11.13). It is a distinctive bluff-forming layer and consists mainly of tan, medium-grained quartz sandstone, slightly ferrugenous, with some secondary enlargement of grains (Fig. 3.52). Two other aspects of the unit are especially noteworthy. It contains scattered rounded quartz and quartzite pebbles up to about 1 inch (2.5 cm) in maximum diameter which are somewhat more abundant in the lower part. Also, though the basal Winslow rests on a wide variety of lithologies, including shale, limestone, calcareous sandstone, and siltstone,

there is no basal conglomerate, and no notable clasts of these lithologies occur in the basal Winslow. What does occur there is perhaps even more significant. Where the base of the lower sandstone unit can be viewed beneath overhanging bluffs, it is often seen to consist of an interwoven mat of vegetal matter 0.5–1 inch thick consisting of Carboniferous scale-bark tree fragments up to 8 inches (12 cm) wide and 3–4 ft (1 m) long. Most notable are pieces of *Calamites* sp. and *Lepidodendron* sp.

The basal Winslow unconformity, once again, is a buried surface of appreciable erosional relief. The transgressive facies is mineralogically very mature, a fact readily traced to an associated humid coastal environment. The latter relation is thoroughly documented by the plant fossil association. Also, the survival of the vegetal material in extensively recognizable form attests to the tranquil, low-energy form of the sea during transgression. Absence of metastable rocks and minerals from the basal Winslow is therefore hardly due to marine erosion or reworking and can therefore be attributed to subaerial environmental controls.

Of the four major unconformities discussed from the Ozark region, three exhibit considerable erosional relief, and at least one of the three combines this with evidence of crustal movement. In each instance there is evidence that the seas were comparatively warm (reefs), that the adjacent lands were humid, and associated sediments are chemically and mineralogically mature quartzose sandstone with evidence for offshore argillite facies. All three of the substrates beneath the unconformities include metastable rock types but none of them survive in fragmental form in superjacent deposits. The single unconformity that displayed essentially no erosional relief or structural movement had associated metastable conglomerates. Related marine deposits include siliceous and probably cold-water facies, and indications are that the land was a coastal desert. In at least these examples, the postulate made in

Fig. 11.12. Unconformity between Winslow Formation and older rocks in the southern Ozark region.

Fig. 11.13. A butte near Heber Springs, Arkansas, capped by the massive basal Winslow sandstone interval. (See also Fig. 3.52.)

the introduction to this chapter would appear to be borne out.* The high-relief unconformities are related to humid conditions, and the low-relief example suggests aridity. Furthermore, there is no apparent lithologic response to local crustal movements in terms of correlative sediments.

EARLY PENNSYLVANIAN GEOMORPHOLOGY OF THE MID-CONTINENT

Let us begin with a geologic enigma of the North American mid-continent region and follow it through a paleogeomorphologic analysis to a possible solution. In the sequence of rock formations and uncon-

formities just discussed there is a basal Pennsylvanian formation known as the Hale in northern Arkansas and Oklahoma (cf. Case, 1954; Frezon and Glick, 1959). Its lower Cane Hill Member, consisting of thin-bedded siltstone and very fine sandstone, overlies the Mississippian-Pennsylvanian unconformity. The upper (Prairie Grove) member is similarly thick [60–80 ft (20–25 m)] but consists of Christmastree-shaped carbonate reef mounds and inter-reef, calcareous sandstone and some minor shale. However, our interest here is in a basal Prairie Grove conglomerate which is associated with a surface having notable erosional irregularity. Though this surface has some of the attributes of an unconformity and may, by some definitions, be one, we omitted it from our

---

* The implied genetic relation between conglomerate occurrence and maturity, unconformity relief, and climate may be found not to hold in all instances, but in a probability sense it seems to be a working hypothesis worth testing further.

earlier consideration for reasons that will shortly be clear.

The basal Prairie Grove conglomerate ranges from a few inches to perhaps 10 ft (3 m) in thickness, but in three, rather sharply defined areas continues upward into the formation as conglomeratic sandstone (Fig. 11.14). Two separate aspects of the basal conglomerate are intriguing. First, the conglomerate contains numerous angular limestone fragments of the type common to the Pitkin formation and older units known to be cropping out on the coastal plain not far to the north. Also present are shards of shale, chert chips, and occasional concretion fragments from the same area. More significantly, there are numerous, rounded quartzite and quartz pebbles averaging about 0.5 inch (1 cm) in diameter, plus larger fragments ranging up to 3–5 inch (7.5–12.5 cm) cobbles of quartzite none of which are apparently of local derivation. NOTE that this is the first appearance of such coarse quartzitic materials in the southern-Ozark Paleozoic section. In other words, the basal Prairie Grove incorporates metastable rock materials which have probably moved less than 5 mi (7 km) and metamorphic and igneous clasts which have moved some distance. Thus the enigma previously mentioned. Two questions immediately suggest themselves. (1) How far did the coarse quartz and quartzite substances travel? (2) Why were both locally derived and foreign materials introduced into the seaway simultaneously?

Quartz and Quartzite Provenances

The quartzite clasts of the basal Prairie Grove include some rose quartz commonly attributed to igneous or hydrothermal vein occurrences, some orthoquartzitic material of possible local derivation, and some metaquartzite. Looking to the north, where strand configurations suggest the land was situated, only three possible areas appear to be so situated as to have contributed the materials in question (Fig. 11.15). At the outset, it should be noted that on two counts we are eliminating from our consideration the St. Francis Mountains of southeastern Missouri where some pre-Cambrian rocks are presently exposed. First, paleogeographic considerations indicate that this area was still covered by an extensive area of older sedimentary rocks including, notably, the Mississippian Boone Formation. Second, quartzite is not a notable constituent of the area in any case.

The accompanying map shows three other areas in which quartzite occurs, apart from regions in the Appalachian sector. These are the Precambrian Sioux Quartzite in northwest Iowa, the Baraboo Quartzite of Wisconsin, and the Precambrian crystalline region of the Northern Great Lakes area and southern Canada. It can be seen that the distances to these areas from the southern Ozarks are respectively 500, 600, and 900 mi (700, 840, and 1260 km). To borrow a phrase from detective novels of yore, "The plot thickens." Now we must add a third question to the ones posed at the end of the previous section, namely, (3) How could the materials move such great distances and still be of such large dimensions?

This question arises because the Pennsylvanian Age quartzitic gravels in the

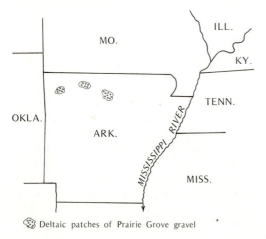

Fig. 11.14. Development of quartzitic deltaic gravels of the Prairie Grove with respect to local geography.

southern Ozarks are of such abundance, dimensions, and clast shapes as to require transport by fluvial means, yet modern rivers of high competency with ready access to similar material cannot transport detritus of such caliber so far. We say "cannot" simply because they "do not." For example, tributary streams to the Mississippi River in the Ozark region have erosional access to coarse relict arid alluvium (cf. Chapters 6 and 7) via such feeder drainages as the White, Black, and Arkansas rivers all some 400–500 mi (560–700 km) from the Mississippi Delta. Yet the materials being deposited in the delta are very fine-grain sand, silt, and clay. In fact, dredges plying their channel-cleaning chores along the main Mississippi River channel in southeastern Arkansas constantly pump up gravel 2–3 inches (5–7.5 cm) in diameter (and an occasional Pleistocene horse tooth) which have been locally exposed by deep current scour less than 400 mi (560 km) from the delta, but none of these materials are being effectively moved downstream by the river. Two things are evident. (1) Factors other than simple hydraulic-sediment interaction (in a slope-volume-velocity-competency relationship) govern the sediment load and deposits of the Mississippi (and presumably similar rivers also). (2) Most of the transport of coarse gravels into the southern Ozark region in Pennsylvanian time was presumably not accomplished by open-channel river flow, but rather by some other mechanism permitting larger clasts to move without undue size attrition.

Space-Time Drainage Relationships

A search for data that might document sediment distribution mechanisms in the North American mid-continent region during early-Pennsylvanian time draws our attention to areas north and east of the present Ozark uplift (cf. Fig. 11.15). It is in these areas that rocks of comparable age are known to occur, and a measure of time control can be established over drain-

CANADA

× Non-quartzitic Precambrian
⬤ Basement quartzite
▱ Prairie Grove Quartzite gravels

Fig. 11.15. Disposition of known quartzite provenances in early Pennsylvanian time relative to the occurrence of derivative gravels.

age and depositional events. The late Mississippian and early Pennsylvanian rocks of the Illinois region are among the most intensively studied in the present context. Swann (1964) has documented a southwest trending drainage system of late Mississippian age (the so-called Michigan River) which brought sediments into the Illinois area from Canada (Fig. 11.16). During the same period, Swan documents 15 major strand line oscillations plus an additional 70 minor fluctuations, a number strongly suggestive of glacio-eustatic sea level influences. But our immediate concerns are with slightly younger rocks.

The youngest Mississippian deposits in the Illinois area are generally conceded to be the Kinkaid Limestone (cf. Swann, 1963), a unit with an upper surface that was extensively incised by a complex of channels (Fig. 11.17) and later buried by early Pennsylvanian "pebbly" sandstones of the Casyville-Tradewater Group (cf. Siever, 1951; Siever and Potter, 1962; Potter and Desborough, 1965). Several aspects of this research appear to be most significant in terms of the Prairie Grove gravels in the southern Ozarks. First, as

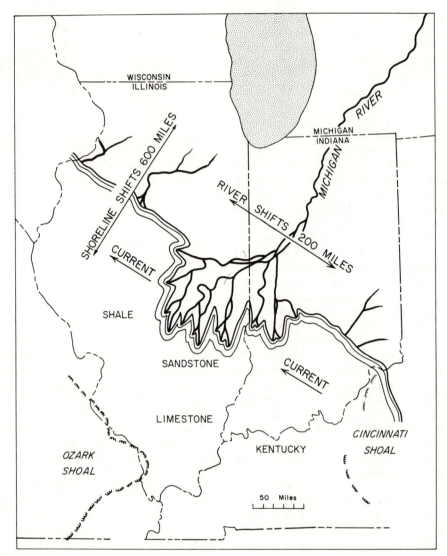

Fig. 11.16. The major Mississippian-Pennsylvanian drainage system of the upper mid-continent region (Michigan River) with suggested shifts in strandline and drainage. (After D. H. Swann.)

illustrated by Potter (1963), the post-Mississippian drainages in Illinois trend northeast-southwest toward the southern Ozark region. Second, the basal Pennsylvanian arenaceous sediments include peanut-size quartz and quartzite pebbles like those that appear in northern Arkansas at the beginning of Prairie Grove time (Fig. 11.18). Third, the drainages illustrated do not have the trunk-stream-tributary configurations of a normal river. Instead, a dendritic flow pattern is confined to the drainage basin margins to the northeast. Farther to the southwest are belts of anastomosing channel systems up to 30 mi (42 km) wide, similar to those encountered on modern riverine plains in Australia and Africa (cf. Figs. 4.23; 4.24) and otherwise developed where extensive arid aggradation has been followed by

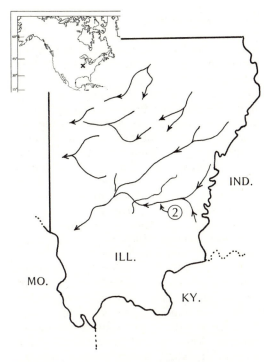

Fig. 11.17. Northeast-southwest trending channels incised into the Mississippian Kincaid Limestone, Illinois. (See also Fig. 11.18.) (After Siever and Potter, 1964.)

humid fluvial conditions as previously discussed in Chapter 7 (cf. Figs. 11.7 and 7.50, 7.51, and 7.52). It will further be noted that the early Pennsylvanian drainages of Illinois develop extensive, complex distributary systems with multiple débouchérs such as the Shari River in Africa.

In his previously cited discussion of the Michigan River, Swann (1964) emphasizes lateral shifts in the débouchér area of up to 200 mi (280 km), and Thornbury (1969) emphasizes the analogy between this type of behavior and the several Hwang Ho River débouchér shifts in modern times to the north and south sides of China's Shantung Peninsula. In the latter instance, however, we are dealing with a river analogous to the Mississippi which sporadically crevasses its natural levees during times of flood and thereby diverts itself. That could have been the reason for the Michigan River adjustments also, but,

as already noted, it is hardly possible to explain the coarse texture of the Arkansas (or Illinois) deposits at all if they were moved by an open-channel, high-velocity river. The early Pennsylvanian drainage configurations documented by Potter (1963), however, are those of an anastomosing channel system (Fig. 11.17) with sporadically braided reaches. Consideration of the Prairie Grove conglomerate distributions (Fig. 11.14) is sufficient to establish that the main concentrations represent débouchér deposits made on an open east-west trending coast in what is now northern Arkansas by a drainage system that probably had two or three mouths operating simultaneously with spacings of 25–50 mi (35–70 km). It seems probable that these débouchérs are the downstream termini of Potter's Illinoian anastomosing channel system (Fig. 11.19).

Thus, it would appear that we may have an answer to one of the three questions posed at the beginning of our discussion of early Pennsylvanian geomorphology, namely, how far did the coarse quartz and quartzite sediments in the southern Ozarks travel? On the basis of correlative drainage configurations, the answer would ap-

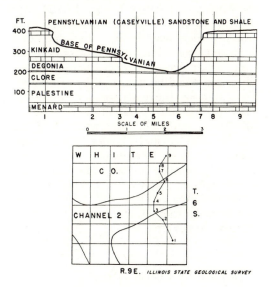

Fig. 11.18. Cross section of channel number 2 of Fig. 11.17. (From Siever, 1961.)

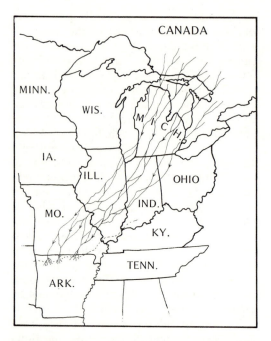

Fig. 11.19. Mid-continental drainage configurations developed intermittently during early Pennsylvanian time and capable of transporting quartzite materials from Canada to Arkansas.

pear to be at least 900 mi (1,250 km) from the Canadian Shield region to the north of the Michigan Basin (cf. Fig. 11.15). Apparently coarse quartzitic materials were working their way to the south and west from this provenance during Mississippian time as intervening depositional basins filled and could be traversed by drainages. For example, the early middle Mississippian (Osagean) Marshall Sandstone of Michigan, studied by the writer and his mentor (Miller and Garner, 1955) contains the same peanut-size quartz and quartzite pebbles in a "millstone grit" facies which slightly later appeared in southern Illinois, and still later in the southern Ozarks after the beginning of Pennsylvanian time.

Environmental Relationships

We have yet to consider why the southern Arkansas Pennsylvanian gravels are so coarse and why materials of both local and foreign derivation were introduced into the seaway there simultaneously. The answers would appear to lie in the realm of environmental space-time relations since, as previously observed, transport by a "normal" humid river involves certain apparent inconsistencies. In addition to the fact that rivers apparently do not readily move coarse detritus great distances during open-channel flow, the order of depositional events in the southern Ozarks is not consistent with the development of Prairie Grove deposits by a river extending itself upslope through headward erosion.

A river system developing adjacent to a coast and gradually extending itself upslope by headward erosion into more distant regions should initially deposit detritus in the seaway eroded from nearby lands and should later add constituents from progressively more and more distant sources. The early Pennsylvanian drainages in the North American mid-continent did not do this. Instead, it is as though water carrying coarse detritus from distant upslope sources crossed a coastal region and acquired an additional sediment load there before entering the sea. This is, in fact, the only explanation that appears to be compatible with all the paleogeomorphologic facts (cf. Fig. 5.52).

The reader will already have noted that the constituents of the Prairie Grove conglomerates are of two contrasting types and derivations. (1) Those substances from the coastal lowland adjacent to the Prairie Grove sea consist mainly of metastable limestone gravels generated in a rigorous and probably arid environment. (2) Those materials from the distant and elevated regions to the northeast are dominantly mature quartzose sediments consistent not only with a humid, chemical weathering environment but also consistant with the great distance traveled which could well eliminate some weaker rock and mineral types. Humidity for the latter source is argued most effectively by the absence of feldspathic detritus which is abundant in parts of the Canadian provenance in question and should have survived the journey along with the quartz if originally it had been created by weathering. Humidity is even more strongly indicated by the fact that the same elevated source area sporadically sponsored the runoff that effected the over-all transport. We are therefore left with a general paleoenvironmental picture of the region in question during late Mississippian-early Pennsylvanian time in which low, dominantly arid coasts extend to elevated, sporadically humid uplands (Fig. 11.20).

These remarks and data perhaps explain the order of introduction of sediments into a southern Ozark seaway in Prairie Grove time in at least partial satisfaction of one of our questions. But the relationship we are considering is not one of fixed geography, unless we confine our discussion to a relatively brief span of time. And even if we restrict our time relation considerably, for example, to the interval during which the Hale Formation was deposited, we are not dealing with fixed environments because even the geography refuses to remain completely stationary. There is abundant evidence that the seas during the interval in question were fluctuating in temperature as well as position. The answer to our third question about the mode of coarse sediment transport is probably involved in related environmental changes.

Climate Change and Sediment Transport

The patterns of sediment accumulation and transport in the North American mid-continent during the later Paleozoic are surprisingly consistent. As already noted in our analysis of several unconformities, coarse detritus (mainly sand) was flushed seaward at a time of warm, reef-strewn seas when there were abundant indications that lands were humid, at least in upland areas. Arid coastal conditions, such as typified early Hale (Cane Hill) time, coincide with an absence of river deposition, probable eolian silt accumulation, siliceously cemented marine sediments,

and probable cold seas. In those instances of most pronounced aridity, it may be safely assumed that arid conditions extended inland and into more elevated areas.

Figure 11.20 shows the apparent paleoenvironmental relations of early Hale time when it is known that the southern Ozark sea coasts were arid (Fig. 11.21). It is further reasoned that the same type of conditions extended northward into Missouri and Illinois so that rivers originating in more humid uplands even farther north were forced to deposit a riverine plain in the form of the Casyville-Tradewater Group of clastic sediments in the desert downslope. Minor amounts of sand were flushed seaward each time it became humid in late Mississippian time, and sand began to appear in southern Ozark deposits then. Deposition of coarse clastic marine sand occurred whenever the seas were warm and contained reefs. The first influx of extensive sand and gravel into the sea did not occur until Prairie Grove

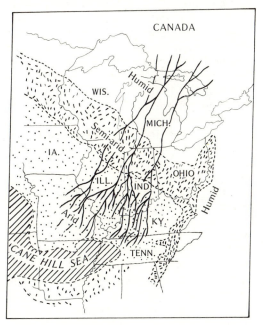

Fig. 11.20. Paleoclimatic and drainage map of the North American mid-continent during Cane Hill (early Pennsylvanian) time showing a restricted and probably cool sea bordered by extensive deserts.

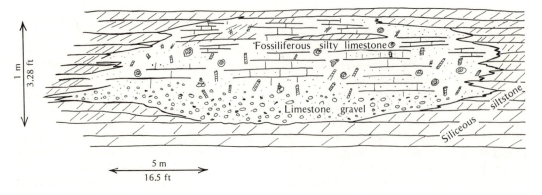

Fig. 11.21.   Section through a gravel pod of the type present in the Cane Hill Member of the Pennsylvanian Hale Formation. The limestone pebble gravel in the pod grades upward into fossiliferous, encrinal limestone.

time when, once again, the seas were warm. The point of this discussion, quite simply, is that the coarse sediments of the Prairie Grove were apparently being prograded to the south from Canada under sporadically arid conditions prior to Prairie Grove time. It has already been noted in Chapter 6 that apparently there is reduced particle attrition in high-density sheetfloods and mudflows in deserts. In other words, fragments can be moved farther with less wear by these agencies than they can by humid-type rivers. Little work has been done on the subject (cf. Schumm and Stevens, 1973) but it may well be that transport within braided stream systems (such as characterize areas of glacial outwash or rivers drying up in deserts) may also involve less particle wear per unit distance of transport than in ordinary open-channel rivers (whether meandering or straight).* A combination of these mechanisms may explain how the Prairie Grove gravels survived the trip from Canada to Arkansas.

Coarse gravels in open river channels apparently tend to be immobilized (possibly through bypassing and/or burial) or are otherwise destroyed while in transit (possibly while under constant bombardment by saltating and suspended sand

grains—*wet blasting*). Certainly, if a by-passed gravel clast, normally entrained mainly by impact or undercutting and moved mainly by rolling, enters a downstream reach of a river with a mud or fine sand bottom, its requisite conditions for transport are altered. The sediment record in deltas suggests that its transport conditions are commonly not met. Possible exceptions occur where the alluviated channels extend essentially to base level (cf. Chapter 10, Arid-Humid Orogenesis).

What we have, in effect, documented in previous paragraphs, is the extension of an alluviated area to base level (by basin filling and riverine plain deposition under sporadically arid conditions). Warming (and slight transgression) of the Prairie Grove sea coincides with the development of an extensive early Pennsylvanian anastomosing channel system on an adjacent alluviated land (Fig. 11.22). These events, with respect to hydrographic and drainage relations previously outlined in some detail (Chapters 4 and 7), suggest that the land adjacent to the Prairie Grove sea became humid essentially from the top down. Runoff related to this followed a similar path. Sedimentologic relations of the late Prairie Grove deposits indicate that even the coast was humid by that time.

* The reasons why this should be true are obscure. But in the late Paleozoic mid-continent we are confronted with coarse detritus which has moved farther than present "rivers" can transport such material. An explanation is clearly needed.

At the start of our discussion of the Prairie Grove Member of the Hale Formation, we were reluctant to call the irregular basal contact with the Cane Hill an unconformity. Our reluctance persists. True, there is an assortment of unconformity criteria including a regionally expressed contact between two contrasting lithic suites, profound local channeling of the Cane Hill, and a distinctive basal conglomerate. What more could we ask? A notable age difference, for one thing. Cephalopod faunas from the two Hale members just across the break are strikingly similar (cf. Quinn, 1963, 1966; Gordon, 1964). Drainage and environmental analysis tells us we are viewing accelerated marine deposition following an arid-to-humid climate change and the basal Prairie Grove channels were cut on the littoral sea floor. If so, we are viewing a *diastem*, a minor break, not an unconformity.

In closing this section, it seems appropriate to observe that the suite of Paleozoic lithologic records we have been discussing constitute a strong documentation for alternating warm, relatively high seas correlated with humid lands and cool or cold, relatively low sea levels correlated with arid lands. The implication that the marine conditions were strongly influencing adjacent land climates is impossible to escape. Also, the implication that these adjustments are, respectively, correlative to inter-glacial and glacial conditions elsewhere (Gondwanaland?) (cf. King, 1958) is again very strong. These notions tend to be reinforced by data presented in the following section.

ALAN B. SHAW AND THE CYCLOTHEMS

The juxtaposition of Dr. Shaw's name with the associated subject is not intended to paraphase a situation implicit in a title such as "Daisy Brown and the Little People." Yet the germ of an idea presented in Shaw's book (1964) provides the basis for

the following synthesis of *cyclothems,* supplemented, of course, by geomorphic concepts set forth in our earlier chapters. A *cyclothem*, as most of our readers will recall, is a cyclically repeated sequence of rock types usually involving a coal deposit (Fig. 11.23). Most relate to the late Paleozoic sedimentary record. The problem of their origins has been widely discussed (cf. Weller, 1930; Moore, 1936; Weller, 1960). Cyclothems are interesting because of the commonly associated economic resource represented by coal; they are also intriguing because of the complex association of stratigraphic, hydrologic, marine, geomorphic, and environmental restrictions they appear to require. It seems best to list these requirements at the beginning of our discussion and to introduce Shaw's concept later.

(1) As illustrated in classic form (Fig.

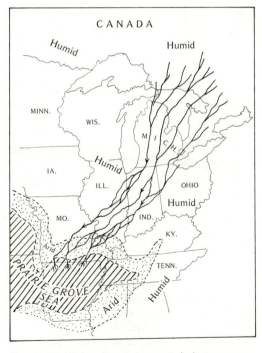

Fig. 11.22. Paleoclimatic map and drainage configurations of the North American mid-continent during Prairie Grove (early Pennsylvanian) time. Shown are a somewhat expanded and probably warm seaway adjacent to a narrowing coastal desert and an expanding humid upland to the northeast.

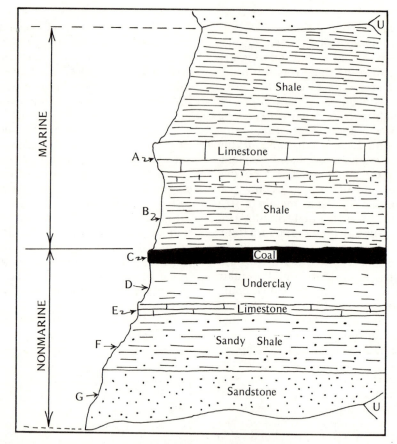

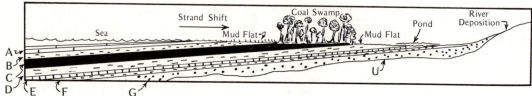

Fig. 11.23. Cross sections of an idealized cyclothem in terms of sequence (stratigraphic), marine and nonmarine conditions, and marine transgressions. (In part after Weller, 1960.)

11.23), a cyclothem is seen to include a marine and a nonmarine interval of sediment, so that continental sedimentation followed by relative changes in sea level are an integral part of the sequence.

(2) In the ideal case, the cyclothem begins and ends with an erosional unconformity of some relief, generally situated at the base of the nonmarine sandstone deposit—because of variable amounts of erosion along the unconformity, various portions of or even most of earlier cyclothems may be missing.

(3) Because sea level (and strandline) shifts are implicit in the cyclothemic sequence, it follows that fluctuations in offshore marine deposits may reflect more landward conditions without, however, actually including nonmarine materials.

(4) Where land environments reflect adjacent marine conditions very closely, as was indicated in the previous section,

it follows that the converse of (3) may also develop—i.e. continental depositional sequences that are cyclic but lack actual marine inclusions such as the bulk of the Pennsylvanian Fountain Formation in Colorado (cf. Chapter 10).

(5) Various lithic units of the usual cyclothem have great lateral persistence, even where thicknesses involved are only a few inches, a relationship that generally requires widely sweeping strands in increments measurable in hundreds of miles and exceedingly low slopes (cf. Swann, 1964).

(6) One of the most outstanding characteristics of cyclothems are their number, those of marine type previously noted from the Mississippian Period totaling at least 20 and possibly as many as 150 to 200 from the Pennsylvanian System in North America—the thickness of resulting sediments generally require more or less constant, isostatically adjusted subsidence, and the frequency of the oscillations essentially eliminate tectonic causes in favor of some glacio-eustatic mechanism.

(7) Finally, and probably most importantly in terms of our forthcoming analysis, there is the matter of the internal lithic consistency of cyclothems. The marine facies always overlies the coal. This raises the question—couched here in teleological terms, "How did the coal know the sea was coming?" This is particularly enigmatic if the plants forming the coal don't "know" anything and are somehow irrevocably tied to the land. Both relations APPEAR to be valid.

The previous point is quite simply that the average transgressing sea at present would not bury a coal forest with its sediments. Along the New Jersey shore a transgression would bury back-beach sands, lagoonal muds, the "pine barrens," and a few tourists. In south Texas a similar transgression would cover barrier beaches, lagoons and coastal plain sands and clays, and some oil wells. In Louisiana there would be backswamp muds, deltaic sands and clays, and fluvatile materials and New Orleans. Only along the southwest coast of Florida where the mangrove swamps hold forth would a transgressive marine facies bury an appreciable amount of vegetation that could conceivably assume the form of an even, contiguous layer of coal many miles in area. But there is a problem. The mangroves are a form of salt-water–adapted vegetation (cf. Chapter 9), and there is no indication whatsoever that the coal forests of the late Paleozoic had any tolerence for salt water. Those Paleozoic forests included a variety of fresh-water dependent plant genera, many of which grew in swamps inhabited by amphibians which similarly required fresh water. This apparent contradiction brings us to the concepts of A. B. Shaw (1964).

The Tideless Sea and the Strand

One of the problems with the Law of Uniformitarianism noted in Chapter 1 centered on the fact that the present is not at all like the past in several respects. For one thing, we now have no epicontinental seas. Present open oceans pound vigorously on our exposed coasts across slopes that are often visibly inclined. The erosive energy expenditure is exceedingly great, tides and currents are effective in exchanging and mixing sea water, and the disposition of particulate sediments generally reflects sorting under more or less turbulent conditions. For many years it was assumed that the ancient epicontinental seas were similar to present oceans. They certainly were not. Shaw (1964) was among the first to emphasize the contrast when he pointed out that bottom slopes for such epicontinental seas were exceedingly low, averaging on the order of 0.1 ft/mi (2 cm/km) with steep examples possibly approaching 0.5 ft/mi (9.0 cm/km). The latter slope (1/180°), as a matter of fact, is presently that of the Mississippi Embayment from southeastern Missouri to the Gulf of Mexico. Present shelf seas have generally steeper bottom slopes (0.5°–1.0°).

Shaw's analysis of epicontinental seas

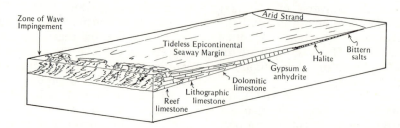

Fig. 11.24. A. B. Shaw's evaporite sequence development landward from the zone of wave impingement, next to an arid land yielding no runoff.

can be related to our cyclothem problem in several additional matters. He (1964) observes that sedimentation would have kept the seas nearly filled, there would be no near-shore surf zone or "beach" in the usual sense of that term, and, "Normal diurnal tides would not have occurred in the interior of epeiric [epicontinental here] seas."* Wave swells developed in deeper water areas could only move landward to the point of wave impingement. If one assumes an average erosional wave base of 30 ft (cf. Chapter 9), the zone of wave impingement in an epicontinental sea would range from 60 mi (84 km) off-shore along steeply inclined bottoms to 300 mi (420 km) off-shore on more gently sloping sea floors. Shaw concludes that the zone of wave impingement in an epicontinental sea would be the locus for carbonate deposition and organic activity. He further notes that increased water salinity should develop landward from that zone in the absence of land runoff (cf. Ginsburg, 1956) and suggests that a landward evaporite sequence should develop (Fig. 11.24).

It should be evident that what we have so far discussed here is sea-floor topography, or rather, an almost complete lack of surface roughness. The combined effects of deposition in broad shallow depressions and erosion in equally broad, gentle up-warps during the lower and middle Paleozoic had generated continental interiors in much of North America and parts of several other continents which were of unsurpassed evenness. Major topographic departures from this scheme took the form of narrow subsiding troughs of some depth, developing orogens such as the Appalachian, Ouachita, or Ural mountains, and barrier reef configurations developing along zones of wave impingement. Within the epicontinental seas, the reefs were "THE" local relief feature. Related relief varied from a few inches around micro-atolls (Fig. 3.53) to many tens of feet next to larger patch reefs in deeper water, and scores or hundreds of feet along major barriers situated along shelf margins (Fig. 11.25).

A few inches or feet of relief does not sound like much, but with the bottom slopes indicated previously it would be a great deal. Stated in terms of a hypothetical would-be swimmer six ft (183 cm) tall entering a Pennsylvanian epicontinental sea with a bottom slope of 0.1 ft/mi (2 cm/km), after he or she has waded seaward about 4.5 mi (7.6 km) the water will be ankle deep (the low land behind would be out of sight). A further wade and some 18 mi (30.5 km) off-shore the water would be knee deep. At the arbitrarily chosen swimming depth of waist deep, our intrepid swimmer will have waded some 30 mi (51 km) off-shore, and the water will be about 40 inches (102 cm) deep.

---

* Evidence presented for closer earth-moon relations and hence stronger tidal forces during the Paleozoic may account for supermaturity of some Cambro-Ordorician sands (cf. Lustig, 1963). However, survival of fragile deposits later may signify the crossing of a friction-energy threshold so that seas, if not tideless, still had inhibited water mixing.

The point of the above discourse is the extreme shallowness of the water along the margins of epicontinental seas. Shaw (1964) visualized these margins as sites of hypersaline water and evaporite deposition, though he states (p. 22), "Runoff fresh waters from the land should modify the basic pattern of increasingly saline waters toward the shore by freshening the waters near shore." Elsewhere, he explains that by freshening he means only a reduction of salinity toward that of normal sea water, and his concerns were mainly with the solubility of the evaporites and the problems of inclusion of the evaporite se-

quence in the time-stratigraphic record. It is at this point that we depart from Shaw's synthesis to consider the role of the epicontinental sea and the cyclothem.

The late Paleozoic deposit and unconformity records already discussed are sufficient to establish that sometimes the lands were arid and shed little runoff, and sometimes they were humid and created major drainage systems and large volumes of fresh water. The margins of epicontinental seas must be viewed in this context and also in the context of climatically controlled sedimentation outlined in conjunction with geomorphic matters in earlier

Fig. 11.25. El Capitan Peak at the southern end of the Guadalupe Mountains, west Texas. Beneath the sliver of cloud, the mountain is capped by the remains of a massive limestone reef of Permian age, probably part of a barrier-shelf complex which may have stood next to a basin 1,000–2,000 ft deep. (National Park Service.)

chapters. Shaw's point (1964, pp. 1–71), which appears to be essentially valid in the context in which it is presented, is that A SHALLOW, TIDELESS, SURFLESS EPICONTINENTAL SEA LACKING APPRECIABLE RUNOFF FROM LAND SHOULD HAVE HYPERSALINE MARGINS, since no dilution by open sea water or mixing can occur. It should also follow that A SHALLOW, TIDELESS, SURFLESS EPICONTINENTAL SEA RECEIVING ABUNDANT RUNOFF FROM A HUMID LAND SHOULD HAVE TRULY FRESH WATER MARGINS, since no pervasive mixing or salinization from the sea can occur (Fig. 11.26).

The last statement involves an extremely important consideration in relation to the cyclothem enigma, and the ideal cyclothem sequence, or an approximate, should thus be re-evaluated (Fig. 11.27). The basal sandstone unit of the usual cyclothem is usually designated nonmarine, though Shaw (p. 53) chooses to consider it as a part of the transgressive facies. It may rest on a comparatively even surface (which previous analyses suggest would develop in deserts, on coasts too close to sea level to permit deep incision,

or both). Alternatively, the basal sandstone may rest on an uneven surface or within channels, but it has too great a lateral continuity in many cases to just represent a channel fill (again, the surface in question, on the basis of earlier syntheses, would presumably reflect stream incision, probably under humid conditions and in relatively elevated land areas, in many cases). The nonmarine sand deposit, itself, is best explained in terms of arid aggradation, possibly in riverine-plain form.

The appearance of layered, sandy, argillaceous materials in the cyclothem sequence, sometimes with a fresh-water limestone and often with plant-bearing muds, signifies an aqueous environment—the type usually postulated for some sort of swamp landward from the beach (cf. Fig. 11.23). If the land becomes humid, however, the shore of the epicontinental sea should be fresh. Furthermore, it has long been known that a wide expanse of ground saturated with fresh water, and even the margins of shallow fresh water bodies within the *photic zone*, quickly

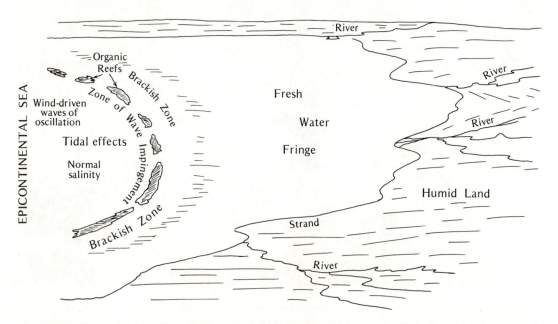

Fig. 11.26. Regional schematic panorama of relations between high-energy midportions of epicontinental seas with normal marine salinity and marginal portions of the same seas with subnormal salinity adjacent to humid lands.

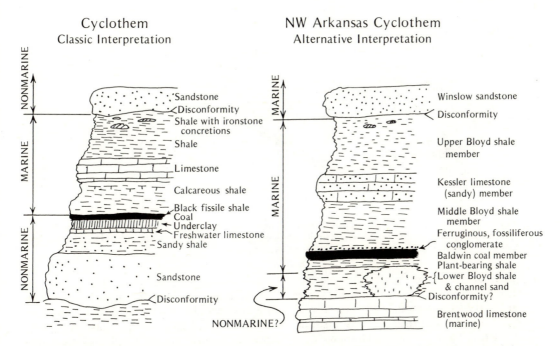

Fig. 11.27. Sections of an idealized Pennsylvanian cyclothem (cf. Moore, 1933) (left) and an actual deposit of similar kind developed in the southern Ozarks (right).

acquire a growth of vegetation—in short, it would become a swamp. And swamps, as is well known, are apparently the sites for coal accumulation. Thus, in answer to the somewhat teleological question posed earlier, the coal "knew" the sea was coming because it probably was formed in the sea's fresh-water fringe. We are clearly discussing a situation in which sea level is rising relatively, and the development of open marine lithofacies, marine fossils, and finally limestone (zone of wave impingement) signifies further inundation of the lands (Fig. 11.28).

The idea of a fresh-water fringe on an otherwise saline sea is unusual, but it is quite compatible with the physical requirements for epicontinental seas set forth by Shaw (1964) and the paleoclimatic and geomorphic frameworks outlined earlier in this work. It would, in fact, appear that the coal portion of a cyclothem probably reflects a particular shallow-water zone along the margins of an epicontinental sea, limited at any one time, on the seaward side, by the extent of salt versus

fresh water and, ultimately, by the lateral extent of marine transgression and low sea bottom slopes. It seems quite probable that most sea level fluctuations during the development of the late Paleozoic cyclothems were glacio-eustatic in nature. If one assumes causally related sea level shifts equivalent to those of the Pleistocene [±350 ft (100 m)], on a bottom slope of 0.1 ft/mi (2 cm/km) a transgression of 3,500 mi (4,900 km) would result—enough to inundate most of central North America. Over a slope of 0.5 ft/mi (5.4 cm/km), a similar sea level rise would shift the strand 700 mi (980 km). This last is the same order of magnitude recorded by Swann (1964) for the Illinois basin during the late Mississippian.

Cyclothemic Landscapes and Environments

A final appraisal of cyclothemic conditions and paleogeomorphic relations of the North American late Paleozoic must attempt to place each of the main interacting geo-

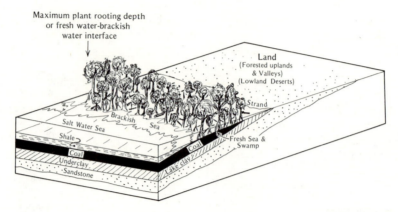

Fig. 11.28. Relations probably developed between freshwater margins of epicontinental seas next to humid lands and coal formations. Seaward limits of plant growth may have related to either plant salinity tolerences or to water-depth rooting limits similar to mangroves.

morphic agencies in some perspective. Several specific sets of relationships are rather well documented. (1) Sea levels were rising and falling, related strands were shifting laterally great distances, and the ocean waters were correlatively changing in temperature—in deeper water areas, probably rather drastically. (2) Both the sea distances from land and the sea temperature changes can drastically alter land moisture conditions (cf. Chapter 4), in general directly—that is, a near, high, warm sea = humid land; a far, low, cold sea = arid land.

In the context of cyclothems and their landscape settings, we therefore offer a synthesis of paleogeomorphic events, with the single additional assumption that the land areas that were the most heavily sedimented would gradually subside isostatically (Fig. 11.29). In this framework, the major features of the ideal cyclothem appear to have the following origins.

(1) The basal unconformity (disconformity) probably developed when sea level was low and waters cold and distant; it is partly an arid planation feature, but there is an increased possibility for fluvial incision under humid conditions in uplands.

(2) The nonmarine sandstone feature is probably a progradational desert deposit spread by desiccating runoff into a lowland which is also an emergent sea floor—

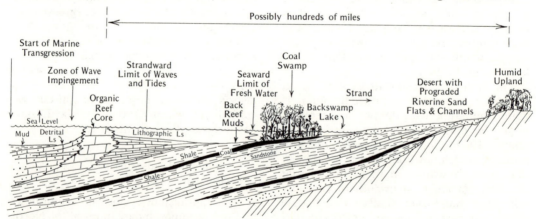

Fig. 11.29. Paleogeomorphic events associated with development of a cyclothem during a marine transgression associated with glacio-eustatic effects; see text for details. The vertical scale is greatly exaggerated.

its youngest portions are seaward and may be marine by virtue of lateral sedimentation; also, it is probably correlative to a glacial episode elsewhere, though one during its later phases because sea levels have begun to rise and waters to warm.

(3) The sandy argillites, fresh-water limestones, and underclay beneath the coal apparently signify the coincidental development of warm rising seas, humid, chemically weathered, vegetated lands, perennial runoff, and an ocean with a developing fresh-water fringe—the "marine" portion of the cyclothem should be indicated as starting at the base of these deposits (Fig. 11.27); much of the material is a sedimentary lag accumulation flushed seaward following an arid-humid climate change rather than classically transgressive.

(4) Humid, upland vegetal types invade now-boggy lowlands and the sea's fresh margins to form coal swamps which are gradually extended landward as seas rise and are terminated seaward as brackish and salty water encroaches—vegetal mats prevent invasion of the coal swamps by fluvial sediment, though streams from the land cut through the coastal swamp at intervals and deposit their loads beyond; seaward sides of coal areas are often topped by a fluvial gravel layer (cf. Fig. 11.27).

(5) Continued submergence and transgression (Fig. 11.29) place open marine facies above coal and nonmarine deposits in that sequence, the arrival of the zone of wave impingement being signaled by the marine limestone deposits that are commonly of a reef facies and may be sandy—the carbonate deposits signify warm seas and are probably the transgressive maximum and turn-around point interglacially and environmentally, the remainder of the cyclothem being most probably deposited by a regressing sea which is becoming cooler adjacent to a land which is, in turn, becoming more arid.

Many aspects of this synopsis are seen in the numerous published analyses of the cyclothem problem. The effort made here was not to be original, but rather to make use of available published information, and particularly of Shaw's work on epicontinental seas which seems to involve some concepts that are basic to the understanding of the cyclothems of the world. With this section we conclude our consideration of paleogeomorphic relationships in cratonic settings. Countless examples of other fossil landscape situations exist, but space does not permit their inclusion here. A great deal of the paleogeomorphic nature of the study of mountain systems has already been brought out in our discussions in Chapter 10, and many of the paleogeomorphic aspects of depositional landforms such as deltas were considered in Chapter 7. Two other geomorphic relationships of ancient character appear to merit at least a brief discussion: the effects of trans-latitudinal continent shifts and the desiccation of oceans. We will consider each now.

PALEOCLIMATOLOGY, TECTONISM, AND TOPOGRAPHY

In this final section dealing with paleogeomorphology, we are mainly concerned with some very subtle and long-term environmental effects that result from two distinct, but sometimes interrelated processes. The first of these effects is evidently a by-product of plate tectonics (cf. Chapter 3) wherein the various continents tend to move from time to time and, in doing so, may migrate across lines of latitude into new subaerial environments. To the extent that different subaerial environments can induce different kinds of landscapes, the movement of a landmass through a sequence of environments in space by continental drift should initiate a sequential series of landscapes on these landmasses through the equivalent interval of time.

The second effect has to do with the disappearance of an ocean or sea by desiccation, possibly with associated impairment of oceanic circulation and usually with drastic geomorphic consequences. We will consider the plate tectonics effect first.

TRANS-LATITUDINAL CONTINENT SHIFTS

In Chapter 4 we outlined some major global climatic belts apparently of a long-term character and disposition, roughly parallel to the Equator, and mainly related to global heat and light distribution as controlled by geostropic winds and ocean current circulation (cf. Fig. 4.5). The main elements include the humid, tropical Doldrum Belt, the zone of Horse Latitude Deserts, the zone of humid, Temperate Forests, and the high latitude boreal areas of cryergic and glacial conditions. In several later chapters (5–9) we discussed in some detail how each of the major subaerial environments may induce specific geomorphic systems and their resulting distinctive landforms and deposits. We have further noted a number of mechanisms that might change climates, but always with respect to some dominant mode established by regional geography, hydrography, and topography. The trans-latitudinal continental movements we are concerned with are those that would establish the long-term environmental "mode" for a region, particularly with reference to latitudinal position.

In the course of the foregoing discussions, we have had occasion to note some of the geomorphic-environmental situations of interest here. For example, it would appear that the Antarctic Continent assumed its polar (and glacial) environment simultaneously during the middle Tertiary. Beneath all that ice, the landscape is certainly dominated by glacial effects. In fact, with the exception of a few minor subaerially exposed terrains (cf. Chapter 8), the Antarctic Continent is in a climax glacial state, geomorphically speaking, largely because of the plate tectonics mechanisms that presumably located the landmass.

In another situation involving trans-latitudinal continent movements, it would appear that North America has moved northward during the Tertiary Period in the course of its westward migration from the mid-Atlantic Ridge. The northward shift in this case may amount to 11°–15° of latitude, which is the equivalent of moving northern Georgia and Alabama from a former position corresponding approximately to that of Cuba. A number of probable effects occur, but perhaps the most notable are the bauxite deposits developed from Cretaceous syenite plugs in Arkansas and the ridge-top bauxite-laterite occurrences in northern Georgia at the same latitude (cf. Bridge, 1950).

We have not the time nor space to examine in depth the many parallel cases of climatically induced landform development during sea-floor spreading. There is one piece of research, however, that is worth a brief synopsis, not only because the subject is germane, but also because the conclusions were reached before the advent of recent plate-tectonic concepts.

Migration and Breakup of Gondwanaland

In 1958, L. C. King attempted to trace the movements of the several southern continents, both with respect to their having been a part of an ancestral southern landmass (Gondwanaland) which eventually broke up, but also to the trans-latitudinal migration of the whole and its eventual parts during the late Paleozoic and Mesozoic (Fig. 11.30). King bases his analysis of the latitudinal positions of these landmasses mainly upon the character of the related deposits, some marine, but most continental. Though one may choose to disagree with some of his detailed conclusions, his basic paleographic interpretations appear to be sound (Fig. 11.31).

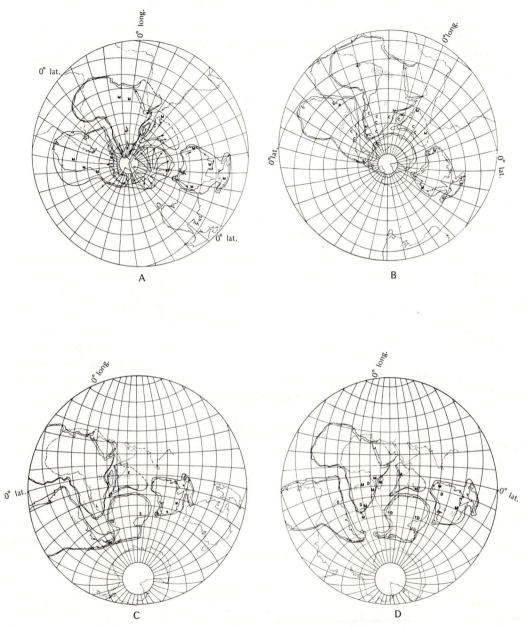

Fig. 11.30. The breakup of Gondwanaland. Shown are (A) Late Carboniferous ice sheet flow pattern with local ice sheets, *L*, peripheral marine deposits, *M*, and North African desert sandstones at the "horse latitudes", 30° S. (B) Continued Permian glaciation in Australia with widespread coldwater shale and arkose, *C*, sometimes with *Glossopteris* cold flora. Warm *Psaronius* flora, *P*, appears in northern Brazil and West Africa. Dots locate desert sandstones. Marine formations remain peripheral, *M*, except in southwest Africa and Brazil. (C) Late Triassic continental environments include desert sandstones (dotted), loess, *L*, equatorial swamp, *E*, seasonal drouth, *S*, and warm pluvial, *W*. (D) Incipient breakup of Gondwanaland during latest Jurassic-earliest Cretaceous with continental facies (dotted), marine facies, *M*, and denudation (= humidity), *D*. (After L. C. King, 1958.)

	Southern Brazil and Uruguay	Southern Africa
Early Cretaceous	Fragmentation of	Gondwanaland
Late Jurassic	?Red sands of the interior	Lualaba Series of Congo
		Marine series in East Africa
Jurassic	"Gondwana"	Cyclic
Rhaetic	Sao Bento basalts	Drakensberg basalts and dolerites
Late Triassic	Botucatu red desert	Stormberg Series ending in "Cave Sandstone" desert
Mid-Triassic	*(Buenos Aires folding)* Santa Maria (reptiles)	*(Cape Province folding)* Upper Beaufort; hiatus in Transvaad and northward
Early Triassic	Hiatus	Middle Beaufort (reptiles)
Late Permian	Grupo Rio do Rasto	Lower Beaufort (green mudstones with *Dicynodon, Taeniopteris*)
	Estrada Nova Series (green mudstones)	Upper Ecca shales (red in South-West Africa)
Early Permian	Estancia Beds (partly red) Irati dark cold-water shales	Middle Ecca coal measures Lower Ecca cold-water shales (rare *Glossopteris*)
	Mesosaurus	*Mesosaurus*
Late Carboniferous	Rio Bonito coal measures	Dwyka tillite with overlying fine dark cold-water shales
	Itarare Series including glacials	
Early Carboniferous	Glacials in western Argentina	

Fig. 11.31. Correlation of Late Paleozoic and Mesozoic Facies in Gondwanaland and global climatic belts related to zones of latitude and continental drift. See also Fig. 11.30.) (After L. C. King, 1958).

Peninsular India	Antarctica	Eastern Australia
nto modern	continental	masses
›alpur (continental) marine ›eds of Cutch Continental		Freshwater beds of Artesian Basin (sauropods and dinosaurs) denudation
ˌjmahal basalts	South Victoria Land dolerites	Tasmanian dolerites
ˌhadevi Series (tropical ›nvironment with saurischia)	Unknown	Wianamatta; beginning with cool-temperate environment, rapidly warming up Hawkesbury
ˌatus	Upper Beacon Sandstone	Narrabeen
ˌnchet Series *(Dicynodon, Taeniopteris)*	Coal beds Lower Beacon Sandstone	Upper coal measures
ˌniganj (coals) ˌrruginous shales ˌrakar Series with coal		Upper marine series (with tillites) Greta coal; lower marine series (with multiple coals)
ˌchir glacial and fluvioglacial	Tillite not yet known	Late Carboniferous glacials and associated beds of southeastern Australia

Simply on the basis that glacial materials are essentially high-latitude deposits, continental deposits are mainly desert accumulations, and similar analytic associations, he traces the following main continental situations.

(1) Late Carboniferous fusion of all southern lands in a near-polar position with extensive continental glaciation—peripheral ice-front phenomena and varved lake deposits detected in South Africa, the Congo, Northern India, and western Australia.

(2) Permo-Triassic migration of the Argentine into the 30° latitude area of deserts, similarly the Nigerian-Lake Rudolf area of Africa—glaciation continued in parts of Australia and Antarctic during early parts of this episode.

(3) Late Triassic aridity of central and south Africa, equatorial humidity in India, and generally widespread arid planation and aggradation throughout much of *Gondwanaland*—widespread plateau basalts marked the end of the period over vast areas of India, Brazil, south Africa, and Antarctica.

(4) Jurassic signs of Gondwanaland breakup and interior seaways; at one stage extensive fresh-water lakes in Australia and a general absence of continental deposits—this would here be taken to indicate humidity and landmass migration into a moist climatic belt, but whether at a low or high latitude is uncertain.

(5) Cretaceous breakup of the megacontinent and gradual migration of the continental fragments into present positions relative to current mid-ocean ridge positions.

OCEAN BASIN DESICCATION

Until very recently, the writer was fond of telling himself that no matter what might happen to the continents, the mountains, the inland epicontinental seas, or, indeed, any of the myriad, obviously transient phenomena that flit across the face of the earth, at least the deep oceans were always

there. Oh, certainly, if continents came together through sea-floor spreading, the ocean basin in between would disappear. But, at least, as long as there was a deep ocean basin, IT WOULD CONTAIN WATER. Right? Wrong!

In late 1972, Hsü published an account of the Mediterranean Basin which is, at once, difficult to accept but impossible to reject totally. Fortunately, the subject is grandiosely geomorphologic. And the discussion is placed here because it is in part, a case of ancient geomorphology. Hsü's report summarizes the findings of many earlier workers in the Mediterranean region who have set forth in a number of published papers the basic data that follows.

(1) The floor of the Mediterranean appears to have a number of salt domes traceable to a salt layer and associated evaporite deposits (anhydrite) signifying evaporation essentially to dryness during the early Pliocene or some six- to ten-million years ago.

(2) Near the Gibraltar area, the drill of the Glomar Challenger (JOIDES Program) brought up fluvatile gravels signifying erosion of the sea floor by running water which eroded only basalt, gypsum, and lithified oceanic ooze.

(3) The ancestral valley of the Rhone River is incised more than 3,000 ft (900 m) below sea level; that of the Nile may be 5,000 ft (1,500 m) deep; neither gorge could have been eroded unless the 9,000- to 10,000-ft sea basin was not filled by water. Other valleys are deeper.

(4) Drilling elsewhere on the same sea floor encountered deposits of wind-blown silts and stromatolitic dolomite, the latter a light-dependent algal deposit, underlying deep-water, normal marine deposits about 5.5 million years old.

It would therefore appear that the floor of the Mediterranean Basin, locally more than 10,000 ft (3,050 m) deep, contains relict eolian material and evaporite deposits of Pliocene age. These materials strongly indicate that the sea there evaporated to dryness then, probably several times before, and possibly since the Pliocene (cf. Zagros Mountain discussion, Chapter 10). Evaporation rates in the region are apparently high enough (1,000 cu mi/year) to cause the desired effect in about 1,000 years, provided that circulatory connections with the Atlantic Ocean were inhibited. This could happen through a northward plate-tectonic shift of Africa (or the converse) or by a glacio-eustatic fall in sea level to diminish inflow into the basin below evaporative loss. It will be recalled that there is excellent evidence for Antarctic glaciation with fluctuations as far back as the mid-Tertiary.

Most impressive among the data presented by Hsü that the Mediterranean did, in fact, dry up, are the valleys of many rivers, including the Rhone and the Nile, which have been found to be incised thousands of feet below sea level (Fig. 11.32). Hsü (1972) also calls attention to the fact that basin refilling probably occurred mainly by a rise in sea level or some similar adjustment permitting a greater water inflow through the Straits of Gibraltar and, further, that this inflow probably took the form of a waterfall to end all waterfalls (Fig. 11.33). Hsü points out that a waterfall through the Straits of Gibraltar which delivered 1,000 cu mi/year (30,000,000 gal/second) would not have countered evaporation losses, and "Gibraltar Falls" must have passed 10,000 cu mi/year to refill the Mediterranean. The fluvial gravels

periodically generated in this enormous fluvial cataract are presumably those evidenced in deposits on the floor of the Mediterranean Basin near Gibraltar as disclosed by the deep-sea drilling efforts of the JOIDES project.

The climatic, isostatic, and geomorphic effects of an ocean-basin desiccation leave room for vast and fascinating speculations. The removal of such a large evaporation source must have drastically affected adjacent, and even distant, land climates including the Zagros Mountains, previously discussed, probably enhancing or inducing aridity in most places. Isostatic depression and uplift should accompany the respective filling and emptying of the basin. The geomorphic ramifications of these environmental and tectonic adjustments are potentially indeed most complex and far beyond the scope of this work. It must suffice for the moment to note that such events apparently can occur. Conceptually, the notion of a depression on the earth's surface 2,000 mi (2,800 km) long, 600 mi (840 km) wide, and 10,000 ft (3,000 m) deep containing only a few sizzling salt flats and tepid saline lakes must go down in the records as one of the major "landforms" of all time. Comparing this with the largest existing desert depression of analogous morphology (the Quattara Depression of Egypt) is like comparing a bathtub with the Grand Canyon.

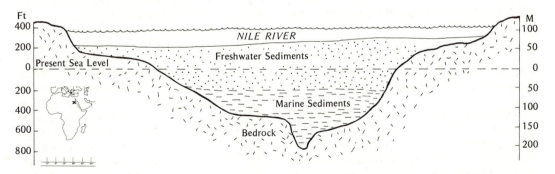

Fig. 11.32. Section showing the buried Nile River valley some 500 mi upstream at the Aswan Dam (location map inset). Closer to the Mediterranean Sea, such valleys are cut thousands of feet below sea level. (After Hsu, 1972, and others.)

Fig. 11.33. Writer's concept of Gibraltar Falls as this gigantic cataract might have appeared from above the coast of Morocco, via satellite, some six million years ago delivering some 10,000 cu mi of water/year into the Mediterranean basin almost 2 mi below. Temperatures over the abyssal salt flats may have exceeded 150° F.

SUMMARY

In retrospect, the study of ancient landscapes seems complex. It requires the use of a deft theoretical touch, to seek out valid analogs in the modern geomorphic scene and to marshall all our wits to what amounts to an archaic landform detective story. To further the analogy, we don't even get to work at the scene of the geomorphologic crime, but rather only with a fragment of it. The stratigrapher tends to curse the unconformity, for it robs his historic record of continuity and forces him to seek the remnants elsewhere. The paleogeomorphologist, however, apart from the occasional channel, reef, or delta, requires unconformities for much of his data. And when one stops to consider that less than half of all geologic time is actually represented by a lithic record, it would pay us to study the gaps.

The application of geomorphic data to the study of the past is esoterically good and of great value to the economy. For the techniques used by the geomorphologist to comprehend fossil landforms and deposits must also be brought to bear, in association with geochemical techniques to locate ore deposits, in association with paleohydrologic techniques, to detect placers, and with data from geophysics, stratigraphy, and sedimentation studies, to discover petroleum. Through the use of isopach, lithofacies, and structure contour maps, the search for an "oil pool" almost invariably involves geomorphology applied to problems of fluid entrapment. The only other requisite is imagination.

Probably the only realm where there is a greater urgency for applied geomorphologic knowledge than in the past is the present. And the potential rewards are at least equal to those that may be gleaned otherwise. On every hand we encounter groups who lay claim to special bodies of information which can be arrayed against the problems of our contemporary environment. Zoologists and botanists know about living things, sociologists know how people live together, engineers know how to build, and so on. Each group can contribute to the betterment of man's environment? What can geomorphologists offer? Our final chapter will briefly consider some of the answers to this question.

REFERENCES

Berkey, C. P. (1906) "Paleogeography of St. Peter Time," *Geol. Soc. Amer. Bull.*, v. 17, pp. 229–250.

Born, K. E. (1940) "Lower Ordovician Sand Zones ('St. Peter') in Middle Tennessee," *Amer. Assoc. Pet. Geologists Bull.*, v. 24, pp. 1641–1662.

Bridge, Josiah (1950) "Bauxite Deposits of the Southeastern United States" [in] *Symposium on Mineral Resources of the Southeast United States*, F. G. Snyder, ed., Univ. of Tennessee Press, Knoxville, esp. pp. 170–201.

Case, J. E. (1956) "Current Ripple Marks in the Lower Pennsylvanian Hale Formation of Northern Arkansas," *Compass*, v. 31, pp. 209–218.

Dake, C. L. (1921) "The Problem of the St. Peter Sandstone," *Mo. Univ. Sch. Min. and Met., Bull.*, v. 6, pp. 1–225.

Easton, W. H. (1942) "Pitkin Limestone of Northern Arkansas," *Ark. Geol. Surv. Bull.*, v. 8, 115 p.

Frezon, S. E. and Glick, E. E. (1959) "Pre-Atoka Rocks of Northern Arkansas," *U. S. Geol. Surv. Prof. Pap.* 314-H, pp. 171–189.

Garner, H. F. (1960) "Upper Paleozoic Stratigraphy and Sedimentation of Northern Arkansas, *Geol. Soc. Amer.* (Abstract), p. 102.

Garner, H. F. (1967) "Moorefield-Batesville Stratigraphy and Sedimentation in Arkansas," *Geol. Soc. Amer. Bull.*, v. 78, pp. 1233–1246.

Giles, A. W. (1935) "Boone Chert," *Geol. Soc. Amer. Bull.*, v. 46, pp. 1815–1878.

Gordon, Mackenzie Jr., (1964) "Carboniferous Cephalopods of Arkansas," *U. S. Geol. Surv. Prof. Pap.* 460, 332 pp.

Hsü, K. J. (1972) "When the Mediterranean Dried Up," *Scientific American*, v. 227, pp. 27–36.

King, L. C. (1958) "Basic Paleogeography of Gondwanaland during the Late Paleozoic and Mesozoic Eras," *Geol. Soc. London Quart. Jour.*, v. 114, pp. 47–70.

Lustig, L. K. (1963) "Uniformitarianism and the Earth-Moon System," *Geol. Soc. Amer. Program.*, N.Y.C.

Martin, Rudolf (1966) "Paleogeomorphology and its Application to Exploration for Oil and Gas (with Examples from Western Canada)," *Amer. Assoc. Pet. Geologists Bull.*, v. 50, pp. 2277–2311.

Miller, A. K. and Garner, H. F. (1955) "Lower Mississippian Cephalopods of Michigan, Part III, Ammonoids and Summary," *Univ. Mich. Mus. Paleont. Contr.*, v. 12, pp. 113–173.

Moore, R. C. (1933) *Historical Geology*, McGraw-Hill Book Co., New York, esp. pp. 281–315.

Moore, R. C. (1936) "Stratigraphic Classification of the Pennsylvanian Rocks of Kansas," *Kans. Geol. Surv. Bull.*, v. 22, esp. pp. 26–35.

Potter, P. E. (1962) "Shape and Distribution Patterns of Pennsylvanian Sand Bodies in Illinois," *Ill. Geol. Surv. Circ.*, 339, 35 p.

Potter, P. E. (1963) "Late Paleozoic Sandstones of the Illinois Basin," *Ill. Geol. Surv. Inves. Rept.* 217, 92 p.

Potter, P. E. and Desborough, G. A. (1965) "Pre-Pennsylvanian Evansville Valley and Caseyville (Pennsylvanian) Sedimentation in Illinois Basin," *Ill. Geol. Surv. Circ.*, 384, 16 p.

Quinn, J. H. (1963) "Stratigraphic Position of Eoasianites globosus, *Okla. Geol. Notes*, v. 23, pp. 26–28.

Quinn, J. H. (1966) "Genus Reticuloceras in America," *Okla. Geol. Notes*, v. 26, pp. 13–20.

Schumm, S. A. and Stevens, M. A. (1973) "Abrasion in Place: A Mechanism for Rounding and Size Reduction of Coarse Sediments in Rivers," *Geology*, v. 1 (1), pp. 37–40.

Shaw, A. B. (1964) *Time in Stratigraphy*, McGraw-Hill Book Co., New York, esp. pp. 1–72.

Siever, Raymond (1951) "The Mississippian-Pennsylvanian Unconformity in Southern Illinois," *Amer. Assoc. Pet. Geologists Bull.*, v. 35, pp. 542–581.

Siever, Raymond and Potter, P. E. (1956) "Sources of Basal Pennsylvanian Sediments in the Eastern Interior Basin, Part II," *Jour. Geol.*, v. 64, pp. 317–335.

Stauffer, C. R. and Theil, G. A. (1941) "Paleozoic and Related Rocks of Southeastern Minnesota," *Minn. Geol. Surv. Bull.*, v. 29, pp. 50–93.

Swann, D. H. (1963) "Classification of Genevievian and Chesterian (Late Mississippian) Rocks of Illinois," *Ill. State. Geol. Surv. Inves. Rept.*, v. 216, 91 p.

Swann, D. H. (1964) "Late Mississippian

Rhythmic Sediments of Mississippi Valley," *Amer. Assoc. Pet. Geologists Bull.*, v. 48, pp. 637–658.

Thornbury, W. H. (1969) *Principles of Geomorphology*, John Wiley and Sons, New York, esp. pp. 521–525.

Trowbridge, A. C. (1917) "Prairie du Chien-St. Peter Unconformity in Iowa," *Iowa Acad. Sci. Proc.*, v. 24, pp. 171–175.

Weller, J. M. (1930) "Cyclical Sedimentation of the Pennsylvanian Period and its Significance," *Jour. Geol.*, v. 38, pp. 97–135.

Weller, J. M. (1960) *Stratigraphic Principles and Practice*, Harper and Bros., New York.

12 Environmental geomorphology

MODERN ENVIRONMENTAL PROBLEMS

Any statement we might make here to the effect that environment and geomorphology are interrelated would be more than redundant. The geomorphologist, however, is in a unique position to help solve some modern environmental problems. That is the point of this short chapter. First geomorphology represents the only science that contributes a long-term historic perspective to the study of natural systems, and the only science that integrates an environmental approach to these systems. One can hardly expect to understand or deal with an environment of obscure origin. Second, geomorphology teaches a special awareness of the role of process-agency interaction as part of a natural system. This book has emphasized the role of natural systems in the genesis of landforms. But, of course, many of the same systems simultaneously comprise habitats for living organisms including man.

In the few pages available to us here, we can only indicate some of the ways in which geomorphic knowledge can help us deal with environmental problems. We will first consider the development of the earth's environments as a matter of increasing complexity, and, then, the main types of environmental problems, particularly as these problems involve interface phenomena, threshold reactions, and equilibrium conditions. When it seems appropriate, we will indicate which remedial actions are being attempted, have been tried, or might be considered. We shall also briefly consider some current work in environmental problem detection and prediction. In each of these areas of interest, our approach will be distinctly geomorphic and problem oriented. For more comprehensive treatments of modern environmental problems that include geomorphic and geologic approaches the reader is referred to Dasmann (1968), A. G. I. Environmental Geology (1970), Flawn (1970), Rumney (1970), Coates *et al.* (1971), Detwyler (1971), Moulton (1972), McKenzie and Utgard (1972), and Tank (1973).

HISTORY OF ENVIRONMENTAL COMPLEXITY

The earth not only has many environments at present, there have been many ancient environments as well. To fully grasp the complexity of contemporary settings it would be well to point out that the earth's first environment between 4,500 and 5,000 million years B.P. was elemental and probably not distinguishable from that of other material in the nascent Solar System. The central Solar System mass was not yet a star. There was no light. Friction between particles probably caused spinning. Later, compaction generated heat where previously the temperature could have been near −273° C (0° A). But the earth's first environment must be conceived as extremely simple—no atmosphere, no free water, no light, no heat.

When the sun began to radiate energy, another earth environment developed—one that by contemporary living standards must have been pretty miserable. Without an atmosphere, solar radiation would have made the surface unbearably hot, not to say deadly, for there was no life. Compaction, heating, volcanism, lithic dewatering, and degassing are believed to have gradually created an atmosphere. Things were getting complicated. Finally atmospheric water saturation reached a threshold where water condensed and, falling as rain, didn't simply re-evaporate. Running water became a reality, as did standing water soon thereafter. Mass wasting and eolian processes were already active, as was a form of weathering. Of course, the creation of living organisms made it a whole new fiesta, particularly when some acquired the ability to generate rock.

In the meantime, some standing water bodies had become saline through leaching of subaerial regolith. Incidentally, this is one of the earliest documented cases of *pollution*. It was natural pollution, of course. Man* has merely invented artificial pollution, pollution, in our definition being an additive to an environment which is cumulative and affects related organisms. "Nature" has been polluting environments for hundreds of millions of years. Perhaps "her" favorite "old standby" is a swamp. But we stray from our point. Each new variable, each new ingredient of the "geosystem" created a new environment, or, if you will, added a new facet to the developing mega-environment.

In this brew of evolving environments, organisms began to evolve and differentiate and in so doing began to occupy developing ecologic niches. The places ON earth that were not converted into habitats by living organisms are hardly worth discussing because there almost aren't any. The aqueous forms alone have tolerances for sub-freezing polar seas and near-boiling thermal springs. In the final analysis, the only biologic measure of success is survival. Extinction must be put down in the book as failure. In general, those creatures that are most adaptable have done best, surviving. The adaptable ones, that is, and the ones that happened upon the most enduring environments. The brachiopod *Lingula* sp., for example, adapted to littoral zones of oceans, and this adaptation has been good for some 500 million years of survival, albeit on a fairly low plane of existence.

But our point is probably made, except for one aspect. The environments of the Earth have progressively become more and more complex. The present array of environmental types is not particularly typical, except for those times where continents were discrete, emergent entities, when there were polar ice sheets, but not fully extended ones, when there were many mountains, etc. As we have heretofore pointed out, most of the earth's subaerial environments are only recently created. In these regions there is often a high degree of environmental disequilibrium, even without any interference by man. In a few regions that qualify as climatic nuclei, there may be a high degree of adjustment to the existing condition, and some form of near-equilibrium may exist. In these two types of situations, man has an obvious choice—if he meddles. He can try his environmental luck with an equilibrium situation, the odds being in favor of some sort of upset. Or, he can try to change "a threshold situation teetering on a environmental precipice." Either way, if he acts in ignorance, he is in trouble (Fig. 12.1).

* The use of the term "man" here is taken to mean all humans of both sexes.

Fig. 12.1. A sinkhole believed to have developed Dec. 2, 1972, possibly due to excessive pumpage of groundwater in Shelby County, Alabama. The depression measures 425 ft (130 m) long, 350 ft (110 m) wide, and 150 ft (45 m) deep and probably represents a disturbance of a pore-pressure balance in the groundwater system of the area. (U. S. Geological Survey.)

In a more or less distinct set of geomorphic circumstances, man also must cope with a variety of long-term geologic environments, incorporating various natural hazards. These include mainly volcanic and tectonic phenomena of the seismic variety. Within the orogenic zones characterized by such phenomena, earthquakes and volcanic eruptions share a number of common attributes. (1) They are inevitable and relatively frequent. (2) Prediction systems for quakes are virtually nonexistant, though some recent findings are promising, and are extremely rudimentary for many volcanos.* (3) Remedial techniques, either in terms of construction techniques or reactions are primitive. In effect, people inhabiting such zones are playing "environmental Russian roulette." This situation places problems related to the Earth's

* Quake prediction techniques in the western United States and Japan are developing, and it is known that fluid injection will alter pore pressures and induce quakes. It would, however, require a brave (or foolhardy) soul who would be willing to "try this out"—one with unlimited funds and a battery of lawyers experienced in damage suits.

subaerial environments in a contrasting perspective. Such problems (1) are essentially predictable with adequate geomorphic study and knowledge of a region's environmental history and (2) can often either be avoided or countered. Since environmental problems relating to earthquakes or volcanic eruptions are as much the province of geologists and geophysicists as of anyone else, we will confine our remarks to those environmental problems more directly associated with the interfaces between land, air, and water.

ENVIRONMENTS, RELICTS, AND DISEQUILIBRIA

We have already had occasion to learn that the Earth is a vast cemetery, one whose caretaker is rather careless and doesn't always cover things up neatly. Moreover, the regions that have undergone recent changes in environments far outnumber those that have not. As a consequence, the vast majority of Earth's inhabitants putter about amidst a dazzling array of largely unrecognized relicts. Naturally, the relicts are changing in their new environments, but the adjustments involved are as yet incomplete. As a result, all sorts of disequilibrium situations exist, many in near-threshold states. Just exactly what it is safe to do—or not to do—depends almost entirely upon the setting, its history of environments, and how long things have been as they are now.

Most of our geomorphic environmental problems can be listed under one of three major categories (1) accelerated erosion, (2) water, and (3) human activity. Any of the three may initiate a situation and either or both of the others may be involved. We are, of course, dealing with a "geosystem," and the parts can interact. In almost every instance there is some relict involvement. We have already seen how changing climates may increase erosion rates. Man alters environments with similar results. Water problems are usually a

matter of occurrence—there is too much in a given place or too little. Even pollution is a reflection of such relationships. If there were enough water to dilute the waste or enough earth to bury it, it could be ignored. For a long time we did the ostrich trick.

Accelerated Erosion Problems

Man has long known that certain events were accompanied by accelerated erosion. Agricultural activities and consequent erosion began in pre-historic times and have continued on down to the present. The problem here is that plowing soil to plant a crop is analogous to changing a climate from humid to arid without diminishing precipitation. The consequences are catastrophic (Fig. 12.2), and, of course, irreplaceable soil resources are lost. In more "progressive" societies, the greatest care is taken to prevent or at least inhibit soil loss. Such methods as elaborate dike systems, paddy farming, strip farming, and plowing on contour are all reasonably effective.

Two forms of accelerated erosion are widely encountered which, though not usually a product of agricultural activity, may affect it. Man may not even be instrumental in the events involved, but he can be affected. We are referring here to slope instabilities and stream bank erosion. A moment's reflection on the content of earlier chapters and especially Chapter 7 is enough to emphasize the probability of climatic effects and of geomorphic relict involvement. Vast areas of almost every continent include formerly desert or glacial regions that are presently humid or semiarid. When water and/or clay contents of relict regoliths increase sufficiently, any disturbance may induce slope instabilities of regolith (Figs. 2.29; 2.30). Possible causes range from excess rainfall to earth tremors to artificial excavations.

Slope Instabilities

At least two strongly contrasting mass-wasting situations exist in different environ-

Fig. 12.2. Severe gullying of Santa soil in Latah County, Idaho, following attempted cultivation in 1941. The horizon is almost completely destroyed, and the field has not now been cultivated for more than 10 years. (H. H. Harris, USDA Soil Conservation Service.)

mental settings. In a climax environment of arid or humid type, slope instabilities are minor because of agency interaction and adjustment. In climax deserts, there are few steep slopes and little water or clay to add to the problem. In climax humid regions (cf. Chapter 5), there tends to be a fine state of adjustment between regolith production via weathering, slope adjustment via creep, and channel incision. Slopes are steep but stable unless disturbed. In the desert, if one disturbs the duricrust and exposes fine-grained clay

and silt the wind apparently will erode temporarily and create a blow-out depression, but sheetwash and gravity will eventually tend to smooth things out. In a humid region, a disturbance of slope and regolith will induce a temporary creep acceleration until the slope is regraded and stable. The effects may be serious, but equilibrium may easily be achieved.

In formerly arid or glacial regions which are humid and which contain abundant relict colluvium, regolith tends to be excessively thick and is easier to erode than most bedrock. Channels tend to be incised quickly, and their banks may be overly steep, particularly in view of developing clay contents and rising water tables. Because of regolith thickness, plant root systems may not anchor debris firmly to bedrock. In time the scene is set for accelerated mass wasting where critical slope-repose relations are exceeded with respect to various lubricants. Landslides and avalanches may be common, often with accentuated effects during severe storms when the water content of soil is elevated or during earthquakes when shock waves alter pore pressure relations, reduce friction, and simultaneously induce incipient shear zones (Fig. 12.3).

In both the foregoing circumstances, a recognition of the climatic condition and the environmental history is half the remedial battle. A correct appraisal of related geologic and engineering factors is the other half. An unstable slope recognized before construction is begun is one thing. The same slope detected from the picture window of one's house as the structure toboggans down a mountain side is "something else." Ultimately, following every major climate change from arid or glacial to humid, a slope adjustment to an equilibrium attitude must occur (Fig. 7.13). Depending upon the region in question, the adjustment may already have occurred, it may be in progress, or the thresholds that should induce it may only just have been reached. This is the sort of thing geomorphologists can work out and construction engineers can allow for, given the proper data in advance.

Stream Channel Manipulation

Man's manipulations of watercourses are not all aimed at problems of accelerated erosion. His purposes are nearly as various as the conditions he manages to induce, some of which are intentional. Most channel manipulation related to accelerated natural erosion has to do with rivers reworking bodies of relict alluvium, usually in the process of meandering. As noted in Chapter 7, the outside banks of bends in watercourses tend to be sites of accelerated erosion in the form of bank undercutting and caving. Any course of action by way of channel manipulation in such a situation should take into account the significance of the meandering habit as a mode of fluvial activity. As noted in previous discussions, meandering may represent one type of equilibrium (least-work) flow form for running water flowing on alluvium. Also, meandering streams are almost the opposite of straight lines. They are "almost" the longest distance between two points—along a channel. Meanders also reflect about the lowest path of slope inclination one could devise in a downstream direction along a given valley flat.

We can only briefly consider some of the things that have been done to meandering streams to inhibit channel shifts and curtail erosion. In passing, we will also touch on similar maneuvers with ostensibly similar techniques, alternative motives, and contrasting results. In general, when dealing with natural situations and the question comes up as to what should be done, the answer probably should always be, "As little as possible." In dealing with meander erosion and cut-off problems along the Missouri River, for example, the U.S. Corps of Engineers used permeable *pile dikes* to diminish erosive force on the outside of meander bends (Fig. 12.4). At the same time, a somewhat sinuous flow route was maintained (cf. Ruhe, 1971), and considerable sedimentation in and around the dikes resulted in progradation of channel banks and increased areas of usable land. The general manipulation reduced channel sinuosity—to produce a straighter

Fig. 12.3. Landslide debris at the base of a landslide scar on the North Coastal Range of Venezuela following torrential rains, early Spring 1951. Similar material passed down steep valleys and crossed the coastal lowland near several towns including La Guaira and Miaquetía causing much loss of life and property (cf. Figs. 2.29 and 2.30).

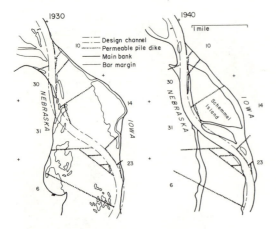

Fig. 12.4. Otoe Bend area, Missouri River Valley, showing artificial channel modifications over the period 1930–40. (From Ruhe *et al.*, 1970.)

channel. For a given reach of valley flat this should increase channel slope somewhat and induce incision. Such effects were not pronounced.

In several instances, watercourses have been straightened out almost completely (cf. Ruhe, 1971; Emerson, 1971; Gillette, 1972). Apparently, in each instance to be considered here, some sort of velocity-slope-sinuosity threshold was crossed. Accelerated erosion ensued. In all instances, the ostensible incentive for manipulation was to prevent flooding along associated bottom lands. In the instance of Crow Creek, which flowed through the east Tennessee Appalachian valleys south into Alabama, a once scenic woodland, teeming with wildlife, is now a barren ditch which in the words of the Philadelphia Academy of Natural Science biologists is an "ecological disaster" (Fig. 12.5). In the case of Willow Creek with flows into the Missouri River from a loess-covered upland in southwest Iowa, we can be somewhat more quantitative.

The effects of straightening Willow Creek began when the main channel was straightened as a form of "engineering improvement," starting in 1906. Consequent changes in width, depth, and shape are striking (Fig. 12.6). Where a tributary

Fig. 12.5. Before (left) and after (right) views of Crow Creek in Tennessee, a pictoral record of the results of channel manipulation. (Aubrey Watson, Tennessee Fish and Game Commission.)

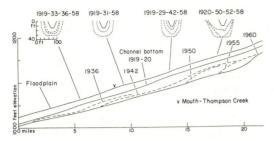

Fig. 12.6. Longitudinal and transverse profiles of Willow River, over the period 1919–58. (After Daniels, 1960.)

stream (Thompson Creek) flows into Willow Creek, channels were deepened some 18 ft (6 m) between 1919 and 1957. Studies by Daniels (1960) indicate canyon-type gullying upstream along Thompson Creek to depths of as much as 40 ft (12 m) during the same interval of time, with related deepening and extension of tributaries (Fig. 12.7). Now, distruction of arable land by gullying is extensive, and many of the vertical-walled gullies are impassable. Ruhe (1971) also notes a resulting depression of the water table so that wells must now be deeper. Flooding has to all intents and purposes stopped, though, purely on a cost-benefits basis, it seems doubtful if the savings related thereto are in any way equivalent to expenses imposed on the surrounding countryside through loss of farm land, water supply, road and bridge construction, and similar related costs.

Too often, it would seem, a natural setting is altered with a particular end in mind but without the necessary overview required to anticipate possible unwanted side effects. In the case of Willow Creek an excessive preoccupation with minor flooding combined with an extreme countermeasure spelled erosional disaster. The same approach produced a similar result on Crow Creek in Tennessee. A more moderate solution to a drainage and erosion problem along a reach of the Missouri River was successful because the flow configuration was imposed with bank stabilization in mind, and it apparently did not exceed critical erosional thresholds. How

close such thresholds were approached, however, is somewhat uncertain. An increasingly accurate appraisal of hydraulic factors is clearly necessary. But such an appraisal should be accompanied by a close look at a drainage's environmental history and developmental stage. We will briefly consider another form of channel manipulation (dam building) in our discussion of water problems.

Coastal Erosion

Accelerated erosion along coasts is most significant where the strand consists mainly of unconsolidated or poorly consolidated materials (cf. El-Ashry, 1971). As in so many of the geomorphic situations discussed previously, coastal erosion problems often come down to matters of systemic budget. Along a given reach of coast the budget is mainly an erosional-sedimentary relation. In other words, does marine erosion strip away material faster than it is brought in by various mechanisms. In the long run, erosional intensity along coasts proceeds at a fairly constant level with a particular land-sea exposure;

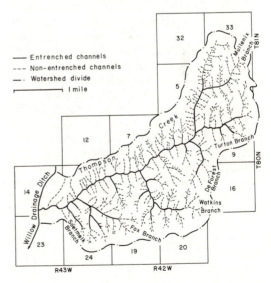

Fig. 12.7. Drainage net of Thompson Creek Watershed shows the extent of extension by tributaries because of deepening of the Willow Creek Drainage Ditch. (From Daniels and Jordan, 1960.)

peak erosional intensities occur during storms. As pointed out in Chapter 9, the over-all tendency along coasts tends to be one of sediment removal, and these effects are intensified on headlands unless countered by organisms (reefs or human reinforcements).

Since the end of the last widespread glacial and arid episodes, sediment influx into oceans has declined along many coasts, and less sediment is being brought in by rivers than the sea removes to ocean deeps. This statement appears to hold for the east coast of the United States but not for the Texas Gulf Coast where major rivers may still maintain a positive sediment budget. For the most part, the mean annual direction of long-shore drift determines the directions from which sediment must come in order to maintain beaches at a particular position. Shore stabilization programs that effectively cut off sediment supplies ultimately ensure strand erosion elsewhere. Thus, groin and jetty construction along the New Jersey coast and at Cape May have cut off sediment into the Sandy Hook area (Fig. 12.8).

In at least one instance, described by Cooper (1950), man has created coastal problems by reducing sediment influx into seaways. The several dams constructed along the Colorado River (Hoover, Parker, Imperial, and Laguna) have acted as sediment traps to such an extent that the Colorado River delta is no longer being depositionally maintained. Coastal erosion at the head of the Gulf of California is accentuated by a tidal bore related to a 33–50 ft (10–15 m) change twice daily. As Cooper points out (1950), the land to the north of the delta in Imperial Valley is below sea level (Fig. 12.9), and, unless remedial action is taken, the strand will ultimately shift northward about 100 mi (140 km) into a valuable piece of real estate.

Water Problems

Appreciable sections of libraries are devoted to the subject at hand, so our re-

marks will be limited to those water problems with the most clear-cut geomorphic affinities. For the most part, "problems" with water exist because for some human kind there is either too much or too little. In turn, there is the implication that the excess or deficiency has to do with a particular place. Deserts generally have too little water—most of the time. But it is immediately apparent that a value judgment is being made involving a specific purpose of man. That is, too little water—for what? To drink? To wash a car? To dilute a pollutant? To water a camel? To wash one's hands? To fill a reservoir? To grow a crop? Alternatively, too much water—for what? To confine within a river channel? To fill one's basement. To contain behind a levee? To permit a particular kind of crop to mature? We will first briefly consider several instances where there is excess water—for one reason or another.

Floods

Even as these words are being placed on paper, one of the all-time great floods is inundating the valleys of the Missouri and Mississippi rivers. In conformance with relations set forth in earlier chapters, where a flood was defined as unconfined runoff, floods must be viewed, assessed, and dealt with in terms of environmental history. This detached view may be somewhat difficult to maintain as one's home or business slowly goes under water. It is nonetheless called for in any effort at long-term flood planning. As was pointed out earlier, floods are a normal runoff form in deserts where planar landforms and alluviated watercourses are the usual consequence of landscape genesis. In climax humid landscapes (selva), floods do not occur because lands not in channel bottoms and banks are in slopes.

Population pressures, the lure of valuable farm lands, and similar effects draw people in large numbers to the floodplains of major drainage systems and low coasts where flood incidence is high. In the case

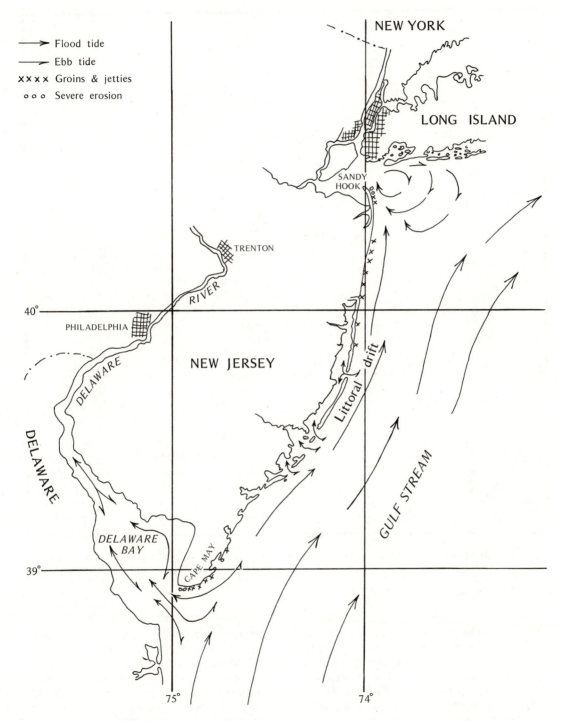

Fig. 12.8.   Major oceanic and tidal circulations along the New Jersey shore. Severe erosion in the Cape May area and to the north in some reaches has led to intensive construction of protective groins and jetties with resultant reduction of sediment supply to Sandy Hook by littoral drift.

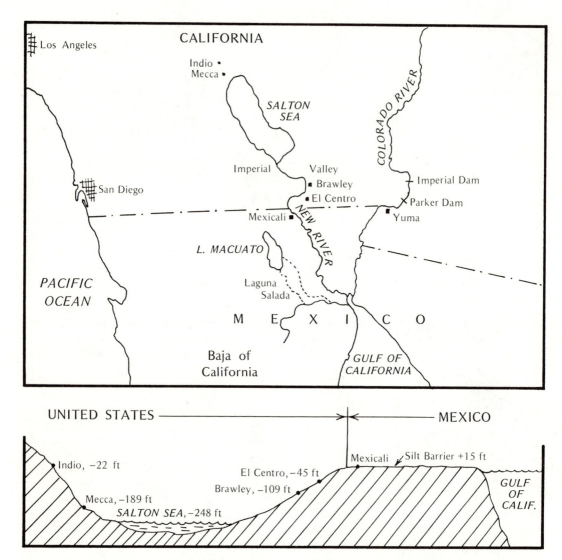

Fig. 12.9. Map and cross section showing the Gulf of California, Salton Sea, and Imperial Valley geography and topography. A tidal bore of great height (commonly 25 ft), operating in the gulf, is slowly removing the silt barrier (deposited by the Colorado River) between the sea and Imperial Valley. Dams on the Colorado prevent maintenance of the barrier by deposition.

of the river floodplain in a humid land, the flat terrain configuration is almost invariably a relict form left by arid, glacial, or similar leveling effects. Floods themselves in such settings reflect the inability of the existing channel cross section to transmit peak discharges without resulting overflow. Severe floods of greatest infrequency are the ones least likely to have been accommodated in any flood-control program. Purely from the standpoint of normal river evolution and a condition of long-term stability without flooding, the channel should be encouraged to deepen and enlarge its cross section. Unfortunately, individuals or groups are not always willing to wait for such natural or artificial adjustments and have insisted upon alternate measures, some of which run counter to natural processes and must ultimately be paid for, one way or another.

Probably the most effective, long-term method of dealing with flooding in humid drainages involves a combination of channel bank stabilization and headwaters dam site construction. The dams are intended to even out flow levels and thus reduce peak discharges. We will consider some of the dam problems shortly. The bank stabilization techniques which were earlier described in some aspects inhibit stream-sediment-load acquisition from lateral sources (mainly banks). This technique can be combined with moderate sinuosity reduction to increase gradient which, in combination with reduced bank caving, should promote bottom erosion and channel deepening. Unless steps are taken to stabilize mouth levels of incoming tributaries, these also will deepen, and tributary erosion will be accentuated as in Thompson Creek discussed earlier. In almost every instance when steps are taken to increase flood runoff, accelerated erosion will result in the absence of adequate countermeasures.

Multi-Purpose Dams

One of the most common ploys for dealing with floods is the multi-purpose dam on a stream or river. This is clearly a form of channel manipulation, but its ramifications are many and complex. At least one federal agency has devoted its energies almost exclusively to the construction of dams across rivers in the United States for several decades. The cry "Dam the rivers" was usually accompanied by a multiple-use public relations package involving some combination of water supply, hydroelectric power, recreation—and, of course, flood control. Viewed more or less strictly from the standpoint of the role that rivers play in nature, the "package" ranges from good to almost totally inappropriate.

The impoundment of water along a drainage line immediately inhibits any developmental tendency toward a stable drainage form, particularly if it is impounded along the middle or lower reaches of a river system. The dam is an artificial knickpoint. As was discovered to almost everyone's surprise when Hoover Dam was constructed across the Colorado River, water leaving the spillways had deposited its load in the lake above and had accentuated erosional energy. Actually, the deposition was anticipated. It is, for essentially every dam. Rivers with solutional loads in humid climax settings of the tropics are perhaps the only ones that might be dammed and not fill with solid sediment. Dams downstream from Hoover Dam are already backing up heavily sedimented shallow lakes and have thus lost much of their water-storage and flood-control capacity. This brings up the point again which may already have been sensed by the reader—many of the "multi" purposes of dams are counter to one another (Fig. 12.10). When one adds this piece of information to the fact that dams counter natural drainage processes, it is difficult to avoid the notion that we are at cross-purposes in these matters—perhaps unbearably.

The realities of population growth and water supply probably require some dam construction for water-supply purposes only. These same realities enhance the possibilities of pollution, but with strict controls, some recreational use of watershed reservoirs seems possible in certain circumstances. Development of nuclear electric power sources may eventually reduce pressures for hydroelectric use of dams. In general, flood-control dams should probably be limited to those headwaters areas of drainages that lack appreciable areas under cultivation upstream. Otherwise, heavy sedimentation of the impoundment will destroy its storage capacity in two or three generations—about long enough for the person who built the structure to be completely beyond reach of most conventional forms of retribution.

	Flood Control	Water Supply	Hydroelectric Power	Recreation
Effects & Manipulations	Lake levels must be dropped prior to anticipated high precipitation or high runoff seasons.	Lake levels should be kept high in case anticipated water fails to develop because of drouth.	Water levels must drop constantly to provide continuous electric power regardless of lake level or season.	Water levels should be high and reasonably constant— otherwise, strand shifts trouble boaters, swimmers and always kill off exposed vegetation and spoil fishing.

Fig. 12.10. Table comparing conflicting aspects of multi-purpose dams and reservoirs.

Irrigation and Related Problems

Often associated with impounded water sources, and invariably practiced in those arid areas of the world that are agriculturally marginal, irrigation merits some discussion from a geomorphic standpoint. Our earlier comments about what happens to water introduced naturally into deserts applies equally well to that introduced artificially. Water that dries up in deserts leaves behind at the ground surface such mineral matter as may have been contained in solution. It has long been known that the precipitation of salts in irrigated soils ultimately tends to poison the soil, particularly in regions where water is otherwise scarce and adequate flushing techniques cannot be practiced. In many areas of the western United States and other desert regions of the world, irrigation, soil poisoning, and land abandonment are common.

Since the amount of fertile land on earth is not infinite, the use of irrigation as the basis of an economy should not be encouraged beyond the point where adequate flushing out of salts is possible. Where irrigation is employed to augment natural precipitation merely as a way to engender steady growth in season, but soil salts are not normally cumulative, there seems to be little problem. Where water is really scarce, irrigation is an absolute form of resource destruction, since potentially good growing surfaces may be permanently ruined. One possible solution to desert irrigation problems involves use of desalinized water which is effectively distilled water with no mineral matter. At present, the costs of such water prohibit appreciable agrarian usage. Fossil groundwater bodies in places like the Sahara are finite and no more than a temporary expedient. One possible alternative is salt water farming, now being attempted in Israel. The technique makes use of the fact that toxic amounts of soil salts tend to be retained in growing media by clay particles in contact with plant rootlets. It appears that clay-free quartz sand provides an almost ideal rooting medium for salt water cultivation. When one considers the vast areas of quartz sand collected into various dune systems and ergs in Africa and elsewhere, the temptation to recommend building a sea water pipeline into the interior is considerable. But then, there may be other reasons for doing the same sort of thing.

Back in the 1920's, an anonymous geomorphologist working in the desert region of the southwestern United States observed and recorded an interesting phenomenon which could have far-reaching implications. It was in the Spring when the region occasionally gets what little precipitation there is. A passing storm dropped enough rain to make a sizable lake in a playa between two desert ranges. Sizable being some 5 mi (8 km) across and maybe 6 inches (15 cm) deep at a maximum. No more weather systems came by after that. But each day as the heat built up, hazy areas could be seen across the lake, clouds formed above it by afternoon and moved away upslope toward the uplands where the water in them condensed through adiabatic cooling and fell as rain. Desert flowers flourished; normally dry and barren stream beds glistened and gurgled. It was a new land.

This rather pleasant scene came to an abrupt end when the lake dried up. So did the streams and the flowers whose life cycle in a desert rarely lasts more than a couple of weeks, in any case. But that is hardly our point. Rather, it seems important to note here that the changed environment observed in the Sonora was due mainly to the temporary existence of a local evaporation source. The atmosphere doing the evaporating was only dependent upon moisture surface area. Water depth was not a factor. Neither was salinity. Salt water in the same setting would presumably have served equally as well. The agency doing the desalinizing in this instance was the atmosphere, and it does the same thing all the time over the oceans. It occurs without expense, moreover, and without a toxic hypersaline brine byproduct. Also, the plant growth was not associated with a buildup of soil salts as long as the rain continued.

The implications of the foregoing environmental event are difficult to miss. We should always be prepared to make the best possible use of available geomorphic knowledge, and it seems wholly appropriate that the final chapter of this work should include an appeal to make a pragmatic use of some of the information presented herein for the betterment of mankind. Population pressures constantly drive us to make use of lands that are ever more marginal. Almost every continent includes a number of interior desert basins where the environments could well be materially altered and improved by the creation of artificial lakes. The floors of many such basins are toxic saline flats, but the surrounding pediment alluvium is a potentially rich growing surface largely uninhabited and certainly untilled. Initial experiments along these lines could be made at comparatively low expense by drilling wells for the initial lake water source, whether saline or fresh wouldn't matter greatly in the long run. Actual water volumes could be comparatively low since area and not depth are what would count. Eventually, if the techniques were successful in pilot runs, consideration could be given to construction of a saltwater pipeline from the nearest ocean to several adjacent desert basins of such a region as the southwestern United States.

It is in situations like the one just discussed that we must be prepared to learn from the past. We propose to reverse the basic precept of the Law of Uniformitarianism by making use of the record of ancient environmental changes to indicate to us the probable consequences of contemporary environmental manipulations. As was noted in Chapter 11, the development of inland water bodies in times past has tended to provide evaporation sources and induce humid climates, particularly if the water is warm. Where the land that remained unsubmerged had appreciable relief, orographic effects were added to other potential causes of precipitation, and, of course, rain-shadow effects would tend to be eliminated in similar modern circumstances. In recent years there has been considerable talk of building an aqueduct from the humid portions of the Canadian northwestern to the arid southwest. In view of the drawbacks of irrigation and potential political complications, a few artificial inland seas in desert basins might prove to be a preferable alternative.

Permafrost Problems

The relict nature of permafrost was described in Chapter 8 as was its fragile character when exposed to certain kinds of treatments. This is no place to exhaustively treat the manifold aspects of the problem. However, in view of continuing efforts to open up oil production in the Arctic regions of Eurasia, as well as North America, rather continuous pressure to construct various pipelines and travel routes in permafrost areas and many other kinds of related activities, a paragraph or two dealing with major aspects of permafrost vulnerability seems most justifiable.

As pointed out by Haugen and Brown (1971), both natural and artificial agents produce the same results in permafrost areas. Unfortunately the results produced by the artificial variety have a habit of being unacceptably accelerated. Critical in the situation is the surface layer of insulat-

ing material, composed mainly of organic material. This layer is thinnest at high latitudes and thickens toward warmer climes. The results of disturbances in the surface layer are most immediate and apparent in the northern areas of ice-rich permafrost such as those of Alaska. Though less immediate and apparent in southern areas with a thicker organic layer, changes there may be no less dangerous.

Disturbances of surface organic layers by fires, bulldozing, passage of tracked vehicles, and building installation directly on the ground are practices that are most definitely to be avoided. They usually re-

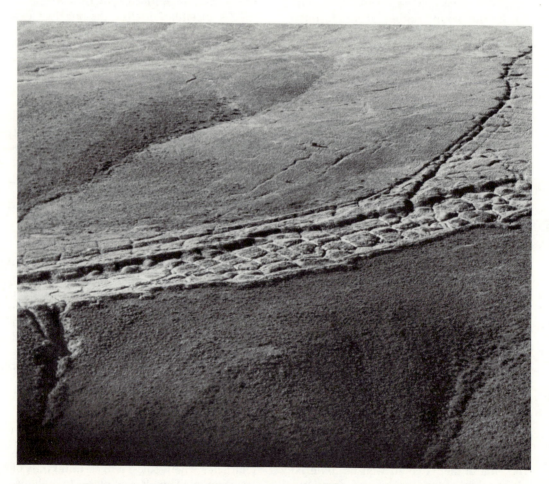

Fig. 12.11. Effects of disturbed permafrost in Alaska near Umiat. Depicted in this aerial view are an entrenched trail and a former airstrip modified by patterned ground. The events in question occurred ca. 1946. (R. K. Haugen, U .S. Army Cold Regions Research and Engineering Laboratory.)

sult in degradation of subjacent permafrost the following year, often permitting excessive depths of thawing. If not followed by equally deep refreezing the following winter, former trails may quickly be converted into watercourses typified by accelerated deep gullying. It is activities such as these that commonly accompany construction projects (Fig. 12.11) and must be strictly guarded against in any programs involving use of terrain underlain by perennially frozen ground. Acceptable engineering techniques for such regions have been outlined by Muller (1947), Linell (1960), and the Corps of Engineers (1950).

SUMMARY

If this chapter is the last and the shortest, it is hardly the least important. All we have tried to do here is touch upon some of the ways in which geomorphic know-how could be brought to bear on the problems of man's environment. We have not nearly exhausted the possible avenues of discussion and interest. But then, the subject matter of this text does not permit us to employ all of the tools required to deal with many problems of the environment. A truly effective approach to such matters would have to be broadly comprehensive in areas of economics, sociology, law, aesthetics, natural resources, engineering, geologic hazards, agrarian needs, and so on.

If this book has had a mission, it was to portray the environments in which landscapes develop, to place in some balance our perspective of the interacting geomorphic elements in both space and time, and to summarize contemporary ideas of gradational processes in a framework of alternating and evolving environments. A threefold value potential emerges where related geomorphic concepts may be applied to the understanding of landform genesis and evolution, to the comprehension of the intricacies of earth history, and to the better coexistence of men and earth.

REFERENCES

A. G. I. (1970) *Environmental Geology, Short Course Lecture Notes,* Amer. Geol. Institute, Milwaukee.

Bryan, Kirk (1923) "Erosion and Sedimentation in the Papago Country, Arizona," *U. S. Geol. Surv. Bull.,* v. 730, pp. 19–90.

Coates, D. R., ed. (1971) "Environmental Geomorphology," *1st Ann. Geomorph. Symposium Proc., Binghampton,* 262 pp.

Cooper, A. M. (1950) "A Cataclysm Threatens California," *Harpers Mag.,* April, pp. 66–69.

Corps of Engineers, U. S. Army (1950) "Comprehensive Report Investigation of Military Construction in Arctic and Subarctic Regions 1945–1948," *St. Paul Dist. ACFEL Tech. Rept.* 28.

Daniels, R. B. (1960) "Entrenchment of the Willow Creek Drainage Ditch, Harrison County, Iowa," *Amer. Jour Sci.,* v. 258, pp. 161–176.

Dasmann, R. R. (1968) *Environmental Conservation,* John Wiley and Sons, New York, esp. pp. 1–25.

Detwyler, T. R. (1961) *Man's Impact on Environment,* McGraw-Hill Book Co., New York, esp. pp. 195–383.

El-Ashry, M. T. (1971) "Causes of Recent Increased Erosion along United States Shorelines," *Geol. Soc. Amer. Bull.,* v. 82, pp. 2033–2038.

Emerson, J. W. (1971) "Channelization: A Case Study," *Science,* v. 173, pp. 325–326.

Flawn, P. T. (1970) *Environmental Geology,* Harper and Row, New York, 313 pp.

Gillett, Robert (1972) "Crow Creek: Case

History of an 'Ecologic Disaster' [in] Stream Channelization; Conflict between Ditchers; Conservationists," *Science*, v. 176, pp. 890–894.

Haugen, R. K. and Brown, Jerry (1971) *Natural and Man-Induced Disturbances of Permafrost Terrain* [in] D. R. Coates, ed., esp. pp. 139–150.

Linell, K. A. (1960) "Frost Action and Permafrost" [in] *Highway Engineering Handbook*, K. B. Woods, ed., McGraw-Hill Book Co., New York, esp. sect. 13.

McKenzie, G. D. and Utgard, B. O. (1972) *Man and His Physical Environment*, Burgess Pub. Co., Minneapolis, 338 pp.

Moulton, Benjamin, ed. (1972) *Readings in Earth Sciences*, Van Nostrand Reinhold Co., New York, 247 pp.

Muller, S. W. (1947) *Permafrost or Permanently Frozen Ground and Related Engineering Problems*, Edwards Bros., Ann Arbor, Mich., 231 pp.

Ruhe, R. V. (1971) "Stream Regimen and Man's Manipulation," [in] D. R. Coates, ed., esp. pp. 9–23.

Rumney, G. R. (1960) *The Geosystem*, Wm. C. Brown Co., Dubuque, Iowa, 135 pp.

Tank, R. W. (1973) *Focus on Environmental Geology*, Oxford Univ. Press, New York, 450 pp.

Glossary

This glossary contains the technical terms used in the body of this volume. In instances where a detailed discussion of the word exists elsewhere, an appropriate citation is given.

ablation—wastage of an ice or snow mass by melting, evaporation, and similar processes.

ablation till—till deposited by ablation, usually from stagnant ice.

abyssal plain—a plain, usually depositional, on the deep ocean floor.

accordant summits (gipfelflur)—a set of hill, mountain, or ridge crests having the same elevation; also called even-crested ridges.

active layer—the annual freeze-thaw layer of permafrost.

adhesion (molecular)—the molecular attraction between surfaces of bodies in contact.

adiabatic gap—a lower zone of the atmosphere, consisting of undersaturated air, from which air must rise before precipitation can occur in a given area.

adiabatic expansion—the expansion of air masses as they rise; usualy accompanied by cooling.

aggradation—buildup of a land surface, usually by alluvium.

aggradation plain—a plain formed by deposition of terrigenous deposits, mainly alluvium.

akle (pl. akeile) dune—crescentic transverse dunes; see Chapter 6.

alluvial plug—a mass of coarse gravel bypassed by runoff in a channel, constituting an obstruction to flow.

alluvial terrace—see terrace and Chapter 7.

alluvium—a more or less stratified deposit made by running water of clay to boulder-size debris.

alpine glacier—a glacier in a mountainous region with a catchment area above the snow line.

anabranch—see distributary.

anastomosing (pattern)—an interwoven pattern of flow lines developed by running water in distinct channels.

anastomosing channel system—a system of interlaced channels formed by consequent runoff passing around pre-existing obstructions of alluvium or bedrock.

andesite—a fine-grained, gray igneous rock of intermediate composition common to island arcs; see Chapter 3.

annular drainage—circular streams on an uplift connected by radial streams.

antecedent drainage—a drainage that maintains its course during development of transverse structure.

arch (sea)—a rocky headland pierced by a sea cave.

arete—sharp, serrate ridge in glaciated mountains, usually between two cirques.

arid (geomorphic environment)—a landform-making environment characterized by discontinuous (usually sparse) vegetation.

arid (geomorphic system)—an association of landform-making agencies and processes in an arid geomorphic environment.

arid fluvial event—the occurrence of precipitation in a desert plus associated evaporation, infiltration, runoff, erosion, and deposition until all runoff has ceased.

aridity (cryergic)—the conditions of a boreal desert, usually characterized by an absence of surface water or groundwater.

aridity (meteorologic)—a dry environment defined on meteorologic bases; often those places with less than 10 inches (25.4 cm) annual precipitation.

argillite—sedimentary rock dominantly silt size or finer; see Chapter 3.

arkose—clastic rock with more than 25% feldspar; see Chapter 3.

arkosic—rock containing 5–25% feldspar; see Chapter 3.

arroyo—a narrow, deep, often steep-walled, flat-bottomed ravine; southwest U.S. usage.

aseismic—those portions of the Earth's crust infrequently subject to earthquakes.

atmosphere—the gaseous envelope surrounding the Earth.

atoll—an encircling organic reef, usually enclosing a lagoon.

attrition—term applied to wear of rock particles in transit.

back swamp—areas of saturated ground behind levees.

backwash—the return of water following the breaking of a wave.

backwash current—current related to backwash; see Chapter 9.

badland (topography)—finely dissected terrain, usually developed on materials of small grain size and unconsolidated.

bar (mid-bay, bay head, bay mouth, offshore)—linear deposit of beach material, usually sand size, mainly along the trend of a weakening current.

barbed tributary—a feeder stream intersecting a trunk drainage line with an acute angle pointing upstream.

barchan dune—cusp-shaped, migratory sand dune common to areas of short sand supply; concave side points down wind.

barrier (beach, island, bar)—partly emergent, elongate ridge of sand or shingle situated somewhat offshore and sub-parallel with the mainland; may be tied to a headland.

basal slippage—movement mode of a glacier; see Chapter 8.

basalt—fine-grained, basic igneous rock common to oceanic extrusions; see Chapter 3.

base level (local)—local limit on downward fluvial erosion established by resistant rock, depositional datum, or similar feature.

base level (regional)—generally, mean sea level; otherwise, lower limit of erosion by a subaerial fluvial agency.

base leveling—reduction of a land surface to base level by erosion.

basin (sedimentary)—depression in which sediment accumulates.

batholith—largest intrusive igneous rock mass; generally granitic and more than 40 sq mi area.

bauxite—high alumina weathered residue; see Chapter 5.

beach—shore at least partly comprised of unconsolidated debris.

beach cusps—regularly spaced U-shaped indentations along a beach separated by prong-like projections.

beach ridge—the raised part of a beach centering on the berm.

bed load—see traction load.

bergschrund—open crevass between a cirque headwall and the glacier ice.

block field—a surface strewn with large, commonly angular rock fragments often related to frost shattering; exfoliation of clasts may produce a boulder field.

blueschist facies—metamorphic rock group associated with collision-type mountains; see Chapter 3.

bolson—a desert depression entirely surrounded by hills or mountains; the lowest part is usually a playa.

botn—see cirque.

box canyon—an arroyo whose upsteam end is a steep-walled cul-de-sac.

braided stream—runoff flowing on alluvium and depositing alluvium as obstructions to produce an interwoven flow pattern within a channel or valley.

breakers (plunging)—waves of translation that pile upon shore and break abruptly and steeply.

breakers (spilling)—waves of translation that spill gradually up on the beach.

bugor—see turf hummock.

butte—usually flat topped, monument-like erosional outlier.

calcrete—a form of duracrust, usually involving $CaCO_3$.

caldera subsidence—collapse of a volcanic vent area into a large, crater-like depression.

caliber (load)—the size of material being moved by a transport agency.

caliche—see calcrete; see also Chapter 6.

capacity—the quantity of sediment an agency can carry.

case hardening—precipitation of mineral matter in a toughened outer layer of a rock being weathered.

cave (sea)—an opening into a sea cliff, developed by wave action.

cavernous weathering—development of pits and depressions in a rock surface during weathering.

cavern (solutional)—underground openings of appreciable extent developed by solution of percolating groundwater.

centrifugal force—the force that tends to make a body moving in a curve go off at a tangent.

chain lakes—a sequence of lakes along a glaciated depression, usually linked by a stream.

channel—an eroded depression through which a stream flows.

channel piracy—a selective incision of one or more channels within a network at the expense of other channels.

chernozem—a dark-colored soil; see Chapter 5.

chert—sedimentary rock composed of cryptocrystalline silica; see Chapter 3.

cinder cone—a cone-shaped eruptive deposit of pyroclastic debris.

circo—see cirque.

cirque—a bowl-shaped depression eroded by ice on an upland or a mountain.

cirque basin—the lower part of a cirque depression.

cirque glacier—a glacier restricted to a cirque.

cirque threshold—the raised lip of a cirque.

cliff (sea)—an escarpment eroded by wave action.

climate—the average weather of a region.

climatic nucleus (pl.nuclei) (humid; arid; glacial)—a region so favored by temperature and hydrography as to retain a particular climate almost indefinitely.

climax—an equilibrium expression of a particular environment.

climax deposit—an accumulation of sediment, reflecting the dynamic equilibrium condition of the parent environment.

climax erosional mode—a least-work erosional mode for a particular agency, usually in a climatic nucleus.

climax flora—an assemblage of plants ecologically adapted to their environment.

climax landform (landscape)—a terrain configuration, reflecting the ultimate morphologic effect of a particular geomorphic system.

climax soil—a soil configuration reflecting an equilibrium adjustment between parent material, environment, and slope.

closed systems (environmental)—a sealed environment into which energy cannot pass, usually artificial.

cloud cover—the per cent of sky covered by clouds per unit time.

coast—the land zone immediately adjacent to a body of standing water.

coastal environment—conditions prevailing along a coast, usually involving interaction of land, air, and water.

coastal geomorphic system—those interacting agencies along a coast that tend to generate landforms.

cohesion—molecular attraction between particles of a body.

col—a narrow pass between two glaciated peaks or ridges.

collision-type mountains—mountains apparently resulting from a collision between two crustal plates.

colluvium—weathered debris including soil which has moved downslope under the force of gravity.

competence—the ability of a geologic agency to move clasts in terms of their size.

composite cone—a volcanic cone of both lava and pyroclastic debris.

compound landscapes—landscapes generated by two or more geomorphic systems.

cone (alluvial)—a conate pile of rock debris, usually at the base of a steep slope.

cone (volcanic)—cone-shaped deposit of volcanic debris, usually centered on a localized vent.

consequent (runoff; drainage)—overland flow of water taking a coarse reflecting original land slope.

continent—one of the several major land subdivisions of earth.

continental accretion—peripheral enlargment of a continent by marginal mountain building, usually involving collision.

continental glacier—an ice sheet occupying a large portion of a continent.

continental rise—the portion of a continental mass that ascends from the deep ocean basins toward the land.

cordilleran-type mountains—mountain system

formed on a continent adjacent to an ocean trench; see Chapter 3.

core stones—more or less intact, rounded remnants of joint blocks generated by spheroidal weathering.

coriolis force—deflective force acting on a horizontally moving mass near the earth's surface; see Chapter 4.

corrasion—wearing of a rock surface by moving rock particles.

corrie—see cirque.

corrosion—solutional erosion, particularly of a stream channel.

countryrock—rock that surrounds igneous intrusions or otherwise comprises bedrock.

covermass—a mass of sediment burying prior bedrock and terrain on which drainage may be initiated.

craig-and-tail (features)—a drumlinoid type glacial deposit made in the lee of a bedrock knob.

craton—the generally aseismic, crystalline portion of a continent; it may have a sedimentary rock veneer.

creep (rock, rock glacier, talus, soil)—very slow, particle-by-particle downslope movement of rock debris.

crevasse (deltaic)—a notch eroded in a levee during flood overflow in a deltaic region.

crevasse (glacial)—an open fracture in glacial ice.

crust—the outer layer of the lithosphere which reacts as a brittle substance when stressed to the point of failure.

cryergic environment—conditions typified by ice development through much or all of each year.

cryosphere—the more or less discontinuous area of cryergic conditions on the Earth's surface.

cuesta—an inclined ramp developed along rock layering, usually terminated up-dip by a scarp.

current—a directionally oriented movement of a gas or liquid.

cut bank—low bench or bluff usually coinciding with a stream margin along which erosion is occurring.

cwm—a steep valley; see cirque.

cyclothem—a cyclic deposit of sediment, including coal and a marine and nonmarine phase.

deaggradation—removal of unconsolidated detritus from an aggraded land surface.

débouchér—a river's mouth.

décollement—structural discordance over which an allochthonous rock mass has moved; see Chapter 3.

deflation—picking up of loose particles by the wind.

deflation depression (blow-out)—an area lowered by wind erosion.

degradation—wearing away of a land by erosion.

delta—a distinct fluvatile deposit made at a river's débouchér.

dendritic drainage (pattern)—a drainage whose tributaries branch in the manner of a plant's boughs.

denudation—to lay bare by erosion.

deposit—an accumulation of rock waste.

depositional coastal segment—a dominantly depositional reach of an irregular coast, usually an embayment.

depositional landform—landforms shaped mainly by an agent of deposition.

depositional plain—planar landform shaped mainly by sediment deposition (see also aggradational plain).

deranged drainage—a drainage whose original flow pattern has been disrupted, commonly by crustal movements, volcanism, glaciation, or aridity.

desert pavement—accretionary surface layer of rock fragments in a desert too coarse to blow away, commonly duricrusted.

desert varnish—surficial precipitate on a rock surface left mainly by evaporation.

diabase—dark-colored igneous rock of intermediate texture and basic composition; see Chapter 3.

diagenesis—the changes that occur in a sediment following accumulation.

diagenetic—a feature of a sedimentary rock developed after accumulation and burial.

diastem—localized break in the stratigraphic sequence signifying erosion but little time duration.

diastrophism—relative movement within the lithosphere.

differential erosion—selective removal of rock material in accordance to erosive susceptibility and agency.

dike—tabular, discordant igneous intrusive.

diorite—igneous rock of intermediate composition and texture; see Chapter 3.

disequilibrium (environmental; agency)—accelerated activity within a system due to a change.

distributary—a channel within a network that takes runoff from a trunk drainage line.

dolomite—sedimentary rock consisting mainly of $MgCO_3$; see Chapter 3.

drainage basin—a depression into which precipitation falls and water gathers and drains; restricted by some to those depressions eroded by drainage nets.

drainage net—a system of connected watercourses usually involving a trunk drainage line with tributaries.

drift—material deposited more or less directly by a glacier.

dripstone—fine textured, dense variety of $CaCO_3$ deposited in thin layers by water, usually in caverns.

drumlin—a whaleback-shaped hill deposited by an ice sheet, usually in groups; see Chapter 8.

dry frozen ground—ground lacking free water below freezing temperature.

dry permafrost—perennially frozen, dry ground.

dune (sand; clay)—mound of clastic material (usually sand) accumulated by wind.

duricrust—ground surface mineral incrustation formed by water solutioning and precipitation, usually in deserts.

dust devil—small spinning vortex of dust-charged air.

dynamic equilibrium (steady state)—a least-work, equal area energy expenditure condition for a system characterized by imperceptible, short-term physical changes.

dynamic relief—relief in an orogenic zone developed because of active lateral stress.

elastic limit—stress limit of a substance which if exceeded results in failure, usually by shear.

elb dune (pl. alab)—symmetric ridges of sand parallel with dominant wind.

elevation—height of a point above mean sea level or other standard datum.

endogenic—formed by forces and agencies acting within the Earth.

endoric—drainage features of desert regions that do not give rise to rivers; interior drainage.

englacial (load; position)—rock material within a glacier mass or that location.

enneri—dry desert wash in the Sahara; see arroyo.

environmental dynamism—pertaining to the theory that environments change in time and space.

environmental extract—that portion of the weathering product that tends to be removed as it is formed by agents of erosion.

environmental residual—the increment of weathered debris that tends to form and remain within a particular environmental area (i.e. soils, alluvium).

embayment (structural)—depressed re-entrant in the margin of a land with a tendency to be submerged and/or subside.

eolian—having to do with the activity and effects of the wind.

epeirogenic (movement)—effectively vertical shifts in continents mainly expressed by relative emergence or submergence; usually for millions of years.

epeirogeny—vertical shift of a continental mass, usually related to crustal thickening or lithic dewatering.

ephemeral (flow)—runoff limited to times of rainfall and shortly thereafter.

equilibrium coastal segment—a portion of a coast more or less straight or curvilinear through the process of erosion, deposition, or organic activity; see Chapter 9.

erg—a sand sea; see Chapter 6.

erosional coastal segment—that part of an irregular terrigenous coast being subjected to essentially continuous wave action.

erosional landform—a landform shaped by the removal of rock material.

erosional phase—a period of erosion equivalent to the duration of an environment in a particular area.

erratic—a clast transported some distance by glacial ice.

esker—an elongate, often sinuous ridge of glacio-fluvial debris, usually deposited under the ice.

eugeosyncline—a deeply subsided, elongate area of the Earth's crust with ocean-trench sedimentary affinities at one stage; it may equal outer continental margin; see Chapter 3.

evaporation—the taking up of moisture from a surface by under-saturated air.

exfoliation—the peeling off of rock in more or less concentric layers during weathering (see spheroidal weathering, sheeting).

exhoric drainage—characteristic of humid regions, water drains beyond the environmental limits.

exhumed (landform; landscape)—a terrain feature, previously buried, which has been re-exposed with little modification.

exogenic—having to do with forces and agencies acting outside the Earth's surface.

extrusion flow—fluid or plastic material escaping or being pushed to the surface from beneath a mass of similar composition, as in valley glaciers and lava flows.

faceted spur—a ridge extending into a valley which has been truncated, commonly by glacial erosion or faulting.

facies (litho)—compositional and textural aspect of a rock, usually sedimentary.

fan (alluvial)—a broad, flattened connate alluvial deposit, usually situated at the base of a steep slope or cliff.

fanglomerate—lithified clastic deposits of an alluvial fan, usually includes much sand and gravel.

fault—a fracture along which there has been relative movement.

felsenmeer—ground surface of coarse, commonly angular blocks so permeable all precipitation soaks in (no runoff).

feral—topography consisting of sharp ridges and steep ravines; see selva.

ferricrete—a form of duricrust with iron and usually some silica.

fetch—the expanse of open water over which wind can blow to generate waves.

firn (névé)—grains of ice developed from snow by repeated freezing and thawing above the snow line.

fissure—an open crack in a rock.

flatiron—flattened, inclined, wedge-shaped rock ramp on a mountain spur formed by erosion of inclined strata.

flood—unconfined runoff.

floodplain—the flattened area along a river subject to periodic overflow.

floral blend—a plant assemblage reflecting two or more distinct environments; cf. cactus and deciduous forest.

flysch—sedimentary facies commonly related to orogenic conditions; see Chapter 3.

foliation—an alignment of mineral grains common to metamorphic rocks.

fosse—an elongate depression between an alpine glacier and the adjacent valley wall.

fracture—any break in a rock.

free face (back slope)—the steep, upper element of a hillslope, often with exposed bare bedrock.

friable—crumbling property of a very loosely cemented, granular sedimentary rock.

fringing reef—an organic reef ranging from slightly offshore to essentially on the strand.

frost cracking—splitting of a rock by low temperature, see Chapter 8.

frost heaving—lifting of blocks of rock by frost action.

frost sorting—separation of regolith into particle size fractions by frost action; see Chapter 8.

frost thrusting—see frost heaving.

frost zone—upper layer of the atmosphere where temperatures are consistently below freezing; intersected by some mountain peaks; descends toward poles.

gabbro—coarse-grained, dark igneous rock; see Chapter 3.

gelifluction—downslope movement of water-saturated earth in association with perennially frozen ground.

geomorphic area—the region affected by a particular geomorphic system.

geomorphic discordance—a contact between two terrains formed under distinct conditions at different times.

geomorphic system—a group of agencies and processes interacting under a particular environment to produce a landscape.

geomorphology—study of the origin of landforms.

geosyncline—elongate, regional depression in the earth's crust in which sediment tends to accumulate, possibly a continental shelf and margin.

geothermal gradient—gradual downward increase in earth's temperature.

giant cusp—see beach cusp; see also Chapter 9.

gibber gravel—see lag gravel.

gibbsite—a form of clay due to chemical weathering; see Chapter 5.

gipfelflur—see accordant summits.

glacial environment—those conditions that most favor the growth, expansion, and wastage of glacial ice, sporadically; the environment is probably cold and moist to begin with.

glacial geomorphic system—the association of glacial and cryergic activities that generate landforms.

glacial steps (stairway)—a series of giant, tread-like risers developed along the floor of a glaciated valley.

glacial surge—relatively rapid advance of glacier ice over a period of a few weeks or months.

glacio-eustatic sea level changes—changes in sea level due to accumulation of water in ice sheets and alternate melting.

glacio-fluvial—activity and effects due to glacial meltwater.

gneiss—foliated, commonly banded, massive metamorphic rock; see Chapter 3.

Gondwanaland—ancestral southern hemisphere landmass; see Chapter 11.

gradation—those processes tending to wear away the land.

graded slope—landslope developed in response to repose of its constituents or through gradational processes.

graded stream—theoretical concept of a stream able to flow and carry its load while neither eroding nor depositing.

granite—coarse-grained, light-colored igneous rock; see Chapter 3.

granodiorite—coarse-grained, light-colored igneous rock; see Chapter 3.

granular disintegration—breakdown of a rock into its mineral constituents by weathering.

gravity flow—movement of a substance down slope by gravity.

gravity waves—wave forms perpetuated by gravity; see Chapter 9.

greenschist (facies)—metamorphosed basic igneous rock suite; see Chapter 3.

groove (glacial)—elongate, rather narrow depression eroded by glacial ice.

ground moraine—planar depositional landform made mainly by wasting continental ice sheets behind terminal moraines.

groundwater body—the body of water that saturates the zone of fracture within the lithosphere.

grus—granular disintegration product of coarse-grained rock, typically a granite.

gully—small, commonly V-shaped watercourse.

guyot—flat-topped sea mount.

gyre—rotating current, especially those currents of the major ocean basins.

hammada (hamada)—slightly undulating plateau of the central Sahara and the Arabian Peninsula, usually capped by lag gravel or other duricrust.

hanging valley—tributary valley whose base is above that of the watercourse it feeds.

hardpan—an impermeable layer near the ground surface which inhibits the infiltration of water.

headwall (cirque)—upslope wall of a cirque.

headwall (trough)—the abrupt rise from a glacial trough to the cirque threshold.

heat budget—relation between the amount of heat entering a given geomorphic system and that dissipated.

heat equator—see thermal equator.

hogback—narrow, elongate rock ridge formed by differential erosion of vertical strata.

horizontal accretion (lateral sedimentation)—outbuilding of sediment by an agency which laterally increases the extent of the deposit, often linearly.

horn—a cirque-shaped peak or the seaward extending portion of a beach cusp.

Horton Number (Q_rK)—see Chapter 7.

hoyo—see cirque.

humid (crygeric)—polar environment where the ground is water saturated but not ice covered in all cases.

humid (geomorphic environment)—an environment capable of inducing particular landforms typified by continuous groundcover of plants.

humid (geomorphic system)—the processes and agencies that interact in a vegetated region to produce landforms.

humid (meteorologic climate)—a relatively moist climate defined on the basis of meteorologic parameters; often a region receiving more than 30 inches (76 cm) of precipitation.

hydration—the chemical bonding of water; especially "during weathering."

hydraulic action—the surge, weight, pressure, and force of water.

ice burst—hydraulically induced explosion of river ice.

ice-contact delta (fan)—glacio-fluvial deposit made, at least in part, in an ice-margin water body.

ice-margin lake—water body impounded between glacier ice and adjacent terrain.

ice-margin stream—a watercourse located by the margin of an ice sheet and associated terrain.

ice-scoured plain—a planar surface developed primarily by glacial erosion.

icing—a condition where ice freezes to the bottom of a frozen-over stream, thereby constricting flow.

ignimbrite—welded tuff; see Chapter 3.

illite—a type of clay, see Chapter 5.

inert (semi-insoluble)—rock or mineral that is stable in a particular environment, i.e. it doesn't dissolve or alter.

infiltration—the process of water soaking into the ground.

inlier—a localized exposure of a rock mass through an otherwise widespread formation.

inselberg (landscape)—an isolated, often rounded erosion remnant, characteristic of some planar landscapes.

interface (geomorphic)—a contact between two media along which landforms develop.

interference (~rhomboidal) ripples—irregular to geometric ripple marks made by crossing currents, usually in the intertidal zone.

intermittent (flow)—runoff occurring a part of each year.

iron formations—sedimentary formations, mainly of Precambrian age; see Chapter 3.

island arc—arcuate disposition of andesite volcanos, usually along an ocean trench.

isohyat—a line connecting points of equal precipitation.

isostasy—a theory that crustal segments are in essentially buoyant repose with elevations commensurate with their thicknesses and densities.

isostatic relief—Earth relief due to different densities and thicknesses of crustal segments.

isotherm—a line connecting points of equal temperature.

joint—a closed fracture in rock.

kame—glacio-fluvial deposit, commonly conate in shape.

kame terrace—deposit of glacial meltwater, commonly between valley glacier and valley wall.

kaolin—clay produced by chemical weathering; see Chapter 5.

kar—see cirque.

karst—topography developed by solution of bedrock; mainly limestone in temperate regions; other rock types respond in the tropics.

kjedel—see cirque.

knickpoint—steepening in longitudinal profile of a stream, commonly due to locally resistant lithology.

laccolith—dome-shaped igneous intrusion.

lacustrine—pertaining to lakes.

lag concentrate—see lag gravel.

lag gravel—coarse surficial sediment residue resulting from loss of fine material by sheetwash and eolian deflation.

laminar flow—low velocity movement by shear between layers of a fluid.

landform—a discrete shape developed over an area of lithosphere.

landscape—an assemblage of landforms.

landslide—downslope movement of a comparatively large mass of rock debris, usually rather suddenly.

landslide surge—large water displacement by a landslide entering standing water.

lapis—solutional enlargement of joints resulting in rounded, often sub-parallel grooves.

lateral moraine—ridge of rock debris deposited by ice along the side of a valley glacier.

laterite—iron-rich soil developed in tropics; see Chapter 5.

lateritic earths—regolith at least partially altered to laterite.

Laurasia—ancient landmass, generally related to the Northern Hemisphere, which later broke up.

lava—moulten rock extruded at the Earth's surface.

lava cave—cave formed in a lava flow by extrusion of a still-liquid part.

lava plateau—elevated, tabular landform developed by widespread lava extrusion, usually of fissure flow basalt (traprock).

leaching—removal of lithic constituents by solutioning of percolating groundwater.

lee—the sheltered side of an object; downwind, downcurrent, downslope.

limestone—sedimentary rock composed mainly of $CaCO_3$; see Chapter 3.

lineament—regional, generally rectilinear fracture in the lithosphere.

lit-par-lit—a mode of igneous intrusion, injection, or replacement into previously layered material (bed-by-bed).

littoral drift—see longshore current.

littoral zone—shore zone of the ocean from storm tide level to a depth of one-third the storm wavelength.

lodgment till—till plastered on a surface by moving ice.

loess—wind-blown dust, mainly silt-size clasts.

longitudinal dune (seif)—elongate sand dune ridge, usually one of several similar parallel examples; see Chapter 6.

longitudinal profile—a surface outline drawn in the long dimension of a feature, e.g. stream channel.

longshore current—major, persistent movement of water parallel to and close to a shore, mainly as a result of waves.

lopolith—a complex laccolith, usually involving sill-like intrusions.

lunette dune—an arcuate dune form, commonly developed just downwind from a blowout or pan.

makatea—a raised ridge around an island formed by an atoll emergence.

mammillated surface—a series of rounded, breast-like hills, usually of glacial origin.

mantle—the outer, non-brittle layer of the earth underlying the crust.

mantle plume—a hot, presumably rising current within the mantle.

marble—metamorphosed carbonate rock; see Chapter 3.

massive (rock)—a rock lacking closely spaced partings.

meander—one of several, regularly arcuate river bends.

meandering (pattern)—stream flow pattern with meanders.

meander scroll—arcuate markings on a flood-plain made by migrating meanders.

medial moraine—ridge of rock debris down medial portions of valley glaciers or deposits thereof.

mega-ripples—very large scale ripples developed by high discharges of floods and geo-stropic currents of oceans.

melange—tectonically modified lithofacies associated with trench orogenesis; see Chapter 3.

mesa—a flat-topped hill or mountain (table mountain) of intermediate dimensions (smaller than a plateau, larger than a butte).

mesogenic—a realm of geomorphic and geologic activity involving internal earth forces that achieve surficial continuation.

metastable—those substances that are comparatively fragile when exposed to subaerial processes and especially weathering of a chemical type.

migamatite—a body of rock altered to granitic texture and composition.

Milankovitch theory—the idea that the Earth's climatic cycles are significantly influenced by astronomic events, particularly the precession of the equinoxes.

mima mounds—see prairie mounds.

miogeosyncline—elongate area of subsidence and sedimentation, characterized by orthoquartzitic sandstone and carbonates.

moisture-depleted air (zone)—upper zone of the atmosphere usually deficient in water because of precipitation at lower elevations.

mollasse—lithofacies of presumed tectonic implications, comprised, in most instances, of arid alluvial covermasses.

monadnock—isolated erosion remnant; especially one of particular resistance.

montmorillonite—a clay produced by chemical weathering; see Chapter 5.

monument—pedestal-shaped erosion remnant; see butte.

moraine—a deposit of glacial debris (see ground, terminal, lateral, medial, recessional).

morphogenesis—landform development, especially as it occurs in a geomorphic system.

morphogenetic region—see geomorphic area.

morphogenetic system—see geomorphic system.

mudflow (gravity)—downslope, often rather rapid movement of a water-saturated mass of colluvium.

mudflow (pluvial)—end result of sheetflood desiccation in the presence of appreciable argillaceous material.

mudlump—mobilized mass of water-saturated mud in deltaic areas which forms a mound-like elevation.

mudstone—a rock composed of clay-size material; see Chapter 3.

nappe—laterally transported, tabular mass of rock, often including recumbent or overturned strata as part of a thrust sheet or gravity allochthon.

natural levee—river bank raised by vertical accretion during flood overflow.

neritic zone—ocean margin from 0 to 100 fathoms depth.

net drift—the increment of water transport related to open-ocean wave movement.

névé—see firn.

nisch—see cirque.

nivation—freeze-thaw processes that create firn from snow.

nunatak—mountain peak protruding through an ice sheet.

ocean basin—one of the several major oceanic depressions.

ocean current gyres—the larger circling water currents in each of the major ocean basins.

open system (environmental)—a natural system through which energy can be transmitted.

ophiolite—a suite of essentially basic igneous rocks believed to relate to mid-ocean ridges; see Chapter 3.

organic coast—a coast whose morphologic configurations are due to organic activity.

organic reef—a mound-like deposit based on a framework of wave-resistant, sediment-binding organisms.

orientation (glacial movement)—having to do with place of origin of glacial ice.

orogeny—mountain-building activity along the margins of crustal plates; see Chapter 3.

orographic rainfall—rainfall induced by air masses forced to move up mountainous slopes with resultant cooling and condensation of moisture.

orography—the description of mountains and their influences.

orthogonal block—a multi-sided block of rock bounded by mainly planar fracture surfaces.

oued—a dry watercourse, mainly in north Africa, often of valley dimensions, usually flat bottomed; see arroyo.

outwash apron—glacio-fluvial debris slope developed outside of a terminal moraine.

outwash gravel—glacio-fluvial gravel spread away from the glacier margins by meltwater.

outwash head—glacio-fluvial debris accumulated between the glacier terminus and the terminal moraine.

ox-bow lake—curved lake formed in an abandoned meander.

oxidation—chemical bonding of a substance with oxygen, especially during weathering.

paleogeomorphology—the study of ancient landscapes.

paleontologic unconformity—a break in the stratigraphic record indicated by fossils.

pan—depression, often abrupt, developed on desert plains and possibly related to eolian deflation; see Chapter 5.

patterned ground—ground surfaces exhibiting regular geometric patterns, particularly as induced by wetting and drying or freezing and thawing.

pedalfer—an aluminum-iron-rich soil; see Chapter 5.

pedestal rock—a tall, slender erosion remnant; see stack, butte, monument.

pediment—a low-inclination, planar landform truncating bedrock structure, commonly steepening upslope where developed on drainage net profiles.

pedimentation—the processes under which pediments develop.

pediplain—a plain developed through coalescence of a series pediments, in theory, at least, under arid conditions.

pediplanation—the processes under which a pediplain forms.

pedocal—a carbonate-rich soil; see Chapter 5.

pedologic—having to do with the development of soil.

peneplain—a theoretical landform presumably resulting from erosion under a humid environment.

perched water table—a groundwater body situated above the main one of a region on an impervious rock layer.

perennial flow—runoff which continues throughout the year in a region.

peridotite—an ultra-basic igneous rock; see Chapter 3.

periglacial—having to do with conditions and events along the margin of a glacier.

perisediment—the sedimentary veneer on the more distal parts of a pediment; it may thicken greatly in closed depressions.

permafrost—perennially frozen ground.

permeable pile dike—a row of spaced pilings in a stream channel intended to influence flow effects.

photic zone—the upper water layer penetrated by light.

phreatic zone—the interval occupied by the groundwater body.

phyllite—low-grade metamorphic rock characterized by weakly developed schistosity; see Chapter 3.

piedmont angle—the angle often created by an abrupt slope steepening at the upslope end of a pediment or plain.

piedmont benchland—a landscape composed of broad planar areas separated by steep scarps; see Chapter 1.

pillow lava—lava in small mound-like masses due to eruption under water.

pingo—a dome-shaped hill apparently uplifted by freezing groundwater in an area of permafrost.

plateau—extensive planar upland of low inclination, some intermontane, some marginal to high mountain areas.

plate tectonics—the array of theories and concepts pertaining to sea-floor spreading, continental drift, and crustal subduction; see Chapter 3.

playa—ephemeral arid lake in a bolson.

plunge pool—erosional depression at the base of a waterfall.

pluton—an intrusive mass of igneous rock.

plutonism—the activity associated with the emplacement of igneous rock.

podzol—a class of soils; see Chapter 5.

point bar—an alluvial deposit on the inside of a stream bend or meander.

pollution—an additive to an environment that is cumulative and affects indigenous organisms.

pool (stream)—area of deep, comparatively quiet water along a stream, usually on a rather straight reach.

porphyroblast—an essentially equidimensional crystal in a schistose rock.

positive infiltration increment—a condition where more precipitation soaks into the ground than is returned to the surface by evaporation, common to humid regions.

pot hole—a circular depression scoured in a stream bed by clasts carried in a rotating current.

prairie mounds—mounds of soil and wind-blown dust, probably accumulated in clumped vegetal areas during aridity.

prairie pimples—see prairie mounds.

profile of equilibrium—idealized longitudinal stream profile, presumably reflecting fluvial energy distribution.

protalus rampart—a ridge of rock debris accumulated at the base of a scarp-base snow drift or cirque glacier.

provenance—the geologic attributes of a sedimentary source area.

pyroclastic—material fragmented during the course of a volcanic eruption.

quartzite—rock consisting of quartz grains bonded by silica; see Chapter 3.

quartz monzonite—light-colored, coarse-grained igneous rock; see Chapter 3.

quasi-sheetflood—an expanse of runoff lacking in confinement and locally interrupted by areas of high ground or vegetation; it may be perennial.

quebrada—Spanish term for a steep-sided gully; see arroyo.

radial drainage—a drainage pattern in which streams diverge away from a central upland.

rain splash erosion—erosion by raindrops falling on bare ground.

reach—any stretch of stream measured between any series of objects or conformations associated with the stream, such as between two curves or two confluence points.

recessional moraine—a moraine behind a terminal moraine signifying an ice re-advance short of prior maxima.

rectangular drainage—a drainage pattern controlled by fractures intersecting at very high angles.

refuge—a place of safety, especially a climatic nucleus where animals or plants survive during unfavorable times elsewhere.

reg—see lag gravel.

regolith—the interval of unconsolidated debris overlying bedrock.

rejuvenation—accentuated stream incision, often attributed to uplift, but requiring perennial runoff in any case and extremely difficult to document.

relegation—pressure-thermal melting of ice at the base of a glacier ice mass.

relict (environmental)—any feature induced by an environment and left in an area by climate change.

relict glacier—a mass of glacier ice that no longer moves.

relief—elevation difference between high and low points of a local area or region.

relief reversal—erosion of rock in an orogen such that structural highs appear as valleys and lows as uplands.

remnant magnetism—magnetism in a rock due to orientation of ferromagnetic particles at the time of lithification.

repose angle—the angle of a slope of loose clasts induced by free downslope particle movement.

rhomboidal ripple marks—see incipient rills.

rhyolite—light-colored, fine-grained igneous rock; see Chapter 3.

ridge (mid-ocean)—a submarine mountain system characterized by high heat flow rates, volcanism, and seismicity; see Chapter 3.

ridge (topographic)—an elongate upland area with a narrow crestal zone, commonly an interfleuve.

ridge-ravine topography—see selva.

riegel—the slightly raised lip of a giant tread in a glacial stairway.

riffle—a shallow area of comparatively fast water in a stream.

rill—a narrow, linear, V-shaped groove cut by running water on a granular surface not previously incised.

rip current—fast-moving seaward trending currents along coasts, frequently off points of land or jetties.

ripplemarks—small, elongate sequences of low ridges induced by air or water currents on unconsolidated sediment.

river—stream of water flowing in a channel as part of a drainage net.

rochés moutonnées—rounded knobs of rock

shaped by glacial ice; from a distance, they look like reclining sheep.

rock cut terrace—low bench paralleling a stream with a rock core; see terrace.

rock flour—finely divided rock chips produced by the grinding action of glacial ice.

rock glacier—lobate mass of large, angular rocks; see block field.

saltation—movement of sediment particles by bouncing.

salt fretting—growth of salt crystals in the outer surface layer of a rock to induce its disruption.

salt wedging—pressure induced by salt crystal growth between other mineral grains.

sand waves—term used by some for horns of beach cusps.

sandstone—sedimentary rock of particles between 2 and 1/16 mm.

sapped cliff—cliff with a relatively resistant rock ledge that has been undercut, often by spring flow or eolian blasting.

saprolite—a thoroughly decomposed, *in situ* weathering residue approximating a climax soil in the tropics; see soil.

scar—a mark left by a previously active erosional agent (e.g. landslide, meander).

scarp (escarpment)—an elongate steep slope, especially one separating lower-inclination land surfaces; see free face.

scarp retreat—recession of an escarpment, usually by processes of pedimentation.

schist—foliated, micaceous metamorphic rock; see Chapter 3.

scree—loose rock debris strewn down a slope; see talus.

sea—a shallow body of water, at least partly saline, broadly connected with the ocean on one side (shelf), land locked (epiric), or extensively inland but connected (epicontinental).

sea-floor spreading—addition of new crustal material at a mid-ocean ridge accompanied by lateral movement of adjacent crustal plates.

sediment—material being moved or deposited by such geologic agencies as wind or running water.

sedimentary bypassing—removal of the more readily shifted detritus (fine?) leaving a residue (coarse?).

sedimentary lag—effectively accelerated removal of a weathered residuum from a geomorphic region by the agencies of a subsequent environment.

seif dune—see longitudinal dune.

selva—topography entirely reduced to slopes between narrow ridges and channel floors at the bottoms of V-shaped valleys, generally under humid conditions unless the materials being eroded are fine grained and poorly consolidated.

semiarid (geomorphic environment)—a transitional condition between humidity and aridity, generally vegetated by grassland (subhumid), where it is an environmental end-member it is geomorphically most important in alternations with aridity.

semiarid (geomorphic system)—landform-making association of processes involving intermittent stream flow, partly exhoric drainage, continuous grassland groundcover, and soil development in association with some chemical activity.

semiarid (meteorologic climate)—a climate type commonly defined as receiving between 10 and 20 inches (25–50 cm) of precipitation; it is characterized by grassland vegetation and termed a steppe by many geographers.

sense of movement (glacial)—the route along which ice moved without directional orientation.

serrate topography—incised topography with narrow, sharp ridges; see selva.

sheet erosion—erosion by water moving in sheets across bare ground.

sheetflood—a broad shallow sheet of water moving over a planar surface; usually ephemeral.

sheeting—the splitting away of thick rock layers, roughly parallel to the surface along fractures commonly due to unloading stresses.

shelf (continental)—the presently submergent continental margin inclined away from the land about 1°–2° and composed mainly of sediment in varying states of induration.

shield—a large area of exposed crystalline rocks; part of a craton.

shield volcano—large volcano with gentle slopes composed largely of lava, usually basaltic.

shingle—coarse beach gravel of flattened clasts.

shooting flow—a form of high-turbulance water flow; see Chapter 7.

sialic—pertaining to rocks high in silica and aluminum and common to continents.

silcrete—a form of duricrust rich in silica.

silk (pl. slouk) dune—a sand dune form of the Arabian region; see Chapter 6.

sill—tabular, concordant igneous intrusion.

siltstone—sedimentary rock of silt size particles, 1/16–1/256 mm, much may be lithified loess.

simatic—rocks rich in silica and magnesium; oceanic crust.

sink—an opening, commonly ovate, into a cavern from the ground surface due to solution.

sinuosity—in a stream channel, the ratio of thalweg length to valley length or flow distance in absence of a valley.

slate—metamorphosed mudstone with pronounced cleavage; see Chapter 3.

slip-off slope—the usually more gentle slope of the inner side of an incised river bend.

slough—low, commonly linear boggy area, usually along a river floodplain, often an abandoned flow route or flood-scour depression.

slump-block—a portion of the crust (commonly regolith) that has subsided under gravity and commonly has rotated along a shear surface.

snow line—lower limit on a mountain peak to which snow cover persists all year.

soil—an "in-place" weathered residue developed primarily through chemical decomposition and organic activity usually under plant cover and often zoned.

soil stripes—elongate zones of fine-grained material trending downslope in areas of permafrost-patterned ground.

soil (tropical)—one of the lateritic or bauxitic weathering residues.

soil zone—one of several possible discrete layers of a soil reflecting pedologic processes.

solifluction—slow, downslope movement of water-saturated regolith, often in association with frost action or wetting and drying.

solute—a dissolved substance.

solution (aqueous)—water in which a substance (solute) is dissolved.

spalling—see exfoliation.

spatter cone—a usually small volcanic cone build up steeply by spattered lava.

spit (hook)—elongate sand deposit extending down-current from a headland; a hook is a curved "spit."

spheroidal weathering—weathering propagated inward from fracture surfaces in jointed bedrock to produce rounded residual centers; see core stones.

stalactite—an elongate dripstone deposit attached to the roof of a cavern.

stalagmite—an elongate dripstone deposit built up from the floor of a cavern.

star (oghurd) dune—ovate dune with many radiating ridges; see Chapter 6.

stock—comparatively small, rounded (plan view), igneous intrusion.

stone circle—repeated rounded regolith pattern related to frost sorting.

stone net—frost-sorted ground in which large clasts are in geometric patterns.

stone polygons—patterned ground in which stone sorting is polygonal.

stone stripes—elongate rocky bands extending downslope in areas of solifluction or gelifluction.

storm surge—the piling up of water on a windward shore during a storm, accentuated by high waves and high tides.

stoss—upflow side of a glaciated knob.

strain—a change in the form or bulk of matter due to stress.

stream—a small river.

stream order—a numbering system for branches of a drainage net; see Chapter 5.

stream piracy—diversion of a drainage line by headward erosion of a stream in another drainage basin.

streamflood—a large discharge of water down a normally dry watercourse in a desert.

streaming flow—a form of fluid movement; see Chapter 7.

stress—a force exerted between two bodies proportional to the strain induced.

striation (glacial)—a scratch made by an ice-borne rock fragment.

striped soil—see soil stripes.

structural plain—a plain in which major parting surfaces are parallel to the land surface.

strudel—surface drainage of fresh water through a hole in sea ice.

subglacial—the basal portion of a glacier; also, rock debris carried in that position.

submarine canyon—submerged notch cut in the margin of a continental rise, some by density currents, others during ocean basin desiccation.

subsequent drainage—drainage developed by headward erosion of tributary streams in a drainage net.

superimposition—drainage initiated on a cover-mass incising downward onto pre-existing rock structures.

supraglacial—a position on top of a glacier; much alpine glacial debris is carried in this manner.

surficial—having to do with the Earth's surface, or, indeed, any other.

swash marks—curved marks made on a beach by a breaker's surge and backwash.

swell—average, day-to-day waves at sea or advancing on a coast.

swelling—see frost heaving.

syenite—light-colored, coarse-grained igneous rock lacking quartz; see Chapter 3.

tableland—large, flat-topped landforms; see mesa, plateau.

taiga—northern coniferous forest; see Chapter 5.

talus—loose rock debris at the base of a slope.

talus slope—a slope developed on a talus.; see footslope.

turbidity current—a subaqueous current of high density because of sediment content (cf. density current).

tarn—a lake in a cirque basin.

tectonic denudation—gravity sliding of large rock masses from a rising orogenic zone.

tectonic relief—Earth relief due to crustal movement.

tectonism—structural and igneous activity; see diastrophism.

terminal (end) moraine—ridge of debris deposited at the line of farthest glacier advance.

terrace (marine)—surface of erosion or deposition formed along a coast by wave erosion and deposition.

terrace (stream)—elevated portions of alluvial fills, rock-cut benches, or other planar features along a stream valley.

terra rossa—residual red-colored soil produced by solutioning of carbonate rock.

terrigenous coast (disequilibrium type)—a coast composed of inorganic materials, for the most part, which is also highly irregular in shape.

terrigenous coast (equilibrium type)—a coast mainly composed of inorganic materials, straight to curvilinear.

thalweg—the deepest part of a stream channel.

thalweg sinuosity—lateral undulations in the thalweg.

thermal (heat) equator—terrestrial zone of greatest solar heat influx.

thermokarst—terrain common to permafrost areas due to freezing and thawing.

thufur—see turf hummock.

tidal bore—a tidal advance into a narrowing inlet expressed on the water surface by turbulence or steepening.

tidal current—current induced by changing tides.

tide—diurnal shifts in water level induced by attraction of extraterrestrial bodies in the solar system.

till—non-sorted and non-stratified ice deposits.

till plain—a plain composed of till, commonly with a swell-swale topography.

time-dependent landform—a landform not in morphologic equilibrium with its environment and hence changing at a measureable rate.

time-independent landform—a landform changing so slowly that measurement of change is sensibly impossible but theoretically probable in terms of occurrence; see climax landform.

toughness—rock resistance to weathering and erosion (particularly physical effects).

traction load—sediment being transported by rolling or bouncing.

trans-environmental river—a river that flows out of its parent environment into another (i.e. humid to arid).

transform fault—a fault developed transverse to a mid-ocean ridge due to crustal emplacement; see Chapter 3.

transpiration—giving off of water to the atmosphere by plants.

transverse drainage—drainage lines flowing across the "grain" of the bedrock in terms of partings or structures.

transverse (cross) profile—the outline drawing of a feature at right angles to its long dimension.

transverse (cross) section—portrayal of the details that would appear in a plane section of a feature.

travertine—a $CaCO_3$ deposit usually formed by streams or springs which are agitated and lose CO_2; the material often exhibits a cellular structure, since the precipitation surface is often vegetal.

trellis (drainage)—a series of streams coming at essentially right angles on either side of a drainage line.

trench (oceanic)—an elongate ocean deep characterized by intense seismicity and andesitic volcanism; see Chapter 3.

troposphere—a lower, meteorologically defined layer of the atmosphere; see Chapter 4.

tsunami—a seismically induced sea wave.

turf hummock—a mossy knoll developed in boggy ground through frost action, fluid filled in summer.

tussock—see turf hummock.

unconformity—a buried erosion surface.

underfit (stream)—the opposite of a flood; a stream which by some criterion is smaller than the one that eroded the enclosing valley.

undertow—colloquial term for rip current or for wave backwash which tugs at a swimmer's feet.

uniformitarianism—the concept that the present is the key to the past; see Chapter 1.

upthrusting—orogeny characterized by high-angle reverse faulting.

upwelling—movement toward the surface of cold, deep, and bottom waters in near-equatorial zones of oceans.

vadose (zone)—the sporadically saturated upper zone of the lithosphere above the water-table.

valley glacier—a glacier occupying a valley; see alpine glacier.

valley train—glacio-fluvial deposits spread down a valley by meltwater.

varves—paired, light- and dark-colored layers of lacustrine sediment common to pro-glacial lakes.

ventifact—a rock faceted by wind-blown sand.

vertical accretion—the building up of a sedimentary deposit.

wadi—dry desert watercourse, particularly in Arabia; see arroyo, oued.

waning slope (toe slope)—the lower segment of a hillslope; see Chapter 2.

ware base (erosional)—the depth to which waves can plane bedrock or build up deposits.

water table—the top of the zone of saturation.

wave refraction—a bending of wave trains where their orbiting water particles encounter bottom.

waves of oscillation—open ocean waves whose particles of water follow uninterrupted orbital paths.

waves of translation—waves of oscillation that acquire lateral motion through water-particle contact with the bottom.

waxing slope (summit; shoulder)—the crest of a hill or escarpment above the free face.

weather—day-to-day conditions of the atmosphere.

weathering—decomposition of the lithosphere due to exposure to the atmosphere.

weathering residual—weathering by-products on essentially horizontal surfaces which are essentially unmoved.

weathering rinds—surficial layers of a rock separated during weathering; see exfoliation, spheroidal weathering.

wet blasting—sand blasting in an aqueous environment.

wetted perimeter—the line of contact between the stream channel wall and bed and the water.

wildflysch—chaotic sedimentary deposit probably developed during tectonic denudation or subduction.

xerophytic—plants common to desert regions capable of surviving on low-moisture budgets.

yardang—grooves and intervening pinnacles eroded in deserts by the wind.

zastruga—grooves eroded in the snow by eolian action.

zone of aeration—upper zone of the lithosphere through which water percolates to reach the water table.

zone of fracture—outer portion of the Earth in which materials yield to stress by fracturing.

zone of saturation—that portion of the zone of fracture which is below the water table.

zone of weathering—that portion of the lithosphere directly subject to atmospheric alteration.

Index

Page numbers in boldface type include illustrations

Abbe, C. J., 570, 585
ablation, 450, 453
ablation till, 488, **489**, 490
Abrahams, A. D., 295, 307
abrasion, 17
Ab'Sáber, A. N., 69, 101, 243
Absaroka Mountains, **525**
abyssal plain, 72, 107, 122
accelerated erosion, 39, 680–686
accordant summits (gipfelflur), 13, 14, 54, **55**, 634
Ackers, P., 413, 414, 444
active geologic substrate, 78–85
active layer, 518
adhesion (molecular), 427
adiabatic (gap), 221, 236
Adirondak Mountains, 147
Adriatic Sea, 281
aerial photos, 47, 48, 59, 69
Afar (Rift) Triangle, 186, **189**, 198, 541, 545
Afghanistan, 169, 170, **171**, 612
Africa, 12, 18, 19, 20, 30, 35, 53, 57, 59, 69, 70, 111, 126, 129, 157, 164, 172, 176, 186, **188**, 199, 201, 221, 223, 240, 241, 242, **243**, 244, 246, 260, 267, **268**, 277, 301, 303, 313, 328, 329, 349, 353, 360, 371, 372, 375, 382, 383, 400, 401, 421, 423, 431, 435, 437, 440, 453, 538, 575, 576, 624, 629, 641, 654, 669, 672, 673, 690
Agades, 329
Agassiz, L., 17
agent (geomorphic), 18, 24, 26
aggradation, 5, 85, 87, 91, 92, 93, 94, 95, 96, 97, 98, 99, 340, 341, 342, 343, 345, 346, 347, 348, 349, 394, 395, 396, 397, 602, 603, 604, 609, 654, 672

aggradational plain, 53, 54, 302–306, 397, 654, 657, 666, 667
A.G.I., 677, 693
air-photo mosaic, 361
akle (dune), 354
Alabama, 425, 668, 679, **684**
Alaska, 463, 471, 481, 492, 495, 519, 526, 552, 597, 692
Alberta, 459
Aletsch Glacier, 456
Aleutian Trench, 122
Algeria, 3, 20, 31, 70, 172, 313, 315, 317, 325, 351, **353**, 355, 358, 359, 360, 369, 370, 371, 372, 384, 564, 601, 632, 641
Alice Springs, 16, 340
allochthon, 185–195, 595, 604, 606, 645
allometric growth, 32, 38
alluvial fan (*see* fan)
alluvial plug, 403, **408**
alluvial terraces (*see* terraces)
alluviated channel behaviors, 414–426
alluvium, 10, 19, 21, 48, 51, 62, 65, 71, 74, 75, 81, 86, 87, 88, 90, 91, 92, 94, 95, 96, 98, 99, 154, 162–165, 311, 396, 399, 400, 602, 605, 606, 608, 610, 615, 682
Alpha (Rise) Ridge, 115, 119
alpine glacial landforms, 471, 615–617
alpine glacier, 471, 615, 617
Alps (Swiss), 79, 178, 189, 194, 449, 457, 459, 470, 471, 526, 609, 617
Altai Mountains, 624
alternating arid-humid geomorphic systems, 381–444
altiplano, 139, 200, 593, 612, 614, 616
Altithermal, 31
Amargosa Desert/Mountains, 190, 318, 602

Amazon River/Basin, 152, 293, **420**, 437, 438, 440, 555, 631
Amer. Comm. Strat. Nomenclature, 95, 97, 101
amphitheater valley head, 301, 303
anabranch (*see* distributary)
Anabunga River, **417**
anastomosing (pattern), 60, 61, **655**
anastomosing channel system, 435, 436, 437, 439, 440, 441, 654, **655**, 656
Ancestral Rockies covermass and drainage, 625–631
ancient landforms and landscapes, 640–674
Ancon Group, 184
Ancon Point, 184
Andel, T. van, 424, 444, 596, 638
Andes Mountains, 46, 89, 91, 92, 93, 138, 176, 177, 192, 195, 199, 200, 222, 234, 239, 243, 246, **254**, 268, 269, 303, 404, 481, 591, 593, 594, 598, 599, 600, 601, 602, 603, 604, 605, 606, 607, 608, 609, 616, 617, 631
andesite, Chapter 3, *see also* pp. 134, 135, 195
Angel Falls, 144
angular unconformity, 643
anhydrite, 85, 169, 172, 672
annular (drainage pattern), 60, 299
Anon., 412, 444
Antarctica, 34, 111, **197**, 216, 238, 315, 452, 453, 461, 471, **515**, 521, 522, 526, 527, 590, 668, 672
Antarctic ice sheet, 450, 458, 461, 516, 521
antecedent (drainage), 84, 617, 619, 621, 632
anticline, 181, 621, 622
Antrim (Thulean) Plateau, 201, 204
Appalachian covermass and drainage, 621–625
Appalachian Mountains, 46, 54, 106, 178, 191, 222, 245, 270, 271, 307, 394, 477, 498, 502, 592, 595, 609, 617, 618, 619, 620, 622, 623, 624, 635, 652, 662, 684
Apure River, 436, 437
Arabia, 1, 2, 18, 20, 126, 349, 353, 354, 372, 375, 648
Araguaya River, 441
Arapin Salt, 68
Arauca River, 436, 437
Arctic (region/climate), 46, 272, 692
Arctic Ocean (Basin), 119, 451, **522**
arete, 476
argillite, Chapter 3, *see also* pp. 596, 599, **601**
arid (meteorologic climate), 13, 15, 17, 18, 20, 21, 32, **85**, 91, 310
arid (cryergic) environment, 91, 517–529
arid (geomorphic) environment/system, 21, 27, **85**, 91, **98**, 310–377, 381, 599, 610, 642, 643, 656–659, 673, 680
arid Geographical Cycle, 13
arid geomorphic parameters, 312
arid geomorphic relicts, 27, 91, 98, 362–377, 399–401, 540, 681
Arid geomorphic systems and landforms, 310–377

arid-humid continuum, 381–385
arid-humid deranged drainages, 426–441
arid-humid orogenesis, 604–606
arid orogenesis, 601–604
arid-to-humid channel redevelopment, 406–414
arid-to-humid semiaridity, 398–399
arid weathering and mass wasting, 316–326
Arizona, 3, 49, 53, 156, 165, 202, 213, 237, 284, 333, 334, 366, 368, 369
Arkansas, 9, 10, 29, 30, 31, 48, 50, 51, 58, 74, 75, 76, 77, 78, 157, 159, 162, 163, 167, 168, 183, 239, 256, 390, 394, 395, 396, 397, 403, 411, 647, 648, 649, 650, 651, 653, 654, 655, 656, 658, 668
Arkansas River, 143, 383, 414, **490**
Arkansas Valley, 74, 75
arkose, 156, 321, 623, 624, 625, 626, 627, 628, 629, 669
arkosic, 19
Arolla Glacier, 479
arroyo, 62, 164
Arroyo de los Frijoles, 65
ash (tephra), 25, 62, 198, 608
Ashraf, A., 556, 587
Asia, 2, 112, 126, 223, 240, 277, 383, 401
Asmari Limestone, 84
Asquith, 422
assimilation, 23
Atacama Desert, 18, 246, 362, 364, 604
Atiu Island, **584**
Atlantic Ocean (Basin), 115, 116, 118, 121, 122, 126, 353, 387, 438, 549, 553, 570, 609, 623, 673
Atlas Mountains, 172, 369, 384, 601, 609, 632
atmosphere, 40, 42
atmosphere–hydrosphere (interaction), 218–225
atmospheric circulation, 217, 218
atmospheric moisture transfer, 220–225
atoll, 199, 582, 583, **584**
attrition, 560
Atwater, G. I., **170**, 205
Atwater, T., 119
Australia, 8, 16, 111, 160, 221, 223, 240, **241**, 260, 277, 325, 333, 339, 340, 343, 345, 357, 364, 374, 389, 397, 400, 401, 417, 426, 431, 435, 440, 466, 537, 545, 575, 580, **582**, 631, 654, 669, 672
Avery Island (salt dome), **172**
Axelrod, D. I., 632, 636
axial valley, 115, 118, 119, 121
Ayres Rock, 16, 339, 340
Azucar Group, 192, 193

back swamp, 411
backwash, 550, 553, **554**
Bader, H., 456, **457**, 531
Badger Wash, 314
Badland National Monument, 337, 338
badland topography, 336, **337**, 338, 605
Bagnold, R. A., 350, 353, 377, 578, 585
bajada, 69, 71

Bajorunas, L., 539, 585
Baker, 188
Baker, B. R., 512, 513, 531
Bakhtiari Conglomerate, 618, 620, 628
Bakhtiari Mountains, 84
Bakhtiari River, 84
Balk, R., 139, 205
Ballard, J. A., 172, 205
Baltic Sea (coast), 545
Bandy, O. L., 35, 43
Bangor, 146
bank (stream channel), 53, 62, 680
bank-full (stream), 417
bar (channel, mid-bay, bay head, bay mouth, off-shore), 65
Baraboo Quartzite, 652
Barbados, 537, 539, 562
barbed tributary, 22, 436
barchane dune, 354, 356
Barinas (state of), 184, 246
Baron, J. G., 34, 44, 110
barrier (beach, island, bar), 3, 572, 573, 574, 575, 576
barrier reef, 582
Bartholomew, B., 385, 444
Bartholomew, J., 229, 248, 441, 444
Barton, D. C., 170, 206
basal slippage (glacial), 457
basalt, Chapter 3, see also pp. 57, 81, 82, 107, 108, 109, 115, 133, 135, 136, 139, 162, 181, 195, 197, 279, 283, 388, 464, 540, 563, 608, 619, 623, 672
base level (local), 7, 331–343, 344–349
base level (regional), 4, 7, 10, 14, 15, 46, 80, 85, 88, 92, 594
base leveling, 7, 88
Basin and Range Province, 48, 162, 186, 187, 190, 199, 201, 330, 366, 368, 376
basin (cirque), 471
basin (drainage) (see drainage)
basin (sedimentary), 69, 94, 152, 154, 598, 599, 600, 601, 605, 606
Basutoland Plateau, 57
Bates, J. D., 537, 588
Batesville, 157
Batesville Formation, 78
batholith, 131
bathymetric map, 62, 63, 120
Baulig, H., 517, 531
bauxite, Chapter 5, see also pp. 154, 275, 277, 279, 668
bay deposits, 568, 569, 570
Bay of Bengal, 632
Bay of Fundy, 555
Bayrock, L. A., 489, 531
beach, 3, 548, 554, 576, 662
beach cusp, 554, 576, 577, 578, 579, 580
Beaufort Sea, 519, 520
Beaumont, P., 328, 333, 377
Beckinsale, R. P., 43, 86
bedding, 65, 79, 483, 484, 640, 641
Bedford Limestone, 27

bed (traction) load, 29, 404
Belo Horizonte, 301, 302
Beloussov, V. V., 185, 206
bench, 34
Benguela Current, 219
Beni Abbés (oasis), 353, 369, 370, 371
Benioff, H., 111, 125, 206
Bennett, H. H., 229, 230, 248, 257, 262, 265, 307, 331, 391, 444
Benue River, 438
Berg, R. R., 153, 206
bergschrund, 471, 473, 474
Berkey, C. P., 646, 675
Bermuda, 270
Bernhardi, A., 17
Berry, L. R., 283, 284, 309
Biehler, S., 114, 209
Bigarella, J. J., 19, 27, 33, 34, 37, 38, 43, 69, 70, 94, 96, 97, 98, 99, 101, 238, 243, 249, 507
Big Pine Creek Canyon, 331
Big Sur (coast), 539
Bird, E. C. F., 426, 429, 444
Bird, J. M., 127, 174, 176, 177, 178, 179, 180, 185, 187, 191, 206
Bishop, B. C., 470, 531
black duster, 387
Blackett, P. M. S., 108, 206
Black, R. F., ix, 47, 83, 90, 101, 102, 462, 516, 517, 518, 522, 527, 531
Black Forest (horst), 186, 188
Black Hills (uplift), 150, 163
Black Volta River, 305
Blackwelder, E., 307, 317, 377, 521, 531
block fields, 524
Bloom, A. L., 543, 569, 585, 634, 636
blow-out, 358
Bloyd Formation, 649–651, 665
Bloyd Mountain, 76
Bloyd-Winslow unconformity, 649–651
Blue Ridge Mountains, 186
blueschist (facies), Chapter 3, see also pp. 176, 185
Boat Mountain, 76
Bolivian Plateau, 8, 593, 612
bolson, 360
Bonatti, E., 186, 206
Boomer Beach, 560, 561
Boone Formation, 78, 162, 648
Bormann, F. H., 265, 309
Born, K. E., 646, 675
Born, S. M., 419, 421, 444
Borneo, 597
Bornhauser, M., 170, 206
Boston Mountain Plateau, 77, 648
Boston Mountain Rise, 58, 74, 77, 78, 89
Boston Mountains, 58, 74, 76, 77, 78, 159
Boswell, K. C., 13, 45
Bosworth, T. O., 377, 602, 636
boulder clay, 279, 604
boulders, 51, 149, 151, 388, 403, 410
Bowen, N. L., 107, 207, 270

Bowman, I., 238, 249, 386, 444, 528, 531, 602, 631, 636
Bowser, C. J., 522, 531
box canyon, 62
Bradley, W. H., 492, 549, 585
Brahmaputra River, 632
braided stream, 19, 60, 61, 410, 415, 435, 655, 658
Branner, J. C., 315, 316, 318, 319, 378
Braun, L. E., 516, 531
Brazil, 19, 22, 98, 99, 186, 244, 301, 302, 315, 316, 319, 436, 441, 669, 672
Brazil Current, 219
breakers (*see* spilling, plunging)
Bretschneider, C. L., 549, 552, 585
Bretz, J. H., 90, 101, 300, 307, 435, 436, 440, 441, 512, 531
Bridge, J., 668, 675
Broecker, 41
Brooks Range, 579
Brosset, D., 354, 378
Brown, J., 692, 694
Brown, J. H., 244
Brückner, E., 489, 490, 533
Bruder, K. F., 519, 534
Brunn, P., 579, 585
Bryan, K., 13, 43, 344, 378, 693
Bryce Canyon, 76
Büdel, J., 35, 43, 86, 87, 90, 101, 238, 249
Buffalo River, 9, 10
Bull, W. B., 341, 343, 378
Bullard, F. M., 81, 101
Bumpus, D. F., 570, 585
buried Paleozoic landscapes of the Ozarks, 646–651
Burma, 606
Butler, B. E., 305, 307, 397, 444
Butler, E. A., 35, 43
butte, 166, 651
Butzer, K. W., 35, 43, 238, 240, 248, 302, 349, 378

Caddo Gap, 183
Cady, J. G., 271, 307, 309
Cahir, J., 225, 250
Cailleux, A., 19, 26, 45, 103
calcrete (caliche), 67, 323
caldera subsidence, 199
caliber (load clast size), 404
caliche (*see* calcrete)
California, 19, 56, 95, 187, 319, 322, 332, 348, 351, 362, 388, 390, 396, 464, 478, 480, 542, 549, 551, 555, 560, 561, 633, 686
California Current, 220
Canada, 80, 81, 129, 130, 133, 148, 149, 167, 267, 272, 458, 459, 460, 473, 477, 498, 499, 500, 516, 520, 522, 528, 545, 572, 652, 653, 656, 658
Canadian Shield, 656
Canary Current, 220
Canary Islands, 133

Cane Hill Member (Hale Formation), 648–649, 658
canga, 324, 339
capacity (sediment load amount), 4, 5, 331
Cape May Point, 289, 553, 554, 686, 687
Caribbean, 38, 41, 73, 537, 632
Carlesbad Caverns, 301
Carlston, C. W., 5, 43, 418
Carol, H., 462, 463, 464, 531
Carolina Bays, 437, 438, 545
Carolina Capes, 424
Carolina Coast, 545, 570, 571, 572, 573
Carrizo Plain, 56
Carroll, Lewis, 343
Cascade Mountains, 478, 617
Case, J. E., 651, 675
case hardening, 319, 320, 321, 521
Caspian Sea (coast), 545
Casgville-Tradewater Group, 653, 655
catastrophism, 1, 2, 39
Cathkin Peak, 57
cavern (*see* karst)
cavernous weathering, 521
Celebes, 597
Cenomanian, 84
centripetal drainage (pattern), 19, 22, 61, 437, 441
Cerame-Vivas, M. J., 570, 586
Cerro Bolivar, 149
chain lakes (*see* alpine glaciation)
chalk, 58
Challenger Knoll, 172
Chamberlin, R. T., 463, 531
Chamberlin, T. C., 462, 463, 468, 489, 531
Champua-Keonjhar Upland, 151
channel (stream), 7, 8, 15, 53, 62, 65, 69, 88, 406, 407, 655, 681, 682, 684, 685, 686
Channeled Scabland, 440, 441, 512, 513, 514
channel manipulation, 682–685
channel piracy, 436, 437
Chapman, 374
Chapman, C., 199, 207
Chapman, R., 199, 207
Chappell, J., 38, 45
character and maintenance of mountainous relief, 592–594
Chari (Shari) River, 241, 303, 431, 438, 655
Charlesworth, J. K., 35, 43, 239, 249, 451, 486, 504, 531
Charlton, F. G., 413, 414, 444
Charpentier, J. de, 17
chemical (decomposition) weathering, 19, 21, 269–281, 318–324
Cherkauer, D., 53, 333, 334, 340, 341, 378
chernozem, Chapter 5, *see also* pp. 274, 275
Cherokee Erosion Surface, 504, 505, 506, 507
chert, Chapter 3, *see also* pp. 58, 75, 154, 161, 162, 163, 175, 181, 183, 319, 320, 394, 395, 647, 648, 649, 652
Chief Mountain, 194
Chien, Chien-Wu, 250

Chile, 223, 239, 246, 481, 545, 601, 604, 607, 631
Chin, A., x, 166, 208
China, 655
Cholley, A., 25, 43
Chorley, R. J., 26, 32, 43, 86, 317, 334, 341
Christiansen, R. L., 102, 208
Cima Dome, 348
Cimmeron River, 414
Cincinnati Arch, 153
cinder cone, 199
circo (*see* cirque)
cirque, 471, 472, 473, 474, 475, 476, 477, 478
cirque glacier, 472, 474
Clarindenfirn Glacier, 457
Clark, T. H., 126, 207, 486, 531, 547
clay, 51, 70, 595, 661, 681, 682, 683, 690
cliff (*see* scarp, sea cliff)
climate, 13, 15, 17, 18, 19, 20, 22, 26, 27, 32, 33, 34, 35, 36, 37, 38, 83, 86, 225
climates, geomorphic environments and interfaces, 225–231
climate zone, 86, 87, 668, 669, 670, 671, 672
climatic accidents, 4, 15, 20, 32, 35, 38
climatic ineffectualness (theory of), 17
climatic map/section, 87, 227, 258, 260, 452, 455
climatic nuclei (humid; arid; glacial), 38, 243, 244, 245, 246, 247, 362–376, 381, 448, 460, 597, 604, 678
climax (systemic expression), 39, 298, 362, 668, 681
climax arid morphogenesis, 362–376
climax deposit, 39, 362
climax erosional mode, 294, 295, 296, 375, 668
climax flora, 39, 266, 267
climax landform (landscape), 39, 42, 298, 362–376
climax soil, Chapter 5, *see also* p. 275
Cloos, H., 139, 188, 207
closed system (environmental), Chapter 1
closed system (sedimentary), 394
cloud cover (*see* climate)
coal, 659, 661, 665, 666, 667
coastal classifications, 541, 542, 543, 544, 545, 546
coastal erosion, 685–686
coastal geomorphic systems, Chapter 9, *see also* pp. 537–541, 643
coastal plain, 73, 606
coastal waves and currents, 546–556
coast ranges (Pacific), 617
coasts, 58, 59, 70, 536–585
Coates, D. R., 575, 586, 677, 693
Coats, R. R., 110, 207
coccoliths, 34
Cohen, S. B., 441, 444
cohesion, 427
col, 476, 477
Cole, F. W., 215, 249
Cole, M. M., 268, 307

Coleman, J. M., 424, 426, 428, 429, 430, 444
Coleman, P. J., 551, 586
Collins, J. B., 628, 638
collision-type mountains, 176–180, 589–639
colluvium, 30, 71, 98, 271, 681
Colombia, 22, 169, 187, 599
Colorado, 143, 150, 158, 165, 182, 259, 261, 269, 284, 314, 315, 364, 387, 479, 490, 625, 627, 628, 632, 642, 661
Colorado Plateau, 4, 8, 105, 165, 612, 614, 615
Colorado River, 4, 9, 165, 303, 425, 686, 689
Colorado River Delta, 686
Columbia Plateau, 201
Columbia River, 303, 510, 514
Columbus, C. 267
columnar jointing, 135, 201, 202, 204, 205, 206
competency (sediment load), 331, 332, 333, 404
composite cone (volcanic), 199, 200
compound landscapes, 39
cone (alluvial), 62, 69
cone (volcanic) 82, 198, 199
conglomerate, 162, 163, 605, 619, 620, 647, 648, 649, 650, 651–655, 656
Congo, 672
Connally, G. G., 400, 444
Connecticut, 569
Conolly, J. R., 198, 209
Conrad, C., 49
consequent (runoff-drainage), 48, 287, 617
contemporary environmental problems, 99, 100, 101
continent, 34, 42, 78, 83, 99, 125, 126–128, 608
continental accretion, 126, 127, 173–179, 180–185
continental (non-marine) deposits, Chapter 6, *see also* pp. 644, 660, 668, 669, 670, 671, 672
continental geomorphology, 125, 126, 127, 128
continental glaciation, 497–517
continental glaciers, 22, 32, 34, 497–517, 641, 669, 670, 672
continental rise, 126, 185
controls (environmental), 26
Cooke, C. W., 570, 586
Cooper, A. M., 686, 693
Cooper, G., 49
Cordillera de la Coasta, 72
Cordilleran Ice Sheet, 453, 512
Cordilleran-type mountains, Chapter 10, *see also* pp. 176, 177, 178, 179
Cordillera Occidental (Andes), 92, 243
Cordillera Oriental (Andes), 91, 244
core stones, 283, 284, 385, 388, 390
coriolis force (*see* Chapter 4)
Corps of Engineers (U.S.), 693
corrasion, 7, 560
corrie (*see* cirque)
Costeau, J., 593
Cotton, C. A., 4, 8, 18, 26, 27, 38, 43, 91, 199,

207, 305, 308, 310, 346, 372, 378, 398, 478, 486, 513, 528, **529**, 531, **541**, 544, **562**, 564, 586, 593, 594, 636
covermass (alluvial), 617, 618, 619, **620**, **621**, **622**, **623**, **624**, **625**, **626**, **627**, **628**, **629**, **630**
Covermasses and transverse drainages, 617
craig-and-tail features, 466, 467
Crandell, D. R., 330, 378
Crary, A. B., 481, 531
Craters of the Moon, 25, 81, 82, **201**
craton, 72, 73, 126, 129, 645, 646
creep (rock, rock glacier, talus, soil), 30, 31, 62, 523, 681
Cressy, G. B., 94, 101
crevass (deltaic-fluvial), 426, 430, 655
crevasse (glacial), 457
Crook, K. A. W., 596, 636
Crosby, W., 461, 531
Crow Creek, 684
crust, 80, 92
cryergic environment, 90, 91, 461, 517–529
cryergic geomorphic systems and landforms, 517–529
cryergic processes, 519–526
cryosphere (*see* cryergic environment)
Cuba, 545
cuesta, 166
Curray, K., 596, 638
cut bank (stream), 10
Cuzco, 138
cwn (*see* cirque)
Cyclothems, 629, 659–667
Cyprus, 119
Czech, H., 13, 45
Czechoslovakia, 41

Dake, C. L., 646, 675
Dakota Glacial Lobe, **508**, 509
Dalmation Coast, 545
Daly, R. A., 423, 444, 582, 586
Damuth, J. E., 19, 38, 43, 238, 243, 249, 451, 452, 531
Dan, J., 270, 308
Dana, J. D., 636
Daniels, R. B., **685**, 693
Dansgaard, W., 33, 34, 43
Darling-Murray (Murrumbigee) River, 241, 303, 305, 333
Dartmoor, 133
Darwin, C., 562, 582, 586
Darwin, G., 107
Da Silva, J. S., 34, 43, 101
Daule Basin, **598**, 608
Dasmann, R. R., 693
Davidson, C., 523, 531
da Vinci, L., 1, 7, 212
Davis, L. C., 31, 43, 239, 249, 267
Davis, M. B., 267, 308
Davis, W. M., 4, 5, 6, 7, 8, 10, 11, 12, 13, 15, 17, 18, 20, 21, 25, 27, 32, 33, 42, 43, 86, 88, 90, 101, 105, 136, 266, 285, 290, 293, 297,

299, 308, 312, 313, 339, 341, 346, 364, 375, 376, 378, 392, 398, 417, 419, 426, 444, 475, 476, 498, 592, 593, 594, 596, 617, 636
deaggradation, 402, 403, 404, **405**, 408, 411, 412, 413, 414
Dean, H. L., 511, 534
de Andrade, G. O., 238, 249
Death Valley, 18, 19, 315, 328, 332, 351
débouchér (multiple), 438
debris slope, 86
Deccan Plateau, 201
décollement, 189 (*see also* allochthon)
Deep Springs Valley, 396
Deffeyes, K. S., 114, 115, 119, 207
deflation, 32, 350, 355, 362
Deford, R. K., 308
De Geer, G., 495, 531
deglaciated coasts, 537, 538
deglaciation, 72
degradation, 402
Deitz, R. S., 108, **124**, 125, 126, 128, 153, 207, 546, **547**, 548
Delaware Bay, 553
Delaware River, 425
Delaware Watergap, **52**, 150
delta, 62, 67, 92, 423, 424, 425, 426, 537, **540**, 608, 653, 686
Demchinski, B., 519, 534
Demorest, M., 462, 464, 532
dendritic (drainage pattern), 60, 73, 288, 289, 654
Denny, C. S., 345, 378
denudation, 402
deposit, 27, 39, 41, 47, **51**
depositional landforms, 34, 48, 50, 66–78, 86
depositional plain (*see* aggradational plain)
deranged drainage, 19, 60, 61, 81, **241**, **242**, 303, **305**, 306, 398, 426–442, 498–500
Derbyshire, E., 471, 475, 477, 532
Desborough, G. A., 653, 675
description and measurement (in geomorphology), 47, 48
desert, Chapter 6, *see also* pp. 13, 18, 31, 38, 83, 597, 607, 641, 644, 645, 650, 664, 671, 681, 691
desert fluvial activity, 327–343
desert geomorphic setting, 311, 312
desert pavement (*see* lag gravel)
desert precipitation–temperature effects, 312–316
desert storms (accounts of), 328–331
desert varnish, 319, 320, 521
Detwyler, T. R., 677, 693
Devils Post Pile National Monument, **464**
Dewey, J. F., 127, **174**, 176, 177, 178, 179, 180, 185, 187, 191, 206
diabase, Chapter 3, *see also* pp. 133, 136, 141, 142, 148, 619
diagenesis, 99, 595
Diamantina Trench, 122
diaper, 170, **171**
diastrophism (*see* tectonism)

diatomite, 604
differential (weathering/erosion), 73
dike, 108, 131, 139, **140**, **141**
diorite, Chapter 3, *see also* pp. 134, 135, **137**
disequilibrium (environmental), 27, **28**, 38, 39, 386–390, 678, 680, 681, 682
distributary (channels), 60, 61, 303, 426, **428**, **429**, 430
Doering, J. A., **424**, 444
Dolan R., 544, 570, 571, 576, 577, 586
Doldrum Belt, 217, 220, 234, 235, 243, 262, 597, 610, 668
dolerite, 151
dolomite, Chapter 3, *see also* pp. 57, 158, 647
Dombrowolski, A. B., 517, 532
Donahue, J., 486, 533
Donn, W. L., 451, 532
Doria-Medina, J. H., 394, 395, 444
Dos Bocas Shale, 604
downwasting, 8
drainage basin, 12, 15, 26, 46, 84, 92, 288, 293, 612, 615
drainage (pattern) configurations, 59, 60, 61, 84, 287, 288, 289, 290, 291, 293, 304, 305, 392, 393, 395, 397, **402**, 404, 407, 411, 412, 413, 414, 415, 421, **425**, **426**, **427**, **428**, **429**, **434**, **435**, **436**, **437**, **439**, **440**, **441**, 492, **500**, **505**, **510**, **511**, **512**, 556, 613, 615, 618, 619, **654**, **655**, **656**, **657**, 684, 685
drainage density, 294
drainage derangement (*see* deranged drainage)
drainage divides (primary, secondary, etc.), 18, 58, **59**, 74, 612
drainage net (system), 2, 18, 19, 27, 46, 48, 59, 62, 80, 83, 88, 241, 285–294, 310, 345, **346**, **392**, **393**, **394**, 395–398, 504, **505**, 506, **510**, **511**, 605, 612, 615, 617–630, 631, 653, **654**, **655**, **656**, **657**, 658, 659, 673
drainage nets and hillslopes, 290–294
drainage patterns, 59, 60, 61, 80
drainage system development, 286–294
Drake, C. L., 174, **175**, 207
Drakensberg Escarpment, **57**, 59
Dresch, J., 91, 101, 528, 593, 601, 636
drift, 81, 94, 483, **484**, 486, 497, 504, **505**, 643
drift lithology and morphology, 485–497
dripstone, 67
drumlin, 502, **503**, **504**
dry frozen ground (*see* permafrost)
Duane, D. B., 539, 585
DuBoys formula, 288
dune (sand; clay), 31, 51, 62, 67, **69**, 94, 353–362, 399, **548**
Dunkard Group, 623
Dunn, A. J., 43
duricrust, 164, 323, 351, 384, 390, 681
Dury, G. H., xvii, 6, 7, 29, 32, 33, 43, 165, 207, 305, 308, 330, 334, 338, 344, **345**, 351, 378, 398, 401, 408, 417, 445, 555, 556, 586
Dust Bowl, 386, **387**
dust devil, 360
Dutton, C. E., 4

dynamic equilibrium (steady state), 26, 27, **28**, 29, 38, 39, **59**, 80, 212
dynamic relief, 79, 105, 106, 591, 592, 593, 594

Eardley, A. J., 129, 207
early orogenic denudation, 594–596
Early Pennsylvanian geomorphology of North American mid-continent, 651–659
earthquake (seismic) activity, 3, 110, 125
East Australian Current, 219
East Indies, 597
Easton, W. H., 649, 675
ecologite, 109
Ecuador, 155, **160**, **161**, 183, **184**, 187, **192**, **193**, 195, 221, 236, **255**, 267, 268, 301, 303, 400, 404, 556, 565, **567**, 584, 593, 597, **599**, **600**, 601, 604, 605, 607, 608, 610, 615
Egypt, 3, **388**, 673
El-Ashry, M. T., 294, 296, 308, 685, 697
elastic limit, 453
elb dune, 354, **361**
elevation, 8, 46, 54, 57, 596
Ellis, C. W., 569, 585
Ellsworth Mountains, **590**
El Niño (current), 220, 243
Elson, J. A., 486, 531
embayment (structural), 168
Emerson, J. W., 684, 693
Emiliani, C., 33, 37, 38, 43, 44
Emmett, W. W., 102, 250, 446
Endhaven (Netherlands), 279
endogenic topographic effects, 104–189
endoric, 311, 327–344, 699
Engeln, O. D. von, 34, 44, 547
englacial (load/position), 468, 489
England, 1, 39, **58**, 133
enneri (*see* oued; wadi)
entrenched meanders, **10**, **11**
environment (parent/formative), 22, 24, 26
environmental blend, 39
environmental climax, (*see* climax)
environmental complexity (history of), 677–680
environmental continuum, 381
environmental dominance, 42
environmental dynamism, 17, 39, 42, 99
environmental evolution, 26, 42, 678
environmental extract, 93
environmental geomorphology, 677–693
environmental passivism, 3, 17
environmental problems (modern), 101, 677–693
environmental residual, 93
environmental sequence, 42, **442**
environments of orogenic uplift, 596–609
environments, relicts, and (modern) disequilibria, 680
eolian, 42, 61, 70, 76, 80, 86, 89, 211, 311, 605, 610, 657, 672, 673
eolian geomorphic effects, 349–362

epeirogenic (submergence/emergence), 646, 647
epeirogenic continental settings, 128–168
epeirogeny, 106, 114, 120, 128
ephemeral (runoff), 19, 53, 295
epicontinental sea, 643, 661, 662, 663, 664, 665
Epstein, S., 34, 44
Equator, 221, 223
equatorial (zone), 215, 450, 668
Equatorial Counter Current, 220
equilibrium coastal segments, 545
equilibrium (stream) profile, 8
erg, 18, 20, 31, 32, 355, 359, 360, 361, 369, 399, 690
Erhart, H., 274, 275, 276, 277, 278, 279, 308
Eriksson, 496
erosion, 4, 71, 83, 680, 685, 686, 687
erosional coastal segments, 559
erosional landforms, 34, 74, 76, 79, 83, 86, 87, 559
erosion remnant, 13, 16, 53, 59, 74, 86, 388, 389
erosion versus deposition (in deserts), 331–343
erratic (glacial), 458
escarpment (see scarp)
esker, 491, 493, 494, 495, 496
Esmeraldas Basin, 598, 599, 600, 601, 608
estuarine (deposits), 595, 604
Ethiopia, 186
eugeosynclinal facies, 175, 176
Eurasia, 126, 260, 305, 439, 509
Eureka Springs Escarpment, 58
Europe, 1, 36, 37, 41, 79, 89, 90, 92, 152, 248, 270, 383, 401, 402, 501, 511, 512
evaporation, 17, 223
Everton Dolomite, 647
Ewing, M., 174, 207, 451, 532
exfoliation, 149, 318, 319, 320, 321, 521
exhoric drainage, 8, 88, 383
exhumed (landform/landscape), 98, 640, 641
exogenic environmental phenomena, 211
exogenic topographic effects, 104, 211
extrusion flow (glacial), 457

faceted spur, 481
facies (litho), 650, 665
Fahnestock, R. K., 332, 378
Faill, R. T., 623, 624, 637
Fair, T. J. D., 53, 101
Fairbridge, R. W., ix, x, 19, 38, 43, 53, 62, 102, 238, 243, 249, 271, 308, 319, 323, 378, 424, 445, 451, 452, 531, 546, 547, 548, 549, 559, 582, 586, 641
fan (alluvial), 48, 51, 62, 66, 69, 599
fanglomerate, 601, 623, 624, 604
Falkland Current, 220
fault, 50, 56, 57, 70, 74, 79, 83, 90, 138, 140, 540, 541, 645
fault-block mountains, 185–189
fault breccia, 645
fault scarp, 53, 140
Fayetteville Shale, 30

Feden, R. H., 172, 205
feedback (mechanisms), 26
felsenmeer, 162, 163
Fenneman, N., 59
Fenner, C. N., 398, 401, 445, 602, 613, 631, 637
feral (topography) (see selva)
Ferm, J. C., 570, 571, 576, 577, 586
ferricrete, 323
fetch (wave), 549
fill (alluvial), 69, 71, 86
Finland, 469, 500, 545
Finsterwalder, R., 463, 532
firn (névé), 449, 456
first- and second-order landforms, 34, 60, 61, 62, 63, 64
first-order climate changes, 35, 36
Fisher, R. L., 122, 207
Fisk, H. N., 426, 445, 492, 494, 532, 574
fissure, 81, 83, 136, 139, 540
fjord, 481, 482
Flandrian, 37
flat (topographic), 13, 50, 89, 345, 594
flatiron, 182
Flawn, P. T., 677, 693
Fleetwood, A. R., 421, 422, 446, 452, 492, 506
Flint, R. F., 35, 44, 94, 208, 238, 239, 249, 267, 268, 270, 271, 277, 280, 450, 451, 454, 455, 458, 459, 465, 466, 467, 468, 469, 471, 475, 481, 485, 486, 488, 492, 493, 494, 496, 502, 505, 509, 511, 512, 518, 532
flood, 3, 405, 406, 407, 655, 684, 686, 688, 689
floodplain, 411, 412, 686
floral blend, 267
Florida, 299, 425, 437, 537, 538, 545, 575, 582, 661
fluvial (stream) erosion/incision, 3 , 5, 15, 31, 46, 50, 53, 70, 73, 80, 85, 86, 88, 93, 211, 335, 336, 337, 338, 339, 598, 603, 610, 615, 631, 632, 633, 655, 666, 672, 673, 674, 681
fluvial relief, 15
fluvial terraces and deltas, 418–426
flysch, Chapter 3, see also pp. 84, 176, 180, 181, 184, 193
folds (structural), 60, 62, 74, 79, 83, 84, 541
foliated rock fabrics, 141–150, 154–167, 179–185
Folk, R. L., 354, 357, 358, 359, 378, 415, 445
Foose, R. M., 153, 207
foraminifera, 34, 41
Forbes, J. D., 318, 378
Forbes, R. H., 304, 308
forest, 80, 401
Forman, M. J., 170, 205
fosse (see kame terrace)
fossils, 33, 34, 70, 80, 86, 87, 93, 649, 650, 653, 665, 669, 671
Fountain Formation, 627, 628
Fox, P. J., 115, 122, 208
fracture, 73, 136, 141, 142, 167
France, 3
Frankel, J. J., 560, 586

free face (backslope), 49, 50, **52**, 86
Frenzel, B., 266, 308
Frey, D. G., 95, 103, 239, 250, 451, 486, 509, 535
Frezon, S. E., 646, 648, 651, 675
friable, 156
Friedman, G. M., 166, 207, 239, 249, 451, 532
fringing reef, **581**
frost cracking, 521
frost heaving, 519, 521
frost sorting, 521
frost thrusting (*see* frost heaving)
frost zone (atmospheric), 19
Froude Number, 433
Frye, J. C., 14, 37, 44, 69, 95, 97, 100, 102, 103, 301, 304, **485**, 517, 532, 535
Fugilister, F. C., 587
Furon, R., 349, 378
Fyles, J. G., 533

gabbro, Chapter 3, *see also* pp. 133, 135, 137
Gage, M., 404, 445
Gagliano, 428, 429, 430
Garbarini, G. S., 207
Gardner, L. R., 390, 445
Garner, H. F., 4, 8, 15, 19, 27, 28, 29, 35, 38, 44, 46, 71, 74, 78, 86, 87, 88, 91, 93, 102, 139, 154, 162, 207, 225, 229, 238, 240, 243, 244, 246, 249, 267, 269, 270, 277, 299, 308, 326, 346, 372, 373, 378, 382, 385, 386, 394, 395, 398, 399, 400, 410, 413, 414, 424, 431, 435, 445, 451, 498, 512, 528, 532, 593, 594, 596, 597, 598, 599, 601–604, 606, 607, 613, 624, 625, 629, 631, 637, 648, 656
Garrels, R., 547
Geiger, R., 216, 249
gelifluction, 524
general atmospheric parameters, 215–217
General Systems Theory, 24, 25, 26
Gentelli, J., 320, 324, 378
geochronology, 33, 34, **35**, 36, 39, 70, 80–94
Geographical Cycle (Davis), 4, 12, 15, 25, 27
geologic cross sections, 74, 75, 108, 109, 110, 112, 119, 121, 122, 129, 131, 170, 171, 174, 175, 177, 178, 179, 180, 185, 187, 188, 194, 594, 601, 605, 606, 611, 621, 630, 643, 644, 645, 648, 649, 650, 660, 665, 673
geologic substrate, 32, 47
geomorphic agency, 211
geomorphic area, 26, 83, 92, 229
geomorphic discordance, 700
geomorphic environment, 211
geomorphic interface, 212, 214
geomorphic maps, 62, 71, 203, 232, 242, 243, 244, 363, 370, 374, 375, 421, 429, 430, 462, 469, 492, 504, 508, 569, 571, 574, 598, 620, 623, 653, 654, 669, 670, 671, 688
geomorphic principles, 41
geomorphic problems and analysis, 46, 47, 80–83
geomorphic space-time concept, 246

geomorphic system, 24, 38, 39, 42, 47, 80, 211, 266, *see also* Chapters 3–10
geomorphic system attributes, 24–29, 42, 83, 86, 90, 231–237
geomorphic theory, Chapter 1, *especially* pp. 32, 33
geomorphic zone, 243
geomorphology, 46, 47
Georgia, 425, 668
geosyncline, 76, 126, 168, 174, 175
geotectonic theory, Chapter 3, *see also* pp. 32, 589–639, 640–674
geothermal gradient, 518
Gerhard, J. E., 172
German, K., 486, 532
Germany, 387
Gettysburg (Battlefield/Basin), 139, **141**, **142**, 624
Geyl, W. F., 556, 586
giant cusps (*see* beach cusps)
Giant's Causeway, 204, **205**
gibber (*see* lag gravel)
Gibbons, J. F., ix, x, 139, 152, 153, 166, 207
gibbsite, 154
gibbsite, 154
Gibralter, 194, **195**
Gignoux, M., 93, 94, 102, 164, 207, 326, 378, 601, 637
Gilbert, G. K., 4, 5, 7, 44, 106, 298, 394, 573
Giles, A. W., 648, 675
Gillberg, G., 488, 532
Gillett, R., 684, 693
Gilliland, W. N., ix, x, 9, 11, 30, 68, 120, **121**, **132**, 139, 166, 173, 182, **202**, 207, 368, **470**, **483**, 487, **493**, **559**, **563**, **565**, **566**, **570**, **643**
Ginsburg, 662
gipfelflur (*see* accordant summits)
Gippsland Lakes, 426, **429**
Girdler, 187
glacial environment, 21, 35, 39, 94, 449–459, 668
glacial episodes (stages), 35, 36, 37, 46, 47, 80, 90, 451–453, 673
glacial erosion-deposition patterns, 459–471
glacial geomorphic effects, 453–517
glacial geomorphic system, 94, 211, 453, 468, 643, 683
glacial geomorphic systems and landforms, 448–517
Glacial Lake Agassiz, 512
Glacial Lake Missoula, 513
glacial landform/deposit, 53, 63, 80, 89, 94, 540, 644
glacial steps (stairway), 478, 479
glacier, 17, 22, 86, 94, 448, 449, 450, 451, 453, **454**, **455**, **456**, **457**, **458**, **459**, **460**, **461**, **462**, **463**, **464**, **465**, **466**, **467**, **468**, **469**, **470**, **471**, **472**, **473**, **474**, **475**, **476**, **477**, **478**, **479**, 480, **481**, **482**, **483**, 488, 489, **493**, 495, 497, **511**, 515, 641
glacier budgets, 449–451
glacier directional indicators, 466–468
glacier erosion, 81, 94, 461–466

glacier map, 449
glacier movement, 453–459
Glacier National Park, 457
glacier surge, 457
glacier transport, 468–471
glacio-eustatic sea level changes, 7, 8, 37, 80, 542, 544, **557**, 646, 653, 659, 665, 666
glacio-fluvial, 460, 491, 615–617, 643, 658
Glass, H. D., 535
Glen, J. W., 459, 488, 532
Glen Eyrie Shale Member (Fountain Fm.), 627
Glen's Flow Law, 459
Glick, E. E., 646, 648, 651, 675
gneiss, Chapter 3, *see also* pp. 80, 81, 138, 146, 147, 465
Goddard, E. N., 273, 308, 627, 637
Goldsmith, V., ix, 546
Gomez, R. S., 309
Gomez-Pompa, A., 265, 268, 308
Gondwanaland, 659, 668, **669**, 670, **671**, 672
Goodlet, J. C., 400, 445
Gorda Axial Valley, 119
Gordon, M., Jr., 649, 659, 675
gorge, 289, 384, 616, 672
Gorycki, M. A., 415, 445
Goudie, A., 378, 453, 532
Gow, J. J., 34, 44
Grabau, A., 104, 106, 207
graben, 187, **188**, **606**, 608
graded slope, 29 (*see also* repose angle)
graded stream, 5, 6, 7, 15, 85, 88
Gradualism (Theory of), 2, 3, 39
Grahm, J. W., 209
Grand Canyon, 4, **156**, 264, 673
Grand Erg Occidental, 313, **359**, 360
Grand Tetons, 191, 472
granite, Chapter 3, *see also* pp. 23, 57, 80, 107, **134**, 143, **151**, 279, 284, 319, **322**, 323, 354, 388, 627
granodiorite, Chapter 3, *see also* pp. 133, **134**, 284
granular disintegration, 318, 321, 323, 521
grassland (steppe/savannah), 38, 401
gravel, 51, 65, 80, **90**, 98, 395, 595, 600, 602, 604, 605, 606, 619, **641**, 649, 652, 653, 654, 655, 656, 657, **658**, 672
Gravenor, C. P., 486, 532
gravity, 42, 83
gravity flow (glacier), 457
gravity tectonics, 184–195
gravity tectonics and topography, 189–195
gravity waves, 551
Gray, I. E., 570, 586
graywacke sandstone, Chapter 3, *see also* pp. 175, **183**, **184**, 608
Great Barrier Reef, 160
Great Dividing Range, 632
Great Lakes, 80, 128, 512, 652
Great Slave Lake, 167
Great Valley (of California), 632
Greece, 2

Green, J., 197, 207
Greenland, 1, 34, **35**, **41**, 46, **232**, 238, 461, 481, 482, 521, 523, 527
Greenland Ice Sheet, 449, 450, 461, 482, 516
greenschist, 142, **186**
Grenoble, 79
Groen, P., 552, 586
groove (glacial), **465**
Gross, D. L., 37, 44
groundcover, 451
ground moraine, 488
groundwater body, 71, **253**, 401
Grove, A. T., 19, 44, 240, **242**, 249, 305, 308, 315, 329, 339, 344, 347, 350, 355, 371, 372, 373, 374, **375**, 378, 399, 431, 438, 453, 532
Groves, G. W., 552, 586
Grüger, E., 266, 308, 516, 532
gruss, 324, 354
Guadalupe Mountains, 663
Guayaquil, 607
Guayas (Basin) Province, 192, 598, **604**, **605**, 606, 608
Guayas River, 608
Guevara, S., 308
Guiana Highlands, 307
Guilcher, A., 553, 576, **577**, 586
Guillier, Y., 517, 532
Guinea, 12, 244
Gulf Coastal Plain, 169, 390, 400, 539, 545, 686
Gulf Coast Embayment (U.S.) (*see* Mississippi Embayment)
Gulf of Aden, 541
Gulf of California, 686, **688**
Gulf of Campeche, 625, **626**
Gulf of Guayaquil, 608
Gulf of Mexico, 169, **172**, 222, 365, 397, 425, 575, 625, 661
Gulf of Suez, 186, 187
Gulf Stream, 39, 220, 538
Gulliver, F. P., 541, 586
gully, 34, 57, 62, 74, 289, 310, 398, 681, 685, 693
Günz, 90
Guppy, H. B., 584, 586
guyot, 107, 198
gyre (ocean current), **454**, **455**, 457, 548

Hack, J. T., 27, 29, 32, 44, 297, 308, 400, 445
Hackettstown, 161
Hadley, R. F., 315, 344, 378
Haefeli, R., 456, 531
Haffer, J., 38, 44, 243, 244, 249
Hale Formation, 648, 649, 651–659
Hammada, 351, 353, 355, 370, 372
Hammock Tidal Marsh, 569
Hammond, A. L., 112, 113, 207
hanging valley, 58, 480
Hansen, 302
hardpan, 274
Harlan, W. B., 486, 532

Harris, K., **601**
Harris, S. A., 32, 44
Hart, P. J., 207
Hartshorn, J. H., 48, 69, 102, 471, 489, 532
Haugen, R. K., 692, 694
Hawaii, 199, 203, 301, 537, 545
Hawaiian Islands, 153, 199
headlands, 59
headwall (cirque), 471, **472**
headward erosion (gully), 19
Heart Mountain Thrust, 193, **194**
Heezen, B. C., 61, 63, 64, 102, 114, 115, **116**, 118, 120, **122**
Heirtzler, J. R., 34, 44, 108, **110**
helical flow, 415, **416**
Hemisphere (Northern), 32, 34, **40**, **41**, 99, 216, 219, 239, **454**, **455**, **462**, 471
Hemisphere (Southern), 34, **40**, **41**, 91, 99, 216, 219, 239, 629
Henry, V. J., Jr., 570, 571, **572**, 586
Henry Mountains, 4
Hershey, O. H., 495, 532
Hess, H. H., 107, 108, 110, 113, 115, 119, 125, 198
Higgins, C. G., 73, 102, 299, 308
high montane environments, 610, 611, 612
High Plains (North American), 154, **164**, 234, 236, 259, 386, 431, 437, 619
hillslope elements, 14, 57
hillslopes, 8, 10, 12, 14, **15**, 27, 29, 30, 31, 48, 49, 50, 51, **52**, 53, **55**, **56**, **57**, 58, 59, 71, **72**, 73, 85, 86, **87**, 88, 97, 253, 254, 290, 291, 292, 396, 397, **400**, 661, 662, 680, 681, 682
Himalayan Mountains, 176, **178**, 591, 606, 612, 617, 624, 632
Hindu Kush, 612
historic predisposition (of environments), 26
Hobbs, H. G., 462, 475, 532
hogback, **85**, **182**
Holden, J. C., 108, 128, 153, 207
Hollister, 121
Holm, D. A., 354, 378
Holms, A., 104, 107, 108, **110**, 126, 188, 197, 203, 208, 270, 274, 277, 279, 281, 301, 553, 555, 559, 560, 564, 573, 576, 586, 591, **592**, 637
Holms, C. D., 486, 488, 491, 533
Hong Kong, 283, **284**
hook (see spit)
Hooke, R. LeB., 341, 379
Hook Mountain, 283
Hopkins, W., 107, 208
Horberg, L., **511**, 533
horizontal accretion, 426
horn (glacial), 71
Horse Latitude (deserts), 218, 315, 362, 668
horst (mountains), 187, **188**, **190**, 608
Horton, R. E., 46, 102, 287, 288, 289, 290, 291, 292, 293, 294, 295, 296, 299, 308, 327, 335, 336, 365, 375, 392, 406, 445, 504
Horton Number, 393
Hough, J. L., 512, 533

Howard, A. D., 24, 29, 44, 59, 102, 296, 411, 419, 445
Howell, B. F., 112, 116, 208
hoyo (see cirque)
Hoyt, J. H., 568, 570, 571, **572**, 573, 576, 586
Hsü, K. J., 619, 637, 672, **673**
Hubert, J. F., 386, 445, 623, 627, 628, 629, 637
Hudson, F. S., 632, 637
Hudson River, 132, 262, 425
Hughes, R., 456, 533
humic (soil zone), 252
humid climax drainage, 294–299
humid cryergic environment, 521–523
humid fluvial erosion, 285–299
humid geomorphic environment, 12, 27, 39, 251–263, 597–601, 656–659, 664
humid geomorphic parameters, 256–302
humid geomorphic relicts, 385, 386
humid geomorphic system, 27, 39, 75, 85, 98, 251–307, 642, 680
humid geomorphic systems and landforms, 251–307
humid meteorologic climate, 13, 15, 17, 18, 20, 21, 27, 30, 32, 39, 664
humid orogenesis, 597–601
humid relicts, 27, 680
humid river, 5, **12**, 75
humid-to-arid disequilibrium, 386–390
humid-to-arid semiaridity, 385
humid weathering, 269–285
Hummel, D., 494, 533
Hunkins, K., 119, 208
Hunt, C. B., 272, 308
hurricane, 3
Hussey, K. M., 324, 379
Hutton, J., 2, 3, 7, 18, 20, 38, 39, 41, 42
Hwang Ho (Yellow) River, 241, 303, 655
hydration, 270, 280
hydraulic action, 560, 561, 563
hydrography, 597, 609, 610, 612, 618, 631

Ibex Mountains, 348
ice, 42, 67, 74, 448, **457**, **458**
icebox effect, 451
ice burst, 306
ice-contact deposits, 483–490
Iceland, 108, **139**, **140**, 198, 199, 450, 483, 487, **493**, 563, 575
ice-margin lakes, 512, 513
ice-margin streams, 509–511
ice-scoured plain, 498, 499, 641, 643
ice sheet, 17, 669, 678
icing (stream), 519
Idaho, 25, 81, 82, 201, 681
igneous (rock/process), 10, 20, 22, 33, 72, 73, 82, 83, 114–120, 123, 125, 130–137, **139**, **140**, 141, 142, 195–206, 652
ignimbrite, 199
Illinesa, 200
Illinois, 37, 100, 387, 504, 653, **654**, 655, 657
illite, 154

Imperial Valley, 686, **688**
India, **55**, 107, **112**, 126, **151**, 201, 223, 325, 326, 339, 401, 466, 619, 624, 672
Indiana, 27, 387, 504
Indian Ocean, 63, 64, **112**, 114, **118**, 186, 438
Indochina, 606
inert (semi-insoluble) rocks and minerals, 270
infiltration, 223
inlet, 3
inlier, 394
inselberg (landscape), 13, 15, 16, 74, 325, 340, 347, 348, 366, 367, 368, 369
instruments (geomorphic/geologic), 48, 71
interface (geomorphic) (*see* geomorphic interface)
interference (rhomboidal) ripples, 289, 577
interglacial episodes (stages), 37, 90
intermittent (runoff), 59, 259, 294
intermontane (basin/deposit/environment), 81, 82, 83, 612, 613, 614, 615, 628
Iowa, 37, 84, 89, 386, 387, 486, 504, **505**, 652
Iran, 46, 83, 84, 85, **181**, 402, 572, 672
Ireland, 204, 205, 503, 504, 545
iron formations, 142
irrigation and related problems, 690–691
Isacks, B., 111, 123, 126, 208
island arc, 122–125
isohyat, Chapter 4, *see also* p. 702
isopach maps, **600**, **605**, 623
isostasy, 78, 79, 80, 107, 673
isostatic rebound, 39, 78, 80, 514, **516**, 540, 673
isostatic relief, 78, 105, 106, **591**, **592**, **593**, **594**
isotherm, Chapter 4, *see also* p. 702
Israel, 186, 270
Italy, 552
Ives, J. D., 495, 533

Jackson, K. C., 69, 102, 136, 147, 154, 155, 190, **252**, 270
Jahn, A., 526, 533
Jamieson, 17
Janda, R. J., **594**, 637
Japan Trench, 123, **124**, **185**
Java Trench, 122
Jefferson, M. S., 576, 586
Jenks, W. F., 601, 602, 607, 637
Jenny, H., 279, 280, 308
Jewell (Interval), 31
Johnson, D. S., 250
Johnson, D. W., 54, 88, 102, 472, 533, 541, 545, 555, **558**, 559, 560, 568, 569, 573, 586, 617, 622
Johnson, H., 208
Johnson, J. W., 548, 586
Johnson, S. J., 34, 43
Johnson, W. D., 93, 102, 341, 379, 601
joint, **12**, 29, 30, 65, 73, 79, 136, **140**, 302, 596
Joly, F., 601, 637
Jones, S. B., 301, 308
Judson, S., 547
Jukes, J. B., 617, 637

Jungfraujoch Glacier, **459**
Juvashytt Glacier, 479

Kahn, H. R., 412, **413**, 414, 446
Kaibab Plateau, **264**
Kalahari Desert, 188, 240, **242**, 343, 362, 364, 373, 374, 375, 382, 399, 438
kames, 47, 491, 496, **497**
kame terrace, **488**
Kansas, 239, 267, 301, 304, 387, 500
kaolinite, 154
kar (*see* cirque)
Karakorum Mountains, 612
karst (topography), 62, 74, 83, 90, 300, 627
Karst Plateau, 281
Kay, M., 126, 169, 174, 175, 208
Kazakstan Steppes, 236, 431
Kehle, R. O., 191, 193, 208
Keller, E. A., 430, 445
Kempton, J. P., 37, 44
Kennedy, C., 208
Kennett, J. P., 35, 44
Kentucky, 267, 407, 412
Kentucky River, **407**
Kermsdec Trench, 111
Kern River Canyon, 633
Kesseli, J. E., 7, 44, 526, 533
Kessler Mountain, 76
Key Minor Island, 541, **545**
Kharas Mountains, 188
Kidson, C., 543, 546, 586
Kilgour, R. H., 522, 523
King, C. A. M., 102, 543, 586, 587
King, L. C., 8, 13, 14, 15, 16, 17, 18, 32, 33, 42, 44, 49, 50, 51, 52, 53, 57, 58, 59, 73, 90, 128, 186, 208, 229, 250, 285, 290, 308, 328, 329, 344, 364, 379, 385, 445, **594**, 659, 668, 669, **670**, **671**, 672
King Ranch, 135
King's River Canyon, 633
Kinkaid Limestone, 653, 655
Kirkland, D. W., 172, 208
kjedel (*see* cirque)
knickpoint, 7, 62, 673, 674
Knik Glacier, 492
Knik River Valley, **492**
Knopf, A., 547
Knopoff, L., 112, 115, 116, 117
Koksoak River, **500**
Komar, P. D., 571, 577, 578, **579**, **580**, 587
Konig, R. H., x, 166, 167, 168, 208
Krakatoa, 199
Krinsley, D. H., 352, 355, 379, 444, 486, 533
Krook, L., 596, 599, 637
Kruit, C., 444
Krynine, P. D., 335, 352, 379, 623, 625, 626, 637
Kuenen, P. H., 123, 208, 352, 355, 379
Kuh-i-Nawak (salt glacier), **171**
Kuiseb River, 342
Kukla, G. J., 40, 41, 44
Kumargoan, **151**

Kummel, B., 154, 208
Kunlun Mountains, 612
Kuril Trench, 124
Kuro Shio (current), 220

Labrador, 481, 498, **500**, 545
Labrador Current, 220
laccolith, 131
Lachenbruch, **518**, **523**
lacustrine, 18, 61, 92, 266, 595
La Fond, E. C., 576, 588
lag concentrate (*see* lag gravel)
lag gravel, 73, **317**, **318**, 350, 351, 353, 390, 607
lahar, 330
Lake Chad, 399, 438
Lake Guilietti, 541
Lake Michigan, 51
Lake Rudolf, 672
Lake Sacandaga, 289
Lake Superior, 81, **147**, **148**, **465**, 467, 539, **562**, 578
Lake Titicaca, 593, 612, 614
Lakota Formation, 158
laminar flow, **432**, **433**
landform (definitions), 17, 21, 27, 39, 40, 46, 47, 48, 53, 54, 57, 59, 60, 62, 65, 66, 67, 69–73, 76, 78, 79, 83
landform histories (chronologies), 34, 83–93, 96, 97
landscape (definitions), 17, 21, 23, 27, 29, 32, 40, 42, 46, 47, 48, 54, 74, 86, 89, 98, etc.
landslide surge, 550, 552
landslide topography, 62, 73, **682**, **683**
Lane, E. W., 413, 414, 445
Langbein, W. B., 262, 263, 265, 308, 418, 445
Langford-Smith, T., 305, 309, 334, 338, 345, 379, 398
Langway, C. C., Jr., 34, 43
lapis, 301, 302, 467
Las Cañas Formation, 604
Las Vegas, 318
lateral moraine, **486**, **487**, 488
lateral stream erosion, 8, 9, 411, 412, 413, 414, 415, 416, 417, 418
laterite, Chapter 5, *see also* pp. 18, 154, 275, 276, **277**, **278**, **279**, 604, 668
lateritic earths, 5, 274, 279
Laurentian Upland, 498
lava, 18, 22, **57**, 62, 82, 195, 196, 197, 199, 201, 540
lava cave, 199, 299
lava plateau, 199, **204**
Law of Accordant Junctions, 2, 46
Law of Superposition, 17
leaching, 264
Leaf River, 500
Lees, G. M., 618, 637
lee side (glacial), 463
Leet, D., 547
Leighley, 433
Leliavsky, S., 415, 446
Le Mehaute, B., 577, 579, 587

Lena River, 305
Leningrad, 274
Leopold, L. B., 3, 5, 6, 7, 33, 44, 48, 62, 65, 102, 229, 250, 256, **258**, 263, 309, 385, 408, 411, 413, 414, **415**, **416**, **417**, **418**, **419**, 427, 445, 505
LePichon, X., 34, 44, 108, **110**, **111**
Leslie, **157**
Levee (natural), **412**, 655
Leverett, F., 512, 533
Lewis, W. V., **474**, **478**, **479**, 533, 568, 587
Libya, 329, 349
Lichty, R. W., 3, 45, 229, 250
limestone, Chapter 3, *see also* pp. 10, 27, 28, 29, 30, 31, 32, 33, 58, 74, 78, 154, 158, 159, 160, 161, 162, 175, 193, 266, 355, 388, 599, 603, 604, 607, 619, 641, 648, 649, 650, 652, 656, 663, 665
Lindgren, W., 632, 637
lineament, 166, 167, 168
linear erosion (*see* fluvial incision)
Linell, K. A., 693, 694
Linton, D. L., 283, 309
Lipman, P. W., 83, 102, 199, 208
Lister, C. R. B., 603, 637
lithology, Chapter 3, *see also* pp. 3, 10, 20, 22, 28, 29, 30, 31, 32, 33, 47, 50, 51, 57, 58, 65, 73, 74, 76, 78, 80, 81, **82**, 83, 85, 98, 104, 109, 113, 114, **121**, 130–138, 141, 142–148, **151**, 154–164, 169–172, 175, 176, 179–186, 190, 191, 195–202, 599–600, 601, 602, 603, 604, 605, 618, 619, 623, 624, 627–629, 646–650, 651–654, 660–663, 669, 672–674
lithostatic subsidence and topography, 169–172
litho-structural drainage effects, 299–302
lit-par-lit, 141
Little Red River, 414
littoral drift (*see* longshore current)
littoral zone, 546
Lituya Bay, 552
Llano Estacado, 437
Llanos, 414, 436
load (stream), 7, 27
Lobeck, A. K., 346, 559, 587, 595
lodgment till, 488, **489**, **490**
loess, 50, 52, 60, 70, 71, 86, 90, 94, 155, 270, 279, 684
Long Island, 484, 545, **575**
longitudinal (seif) dune, 438
longitudinal profile (stream), 7
longshore current, 544, 553, 686, **687**
Longuet-Higgins, M. S., 576, 578, 579, 587
Longwell, C. R., 35, 44, 197, 208, 547
lopolith, 131
Los Angeles Basin, 187
Louisiana, **170**, **171**, 167, 382, 401, 661
Lower Fars Formation, 85
low-level planation and aggradation, 609–610
Lucke, J. B., 546, 587
Lugn, A. L., 18, 44, 69, 71, 94, 102, 250
lunette dune, 375, 438

Lusby, G. C., 328, 342, 379, 413
Lustig, L. K., ix, x, 48, 69, 95, 102, 162, 220, 229, 240, 250, 268, 313, 314, 330, 331, 332, 333, 334, 336, 339, 341, 342, 345, 347, 362, 373, 374, 379, 396, 459
Lyell, C., 2, 20, 39, 44, 483, 533
Lykins Formation, 628
Lyons Formation, 628, 630

Mabbutt, J. A., 329, 379, 601, 632, 637
Macdonnell Ranges, 631
Machakandana River, 151
MacKay, J. R., 461, 528, 535
Mackenzie River, 305, 466, 492
Mackin, J. H., 6, 44, 418, 446, 594, 612, 624, 638
Madagascar, 276, 277, 278, 279
Maddock, T., 5, 7, 44, 418, 445
Madeira Island, 562
Magdelena River, 169
magnetic map, 110
magnetic reversals, 34, 108, 110
Maher, J. C., 628, 638
Maine, 539, 545
makatea, 584
Malaspina Glacier, 459
Maloney, N. J., 598, 638
mamillated surface, 498
Mammoth Hot Springs (Yellowstone), 67
mangrove coasts, 538, 582, 584
Manila Trench, 185
Manitoba, 80, 511
Manitou Dolomite, 629
mantle plume (hot spot), 153
maps (*see* specific types, structural, geomorphic, etc.)
Marañon River, 631
marble, Chapter 3, *see also* p. 146
Margolis, S. V., 35, 44
Mariana Trench, 122
marine deposits, 20, 62, 92, 669
marine geomorphic evolution, 556–584
marine plain, 39, 53
marine transgressions, 644, 646, 647, 648, 649, 650, 653, 665, 666, 667
Mark, D. M., 488, 533
Marks, P. L., 265, 309
marl, 85, 595
Marmer, H. A., 574, 587
Maroon Formation, 628
Marshall Formation (Sandstone), 656
Martin, R., 640, 675
Martinsburg Shale, 191
Maryland, 437
Masjed-i-Sulaiman (oil field), 85
Massachusetts, 137
massive rock fabrics, 130–140
mass wasting, 3, 8, 12, 46, 70, 71, 72, 264, 310, 316, 317, 318, 319, 320, 321, 322, 323, 324, 325
Mathieu, G., 208
Matterhorn, 470

Matthes, F., 415, 471, **478**, 480, 533
Matthews, P. M., 108
maturity (weathering/mineralogic), 4, **5**, 8, 10, 48, 648, 650, 656
Mauna Loa, 199
Mauritania, 545
Maxwell, J. C., 131, 133, 141, 190, 191, 207
Mazer, J., 448
McBride Canyon, 164
McClures Beach, **555**
McDonald Lake, 167
McGee, W. J., 13, 44, 315, 316, 317, 328, 329, 330, 334, 335, 341, 344, 366, 379, 431
McGill, J. T., 542, 587
McIntire, W. G., 587
McKee, E. D., 574, 587
McKenzie, G. D., 677, 694
McLaughlin, D. H., 528, 533
McMaster, R. L., 556, 587
McNeil, F. S., 580, 583, 587
meander, 8, 9, **10**, **11**, 48, **411**, 412, 413, 415, 416
meandering pattern (stream), 10, 60, 289, 418, 419, 658, 682, **684**, **685**
meander scar, 48, 411, 412
mechanical erosion, 22
mechanical weathering, 19, 21, 29, 30, 31, 83, 316–327, 468, 521–523, 610–617
medial moraine (*see* moraine)
Mediterranean (Sea) Basin, 170, 199, 369, 401, 564, 619, 672, 673, 674
mega-ripples, 440, 512, **514**
Meier, M. F., 459, 533
Meinesz, V., 123
melange, 181
Melloni, 316
Melton, M. A., 295, 309
Menard, H. W., 115, 122, 197, 209, 332, 379
Mendeleyev Ridge, 119
Mercer, J. H., 239, 250
mesa, 142, 166
Mesa de Jajo, 184, **246**
mesogenic topographic effects, 104, 189–206
metamorphic (high-temperature facies), 185
metamorphic (rock/process), 10, 20, 22, 141–146, 176, 177, 178, 179, 183, 185, 652
metastable (rock/mineral), 132, 619, **620**, **623**, 624, 626, 627, 628, 629, 630, 650, 652
meteorite (scar), 212, 213
meteorologic (map/section), **216**, **217**, 221, 223, 224, **225**, 226, 258, 452, 454, 456, 613
Mexican Hat, 202
Mexico, 7, 161, 316, 328, 390, 394, 540, 545, 625, 626
Meyerhoff, A. A., 107, 110, 111, 121, 209
Meyerhoff, H. A., 110, 111, 121, 209, 617, 624
Michigan Basin, 152, 387, 656
Michigan River, 653, **654**, 655
microclimate, 31, 66
Middle Pinnacle Creek, 345
mid-ocean ridge, 72, 83, 108, **109**, 110, 111, 112, 113–120, **121**, **122**, 197, **541**, 623, 668, 672

mid-ocean ridge orogenesis, 114, 115, 116, 117, 118, 119
migmatite, 130
Milankovitch Theory, 38
Miller, A. K., 656, 675
Miller, D. J., 552, 587
Miller, J. P., 3, 44, 258, 309, 394, 404, 416, 419, **427**, 445, 446, 505
mima mounds (*see* prairie mounds)
Minard, J. P., 400, 446
Mindel, 90
Minnesota, 134, 500, 646
Miocene, **85**
miogeosynclinal facies, 175
Mississippi, 401, 494
Mississippi-Pennsylvanian Unconformity, 648–649
Mississippian (period-system), 78, 648, 649
Mississippi Delta, 426, **428, 429**, 430, 653
Mississippi Embayment, 152, 169, 661
Mississippi River Valley, 8, **52**, 169, 170, 267, 293, 397, 418, 421, **424**, 492, 493, 494, 510, 653, 655, 686
Missouri, 150, **153**, 166, 167, 187, 257, 267, 280, 354, 652, 657, 661
Missouri River, 71, 492, 510, 682, 684, 686
Mitchell, J. G., 628, 638
Mitchell Delta, **429**
Moberley, R., 596, 638
moisture-depleted air (zone), **221**
Mojave Desert, 319, 320, 325, 366, 388, **389**
molasse, Chapter 3, *see also* pp. 176, 181, 184, 619, 620, 623, 624, 630
Møller, B., 34, 43
Molopo River, 343, **374**
Moluccas Archipelago, **541**
monadnock, 55, 74, 136
monsoonal conditions, 261, 262, 632
Montana, 342, 513
montane geomorphic asymmetry, 631–633
montane plateaus, 612–615
Montauk Point, **484**
Monteregian Hills, 133
Montgomery, A., 446
montmorillonite, 154
Montreal, 133
monument (*see* erosion remnant)
Monument Valley, 165, 366, 368, **369**
Moore, **422**
Moore, R. C., 659, 675
moraine, 24, 41, 67, 468, **470**, 477, 484, 487, 488, 490, 493, **501, 502, 508**
Moramanga, 277
Moreau, R. E., 267, **268**, 309
Morgan, 111
Morgan, J. P., 426, 446
Morgan, R. S., 546, 588
Morisawa, M., 408, 432, 433, 446
Morocco, 30, 133, **570**, 601, 632, 674
morphogenesis, 29, 83
morphogenetic region (*see* geomorphic region)
morphometric analysis, 57

morphostratigraphic terms, 97
Morrison, 182
Morrison, R. B., 17, 37, 45, 238, 250
Moss, J. H., 524, 534
Moulton, B., 677, 694
Mountain Fork Creek, 74, 75, 394, **395**, 396, 397, **411**
mountain roots, 633–635
mountains, Chapter 10, *see also* pp. 34, 46, 47, 48, 62, 73, 74, 79, 95, 176–206, 384, 678
Mount Asmara, 198
Mount Assiniboine, 473, **477**
Mount Chimborazo, 199, **450**
Mount Collon, 479
Mount Cotapaxi, 199, **200**
Mount El Misti, 631
Mount Everest, 122
Mount Gardner, **590**
Mount Lofty Range, 133
Mount McKinley, **463**
Mount Monadnock, 55, **136**
Mount Pelee, 199
Mount Pichu Pichu, 631
Mount St. Hiliare, 133
Mount Wilson, **479**
Mousinho (de Meis), M. R., 34, 43, 101, 243, 249, 250, 453, 533
moving water and sediment, 401–441
Mozambique-Agulhas Current, 219
mudflow (fluvatile), 62, 66, 330, **331**, 332, 333, 334, 335, 404
mudflow (gravity), 330
Mudie, 119
mudlump, 426
mudstone, 154, 155, **157**, 595
Muller, J., 424, 446
Muller, S. W., 518, 519, 533, 693, 694
multi-purpose dams, 689
Musgrave, 261
Musinia (fault blocks), 188
Myrick, R. M., 102, 250, 446

Nagelfluh Molasse, 618–619
Namib Desert, **69**, 246, 342, 354, 362, 364, 374
Nansen Fracture, 120
Nantucket Island, 579
nappe, 189
Nashville Dome, 153
Natal, **53**, 57
natural levee, **411**, 412
Navajo Uplift, 182
Nebraska, 18, 71, 239, 399, 500
Nebraska Sand Hills, 18, 71, 399
Neff, G. E., 444, 531
Neiburger, M., 225, 250
net drift, 548, 553
Netherlands, 279
Neuberger, H., 225, 250
Nevada, 190, 318, 419, 421, **602**
névé (*see* firn)
Newark Group, 623, 624
New Brunswick, 555

Newell, N. D., 602, 612, 638
New England, **55**, 262
New Guinea, 176, **417**, 597, 606
New Hampshire, 55, 387
New Jersey, 13, 24, **52**, 111, 132, 135, 148, **150**, 283, 289, 321, 382, 400, 538, 553, **569**, 575, 576, 623, 624, 661, 686, 687
New Mexico, 6, 65, 150, 300, 301, 356, 387, 437, 625, 628
New South Wales, 333
Newton, I., 42
New York, 132, 262, 289, 387, 400, **484**, 491, 502, **575**, 624
New York City, 387
New Zealand, 239, 404, 453, 545, 572, **573**, 615
Neyes Provincial Park, 465, 467
Ngamiland, 242
Nickelsen, D. W., 209
Nigeria, 18, 281, **305**, 361, 367, 389, 672
Niger River, 303, **305**, 383, 438
Nile Delta, **423**, 426
Nile River (White), 241, 303, 371, 673
Nipigon, 148
nisch (*see* cirque)
nivation, 471
Noah, 1
Nogales, 329
normal erosion cycle, 4
North America, 1, 4, 9, 10, 13, 14, 18, 19, 22, **24**, **25**, 31, 35, 36, 37, 49, 50, 51, **52**, 53, 55, **56**, 65, 67, 68, 71, 74, 75, 76, 79, 80, 81, 82, 88, 89, 92, 95, 100, 126, 127, 128, **129**, 150, **152**, 153, 161, 164, 169, 170, 174, **175**, 176, 185, 187, 201, 222, 234, 235, 236, 238, 239, 240, 244, 248, **256**, 259, **260**, 266, **267**, **272**, 273, 277, 303, 306, 320, 349, 365, 367, 383, 386, 394, 400, 401, 414, 418, 421, 431, 436, 439, 453, 457, 500, 501, **502**, 509, 510, **511**, 512, 517, 519, 537, 538, 617, 624, 633, 646, 651, 656, **657**, **659**, 662, 665, 668, 692
North Carolina, 438
North Dakota, 500, 511
North Sea, 537
Northwest Territories, 520, **528**
Norway, **474**, 479, 481, **493**, 545
Nova Scotia, 555
nunatak, 461
Nye, F., 472, 533

Oahu, 301
oasis, 313, **353**
Oberlander, T., 46, 83, 84, 102, 325, 328, 333, 345, 377, 379, 388, 398, 446, 596, 601, 613, 617, **618**, **619**, **620**, **621**, 634
Ob River, 305
Ob Trench, 122
ocean, 42, 62
ocean basin, 34, 114–125, 672–674
ocean basin desiccation, 668, 672–674
ocean current gyres, 548
ocean current map, 219, **687**

ocean ridge (*see* mid-ocean ridge)
ocean trenches, 72, 107, 110, **111**, 114, 122–125, 603, 608
ocean trench orogenesis, 122, 123, 124, 125
oghurd dune (*see* star dune)
Ohio, 267, 500, 623
Ohio River, **412**, 492, 510
Okavango Delta, 240, **242**, 431, 438
Oklahoma, 237, 239, 257, 267, 387, 390, 394, 401, 651
Oklahoma (Anadarko) Basin, **152**
old age (Davisian topographic assessment), 4, 5, 10
Oliver, J., 111, **123**, 208
Olmstead, E. W., 617, 638
Ontario, 80, 81, 148, 511, **562**
open system (environmental), 28, 32
open system (sedimentary), 394
ophiolite (igneous lithofacies), 131, 133
Ordai River, 326
Oregon, 201, 223, 390, 541
organic coasts, 580–584
organic (vegetal) decay, 33, 373
organic reef, 67, 76, 78, 159, 160, 162, 199, 538, 560, 580, 581, **582**, **583**, 648, 650, 651, 657, 662, **663**, 686
orientation of glacier ice motion (*see* glacier movement)
Orinoco River, 149, 424, **425**, **432**, 437, 438, 440
Orme, A., 575
orogenic belts, 72, 78, 79, 80, 81, 173–179, 180–185, 589–636
orogenic continental settings, 168–206
orogenic elevation and relief, 591, 592
orogenic landscape case studies, 617
orogeny, 83, 106, 114, 173–179, 180–185, 597–609
orographic rainfall, 312–314, 338–340, 597–601
Orowan, E., 472, 533
orthogonal block, 139, 143
orthophoto map, 49
Ostland, G., 206
Otvos, E. G., 400, 446, 576, 587
Ouachita Geosyncline, 76
Ouachita Mountains, 106, 166, **183**, 190, 191, 662
Ouachita River, 492
oued, 31, 369, **317**, **353**, 355, 369, 372
Oued Saoura, **317**, **353**, 355, 369, 372
outwash deposits (gravel), 86, 461, 491, 493
outwash head, 493
Overbeck, R. M., 308
Owen Fracture, 120
Owens Valley, **95**
ox-bow lake, 9, **411**
oxidation, 270, 280
Ozark Uplift, 74, 75, 76, 77, 78, 88, 90, 150, 153, 157, 161, 162, 163, 165, 167, 187, 259, 261, 394, 401, 402, 405, **408**, 529, 646, 647, **648**, **650**, 652, 653, 654, 655, 657, **665**

Pacific Ocean (Basin), 107, **118**, 122, **124**, 126, 200, 243, 550, 551, 593, 603, 632
Padre Island, 574
paleoclimatic map, 240, **244**, **452**, **454**, **456**, **657**, **659**
paleoclimatic tectonism, 667–672
paleoclimatology, 33–41
paleogeomorphology, 98, 99, *see also* Chapter 11
paleomagnetism, 34, 35, 36, 108, **110**
paleontologic unconformity, 641
paleopavement, 98, 507, **509**
paleosol, **25**
paleotemperatures, 33, 34, 37, 38, **40**, **41**
Paleozoic, 72, 153–162, 625–631, 641, 646–667
Palisades Intrusive, 132, 139
Pallister, J. W., 344, 379
Palynology, 34, **41**, 70
pan, 61, **70**, 359, 362, 374, 375, 437, 438
Panamint Mountains, 348
Panamint Valley, 319
Pannett, D. J., 445, 586
Papua, 606
Paraguay River, 438
parallel (drainage pattern), **60**
parallel scarp retreat, 12
Pardee, J. T., 513, 533
Paris Basin, **152**
Parkin, D. W., 576, 587
Passarge, S., 364, 379
passive geologic substrate, 72
Patagonian Desert, 362
patterned ground, 523, 526, **527**, 528, **529**
Patterson, W. S. B., 457, 533
Peabody, F. E., 301, 309, 382, 383, 446
Pecos River, 300
pedalfer soil, Chapter 5, *see also* pp. 273, **274**, 384, 485
pedestal rock (*see* butte)
pediment, 12, 14, 15, 16, 49, 52, 53, 71, 74, 86, **326**, **338**, 343, 344, 345, **346**, 347, 348, 396, 605, 608, 612, **644**, 691
pedimentation, 14, 15, 16, 86, 346, 347, 348, 608, 641
pediments and planation, 343–349
pediplain, 15, 16, 54, 340, 605, **607**
pediplanation, 8, 15, 33, 90, 609
pedocal soil, Chapter 5, *see also* pp. 274, **275**, 384, 485
pedology, 22, 69, 271–281
Peel, R. F., 315, 329
Penck, A., 7, 54, 103, 489, **490**
Penck, W., 8, 12, 13, 14, **15**, 16, 17, 21, 32, 45, 285, 309, 364, 379, 594
peneplain, 4, 7, 11, 12, 54
Pennsylvania, 3, **52**, 139, **141**, 142, 146, 265, 293, 387, 502, **524**, 623
Pennsylvanian (period/system), 76, 625–631, 648–667
Pennyback Creek, 293
Pequest River, 24

perched water table, 155
perennial runoff, 294
peridotite, 133
periglacial, 515, 516
perisediment, 345, 346
permafrost, 62, 90, 461, 462, 517, **518**, 519, 523, **528**, 691, **692**, 693
permeable pile dike, 682
Persia, 170
Persian Gulf, 541, 545, 619
Peru, 54, 91, 92, 187, 192, 195, 243, **245**, 246, 254, 267, 364, **420**, 545, **548**, 593, 601, 602, 603, 604, 607, 610, 616, 631, 632
Peru–Chile Trench, 122, 126, 199, 603
Peru Current, 243
Perutz, M., 456, 458, 459, 533
Pettijohn, F. J., 93, 99, 103, 180, 181, 209, 404, 446
Philippines, 606
Philippine Trench, **124**
photic zone, 664
phreatic zone, 300
phyllite, 98, 145, 149
physiographic diagram, 64, **116**, 117, **611**
Picard, D., 187
piedmont, 12, 83, 608, 628
piedmont angle, 325, 326
piedmont benchland, 13, **15**
Pierce, W. G., 193, 194, 209
Pikes Peak Granite, 627
Pillewitzer, W., 526, 533
pillow lava, **197**
pingo, 519, **520**, 521
pinnate drainage pattern, **60**
Pissart, A., 519, **521**, 534
Pitkin Limestone, 29, 648, 649
Pitty, A. F., 47, 103
plain (planar landform), 4, 7, 11, 12, 15, 16, 53, 54, **55**, 56, 57, 62, 67, 73, 81, **82**, 83, 85, 87, **88**, **89**, 90, 91, 92, 93, 96, 99, 313, 317, 318, 326, 340, 341, 342, 343, 344, 345, 346, 347, 348, 364–376, 390, 537, 594, 612
planar erosion, 50, 88, 609, 672
plant (vegetal) cover, 15, 18, 19, 21, 29, 38, 39, **41**, 42, 59, 65, 71, 80, 81, **82**, 83, 93, **95**, 237, 251, **255**, 257, 259, 260, 263, **264**, 266, 267, 268, 269
plateau (intermontane, marginal), 612, 613, 614, **615**, 672
plate tectonic concepts, 106, 107, 108
plate tectonics, 23, 83, 106–125, 173–195, 667, 668, **669**, 670, **671**, 672
plate tectonics and geomorphology, 111, 112, 113
Platte River, 414
playa, 61, 62, 67, **347**, 359, 690, 691, 704
Playfair, J., 2, 17, 45, 46
Pleistocene (time/effects), 32, 35, 37, 83, 86, 90
Pliocene, 90
plunge pool, 62
plunging breakers, 576, **578**

pluton, 130, 131
plutonism, 104
pluvial, 19
podzol, Chapter 5, *see also* pp. 273, **274**
point bar deposit, 9, 62, 67, 411, **412**
Polach, H. A., 398, 447
polar (regions), 215, 217, 448, 452
pollution, 678
polycyclic mountain erosion theory, 592, 593
polygenetic landscapes (coasts), 536–585
polygenetic landscapes (mountains), 589–636
pool (channel), **404, 408**
pore pressure (disturbances), 682
porphyroblast, 145, **147**
Porter, S. C., ix, 506
post-Boone unconformity, 648
Postma, H., 444
pothole, 62
Potter, N., **524, 525**, 534
Potter, P. E., 412, 653, 654, **655**, 675
Powell, J. W., 4, 7, 45, 617, 634, 638
prairie mounds, 390, **391**
prairie pimples (*see* prairie mounds)
Precambrian, 35, 42, 80, 130–154
precipitation, 18, 29, 31, 95, **258**, 260, **261, 262**, 312, 313, 314, 315, 316, 381, 382–385, 449–453, 596–609, 610–617
precipitation and temperature, 256–263
Prescott, J. R. U., 240, 250
Prest, V. K., 495, 534
Price, W. A., 437, 438, 446, 556, 571, 573, 574, 577, 579, 587
process (geomorphic), 15, 18, 24, 26, 32, 42, 46, 62
product (geomorphic), 24, *see also* landforms and deposits
profile of equilibrium, 6, 85, 88
Progresso Formation, 604
Prostka, H. J., 102, 208
protalus rampart, 473, **475**
Prouty, W. F., 437, 438, 446
provenance, 69, 99, 174
Prucha, J. J., 153, 209
Prus-Chacinski, T. M., 415, 446
Puerto Rico, 581
Purington, W., 518, 534
Pyramid Lake, 420
Pyrenees Mountains, 404
pyroclastic, 25, 82, 195, 198, 199, 540

quantification (geomorphic), 47, 48
quartz, 652, 654, 655, 656, 690
quartzite, Chapter 3, *see also* pp. 57, 146, **148**, **150**, 281, 652, 654, 655, 656
quartz monzonite, 133, **135**, 388
quasi-sheetflood, 431, 432, **433, 434**
Quaternary, 31, 32, 33, 34, 35, 36, 89, 92, 162–165, 169–172, 302–306, 362–376, 448–530
Quattara Depression, 673
Quebec, 458, 460, 486, 500
quebrada (*see* arroyo)

Quinn, J. H., 59, 74, 88, 103, 240, 250, 364, 379, 390, 394, 404, 411, 445, 446, 659

radial drainage pattern, **61**, 299
radiometric ages, 33, 34, 37, 41
Rahn, P. H., 51, 103, 496, 534
rainfall (incidence, amount), 17, 72, 73, 83, 680
rain forest, 21, 38, 254, **255**, 260, 268, 277, **278**, 599
rain shadow, 613, 631
rain splash erosion, 314
Ramana Rao, K. L. V., x, 55, 323, 324, 325, 326, 333, 339
Rampton, V. N., 533
Ramsey, A. C., 471
ravine (*see* gully)
Reading Prong (New England Precambrian Province), **135**, 148
recent geomorphic system displacements, 237–246
recessional moraine (*see* moraine)
rectangular drainage pattern, **12, 60**, 299
Red Sea, 186, **187**, 189, 198, 545
reef-buttressed coasts, 580, **581, 582, 583**, 584
refuge (organic), 239–245, *see also* climatic nuclei
regolith, 16, 29, 71, **72**, 211, 264, 270, 383, 392, 400, 401, 604, 681, 682
Reich, B. M., 260, 309
Reimnitz, E., 519, **522**, 534
rejuvenation, 10, 11, 27
relative geomorphic sequences (*see* landform chronologies)
relegation, 458
relict (environmental), 3, 17, 18, **19**, 20, 21, 22, 23, 26, 28, 29, 30, 31, 32, 38, 39, 47, 48, 62, 71, 80, 94, 99, 105, 396, 399–401, 513, 615, 680, 681, 682
relict destruction, 27–32, 38, 39, 381–444
relict glacier, 453
relict problems, 18
relief, 8, 32, 34, 46, 80, 104–106, 113, 114, 115–119, 122–125, 173, 185–189, 190–206, 596, 650, 651
relief inversion (reversal), 84, 594, **595**
remnant magnetism, 108
remnant planar landforms, 13, **15, 16**, 53, 54, 55, 57, 62, 77, 136, 325, 340, 347, 348, 366, 367, 368, 369
repose (angle of slope), 50
repose angle, 291, 292, 400, 680–682
residual mountains, 74
Reykjanes Ridge, **110**
Reynolds Number, 433
Rhine Valley (Graben), 186, **188**
Rhode Island, 566
rhomboidal ripple marks, 553, **577**, *see also* rilling
Rhone River, 672
rhyolite, 134, **135**
ria (river-incised) coast, **559**

Richardson, 431
Richmond, G. M., 280, 309
ridge-ravine topography (*see* selva)
Ridley, H. N., 584, 587
riegel, 478, 479
riffel (stream), 408
Riley, C., 464, 534
rill, 15, 62, **287**, **288**, **289**, **290**, **291**, 540
Rio Caroni, 144, 145, 431, **432**, **434**, **435**, **436**, 437, **439**
Rio de Janerio (state), 98
Rio Grande, 303, 383, 425
Rio Guayas, 555
Rio Javita, 255
Rio Mototan, 184, **246**
Rio Negro, 22
Rio Pao, 414, **415**
Rio Yaguachi, 404, **405**
rip currents, 553, 554, **555**, **579**, 580
ripplemarks, 359, 427
Ritter, D. F., 419, **421**, **444**, 594, 638
riverine plain (*see* aggradation plain)
rivers (*see* specific names)
river system, 3, 14, 18, 26, 653, 658, *see also* drainage net
Roach, C. B., 209
Robinson, A. R., 587
rochés moutonnées, 363, **364**, **367**
rock cut terrace (*see* terrace)
rock cycle, 22, 23
rock flour, 706, *see also* Chapter 8
rock glacier, 524, **525**
rocks and topography, 113
rock stratigraphic (terms), 97
rock toughness, 130
Rocky Mountains, 14, 153, 172, 178, **182**, 187, 473, 477, 592, 594, 606, 610, 624, 625, **627**, 628, **629**, 630, 632, 645
Rodgers, N. D., 617, 638
Rodriques Fracture, 120
Romanche Fracture, **116**, **117**
Romania, 169
rooting media (for plants), 265
Roraima Series, 142, **144**
Roswell, 300
Rouse, H., 332, 379
Rove, O. N., 308
Roy, C. J., 324, 379
Royal Gorge, 14
Royo y Gómez, J., 599, 638
Rub al-Khali, 354
Rubin, M., 83, 90, 102
Ruhe, R. V., 31, 33, 34, 37, 45, 47, 49, 52, 69, 70, 86, 89, 94, 97, 103, 273, 277, 309, 504, **505**, 506, 507, 682, **684**, **685**, 694
Rumney, G. R., 215, 216, 217, 218, 220, 222, 224, **225**, 250, 677, 694
Runcorn, S. K., 108, 209
runoff, 16, 31, 56, 81, 83, 229, **230**, **231**, 257, **258**, 259, 260, **261**, **262**, **263**, 285–307, 327–349, 390–398, 401–444, 498–500, 504–513, 527–530, 597–601, 604–606, 607–609, 610, 612–617, 618–636, 640–668, 672–674, 680–685, 686, 688, 689
Russell, I. C., 17, 45, 462, 534
Russell, R. J., 259, 309, 421, 426, 440, 537, 556, 576, 587
Russia (U.S.S.R.), 236, **274**, **275**, 279, 501
Ruth Glacier, 463
Ruxton, B. P., 283, **284**, 309, 325, 329, 597, 606
Rydell, H., 206
Ryder, R. T., 624, 638

Sahara, 18, 19, **70**, 172, 218, 236, 240, **241**, 246, 305, 313, 314, 315, 317, 325, 329, 339, 349, 351, 352, **353**, 355, 358, 359, 361, 364, 368, **370**, 371, 372, 373, 399, 438, 441, 466, 641, 648
Saidmarreh (Kuh-i-Kailan) Anticline, **181**
Salida Basin, **490**
Salisbury, R. D., 462, 468, 531
salt, 51, 68, 85, 169, 170, 672, 673, 674, 690, 691
saltation (*see* bed load)
salt dome, 170, **171**, 172, 541, 672
salt fretting, 522
salt glacier, 170, **171**
Salton Sea (Coast), 545, 688
salt wedging, 521
San Andreas Fault, 56, 187
San Benedicto Island, 540
sand, 51, 65, 94, 595, 599, 600, 602, 610, 641, 653, 657, 661
Sand Hills (*see* Nebraska Sand Hills)
sand sea (*see* erg)
sandstone, 10, 30, 58, 74, **78**, 154, 156, 175, 266, 394, 395, 604, 641, **642**, 646, 650, **651**, 653, 660, 664, 666, 669
sand waves (*see* beach cusps)
Sandy Hook, 569, 659, 686
San Eduardo Limestone, 255
Sangre de Cristo Mountains, 614
San Juan River, 9, **11**
Sanpete-Sevier Valley Anticline, **173**
Santa Catrina (state of), 98
Santa Elena Peninsula, 193
sapped cliff, 62
Sapper, K., 597, 638
saprolite (*see* soil)
Saratov, 275
Saskatchewan, 511
Saskatchewan Glacier, 459
Saucier, R. T., 420, 422, 446, 452, 492, 506
Saunders, J. E., 35, 44, 208, 239, 249, 451, 523
savannah, 267, **268**, 401
Savegear, 315, 329
Scandinavia, 460, 469, 474, 496, 498, 500, 514
scarp (escarpment), 12, 14, 15, 50, 51, 56, 57, 59, 71, 73, 76, 86, 88, 89, 90, 91, 92, 93, **166**, **167**, 303, 345, 594, 607, 644
scarp retreat (parallel), 12, 13, 14, 15, 53, 59,

86, 87, 88, 91, 92, 93, 303, 343–349, 601–604, 609, 610, 613–617
scars (meander, landslide), 9, 62, 72, 73
Schaffer, J. P., 69, 103
Scheidegger, A. E., 33
schist, 80, 145, 147, 149
Schmitz, W. J., 548, 587
Schoff, S. L., 301, 304, 308
Schofield, J. C., 579, 587, 588
Scholl, D. W., 584, 588
Scholtem, R., 624, 638
Schooley Surface (peneplain of W. M. Davis), 13, 54, 55
Schuchert, C., 126, 209
Schumm, S. A., 3, 39, 45, 78, 103, 229, 250, 262, 263, 265, 290, 305, 308, 309, 314, 317, 328, 334, 336, 341, 342, 346, 379, 397, 412, 413, 414, 416, 417, 446, 594, 638, 658, 675
Schwelnus, C. M., 188
scientific technique (in geomorpohology), 46, 47
Scotland, 561
Scottish Highlands, 178
Scott, T., 601
scree, 25, 62, see also talus
sea arch, 564
sea cave, 562, 563
sea cliff, 53, 58, 59, 184, 217, 484, 539, 540, 542, 548, 551, 562, 563, 564, 566, 568
sea-floor epeirogenesis, 120, 121, 122
sea-floor spreading, 111, 541
sea stack, 562, 563
Seca Shale, 184, 604
second-order climate changes, 37
sedimentary bypassing, 404, 658
sedimentary lag, 625
sedimentary rock, 22, see also Chapters 3, 10, and 11
sedimentology, 69
seif dune (see longitudinal dune)
Selby, M. J., 513, 534
Seligman, G., 456, 457, 533
selva (topography), 260, 297, 599, 604, 606, 627
semiarid (geomorphic environment), 385
semiarid (meteorologic climate) 13, 15, 17, 18, 385
semi-insoluble (matter), 4, 5, 8, 10, 48, 132, 648
Senegal River, 12, 281, 294, 295, 296, 303
Senstius, M. W., 33, 39, 45, 273, 275, 277, 309
Serangeti Plains, 386
Severn River, 417, 555
Seychelles Islands, 133
shale, 10, 29, 30, 31, 57, 58, 74, 76, 78, 154, 156, 157, 158, 159, 169, 172, 173, 175, 181, 183, 256, 261, 394, 600, 604, 608, 649, 651, 652, 669
shallow neritic zone, 564
Shand, S. J., 136, 209
Shantung Peninsula, 655
Shari (see Chari)

Sharp, C. F. S., 526, 529, 534
Sharp, R. P., 34, 44, 457, 458, 459, 534
Shaw, A. B., 537, 588, 659, 661, 662, 663, 664, 665, 675
sheet (flow-wash) erosion, 15, 16, 76, 394
sheetflood, 62, 66, 328, 329, 330, 392, 430, 431, 658
sheeting, 139, 320, 322
shelf (continental), 168, 169, 170–175
Shelton, J. S., 512, 514, 534
Shenandoah National Park, 186
Shepard, F. P., 423, 446, 542, 543, 545, 557, 573, 576, 577, 588
Shepard, R. G., 11, 45
Sherman, 261
shield, 72, 73, 129, 130–150
shield volcano, 199, 203
Shilts, W. W., 94, 103
shingle (beach), 560, 562
shooting flow, 433, 434
Short, N., 197, 207
Shrewsbury Plain, 417
sialic, 108
Siberia, 401, 519, 526
Sierra del Norte, 597, 598, 600, 606, 632
Sierra Madre Oriental, 625
Sierra Nevada, 185, 187, 322, 323, 331, 365, 478, 480, 592, 609, 617, 632, 633, 634
Siever, R., 653, 655, 675
silcrete, 323
silk dunes, 354, 357
sill, 120, 122, 131, 139, 141, 148
silt (sediment grain size), 50, 51, 70, 672
siltstone, Chapter 3, see also pp. 30, 154, 155, 156, 157, 394, 395, 604, 619, 649, 650, 657
Silvester, R., 550, 553, 588
simatic, 108
Simonett, D. S., 606, 638
Simons, 431
Simpson Desert, 357
Singewald, J. T., Jr., 308
sink, 30, 300, 679
Sinker, C. A., 445, 586
sinter, 66, 67
sinuosity, 412, 416
Sioux Quartzite, 652
Sirispal, 151
Sirkin, L. A., 444
Siwalik Conglomerate, 619
Skauthoe Glacier, 479
slate, 145, 146, 149
Sleep, N. H., 114, 209
slip-off slope, 9, 10, 62
slope (see hillslope)
slope instabilities, 680–682
slough, 9, 62, 411
slump block, 62
Smally, I. J., 24, 45, 352, 355, 379, 502, 534
Smith, B. L., 208
Smith, H. T. U., 18, 45, 444, 513, 516, 531, 534
Smith, N. D., 418, 430

snails, 34
Snake River, 303
snow line, 450
Snyder, 498
Socorro Sandstone, 184, 604
soil, 12, 18, **21**, 22, 24, 25, 27, 29, 30, 32, 37, 41, **52**, 62, 65, 69, 86, 89, 93, 143, 154, 155, 252, 254, 270, **271, 272, 273, 274, 275, 276, 277, 278, 279, 280,** 281, **282,** 283, 284, 294, 383, 388, 389, 390, 398, 400, 485, 644, 680, 691
soil (tropical), 21, **274, 275, 276, 277, 278,** 279, **282**
soil development, 271–285
soil profiles, 272, 274, 275, 276, 277, 278, 279, 280, 282, 284
soil stratigraphic (terms), 97
soil stripes (*see* patterned ground)
solar insolation, 38, 40, 41, 42
Solar System, 678
Solheimajökull Glacier, 483, 487, 493
solifluction, 30, 62, 90, 523
solution (solute), 596, 599, 689
Sonora Desert, 316, 328, 362, 366
South America, 21, 22, 46, 54, 72, 73, **91, 92, 93,** 126, **152,** 187, **192,** 221, 222, 223, 239, 240, 243, 244, 254, **255,** 260, 268, 277, 297, **298,** 301, 421, 435, 438, 440, 452, 512, 538, 541, 597–609
South Dakota, 150, 336, 337, 338, 500, **501,** 507, **508, 510**
South Pole, 216
space-time continuum (environmental), 42, **762**
Spain, 674
spalling (*see* spheroidal weathering)
spatter cone, 199
Speight, J. G., **417,** 447
Spencer, E. W., 167, 209
spheroidal weathering, 283
Spieker, 188
spilling breakers, 576, 578
spit, 3, 567, 569
Spitsbergen, 111, **529**
Springfield Plateau, 90
stage (geomorphic developmental), 15, 42, 442
stagnant ice features, 47, **470, 488, 489, 494, 497**
stalactite, 68, **301**
stalagmite, 301
Stanley, K. O., 153, 209
star (oghurd) dune, 69, 354, **355, 358, 359**
Stauffer, C. R., 646, 675
St. Cloud, 134
steady state (*see* dynamic equilibrium)
Stearn, C. W., 126, 207, 574
Steers, J. A., 568, 588
Stein, G., 46
Steno, N., 17
Stephen, T., 464, 534
Sterett, T. S., 574, 587
Stevens, M. A., 658, 675

St. Francis Mountains, **280,** 354, 652
St. Lawrence River Valley, **486,** 500
St. Louis, 397
stock, 131
Stokes, W. L., 163, 209, 330, 380
Stokes Law, 410
Stone, R., 315, 380
stone circle (*see* patterned ground)
stone net, 526, *see also* patterned ground
stone polygons, 526, **529**
stone stripes, 526, **529**
storm surge, 3, 54, 550, 552
St. Peter Sandstone, 646, 647, 648
Strahler, A. N., 25, 26, 33, 45, 57, 103, 336, 380, 393, 447, 617, 638
strain, 707
Straits of Gibralter, 673, **674**
Strangford Lough, **503,** 545
stratification (*see* bedding)
stratigraphic nomenclature, 92, 95, 97
stratigraphy (Quaternary), 34, 36, 37, 69, 90–100
stream, 14, 15, 56
stream and pediment profiles, 6, 7, 53, 54, 326, 333, 345, 346, **685**
streamflood, 391, 392, 394
streamfloods and stripped surfaces, 390–398
stream incision (*see* fluvial incision)
streaming flow, 66, 433, 434
stream order, 46, 294–296
stream piracy (*see* channel piracy; deranged drainage)
Streiff-Becker, R., 457, 458, 534
stress, 707
striation (glacial), 465
striped soil (*see* patterned ground)
stripped plain, 317
structural geomorphology, 86
structural (tectonic) landforms, 34, 61, 62, 79, 83, 104, 111, 114, 122–125, 150–154, 165–172, 173–179, 185–189, 190–195, *see also* Chapter 10
structural plain, 54, 67, 165
structural-tectonic stability, 85, 104–111, 114, 116–122, 128–130, 168, 169–172, 173–179, 185–189, 190–206
structure (rock), 15, 42, 46, 59, 75, 104–111, 122–125, 136–140, 141–150, 165–168, 169–172, 179–185, 189–195, 196–206
strudel, 519, **522**
Stuvier, M., 569, 585
subaerial (environment), 22, 23, 32, 34, 38, 39, 47, 73, 83, 89, 92, *see also* Chapters 4–10
subduction, 23, 110, 113, **125,** 175, 176, 608
subglacial (load/environment), 86, 468, 489
sublimation, 449
submarine canyon, 107, **422,** 423
subsequent drainage, 84, 401–444, 617
Sumatra, 597
Sumgin, M., 519, 534
Sundborg, A., 408, **409,** 432, 433, 447

superimposition (drainage), 4, 617, **618**, 619, **620**, 621, **622**, 623, 625, 627, 628, 629 630
supraglacial (load/environment), 468, **489**
surf (*see* breakers)
surficial, 2, 42
surficial geomorphic systems, 211–250
Surtsey, 199
Sussex, **58**, 545, **568**
Sutherland, P. K., 446
Sutton, G. H., 174, 207
swamp flora, 265
Swann, D. H., 653, **654**, 655, 665, 675
swash marks, 550
Sweden, 496, 500
Sweetwater Uplift, **132**
swell, 552, 553
swelling (*see* frost heaving)
Switzerland, 456, **470**, 479, 619
syenite, 134, 135, **136**, 668
Sykes, J. L., 111, **123**, 174, 208
system (*see* open; closed)

Tablazo Limestone, **193**, 604
tableland, 165, 166 (*see also* plateau; mesa)
Tahrit Oasis, 313
taiga (*see* plant cover)
Talara, 192, **548**, 598, 607
Talara Basin, **598**
Tallahatchie River, **261**
talus (debris slope/footslope), 29, 49, 50, 51, **52**, 62, 71, 94, **150**, 325, 369, 372, **602**
Talwani, 109
Tank, R. W., 677, 694
Tanner, W. F., 10, 46, 57, 103, 537, 588
tarn, 471
Tator, B. A., 344, 680
Tauber, W., 33, 43
Taylor, F. B., 512, 533
Tazieff, H., 186, 189, **198**, 209
tectonic denudation, 189–195, 594–596, 609
tectonic geomorphic systems, Chapters 3, 10, *see also* p. 644
tectonic landscape elements in buried cratons, 150–168
tectonic landscape elements in shields, 130–150
tectonic maps, **75**, 111, **123**, 127, **152**, **153**, 192, **196**, 598, **627**
tectonic relief, Chapters 3, 10, *see also* pp. 46, 80, 83, 105
tectonic topographic elements, 104
tectonic uplift, 4, 14, 15, 27, 39, **75**, 79, 80, 83, 85, 105
tectonism, 32, 47, 72, 73, **75**, 78, 79, 80, 83, 104
Teeter, S., 208
Tennessee, 52, 267, 281, 684, 685
tephrochronology, 35
Te Punga, M. T., 528, 531
terminal (end) moraine, 24, **191**, 487
terminal orogenic topography, 633

terraces (marine), 499, 541, **542**
terraces (stream), 12, 34, 69, 71, 90, 98, 191, 411, 418, **419**, **420**, 421
terra rossa, 30, 281
terrigenous coasts of disequilibrium, 545
terrigenous coasts of equilibrium, 545
terrigenous coasts of primary morphologic disequilibrium, 558–571
terrigenous coasts of primary morphologic equilibrium, 571
Tertiary, 35, **91**
Texas, 135, 164, 387, 401, 425, 437, 539, 572, **574**, 575, 661, 663, 686
Thailand, 606
thalweg, 62, 408, 413, 414
Thar Desert, 364
Tharp, M., 61, 62, 63, 64, 102, 114, 116, 120, 122
Theil, G. A., 646, 675
theory of pediplanation, 12, 30
theory of peneplanation, 4
theory of plate tectonics, 106
thermal equator, 216
thermokarst, 517
Thingrellic Rift Valley, **140**
third- and fourth-order landforms, 34, 62, **65**
Thompson, H. D., 617, 638
Thornbury, W. D., 4, 10, 39, 45, 466, 475, 478, 480, 487, 496, 498, **511**, 512, 517, 534, 542, 549, 552, 553, 560, 573, 588, 624, 655, 675
Thornthwait, C. W., 226, 229, 250, 257, 264, 309
threshold (systemic), 26, 39, 685
thufur, 526
Tibet, 364, 592, 612, 615
Tibisti, 371, 441
tidal bore, 555, 556, 686, 688
tidal current, 554, 555, 556
tidal estuary, **556**
tidal marsh, 569
tide, 42
tideless seas, 661–665
Tien Shan Mountains, 624
till, 22, 24, 37, **483**, 485, 489, 491, 501, 509, **510**
till fabric, 69
till plain, 67, **501**
time-dependent landforms, 27, 29
time-independent landforms, 26, 27, 29
time-stratigraphic terms, 97
Tindouf Basin, **152**
Tinkler, K. J., 62, 103, 404, 408, 447
Todd, J. E., 507, **508**, **509**, 511, 534
Toit, A. L. du., 107, 209
Tomales Bay, **551**
Tomales Point, **555**
Tompoketsa, 279
Tonga Trench, 111, 123
topographic maps, 48, **50**, **51**, 54, 59, 62, 136, 183, 342
topographic profiles, **55**, 118, 124, 591, 592, 594, 611, 634

topography of rifted continental rises, 185–189
traction load (stream) (see bed load)
Trade Winds Belt, 214, 217, 218, 222
trans-environmental rivers, 302–306
transform faults, 72, 109
trans-Pacific Ridge, 119
transpiration, 223, 226
transverse drainage, 46, 84, 617, 618, 619, 620, 621, 622, 630
Trask, P. D., 308, 596, 603, 638
travertine, 66, 301
trellis drainage pattern, 60, 299
trench (see oceanic trench)
Treppen Concept, 12, 13
Trewartha, G. G., 216, 250
tributary, 2
Tricart, J., 19, 26, 45, 69, 103, 281, 294, 295, 296, 301, 302, 309, 404, 447
Tricker, R. A. R., 554
Tripoli, 315
Trona, 319
tropical soil (see soil)
troposphere, 215
Trowbridge, A. C., 51, 87, 88, 103, 646, 676
Truckee River, 419, 421
Trumpy, R., 181, 189, 209
trunk stream, 2
tsunami, 550, 551
Tumbez Desert (basin), 598, 602, 603
Tunesia, 369, 400
Tuolumne River Canyon, 633
Tuomey, M., 570, 588
turbidity (density) currents, 8, 422
turbulent flow, 433, 434
turf hummock, 526
Turner, F. J., 136, 209
Turner, H. W., 632, 637
tussock ring, 526
Twain, M. (S. L. Clemens), 5
Twenhofel, W. H., 310, 349, 350, 380
Twidale, C. R., 32, 44, 130, 133, 139, 140, 149, 209, 325, 344, 347, 451

Ucayali River, 631
Uinta Mountains, 475
unconformity, 78, 98, 640–667
unconformity (basal St. Peter Sandstone), 646–648
unconformity analysis, 640
underfit stream, 417
undertow, 553
Ungava Bay, 500
Uniformitarianism, 2, 17, 18, 24, 38, 39, 40, 41, 42, 99, 661
United States, 75, 89, 129, 150, 154, 156, 157, 163, 164, 169, 185, 186, 223, 246, 267, 272, 280, 328, 347, 356, 386, 387, 417, 452, 463, 517, 524, 549, 573, 601, 619, 646, 689, 690
Unwin, D. J., 502, 534
uplift morphogenesis, 609–617
upthrusting, 78, 153, 187, 191

upwelling, 220
Ural Mountains, 106, 662
Urubamba River, 54, 245, 254, 420, 481, 616
Utah, 11, 68, 76, 173, 182, 188, 336, 337, 366, 475, 643
Utgard, B. O., 677, 694

vadose (groundwater zone), 300
Vaidyanadhan, R., 323, 379
Vaiont Canyon, 552
Valentin, H., 542, 545, 588
valley flat, 10, 11, 48, 684
valley glacier (see alpine glacier)
valley glaciation, 478–485
valleys, 18, 29, 48, 69, 73, 75, 77, 79, 88, 90, 184, 289, 290, 292, 419, 605, 672, 673
valley train, 491, 492
valley wall, 53, 75, 86
Van Donk, 41
Van Dorn, W. G., 550, 588
Vanzolini, P. E., 239, 243, 244, 250
varves, 461
Vatnajökull Glacier, 450
Vázquez-Yanes, C., 308
Veeh, H. H., 38, 45
vegetated areas (see plant cover)
vegetated setting, 251–256
vegetation maps, 228, 252, 267, 268
Vema Fracture, 122
Venetz, 17
Venezuela, 19, 21, 72, 73, 142, 144, 145, 149, 184, 260, 268, 297, 298, 389, 414, 415, 432, 434, 435, 545, 559, 563, 565, 566, 597, 598, 599, 600, 606, 632, 682
Ventauri River Delta, 425
ventifact, 318, 320, 351, 352, 486, 607
Verhoogen, J., 136, 209
Vermont, 178
vertebrates, 33, 34, 36, 37
vertical accretion, 285
vertical stream incision, 8, see also fluvial incision
Victoria Land, 471, 522
Vigil site, 3
Vine, F. J., 34, 45, 108, 115
Virginia, 186, 545, 546
Virkkala, 469
viscosity, 432
Vita, finzi, C., 24, 45
Vogt, P. R., 198, 209
volcanic deposits, 18, 22, 25, 33, 35, 53, 57, 62, 80, 81, 82, 91, 92, 175, 196–206, 450, 605, 607, 631
volcanic eruptions, 3, 47, 83, 113–120, 122–125, 196, 623, 679
volcanic geomorphic mechanisms, 79, 81, 82, 195–206
volcanic neck, 202
volcanic terrain, 61, 72, 78, 81, 82, 113–120, 197–206, 450
Volga River Valley, 492
von Richthofen, F. F., 541, 587

Vuilleumier, B. J., 239, 243, 250

Wachung (Basalt) Mountains, 111, 135
wadi, 62
Wadi Taffasassett, 641
Wahlstrom, E. E., 136, 209, 284, 309, 625, 639
Wahrhaftig, C., 632, 634, 639
Wales, 417
Walker, 52
Wallcott Cirque, 471
waning slope (toe slope), 49, 51, 52, 53
Wanless, H. R., 3, 45
War Eagle River, 403
Warran, A., 240, 355, 378
Warren, C. R., 511, 534
Wase Rock, 367
Washburn, A. L., 3, 45, 46, 103, 521, 523, 526, 528, 529, 534
Washington, 201, 223, 391, 440, 441, 512, 513, 514, 515
watercourse, 71, 253, see also channel
waterfall (see knickpoint)
water problems (modern), 686–691
Waters, R. S., 133, 209
water table, 295
wave base, 546, 547, 549, 565, 567
wave-cut (bench/terrace/platform), 564, 565, 567
wave refraction, 549, 550, 551
waves of oscillation, 54, 548
waves of translation, 547, 549
wave-splash fretwork, 566
wave train, 549
waxing slope (summit/shoulder), 49, 50, 52
Wayne, W. J., 153, 209, 511, 534
weather, 709
weathering, 3, 22, 42, 46, 71, 76, 79, 269, 270, 271, 272, 273, 274, 386
weathering maturity, 269, 275, 386, see also maturity
weathering residual, 93, 269–279
weathering rinds (see spheroidal weathering)
Weeks Island salt dome, 170
Wegener, A., 107, 209
Wegman, E., 596, 639
Weller, J. M., 154, 209, 659, 660, 676
Wellman, H. W., 522, 534
Wells, P. V., 266, 309
Welter, O., 593, 639
Wendorf, F., 394, 446
Werner, A. G., 2
West Australian Current, 219
West Virginia, 623
West Wind Drift, 219, 456
wet blasting, 658
wetted perimeter, 5, 16
Wheeler, H. E., 8, 45
White, H. P., 240, 250
White, W. A., 512, 535, 571, 588
White Mountain, 76
White River, 48, 50, 51
White Sands National Monument, 356
Wiegel, R. L., 549, 550, 588

wildflysch, 181, 193
Wildwood Beach, 553
Wiles, W. W., ix
Williams, E. E., 239, 243, 244, 250
Williams, G. E., 398, 447
Willman, H. B., 69, 95, 97, 100, 102, 103, 485, 486, 517, 532, 535
Willow River, 684, 685
Wilson, A. T., 513, 522, 534
Wilson, J. T., 109, 112, 115, 120, 127
Wilson, L., ix, 32
wind gap, 283
Winslow Formation, 649–651
Winterer, E. L., 194, 210
Wisconsin (glacial stage), 31, 40, 80
Wisconsin (state of), 500, 501, 652
Wise, D. U., 153, 207, 210
Wolff, J. E., 136
Wolman, M. G., 3, 5, 7, 44, 45, 258, 309, 413, 414, 416, 419, 427, 445, 505
Wood, A., 49, 103
Woodworth Glacier, 495
Wooldridge, S. W., 546, 588
Wooley, R. R., 330, 380
Woollard, G. P., 129, 130, 210, 592, 639
Woolnough, W. G., 323, 380
Wright, H. E. Jr., 95, 103, 239, 250, 451, 486, 509, 535
Wright, R. C., 35, 43, 103
Wright Dry Valley, 515, 522, 527
Würm, 90
Wyoming, 9, 67, 132, 150, 194, 206, 472, 628

xerophytic (plants), 81, 231, 267, 268, 311

Yaalon, D. H., 29, 45, 270, 272, 308
yardang, 62
Yasso, W., ix, 550
Yellowstone River, 9
Yellowstone River Canyon, 206
Yenisei River, 305
Yoe Lake, 168
Yosemite Valley, 478, 480, 633
youth (Davisian geomorphic age), 4, 5
Yucca Mountains, 602
Yugoslavia, 281, 545

Zagros covermass and drainage, 618–621
Zagros Mountains, 46, 83, 84, 85, 178, 181, 402, 601, 609, 610, 617, 618, 620, 621, 622, 628, 673
Zambezi River, 240, 242, 373, 438
zastruga, 62
Zenkovitch, V. P., 546, 556, 570, 577, 588
zone of aeration, 253
zone of fracture, 253
zone of moisture-depleted air, 612
zone of prevailing westerly winds, 218, 219
zone of saturation, 253
zone of triggered precipitation, 236
zone of wave impingement, 662
zone of weathering, 271
Zululand, 575